Birkhäuser

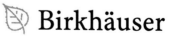

Progress in Mathematics

Volume 296

For further volumes:
http://www.springer.com/series/4848

Ilia Itenberg • Burglind Jöricke • Mikael Passare
Editors

Perspectives in Analysis, Geometry, and Topology

On the Occasion of the 60th Birthday of Oleg Viro

Birkhäuser

Editors

Ilia Itenberg
Université Pierre et Marie Curie
and Institut Universitaire de France
Institut de Mathématiques de Jussieu
4 place Jussieu
75005 Paris
France
itenberg@math.jussieu.fr

Burglind Jöricke
Institut des Hautes Études Scientifiques
Le Bois-Marie
35, route de Chartres
Bures-sur-Yvette 91440
France
joericke@googlemail.com

Mikael Passare
Department of Mathematics
Stockholm University
Stockholm SE-10691
Sweden
passare@math.su.se

ISBN 978-0-8176-8276-7 e-ISBN 978-0-8176-8277-4
DOI 10.1007/978-0-8176-8277-4
Springer New York Dordrecht Heidelberg London

Library of Congress Control Number: 2011939742

Mathematics Subject Classification (2010): 11H06, 14P25, 14T05, 14H99, 20F36, 31A25, 32U05, 53D05, 53D10, 53D40, 53D42, 57Mxx, 57M25, 57M27

Printed on acid-free paper

Springer is part of Springer Science+Business Media (www.birkhauser-science.com)

This volume is dedicated to Oleg Viro on the occasion of his 60th birthday.

Contents

Preface

Encounters between the fields of analysis, geometry, and topology are widespread and often provide major impetus for breakthroughs in these domains. Recent developments in low-dimensional topology, algebraic geometry, symplectic geometry, complex analysis, and tropical geometry are so rich with impressive examples that any attempt to reflect a major portion of them in a single volume would be too ambitious.

For example, recent exciting progress in low-dimensional topology and geometry would have been unthinkable without powerful techniques from analysis; new achievements in topology and symplectic geometry have been influencing complex analysis; methods of tropical geometry are being used in algebraic geometry, statistical physics, and complex analysis; differential equations coming from complex analysis have applications in algebraic geometry.

The variety of topics presented by leading experts in this volume should nonetheless give some sense of recent developments in these fields and their interactions with one another.

A Marcus Wallenberg Symposium on Perspectives in Analysis, Geometry, and Topology was held at Stockholm University in May 2008. The choice of subjects for this symposium and this volume was motivated by the work and the mathematical interests of Oleg Viro, to whom the symposium and this collection are dedicated. As a mathematician with broad education and interests, Viro has a deep feeling for the unity of mathematics. Viro is famous for fundamental results in several areas of geometry and topology. As a professor at Uppsala University, Viro has made invaluable contributions to Swedish research by complementing the country's longstanding tradition in analysis with his own renowned expertise in geometry and topology.

A Glimpse into Viro's Work

Oleg Viro graduated from the Leningrad (now St. Petersburg) school of topology, one of the strongest that ever existed. He defended his PhD thesis in 1974 under the supervision of V. A. Rokhlin, the founder and head of this school.

Viro's first publications date back to his undergraduate years. They are devoted to branched coverings and braids. He proved that any closed orientable 3-manifold of genus two is a two-fold branched covering of the three-dimensional sphere branched over a link with three bridges (this was proved independently by J. Birman and H. Hilden).

Viro's results on branched coverings also include interpretations of the signature invariants of a link of codimension 2 as signature invariants of cyclic branched covering spaces of a ball, and estimates for the slice genus of links and the genus of non-locally-flat surfaces in 4-manifolds.

During the year of Viro's thesis defense, Rokhlin was stricken with a severe heart attack, and Viro, then 26 years old, took over Rokhlin's course in topology for second-year undergraduates. Teaching Rokhlin's course was no easy task. On one hand the brilliance of Rokhlin's lectures was unique, meeting his standard of lecturing seemed to be out of reach. On the other hand, the novelty and level of abstractness provided difficulties for many students, which was reflected in examination scores that displeased the University administration.

As a result of increasing antisemitism and suppression of intellectual freedom in the Soviet Union during the 1970s, Viro was at that time the only graduate of Rokhlin's extremely strong school of topology to be hired by the department. He took on himself the responsibility for topology as a research subject at the department and for the education of students in this field. In the years 1975–1985 he became a central figure in the scientific and pedagogical life at the Mathematical–Mechanical Department of Leningrad University.

A. Vershik has communicated to us that in 1979, Rokhlin remarked to him, "Do you know why it is worthwhile for a student to study in this department? Because Oleg Viro is working there." Anyone who is aware of Rokhlin's high standards and restrained praise can imagine the value of these words of appreciation.

In this period, Oleg Viro introduced the *patchworking technique*, his best-known contribution to *real algebraic geometry*. Viro's investigation followed a breakthrough made in topology of real algebraic varieties in the late 1960s and 1970s. The work of Gudkov, Arnold, Rokhlin, Utkin, and Kharlamov provided answers to questions formulated in the first part of Hilbert's 16th problem (isotopy classification of nonsingular curves of degree 6 in the real projective plane $\mathbb{R}P^2$ and topological classification of nonsingular quartic surfaces in the three-dimensional real projective space) and gave a conceptional understanding of many general phenomena concerning the topology of real algebraic varieties.

For curves of higher degree, an isotopy classification was not accessible by the previous techniques. The patchworking construction allowed Oleg Viro to complete the isotopy classification of nonsingular curves of degree 7 in $\mathbb{R}P^2$ (the largest degree for which such a classification is available even today) and to disprove Ragsdale's conjecture from 1906 concerning curves of even degree.

The impact of the patchworking technique is much wider. Patchworking revolutionized real algebraic geometry by linking together real algebraic geometry, toric geometry, and combinatorics of convex polytopes. It results in a powerful construction of real algebraic varieties with prescribed topology. Almost all recent constructions of real algebraic varieties use Viro's theorem. Patchworking was a fundamental idea in motivating the appearance of tropical geometry. In fact, Viro established a relation between patchworking and Maslov's dequantization of positive real numbers that was one of the starting points of tropical geometry.

A combinatorial version of Viro's theorem is as follows. Let n be a positive integer and Δ a convex lattice polytope in $\mathbb{R}^n$. Consider a lattice triangulation of Δ and a distribution of signs at the vertices of the triangulation. In this situation, Viro constructs a piecewise-linear hypersurface H in the real part $\mathbb{R}T(\Delta)$ of the toric variety $T(\Delta)$ associated with Δ. His theorem then produces, under a certain condition on the triangulation, a real algebraic hypersurface in $\mathbb{R}T(\Delta)$ that is isotopic to H and defined by a polynomial of Newton polytope Δ. This version of Viro's theorem is called *combinatorial patchworking* and provides a rich collection of real algebraic hypersurfaces. The piecewise-linear hypersurfaces appearing in this construction can be seen as real tropical hypersurfaces.

The achievements of Viro in real algebraic geometry also include important restrictions on the topology of real algebraic curves and a notion of complex orientations of real algebraic varieties of dimension ≥ 2 (generalizing the notion of complex orientations of a real algebraic curve dividing its complexification).

Viro was invited to the International Congress of Mathematicians in Warsaw in 1983. His article on real algebraic varieties was published in the proceedings, but he was forbidden by the administration of Leningrad University to attend the congress. The reason behind the denial was the following story. In a wave of antisemitism and personal attacks against Rokhlin, whose independence and nonconformity were taken by the authorities as a provocation, the administrators of the Mathematics Department tried to force Rokhlin, then 60 years old and mathematically active, into early retirement. One of the measures was an attempt to prevent Rokhlin from supervising his graduate student Mikhail Goussarov and to assign Goussarov to

Oleg Viro. Viro declined, and Goussarov remained Rokhlin's student. Viro was thus punished for his display of loyalty and human decency. He never lost the remarkable personality trait that emerged on this occasion: a disposition to protect others, to ease their burdens, and to expose himself to the slings and arrows of outrageous fortune. His empathy and his courage to speak the truth and take a principled stand have always been unquestioned.

Viro's best-known contribution to *low-dimensional topology* is his joint work with Vladimir Turaev that defines "quantum invariants" of 3-manifolds as a certain state sum over an arbitrary triangulation. The state sum is based on so-called quantum 6j-symbols related to the representation theory of the quantum group $U_q\mathfrak{sl}(2)$ for a root of unity q. Notably, the construction leads to a $(2+1)$-dimensional topological quantum field theory – the first rigorous realization of an approach by the physicists Ponzano and Regge to $(2+1)$-dimensional quantum gravity. The paper contributes to the relationship – for mathematicians still largely mysterious – between topology and mathematical physics. Remarkable parallels were revealed concerning the relationship between analysis and theoretical physics: for instance, the interest of Sweden's world-renowned analyst Lennart Carleson in two-dimensional conformal field theory paved the way for Stanislav Smirnov to prove a formula for critical site percolation predicted by the physicist Cardy.

We also mention the following articles among the numerous papers of Viro in *topology*. The volume of Lecture Notes in Mathematics dedicated to the memory of V. A. Rokhlin contains a paper that caused McPherson to refer jokingly to Viro as the guy "who is able to integrate the Euler characteristic." In this paper Viro discussed one of the manifestations of certain "universality" of the Euler characteristic. The same volume contains a construction of the first infinite series of surfaces smoothly embedded in the four-dimensional sphere, which are pairwise ambiently homeomorphic but not diffeomorphic (joint work with S. Finashin and M. Kreck). A joint work of Viro with M. Goussarov and M. Polyak treats diagrammatic formulas for finite-type knot invariants for classical and virtual knots.

Viro has spent a great deal of time and energy with his graduate students. His former students, many of them now renowned professors of mathematics, represent diverse areas of geometry and topology. Viro's principle is to put high demands on his students but to be generous and supportive of them. He believes that education of the next generation of mathematicians cannot be reduced to imparting knowledge and skills. It is also important to mold responsible personalities capable of mastering the challenges of the age, which is often accomplished less through words than by example, through nonverbal messages conveyed by the personality of the teacher, as Viro experienced himself as a student of Rokhlin.

Viro's activities are not limited to research in topology and geometry and the education of graduate students. He is concerned as well about the role of mathematics as an irreplaceable part of human culture and about its survival – in his view a very complex task. This includes an understanding of the development of mathematics as a whole and of its interaction with other sciences, as well as an open mind for promising areas of mathematics, a careful choice of research subjects, and above all, the education of the next generation beginning with the foundations of mathematics. Viro has striven to return geometry to its important place in the educational program. While still in St. Petersburg, he participated in a project to provide students their first approach to mathematics and mathematical rigor through developing geometric intuition rather than through developing the ability to formally manipulate mathematical symbols. In Uppsala, Viro initiated a special program for strong students following similar lines and offered a newly developed course on basic geometry.

Viro taught at Uppsala University until his forced resignation in 2007. Now he is professor at State University of New York, Stony Brook.

Despite all the difficulties he has experienced, Viro has maintained his ability to meet people with a delightfully warm and sincere smile.

Paris, France — Ilia Itenberg
Bures-sur-Yvette, France — Burglind Jöricke
Stockholm, Sweden — Mikael Passare

On the Content of This Volume

The papers of the volume represent various topics in geometry, topology, and analysis. To give an impression of the diversity of these papers, we have divided them provisionally into groups and have included a short description of each. A complete list of the symposium talks appears after the description of the content of this volume.

Algebraic Geometry

K. Kaveh and A. Khovanskii: *Algebraic Equations and Convex Bodies*. This paper is a summary of the authors' recent work on intriguing relationships between algebraic geometry on the one hand, and convex bodies and their combinatorics on the other. Their results can be seen as far-reaching generalizations of the Bernstein–Kushnirenko theorem, which relates algebraic geometry and the theory of mixed volumes.

N. Mok: *Projective Algebraicity of Minimal Compactifications of Complex-Hyperbolic Space Forms of Finite Volume*. This paper considers quotients of the unit ball in complex affine space by lattices (i.e., by discrete subgroups of the holomorphic automorphism group with quotients of finite Bergman volume). The author shows that for (possibly nonarithmetic) quotients, the minimal compactification (obtained by adding a finite number of points to the cusps) is projective algebraic.

E. Shustin: *Tropical and Algebraic Curves with Multiple Points*. This paper is devoted to a new patchworking theorem allowing one to construct algebraic curves with multiple points. This theorem can be seen as a generalization of Viro's patchworking construction and has important applications. It can be used to obtain a new correspondence theorem between the complex algebraic and tropical worlds.

Real Algebraic Geometry

The topics of this group are intimately related to the work of Oleg Viro in topology of real algebraic varieties.

A. Degtyarev: *Toward a Generalized Shapiro and Shapiro Conjecture*. The Shapiro and Shapiro conjecture, proposed in 1993, is the following: *if all flattening points of a rational curve $\mathbb{C}P^1 \to \mathbb{C}P^n$ lie on the real line $\mathbb{R}P^1 \subset \mathbb{C}P^1$, then the curve is conjugate to a real algebraic curve under an appropriate projective automorphism of $\mathbb{C}P^n$*. The conjecture was proved in 2005 by E. Mukhin, V. Tarasov, and A. Varchenko. The present paper is devoted to a generalization of the conjecture to curves of arbitrary genus (this generalization was proposed by T. Ekedahl, B. Shapiro, and M. Shapiro). A new, asymptotically better, bound on the genus of a curve that may violate the generalized conjecture is obtained.

A. Degtyarev, I. Itenberg, and V. Kharlamov: *On the Number of Components of a Complete Intersection of Real Quadrics*. This paper concerns the topology of complete intersections of three real quadrics. The main result is the following: The maximal possible number $B_2^0(N)$ of connected components of a regular complete intersection of three real quadrics in $\mathbb{P}^N$ differs by at most one from the maximal number of ovals of the submaximal depth $[(N-1)/2]$ of a real plane projective curve of degree $d = N+1$.

S. Orevkov: *Some Examples of Real Algebraic and Real Pseudoholomorphic Curves*. The paper contains several results concerning the embedded topology of algebraic and pseudoholomorphic curves in the real projective plane. The comparison between real algebraic and real pseudoholomorphic curves can be seen as a natural continuation of the study of *flexible curves* which was initiated by Viro.

E. Shustin: *Tropical and Algebraic Curves with Multiple Points* (see under *Algebraic Geometry*).

Differential Geometry and Differential Equations

Y. Eliashberg and N. Mishachev: *Topology of Spaces of S-immersions*. This paper discusses the *h*-principle for maps between equidimensional manifolds whose only singularities are prescribed folds.

Symplectic and Contact Geometry

K. Baker and J. Etnyre: *Rational Linking and Contact Geometry*. The authors extend the self-linking number of transverse knots and the Thurston–Bennequin invariant and rotation number of Legendrian knots to the case of rationally null-homologous knots. The paper contains a generalization of Bennequins inequality for these knots, a study of sharpness of the Bennequin bound for fibered knot types, and a proof of the fact that rational unknots in tight contact structures on lens spaces are weakly transversely simple and Legendrian simple.

T. Ekholm: *Rational SFT, Linearized Legendrian Contact Homology, and Lagrangian Floer Cohomology*. This paper relates the version of rational symplectic field theory for exact Lagrangian cobordisms introduced in a previous article of the author with linearized Legendrian contact homology.

M. Entov, L. Polterovich, and P. Py: *On Continuity of Quasimorphisms for Symplectic Maps. With an appendix by Michael Khanevsky*. This paper is devoted to C^0-continuous homogeneous quasimorphisms on the identity component of the group of compactly supported symplectomorphisms of a symplectic manifold. The authors give a topological characterization of such quasimorphisms in the case of surfaces and show that for standard symplectic balls of any dimension and for compact oriented surfaces other than the sphere, the space of such quasimorphisms is infinite-dimensional.

D. Gay and A. Stipsicz: *On Symplectic Caps*. This paper gives an explicit construction of symplectic caps (concave fillings) of links of minimal rational singularities. As an application, new examples of surface singularities are obtained that do not admit rational homology disk smoothing.

C. Manolescu and Ch. Woodward: *Floer Homology on the Extended Moduli Space*. The goal of this paper is to approach the Atiyah–Floer conjecture by suitably modifying its symplectic side. The Atiyah–Floer conjecture relates Floer's instanton homology and his Lagrangian homology. Given a Heegaard splitting of a 3-manifold, the authors construct, using Lagrangian Floer homology, a relatively $\mathbb{Z}/8\mathbb{Z}$-graded abelian group that is conjecturally a 3-manifold invariant.

S. Orevkov: *Some Examples of Real Algebraic and Real Pseudoholomorphic Curves* (see under *Real Algebraic Geometry*).

Complex Analysis

R. Berman and J.-P. Demailly: *Regularity of Plurisubharmonic Upper Envelopes in Big Cohomology Classes*. The subject of this paper originates from constructions of Hermitian metrics with minimal singularities on a big line bundle over a compact complex manifold. The developed analytic tool, the proof of a certain regularity of an upper envelope of "quasiplurisubharmonic" functions, is applied to study the Dirichlet problem for a degenerate Monge–Ampère operator as well as geodesics in the space of Kähler metrics.

G. Henkin: *Cauchy–Pompeiu-Type Formulas for $\bar\partial$ on Affine Algebraic Riemann Surfaces and Some Applications*. The author gives explicit formulas for solving $\bar\partial$- and perturbed ∂-problems on complex affine algebraic curves in complex two-space and applies them to the two-dimensional inverse conductivity problem on surfaces with boundary. This is related to electric impedance tomography, geophysics, and other topics.

N. Mok: *Projective Algebraicity of Minimal Compactifications of Complex-Hyperbolic Space Forms of Finite Volume* (see under *Algebraic Geometry*).

Three- and Four-Dimensional Manifolds, Invariants of Links

S. Akbulut: *Exotic Structures on Smooth Four-Manifolds*. This survey treats topological methods of constructions of exotic copies of 4-manifolds (smooth manifolds that are homeomorphic but not diffeomorphic to the original one). One of the methods (gluing "corks") consists of the following. Cut off a smoothly bounded domain that admits the structure of a relatively compact strictly pseudovonvex domain in a Stein manifold. Glue it back by an involution of the boundary that extends to the domain homeomorphically but not diffeomorphically.

L. Kauffman: *Remarks on Khovanov Homology and the Potts Model*. This paper explores relationships between the Potts model in statistical mechanics and Khovanov homology. The author proves in particular that the Euler characteristics of Khovanov homology (precisely, of its subcomplexes for fixed quantum grading) figure in the computation of the Potts model at certain imaginary temperatures.

N. Reshetikhin, C. Stroppel, and B. Webster: *Schur-Weyl-Type Duality for Quantized* $\mathfrak{gl}(1|1)$, *the Burau Representation of Braid Groups, and Invariants of Tangled Graphs*. This paper is a concise survey of the authors' results in the subject. The paper clarifies the relationship between reduced Burau representations of braid groups, nonreduced Burau representations, and the representation of the braid group defined by R-matrices related to $U_q(\mathfrak{gl}(1|1))$.

A. Shumakovich: *Khovanov Homology Theories and Their Applications*. This survey presents various versions of Khovanov homology theories. The author discusses relationships between these theories, as well as their properties and applications to other areas of knot theory and low-dimensional topology.

Ilia Itenberg
Burglind Jöricke
Mikael Passare

Acknowledgments The editors want to express deep gratitude to many friends and colleagues of Oleg Viro who generously provided information and assistance. The photographs are courtesy of his family, friends, and colleagues. The computing staff at MSRI and IHES provided help with editing photographs. We would like to thank them all. We are grateful to all speakers at the Marcus Wallenberg Symposium on Perspectives in Analysis, Geometry, and Topology, to all colleagues who contributed to the present volume, and to the anonymous referees of the papers. Lennart Carleson kindly granted permission to include the transcript of his opening remarks. His words have not been edited, and hence some of the charm of a vivid speech has been retained. The symposium was made possible by the generous support of the Marcus Wallenberg Foundation for International Scientific Cooperation, from Stockholm University, and from the Mittag–Leffler Institute.

In memoriam notice On September 15, 2011 our friend, colleague and co-editor Mikael Passare unexpectedly passed away in the result of an accident. He has been widely appreciated as a mathematician, teacher, supervisor and unresting organizer on behalf of mathematics. The symposium would hardly have taken place without his energy and organizational skills. He is missed sadly.

Ilia Itenberg
Burglind Jöricke

Oleg Viro 1976

Oleg Viro around 1986

Lennart Carleson and Oleg Viro 2008

Oleg Viro 2002

The poster of the Marcus Wallenberg Symposium on Perspectives in Analysis, Geometry, and Toplogy

Program of the Symposium

Monday, May 19:

10.00–10.45	Opening, with a welcoming address by Vice-Chancellor Kåre Bremer and a speech by Lennart Carleson on the "Unity of Mathematics"
11.00–12.00	Louis Kauffman: "An Extended Bracket State Summation for Virtual Knots and Links"
LUNCH	
14.00–15.00	Michael Polyak: "Enumerative Geometry and Finite Type Invariants"
15.30–16.30	Alexander Degtyarev: "Toward a Generalized Shapiro and Shapiro Conjecture"
17.00–18.00	Ngaiming Mok: "Geometric Structures on Fano Manifolds of Picard Number 1"

Tuesday, May 20:

09.30–10.30	Gang Tian: "Ricci Flow and Projective Manifolds"
11.00-12.00	Ludwig Faddeev: "Discrete Series of Representations for the Modular Double of $U_q(\mathfrak{sl}(2,R))$"
LUNCH	
14.00–15.00	Tobias Ekholm: "A Surgery Exact Sequence in Linearized Contact Homology"
15.30–16.30	Paolo Lisca: "Heegaard Floer Invariants of Legendrian and Transverse Knots"
17.00–18.00	John Etnyre: "Fibered Knots and the Bennequin Bound"

Wednesday, May 21:

09.30–10.30	Gennadi Henkin: "Cauchy–Pompeiu-Type Formulas for $\overline{\partial}$ on Affine Algebraic Riemann Surfaces and Some Applications"
11.00–12.00	Charles Epstein: "Solving Maxwell's Equations in Exterior Domains"

LUNCH
14.00–15.00 Elliott Lieb: "Four Decades of 'Stability of Matter' and Analytic Inequalities"
15.30–16.30 Eric Bedford: "Dynamics of Rational Surface Automorphisms of Positive Entropy"
17.00–18.00 Mikhail Lyubich: "Yang–Lee Zeros for Diamond Lattices and 2D Rational Dynamics"

Thursday, May 22:
09.00–10.00 Viatcheslav Kharlamov: "On the Number of Connected Components in the Intersection of Three Real Quadrics"
10.30–11.30 Grigory Mikhalkin: "Patchworking of Real Algebraic Knots and Links"
12.00–13.00 Stepan Orevkov: "Classification of Algebraic Links in RP^3 of Degree 5 and 6"
LUNCH
14.00–18.30 Excursion
19.00 Reception at Town Hall

Friday, May 23:
09.30–10.30 Selman Akbulut: "On Exotic Structures on 4-Manifiolds"
11.00–12.00 Nicolai Reshetikhin: "Invariants of Links and Quantum Groups at Roots of Unity"
LUNCH
14.00–15.00 Robert MacPherson: "The Geometry of Grains"
15.30–16.30 Laszlo Lempert: "Two Examples of Complex Manifolds, Motivated by (Theoretical) Physics"
07.00-18.00 Boris Khesin: "A Nonholonomic Moser Theorem and Diffeomorphism Groups"
19.00 Banquet

Saturday, May 24:
09.30–10.30 Tomasz Mrowka: "Knot Invariants from Instantons"
11.00–12.00 Askold Khovanskii: "Algebraic Equations, Convex Bodies, and Bernstein Theorem for Some Spherical Varieties"
LUNCH
14.00–15.00 Ronald Fintushel: "Constructions of 4-Manifolds"
15.30–16.30 Stefan Nemirovskii: "Lagrangian Embeddings and Complex Analysis"
17.00–18.00 Stanislav Smirnov: "Conformal Invariance and Universality in 2D Ising Model"

Sunday, May 25:

09.30–10.30	Eugenii Shustin: "Computing Real Algebraic and Tropical Enumerative Invariants"
11.00–12.00	Alexander Shumakovitch: "Khovanov Homology, Its Properties and Applications"
12.00	Closure

On the Unity of Mathematics

Opening speech by Lennart Carleson
Transcript by the Editors

It is a great pleasure for me to have this opportunity to welcome you all to Stockholm (and on such a beautiful day as we have today—we are not always so lucky!), and to this conference on "Perspectives in Analysis, Geometry, and Topology."

There is, of course, also a very special reason for such a conference at this time and in this place, namely Sweden, and this is that it gives us an opportunity to congratulate Oleg Viro on his 60th birthday, which will happen this year—and may have happened. He had a great influence on the development of mathematics in Uppsala, and on the infusion of geometry and topology in this land of analysis and algebra. Therefore his work is very much in the spirit of this conference, and it is rightly dedicated to him.

Many mathematicians, like myself, are very upset by the way that he was treated by the administration in the University of Uppsala, and by the very unconstructive and exceptional ways in which the administration decided to handle the difficulties there. We were not able to reverse anything, but you should think of this conference as a way of expressing our sentiments to what has happened here.

The subject of the conference defines a subset of the greater issue of the unity of Mathematics. This problem arose really during the 1800s in two different ways. It started with the introduction of the notion of applied mathematics. Gauss might be considered the last universal mathematician in the sense of also including applied mathematics, and what happened during the nineteenth century is the introduction of serious rigor in what we now call pure mathematics, and also the separation of the applied fields, first into different kinds of sciences, and later now also into all kinds of biology and social sciences. And this is a development that still goes on.

This is of course a great issue, which I am not really going to talk about, because the conference here belongs to the second movement, namely the separation of smaller areas of what we now call pure mathematics. This fragmentation I think is a serious business that has been going on for one hundred years now. You can probably say that Hilbert and Poincaré were the last generation of people who could somehow be considered to represent all of mathematics.

Then the fragmentation has continued, and this is seen in the names of professorships and the creation of journals and I don't know what. We have journals in, for example, approximation, in semigroups, or in inequalities. This specialization has really accelerated with the introduction of the electronic way of searching for information, which means that you can now define what you are looking for so well that you are sure that you don't see anything else. It is really difficult to have general colloquia and information outside the specialties.

I remember when I was a student and I was looking for information, I went to the library and found the book which contained the article in question, and I was reading that. Often I took it home and and kept it there, and I looked at the other articles in the journal. Very often you found something that interested you, and very often (mostly, I would say) you found something you couldn't understand. But still you got some information, and you got some impression of the field and of what you had around you.

I think that this possibility of finding unexpected information is what corresponds really to what one has in experimental sciences by making mistakes. There are very famous examples here, for instance the Curies, who found radioactivity by putting a stone on a photographic plate, or Fleming, who found penicillin by similarly having contamination of bacteria.

I think the closest we can get here is that we get information that we didn't really expect about something that we didn't know in advance what it was. I have been trying to find an example of this, but I can't really remember any well-known problem being solved by somebody by mistake reading an article on something else. The most well known story that I can recall (probably untrue, but still) is the story about Heisenberg going to the wrong lecture and learning about matrices. But it could be true!

This conference is, of course, an example of this, but I should like to make a personal illustration of something that happened recently to me. I was in Helsinki to give a description of Lars Ahlfors's work, so I had a reason to look at what he had done, and I looked at his famous 1935 paper on Riemann surfaces. It occurred to me that his approach to Nevanlinna theory, which is purely geometric and doesn't contain any formulas or any integration, but still gives a complete description of the value distribution of a meromorphic function, could be the way to understand Vojta's observations on the formal similarity between valuations and Nevanlinna theory, which nobody understands today. I think the reason one doesn't understand it is that one has the wrong model, and Ahlfors's model is of course completely different from the Nevanlinna model. So if I could suggest a good subject where infusion of ideas from a different field could give something, I would suggest that one look more carefully at this.

So I think it is very important that we take actions in the organization of education which make us all and future students more aware of fields outside their specialty, and that we avoid specialization in the narrow way that has been done here. What you could do is to make sure that the programs of graduate studies are more varied and that specialization comes later.

In terms of journals I think the development accelerates now with the electronic journals, and I should think it is very important that we keep a number of the really best journals, so that they still get printed, and so that people still can go to the library and by mistake read an article on something that they didn't really expect. I think it is a great risk that we make everything computerized, and by doing that, as I said, you make sure that you don't learn anything unexpected.

One more and final point would be that one organizes conferences, not only like this one, but also makes sure that the International Congresses are kept. There is a great risk that people criticize them and say that they don't give anything, it is too wide, there are too many people, and so on. One should resist that. It is very important that we keep the subject together.

And then I have avoided the greatest issue, that is, the relation to applied mathematics, which I think takes a very serious reorganization of mathematics. But this is not the time and place for that.

Thank you!

Exotic Structures on Smooth Four-Manifolds

Selman Akbulut

Dedicated to Oleg Viro on the occasion of his 60th birthday

Abstract A short survey of exotic smooth structures on 4-manifolds is given with a special emphasis on the corresponding cork structures. Along the way we discuss some of the more recent results in this direction, obtained jointly with R. Matveyev, B. Ozbagci, C. Karakurt, and K. Yasui.

Keywords Four-manifold • Exotic structure • Lefschetz fibration • Cork

1 Corks

Let M be a smooth closed simply connected four-manifold, and M' an exotic copy of M (a smooth manifold homeomorphic but not diffeomorphic to M). Then we can find a compact contractible codimension-zero submanifold $W \subset M$

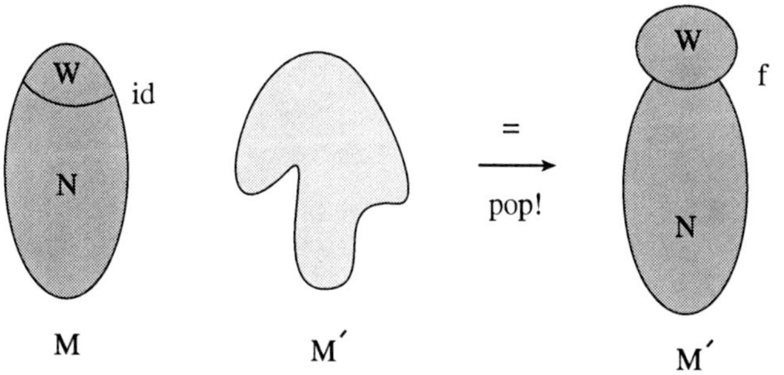

Fig. 1 Popping a cork

S. Akbulut (✉)
Department of Mathematics, Michigan State University, MI 48824, USA
e-mail: akbulut@math.msu.edu

I. Itenberg et al. (eds.), *Perspectives in Analysis, Geometry, and Topology: On the Occasion of the 60th Birthday of Oleg Viro*, Progress in Mathematics 296,
DOI 10.1007/978-0-8176-8277-4_1,

with complement N, and an involution $f : \partial W \to \partial W$ giving decompositions (identifications are by diffeomorphisms)

$$M = N \cup_{\mathrm{id}} W, \quad M' = N \cup_{\mathrm{f}} W \tag{1}$$

The existence of this structure was first observed in an example in [A1]. Then in [M] and [CFHS], this was generalized to the general form discussed above. Also, the 5-dimensional h-cobordisms induced by corks were studied in [K]. Since then, the contractible pieces W appearing in this decomposition have come to be known as *corks*.

Recall that a properly embedded complex submanifold of an affine space $X \subset \mathbf{C}^N$ is called a *Stein manifold*. Also, in topology, a smooth submanifold $M \subset X$ is called a *compact Stein* manifold if it is cut out from X by $f \leq c$, where $f : X \to \mathbf{R}$ is a strictly plurisubharmonic (proper) Morse function and c is a regular value. In particular, M is a symplectic manifold with convex boundary and the symplectic form $\omega = \frac{1}{2}\partial\bar{\partial} f$. The form ω induces a contact structure ξ on the boundary ∂M. We call (M, ω) a *Stein filling* of the boundary contact manifold $(\partial M, \xi)$. Stein manifolds have been a useful tool for studying smooth four-manifolds. In this paper, we will sometimes for the sake of brevity abuse conventions and call compact Stein manifolds just Stein manifolds.

By [AM], in the cork decomposition (1), each W and each N piece can be made Stein. This is achieved by a useful technique (called "creating positrons" in [AM]) that amounts to moving the common boundary $\Sigma = \partial W = \partial N$ in M by a convenient homotopy: First, by handle exchanges we can assume that each W and N side has only 1- and 2-handles. Eliashberg's criterion (cf. [G]) says that manifolds with 1- and 2-handles are Stein if the attaching framings of the 2-handles are sufficiently negative (let us call these *admissible*). This means that any 2-handle H has to be attached along a knot K with framing less than the Thurston–Bennequin framing $\mathrm{tb}(K)$ of any Legendrian representative of K (i.e., K is tangent to ξ, and $\mathrm{tb}(K)$ is the framing induced by ξ). The idea is that when the attaching framing is bigger than $\mathrm{tb}(K) - 1$, by local handle exchanges near H (but away from H) to alter $\Sigma \rightsquigarrow \Sigma'$, which results in an increase in the Thurston–Bennequin numbers $\mathrm{tb}(K) \rightsquigarrow \mathrm{tb}(K') = \mathrm{tb}(K) + 3$.

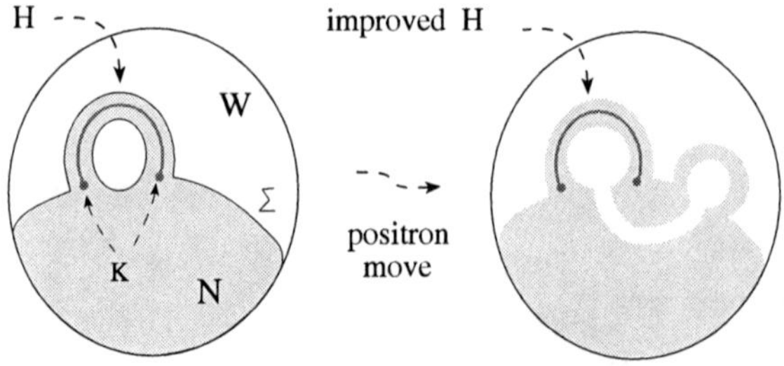

Fig. 2 Carving to make Stein

For example, as indicated in Fig. 3, carving out a tubular neighborhood of a properly embedded 2-disk **a** from the interior of N increases $\mathrm{tb}(K)$ by 3. Carving in the N side corresponds to attaching a 2-handle **A** from the W side, which itself might be attached with a "bad" framing. To prevent this, we also attach a 2-handle **B** to N near **a**, which corresponds to carving out a 2-disk from the W side. This makes the framing of the 2-handle **A** in the W side admissible. Furthermore, **B** itself is admissible. So by carving a 2-disk **a** and attaching a 2-handle **B**, we have improved the attaching framing of the 2-handle H without changing other handles (we changed Σ by a homotopy). This technique gives the following result.

Theorem 1. *([AM]) Given any decomposition of a closed smooth four-manifold $M = N_1 \cup_\partial N_2$ by codimension-zero submanifolds, with each piece consisting of 1- and 2-handles, after altering pieces by a homotopy, we can obtain a similar decomposition $M = N_1' \cup N_2'$, where both pieces N_1' and N_2' are Stein manifolds.*

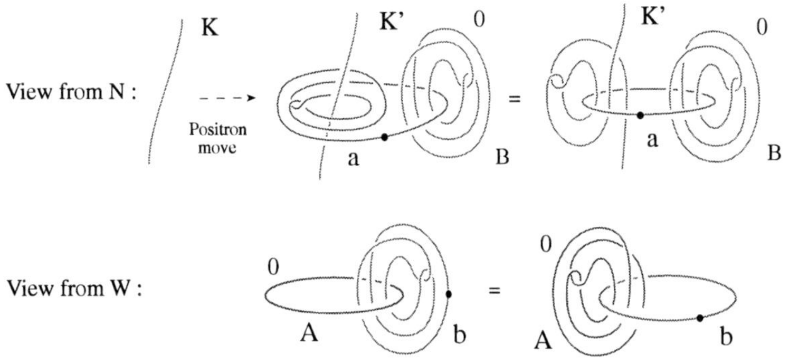

Fig. 3 Making W and N Stein

Using [Gi], one can also assume that the two open books on the common contact three-manifold boundaries match, but with the wrong orientation [B1].

Definition 1. A cork is a pair (W, f), where W is a compact Stein manifold and $f : \partial W \to \partial W$ is an involution that extends to a self-homeomorphism of W but does not extend to a self-diffeomorphism of W. We say that (W, f) is a cork of M if we have the decomposition (1) for some exotic copy M' of M.

In particular, a cork is a fake copy of itself. There are some natural families of corks W_n, $n = 1, 2, \ldots$, which are generalizations of the Mazur manifold W used in [A1], and $\overline{W}_n$, $n = 1, 2, \ldots$ (Fig. 4), which were introduced in [AM] (so-called positrons). Recently, in [AY1], all these infinite families were shown to be corks. Now it is a natural question to ask whether these small standard corks are sufficient to explain all exotic smooth structures on four-manifolds? (Fig. 5). For example, we know that W_1 is a cork of the blown-up Kummer surface $E(2)\#\overline{\mathbb{CP}}^2$ [A1], and $\overline{W}_1$ is a cork of the Dolgachev surface $E(1)_{2,3}$ [A2], where $E(n)$ is the elliptic surface of signature $-8n$.

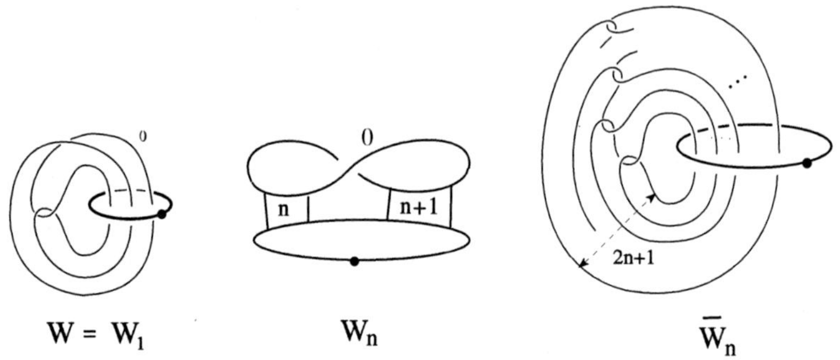

Fig. 4 Variety of corks

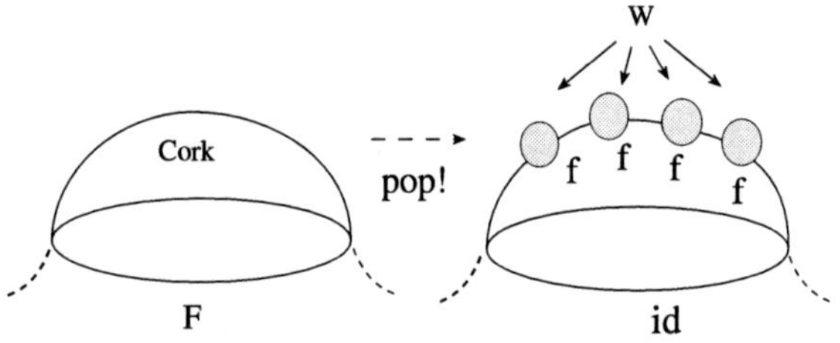

Fig. 5 Do all corks decompose into standard corks?

We require corks to be Stein manifolds in order to rule out trivial examples, as well as to introduce rigidity in their structures. For example, a theorem of Eliashberg says that if a Stein manifold has boundary S^3 or $S^1 \times S^2$, then it has to be a B^4 or $S^1 \times B^3$, respectively.

1.1 How to Recognize a Cork

In general, it is hard to recognize when a codimension-zero contractible submanifold $W \subset M^4$ is a cork of M. In fact, all the corks obtained in the general cork decomposition theorem of [M] have the property that $W \cup -W = S^4$ and $W \cup_f -W = S^4$. So it is easy to embed W's into charts of M without being corks of M. One quick way of showing that (W, f) is a cork of a manifold M with nontrivial Seiberg–Witten invariants is to show that the change $M \rightsquigarrow M'$ in (1) gives a split manifold M', implying zero (or different) Seiberg–Witten invariants. In [AY1] and [A2], many interesting corks were located using this strategy.

There are also some hard-to-calculate algebraic ways of checking whether $W \subset M$ is a cork, provided that we know the Heegard–Floer homology groups of the boundary of W [OS]. This follows from the computation of the Ozsváth–Szabó 4-manifold invariant, i.e., by first removing two B^4's from M as shown in Fig. 6, and computing a certain trace of the induced map on the Floer homology of the two S^3 boundary components (induced from the cobordism). For example, we have the following theorem.

Theorem 2. *([AD]): Let $M = N \cup_{\partial} W$ be a cork decomposition of a smooth closed four-manifold, where W is the Mazur manifold and $b_2^+(M) > 1$ (the union is along the common boundary Σ). Let N_0 be the cobordism from S^3 and Σ obtained from N by removing a B^4 from its interior. Then $Q' = N \cup_{f} W$ is a fake copy of Q if the image of the "mix map" $F^{\mathrm{mix}}_{(N_0,s)}$ (defined in [OS]) lies in T_0^+ for some* Spin^c *structure s.*

$$F^{\mathrm{mix}}_{(N_0,s)} : HF^-(S^3) \to HF^+(\Sigma) \cong T_0^+ \oplus \mathbf{Z}_{(0)} \oplus \mathbf{Z}_{(0)}.$$

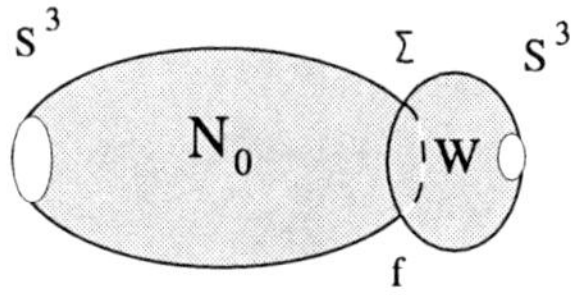

Fig. 6 Cobordism to compute Ozsvath-Szabo invariant

1.2 *Constructing Exotic Manifolds from Corks*

By thickening a cork in two different ways one can obtain absolutely exotic manifold pairs (i.e., homeomorphic but not diffeomorphic manifolds). Here is a quick review of [A3]: Let (W, f) be the Mazur cork. Then its involution $f : \partial W \to \partial W$ has an amazing property: There is a pair of loops α, β with the following properties:

- $f(\alpha) = \beta$.
- $M := W +$ (2-handle to α with -1 framing) is a Stein manifold.
- β is slice in W; hence $M' := W +$ (2-handle to β with -1 framing) contains an embedded (-1)-sphere.

So M' is an absolutely exotic copy of M; if not, M' would be a Stein manifold also, but any Stein manifold compactifies into an irreducible symplectic manifold [LM], contradicting the existence of the smoothly embedded (-1)-sphere (Fig. 7).

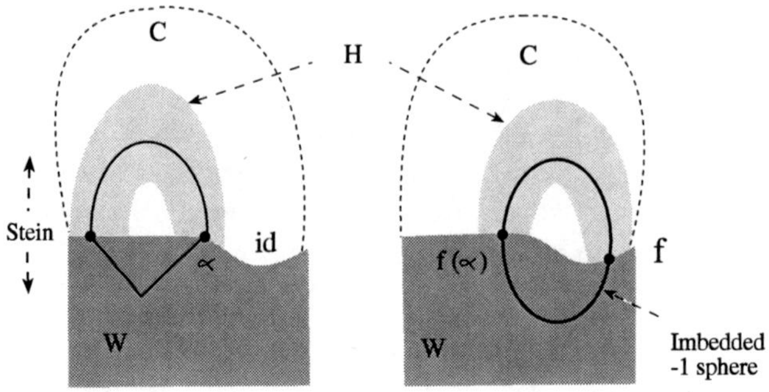

Fig. 7 Inflating a cork to exotic manifold pairs

Interestingly, by handle slides one can show that each of M, M' is obtained by attaching a 2-handle to B^4 along a knot, as shown in Fig. 8 [A3]. The reader should contrast this with [A5], where examples of other knot pairs $K, L \subset S^3$ are given (one is a slice; the other is not a slice) such that attaching 2-handles to B^4 along K, L gives diffeomorphic four-manifolds, as in Fig. 9.

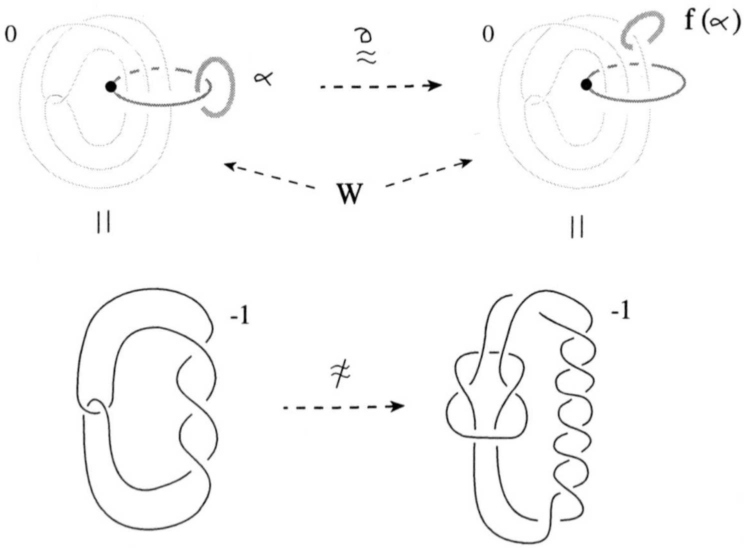

Fig. 8 Exotic manifold pairs

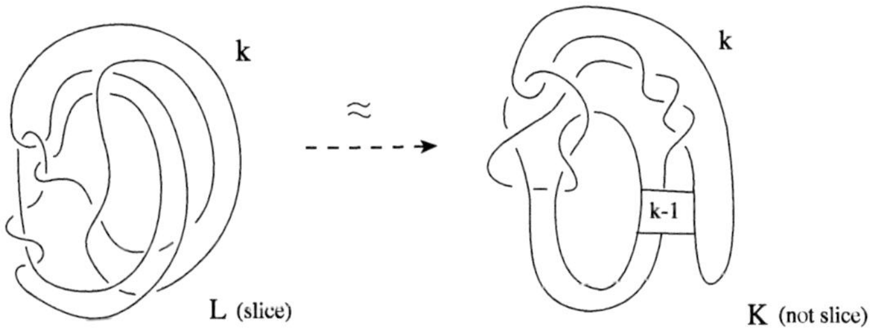

Fig. 9 Diffeomorphic manifold pairs

2 PALFs

It turns out that Stein manifolds admit finer structures as primitives. They are "positive allowable Lefschetz fibrations" over the 2-disk, where the regular fibers are surfaces F with boundaries; in [AO1], we called them PALFs. Here "allowable" means that the monodromies of Lefschetz singularities over the singular points are products of positive Dehn twists along nonseparating loops (this last condition is not a restriction; it comes for free from the proofs).

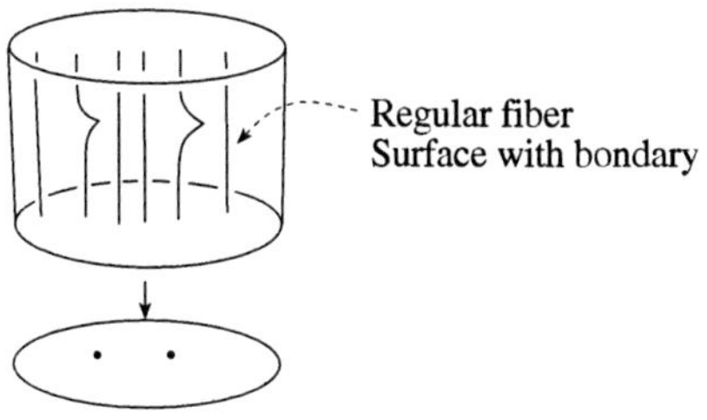

Fig. 10 PALF

So PALFs are certain topological structures underlying a Stein manifold: Stein = |PALF|. Existence of these structure on Stein manifolds was first proven in [LP]; later, in [AO1], a constructive topological proof along with its converse was given, thereby establishing the following result.

Theorem 3. *There is a surjection*

$$\{\text{PALFs}\} \Longrightarrow \{\text{Stein manifolds}\}.$$

Here, by Eliashberg's characterization [G], a *Stein manifold* means a handlebody consisting of 1- and 2-handles, where the 2-handles are attached along a Legendrian framed link, with each of its components K framed with $\text{tb}(K) - 1$ framing.

The proof that a PALF $\mathcal{F}$ gives a Stein manifold $|\mathcal{F}|$ goes as follows: By [Ka], $\mathcal{F}$ is obtained by starting with the trivial fibration $X_0 = F \times B^2 \to B^2$ and attaching a sequence of 2-handles to the curves $k_i \subset F$, $i = 1,2,\ldots$, on the fibers, with framing one less than the page framing: $X_0 \rightsquigarrow X_1 \rightsquigarrow \cdots \rightsquigarrow X_n = \mathcal{F}$. On an $F \times B^2$, we start with the standard Stein structure and assume that to the contact boundary F there is a convex surface with the "dividing set" ∂F [T]. Then by applying the "Legendrian realization principle" of [H], after an isotopy we make the surface framings of $k_i \subset F$ into the Thurston–Bennequin framings, and then the result follows from Eliashberg's theorem.

Conversely, to show that a Stein manifold W admits a PALF (here we indicate the proof only for the case in which there are no 1-handles), we isotope the Legendrian framed link to square bridge position (by turning each component counterclockwise 45°), and put the framed link on a fiber F of the (p,q) torus knot L as indicated in Fig. 11. This gives a PALF structure on $B^4 = |\mathcal{F}|$. Attaching handles to this framed link has the effect of enhancing the monodromy of the (p,q) torus knot by the Dehn twist along them, resulting in a bigger PALF. An improved version of this theorem is given in [Ar].

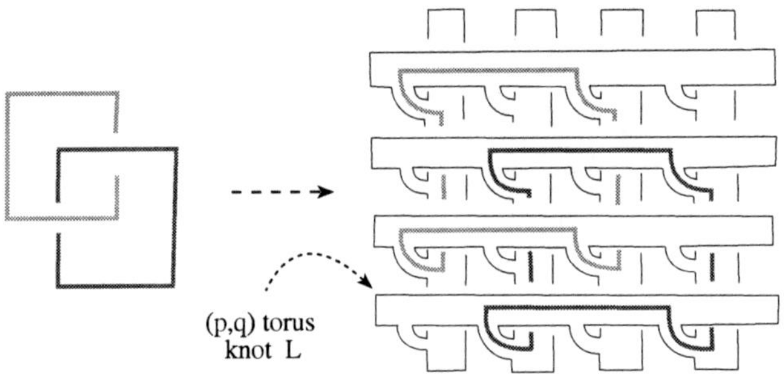

Fig. 11 Surgery on a framed link induces Dehn twists on the page

A PALF structure $\mathcal{F}$, like a triangulation or handlebody structure on a smooth manifold, should be viewed as an auxiliary topological structure on a Stein manifold $X = |\mathcal{F}|$. On the boundary, a PALF gives an open book compatible with the induced contact manifold $\partial X = |\partial \mathcal{F}|$. Usually, geometric structures come as primitives of topological structures: topology = |geometry|, such as the real algebraic structures or complex structures on a smooth manifold; but surprisingly, in this case the roles are reversed: geometry = |topology|. For example, B^4 has a unique Stein structure, whereas it has infinitely many PALF structures corresponding to fibered links.

Choosing an underlying PALF is often useful for solving problems in Stein manifolds. A striking application of this principle was the approach in [AO2] to the compactification problem of Stein manifolds, which was later strengthened by [E] and [Et]. The problem of compactifying a Stein manifold W into a closed symplectic manifold was first solved in [LM]. Then in [AO2], an algorithmic solution was given using PALFs. An analogous case involves compactifying the interior of a compact smooth manifold to a closed manifold by first choosing a handlebody on W, then canonically closing it up by attaching dual handles (doubling).

In the symplectic case, we first choose a PALF on $W = |\mathcal{F}|$, then attach a 2-handle to the binding of the open book on the boundary (Fig. 12), thereby obtaining a closed surface F^*-bundle over the 2-disk with monodromy a product of positive Dehn twists $\alpha_1 \cdot \alpha_2 \cdots \alpha_k$. We then extend this fibration $\mathcal{F}$ by doubling monodromies $\alpha_1 \cdot \alpha_2 \cdots \alpha_k \cdots \alpha_k^{-1} \cdots \alpha_1^{-1}$ (i.e., attaching corresponding 2-handles) and capping off with $F^* \times B^2$ on the other side. We do this after converting each negative Dehn twist α_i^{-1} in this expression into products of positive Dehn twists using the relation $(a_1 b_1 \cdots a_g b_g)^{4g+2} = 1$ among the standard Dehn twist generators of the surface F^* of genus g (cf. [AO2]).

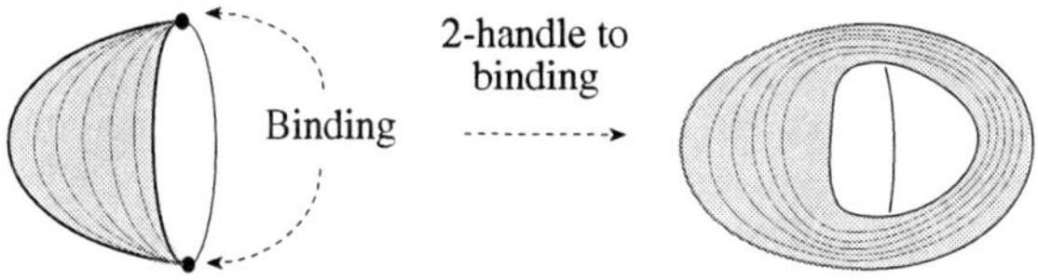

Fig. 12 Adding a 2-handle to a binding

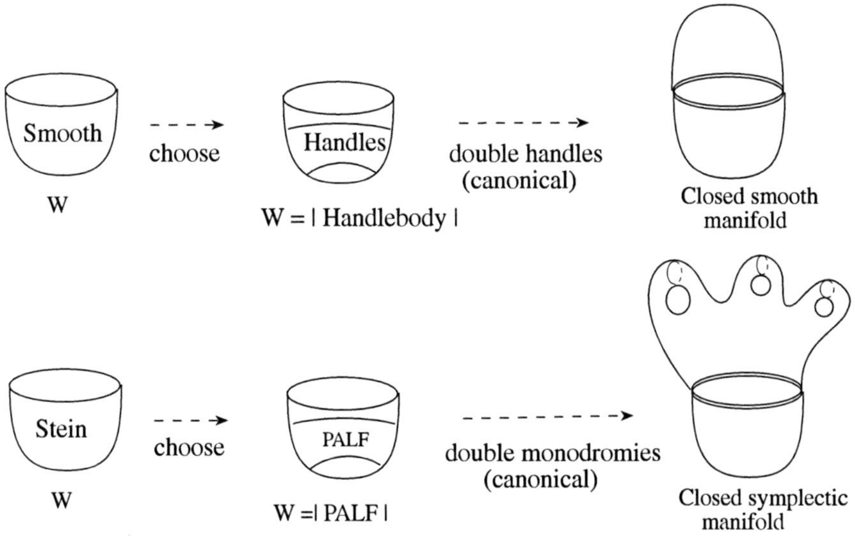

Fig. 13 Natural compactification after choosing an auxiliary structure

Choosing a PALF on W makes this an algorithmic canonical process. Even in the case of the Stein ball B^4, using different PALF structures on B^4 we get a variety of different symplectic compactifications of B^4. For example, $B^4 = |\mathcal{U}| \rightsquigarrow S^2 \times S^2$, where $\mathcal{U}$ is the trivial PALF $D^2 \times D^2 \to D^2$, whereas $B^4 = |\mathcal{T}| \rightsquigarrow K3$ Surface, where $\mathcal{T}$ is the PALF induced by the trefoil knot.

3 BLFs

Lack of a uniqueness result in the "cork decompositions" and the differing orientations of the two Stein pieces obtained from Theorem 1 make it hard to define four-manifold invariants. In [ADK], a more general version of the Lefschetz fibration (or pencil) structure on four-manifolds is introduced, namely the "broken Lefschetz fibration" BLF (or broken Lefschetz pencil BLP), where Lefschetz fibrations or pencils $\pi : X^4 \to S^2$ (with closed surfaces as regular fibers) are allowed to have circle singularities; that is, on a neighborhood of some circles, π can look like a map $S^1 \times B^3 \to \mathbb{R}^2$ given by $(t, x_1, x_2, x_3) \mapsto (t, x_1^2 + x_2^2 - x_3^2)$ (otherwise, it is a Lefschetz fibration). In [ADK], using analytic techniques, it was shown that every four-manifold X with $b_+^2 > 0$ is a BLP. Also, after a useful partial result in [GK], in [L] and [AK] two independent proofs that all four-manifolds are BLFs were given; the first proof uses singularity theory, and the second uses handlebody theory (in [B2], another singularity approach is employed, resulting in a weaker version of [L]). In [P] there is an approach using BLFs to construct four-manifold invariants.

The proof in [AK] proceeds along the lines of Theorems 1 and 3, discussed earlier. Roughly, it goes as follows: First define ALFs, which are weaker versions of PALFs. They are "achiral Lefschetz fibrations" over the 2-disk with bounded fibers, where we allow Lefschetz singularities to have monodromies that are negative Dehn twists. From the proof in [AO1], we get the surjection

$$\{\text{ALFs}\} \Longrightarrow \{\text{almost Stein manifolds}\}.$$

Here an *almost Stein manifold* means a handlebody consisting of 1- and 2-handles, where the 2-handles are attached to a Legendrian framed link, with each component K framed with $\mathrm{tb}(K) \pm 1$ framing (it turns out that every 4-dimensional handlebody consisting of 1- and 2-handles has this nice structure).

First, we make a tubular neighborhood X_2 of any embedded surface in X a "concave BLF" (e.g., [GK]). A *concave BLF* means a BLF with 2-handles attached to circles transversal to the pages on the boundary (open book), as indicated in Fig. 15 (so a concave BLF fibers over the whole S^2, with closed-surface regular fibers on one hemisphere, and 2-disk regular fibers on the other hemisphere). Also we make sure that the complement $X_1 = X - X_2$ has only 1- and 2-handles; hence it is an ALF. Applying [Gi], we ensure that the boundary open books induced from each side X_i, $i = 1, 2$, match. So we have a (matching) union $X = X_1 \cup X_2$ consisting of an ALF X_1 and concave BLF X_2. This would have made X a BLF had X_1 been a PALF.

Now comes the crucial point. In analogy to obtaining a butterfly by drilling into its cocoon, we will turn the ALF X_1 into a PALF by removing a disk from it, i.e., we will apply the positron move of Theorem 1: For each framed knot representing a 2-handle of X_1, we pick an unknot (K in Fig. 14) with the properties that (1) it lies on a page and (2) it links that framed knot twice, as in Fig. 14. These conditions allow us to isotope K to the boundary of a properly embedded 2-disk in X_1 meeting each fiber of the ALF once (here K is isotoped to the meridian of the binding curve). Therefore, carving out the tubular neighborhood of this disk from X_1 preserves the ALF structure (this is indicated by putting a dot in K in the figure, a notation from [A5]), where the new fibers are obtained by puncturing the old ones. But this magically changes the framings of each framed knot by $\text{tb}(K)+1 \rightsquigarrow \text{tb}(K)-1$; hence after carving out X_1, we end up with a PALF. On the other, X_2, side, this carving corresponds to attaching 2-handles to X_2 to circles transverse to fibers, so the enlarged X_2 is still a concave BLF, and so the open books of each side still match.

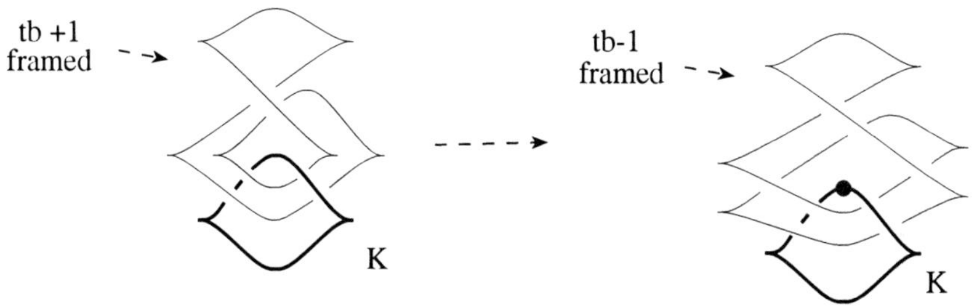

Fig. 14 Turning an ALF into a PALF by carving

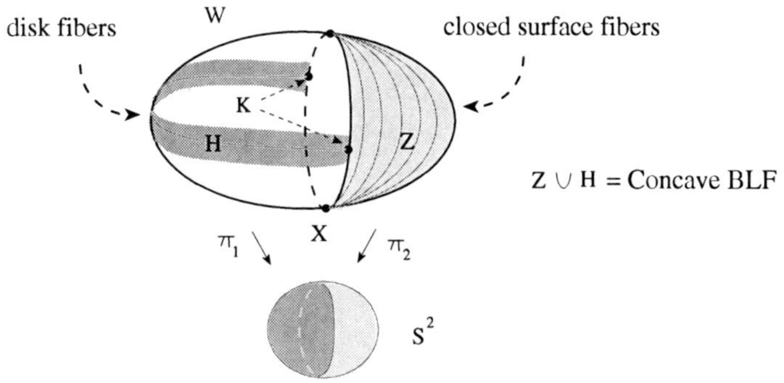

Fig. 15 Extending to concave BLF

4 Plugs

To understand exotic structures of four-manifolds better, recently in [AY1], Yasui and I started to search for corks in the known examples of exotic four-manifolds, since we knew theoretically that they exist [AY1, Sect. 0]. From this endeavor we learned two important lessons: First, as in the case of corks, there are differently behaving codimension-zero submanifolds that are also responsible for the exoticness of four-manifolds (we named them *plugs*). Second, the position of corks and plugs in four-manifolds plays an important role. For example, it helps us to construct exotic Stein manifolds in Theorem 4, whose existence had eluded us for a long time. In some sense, plugs generalize the *Gluck twisting* operation, just as corks generalize the Mazur manifold. The rest of this section is a brief summary of [AY1, AY2].

Definition 2. A plug is a pair (W, f), where W is a compact Stein manifold and $f : \partial W \to \partial W$ is an involution that does not extend to a self-homeomorphism of W and such that there is the decomposition (1) for some exotic copy M' of M.

Plugs might be deformations of corks (to deform corks into each other, we might have to go through plugs). We can think of corks and plugs as freely moving particles in four-manifolds (Fig. 21), like fermions and bosons in physics, functioning like little knobs on a wall to turn on and off the ambient exotic lights.

An example of a plug that frequently appears in four-manifolds is $W_{m,n}$, where $m \geq 1, n \geq 2$ (Fig. 16). By canceling the 1-handle and the $-m$-framed 2-handle, we see that $W_{m,n}$ is obtained from B^4 by attaching a 2-handle to a knot with $-2n - n^2m^2$ framing. The involution f is induced from the symmetric link.

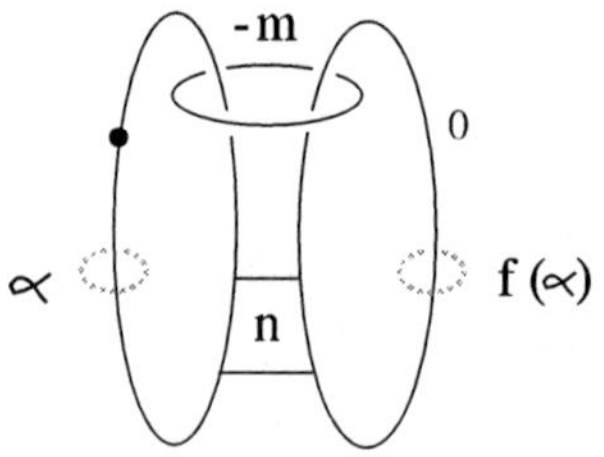

Fig. 16 $W_{m,n}$

Here the degenerate case is also interesting. Removing $W_{1,0}$ and gluing with f corresponds to the Gluck twisting operation. Previously, we knew only one example of an exotic manifold that is obtained from the standard one by the Gluck operation, and that manifold is nonorientable [A4].

Observe that if this obvious involution $f : \partial W_{m,n} \to \partial W_{m,n}$ extended to a homeomorphism, we would get homeomorphic manifolds $W^1_{m,n}$ and $W^2_{m,n}$, obtained by attaching 2-handles to α and $f(\alpha)$ with -1 framings, respectively. But $W^2_{m,n}$ and $W^1_{m,n}$ have the following nonisomorphic intersection forms, which is a contradiction:

$$\begin{pmatrix} -2n - mn^2 & 1 \\ 1 & -1 \end{pmatrix}, \quad \begin{pmatrix} -2n - mn^2 & -1 - mn \\ -1 - mn & -1 - m \end{pmatrix}.$$

The following theorem implies that $(W_{m,n}, f)$ is a plug, and also it says that this plug can be inflated to exotic Stein manifold pairs.

Theorem 4. *([AY2]) The simply connected Stein manifolds shown in Fig. 17 are exotic copies of each other.*

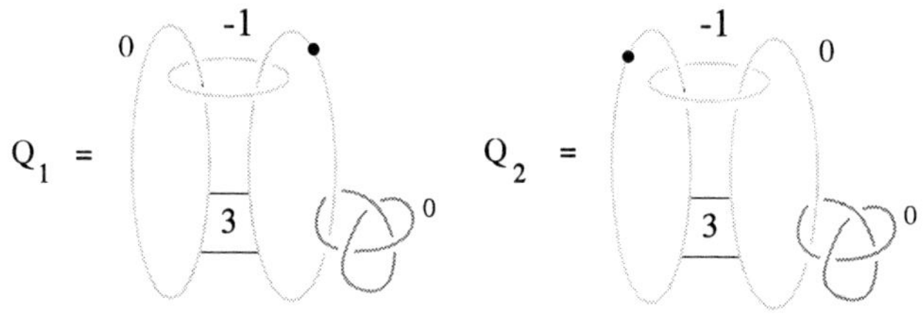

Fig. 17 Inflating a plug to an exotic Stein manifold pair

Notice that the transformation $Q_1 \rightsquigarrow Q_2$ is obtained by twisting along the plug $(W_{1,3}, f)$ inside. It is easy to check that both are Stein manifolds, and clearly the boundaries of Q_1 and Q_2 are diffeomorphic, and the boundary diffeomorphism extends to a homotopy equivalence inside, since they have isomorphic intersection forms $(-1) \oplus (1)$; cf. [Bo]. Hence by Freedman's theorem they are homeomorphic. The fact that they are not diffeomorphic follows from an interesting embedding $Q_1 \subset E(2)\#2\overline{\mathbb{CP}}^2$ (so the positions of plugs are important!), where the two homology generators $\langle x_1, x_2 \rangle$ of $H_2(Q_1)$ intersect the basic class $K = \pm e_1 \pm e_2$ of $E(2)\#2\overline{\mathbb{CP}}^2$ with $x_i \cdot e_j = \delta_{ij}$, (here e_j are the two $\overline{\mathbb{CP}}^1$ factors). Now by applying the adjunction inequality we see that there is no embedded torus of self-intersection zero in Q_1, whereas there is one in Q_2.

We can also inflate corks to exotic Stein manifold pairs (compare these examples to the construction in Sect. 1.2).

Theorem 5. *([AY2]) The simply connected Stein manifolds shown in Fig. 18 are exotic copies of each other.*

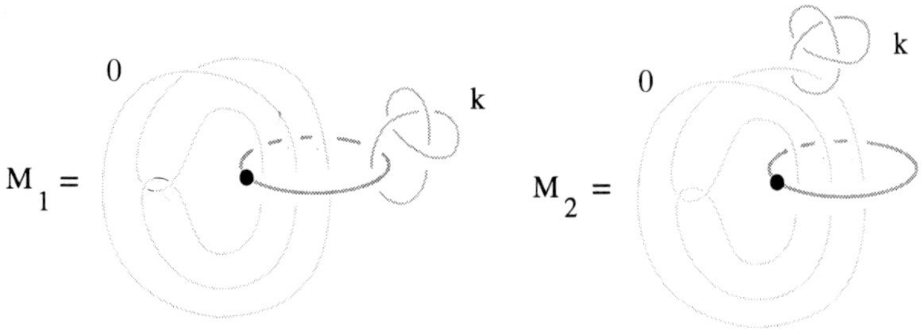

Fig. 18 Inflating a cork to an exotic Stein manifold pair ($k \leq 0$)

The proof of this is similar to the previous theorem. The crucial point is finding an embedding of M_1 into a useful closed four-manifold with nontrivial Seiberg–Witten invariant. Note that by a result of Eliashberg, the only Stein filling of S^3 is B^4. The existence of simply connected exotic Stein manifold pairs was established recently in [AEMS] using the technique of knot surgery. Now a natural question is this: Are all exotic structures on four-manifolds induced from the corks W_n, $\bar{W}_n$ and plugs $W_{m,n}$? Here is a result of some recent searches.

Theorem 6. *([AY1]) For $k,r \geq 1$, $n,p,q \geq 2$, and $\gcd(p,q) = 1$, we have the following:*

- $E(2k)\#\bar{\mathbf{CP}}^2$ *has corks* (W_{2k-1}, f_{2k-1}) *and* (W_{2k}, f_{2k})*;*
- $E(2k)\#r\bar{\mathbf{CP}}^2$ *has plugs* $(W_{r,2k}, f_{r,2k})$ *and* $(W_{r,2k+1}, f_{r,2k+1})$*;*
- $E(n)_{p,q}\#\bar{\mathbf{CP}}^2$ *has cork* W_1*, and plug* $W_{1,3}$*;*
- $E(n)_K\#\bar{\mathbf{CP}}^2$ *has cork* W_1*, and plug* $W_{1,3}$*;*
- *Yasui's exotic* $E(1)\#\bar{\mathbf{CP}}^2$ *in [Y] has cork* W_1*.*

An interesting question is whether any two cork embeddings $(W,f) \subset M$ are isotopic to each other. Put another way, can you knot corks inside of four-manifolds? It turns out that there are indeed such knotted corks [AY1]. For example, there are two nonisotopic cork embeddings $(W_4, f) \subset M = \mathbb{CP}^2\#14\bar{\mathbb{CP}}^2$. It is also possible to knot some corks in infinitely many different ways [AY3]. This is proved by calculating the change in the Seiberg–Witten invariants of the two manifolds obtained by twisting M along the two embedded corks and getting different values. This calculation uses the techniques of [Y].

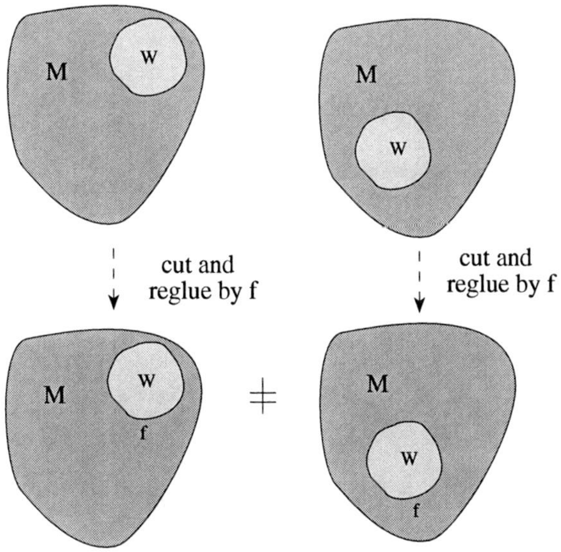

Fig. 19 Nonisotopic corks

Another natural question arises: Since every exotic copy of a closed 4-manifold can be explained by cork twisting, and there are many ways of constructing exotic copies of four-manifolds, e.g., logarithmic transform, rational blowing down, knot surgery operations [FS1, FS2, GS], are there ways of linking all these constructions to corks? In some cases this can be done for the rational blowing-down operation $X \rightsquigarrow X_{(p)}$, by showing that $X_{(p)}\#(p-1)\bar{\mathbf{CP}}^2$ is obtained from X by a cork twisting along some $W_n \subset X$ [AY1]. It is already known that there is a similar relation between logarithmic transforms and the rational blowings down. The difficult remaining case seems to be the problem of relating a general knot surgery operation to cork twisting.

Remark 1. We don't know whether an exotic copy of a manifold with boundary differs from its standard copy by a cork. It is likely that there is a relative version of the cork theorem. Perhaps the most interesting example to check is the exotic cusp of [A6], which is the smallest example of a simply connected exotic smooth manifold we know that requires 1- or 3-handles in any handlebody decomposition, whereas its standard copy has only 2-handles [AY1].

Acknowledgments The author is partially supported by NSF grant DMS 9971440 and IMBM.

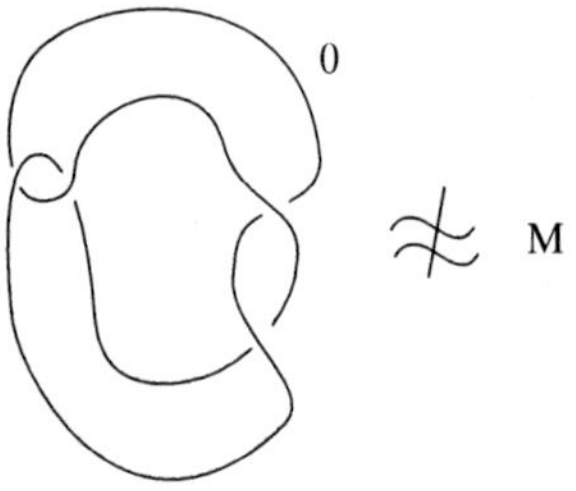

Fig. 20 The cusp and its fake copy

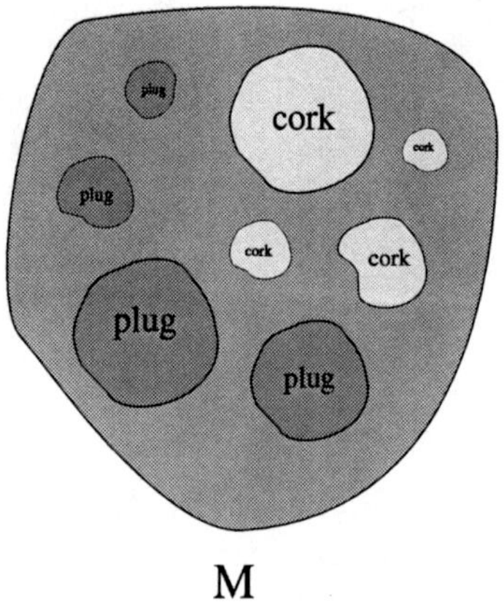

Fig. 21 Zoo of corks and plugs in a four-manifold

References

[A1] S. Akbulut, A Fake compact contractible 4-manifold, *Journ. of Diff. Geom.* 33 (1991), 335–356.

[A2] S. Akbulut, The Dolgachev Surface, arXiv:0805.1524.

[A3] S. Akbulut, An exotic 4-manifold, *Journ. of Diff. Geom.* 33 (1991), 357–361.

[A4] S. Akbulut, Constructing a fake 4-manifold by Gluck construction to a standard 4-manifold, *Topology* 27:2 (1988), 239–243.

[A5] S. Akbulut, On 2-dimensional homology classes of 4-manifolds, *Math. Proc. Camb. Phil. Soc.* 82 (1977), 99–106

[A6] S. Akbulut, A fake cusp and a fishtail, *Turkish Jour. of Math.* 1 (1999), 19–31, arXiv:math.GT/9904058.

[AD] S. Akbulut and S. Durusoy, An involution acting nontrivially on Heegaard–Floer homology, in *Geometry and Topology of Manifolds* Fields Inst. Commun. 47, , pp. 1–9. Amer. Math. Soc., Providence, RI, 2005.

[AK] S. Akbulut and C. Karakurt, Every 4-manifold is BLF, *Jour. of GGT* 2 (2008), 83–106, arXiv:0803.2297.

[AM] S. Akbulut and R. Matveyev, A convex decomposition theorem for 4-manifolds, *Internat. Math. Res. Notices* (1998), no. 7, 371–381.

[AO1] S. Akbulut and B. Ozbagci, Lefschetz fibrations on compact Stein surfaces, *Geometry and Topology* 5 (2001), 319–334.

[AO2] S. Akbulut and B. Ozbagci, On the topology of compact Stein surfaces, *IMRN* 15 (2002), 769–782.

[AY1] S. Akbulut and K. Yasui, Corks, plugs and exotic structures, *Jour. of GGT* 2 (2008), 40–82, arXiv:0806.3010v2.

[AY2] S. Akbulut and K. Yasui, Small exotic Stein manifolds, Comm. Math. Helvetici, 85(3) (2010), 705–721.

[AY3] S. Akbulut and K. Yasui Knotting corks, Jour. of Topology, 2(4) (2009), 823–839.

[ADK] D. Auroux, S. K. Donaldson, and L.Katzarkov, Singular Lefschetz pencils, *GT* 9 (2005), 1043–1114.

[AEMS] A. Akhmedov, J. B. Etnyre, T. E. Mark, and I. Smith, A note on Stein fillings of contact manifolds, arXiv:0712.3932.

[Ar] M. F. Arikan, On support genus of a contact structure, *Journal of Gökova Geometry Topology*, 1 (2007), 96–115.

[B1] I. Baykur, Kähler decomposition of 4-manifolds, arXiv:0601396v2

[B2] I. Baykur, Existance of broken Lefschetz fibrations, arXiv: 0801.3139.

[Bo] S. Boyer, Simply-connected 4-manifolds with a given boundary, *Trans. Amer. Math. Soc.* 298:1 (1986), 331–357.

[CFHS] C. L. Curtis, M. H. Freedman, W. C. Hsiang, and R. Stong, A decomposition theorem for h-cobordant smooth simply-connected compact 4-manifolds, *Invent. Math.* 123:2 (1996), 343–348.

[E] Y. Eliashberg, Few remarks about symplectic filling, arXiv:math/0311459.

[Et] J. Etnyre, On symplectic fillings, arXiv:math.SG/0312091.pdf (2003).

[FS1] R. Fintushel and R. J. Stern, Rational blowdowns of smooth 4-manifolds, *J. Differential Geom.* 46:2 (1997), 181–235.

[FS2] R. Fintushel and R. J. Stern, Knots, Links, and 4-Manifolds, *Invent. Math.* 134:2 (1998), 363–400.

[GK] D. Gay and R. Kirby, Constructing Lefschetz-type fibrations on four-manifolds, *Geom. Topol.* 11 (2007), 2075–2115.

[G] R. Gompf, Handlebody contruction of Stein surfaces, *Ann. of Math.* 148, (1998), 619–693.

[Gi] E. Giroux, Contact geometry: from dimension three to higher dimensions, in *Proceedings of the International Congress of Mathematicians (Beijing 2002)*, pp. 405–414.

[GS] R. E. Gompf and A. I. Stipsicz, *4-Manifolds and Kirby Calculus*, Graduate Studies in Mathematics 20. American Mathematical Society, 1999.

[H] K. Honda, On the classification of tight contact structures I, *Geometry and Topology* 4 (2000), 309–368.

[Ka] A. Kas, On the handlebody decomposition associated to a Lefschetz fibration, *Pacific J. Math.* 89 (1980), 89–104.

[K] R. Kirby, Akbulut's corks and h-cobordisms of smooth, simply connected 4-manifolds, *Turkish J. Math.* 20:1 (1996), 85–93.

[L] Y. Lekili, Wrinkled fibrations on near-symplectic manifolds, arXiv:0712.2202.

[LM] P. Lisca and G. Matic, Tight contact structures and Seiberg–Witten invariants, *Invent. Math.* 129 (1997), 509–525.

[LP] A. Loi and R. Piergallini, Compact Stein surfaces with boundary as branched covers of B^4, *Invent. Math.* 143 (2001) 325–348, math.GT/0002042.

[M] R. Matveyev A decomposition of smooth simply-connected h-cobordant 4-manifolds, *J. Differential Geom.* 44:3 (1996), 571–582.

[OS] I. Ozsváth and Z. Szabó, Holomorphic triangles and invariants for smooth four-manifolds, arXiv:math/0110169v2.

[P] T. Perutz, Lagrangian matching invariants for fibred four-manifolds: I, arXiv:0606061v2.

[T] I. Torisu, Convex contact structures and fibered links in 3-manifolds, *Int. Math. Research Notices* 9 (2000), 441–454.

[Y] K. Yasui, Exotic rational elliptic surfaces without 1-handles, *Algebraic and Geometric Topology* 8 (2008) 971–996.

Rational Linking and Contact Geometry

Kenneth Baker and John Etnyre

This paper is dedicated to Oleg Viro on the occasion of his 60th birthday

Abstract In the note we study Legendrian and transverse knots in rationally null-homologous knot types. In particular, we generalize the standard definitions of self-linking number, Thurston–Bennequin invariant, and rotation number. We then prove a version of Bennequin's inequality for these knots and classify precisely when the Bennequin bound is sharp for fibered knot types. Finally, we study rational unknots and show that they are weakly Legendrian and transversely simple.

Keywords Legendrian knot • Transverse knot • Contact geometry • Self-linking • Open book

In this note we extend the self-linking number of transverse knots and the Thurston–Bennequin invariant and rotation number of Legendrian knots to the case of rationally null-homologous knots. This allows us to generalize many of the classical theorems concerning Legendrian and transverse knots (such as the Bennequin inequality) as well as to put other theorems in a more naturalcontext (such as the result in [10] concerning exactness in the Bennequin bound). Moreover, due

K. Baker
Department of Mathematics, University of Miami, PO Box 249085, Coral Gables, FL 33124-4250, USA
e-mail: k.baker@math.miami.edu

J. Etnyre (✉)
School of Mathematics, Georgia Institute of Technology, 686 Cherry St., Atlanta, GA 30332-0160, USA
e-mail: etnyre@math.gatech.edu

I. Itenberg et al. (eds.), *Perspectives in Analysis, Geometry, and Topology: On the Occasion of the 60th Birthday of Oleg Viro*, Progress in Mathematics 296, DOI 10.1007/978-0-8176-8277-4_2,

to recent work on the Berge conjecture [3] and surgery problems in general, it has become clear that one should consider rationally null-homologous knots even when studying classical questions about Dehn surgery on knots in S^3. Indeed, the Thurston–Bennequin number of Legendrian rationally null-homologous knots in lens spaces has been examined in [2]. There is also a version of the rational Thurston–Bennequin invariants for links in rational homology spheres that was previously defined and studied in [13].

We note that there has been work on relative versions of the self-linking number (and other classical invariants) in the case of general (even non-null-homologous) knots; cf. [4]. While these relative invariants are interesting and useful, many of the results considered here do not have analogous statements. So rationally null-homologous knots seems to be one of the largest classes of knots to which one can generalize classical results in a straightforward manner.

There is a well-known way to generalize the linking number between two null-homologous knots to rationally null-homologous knots; see, for example, [11]. We recall this definition of a rational linking number in Sect. 1 and then proceed to define the rational self-liking number $\mathrm{sl}_{\mathbb{Q}}(K)$ of a transverse knot K and the rational Thurston–Bennequin invariant $\mathrm{tb}_{\mathbb{Q}}(L)$ and rational rotation number $\mathrm{rot}_{\mathbb{Q}}(L)$ of a Legendrian knot L in a rationally null-homologous knot type. We also show the expected relation between these invariants of the transverse pushoff of a Legendrian knot and those of stabilizations of Legendrian and transverse knots. This leads to one of our main observations, a generalization of Bennequin's inequality.

Theorem 2.1. *Let (M,ξ) be a tight contact manifold and suppose K is a transverse knot in it of order $r > 0$ in homology. Further suppose that Σ is a rational Seifert surface of K. Then*

$$\mathrm{sl}_{\mathbb{Q}}(K) \leq -\frac{1}{r}\chi(\Sigma).$$

Moreover, if K is Legendrian, then

$$\mathrm{tb}_{\mathbb{Q}}(K) + |\mathrm{rot}_{\mathbb{Q}}(K)| \leq -\frac{1}{r}\chi(\Sigma).$$

In [10], bindings of open book decompositions that satisfy equality in the Bennequin inequality were classified. We generalize that result to the following.

Theorem 4.2. *Let K be a rationally null-homologous fibered transverse knot in a contact 3-manifold (M,ξ) such that ξ is tight when restricted to the complement of K. Denote by Σ a fiber in the fibration of $M - K$ and let r be the order of K. Then $r\,\mathrm{sl}^{\xi}_{\mathbb{Q}}(K,\Sigma) = -\chi(\Sigma)$ if and only if either ξ agrees with the contact structure supported by the rational open book determined by K or it is obtained from it by adding Giroux torsion along tori that are incompressible in the complement of K.*

A rational unknot in a manifold M is a knot K with a disk as a rational Seifert surface. One may easily check that if M is irreducible, then for M to admit a rational unknot (that is not actually an unknot), it must be diffeomorphic to a lens space.

Theorem 5.1. *Rational unknots in tight contact structures on lens spaces are weakly transversely simple and Legendrian simple.*

In Sect. 5 we also given an example of the classification of Legendrian rational unknots (and hence transverse rational unknots) in $L(p,1)$ when p is odd. The classification of Legendrian and transverse rational unknots in a general lens space can easily be worked out in terms of the classification of tight contact structures on the given lens space. The example we give illustrates this.

In Sect. 6, we briefly discuss the generalization of our results to the case of links.

1 Rational Linking and Transverse and Legendrian Knots

Let K be an oriented knot of $\mathbb{Z}$-homological order $r > 0$ in a 3-manifold M and denote a tubular neighborhood of it by $N(K)$. By $X(K)$ denote the knot exterior $\overline{M \setminus N(K)}$. We fix a framing on $N(K)$. We know that half the $\mathbb{Z}$-homology of $\partial X(K)$ dies when included in the $\mathbb{Z}$-homology of $X(K)$. Since K has order r, it is easy to see that there is an embedded (r,s)-curve on $\partial X(K)$ that bounds an oriented connected surface Σ° in $X(K)$. We can radially cone $\partial\Sigma^\circ \subset \partial X(K) = \partial N(K)$ in $N(K)$ to get a surface Σ in M whose interior is embedded in M and whose boundary wraps r times around K. Such a surface Σ will be called a *rational Seifert surface* for K, and we say that K *r-bounds* Σ. We also sometime say that Σ is of *order r* along K. We also call $\Sigma \cap \partial N(K)$ the *Seifert cable of* K. Notice that Σ may have more than one boundary component. Specifically, Σ will have $\gcd(r,s)$ boundary components. We call the number of boundary components of Σ the *multiplicity* of K. Notice that Σ defines a $\mathbb{Z}$-homology chain Σ and $\partial\Sigma = rK$ in the homology 1-chains. In particular, as $\mathbb{Q}$-homology chains, $\partial(\frac{1}{r}\Sigma) = K$.

We now define the *rational linking number* of another oriented knot K' with K (and Seifert surface Σ) to be

$$\mathrm{lk}_{\mathbb{Q}}(K,K') = \frac{1}{r}\Sigma \cdot K',$$

where $\cdot$ denotes the algebraic intersection of Σ and K'. It is not hard to check that $\mathrm{lk}_{\mathbb{Q}}$ is well defined given the choice of $[\Sigma] \in H_2(X(K), \partial X(K))$. Choosing another rational Seifert surface for K representing a different relative second homology class in $X(K)$ may change this rational linking number by a multiple of $\frac{1}{r}$. To emphasize this, one may prefer to write $\mathrm{lk}_{\mathbb{Q}}((K,[\Sigma]),K')$. Notice that if there exist rational Seifert surfaces Σ_1 and Σ_2 for which $\mathrm{lk}_{\mathbb{Q}}((K,[\Sigma_1]),K') \neq \mathrm{lk}_{\mathbb{Q}}((K,[\Sigma_2]),K')$, then K' is not rationally null-homologous.

Moreover, if K' is also rationally null-homologous, then it r'-bounds a rational Seifert surface Σ'. In $M \times [0,1]$ with Σ and Σ' thought of as subsets of $M \times \{1\}$, we can perturb them relative to the boundary to make them transverse. Then one may also check that

$$\mathrm{lk}_{\mathbb{Q}}(K,K') = \frac{1}{rr'}\Sigma \cdot \Sigma'.$$

From this one readily sees that the rational linking number of rationally null-homologous links is symmetric.

1.1 Transverse Knots

Let (M,ξ) be a contact 3-manifold (with orientable contact structure ξ) and K a (positively) transverse knot. Given a rational Seifert surface Σ for K with $\partial\Sigma = rK$, we can trivialize ξ along Σ. More precisely, we can trivialize the pullback $i^*\xi$ to Σ, where $i : \Sigma \to M$ is the inclusion map. Notice that the inclusion map restricted to $\partial\Sigma$ is an r-fold covering map of $\partial\Sigma$ to K. We can use the exponential map to identify a neighborhood of the zero section of $i^*\xi|_{\partial\Sigma}$ with an r-fold cover of a tubular neighborhood of K. Let v be a nonzero section of $i^*\xi$. By choosing v generically and suitably small, the image of $v|_{\partial\Sigma}$ gives an embedded knot K' in a neighborhood of K that is disjoint from K. We define the *rational self-linking number* to be

$$\mathrm{sl}_{\mathbb{Q}}(K) = \mathrm{lk}_{\mathbb{Q}}(K,K').$$

It is standard to check that $\mathrm{sl}_{\mathbb{Q}}$ is independent of the trivialization of $i^*\xi$ and the section v. Moreover, the rational self-linking number depends only on the relative homology class of Σ. When this dependence is important to note, we denote the rational self-linking number by

$$\mathrm{sl}_{\mathbb{Q}}(K,[\Sigma]).$$

Just as in the case of the self-linking number one can compute it by considering the characteristic foliation on Σ. To this end we can always isotop Σ so that its characteristic foliation Σ_ξ is generic (in particular has only elliptic and hyperbolic singularities) and we denote by $e_\pm$ the number of $\pm$-elliptic singular points, and similarly, $h_\pm$ denotes the number of $\pm$-hyperbolic points.

Lemma 1.1. *Suppose K is a transverse knot in a contact manifold (M,ξ) that r-bounds the rational Seifert surface Σ. Then*

$$\mathrm{sl}_{\mathbb{Q}}(K,[\Sigma]) = \frac{1}{r}\left((e_- - h_-) - (e_+ - h_+)\right). \tag{1}$$

Proof. We begin by constructing a nice neighborhood of Σ in (M,ξ). To this end, notice that for suitably small ϵ, K has a neighborhood N that is contactomorphic to the image C_ϵ of $\{(r,\theta,z) : r \leq -\epsilon\}$ in $(\mathbb{R}^3, \ker(dz + r^2\, d\theta))$ modulo the action $z \mapsto z+1$. Let C' be the r-fold cover of C_ϵ. Taking ϵ sufficiently small, we can assume that $\Sigma \cap \partial N$ is a transverse curve T. Thinking of T as sitting in C_ϵ, we can take its lift T' to C'. Let N' be a small neighborhood of $\overline{\Sigma - (N \cap \Sigma)}$. We can glue N' to C_ϵ along a neighborhood of T to get a model neighborhood U for Σ in M. Moreover, we can glue N' to C' along a neighborhood of T' to get a contact manifold U' that will map onto U so that C' r-fold covers C_ϵ and N' in U' maps diffeomorphically to N' in U. Inside U' we have $K' = \partial\Sigma$, which r-fold covers K in U. The transverse knot K' is a null-homologous knot in U'. According to a well-known formula that easily follows by interpreting $\mathrm{sl}(K')$ as a relative Euler class, see [5], we have that

$$\mathrm{sl}(K') = (e_- - h_-) - (e_+ - h_+),$$

where $e_\pm$ and $h_\pm$ are as in the statement of the theorem. Now one easily sees that $\mathrm{sl}_{\mathbb{Q}}(K) = \frac{1}{r}\,\mathrm{sl}(K')$, from which the lemma follows. □

1.2 Legendrian Knots

Let (M,ξ) be a contact 3-manifold (with orientable contact structure ξ) and K a Legendrian knot. Choose a framing on K. Given a rational Seifert surface Σ for K, the Seifert cable of K is $K_{(r,s)}$.

The restriction $\xi|_K$ induces a framing on the normal bundle of K. Define the *(rational) Thurston–Bennequin number* of the Legendrian knot K to be

$$\mathrm{tb}_{\mathbb{Q}}(K) = \mathrm{lk}_{\mathbb{Q}}(K, K'),$$

where K' is a copy of K obtained by pushing off using the framing coming from ξ.

We now assume that K is oriented. Recall that the inclusion $i\colon \Sigma \hookrightarrow M$ is an embedding on the interior of Σ and an r-to-1 cover $\partial\Sigma \to K$. As above, we can trivialize ξ along Σ. That is, we can trivialize the pullback $i^*\xi$ to Σ. The oriented tangent vectors T_K give a section of $\xi|_K$. Thus i^*T_K gives a section of $\mathbb{R}^2 \times \partial\Sigma$. Define the *rational rotation number* of the Legendrian knot K to be the winding number of i^*T_K in $\mathbb{R}^2$ divided by r:

$$\mathrm{rot}_{\mathbb{Q}}(K) = \frac{1}{r}\mathrm{winding}(i^*T_K, \mathbb{R}^2).$$

Recall [8] that given a Legendrian knot K, we can always form the *(positive) transverse pushoff of* K, denoted by $T(K)$, as follows: the knot K has a neighborhood contactomorphic to the image of the x-axis in $(\mathbb{R}^3, \ker(dz - y\,dx))$ modulo the action $x \mapsto x+1$ such that the orientation on the knot points toward increasing x-values. The curve $\{(x,\epsilon,0)\}$ for $\epsilon > 0$ small enough will give the transverse pushoff of K.

Lemma 1.2. *If K is a rationally null-homologous Legendrian knot in a contact manifold (M,ξ), then*

$$\mathrm{sl}_{\mathbb{Q}}(T(K)) = \mathrm{tb}_{\mathbb{Q}}(K) - \mathrm{rot}_{\mathbb{Q}}(K).$$

Proof. Notice that in pulling K back to a cover U' similar to the one constructed in the proof of Lemma 1.1, we get a null-homologous Legendrian knot K'. Here we have the well-known formula (see [8])

$$\mathrm{sl}(T(K')) = \mathrm{tb}(K') - \mathrm{rot}(K').$$

One easily computes that $r\,\mathrm{sl}(T(K')) = \mathrm{sl}_{\mathbb{Q}}(T(K)), r\mathrm{tb}(K') = \mathrm{tb}_{\mathbb{Q}}(K')$ and $r\mathrm{rot}(K') = \mathrm{rot}(K)$. The lemma follows. □

We can also construct a Legendrian knot from a transverse knot. Given a transverse knot K, it has a neighborhood as constructed in the proof of Lemma 1.1. It is clear that the boundary of a sufficiently small closed neighborhood of K of the appropriate size will have a linear characteristic foliation by longitudes of K. One of the leaves in this characteristic foliation will be called a *Legendrian pushoff of K*. We note that this pushoff is not unique, but that different Legendrian pushoffs are related by negative stabilizations; see [9].

1.3 Stabilization

Recall that stabilization of a transverse and Legendrian knot is a local procedure near a point on the knot, so it can be performed on any transverse or Legendrian knot whether null-homologous or not.

There are two types of stabilization of a Legendrian knot K: positive and negative stabilization, denoted by $S_+(K)$ and $S_-(K)$, respectively. Recall that if one identifies a neighborhood of a point on a Legendrian knot with a neighborhood of the origin in $(\mathbb{R}^3, \ker(dz - y\,dx))$ so that the Legendrian knot is mapped to a segment of the x-axis and the orientation induced on the x-axis from K is toward increasing x-values, then $S_+(K)$, respectively $S_-(K)$, is obtained by replacing the segment of the x-axis by a "downward zigzag," respectively "upward zigzag"; see [8, Fig. 19]. One may similarly define stabilization of a transverse knot K, and we denote it by $S(K)$. Stabilizations have the same effect on the rationally null-homologous knots as they have on null-homologous ones.

Lemma 1.3. *Let K be a rationally null-homologous Legendrian knot in a contact manifold. Then*

$$\mathrm{tb}_{\mathbb{Q}}(S_\pm(K)) = \mathrm{tb}_{\mathbb{Q}}(K) - 1 \quad \textit{and} \quad \mathrm{rot}_{\mathbb{Q}}(S_\pm(K)) = \mathrm{rot}_{\mathbb{Q}}(K) \pm 1.$$

Let K be a rationally null-homologous transverse knot in a contact manifold. Then

$$\mathrm{sl}_{\mathbb{Q}}(S(K)) = \mathrm{sl}_{\mathbb{Q}}(K) - 2.$$

Proof. One may check that if K' is a pushoff of K by some framing $\mathcal{F}$ and K'' is the pushoff of K by a framing $\mathcal{F}''$ such that the difference between $\mathcal{F}$ and $\mathcal{F}'$ is -1, then

$$\mathrm{lk}_{\mathbb{Q}}(K, K'') = \mathrm{lk}_{\mathbb{Q}}(K, K') - 1.$$

Indeed, by noting that $r\,\mathrm{lk}_{\mathbb{Q}}(K, K')$ can easily be computed by intersecting the Seifert cable of K on the boundary of a neighborhood of K, $T^2 = \partial N(K)$, with the curve $K' \subset T^2$, the result easily follows. From this one obtains the change in $\mathrm{tb}_{\mathbb{Q}}$.

Given a rational Seifert surface Σ that is r-bounded by K, a small Darboux neighborhood N of a point $p \in K$ intersects Σ in r disjoint disks. Since the stabilization can be performed in N, it is easy to see that Σ is altered by adding r small disks, each containing a positive elliptic point and negative hyperbolic point (see [8]). The result for $\mathrm{sl}_{\mathbb{Q}}$ follows.

Finally, the result for $\mathrm{rot}_{\mathbb{Q}}$ follows by a similar argument or from the previous two results, Lemma 1.2, and the following lemma (whose proof does not explicitly use the rotation number results from this lemma). □

The proof of the following lemma is given in [9].

Lemma 1.4. *Two transverse knots in a contact manifold are transversely isotopic if and only if they have Legendrian pushoffs that are Legendrian isotopic after each has been negatively stabilized some number of times. The same statement is true with "transversely isotopic" and "Legendrian isotopic" both replaced by "contactomorphic."* □

We similarly have the following result.

Lemma 1.5. *Two Legendrian knots representing the same topological knot type are Legendrian isotopic after each has been positively and negatively stabilized some number of times.* □

While this is an interesting result in its own right, it clarifies the range of possible values for $\mathrm{tb}_{\mathbb{Q}}$. More precisely, the following result is an immediate corollary.

Corollary 1.6. *If two Legendrian knots represent the same topological knot type, then the difference in their rational Thurston–Bennequin invariants is an integer.*

2 The Bennequin Bound

Recall that in a tight contact structure, the self-linking number of a null-homologous knot K satisfies the well-known Bennequin bound

$$\mathrm{sl}(K) \leq -\chi(\Sigma)$$

for any Seifert surface Σ for K; see [6]. We have the analogous result for rationally null-homologous knots.

Theorem 2.1. *Suppose K is a transverse knot in a tight contact manifold (M,ξ) that r-bounds the rational Seifert surface Σ. Then*

$$\mathrm{sl}_{\mathbb{Q}}(K,[\Sigma]) \leq -\frac{1}{r}\chi(\Sigma). \tag{1}$$

If K is a Legendrian knot, then

$$\mathrm{tb}_{\mathbb{Q}}(K)+|\mathrm{rot}_{\mathbb{Q}}(K)| \leq -\frac{1}{r}\chi(\Sigma).$$

Proof. The proof is essentially the same as the one given in [6]; see also [7]. The first thing we observe is that if v is a vector field that directs Σ_ξ, that is, v is zero only at the singularities of Σ_ξ and points in the direction of the orientation of the nonsingular leaves of Σ_ξ, then v is a generic section of the tangent bundle of Σ and points out of Σ along $\partial\Sigma$. Thus the Poincaré–Hopf theorem implies

$$\chi(\Sigma) = (e_+ - h_+) + (e_- - h_-).$$

Adding this equality to r times equation (1) gives

$$r\,\mathrm{sl}_{\mathbb{Q}}(K,[\Sigma]) + \chi(\Sigma) = 2(e_- - h_-).$$

So if we can isotop Σ relative to the boundary so that $e_- = 0$, then we clearly have the desired inequality. Recall that if an elliptic point and a hyperbolic point of the same sign are connected by a leaf in the characteristic foliation, then they may be canceled (without introducing any further singular points).

Thus we are left to show that for every negative elliptic point we can find a negative hyperbolic point that cancels it. To this end, given a negative elliptic point p, consider the *basin* of p, that is, the closure of the set of points in Σ that limit under the flow of v in backward time to p. Denote this set by B_p. Since the flow of v goes out the boundary of Σ, it is clear that B_p is contained in the interior of Σ. Thus we may analyze B_p exactly as in [6,7] to find the desired negative hyperbolic point.

We briefly recall the main points of this argument. First, if there are repelling periodic orbits in the characteristic foliation, then add canceling pairs of positive elliptic and hyperbolic singularities to eliminate them. This prevents any periodic orbits in B_p, and thus one can show that B_p is the immersed image of a polygon that is an embedding on its interior. If B_p is the image of an embedding, then the boundary consists of positive elliptic singularities and hyperbolic singularities of either sign and flow lines between these singularities. If one of the hyperbolic singularities is negative, then we are done, since it is connected to B_p by a flow line. If none of the hyperbolic points are negative, then we can cancel them all with the positive elliptic singularities in ∂B_p, so that ∂B_p becomes a periodic orbit in the characteristic foliation and, more to the point, the boundary of an overtwisted disk. In the case that B_p is an immersed polygon, one may argue similarly; see [6, 7].

The inequality for Legendrian K clearly follows from considering the positive transverse pushoff of K and $-K$ and Lemma 1.2 together with the inequality in the transverse case. □

3 Rational Open Book Decompositions and Cabling

A *rational open book decomposition* of a manifold M is a pair (L,π) consisting of

- an oriented link L in M and
- a fibration $\pi\colon (M \setminus L) \to S^1$

such that no component of $\pi^{-1}(\theta)$ meets a component of L meridionally for any $\theta \in S^1$. We note that $\pi^{-1}(\theta)$ is a rational Seifert surface for the link L. If $\pi^{-1}(\theta)$ is actually a Seifert surface for L, then we say that (L,π) is an *open book decomposition* of M (or sometimes we will say an *integral* or *honest* open book decomposition of M). We call L the *binding* of the open book decomposition and $\pi^{-1}(\theta)$ a *page*.

The rational open book decomposition (L,π) of M *supports* a contact structure ξ if there is a contact form α for ξ such that

- $\alpha(v) > 0$ for all positively pointing tangent vectors $v \in TL$ and
- $d\alpha$ is a volume form when restricted to (the interior of) each page of the open book.

Generalizing the work of Thurston and Winkelnkemper [14], the authors of this paper in work with Van Horn-Morris proved the following result.

Theorem 3.1 (Baker et al. 2008 [1]). *Let (L,π) be any rational open book decomposition of M. Then there exists a unique contact structure $\xi_{(L,\pi)}$ that is supported by (L,π).*

It is frequently useful to deal with only honest open book decompositions. One may easily pass from a rational open book decomposition to an honest one using cables, as we now demonstrate.

Given any knot K, let $N(K)$ be a tubular neighborhood of K, choose an orientation on K and an oriented meridian μ linking K positively once, and choose some oriented framing (i.e., longitude) λ on K such that $\{\lambda,\mu\}$ give longitude-meridian coordinates on $\partial N(K)$. The (p,q)*-cable* of K is the embedded curve (or collection of curves if p and q are not relatively prime) on $\partial N(K)$ in the homology class $p\lambda + q\mu$. Denote this curve (these curves) by $K_{p,q}$. We say that a cabling of K is *positive* if the cabling coefficients have slope greater than the Seifert slope of K. (The *slope* of the homology class $p\lambda + q\mu$ is q/p.)

If K is also a transverse knot with respect to a contact structure on M, then using the contactomorphism in the proof of Lemma 1.1 between the neighborhood

$N = N(K)$ and C_ϵ for sufficiently small ϵ, we may assume that the cable $K_{p,q}$ on ∂N is also transverse. As such, we call $K_{p,q}$ the *transverse (p,q)-cable*.

If $L = K_1 \cup \cdots \cup K_n$ is a link, then we can fix framings on each component of L and choose n pairs of integers (p_i, q_i). Then after setting $(\mathbf{p},\mathbf{q}) = ((p_1,q_1),\ldots,(p_n,q_n))$, we denote by $L_{(\mathbf{p},\mathbf{q})}$ the result of (p_i,q_i)-cabling K_i for each i. It is easy to check (see, for example, [1]) that if L is the binding of a rational open book decomposition of M, then so is $L_{(\mathbf{p},\mathbf{q})}$, unless a component K_i of L is nontrivially cabled by curves of the fibration's restriction to $\partial N(K_i)$.

The following lemma says how the Euler characteristic of the fiber changes under cabling as well as the multiplicity and order of a knot.

Lemma 3.2. *Let L be a (rationally null-homologous) fibered link in M. Suppose K is a component of L for which the fibers in the fibration approach as (r,s)-curves (in some framing on K). Let L' be the link formed from L by replacing K by the (p,q)-cable of K, where $p \neq \pm 1, 0$ and $(p,q) \neq (kr,ks)$ for any $k \in \mathbb{Q}$. Then L' is fibered. Moreover, the Euler characteristic of the new fiber is*

$$\chi(\Sigma_{\text{new}}) = \frac{1}{\gcd(p,r)}\left(|p|\chi(\Sigma_{\text{old}}) + |ps - qr|(1-|p|)\right),$$

where Σ_{new} is the fiber of L' and Σ_{old} is the fiber of L. The multiplicity of each component of the cable of K is

$$\gcd\left(\frac{r}{\gcd(p,r)}, \frac{p(rq-sp)}{\gcd(p,r)\gcd(p,q)}\right)$$

and the order of Σ_{new} along each component of the cable of K is

$$\frac{r}{\gcd(p,r)}.$$

□

The proof of this lemma may be found in [1], but it follows easily by observing that one may construct Σ_{new} by taking $|\frac{p}{\gcd(p,r)}|$ copies of Σ_{old} and $|\frac{rq-sp}{\gcd(p,r)}|$ copies of meridional disks to K and connecting them via $|\frac{p(rq-sp)}{\gcd(p,r)}|$ half-twisted bands.

Now suppose we are given a rational open book decomposition (L,π) of M. Suppose K is a rational binding component of an open book (L,π) whose page approaches K in an (r,s)-curve with respect to some framing on K. (Note that $r \neq 1$ and that r is not necessarily coprime to s.) For any $l \neq s$, replacing K in L by the (r,l)-cable of K gives a new link $L_{K_{(r,l)}}$ that by Lemma 3.2 is the binding of a (possibly rational) open book for M and has $\gcd(r,l)$ new components each having order and multiplicity 1. This is called the *(r,l)-resolution of L along K*. In the resolution, the new fiber is created using just one copy of the old fiber following the construction of the previous paragraph. Thus after resolving L along the other rational binding components, we have a new fibered link L' that is the binding of an integral open book (L',π'). This is called an *integral resolution of L*.

If we always choose the cabling coefficients (r,l) to have slope greater than the original coefficients (r,s), then we say that we have constructed a *positive (integral) resolution of L*.

Theorem 3.3 (Baker et al. 2008 [1]). *Let (L,π) be a rational open book for M supporting the contact structure ξ. If L' is a positive resolution of L, then L' is the binding of an integral open book decomposition for M that also supports ξ.*

4 Fibered Knots and the Bennequin Bound

Recall that in [10], null-homologous (nicely) fibered links satisfying the Bennequin bound were classified. In particular, the following theorem was proven.

Theorem 4.1 (Etnyre and Van Horn-Morris, 2008 [10]). *Let L be a fibered transverse link in a contact 3-manifold (M,ξ) and assume that ξ is tight when restricted to $M \setminus L$. Moreover, assume that L is the binding of an (integral) open book decomposition of M with page Σ. Then $\mathrm{sl}_\xi(L,\Sigma) = -\chi(\Sigma)$ if and only if either*

1. *ξ is supported by (L,Σ) or*
2. *ξ is obtained from $\xi_{(L,\Sigma)}$ by adding Giroux torsion along tori that are incompressible in the complement of L.*

In this section we generalize this theorem to allow for any rationally null-homologous knots. The link case will be dealt with in Sect. 6.

Theorem 4.2. *Let K be a rationally null-homologous fibered transverse knot in a contact 3-manifold (M,ξ) such that ξ is tight when restricted to the complement of K. Denote by Σ a fiber in the fibration of $M - K$ and let r be the order of K. Then $r\,\mathrm{sl}_{\mathbb{Q}}^{\xi}(K,\Sigma) = -\chi(\Sigma)$ if and only if either*

1. *ξ agrees with the contact structure supported by the rational open book determined by K or*
2. *ξ is obtained from the contact structure by adding Giroux torsion along tori that are incompressible in the complement of K.*

Proof. Let K' be a positive integral resolution of K. Then from Theorem 3.3 we know that K' and K support the same contact structure. In addition, the following lemma (with Lemma 3.2) implies that if $r\,\mathrm{sl}_{\mathbb{Q}}^{\xi}(K,\Sigma) = -\chi(\Sigma)$, then $\mathrm{sl}_\xi(K',\Sigma') = -\chi(\Sigma')$, where Σ' is a fiber in the fibration of $M - K'$. Thus the proof is finished by Theorem 4.1.

The other implication is obvious from the generalization of the Thurston–Winkelnkemper construction in [1] and (1), since the characteristic foliation on the page of a rational open book contains only positive singularities and while adding Giroux torsion adds negative singularities, they cancel in the computation of the rational self-linking number and the Euler characteristic of Σ. □

Lemma 4.3. *Let K be a rationally null-homologous transverse knot of order r in a contact 3-manifold. Fix some framing on K and suppose a rational Seifert surface Σ approaches K as a cone on an (r,s)-knot. Let K' be a (p,q)-cable of K that is positive and transverse in the sense described before Lemma 3.2 and let Σ' be the Seifert surface for K' constructed from Σ as in the previous section. Then*

$$\mathrm{sl}(K',[\Sigma']) = \frac{1}{\gcd(r,p)}\left(|p|r\,\mathrm{sl}_{\mathbb{Q}}(K,[\Sigma]) + |rq-sp|(-1+|p|)\right).$$

Proof. For each singular point in the characteristic foliation of Σ there are $\frac{|p|}{\gcd(p,r)}$ corresponding singular points on Σ' (coming from the $\frac{|p|}{\gcd(p,r)}$ copies of Σ used in the construction of Σ'). For each of the $\frac{|rq-sp|}{\gcd(p,r)}$ meridional disks used to construct Σ' we get one positive elliptic point in the characteristic foliation of Σ'. Finally, since cabling was positive, the $|\frac{p(rq-sp)}{\gcd(p,r)}|$ half-twisted bands added to create Σ' each have a single positive hyperbolic singularity in their characteristic foliation. (It is easy to check that the characteristic foliation is as described, since the construction takes place mainly in a solid torus neighborhood of K where we can write an explicit model for this construction.) The lemma follows now from Lemma 1.1. □

5 Rational Unknots

A knot K in a manifold M is called a *rational unknot* if a rational Seifert surface D for K is a disk. Notice that the union of a neighborhood of K and a neighborhood of D is a punctured lens space. Thus the only manifold to have rational unknots (that are not actual unknots) are manifolds with a lens space summand. In particular, the only irreducible manifolds with rational unknots (that are not actual unknots) are lens spaces. So we restrict our attention to lens spaces in this section.

A knot K in a lens space is a rational unknot if and only if the complement of a tubular neighborhood of K is diffeomorphic to a solid torus. This, of course, implies that the rational unknots in $L(p,q)$ are precisely the cores of the Heegaard tori.

Theorem 5.1. *Rational unknots in tight contact structures on lens spaces are weakly transversely simple and Legendrian simple.*

A knot type is *weakly transversely simple* if it is up to contactomorphism (topologically) isotopic to the identity by its knot type and (rational) self-lining number. We have the analogous definition for *weakly Legendrian simple*. We will prove this theorem in the standard way. That is, we identify the maximal value for the rational Thurston–Bennequin invariant, show that there is a unique Legendrian knot with that rational Thurston–Bennequin invariant, and finally, show that any transverse unknot with nonmaximal rational Thurston–Bennequin invariant can be destabilized. The transverse result follows from the Legendrian result, as Lemma 1.4 shows.

5.1 Topological Rational Unknots

We explicitly describe $L(p,q)$ as follows: fix $p > q > 0$ and set

$$L(p,q) = V_0 \cup_\phi V_1,$$

where $V_i = S^1 \times D^2$ and we are thinking of S^1 and D^2 as the unit complex circle and disk, respectively. In addition the gluing map $\phi : \partial V_1 \to \partial V_0$ is given in standard longitude-meridian coordinates on the torus by the matrix

$$\begin{pmatrix} -p' & p \\ q' & -q \end{pmatrix},$$

where p' and q' satisfy $pq' - p'q = -1$ and $p > p' > 0$, $q \geq q' > 0$. We can find such p', q' by taking a continued fraction expansion of $-\frac{p}{q}$,

$$-\frac{p}{q} = a_0 - \cfrac{1}{a_1 - \cdots \cfrac{1}{a_{k-1} - \frac{1}{a_k}}},$$

with each $a_i \geq 2$ and then defining

$$-\frac{p'}{q'} = a_0 - \cfrac{1}{a_1 - \cdots \cfrac{1}{a_{k-1} - \frac{1}{a_k + 1}}}.$$

Since we have seen that a rational unknot must be isotopic to the core of a Heegaard torus, we clearly have four possible (oriented) rational unknots: $K_0, -K_0, K_1, -K_1$, where $K_i = S^1 \times \{pt\} \subset V_i$. We notice that K_0 represents a generator in the homology of $L(p,q)$, and $-K_0$ is the negative of that generator. So except in $L(2,1)$, the knots K_0 and $-K_0$ are not isotopic or homeomorphic via a homeomorphism isotopic to the identity. Similarly for K_1 and $-K_1$. Moreover, in homology, $q[K_0] = [K_1]$. So if $q \neq 1$ or $p-1$, then K_1 is not homeomorphic via a homeomorphism isotopic to the identity to K_0 or $-K_0$. We have established most of the following lemma.

Lemma 5.2. *The set of rational unknots up to homeomorphism isotopic to the identity in $L(p,q)$ is given by*

$$\{\text{rational unknots in } L(p,q)\} = \begin{cases} \{K_1\}, & p = 2, \\ \{K_1, -K_1\}, & p \neq 2, q = 1 \text{ or } p-1, \\ \{K_0, -K_0, K_1, -K_1\}, & q \neq 1 \text{ or } p-1. \end{cases}$$

Proof. Recall that $L(p,q)$ is an S^1-bundle over S^2 if and only if $q = 1$ or $p-1$. In this case, K_0 and K_1 are both fibers in this fibration and hence are isotopic. We are left to see that K_0 and $-K_0$ are isotopic in $L(2,1) = \mathbb{R}P^3$. To this end, notice that K_0 can be thought of as an $\mathbb{R}P^1$. In addition, we have the natural inclusions $\mathbb{R}P^1 \subset \mathbb{R}P^2 \subset \mathbb{R}P^3$. It is easy to find an isotopy of $\mathbb{R}P^1 = K_0$ in $\mathbb{R}P^2$ that reverses the orientation. This isotopy easily extends to $\mathbb{R}P^3$. □

5.2 *Legendrian Rational Unknots*

For results concerning convex surfaces and standard neighborhoods of Legendrian knots we refer the reader to [9].

Recall that in the classification of tight contact structures on $L(p,q)$ given in [12], the following lemma was proven as part of Proposition 4.17.

Lemma 5.3. *Let N be a standard neighborhood of a Legendrian knot isotopic to the rational unknot K_1 in a tight contact structure on $L(p,q)$. Then there is another neighborhood with convex boundary N' such that $N \subset N'$ and $\partial N'$ has two dividing curves parallel to the longitude of V_1. Moreover, any two such solid tori with convex boundary each having two dividing curves of infinite slope have contactomorphic complements.*

We note that N' from this lemma is the standard neighborhood of a Legendrian knot L topologically isotopic to K_1. Moreover, one easily checks that

$$\mathrm{tb}_{\mathbb{Q}}(L) = -\frac{p'}{p},$$

where $p' < p$ is defined as in the previous sections. The next possible larger value for $\mathrm{tb}_{\mathbb{Q}}$ is $-\frac{p'}{p} + 1 > -\frac{1}{p}$, which violates the Bennequin bound.

Theorem 5.4. *The maximum possible value for the rational Thurston–Bennequin invariant for a Legendrian knot isotopic to K_1 is $-\frac{p'}{p}$, and it is uniquely realized, up to contactomorphism isotopic to the identity. Moreover, any Legendrian knot isotopic to K_1 with nonmaximal rational Thurston–Bennequin invariant destabilizes.*

Proof. The uniqueness follows from the last sentence in Lemma 5.3. The first part of the same lemma also establishes the destabilization result, since it is well known (see [9]) that if the standard neighborhood of a Legendrian knot is contained in the standard neighborhood of another Legendrian knot, then the first is a stabilization of the second. □

To finish the classification of Legendrian knots in the knot type of K_1, we need to identify the rational rotation number of the Legendrian knot L in the knot type of K_1 with maximal rational Thurston–Bennequin invariant. To this end, notice that if we fix the neighborhood N' from Lemma 5.3 as the standard neighborhood of the maximal rational Thurston–Bennequin invariant Legendrian knot L, then we can choose a nonzero section s of $\xi|_{\partial N'}$. This allows us to define a relative Euler class for $\xi|_{N'}$ and $\xi|_C$, where $C = L(p,q) - N'$. One easily sees that the Euler class of $\xi|_{N'}$ vanishes and the Euler class $e(\xi)$ is determined by its restriction to the solid torus C. In particular, $\xi|_C$ is determined by

$$e(\xi)(D) = e(\xi|_C, s)(D) \mod p,$$

where D is the meridional disk of C and the generator of two-chains in $L(p,q)$.

Thinking of D as the rational Seifert surface for L, we can arrange the foliation near the boundary to be by Legendrian curves parallel to the boundary (see [12, Fig. 1]). From this we see that we can take a Seifert cable L_c of L to be Legendrian and satisfy

$$\mathrm{rot}_{\mathbb{Q}}(L) = \frac{1}{p}\,\mathrm{rot}(L_c).$$

By taking the foliation on $\partial N' = \partial C$ to be such that $D \cap \partial C$ is a ruling curve, we see that

$$\mathrm{rot}_{\mathbb{Q}}(L) = \frac{1}{p}\,\mathrm{rot}(L_c) = \frac{1}{p}\,e(\xi|_C, s)(D).$$

By the classification of tight contact structures on solid tori [12], we see that the number $e(\xi|_C, s)(D)$ is always a subset of $\{p' - 1 - 2k : k = 0, 1, \ldots, p' - 1\}$ and determined by the Euler class of ξ. To give a more precise classification we need to know the range of possible values for the Euler class of tight ξ on $L(p,q)$. This is in principal known, but difficult to state in general. We consider several cases in the next subsection.

We clearly have the analog of Theorem 5.4 for $-K_1$. That is all Legendrian knots in the knot type $-K_1$ destabilize to the unique maximal representative L with $\mathrm{tb}_{\mathbb{Q}}(L) = -\frac{p'}{p}$ and rotation number the negative of the rotation number for the maximal Legendrian representative of K_1.

Proof (Proof of Theorem 5.1). Notice that if $q^2 \equiv \pm 1 \mod p$, we have a diffeomorphism $\psi : L(p,q) \to L(p,q)$ that exchanges the Heegaard tori, and if $q = 1$ or $p - 1$, then this diffeomorphism is isotopic to the identity. Thus when $p \neq 2$ and $q = 1$ or $p - 1$, we have competed the proof of Theorem 5.1. Note also that we always have the diffeomorphism $\psi' : L(p,q) \to L(p,q)$ that preserves each of the Heegaard tori but acts by complex conjugation on each factor of each Heegaard torus (recall that the Heegaard tori are $V_i = S^1 \times D^2$, where S^1 and D^2 are a unit circle and disk in the complex plane, respectively). If $p = 2$, then this diffeomorphism is also isotopic to the identity. Thus we have finished the proof of Theorem 5.1 in this case.

We are left to consider the case that $q \neq 1$ or $p-1$. In this case we can understand K_0 and $-K_0$ by reversing the roles of V_0 and V_1. That is, we consider using the gluing map

$$\phi^{-1} = \begin{pmatrix} q & p \\ q' & p' \end{pmatrix}$$

to glue ∂V_0 to ∂V_1. □

5.3 Classification Results

To give some specific classification results we recall that for the lens space $L(p,1)$, p odd, there is a unique tight contact structure for any given Euler class not equal to the zero class in $H_2(L(p,q);\mathbb{Z})$. From this, the fact that $p' = p-1$ in this case, and the discussion in the previous subsection we obtain the following theorem.

Theorem 5.5. *For p odd and any integer $l \in \{p-2-2k : k = 0,1,\dots,p-2\}$, there is a unique tight contact structure ξ_l on $L(p,1)$ with $e(\xi_l)(D) = l$ (here D is again the 2-cell in the CW-decomposition of $L(p,1)$ given in the last subsection). In this contact structure, the knot types K_1 and $-K_1$ are weakly Legendrian simple and transversely simple. Moreover, the rational Thurston–Bennequin invariants realized by Legendrian knots in the knot type K_1 are*

$$\left\{ -\frac{p-1}{p} - k : k \text{ a nonpositive integer} \right\}.$$

The range for Legendrian knots in the knot type $-K_1$ is the same. The range of rotation numbers realized for a Legendrian knot in the knot type K_1 with rational Thurston–Bennequin invariant $-\frac{p-1}{p} - k$ is

$$\left\{ \frac{l}{p} + k - 2m : m = 0,\dots,k \right\},$$

and for $-K_1$, the range is

$$\left\{ \frac{-l}{p} + k - 2m : m = 0,\dots,k \right\}.$$

The range of possible rational self-linking numbers for transverse knots in the knot type K_1 is

$$\left\{ -\frac{p+l-1}{p} - k : k \text{ a nonpositive integer} \right\},$$

and in the knot type $-K_1$ *is*

$$\left\{-\frac{p-l-1}{p}-k : k \text{ a nonpositive integer}\right\}.$$

Results for other $L(p,q)$ can easily be written down after the range of Euler classes for the tight contact structures is determined.

6 Rationally Null-Homologous Links and Uniform Seifert Surfaces

Much of our previous discussion for rational knots also applies to links, but many of the statements are a bit more awkward (or even uncertain) if we do not restrict to certain kinds of rational Seifert surfaces.

Let $L = K_1 \cup \cdots \cup K_n$ be an oriented link of $\mathbb{Z}$-homological order $r > 0$ in a 3-manifold M, and let us denote a tubular neighborhood of L by $N(L) = N(K_1) \cup \cdots \cup N(K_n)$. By $X(L)$ we denote the link exterior $\overline{M \setminus N(L)}$. Fix a framing for each $N(K_i)$. Since L has order r, there is an embedded (r, s_i)-curve on $\partial N(K_i)$ for each i, and together, they bound an oriented surface Σ° in $X(L)$. Radially coning $\partial \Sigma^\circ \subset N(L)$ to L gives a surface Σ in M whose interior is embedded and for which $\partial \Sigma|_{K_i}$ wraps r times around K_i. By tubing if needed, we may take Σ to be connected. Such a surface Σ will be called a *uniform rational Seifert surface* for L, and we say that L *r-bounds* Σ.

Notice that as $\mathbb{Z}$-homology chains, $\partial \Sigma = rL = 0$. Since as 1-chains there may exist varying integers r_i such that $r_1 K_1 + \cdots + r_n K_n = 0$, the link L may have other rational Seifert surfaces that are not uniform. However, only for a uniform rational Seifert surface Σ do we have that $\partial(\frac{1}{r}\Sigma) = L$ as $\mathbb{Q}$-homology chains.

With respect to uniform rational Seifert surfaces, the definition of rational linking number for rationally null-homologous links extends directly: If L is an oriented link that r-bounds Σ and L' is another oriented link, then

$$\mathrm{lk}_{\mathbb{Q}}(L, L') = \frac{1}{r} \Sigma \cdot L'$$

with respect to $[\Sigma]$. If L' is rationally null-homologous and r'-bounds Σ', then this linking number is symmetric and independent of choice of Σ and Σ'.

It now follows that the entire content of Sects. 1 and 2 extends in a straightforward manner to transverse/Legendrian links L that r-bound a uniform rational Seifert surface Σ in a contact manifold. The generalization of Theorem 4.2 is straightforward as well, but relies on the generalized statements of Lemmas 3.2 and 4.3. Rather than record the general statements of these lemmas (which becomes cumbersome for arbitrary cables), we present them only for integral resolutions of rationally null-homologous links with uniform rational Seifert surfaces.

Lemma 6.1. *Let L be a link in M that r-bounds a uniform rational Seifert surface Σ for $r > 0$. Choose a framing on each component K_i of L, $i = 1, \dots, n$, such that Σ approaches K_i as (r, s_i)-curves. Let L' be the link formed by replacing each K_i by its (r, q_i)-cable, where $q_i \neq s_i$. If L is a rationally fibered link with fiber Σ, then L' is a (null-homologous) fibered link bounding a fiber Σ' with*

$$\chi(\Sigma') = \chi(\Sigma) + (1 - r)\sum_{i=1}^{n} |s_i - q_i|.$$

Furthermore, assume that M is endowed with a contact structure ξ and that L is a transverse link. If the integral resolution L' of L is positive and transverse, then

$$\mathrm{sl}(L', [\Sigma']) = r\,\mathrm{sl}_{\mathbb{Q}}(K, [\Sigma]) + (-1 + r)\sum_{i=1}^{n} |s_i - q_i|.$$

Proof. The construction of Σ' is done by attaching $|s_i - q_i|$ copies of meridional disks of $N(K_i)$ with $r|s_i - q_i|$ half-twisted bands to Σ for each i. Now follow the proof of Lemma 4.3. □

Theorem 6.2. *Let L be a rationally null-homologous fibered transverse link in a contact 3-manifold (M, ξ) such that ξ is tight when restricted to the complement of L. Suppose L r-bounds the fibers of the fibration of $M - L$ and let Σ be a fiber. Then $r\,\mathrm{sl}_{\mathbb{Q}}^{\xi}(L, \Sigma) = -\chi(\Sigma)$ if and only if either ξ agrees with the contact structure supported by the rational open book determined by L and Σ or ξ is obtained from the contact structure by adding Giroux torsion along tori that are incompressible in the complement of L.*

Proof. Follow the proof of Theorem 4.2 using Lemma 6.1 instead of Lemmas 3.2 and 4.3. □

Acknowledgments The first author was partially supported by NSF Grant DMS-0239600. The second author was partially supported by NSF Grants DMS-0239600 and DMS-0804820.

References

1. Kenneth L. Baker, John B. Etnyre, and Jeremy Van Horn-Morris. Cabling, contact structures, and mapping class monoids. *J. Differ. Geom.* to appear.
2. Kenneth L. Baker and J. Elisenda Grigsby. Grid diagrams and Legendrian lens space links. J. Symplectic. Geom. 7(4):415–448, 2009.
3. Kenneth L. Baker, J. Elisenda Grigsby, and Matthew Hedden. Grid diagrams for lens spaces and combinatorial knot Floer homology. *Int. Math. Res. Not. IMRN*, (10):Art. ID rnm024, 39, 2008.
4. Vladimir Chernov. Framed knots in 3-manifolds and affine self-linking numbers. *J. Knot Theory Ramifications*, 14(6):791–818, 2005.

5. Yakov Eliashberg. Filling by holomorphic discs and its applications. In *Geometry of low-dimensional manifolds, 2 (Durham, 1989)*, volume 151 of *London Math. Soc. Lecture Note Ser.*, pages 45–67. Cambridge Univ. Press, Cambridge, 1990.
6. Yakov Eliashberg. Legendrian and transversal knots in tight contact 3-manifolds. In *Topological methods in modern mathematics (Stony Brook, NY, 1991)*, pages 171–193. Publish or Perish, Houston, TX, 1993.
7. John B. Etnyre. Introductory lectures on contact geometry. In *Topology and geometry of manifolds (Athens, GA, 2001)*, volume 71 of *Proc. Sympos. Pure Math.*, pages 81–107. Amer. Math. Soc., Providence, RI, 2003.
8. John B. Etnyre. Legendrian and transversal knots. In *Handbook of knot theory*, pages 105–185. Elsevier B. V., Amsterdam, 2005.
9. John B. Etnyre and Ko Honda. Knots and contact geometry. I. Torus knots and the figure eight knot. *J. Symplectic Geom.*, 1(1):63–120, 2001.
10. John B. Etnyre and Jeremy Van Horn-Morris. Fibered transverse knots and the Bennequin bound. IMRN 2011:1483–1509, 2011.
11. Robert E. Gompf and András I. Stipsicz. *4-manifolds and Kirby calculus*, volume 20 of *Graduate Studies in Mathematics*. American Mathematical Society, Providence, RI, 1999.
12. Ko Honda. On the classification of tight contact structures. I. *Geom. Topol.*, 4:309–368 (electronic), 2000.
13. Ferit Öztürk. Generalised Thurston-Bennequin invariants for real algebraic surface singularities. *Manuscripta Math.*, 117(3):273–298, 2005.
14. W. P. Thurston and H. E. Winkelnkemper. On the existence of contact forms. *Proc. Amer. Math. Soc.*, 52:345–347, 1975.

Regularity of Plurisubharmonic Upper Envelopes in Big Cohomology Classes

Robert Berman and Jean-Pierre Demailly

Dedicated to Professor Oleg Viro for his deep contributions to mathematics

Abstract The goal of this work is to prove the regularity of certain quasi-plurisubharmonic upper envelopes. Such envelopes appear in a natural way in the construction of Hermitian metrics with minimal singularities on a big line bundle over a compact complex manifold. We prove that the complex Hessian forms of these envelopes are locally bounded outside an analytic set of singularities. It is furthermore shown that a parametrized version of this result yields a priori inequalities for the solution of the Dirichlet problem for a degenerate Monge–Ampère operator; applications to geodesics in the space of Kähler metrics are discussed. A similar technique provides a logarithmic modulus of continuity for Tsuji's "supercanonical" metrics, that generalize a well-known construction of Narasimhan and Simha.

Keywords Plurisubharmonic function • Upper envelope • Hermitian line bundle • Singular metric • Logarithmic poles • Legendre–Kiselman transform • Pseudoeffective cone • Volume • Monge–Ampère measure • Supercanonical metric • Ohsawa–Takegoshi theorem

R. Berman
Department of Mathematics, Chalmers University of Technology, Eklandag. 86, SE-412 96 Göteborg, Sweden
e-mail: robertb@chalmers.se

J.-P. Demailly (✉)
Université de Grenoble I, Départ. de Mathématiques, Institut Fourier, BP74, 38402 Saint-Martin d'Hères, France
e-mail: demailly@fourier.ujf-grenoble.fr

I. Itenberg et al. (eds.), *Perspectives in Analysis, Geometry, and Topology: On the Occasion of the 60th Birthday of Oleg Viro*, Progress in Mathematics 296,
DOI 10.1007/978-0-8176-8277-4_3, © Springer Science+Business Media, LLC 2012

1 Main Regularity Theorem

Let X be a compact complex manifold and ω a Hermitian metric on X, viewed as a smooth positive $(1,1)$-form. As usual, we put $d^c = \frac{1}{4i\pi}(\partial - \overline{\partial})$, so that $dd^c = \frac{1}{2i\pi}\partial\overline{\partial}$. Consider the dd^c-cohomology class $\{\alpha\}$ of a smooth real d-closed form α of type $(1,1)$ on X. (In general, one has to consider the Bott–Chern cohomology group, for that boundaries are dd^c-exact $(1,1)$-forms $dd^c\varphi$, but in the case in that X is Kähler, this group is isomorphic to the Dolbeault cohomology group $H^{1,1}(X)$.)

Recall that a function ψ is said to be quasiplurisubharmonic (or quasi-psh) if $idd^c\psi$ is locally bounded from below, or equivalently, if it can be written locally as a sum $\psi = \varphi + u$ of a psh function φ and a smooth function u. More precisely, it is said to be α-plurisubharmonic (or α-psh) if $\alpha + dd^c\psi \geq 0$. We denote by $\mathrm{PSH}(X,\alpha)$ the set of α-psh functions on X.

Definition 1.1. The class $\{\alpha\} \in H^{1,1}(X,\mathbb{R})$ is said to be pseudoeffective if it contains a closed (semi)positive current $T = \alpha + dd^c\psi \geq 0$, and big if it contains a closed "Kähler current" $T = \alpha + dd^c\psi$ such that $T \geq \varepsilon\omega > 0$ for some $\varepsilon > 0$.

From now on in this section, we assume that $\{\alpha\}$ is *big*. We know by [Dem92] that we can then find $T_0 \in \{\alpha\}$ of the form

$$T_0 = \alpha + dd^c\psi_0 \geq \varepsilon_0\omega \tag{1.2}$$

with a possibly slightly smaller $\varepsilon_0 > 0$ than the ε in the definition, and ψ_0 a quasi-psh function with analytic singularities, i.e., locally

$$\psi_0 = c\log\sum|g_j|^2 + u, \quad \text{where } c > 0,\, u \in C^\infty,\, g_j \text{ holomorphic.} \tag{1.3}$$

By [DP04], X carries such a class $\{\alpha\}$ if and only if X is in the Fujiki class $\mathcal{C}$ of smooth varieties that are bimeromorphic to compact Kähler manifolds. Our main result is the following.

Theorem 1.4. *Let X be a compact complex manifold in the Fujiki class $\mathcal{C}$, and let α be a smooth closed form of type $(1,1)$ on X such that the cohomology class $\{\alpha\}$ is big. Pick $T_0 = \alpha + dd^c\psi_0 \in \{\alpha\}$ satisfying (1.2) and (1.3) for some Hermitian metric ω on X, and let Z_0 be the analytic set $Z_0 = \psi_0^{-1}(-\infty)$. Then the upper envelope*

$$\varphi := \sup\{\psi \leq 0,\ \psi\ \alpha\text{-psh}\}$$

is a quasiplurisubharmonic function that has locally bounded second-order derivatives $\partial^2\varphi/\partial z_j\partial\overline{z}_k$ on $X \smallsetminus Z_0$, and moreover, for suitable constants $C,B > 0$, there is a global bound

$$|dd^c\varphi|_\omega \leq C(|\psi_0|+1)^2 e^{B|\psi_0|}$$

that explains how these derivatives blow up near Z_0. In particular, φ is $C^{1,1-\delta}$ on $X \smallsetminus Z_0$ for every $\delta > 0$, and the second derivatives $D^2\varphi$ are in $L^p_{\mathrm{loc}}(X \smallsetminus Z_0)$ for every $p > 0$.

An important special case is the situation in that we have a Hermitian line bundle (L,h_L) *and* $\alpha = \Theta_{L,h_L}$, *with the assumption that L is big, i.e., that there exists a singular Hermitian* $h_0 = h_L e^{-\psi_0}$ *that has analytic singularities and a curvature current* $\Theta_{L,h_0} = \alpha + dd^c\psi_0 \geq \varepsilon_0\omega$. *We then infer that the metric with minimal singularities* $h_{\min} = h_L e^{-\varphi}$ *has the regularity properties prescribed by Theorem 4.1 outside of the analytic set* $Z_0 = \psi_0^{-1}(-\infty)$. *In fact, [Ber07, Theorem 3.4 (a)] proves in this case the slightly stronger result that* φ *in* $C^{1,1}$ *on* $X \setminus Z_0$ *(using the fact that X is then Moishezon and that the total space of* L^* *has many holomorphic vector fields). The present approach is by necessity different, since we can no longer rely on the existence of vector fields when X is not algebraic. Even then, our proof will be in fact somewhat simpler.*

Proof. Notice that in order to get a quasi-psh function φ, we should a priori replace φ by its upper semicontinuous regularization $\varphi^*(z) = \limsup_{\zeta\to z}\varphi(\zeta)$, but since $\varphi^* \leq 0$ and φ^* is α-psh as well, $\psi = \varphi^*$ contributes to the envelope, and therefore $\varphi = \varphi^*$. Without loss of generality, after subtracting a constant from ψ_0, we may assume that $\psi_0 \leq 0$. Then ψ_0 contributes to the upper envelope, and therefore $\varphi \geq \psi_0$. This already implies that φ is locally bounded on $X \setminus Z_0$. Following [Dem94], for every $\delta > 0$, we consider the regularization operator

$$\psi \mapsto \rho_\delta\psi \tag{1.5}$$

defined by $\rho_\delta\psi(z) = \Psi(z,\delta)$ and

$$\Psi(z,w) = \int_{\zeta\in T_{X,z}} \psi\big(\mathrm{exph}_z(w\zeta)\big)\,\chi(|\zeta|^2)\,\mathrm{d}V_\omega(\zeta), \qquad (z,w)\in X\times\mathbb{C}, \tag{1.6}$$

where $\mathrm{exph}: T_X \to X$, $T_{X,z} \ni \zeta \mapsto \mathrm{exph}_z(\zeta)$, is the formal holomorphic part of the Taylor expansion of the exponential map of the Chern connection on T_X associated with the metric ω, and $\chi : \mathbb{R} \to \mathbb{R}_+$ is a smooth function with support in $]-\infty, 1]$ defined by

$$\chi(t) = \frac{C}{(1-t)^2}\exp\frac{1}{t-1} \quad \text{for } t<1, \qquad \chi(t) = 0 \quad \text{for } t \geq 1,$$

with $C > 0$ adjusted so that $\int_{|x|\leq 1}\chi(|x|^2)\,\mathrm{d}x = 1$ with respect to the Lebesgue measure $\mathrm{d}x$ on $\mathbb{C}^n$.

Also, $\mathrm{d}V_\omega(\zeta)$ denotes the standard Hermitian Lebesgue measure on (T_X,ω). Clearly, $\Psi(z,w)$ depends only on $|w|$. With the relevant change of notation, the estimates proved in Sects. 3 and 4 of [Dem94] (see especially Theorem 4.1 and estimates (4.3), (4.5) therein) show that if one assumes $\alpha + dd^c\psi \geq 0$, then there are constants $\delta_0, K > 0$ such that for $(z,w) \in X\times\mathbb{C}$,

$$[0,\delta_0] \ni t \mapsto \Psi(z,t) + Kt^2 \quad \text{is increasing,} \tag{1.7}$$

$$\alpha(z) + dd^c\Psi(z,w) \geq -A\lambda(z,|w|)|\mathrm{d}z|^2 - K\big(|w|^2|\mathrm{d}z|^2 + |\mathrm{d}z||\mathrm{d}w| + |\mathrm{d}w|^2\big), \tag{1.8}$$

where $A = \sup_{|\zeta|\le 1, |\xi|\le 1}\{-c_{jk\ell m}\zeta_j\overline{\zeta}_k\xi_\ell\overline{\xi}_m\}$ is a bound for the negative part of the curvature tensor $(c_{jk\ell m})$ of (T_X, ω) and

$$\lambda(z,t) = \frac{\mathrm{d}}{d\log t}(\Psi(z,t) + Kt^2) \underset{t\to 0_+}{\longrightarrow} \nu(\psi, z) \qquad \text{(Lelong number)}. \tag{1.9}$$

In fact, this is clear from [Dem94] if $\alpha = 0$, and otherwise we simply apply the above estimates (1.7)–(1.9) locally to $u + \psi$, where u is a local potential of α, and then subtract the resulting regularization $U(z,w)$ of u, that is such that

$$dd^c(U(z,w) - u(z)) = O(|w|^2|\mathrm{d}z|^2 + |w||\mathrm{d}z||\mathrm{d}w| + |\mathrm{d}w|^2), \tag{1.10}$$

because the left-hand side is smooth and $U(z,w) - u(z) = O(|w|^2)$.

As a consequence, the regularization operator ρ_δ transforms quasi-psh functions into quasi-psh functions, while providing very good control on the complex Hessian. We exploit this, again quite similarly as in [Dem94], by introducing the Kiselman–Legendre transform (cf. [Kis78, Kis94])

$$\psi_{c,\delta}(z) = \inf_{t\in]0,\delta]} \rho_t\psi(z) + Kt^2 - K\delta^2 - c\log\frac{t}{\delta}, \quad c > 0,\ \delta \in\,]0,\delta_0]. \tag{1.11}$$

We need the following basic lower bound on the Hessian form.

Lemma 1.12. *For all $c > 0$ and $\delta \in\,]0,\delta_0]$, we have*

$$\alpha + dd^c\psi_{c,\delta} \ge -\big(A\min\big(c, \lambda(z,\delta)\big) + K\delta^2\big)\omega.$$

Proof of lemma. In general, an infimum $\inf_{\eta\in E} u(z,\eta)$ of psh functions $z \mapsto u(z,\eta)$ is not psh, but this is the case if $u(z,\eta)$ is psh with respect to (z,η) and $u(z,\eta)$ depends only on $\mathrm{Re}\,\eta$, in that case it is actually a convex function of $\mathrm{Re}\,\eta$. This fundamental fact is known as Kiselman's infimum principle. We apply it here by putting $w = \mathrm{e}^\eta$ and $t = |w| = \mathrm{e}^{\mathrm{Re}\,\eta}$. At all points of $E_c(\psi) = \{z \in X\,;\ \nu(\psi,z) \ge c\}$, the infimum occurring in (1.11) is attained at $t = 0$. However, for $z \in X \smallsetminus E_c(\psi)$ it is attained for $t = t_{\min}$, where

$$\begin{cases} t_{\min} = \delta & if \lambda(z,\delta) \le c, \\ t_{\min} < \delta & such\ that\ c = \lambda(z,t_{\min}) = \frac{\mathrm{d}}{dt}(\Psi(z,t) + Kt^2)_{t=t_{\min}} if \lambda(z,\delta) > c. \end{cases}$$

In a neighborhood of such a point $z \in X \smallsetminus E_c(\psi)$, the infimum coincides with the infimum taken for t close to $t_{\min}$, and all functions involved have (modulo addition of α) a Hessian form bounded below by $-(A\lambda(z,t_{\min}) + K\delta^2)\omega$ by (1.8). Since $\lambda(z,t_{\min}) \le \min(c, \lambda(z,\delta))$, we get the desired estimate on the dense open set $X \smallsetminus E_c(\psi)$ by Kiselman's infimum principle. However, $\psi_{c,\delta}$ is quasi-psh on X, and $E_c(\psi)$ is of measure zero, so the estimate is in fact valid on all of X, in the sense of currents. □

We now proceed to complete the proof or Theorem 4.1. Lemma 1 implies the more brutal estimate

$$\alpha + dd^c \psi_{c,\delta} \geq -(Ac + K\delta^2)\,\omega \qquad \text{for } \delta \in \,]0,\delta_0]. \tag{1.13}$$

Consider the convex linear combination

$$\theta = \frac{Ac + K\delta^2}{\varepsilon_0}\psi_0 + \left(1 - \frac{Ac + K\delta^2}{\varepsilon_0}\right)\varphi_{c,\delta},$$

where φ is the upper envelope of all α-psh functions $\psi \leq 0$. Since $\alpha + dd^c\varphi \geq 0$, (1.2) and (1.13) imply

$$\alpha + dd^c\theta \geq (Ac + K\delta^2)\,\omega - \left(1 - \frac{Ac + K\delta^2}{\varepsilon_0}\right)(Ac + K\delta^2)\,\omega \geq 0.$$

Also $\varphi \leq 0$, and therefore $\varphi_{c,\delta} \leq \rho_\delta\varphi \leq 0$ and $\theta \leq 0$ likewise. In particular θ contributes to the envelope, and as a consequence we get $\varphi \geq \theta$.

Returning to the definition of $\varphi_{c,\delta}$, we infer that for every point $z \in X \smallsetminus Z_0$ and every $\delta > 0$, there exists $t \in \,]0,\delta]$ such that

$$\begin{aligned}\varphi(z) &\geq \frac{Ac + K\delta^2}{\varepsilon_0}\psi_0(z) + \left(1 - \frac{Ac + K\delta^2}{\varepsilon_0}\right)(\rho_t\varphi(z) + Kt^2 - K\delta^2 - c\log t/\delta)\\ &\geq \frac{Ac + K\delta^2}{\varepsilon_0}\psi_0(z) + (\rho_t\varphi(z) + Kt^2 - K\delta^2 - c\log t/\delta)\end{aligned}$$

(using the fact that the infimum is ≤ 0 and reached for some $t \in \,]0,\delta]$, since $t \mapsto \rho_t\varphi(z)$ is bounded for $z \in X \smallsetminus Z_0$). Therefore, we get

$$\rho_t\varphi(z) + Kt^2 \leq \varphi(z) + K\delta^2 - (Ac + K\delta^2)\varepsilon_0^{-1}\psi_0(z) + c\log\frac{t}{\delta}\,. \tag{1.14}$$

Since $t \mapsto \rho_t\varphi(z) + Kt^2$ is increasing and equal to $\varphi(z)$ for $t = 0$, we infer that

$$K\delta^2 - (Ac + K\delta^2)\varepsilon_0^{-1}\psi_0(z) + c\log\frac{t}{\delta} \geq 0,$$

or equivalently, since $\psi_0 \leq 0$,

$$t \geq \delta\exp\big(-(A + K\delta^2/c)\varepsilon_0^{-1}|\psi_0(z)| - K\delta^2/c\big).$$

Now (1.14) implies the weaker estimate

$$\rho_t\varphi(z) \leq \varphi(z) + K\delta^2 + (Ac + K\delta^2)\varepsilon_0^{-1}|\psi_0(z)|;$$

hence, by combining the last two inequalities, we get

$$\begin{aligned}&\frac{\rho_t\varphi(z)-\varphi(z)}{t^2}\\&\leq K\left(1+\Big(\frac{Ac}{K\delta^2}+1\Big)\varepsilon_0^{-1}|\psi_0(z)|\right)\exp\left(2\Big(A+K\frac{\delta^2}{c}\Big)\varepsilon_0^{-1}|\psi_0(z)|+2K\frac{\delta^2}{c}\right).\end{aligned}$$

We exploit this by letting $0<t\leq\delta$ and c tend to 0 in such a way that $Ac/K\delta^2$ converges to a positive limit ℓ (if $A=0$, just enlarge A slightly and then let $A\to 0$). In this way, we get for every $\ell>0$,

$$\begin{aligned}&\liminf_{t\to 0_+}\frac{\rho_t\varphi(z)-\varphi(z)}{t^2}\\&\qquad\leq K(1+(\ell+1)\varepsilon_0^{-1}|\psi_0(z)|)\exp\Big(2A((1+\ell^{-1})\varepsilon_0^{-1}|\psi_0(z)|+\ell^{-1})\Big).\end{aligned}$$

The special (essentially optimal) choice $\ell=\varepsilon_0^{-1}|\psi_0(z)|+1$ yields

$$\liminf_{t\to 0_+}\frac{\rho_t\varphi(z)-\varphi(z)}{t^2}\leq K(\varepsilon_0^{-1}|\psi_0(z)|+1)^2\exp\big(2A(\varepsilon_0^{-1}|\psi_0(z)|+1)\big).\tag{1.15}$$

Now, putting as usual $\nu(\varphi,z,r)=\frac{1}{\pi^{n-1}r^{2n-2}/(n-1)!}\int_{B(z,r)}\Delta\varphi(\zeta)\,\mathrm{d}\zeta$, we infer from estimate (4.5) of [Dem94] the Lelong–Jensen-like inequality

$$\begin{aligned}\rho_t\varphi(z)-\varphi(z)&=\int_0^t\frac{\mathrm{d}}{\mathrm{d}\tau}\Phi(z,\tau)\,\mathrm{d}\tau\\&\geq\int_0^t\frac{\mathrm{d}\tau}{\tau}\left(\int_{B(0,1)}\nu(\varphi,z,\tau|\zeta|)\,\chi(|\zeta|^2)\,\mathrm{d}\zeta-O(\tau^2)\right)\\&\geq c(a)\,\nu(\varphi,z,at)-C_2t^2\quad[\text{where } a<1,\ c(a)>0 \text{ and } C_2\gg 1]\\&=\frac{c'(a)}{t^{2n-2}}\int_{B(z,at)}\Delta\varphi(\zeta)\,\mathrm{d}\zeta-C_2t^2,\end{aligned}\tag{1.16}$$

where the third line is obtained by integrating for $\tau\in[a^{1/2}t,t]$ and for ζ in the corona $a^{1/2}<|\zeta|<a^{1/4}$ (here we assume that χ is taken to be decreasing with $\chi(t)>0$ for all $t<1$, and we compute the Laplacian Δ in normalized coordinates at z given by $\zeta\mapsto\mathrm{exph}_z(\zeta)$).

Hence by Lebesgue's theorem on the existence almost everywhere of the density of a positive measure (see, e.g., [Rud66, 7.14]), we obtain

$$\lim_{t\to 0_+}\frac{1}{t^2}(\rho_t\varphi(z)-\varphi(z))\geq c''(\Delta_\omega\varphi)_{\mathrm{ac}}(z)-C_2\quad\text{a.e. on } X,\tag{1.17}$$

where the subscript "ac" means the absolutely continuous part of the measure $\Delta_\omega \varphi$. By combining (1.15) and (1.17) and using the quasiplurisubharmonicity of φ, we conclude that

$$|dd^c \varphi|_\omega \leq \Delta_\omega \varphi + C_3 \leq C(|\psi_0|+1)^2 e^{2A\varepsilon_0^{-1}\psi_0(z)} \quad \text{a.e. on } X \smallsetminus Z_0$$

for some constant $C > 0$. There cannot be any singular measure part μ in $\Delta_\omega \varphi$ either, since we know that the Lebesgue density would then be equal to $+\infty$ μ-a.e. [Rud66, 7.15], in contradiction to (1.15). This gives the required estimates for the complex derivatives $\partial^2\varphi/\partial z_j \partial \bar{z}_k$. The other real derivatives $\partial^2\varphi/\partial x_i \partial x_j$ are obtained from $\Delta\varphi = \sum_k \partial^2\varphi/\partial z_k \partial \bar{z}_k$ via singular integral operators, and it is well known that these operate boundedly on L^p for all $p < \infty$. Theorem (1.4) follows. □

Remark 1.18. The proof gave us in fact the very explicit value $B = 2A\varepsilon_0^{-1}$, where A is an upper bound of the negative part of the curvature of (T_X, ω). The slightly more refined estimates obtained in [Dem94] show that we could even replace B by the possibly smaller constant $B_\eta = 2(A' + \eta)\varepsilon_0^{-1}$, where

$$A' = \sup_{|\zeta|=1,\, |\xi|=1,\, \zeta \perp \xi} -c_{jk\ell m} \zeta_j \overline{\zeta}_k \xi_\ell \overline{\xi}_m,$$

and the dependence of the other constants on η could then be made explicit.

Remark 1.19. In Theorem (1.4), one can replace the assumption that α is smooth by the assumption that α has L^∞ coefficients. In fact, we used the smoothness of α only as a cheap argument to get the validity of estimate (1.10) for the local potentials u of α. However, the results of [Dem94] easily imply the same estimates when α is L^∞, since both u and $-u$ are then quasi-psh; this follows, for instance, from (1.8) applied with respect to a smooth α_∞ and $\psi = \pm u$ if we observe that $\lambda(z, |w|) = O(|w|^2)$ when $|dd^c\psi|_\omega$ is bounded. Therefore, only the constant K will be affected in the proof.

2 Applications to Volume and Monge–Ampère Measures

Recall that the *volume* of a big class $\{\alpha\}$ is defined, in the work [Bou02] of S. Boucksom, as

$$\mathrm{Vol}(\{\alpha\}) = \sup_T \int_{X \smallsetminus \mathrm{sing}(T)} T^n, \tag{2.1}$$

with T ranging over all positive currents in the class $\{\alpha\}$ with *analytic singularities*, whose locus is denoted by $\mathrm{sing}(T)$. If the class is not big, then the volume is defined to be zero. With this definition, it is clear that $\{\alpha\}$ is big precisely when $\mathrm{Vol}(\{\alpha\}) > 0$.

Now fix a *smooth* representative α in a *pseudoeffective* class $\{\alpha\}$. We then obtain a uniquely defined α-plurisubharmonic function $\varphi = \psi_{\min} \geq 0$ with minimal singularities defined as in Theorem (1.4) by

$$\varphi := \sup\left\{\psi \leq 0,\ \psi\ \alpha\text{-psh}\right\}; \tag{2.2}$$

notice that the supremum is nonempty by our assumption that $\{\alpha\}$ is pseudoeffective. If $\{\alpha\}$ is big and ψ is α-psh and locally bounded in the complement of an analytic $Z \subset X$, one can define the Monge–Ampère measure $\mathrm{MA}_\alpha(\psi)$ by

$$\mathrm{MA}_\alpha(\psi) := \mathbf{1}_{X\smallsetminus Z}(\alpha + dd^c\psi)^n, \tag{2.3}$$

as follows from the work of Bedford and Taylor [BT76, BT82]. In particular, if $\{\alpha\}$ is big, there is a well-defined positive measure on $\mathrm{MA}_\alpha(\varphi) = \mathrm{MA}_\alpha(\psi_{\min})$ on X; its total mass coincides with $\mathrm{Vol}(\{\alpha\})$, i.e.,

$$\mathrm{Vol}(\{\alpha\}) = \int_X \mathrm{MA}_\alpha(\varphi)$$

(this follows from the comparison theorem and the fact that Monge–Ampère measures of locally bounded psh functions do not carry mass on analytic sets; see, e.g., [BEGZ08]). Next, notice that in general, the α-psh envelope $\varphi = \psi_{\min}$ corresponds *canonically* to α, so we may associate to α the following subset of X:

$$D = \{\varphi = 0\}. \tag{2.4}$$

Since φ is upper semicontinuous, the set D is compact. Moreover, a simple application of the maximum principle shows that $\alpha \geq 0$ pointwise on D (precisely as in Proposition 3.1 of [Ber07]: at any point z_0 where α is not semipositive, we can find complex coordinates and a small $\varepsilon > 0$ such that $\varphi(z) - \varepsilon|z - z_0|^2$ is subharmonic near z_0, using the fact that $dd^c\varphi \geq -\alpha$ or rather the induced inequality between traces, and so integrating over a small ball B_δ centered at z_0 gives $\varphi(z_0) - 0 \leq \int_{B_\delta} \varphi(z) - \varepsilon|z - z_0|^2 < 0$, showing that z_0 is not in D).

In particular, $\mathbf{1}_D\alpha$ is a positive $(1,1)$-form on X. From Theorem (1.4) we infer the following.

Corollary 2.5. *Assume that X is a Kähler manifold. For any smooth closed form α of type $(1,1)$ in a pseudoeffective class and $\varphi \leq 0$ the α-psh upper envelope, we have*

$$\mathrm{MA}_\alpha(\varphi) = \mathbf{1}_D\alpha^n, \qquad D = \{\varphi = 0\}, \tag{2.6}$$

as measures on X (provided the left-hand side is interpreted as a suitable weak limit) and

$$\mathrm{Vol}(\{\alpha\}) = \int_D \alpha^n \geq 0. \tag{2.7}$$

In particular, $\{\alpha\}$ is big if and only if $\int_D \alpha^n > 0$.

Proof. Let ω be a Kähler metric on X. First assume that the class $\{\alpha\}$ is big and let Z_0 be the singularity set of some strictly positive representative $\alpha + dd^c\psi_0 \geq \varepsilon\omega$ with analytic singularities. By Theorem (1.4), $\alpha + dd^c\varphi$ is in $L^\infty_{\text{loc}}(X \setminus Z_0)$. In particular (see [Dem89]), the Monge–Ampère measure $(\alpha + dd^c\varphi)^n$ has a locally bounded density on $X \setminus Z_0$ with respect to ω^n. Since by definition, the Monge–Ampère measure puts no mass on Z_0, it is enough to prove the identity (2.6) pointwise almost everywhere on X.

To this end, one argues essentially as in [Ber07] (where the class was assumed to be integral). First, a well-known local argument based on the solution of the Dirichlet problem for $(dd^c)^n$ (see, e.g., [BT76, BT82], and also Proposition 1.10 in [BB08]) proves that the Monge–Ampère measure $(\alpha + dd^c\varphi)^n$ of the envelope φ vanishes on the open set $(X \setminus Z_0) \setminus D$ (this uses only the fact that α has continuous potentials and the continuity of φ on $X \setminus Z_0$). Moreover, Theorem (1.4) implies that $\varphi \in C^1(X \setminus Z_0)$ and

$$\frac{\partial^2 \varphi}{\partial x_i \partial x_j} \in L^p_{\text{loc}} \tag{2.8}$$

for any $p \in \,]1,\infty[$ and $i, j \in [1, 2n]$. Even if this is slightly weaker than the situation in [Ber07], where it was shown that one can take $p = \infty$, the argument given in [Ber07] still goes through. Indeed, by well-known properties of measurable sets, D has Lebesgue density $\lim_{r\to 0} \lambda(D \cap B(x,r))/\lambda(B(x,r) = 1$ at almost every point $x \in D$, and since $\varphi = 0$ on D, we conclude that $\partial\varphi/\partial x_i = 0$ at those points (if the density is 1, no open cone of vertex x can be omitted and thus we can approach x from any direction by a sequence $x_\nu \to x$).

But the first derivative is Hölder continuous on $D \setminus Z_0$; hence $\partial\varphi/\partial x_i = 0$ everywhere on $D \setminus Z_0$. By repeating the argument for $\partial\varphi/\partial x_i$, that has a derivative in L^p (L^1 would even be enough), we conclude from Lebesgue's theorem that $\partial^2\varphi/\partial x_i \partial x_j = 0$ a.e. on $D \setminus Z_0$, hence that $\alpha + dd^c\varphi = \alpha$ on $D \setminus E$, where the set E has measure zero with respect to ω^n. This proves formula (2.6) in the case of a big class.

Finally, assume that $\{\alpha\}$ is pseudoeffective but not big. For any given positive number ε, we let $\alpha_\varepsilon = \alpha + \varepsilon\omega$ and denote by D_ε the corresponding set (2.4). Clearly α_ε represents a big class. Moreover, by the continuity of the volume function up to the boundary of the big cone [Bou02],

$$\text{Vol}(\{\alpha_\varepsilon\}) \to \text{Vol}(\{\alpha\}) \quad (= 0) \tag{2.9}$$

as ε tends to zero. Now observe that $D \subset D_\varepsilon$ (there are more $(\alpha + \varepsilon\omega)$-psh functions than α-psh functions, and so $\varphi \leq \varphi_\varepsilon \leq 0$; clearly, φ_ε increases with ε and $\varphi = \lim_{\varepsilon\to 0} \varphi_\varepsilon$; compare with Proposition 3.3 in [Ber07]). Therefore

$$\int_D \alpha^n \leq \int_{D_\varepsilon} \alpha^n \leq \int_{D_\varepsilon} \alpha_\varepsilon^n,$$

where we used that $\alpha \leq \alpha_\varepsilon$ in the second step.

Finally, since by the big case treated above, the right-hand side above is precisely $\mathrm{Vol}(\{\alpha_\varepsilon\})$, letting ε tend to zero and using (2.9) proves that $\int_D \alpha^n = 0 = \mathrm{Vol}(\{\alpha\})$ (and that $\mathrm{MA}_\alpha(\varphi) = 0$ if we interpret it as the limit of $\mathrm{MA}_{\alpha_\varepsilon}(\varphi_\varepsilon)$). This concludes the proof. □

In the case that $\{\alpha\}$ is an integer class, i.e., when it is the first Chern class $c_1(L)$ of a holomorphic line bundle L over X, the result of the corollary was obtained in [Ber07] under the additional assumption that X is a *projective* manifold; it was conjectured there that the result was also valid for integral classes over a nonprojective Kähler manifold.

Remark 2.10. In particular, the corollary shows that if $\{\alpha\}$ is big, there is always an α-plurisubharmonic function φ with minimal singularities such that $\mathrm{MA}_\alpha(\varphi)$ has an L^∞-density with respect to ω^n. This is a very useful fact when one is dealing with big classes that are not Kähler (see, for example, [BBGZ09]).

3 Application to Regularity of a Boundary Value Problem and a Variational Principle

In this section we will see how the main theorem may be interpreted as a regularity result for (1) a free boundary value problem for the Monge–Ampère operator and (2) a variational principle. For simplicity we consider only the case of a Kähler class.

3.1 A Free Boundary Value Problem for the Monge–Ampère Operator

Let (X,ω) be a Kähler manifold. Given a function $f \in \mathcal{C}^2(X)$, consider the following *free* boundary value problem:

$$\begin{cases} \mathrm{MA}_\omega(u) = 0 & \text{on } \Omega, \\ u = f & \text{on } \partial\Omega, \\ du = df & \end{cases}$$

for a pair (u,Ω), where u is an ω-psh function on $\overline{\Omega}$ that is in $\mathcal{C}^1(\overline{\Omega})$, and Ω is an open set in X. We have used the notation $\partial\Omega := \overline{\Omega} \smallsetminus \Omega$, but no regularity of the boundary is assumed. The reason that the set Ω is assumed to be part of the solution is that for a fixed Ω, the equations are overdetermined. Setting $u := \varphi + f$ and $\Omega := X \smallsetminus D$, where φ is the upper envelope with respect to $\alpha := dd^c f + \omega$, yields a solution. In fact, by Theorem (1.4), $u \in \mathcal{C}^{1,1-\delta}(\overline{\Omega})$ for any $\delta > 0$.

3.2 A Variational Principle

Fix a form α in a Kähler class $\{\alpha\}$ possessing continuous potentials. Consider the following energy functional defined on the convex space $\mathrm{PSH}(X,\alpha)\cap L^\infty$ of all α-psh functions that are bounded on X:

$$\mathcal{E}[\psi] := \frac{1}{n+1}\sum_{j=0}^{n}\int_X \psi(\alpha+dd^c\psi)^j\wedge\alpha^{n-j}. \tag{3.2.1}$$

This functional seems to first have appeared, independently, in the work of Aubin and Mabuchi on Kähler–Einstein geometry (in the case that α is a Kähler form). More geometrically, up to an additive constant, $\mathcal{E}$ can be defined as a primitive of the one-form on $\mathrm{PSH}(X,\alpha)\cap L^\infty$ defined by the measure-valued operator $\psi\mapsto \mathrm{MA}_\alpha(\psi)$.

As shown in [BB08] (version 1), the following variational characterization of the envelope φ holds:

Proposition 3.2.2. *The functional*

$$\psi\mapsto\mathcal{E}[\psi]-\int_X\psi(\alpha+dd^c\psi)^n$$

achieves its minimum value on the space $\mathrm{PSH}(X,\alpha)\cap L^\infty$ *precisely when* ψ *is equal to the envelope* φ *(defined with respect to* α*). Moreover, the minimum is achieved only at* φ*, up to an additive constant.*

Hence, the main theorem above can be interpreted as a regularity result for the functions in $\mathrm{PSH}(X,\alpha)\cap L^\infty$ minimizing the functional (3.2.1) in the case that α is assumed to have L^∞_{loc} coefficients. More generally, a similar variational characterization of φ can be given in the case of a *big* class $[\alpha]$ [BBGZ09].

4 Degenerate Monge–Ampère Equations and Geodesics in the Space of Kähler Metrics

Assume that (X,ω) is a compact Kähler manifold and that Σ is a Stein manifold with strictly pseudoconvex boundary, i.e., Σ admits a smooth strictly psh nonpositive function η_Σ that vanishes precisely on $\partial\Sigma$. The corresponding product manifold will be denoted by $M := \Sigma\times X$. By taking pullbacks, we identify η_Σ with a function on M and ω with a semipositive form on M. In this way, we obtain a Kähler form $\omega_M := \omega+dd^c\eta_\Sigma$ on M. Given a function f on M and a point s in Σ, we use the notation $f_s := f(s,\cdot)$ for the induced function on X.

Further, given a closed $(1,1)$-form α on M with bounded coefficients and a continuous function f on ∂M, we define the upper envelope

$$\varphi_{\alpha,f} := \sup\{\psi : \psi \in \mathrm{PSH}(M,\alpha)\cap C^0(M),\ \psi_{\partial M} \leq f\}. \tag{4.1}$$

Note that when Σ is a point and $f = 0$, this definition coincides with the one introduced in Sect. 1. Also, when F is a smooth function on the whole of M, the obvious translation $\psi \mapsto \psi' = \psi - F$ yields the relation

$$\varphi_{\beta,f-F} = \varphi_{\alpha,f} - F, \quad \text{where } \beta = \alpha + dd^cF. \tag{4.2}$$

The proof of the following lemma is a straightforward adaptation of the proof of Bedford–Taylor [BT76] in the case that M is a strictly pseudoconvex domain in $\mathbb{C}^n$.

Lemma 4.3. *Let α be a closed real $(1,1)$-form on M with bounded coefficients, such that $\alpha_{|\{s\}\times X} \geq \varepsilon_0\omega$ is positive definite for all $s \in \Sigma$. Then the corresponding envelope $\varphi = \varphi_{\alpha,0}$ vanishes on the boundary of M and is continuous on M. Moreover, $MA_\alpha(\varphi)$ vanishes in the interior of M.*

Proof. By (4.2), we have $\varphi_{\alpha,0} = \varphi_{\beta,0} + C\eta_\Sigma$, where $\beta = \alpha + Cdd^c\eta_\Sigma$ can be taken to be positive definite on M for $C \gg 1$, as is easily seen from the Cauchy–Schwarz inequality and the hypotheses on α. Therefore, we can assume without loss of generality that α is positive definite on M. Since 0 is a candidate for the supremum defining φ, it follows immediately that $0 \leq \varphi$ and hence $\varphi_{\partial M} = 0$. To see that φ is continuous on ∂M (from the inside), take an arbitrary candidate ψ for the sup and observe that

$$\psi \leq -C\eta_\Sigma$$

for $C \gg 1$, independent of ψ.

Indeed, since $dd^c\psi \geq -\alpha$, there is a large positive constant C such that the function $\psi + C\eta_\Sigma$ is strictly plurisubharmonic on $\Sigma \times \{x\}$ for all x. Thus the inequality above follows from the maximum principle applied to all slices $\Sigma \times \{x\}$. All in all, taking the sup over all such ψ gives

$$0 \leq \varphi \leq -C\eta_\Sigma.$$

But since $\eta_{\Sigma|\partial M} = 0$ and η_Σ is continuous, it follows that $\varphi(x_i) \to 0 = \varphi(x)$ when $x_i \to x \in \partial M$.

Next, fix a compact subset K in the interior of M and $\varepsilon > 0$. Let $M_\delta := \{\eta_\Sigma < -\delta\}$, where δ is sufficiently small to ensure that K is contained in $M_{4\delta}$. By the regularization results in [Dem92] or [Dem94], there is a sequence φ_j in $\mathrm{PSH}(M, \alpha - 2^{-j}\alpha) \cap C^0(M_{\delta/2})$ decreasing to the upper semicontinuous regularization φ^*. By replacing φ_j with $(1-2^{-j})^{-1}\varphi_j$, we can even assume $\varphi_j \in \mathrm{PSH}(M,\alpha) \cap C^0(M_{\delta/2})$. Put

$$\varphi_j' := \max\{\varphi_j - \varepsilon, C\eta_\Sigma\} \text{ on } M_\delta, \quad \text{and} \quad \varphi_j' := C\eta_\Sigma \text{ on } M \smallsetminus M_\delta.$$

On ∂M_δ we have $C\eta_\Sigma = -C\delta$, and we can take j so large that

$$\varphi_j < -C\eta_\Sigma + \varepsilon/2 = C\delta + \varepsilon/2,$$

so we will have $\varphi_j - \varepsilon < C\eta_\Sigma$ as soon as $2C\delta \le \varepsilon/2$. We simply take $\varepsilon = 4C\delta$. Then φ_j' is a well-defined continuous α-psh function on M, and φ_j' is equal to $\varphi_j - \varepsilon$ on $K \subset M_{4\delta}$, since $C\eta_\Sigma \le -4C\delta \le -\varepsilon \le \varphi_j - \varepsilon$ there. In particular, φ_j' is a candidate for the sup defining φ; hence $\varphi_j' \le \varphi \le \varphi^*$, and so

$$\varphi^* \le \varphi_j \le \varphi_j' + \varepsilon \le \varphi^* + \varepsilon$$

on K. This means that φ_j converges to φ uniformly on K, and therefore φ is continuous on K.

All in all this shows that $\varphi \in C^0(M)$. The last statement of the proposition follows from standard local considerations for envelopes due to Bedford–Taylor [BT76] (see also the exposition in [Dem89] and [Dem93]). □

Theorem 4.4. *Let α be a closed real $(1,1)$-form on M with bounded coefficients such that $\alpha_{|\{s\}\times X} \ge \varepsilon_0\omega$ is positive definite for all $s \in \Sigma$. Consider a continuous function f on ∂M such that $f_s \in \mathrm{PSH}(X,\alpha_s)$ for all $s \in \partial\Sigma$. Then the upper envelope $\varphi = \varphi_{\alpha,f}$ is the unique α-psh continuous solution of the Dirichlet problem*

$$\varphi = f \text{ on } \partial M, \qquad (dd^c u + \alpha)^{\dim M} = 0 \text{ on the interior } M^\circ. \tag{4.5}$$

Moreover, if f is $C^{1,1}$ on ∂M, then for any s in Σ, the restriction φ_s of φ on $\{s\}\times X$ has a dd^c in L^∞_{loc}. More precisely, we have a uniform bound $|dd^c\varphi_s|_\omega \le C$ a.e. on X, where C is a constant independent of s.

Proof. Without loss of generality, we may assume as in Lemma (4.3) that α is positive definite on M. Also, after adding a positive constant to f, that has only the effect of adding the same constant to $\varphi = \varphi_{\alpha,f}$, we may suppose that $\sup_{\partial M} f > 0$ (this will simplify a little bit the arguments below).

Continuity. Let us first prove the continuity statement in the theorem. In the case that f extends to a smooth function F in $\mathrm{PSH}(M,(1-\varepsilon)\alpha)$, the statement follows immediately from (4.2) and Lemma (4.3), since

$$f - F = 0 \text{ on } \partial M \text{ and } \beta = \alpha + dd^c F \ge \varepsilon\alpha \ge \varepsilon\varepsilon_0\omega.$$

Next, assume that f is *smooth* on ∂M and that $f_s \in \mathrm{PSH}(X,(1-\varepsilon)\alpha_s)$ for all $s \in \partial\Sigma$. If we take a smooth extension $\widetilde{f}$ of f to M and $C \gg 1$, we will get

$$\alpha + dd^c(\widetilde{f}(x,s) + C\eta_\Sigma(s)) \ge (\varepsilon/2)\alpha$$

on a sufficiently small neighborhood V of ∂M (again using Cauchy–Schwarz). Therefore, after enlarging C if necessary, we can define

$$F(x,s) = \max_\varepsilon(\widetilde{f}(x,s) + C\eta_\Sigma(s), 0)$$

with a regularized max function $\max_\varepsilon$ in such a way that the maximum is equal to 0 on a neighborhood of $M \smallsetminus V$ ($C \gg 1$ being used to ensure that $\widetilde{f} + C\eta_\Sigma < 0$ on $M \smallsetminus V$). Then F equals f on ∂M and satisfies

$$\alpha + dd^c F \geq (\varepsilon/2)\alpha \geq (\varepsilon\varepsilon_0/2)\omega$$

on M, and we can argue as previously. Finally, to handle the general case in that f is continuous with $f_s \in \mathrm{PSH}(X, \alpha_s)$ for every $s \in \Sigma$, we may, by a parametrized version of Richberg's regularization theorem applied to $(1-2^{-\nu})f + C2^{-\nu}$ (see, e.g., [Dem91]), write f as a decreasing uniform limit of smooth functions f_ν on ∂M satisfying $f_{\nu,s} \in \mathrm{PSH}(X, (1-2^{-\nu-1})\alpha_s)$ for every $s \in \partial\Sigma$. Then $\varphi_{\omega,f}$ is a decreasing uniform limit on M of the continuous functions φ_{ω,f_ν}, (as follows easily from the definition of $\varphi_{\omega,f}$ as an upper envelope).

Observe also that the uniqueness of a continuous solution of the Dirichlet problem (4.5) results from a standard application of the maximum principle for the Monge–Ampère operator. This proves the general case of the continuity statement.

Smoothness. Next, we turn to the proof of the smoothness statement. Since the proof is a straightforward adaptation of the proof of the main regularity result above, we will just briefly indicate the relevant modification. Quite similarly to what we did in Sect. 1, we consider an α-psh function ψ with $\psi \leq f$ on ∂M, and introduce the fiberwise transform Ψ_s of ψ_s on each $\{s\} \times X$, that is defined in terms of the exponential map $\mathrm{exph} : T_X \to X$, and we put

$$\Psi(z,s,t) = \Psi_s(z,t).$$

Then essentially the same calculations as in the previous case show that all properties of Ψ are still valid with the constant K depending on the $C^{1,1}$-norm of the local potentials $u(z,s)$ of α, the constant A depending only on ω and with

$$\partial\Psi(z,s,t)/\partial(\log t) := \lambda(z,s,t) \to \nu(\psi_s),$$

as $t \to 0^+$, where $\nu(\psi_s)$ is the Lelong number of the function ψ_s on X at z.

Moreover, the local vector-valued differential $\mathrm{d}z$ should be replaced by the differential $\mathrm{d}(z,s) = \mathrm{d}z + \mathrm{d}s$ in the previous formulas. Next, performing a Kiselman–Legendre transform fiberwise, we let

$$\psi_{c,\delta}(z,s) := (\psi_s)_{c,\delta}(z).$$

Then, using a parametrized version of the estimates of [Dem94] and the properties of $\Psi(z,s,w)$ as in Sect. 1, arguments derived from Kiselman's infimum principle show that

$$\alpha + dd^c \psi_{c,\delta} \geq (-A\min(c, \lambda(z,s,\delta)) - K\delta^2)\omega_M \geq -(Ac + K\delta^2)\omega_M, \qquad (4.6)$$

where ω_M is the Kähler form on M.

In addition to this, we have $|\psi_{c,\delta} - f| \leq K'\delta^2$ on ∂M by the hypothesis that f is $C^{1,1}$. For a sufficiently large constant C_1, we infer from this that $\theta = (1 - C_1(Ac + K\delta^2))\psi_{c,\delta}$ satisfies $\theta \leq f$ on ∂M (here we use the fact that $f > 0$ and hence that $\psi_0 \equiv 0$ is a candidate for the upper envelope). Moreover, $\alpha + dd^c\theta \geq 0$ on M thanks to (4.6) and the positivity of α. Therefore, θ is a candidate for the upper envelope, and so $\theta \leq \varphi = \varphi_{f,\alpha}$.

Repeating the arguments of Sect. 1 almost word for word, we obtain for $(\rho_t\varphi)(z,s) := \Phi(z,s,t)$ the analogue of estimate (1.15), that reduces simply to

$$\liminf_{t \to 0_+} \frac{\rho_t\varphi(z,s) - \varphi(z,s)}{t^2} \leq C_2,$$

since $\psi_0 \equiv 0$ in the present situation. The final conclusion follows from (1.16) and the related arguments already explained. □

In connection to the study of Wess–Zumino–Witten-type equations [Don99], [Don02] and geodesics in the space of Kähler metrics [Don99], [Don02], [Che00], it is useful to formulate the result of the previous theorem as an extension problem from $\partial\Sigma$, in the case that $\alpha(z,s) = \omega(z)$ does not depend on s.

To this end, let $F : \partial\Sigma \to \mathrm{PSH}(X,\omega)$ be the map defined by $F(s) = f_s$. Then the previous theorem gives a continuous "maximal plurisubharmonic" extension U of F to Σ, where $U(s) := u_s$, so that $U : \partial\Sigma \to \mathrm{PSH}(X,\omega)$.

Let us next specialize to the case in that $\Sigma := A$ is an annulus $R_1 < |s| < R_2$ in $\mathbb{C}$ and the boundary datum $f(x,s)$ is invariant under rotations $s \mapsto s\mathrm{e}^{i\theta}$. Denote by f^0 and f^1 the elements in $\mathrm{PSH}(X,\omega)$ corresponding to the two boundary circles of A. Then the previous theorem furnishes a continuous path f^t in $\mathrm{PSH}(X,\omega)$ if we put $t = \log|s|$, or rather $t = \log(|s|/R_1)/\log(R_2/R_1)$, to be precise. Following [PS08], the corresponding path of semipositive forms $\omega^t := \omega + dd^c f^t$ will be called a *(generalized) geodesic* in $\mathrm{PSH}(X,\omega)$ (compare also with Remark 4.8).

Corollary 4.7. *Assume that the semipositive closed $(1,1)$-forms ω^0 and ω^1 belong to the same Kähler class $\{\omega\}$ and have bounded coefficients. Then the geodesic ω^t connecting ω^0 and ω^1 is continuous on $[0,1] \times X$, and there is a constant C such that $\omega^t \leq C\omega$ on X, i.e., ω^t has uniformly bounded coefficients.*

In particular, the previous corollary shows that the space of all semipositive forms with bounded coefficients in a given Kähler class is "geodesically convex."

Remark 4.8. As shown in the work of Semmes, Mabuchi, and Donaldson, the space of Kähler metrics $\mathcal{H}_\omega$ in a given Kähler class $\{\omega\}$ admits a natural Riemannian structure defined in the following way (see [Che00] and references therein). First note that the map $u \mapsto \omega + dd^c u$ identifies $\mathcal{H}_\omega$ with the space of all smooth and strictly ω-psh functions, modulo constants. Now with the tangent space of $\mathcal{H}_\omega$ identified at the point $\omega + dd^c u \in \mathcal{H}_\omega$ with $C^\infty(X)/\mathbb{R}$, the squared norm of a tangent vector v at the point u is defined as

$$\int_X v^2(\omega + dd^c u)^n/n!.$$

Then the potentials f^t of any given geodesic ω^t in $\mathcal{H}_\omega$ are in fact solutions of the Dirichlet problem (4.5) above, with Σ an annulus and $t := \log|s|$; see [Che00].

However, the *existence* of a geodesic u_t in $\mathcal{H}_\omega$ connecting any given points u_0 and u_1 is an open and even dubious problem. In the case that Σ is a Riemann surface and the boundary datum f is smooth with $\alpha_s + dd^c f_s > 0$ on X for $s \in \partial\Sigma$, it was shown in [Che00] that the solution φ of the Dirichlet problem (4.5) has a total Laplacian that is bounded on M. See also [Blo09] for a detailed analysis of the proof in [Che00] and some refinements.

On the other hand, it is not known whether $\alpha_s + dd^c\varphi_s > 0$ for all $s \in \Sigma$, even under the assumption of rotational invariance, that appears in the case of geodesics as above. See [CT08], however, for results in this direction. A case similar to the degenerate setting in the previous corollary was also considered very recently in [PS08], building on [Blo09].

Remark 4.9. Note that the assumption $f \in C^2(\partial M)$ is not sufficient to obtain uniform estimates on the total Laplacian on M with respect to ω_M of the envelope u up to the boundary. To see this, let Σ be the unit ball in $\mathbb{C}^2$ and write $s = (s_1, s_2) \in \mathbb{C}^2$. Then $f(s) := (1 + \mathrm{Re}\, s_1)^{2-\varepsilon}$ is in $C^{4-2\varepsilon}(\partial M)$, and $u(x,s) := f(s)$ is the continuous solution of the Dirichlet problem (4.5). However, u is not in $C^{1,1}(M)$ at $(x; -1, 0) \in \partial M$ for any $x \in X$. Note that this example is the trivial extension of the example in [CNS86] for the real Monge–Ampère equation on the disk.

5 Regularity of "Supercanonical" Metrics

Let X be a compact complex manifold and $(L, h_{L,\gamma})$ a holomorphic line bundle over X equipped with a singular Hermitian metric $h_{L,\gamma} = \mathrm{e}^{-\gamma} h_L$ that satisfies $\int \mathrm{e}^{-\gamma} < +\infty$ locally on X, where h_L is a smooth metric on L. In fact, we can more generally consider the case in that $(L, h_{L,\gamma})$ is a "Hermitian $\mathbb{R}$-line bundle"; by this we mean that we have chosen a smooth real d-closed $(1,1)$-form α_L on X (whose dd^c cohomology class is equal to $c_1(L)$), and a specific current $T_{L,\gamma}$ representing it, namely $T_{L,\gamma} = \alpha_L + dd^c\gamma$, such that γ is a locally integrable function satisfying $\int \mathrm{e}^{-\gamma} < +\infty$.

An important special case is obtained by considering a klt (Kawamata log terminal) effective divisor Δ. In this situation, $\Delta = \sum c_j \Delta_j$ with $c_j \in \mathbb{R}$, and if g_j is a local generator of the ideal sheaf $\mathcal{O}(-\Delta_j)$ identifying it with the trivial invertible sheaf $g_j\mathcal{O}$, we take $\gamma = \sum c_j \log|g_j|^2$, $T_{L,\gamma} = \sum c_j[\Delta_j]$ (current of integration on Δ) and α_L given by any smooth representative of the same dd^c-cohomology class; the klt condition means precisely that

$$\int_V \mathrm{e}^{-\gamma} = \int_V \prod |g_j|^{-2c_j} < +\infty \tag{5.1}$$

on a small neighborhood V of any point in the support $|\Delta| = \bigcup \Delta_j$. (Condition (5.1) implies $c_j < 1$ for every j, and this in turn is sufficient to imply Δ klt if Δ is a normal crossing divisor; the line bundle L is then the real line bundle $\mathcal{O}(\Delta)$, that makes sense as a genuine line bundle only if $c_j \in \mathbb{Z}$.)

For each klt pair (X, Δ) such that $K_X + \Delta$ is pseudoeffective, Tsuji [Ts07a,Ts07b] has introduced a "supercanonical metric" that generalizes the metric introduced by Narasimhan and Simha [NS68] for projective algebraic varieties with ample canonical divisor. We take the opportunity to present here a simpler, more direct, and more general approach.

We assume from now on that $K_X + L$ is *pseudoeffective*, i.e., that the class $c_1(K_X) + \{\alpha_L\}$ is pseudoeffective, and under this condition, we are going to define a "supercanonical metric" on $K_X + L$. Select an arbitrary smooth Hermitian metric ω on X. We then find induced Hermitian metrics h_{K_X} on K_X and $h_{K_X+L} = h_{K_X} h_L$ on $K_X + L$ whose curvature is the smooth real $(1,1)$-form

$$\alpha = \Theta_{K_X+L, h_{K_X+L}} = \Theta_{K_X,\omega} + \alpha_L.$$

A singular Hermitian metric on $K_X + L$ is a metric of the form $h_{K_X+L,\varphi} = \mathrm{e}^{-\varphi} h_{K_X+L}$, where φ is locally integrable, and by the pseudoeffectivity assumption, we can find quasi-psh functions φ such that $\alpha + dd^c\varphi \geq 0$.

The metrics on L and $K_X + L$ can now be "subtracted" to give rise to a metric

$$h_{L,\gamma} h^{-1}_{K_X+L,\varphi} = \mathrm{e}^{\varphi-\gamma} h_L h^{-1}_{K_X+L} = \mathrm{e}^{\varphi-\gamma} h^{-1}_{K_X} = \mathrm{e}^{\varphi-\gamma} \mathrm{d}V_\omega$$

on $K_X^{-1} = \Lambda^n T_X$, since $h^{-1}_{K_X} = \mathrm{d}V_\omega$ is just the Hermitian (n,n) volume form on X. Therefore the integral $\int_X h_{L,\gamma} h^{-1}_{K_X+L,\varphi}$ has an intrinsic meaning, and it makes sense to require that

$$\int_X h_{L,\gamma} h^{-1}_{K_X+L,\varphi} = \int_X \mathrm{e}^{\varphi-\gamma} \mathrm{d}V_\omega \leq 1, \tag{5.2}$$

in view of the fact that φ is locally bounded from above and because of the assumption $\int \mathrm{e}^{-\gamma} < +\infty$. Observe that condition (5.2) can always be achieved by subtracting a constant from φ. We can now generalize Tsuji's supercanonical metrics on klt pairs (cf. [Ts07b]) as follows.

Definition 5.3. Let X be a compact complex manifold and let (L, h_L) be a Hermitian $\mathbb{R}$-line bundle on X associated with a smooth, real, closed $(1,1)$-form α_L. Assume that $K_X + L$ is pseudoeffective and that L is equipped with a singular Hermitian metric $h_{L,\gamma} = e^{-\gamma} h_L$ such that $\int e^{-\gamma} < +\infty$ locally on X. Take a Hermitian metric ω on X and define $\alpha = \Theta_{K_X+L,h_{K_X+L}} = \Theta_{K_X,\omega} + \alpha_L$. Then we define the supercanonical metric h_{can} of $K_X + L$ to be

$$h_{K_X+L,\mathrm{can}} = \inf_{\varphi} h_{K_X+L,\varphi} \quad \text{i.e.} \quad h_{K_X+L,\mathrm{can}} = e^{-\varphi_{\mathrm{can}}} h_{K_X+L}, \quad \text{where}$$

$$\varphi_{\mathrm{can}}(x) = \sup_{\varphi} \varphi(x) \ \text{ for all } \varphi \text{ with } \alpha + dd^c\varphi \geq 0, \ \int_X e^{\varphi-\gamma} dV_\omega \leq 1.$$

In particular, this gives a definition of the supercanonical metric on $K_X + \Delta$ for every klt pair (X, Δ) such that $K_X + \Delta$ is pseudoeffective, and as an even more special case, a supercanonical metric on K_X when K_X is pseudoeffective.

In the sequel, we assume that γ has analytic singularities, for otherwise, not much can be said. The mean value inequality then immediately shows that the quasi-psh functions φ involved in Definition (5.3) are globally uniformly bounded outside of the poles of γ, and therefore everywhere on X. Hence the envelopes $\varphi_{\mathrm{can}} = \sup_\varphi \varphi$ are indeed well defined and bounded above. As a consequence, we get a "supercanonical" current $T_{\mathrm{can}} = \alpha + dd^c\varphi_{\mathrm{can}} \geq 0$, and $h_{K_X+L,\mathrm{can}}$ satisfies

$$\int_X h_{L,\gamma} h^{-1}_{K_X+L,\mathrm{can}} = \int_X e^{\varphi_{\mathrm{can}}-\gamma} dV_\omega < +\infty. \tag{5.4}$$

It is easy to see that in Definition (5.3) the supremum is a maximum and that $\varphi_{\mathrm{can}} = (\varphi_{\mathrm{can}})^*$ everywhere, so that taking the upper semicontinuous regularization is not needed.

In fact, if $x_0 \in X$ is given and we write

$$(\varphi_{\mathrm{can}})^*(x_0) = \limsup_{x \to x_0} \varphi_{\mathrm{can}}(x) = \lim_{\nu \to +\infty} \varphi_{\mathrm{can}}(x_\nu) = \lim_{\nu \to +\infty} \varphi_\nu(x_\nu)$$

with suitable sequences $x_\nu \to x_0$ and (φ_ν) such that $\int_X e^{\varphi_\nu - \gamma} dV_\omega \leq 1$, the well-known weak compactness properties of quasi-psh functions in the L^1 topology imply the existence of a subsequence of (φ_ν) converging in L^1 and almost everywhere to a quasi-psh limit φ. Since $\int_X e^{\varphi_\nu - \gamma} dV_\omega \leq 1$ holds for every ν, Fatou's lemma implies that we have $\int_X e^{\varphi - \gamma} dV_\omega \leq 1$ in the limit. By taking a subsequence, we can assume that $\varphi_\nu \to \varphi$ in $L^1(X)$. Then for every $\varepsilon > 0$, the mean value $\fint_{B(x_\nu,\varepsilon)} \varphi_\nu$ satisfies

$$\fint_{B(x_0,\varepsilon)} \varphi = \lim_{\nu \to +\infty} \fint_{B(x_\nu,\varepsilon)} \varphi_\nu \geq \lim_{\nu \to +\infty} \varphi_\nu(x_\nu) = (\varphi_{\mathrm{can}})^*(x_0),$$

and hence we get $\varphi(x_0) = \lim_{\varepsilon \to 0} \fint_{B(x_0,\varepsilon)} \varphi \geq (\varphi_{\mathrm{can}})^*(x_0) \geq \varphi_{\mathrm{can}}(x_0)$, and therefore the sup is a maximum and $\varphi_{\mathrm{can}} = \varphi^*_{\mathrm{can}}$.

By elaborating on this argument, we can infer certain regularity properties of the envelope. However, there is no reason why the integral occurring in (5.4) should be equal to 1 when we take the upper envelope. As a consequence, neither the upper envelope nor its regularizations participate in the family of admissible metrics. This is why the estimates that we will be able to obtain are much weaker than in the case of envelopes normalized by a condition $\varphi \le 0$.

Theorem 5.5. *Let X be a compact complex manifold and (L, h_L) a holomorphic $\mathbb{R}$-line bundle such that $K_X + L$ is big. Assume that L is equipped with a singular Hermitian metric $h_{L,\gamma} = \mathrm{e}^{-\gamma} h_L$ with analytic singularities such that $\int \mathrm{e}^{-\gamma} < +\infty$ (klt condition). Denote by Z_0 the set of poles of a singular metric $h_0 = \mathrm{e}^{-\psi_0} h_{K_X+L}$ with analytic singularities on $K_X + L$ and by Z_γ the poles of γ (assumed analytic). Then the associated supercanonical metric h_{can} is continuous on $X \smallsetminus (Z_0 \cup Z_\gamma)$ and possesses some computable logarithmic modulus of continuity.*

Proof. With the notation already introduced, let $h_{K_X+L,\varphi} = \mathrm{e}^{-\varphi} h_{K_X+L}$ be a singular Hermitian metric such that its curvature satisfies $\alpha + dd^c\varphi \ge 0$ and $\int_X e^{\varphi-\gamma} \mathrm{d}V_\omega \le 1$. We apply to φ the regularization procedure defined in (1.6). Jensen's inequality implies

$$\mathrm{e}^{\Phi(z,w)} \le \int_{\zeta \in T_{X,z}} \mathrm{e}^{\varphi(\mathrm{exph}_z(w\zeta))} \chi(|\zeta|^2)\,\mathrm{d}V_\omega(\zeta).$$

If we change variables by putting $u = \mathrm{exph}_z(w\zeta)$, then in a neighborhood of the diagonal of $X \times X$ we have an inverse map $\mathrm{logh} : X \times X \to T_X$ such that $\mathrm{exph}_z(\mathrm{logh}(z,u)) = u$, and we obtain for w small enough,

$$\begin{aligned}
&\int_X \mathrm{e}^{\Phi(z,w)-\gamma(z)} \mathrm{d}V_\omega(z) \\
&\quad \le \int_{z\in X} \left(\int_{u \in X} \mathrm{e}^{\varphi(u)-\gamma(z)} \chi\left(\frac{|\mathrm{logh}(z,u)|^2}{|w|^2} \right) \frac{1}{|w|^{2n}} \mathrm{d}V_\omega(\mathrm{logh}(z,u)) \right) \mathrm{d}V_\omega(z) \\
&\quad = \int_{u \in X} P(u,w)\,\mathrm{e}^{\varphi(u)-\gamma(u)} \mathrm{d}V_\omega(u),
\end{aligned}$$

where P is a kernel on $X \times D(0,\delta_0)$ such that

$$P(u,w) = \int_{z \in X} \frac{1}{|w|^{2n}} \chi\left(\frac{|\mathrm{logh}(z,u)|^2}{|w|^2} \right) \frac{\mathrm{e}^{\gamma(u)-\gamma(z)} \mathrm{d}V_\omega(\mathrm{logh}(z,u))}{\mathrm{d}V_\omega(u)} \mathrm{d}V_\omega(z).$$

Let us first assume that γ is smooth (the case in that γ has logarithmic poles will be considered later). Then a change of variable $\zeta = \frac{1}{w} \mathrm{logh}(z,u)$ shows that P is smooth, and we have $P(u,0) = 1$. Since $P(u,w)$ depends only on $|w|$, we infer

$$P(u,w) \le 1 + C_0 |w|^2$$

for w small. This shows that the integral of $z \mapsto \mathrm{e}^{\Phi(z,w) - C_0|w|^2}$ will be at most equal to 1, and therefore if we define

$$\varphi_{c,\delta}(z) = \inf_{t \in]0,\delta]} \Phi(z,t) + Kt^2 - K\delta^2 - c \log \frac{t}{\delta} \tag{5.6}$$

as in (1.10), the function $\varphi_{c,\delta}(z) \le \Phi(z,\delta)$ will also satisfy

$$\int_X e^{\varphi_{c,\delta}(z) - C_0\delta^2 - \gamma(z)} dV_\omega \le 1. \tag{5.7}$$

Now, thanks to the assumption that $K_X + L$ is big, there exists a quasi-psh function ψ_0 with analytic singularities such that $\alpha + dd^c\psi_0 \ge \varepsilon_0\omega$. We can assume $\int_X e^{\psi_0 - \gamma} dV_\omega = 1$ after adjusting ψ_0 with a suitable constant. Consider a pair of points $x, y \in X$. We take φ such that $\varphi(x) = \varphi_{\mathrm{can}}(x)$ (this is possible by the above discussion). We define

$$\varphi_\lambda = \log\left(\lambda e^{\psi_0} + (1-\lambda)e^{\varphi}\right) \tag{5.8}$$

with a suitable constant $\lambda \in [0, 1/2]$, that will be fixed later, and obtain in this way regularized functions $\Phi_\lambda(z,w)$ and $\varphi_{\lambda,c,\delta}(z)$. This is obviously a compact family, and therefore the associated constants K needed in (5.6) are uniform in λ. Also, as in Sect. 1, we have

$$\alpha + dd^c\varphi_{\lambda,c,\delta} \ge -(Ac + K\delta^2)\,\omega \qquad \text{for all } \delta \in \,]0, \delta_0]. \tag{5.9}$$

Finally, we consider the linear combination

$$\theta = \frac{Ac + K\delta^2}{\varepsilon_0}\psi_0 + \left(1 - \frac{Ac + K\delta^2}{\varepsilon_0}\right)(\varphi_{\lambda,c,\delta} - C_0\delta^2). \tag{5.10}$$

Clearly, $\int_X e^{\varphi_\lambda - \gamma} dV_\omega \le 1$, and therefore θ also satisfies $\int_X e^{\theta - \gamma} dV_\omega \le 1$ by Hölder's inequality. Our linear combination is precisely taken so that $\alpha + dd^c\theta \ge 0$. Therefore, by definition of φ_{can}, we find that

$$\varphi_{\mathrm{can}} \ge \theta = \frac{Ac + K\delta^2}{\varepsilon_0}\psi_0 + \left(1 - \frac{Ac + K\delta^2}{\varepsilon_0}\right)(\varphi_{\lambda,c,\delta} - C_0\delta^2). \tag{5.11}$$

Assume $x \in X \smallsetminus Z_0$, so that $\varphi_\lambda(x) > -\infty$ and $\nu(\varphi_\lambda, x) = 0$. In (5.6), the infimum is reached either for $t = \delta$ or for t such that $c = t\frac{d}{dt}(\Phi_\lambda(z,t) + Kt^2)$. The function $t \mapsto \Phi_\lambda(z,t) + Kt^2$ is convex increasing in $\log t$ and tends to $\varphi_\lambda(z)$ as $t \to 0$. By convexity, this implies

$$\begin{aligned} c = t\frac{d}{dt}(\Phi_\lambda(z,t) + Kt^2) &\le \frac{(\Phi_\lambda(x,\delta_0) + K\delta_0^2) - (\Phi_\lambda(z,t) + Kt^2)}{\log(\delta_0/t)} \\ &\le \frac{C_1 - \varphi_\lambda(x)}{\log(\delta_0/t)} \le \frac{C_1 + |\psi_0(z)| + \log(1/\lambda)}{\log(\delta_0/t)}, \end{aligned}$$

and hence

$$\frac{1}{t} \le \max\left(\frac{1}{\delta}, \frac{1}{\delta_0}\exp\left(\frac{C_1 + |\psi_0(z)| + \log(1/\lambda)}{c}\right)\right). \tag{5.12}$$

This shows that t cannot be too small when the infimum is reached.

When t is taken equal to the value that achieves the infimum for $z = y$, we find that

$$\varphi_{\lambda,c,\delta}(y) = \Phi_\lambda(y,t) + Kt^2 - K\delta^2 - c\log\frac{t}{\delta} \geq \Phi_\lambda(y,t) + Kt^2 - K\delta^2. \qquad (5.13)$$

Since $z \mapsto \Phi_\lambda(z,t)$ is a convolution of φ_λ, we get a bound of the first-order derivative

$$|D_z\Phi_\lambda(z,t)| \leq \|\varphi_\lambda\|_{L^1(X)}\frac{C_2}{t} \leq \frac{C_3}{t},$$

and with respect to the geodesic distance $d(x,y)$ we infer from this that

$$\Phi_\lambda(y,t) \geq \Phi_\lambda(x,t) - \frac{C_3}{t}d(x,y). \qquad (5.14)$$

A combination of (5.11), (5.13), and (5.14) yields

$$\begin{aligned}
\varphi_{\mathrm{can}}(y) &\geq \frac{Ac+K\delta^2}{\varepsilon_0}\psi_0(y) + \left(1-\frac{Ac+K\delta^2}{\varepsilon_0}\right)\left(\Phi_\lambda(x,t)+Kt^2-K\delta^2-\frac{C_3}{t}d(x,y)\right)\\
&\geq \frac{Ac+K\delta^2}{\varepsilon_0}\psi_0(y) + \left(1-\frac{Ac+K\delta^2}{\varepsilon_0}\right)\left(\varphi_\lambda(x) - K\delta^2 - \frac{C_3}{t}d(x,y)\right)\\
&\geq \log\left(\lambda e^{\psi_0(x)}+(1-\lambda)e^{\varphi(x)}\right) - C_4\left((c+\delta^2)(|\psi_0(y)|+1)+\frac{1}{t}d(x,y)\right)\\
&\geq \varphi_{\mathrm{can}}(x) - C_5\left(\lambda+(c+\delta^2)(|\psi_0(y)|+1)+\frac{1}{t}d(x,y)\right),
\end{aligned}$$

if we use the fact that $\varphi_\lambda(x) \leq C_6$, $\varphi(x) = \varphi_{\mathrm{can}}(x)$, and $\log(1-\lambda) \geq -(2\log 2)\lambda$ for all $\lambda \in [0,1/2]$.

By exchanging the roles of x,y and using (5.12), we see that for all $c > 0$, $\delta \in]0,\delta_0]$, and $\lambda \in]0,1/2]$, there is an inequality

$$\left|\varphi_{\mathrm{can}}(y) - \varphi_{\mathrm{can}}(x)\right| \leq C_5\left(\lambda + (c+\delta^2)\left(\max(|\psi_0(x)|,|\psi_0(y)|)+1\right) + \frac{1}{t}d(x,y)\right), \qquad (5.15)$$

where

$$\frac{1}{t} \leq \max\left(\frac{1}{\delta}, \frac{1}{\delta_0}\exp\left(\frac{C_1+\max(|\psi_0(x)|,|\psi_0(y)|)+\log(1/\lambda)}{c}\right)\right). \qquad (5.16)$$

By taking c, δ, and λ small, one easily sees that this implies the continuity of φ_{can} on $X \smallsetminus Z_0$. More precisely, if we choose

$$\delta = d(x,y)^{1/2}, \qquad \lambda = \frac{1}{|\log d(x,y)|},$$
$$c = \frac{C_1+\max(|\psi_0(x)|,|\psi_0(y)|)+\left|\log|\log d(x,y)|\right|}{\log\delta_0/d(x,y)^{1/2}}$$

with $d(x,y) < \delta_0^2 < 1$, we get $\frac{1}{t} \le d(x,y)^{-1/2}$, whence an explicit (but certainly not optimal) modulus of continuity of the form

$$\left|\varphi_{\rm can}(y) - \varphi_{\rm can}(x)\right| \le C_7\big(\max(|\psi_0(x)|,|\psi_0(y)|)+1\big)^2 \frac{\big|\log|\log d(x,y)|\big|+1}{|\log d(x,y)|+1}.$$

When the weight γ has analytic singularities, the kernel $P(u,w)$ is no longer smooth and the volume estimate (5.7). In this case, we use a modification $\mu : \widehat{X} \to X$ in such a way that the singularities of $\gamma \circ \mu$ are divisorial, given by a divisor with normal crossings. If we put

$$\widehat{L} = \mu^*L - K_{\widehat{X}/X} = \mu^*L - E$$

(E the exceptional divisor), then we get an induced singular metric on $\widehat{L}$ that still satisfies the klt condition, and the corresponding supercanonical metric on $K_{\widehat{X}} + \widehat{L}$ is just the pullback by μ of the supercanonical metric on $K_X + L$. This shows that we may assume from the start that the singularities of γ are divisorial and given by a klt divisor Δ. In this case, a solution to the problem is to introduce a complete Hermitian metric $\hat{\omega}$ of uniformly bounded curvature on $X \smallsetminus |\Delta|$ using the Poincaré metric on the punctured disk as a local model transversal to the components of Δ. The Poincaré metric on the punctured unit disk is given by

$$\frac{|dz|^2}{|z|^2(\log|z|)^2},$$

and the singularity of $\hat{\omega}$ along the component $\Delta_j = \{g_j(z) = 0\}$ of Δ is given by

$$\hat{\omega} = \sum -dd^c \log|\log|g_j|| \qquad \mathrm{mod}\ C^\infty.$$

Since such a metric has bounded geometry and this is all that we need for the calculations of [Dem94] to work, the estimates that we have made here are still valid, especially the crucial lower bound $\alpha + dd^c\varphi_{\lambda,c,\delta} \ge -(Ac + K\delta^2)\,\hat{\omega}$. In order to compensate this loss of positivity, we need a quasi-psh function $\hat{\psi}_0$ such that $\alpha + dd^c\hat{\psi}_0 \ge \varepsilon_0\hat{\omega}$, but such a lower bound is possible by adding terms of the form $-\varepsilon_1 \log|\log|g_j||$ to our previous quasi-psh function ψ_0.

With respect to the Poincaré metric, a δ-ball of center z_0 in the punctured disk is contained in the corona

$$|z_0|^{e^{-\delta}} < |z| < |z_0|^{e^{\delta}},$$

and it is easy to see from this that the mean value of $|z|^{-2a}$ on a δ-ball of center z_0 is multiplied by at most $|z_0|^{-2a\delta}$. This implies that a function of the form $\hat{\varphi}_{c,\delta} = \varphi_{c,\delta} + C_9\delta\sum\log|g_j|$ will actually give rise to an integral $\int_X e^{\hat{\varphi}_{c,\delta}-\gamma}dV_\omega \le 1$. We see that the term δ^2 in (5.15) has to be replaced by a term of the form

$$\delta\sum\max\big(|\log|g_j(x)||,|\log|g_j(x)||\big).$$

This is enough to obtain the continuity of $\varphi_{\rm can}$ on $X \smallsetminus (Z_0 \cup |\Delta|)$, as well as an explicit logarithmic modulus of continuity. □

Algebraic version 5.17. Since the klt condition is open and $K_X + L$ is assumed to be big, we can always perturb L a little bit, and after blowing up X, assume that X is projective and that $(L, h_{L,\gamma})$ is obtained as a sum of $\mathbb{Q}$-divisors

$$L = G + \Delta,$$

where Δ is klt and G is equipped with a smooth metric h_G (from that $h_{L,\gamma}$ is inferred, with Δ as its poles, so that $\Theta_{L,h_{L,\gamma}} = \Theta_{G,L_G} + [\Delta]$). Clearly this situation is "dense" in what we have been considering before, just as $\mathbb{Q}$ is dense in $\mathbb{R}$. In this case, it is possible to give a more algebraic definition of the supercanonical metric φ_{can}, following the original idea of Narasimhan–Simha [NS68] (see also Tsuji [Ts07a]) – the case considered by these authors is the special situation in that $G = 0$, $h_G = 1$ (and moreover, $\Delta = 0$ and K_X ample, for [NS68]).

In fact, if m is a large integer that is a multiple of the denominators involved in G and Δ, we can consider sections

$$\sigma \in H^0(X, m(K_X + G + \Delta)).$$

We view them rather as sections of $m(K_X + G)$ with poles along the support $|\Delta|$ of our divisor. Then $(\sigma \wedge \overline{\sigma})^{1/m} h_G$ is a volume form with integrable poles along $|\Delta|$ (this is the klt condition for Δ). Therefore one can normalize σ by requiring that

$$\int_X (\sigma \wedge \overline{\sigma})^{1/m} h_G = 1.$$

Each of these sections defines a singular Hermitian metric on $K_X + L = K_X + G + \Delta$, and we can take the regularized upper envelope

$$\varphi_{\mathrm{can}}^{\mathrm{alg}} = \left(\sup_{m,\sigma} \frac{1}{m} \log |\sigma|^2_{h^m_{K_X+L}} \right)^* \tag{5.18}$$

of the weights associated with a smooth metric h_{K_X+L}. It is clear that $\varphi_{\mathrm{can}}^{\mathrm{alg}} \le \varphi_{\mathrm{can}}$, since the supremum is taken on the smaller set of weights $\varphi = \frac{1}{m} \log |\sigma|^2_{h^m_{K_X+L}}$, and the equalities

$$\begin{aligned} e^{\varphi-\gamma} dV_\omega &= |\sigma|^{2/m}_{h^m_{K_X+L}} e^{-\gamma} dV_\omega \\ &= (\sigma \wedge \overline{\sigma})^{1/m} e^{-\gamma} h_L = (\sigma \wedge \overline{\sigma})^{1/m} h_{L,\gamma} = (\sigma \wedge \overline{\sigma})^{1/m} h_G \end{aligned}$$

imply $\int_X e^{\varphi-\gamma} dV_\omega \le 1$.

We claim that the inequality $\varphi_{\mathrm{can}}^{\mathrm{alg}} \le \varphi_{\mathrm{can}}$ is an equality. The proof is an immediate consequence of the following statement, based in turn on the Ohsawa–Takegoshi theorem [OhT87] and the approximation technique of [Dem92].

Proposition 5.19. *With $L = G + \Delta$, ω, $\alpha = \Theta_{K_X+L,h_{K_X+L}}$, γ as above, and $K_X + L$ assumed to be big, fix a singular Hermitian metric $e^{-\varphi}h_{K_X+L}$ of curvature $\alpha + dd^c\varphi \geq 0$ such that $\int_X e^{\varphi-\gamma}dV_\omega \leq 1$. Then φ is equal to a regularized limit*

$$\varphi = \left(\limsup_{m\to+\infty} \frac{1}{m}\log|\sigma_m|^2_{h^m_{K_X+L}}\right)^*$$

for a suitable sequence of sections $\sigma_m \in H^0(X, m(K_X + G + \Delta))$ with $\int_X(\sigma_m \wedge \overline{\sigma}_m)^{1/m}h_G \leq 1$.

Proof. By our assumption, there exists a quasi-psh function ψ_0 with analytic singularity set Z_0 such that

$$\alpha + dd^c\psi_0 \geq \varepsilon_0\omega > 0,$$

and we can assume $\int_C e^{\psi_0-\gamma}dV_\omega < 1$ (the strict inequality will be useful later). For $m \geq p \geq 1$, this defines a singular metric $\exp(-(m-p)\varphi - p\psi_0)h^m_{K_X+L}$ on $m(K_X + L)$ with curvature greater than or equal to $p\varepsilon_0\omega$, and therefore a singular metric

$$h_{L'} = \exp(-(m-p)\varphi - p\psi_0)h^m_{K_X+L}h^{-1}_{K_X}$$

on $L' = (m-1)K_X + mL$ whose curvature $\Theta_{L',h_{L'}} \geq (p\varepsilon_0 - C_0)\omega$ is arbitrarily large if p is large enough.

Let us fix a finite covering of X by coordinate balls. Pick a point x_0 and one of the coordinate balls B containing x_0. By the Ohsawa–Takegoshi extension theorem applied to the ball B, we can find a section σ_B of $K_X + L' = m(K_X + L)$ that has norm 1 at x_0 with respect to the metric $h_{K_X+L'}$ and $\int_B |\sigma_B|^2_{h_{K_X+L'}}dV_\omega \leq C_1$ for some uniform constant C_1 depending on the finite covering, but independent of m, p, x_0.

Now we use a cutoff function $\theta(x)$ with $\theta(x) = 1$ near x_0 to truncate σ_B and solve a $\overline{\partial}$-equation for $(n,1)$-forms with values in L to get a global section σ on X with $|\sigma(x_0)|_{h_{K_X+L'}} = 1$. For this we need to multiply our metric by a truncated factor $\exp(-2n\theta(x)\log|x - x_0|)$ so as to get solutions of $\overline{\partial}$ vanishing at x_0. However, this perturbs the curvature by bounded terms, and we can absorb them again by taking p larger. In this way, we obtain

$$\int_X |\sigma|^2_{h_{K_X+L'}}dV_\omega = \int_X |\sigma|^2_{h^m_{K_X+L}} e^{-(m-p)\varphi - p\psi_0}dV_\omega \leq C_2. \tag{5.20}$$

Taking $p > 1$, the Hölder inequality for conjugate exponents m, $\frac{m}{m-1}$ implies

$$\int_X (\sigma \wedge \overline{\sigma})^{\frac{1}{m}}h_G = \int_X |\sigma|^{2/m}_{h^m_{K_X+L}} e^{-\gamma}dV_\omega$$

$$= \int_X \left(|\sigma|^2_{h^m_{K_X+L}} e^{-(m-p)\varphi - p\psi_0}\right)^{\frac{1}{m}} \left(e^{(1-\frac{p}{m})\varphi + \frac{p}{m}\psi_0 - \gamma}\right)dV_\omega$$

$$\leq C_2^{\frac{1}{m}} \left(\int_X \left(e^{(1-\frac{p}{m})\varphi + \frac{p}{m}\psi_0 - \gamma} \right)^{\frac{m}{m-1}} dV_\omega \right)^{\frac{m-1}{m}}$$

$$\leq C_2^{\frac{1}{m}} \left(\int_X \left(e^{\varphi-\gamma} \right)^{\frac{m-p}{m-1}} \left(e^{\frac{p}{p-1}(\psi_0-\gamma)} \right)^{\frac{p-1}{m-1}} dV_\omega \right)^{\frac{m-1}{m}}$$

$$\leq C_2^{\frac{1}{m}} \left(\int_X e^{\frac{p}{p-1}(\psi_0-\gamma)} dV_\omega \right)^{\frac{p-1}{m}}$$

using the hypothesis $\int_X e^{\varphi-\gamma} dV_\omega \leq 1$ and another application of Hölder's inequality. Since klt is an open condition and $\lim_{p\to+\infty} \int_X e^{\frac{p}{p-1}(\psi_0-\gamma)} dV_\omega = \int_X e^{\psi_0-\gamma} dV_\omega < 1$, we can take p large enough to ensure that

$$\int_X e^{\frac{p}{p-1}(\psi_0-\gamma)} dV_\omega \leq C_3 < 1.$$

Therefore, we see that

$$\int_X (\sigma \wedge \overline{\sigma})^{\frac{1}{m}} h_G \leq C_2^{\frac{1}{m}} C_3^{\frac{p-1}{m}} \leq 1$$

for p large enough. On the other hand,

$$|\sigma(x_0)|^2_{h_{K_X+L'}} = |\sigma(x_0)|^2_{h^m_{K_X+L}} e^{-(m-p)\varphi(x_0)-p\psi_0(x_0)} = 1,$$

and thus

$$\frac{1}{m} \log |\sigma(x_0)|^2_{h^m_{K_X+L}} = \left(1 - \frac{p}{m}\right) \varphi(x_0) + \frac{p}{m} \psi_0(x_0), \tag{5.21}$$

and as a consequence,

$$\frac{1}{m} \log |\sigma(x_0)|^2_{h^m_{K_X+L}} \longrightarrow \varphi(x_0)$$

whenever $m \to +\infty$, $\frac{p}{m} \to 0$, as long as $\psi_0(x_0) > -\infty$.

In the above argument, we can in fact interpolate in finitely many points $x_1, x_2, \ldots, x_q$, provided that $p \geq C_4 q$. Therefore, if we take a suitable dense subset $\{x_q\}$ and a "diagonal" sequence associated with sections $\sigma_m \in H^0(X, m(K_X + L))$ with $m \gg p = p_m \gg q = q_m \to +\infty$, we infer that

$$\left(\limsup_{m\to+\infty} \frac{1}{m} \log |\sigma_m(x)|^2_{h^m_{K_X+L}} \right)^* \geq \limsup_{x_q \to x} \varphi(x_q) = \varphi(x) \tag{5.22}$$

(the latter equality occurring if $\{x_q\}$ is suitably chosen with respect to φ). In the other direction, (5.20) implies a mean value estimate

$$\frac{1}{\pi^n r^{2n}/n!} \int_{B(x,r)} |\sigma(z)|^2_{h^m_{K_X+L}} dz \leq \frac{C_5}{r^{2n}} \sup_{B(x,r)} e^{(m-p)\varphi + p\psi_0}$$

on every coordinate ball $B(x,r) \subset X$. The function $|\sigma_m|^2_{h^m_{K_X+L}}$ is plurisubharmonic after we correct the not necessarily positively curved smooth metric h_{K_X+L} by a factor of the form $\exp(C_6|z-x|^2)$. Hence the mean value inequality shows that

$$\frac{1}{m}\log|\sigma_m(x)|^2_{h^m_{K_X+L}} \le \frac{1}{m}\log\frac{C_5}{r^{2n}} + C_6 r^2 + \sup_{B(x,r)}\Big(1-\frac{p_m}{m}\Big)\varphi + \frac{p_m}{m}\psi_0.$$

By taking in particular $r = 1/m$ and letting $m \to +\infty$, $p_m/m \to 0$, we see that the opposite of inequality (5.22) also holds. □

Remark 5.23. We can rephrase our results in slightly different terms. In fact, let us put

$$\varphi_m^{\mathrm{alg}} = \sup_{\sigma}\frac{1}{m}\log|\sigma|^2_{h^m_{K_X+L}}, \qquad \sigma \in H^0(X, m(K_X+G+\Delta)),$$

with normalized sections σ such that $\int_X (\sigma\wedge\overline{\sigma})^{1/m} h_G = 1$. Then φ_m^{alg} is quasi-psh (the supremum is taken over a compact set in a finite-dimensional vector space), and by passing to the regularized supremum over all σ and all φ in (5.21), we get

$$\varphi_{\mathrm{can}} \ge \varphi_m^{\mathrm{alg}} \ge \Big(1-\frac{p}{m}\Big)\varphi_{\mathrm{can}}(x) + \frac{p}{m}\psi_0(x).$$

Since φ_{can} is bounded from above, we find in particular that

$$0 \le \varphi_{\mathrm{can}} - \varphi_m^{\mathrm{alg}} \le \frac{C}{m}(|\psi_0(x)|+1).$$

This implies that $(\varphi_m^{\mathrm{alg}})$ converges uniformly to φ_{can} on every compact subset of $X \subset Z_0$, and in this way we infer again (in a purely qualitative manner) that φ_{can} is continuous on $X \smallsetminus Z_0$. Moreover, we also see that in (5.18), the upper semicontinuous regularization is not needed on $X \smallsetminus Z_0$; in case K_X+L is ample, it is not needed at all, and we have uniform convergence of $(\varphi_m^{\mathrm{alg}})$ to φ_{can} on the whole of X. Obtaining such a uniform convergence when K_X+L is just big looks like a more delicate question, related, for instance, to abundance of K_X+L on those subvarieties Y where the restriction $(K_X+L)_{|Y}$ would be, for example, nef but not big.

Generalization 5.24. In the general case that L is a $\mathbb{R}$-line bundle and K_X+L is merely pseudo-effective, a similar algebraic approximation can be obtained. We take instead sections

$$\sigma \in H^0(X, mK_X + \lfloor mG\rfloor + \lfloor m\Delta\rfloor + p_m A)$$

where (A, h_A) is a positive line bundle, $\Theta_{A,h_A} \ge \varepsilon_0\omega$, and replace the definition of $\varphi_{\mathrm{can}}^{\mathrm{alg}}$ by

$$\varphi_{\mathrm{can}}^{\mathrm{alg}} = \Big(\limsup_{m\to+\infty}\sup_{\sigma}\frac{1}{m}\log|\sigma|^2_{h_{mK_X+\lfloor mG\rfloor+p_m A}}\Big)^*, \tag{5.25}$$

$$\int_X (\sigma \wedge \overline{\sigma})^{\frac{2}{m}} h^{\frac{1}{m}}_{\lfloor mG \rfloor + p_m A} \leq 1, \tag{5.26}$$

where $m \gg p_m \gg 1$ and $h^{1/m}_{\lfloor mG \rfloor}$ is chosen to converge uniformly to h_G.

We then find again $\varphi_{\mathrm{can}} = \varphi^{\mathrm{alg}}_{\mathrm{can}}$, with an almost identical proof, though we no longer have a sup in the envelope, but just a lim sup. The analogue of Proposition (5.19) also holds in this context, with an appropriate sequence of sections $\sigma_m \in H^0(X, mK_X + \lfloor mG \rfloor + \lfloor m\Delta \rfloor + p_m A)$.

Remark 5.27. The envelopes considered in Sect. 1 are envelopes constrained by an L^∞ condition, while the present ones are constrained by an L^1 condition. It is possible to interpolate and to consider envelopes constrained by an L^p condition. More precisely, assuming that $\frac{1}{p}K_X + L$ is pseudoeffective, we look at metrics $\mathrm{e}^{-\varphi} h_{\frac{1}{p}K_X+L}$ and normalize them with the L^p condition

$$\int_X \mathrm{e}^{p\varphi-\gamma} \mathrm{d}V_\omega \leq 1.$$

This is actually an L^1 condition for the induced metric on pL, and therefore we can just apply the above after replacing L by pL. If we assume, moreover, that L is pseudoeffective, it is clear that the L^p condition converges to the L^∞ condition $\varphi \leq 0$ if we normalize γ by requiring $\int_X \mathrm{e}^{-\gamma} \mathrm{d}V_\omega = 1$.

Remark 5.28. It would be nice to have a better understanding of the supercanonical metrics. In case X is a curve, this should be easier. In fact, X then has a Hermitian metric ω with constant curvature, that we normalize by requiring that $\int_X \omega = 1$, and we can also suppose $\int_X \mathrm{e}^{-\gamma}\omega = 1$. The class $\lambda = c_1(K_X + L) \geq 0$ is a number, and we take $\alpha = \lambda\omega$. Our envelope is $\varphi_{\mathrm{can}} = \sup \varphi$, where $\lambda\omega + dd^c\varphi \geq 0$ and $\int_X \mathrm{e}^{\varphi-\gamma}\omega \leq 1$.

If $\lambda = 0$, then φ must be constant, and clearly $\varphi_{\mathrm{can}} = 0$. Otherwise, if $G(z,a)$ denotes the Green function such that $\int_X G(z,a)\,\omega(z) = 0$ and $dd^c G(z,a) = \delta_a - \omega(z)$, we obtain

$$\varphi_{\mathrm{can}}(z) \geq \sup_{a \in X} \left(\lambda G(z,a) - \log \int_{z \in X} \mathrm{e}^{\lambda G(z,a) - \gamma(z)} \omega(z) \right)$$

by taking the envelope already over $\varphi(z) = \lambda G(z,a) - \text{const}$. It is natural to ask whether this is always an equality, i.e., whether the extremal functions are always given by one of the Green functions, especially when $\gamma = 0$.

References

[BT76] Bedford, E. and Taylor, B.A.: *The Dirichlet problem for the complex Monge–Ampére equation*; Invent. Math. **37** (1976) 1–44

[BT82] Bedford, E. and Taylor, B.A.: *A new capacity for plurisubharmonic functions*; Acta Math. **149** (1982) 1–41

[Ber07] Berman, R.: *Bergman kernels and equilibrium measures for line bundles over projective manifolds*; Amer. J. Math. **131** (2009) no. 5, 1485–1524

[BB08] Berman, R., Boucksom, S.: *Growth of balls of holomorphic sections and energy at equilibrium*; Invent. Math. **181** (2010) no. 2, 337–394

[BBGZ09] Berman, R., Boucksom, S., Guedj, V., Zeriahi, A.: *A variational approach to solving complex Monge–Ampère equations*; in preparation

[Blo09] Błocki, Z: *On geodesics in the space of Kähler metrics*; Preprint 2009 available at http://gamma.im.uj.edu.pl/~blocki/publ/

[Bou02] Boucksom, S.: *On the volume of a line bundle*; Internat. J. Math **13** (2002), 1043–1063

[BEGZ08] Boucksom, S., Eyssidieux, P., Guedj, V., Zeriahi, A.: *Monge–Ampère equations in big cohomology classes*; Acta Math. **205** (2010) no. 2, 199–262

[CNS86] Caffarelli, L., Nirenberg, L., Spruck, J.: *The Dirichlet problem for the degenerate Monge–Ampère equation*; Rev. Mat. Iberoamericana **2** (1986), 19–27

[Che00] Chen, X.: *The space of Kähler metrics;* J. Differential Geom. **56** (2000), 189–234

[CT08] Chen, X. X., Tian, G.: *Geometry of Kähler metrics and foliations by holomorphic discs;* Publ. Math. Inst. Hautes Études Sci. **107** (2008), 1–107

[Dem89] Demailly, J.-P.: *Potential Theory in Several Complex Variables*; Manuscript available at www-fourier.ujf-grenoble.fr/~demailly/books.html, [D3]

[Dem91] Demailly, J.-P.: *Complex analytic and algebraic geometry;* manuscript Institut Fourier, first edition 1991, available online at http://www-fourier.ujf-grenoble.fr/~demailly/books.html

[Dem92] Demailly, J.-P.: *Regularization of closed positive currents and Intersection Theory*; J. Alg. Geom. **1** (1992), 361–409

[Dem93] Demailly, J.-P.: *Monge–Ampère operators, Lelong numbers and intersection theory*; Complex Analysis and Geometry, Univ. Series in Math., edited by V. Ancona and A. Silva, Plenum Press, New-York, 1993

[Dem94] Demailly, J.-P.: *Regularization of closed positive currents of type* $(1,1)$ *by the flow of a Chern connection;* Actes du Colloque en l'honneur de P. Dolbeault (Juin 1992), édité par H. Skoda et J.-M. Trépreau, Aspects of Mathematics, Vol. E 26, Vieweg, 1994, 105–126

[Don99] Donaldson, S.K.: *Symmetric spaces, Kähler geometry and Hamiltonian dynamics;* Northern California Symplectic Geometry Seminar, 13–33, Amer. Math. Soc. Transl. Ser. 2, 196, Amer. Math. Soc., Providence, RI, 1999

[Don02] Donaldson, S.K.: *Holomorphic discs and the complex Monge–Ampère equation;* J. Symplectic Geom. **1** (2002), 171–196

[Kis78] Kiselman, C. O.: *The partial Legendre transformation for plurisubharmonic functions;* Inventiones Math. **49** (1978) 137–148

[Kis94] Kiselman, C. O.: *Attenuating the singularities of plurisubharmonic functions;* Ann. Polonici Mathematici **60** (1994) 173–197

[NS68] Narasimhan, M.S., Simha, R.R.: *Manifolds with ample canonical class;* Inventiones Math. **5** (1968) 120–128

[OhT87] Ohsawa, T., Takegoshi, K.: *On the extension of* L^2 *holomorphic functions;* Math. Zeitschrift **195** (1987) 197–204

[PS08] Phong, D.H., Sturm, J.: *The Dirichlet problem for degenerate complex Monge–Ampère equations;* Comm. Anal. Geom. **18** (2010) no. 1, 145–170

[Rud66] Rudin, W.: *Real and complex analysis;* McGraw-Hill, 1966, third edition 1987

[Ts07a] Tsuji, H.: *Canonical singular Hermitian metrics on relative canonical bundles;* arXiv:0704.0566

[Ts07b] Tsuji, H.: *Canonical volume forms on compact Kähler manifolds;* arXiv: 0707.0111

Toward a Generalized Shapiro and Shapiro Conjecture

Alex Degtyarev

To my teacher Oleg Viro on his 60th birthday

Abstract We obtain a new, asymptotically better, bound $g \leqslant \frac{1}{4}d^2 + O(d)$ on the genus of a curve that may violate the generalized total reality conjecture. The bound covers all known cases except $g = 0$ (the original conjecture).

Keywords Shapiro and Shapiro conjecture • Real variety • Discriminant form • Alexander module

1 Introduction

The original (rational) total reality conjecture suggested by B. and M. Shapiro in 1993 states that if all flattening points of a regular curve $\mathbb{P}^1 \to \mathbb{P}^n$ belong to the real line $\mathbb{P}^1_{\mathbb{R}} \subset \mathbb{P}^1$, then the curve can be made real by an appropriate projective transformation of $\mathbb{P}^n$. (The *flattening points* are the points in the source $\mathbb{P}^1$ where the first n derivatives of the map are linearly dependent. In the case $n = 1$, a curve is a meromorphic function, and the flattening points are its critical points.) There are quite a few interesting and not always straightforward restatements of this conjecture, in terms of the Wronsky map, Schubert calculus, dynamical systems, etc.

Although supported by extensive numerical evidence, the conjecture proved extremely difficult to settle. It was not before 2002 that the first result appeared, due to Eremenko and Gabrielov [4], settling the case $n = 1$, i.e., meromorphic functions on $\mathbb{P}^1$. Later, a number of sporadic results were announced, and the conjecture was

A. Degtyarev (✉)
Department of Mathematics, Bilkent University, 06800 Ankara, Turkey
e-mail: degt@fen.bilkent.edu.tr

I. Itenberg et al. (eds.), *Perspectives in Analysis, Geometry, and Topology: On the Occasion of the 60th Birthday of Oleg Viro*, Progress in Mathematics 296,
DOI 10.1007/978-0-8176-8277-4_4,

proved in full generality in 2005 by Mukhin et al.; see [6]. The proof, revealing a deep connection between Schubert calculus and the theory of integrable systems, is based on the Bethe ansatz method in the Gaudin model.

In the meanwhile, a number of generalizations of the conjecture were suggested. In this paper, we deal with one of them, see [3] and Problem 1.1 below, replacing the source $\mathbb{P}^1$ with an arbitrary compact complex curve (however, restricting n to 1, i.e., to the case of meromorphic functions). Due to the lack of evidence, the authors chose to state the assertion as a problem rather than a conjecture.

Recall that a *real variety* is a complex algebraic (analytic) variety X supplied with a *real structure*, i.e., an antiholomorphic involution $c\colon X \to X$. Given two real varieties (X,c) and (Y,c'), a regular map $f\colon X \to Y$ is called *real* if it commutes with the real structures: $f \circ c = c' \circ f$.

Problem 1.1 (see [3]). Let (C,c) be a real curve and let $f\colon C \to \mathbb{P}^1$ be a regular map such that

1. All critical points and critical values of f are distinct;
2. All critical points of f are real.

Is it true that f is real with respect to an appropriate real structure in $\mathbb{P}^1$?

The condition that the critical points of f be distinct includes, in particular, the requirement that each critical point be simple, i.e., have ramification index 2.

A pair of integers $g \geqslant 0$, $d \geqslant 1$ is said to have the *total reality property* if the answer to Problem 1.1 is affirmative for any curve C of genus g and map f of degree d. At present, the total reality property is known for the following pairs (g,d):

- $(0,d)$ for any $d \geqslant 1$ (the original conjecture; see [4]);
- (g,d) for any $d \geqslant 1$ and $g > G_1(d) := \frac{1}{3}(d^2 - 4d + 3)$; see [3];
- (g,d) for any $g \geqslant 0$ and $d \leqslant 4$; see [3] and [1].

The principal result of the present paper is the following theorem.

Theorem 1.2. *Any pair (g,d) with $d \geqslant 1$ and g satisfying the inequality*

$$g > G_0(d) := \begin{cases} k^2 - 2k, & \text{if } d = 2k \text{ is even,} \\ k^2 - \dfrac{10}{3}k + \dfrac{7}{3}, & \text{if } d = 2k-1 \text{ is odd} \end{cases}$$

has the total reality property.

Remark 1.3. Note that one has $G_0(d) - G_1(d) \leqslant -\frac{1}{3}(k-1)^2 \leqslant 0$, where $k = [\frac{1}{2}(d+1)]$. Theorem 1.2 covers the values $d = 2,3$ and leaves only $g = 0$ for $d = 4$, reducing the generalized conjecture to the classical one. The new bound is also asymptotically better: $G_0(d) = \frac{1}{4}d^2 + O(d) < G_1(d) = \frac{1}{3}d^2 + O(d)$.

1.1 Content of the Paper

In Sect. 2, we outline the reduction of Problem 1.1 to the question of existence of certain real curves on the ellipsoid and restate Theorem 1.2 in the new terms; see Theorem 2.4. In Sect. 3, we briefly recall V. V. Nikulin's theory of discriminant forms and lattice extensions. In Sect. 4, we introduce a version of the Alexander module of a plane curve suited to the study of the resolution lattice in the homology of the double covering of the plane ramified at the curve. Finally, in Sect. 5, we prove Theorem 2.4 and hence Theorem 1.2.

2 The Reduction

We briefly recall the reduction of Problem 1.1 to the problem of existence of a certain real curve on the ellipsoid. Details can be found in [3].

2.1 The Map Φ

Denote by $\operatorname{conj}\colon z \mapsto \bar{z}$ the standard real structure on $\mathbb{P}^1 = \mathbb{C} \cup \infty$. The *ellipsoid* $\mathbf{E}$ is the quadric $\mathbb{P}^1 \times \mathbb{P}^1$ with the real structure $(z,w) \mapsto (\operatorname{conj} w, \operatorname{conj} z)$. (It is indeed the real structure whose real part is homeomorphic to the 2-sphere.)

Let (C,c) be a real curve and let $f\colon C \to \mathbb{P}^1$ be a holomorphic map. Consider the *conjugate map* $\bar{f} = \operatorname{conj} \circ f \circ c\colon C \to \mathbb{P}^1$ and let

$$\Phi = (f, \bar{f})\colon C \to \mathbf{E}.$$

It is straightforward that Φ is holomorphic and real (with respect to the above real structure on $\mathbf{E}$). Hence, the image $\Phi(C)$ is a real algebraic curve in $\mathbf{E}$. (We exclude the possibility that $\Phi(C)$ is a point, for we assume $f \neq \text{const}$; cf. Condition 1.1(1).) In particular, the image $\Phi(C)$ has bidegree (d', d') for some $d' \geqslant 1$.

Lemma 2.1 (see [3]). *A holomorphic map $f\colon C \to \mathbb{P}^1$ is real with respect to some real structure on $\mathbb{P}^1$ if and only if there is a Möbius transformation $\varphi\colon \mathbb{P}^1 \to \mathbb{P}^1$ such that $\bar{f} = \varphi \circ f$.* □

Corollary 2.2 (see [3]). *A holomorphic map $f\colon C \to \mathbb{P}^1$ is real with respect to some real structure on $\mathbb{P}^1$ if and only if the image $\Phi(C) \subset \mathbf{E}$ (see above) is a curve of bidegree $(1,1)$.* □

2.2 The Principal Reduction

Let $p\colon \mathbf{E} \to \mathbb{P}^1$ be the projection to the first factor. In general, the map Φ as above splits into a ramified covering α and a generically one-to-one map β,

$$\Phi\colon C \xrightarrow{\alpha} C' \xrightarrow{\beta} \mathbf{E},$$

so that $d = \deg f = d' \deg \alpha$, where $d' = \deg(p \circ \beta)$, or alternatively, (d', d') is the bidegree of the image $\Phi(C) = \beta(C')$. Then f itself splits into α and $p \circ \beta$. Hence the critical values of f are those of $p \circ \beta$ and the images under $p \circ \beta$ of the ramification points of α. Thus, if f satisfies Condition 1.1(1), the splitting cannot be proper, i.e., either $d = \deg \alpha$ and $d' = 1$ or $\deg \alpha = 1$ and $d = d'$. In the former case, f is real with respect to some real structure on $\mathbb{P}^1$; see Corollary 2.2. In the latter case, assuming that the critical points of f are real, Condition 1.1(2), the image $B = \Phi(C)$ is a curve of genus g with $2g + 2d - 2$ real ordinary cusps (type $\mathbf{A}_2$ singular points, the images of the critical points of f) and all other singularities with smooth branches.

Conversely, let $B \subset \mathbf{E}$ be a real curve of bidegree (d, d), $d > 1$, and genus g with $2g + 2d - 2$ real ordinary cusps and all other singularities with smooth branches, and let $\rho\colon \tilde{B} \to B$ be the normalization of B. Then $f = p \circ \rho\colon \tilde{B} \to \mathbb{P}^1$ is a map that satisfies Conditions 1.1(1) and (2) but is not real with respect to any real structure on $\mathbb{P}^1$; hence, the pair (g, d) does not have the total reality property.

As a consequence, we obtain the following statement.

Theorem 2.3 (see [3]). *A pair (g, d) has the total reality property if and only if there does* not *exist a real curve $B \subset \mathbf{E}$ of degree d and genus g with $2g + 2d - 2$ real ordinary cusps and all other singularities with smooth branches.* □

Thus, Theorem 1.2 is equivalent to the following statement, which is actually proved in the paper.

Theorem 2.4. *Let $\mathbf{E}$ be the ellipsoid, and let $B \subset \mathbf{E}$ be a real curve of bidegree (d, d) and genus g with $c = 2d + 2g - 2$ real ordinary cusps and other singularities with smooth branches. Then $g \leqslant G_0(d)$; see Theorem 1.2.*

Remark 2.5. It is worth mentioning that the bound $g > G_1(d)$ mentioned in the introduction is purely complex: it is derived from the adjunction formula for the virtual genus of a curve $B \subset \mathbf{E}$ as in Theorem 2.3. In contrast, the proof of the conjecture for the case $(g, d) = (1, 4)$ found in [1] makes essential use of the real structure, since an elliptic curve with eight ordinary cusps in $\mathbb{P}^1 \times \mathbb{P}^1$ does in fact exist! Our proof of Theorem 2.4 also uses the assumption that all cusps are real.

2.3 Reduction to Nodes and Cusps Only

In general, a curve B as in Theorem 2.4 may have rather complicated singularities. However, since the proof below is essentially topological, we follow Yu. Orevkov [9] and perturb B to a real *pseudoholomorphic* curve with ordinary nodes (type $\mathbf{A}_1$) and ordinary cusps (type $\mathbf{A}_2$) only. By the genus formula, the number of nodes of such a curve is

$$n = (d-1)^2 - g - c = d^2 - 4d - 1 - 3g. \tag{1}$$

3 Discriminant Forms

In this section, we cite the techniques and a few results of Nikulin [8]. Most proofs can be found in [8]; they are omitted.

3.1 Lattices

A *lattice* is a finitely generated free abelian group L equipped with a symmetric bilinear form $b\colon L \otimes L \to \mathbb{Z}$. We abbreviate $b(x,y) = x \cdot y$ and $b(x,x) = x^2$. Since the transition matrix between two integral bases has determinant ± 1, the determinant $\det L \in \mathbb{Z}$ (i.e., the determinant of the Gram matrix of b in any basis of L) is well defined. A lattice L is called *nondegenerate* if $\det L \neq 0$; it is called *unimodular* if $\det L = \pm 1$ and *p-unimodular* if $\det L$ is prime to p (where p is a prime).

To fix the notation, we use $\sigma_+(L)$, $\sigma_-(L)$, and $\sigma(L) = \sigma_+(L) - \sigma_-(L)$ for, respectively, the positive and negative inertia indices and the signature of a lattice L.

3.2 The Discriminant Group

Given a lattice L, the bilinear form extends to $L \otimes \mathbb{Q}$. If L is nondegenerate, the dual group $L^* = \operatorname{Hom}(L, \mathbb{Z})$ can be regarded as the subgroup

$$\{x \in L \otimes \mathbb{Q} \mid x \cdot y \in \mathbb{Z} \text{ for all } x \in L\}.$$

In particular, $L \subset L^*$, and the quotient L^*/L is a finite group; it is called the *discriminant group* of L and is denoted by $\operatorname{discr} L$ or $\mathcal{L}$. The group $\mathcal{L}$ inherits from $L \otimes \mathbb{Q}$ a symmetric bilinear form $\mathcal{L} \otimes \mathcal{L} \to \mathbb{Q}/\mathbb{Z}$, called the *discriminant form*; when speaking about the discriminant groups, their (anti-)isomorphisms, etc., we always assume that the discriminant form is taken into account. The following properties are straightforward:

1. The discriminant form is nondegenerate, i.e., the associated homomorphism $\mathcal{L} \to \operatorname{Hom}(\mathcal{L}, \mathbb{Q}/Z)$ is an isomorphism;
2. One has $\#\mathcal{L} = |\det L|$;
3. In particular, $\mathcal{L} = 0$ if and only if L is unimodular.

Following Nikulin, we denote by $\ell(\mathcal{L})$ the minimal number of generators of a finite abelian group $\mathcal{L}$. For a prime p, we denote by $\mathcal{L}_p$ the p-primary part of $\mathcal{L}$ and let $\ell_p(\mathcal{L}) = \ell(\mathcal{L}_p)$. Clearly, for a lattice L one has

4. $\operatorname{rk} L \geqslant \ell(\mathcal{L}) \geqslant \ell_p(\mathcal{L})$ (for any prime p);
5. L is p-unimodular if and only if $\mathcal{L}_p = 0$.

3.3 Extensions

An *extension* of a lattice S is another lattice M containing L. All lattices below are assumed nondegenerate.

Let $M \supset S$ be a finite index extension of a lattice S. Since M is also a lattice, one has monomorphisms $S \hookrightarrow M \hookrightarrow M^* \hookrightarrow S^*$. Hence, the quotient $\mathcal{K} = M/S$ can be regarded as a subgroup of the discriminant $\mathcal{S} = \operatorname{discr} S$; it is called the *kernel* of the extension $M \supset S$. The kernel is an isotropic subgroup, i.e., $\mathcal{K} \subset \mathcal{K}^\perp$, and one has $\mathcal{M} = \mathcal{K}^\perp/\mathcal{K}$. In particular, in view of Sect. 3.2(1), for any prime p one has

$$\ell_p(\mathcal{M}) \geqslant \ell_p(\mathcal{L}) - 2\ell_p(\mathcal{K}).$$

Now assume that $M \supset S$ is a *primitive* extension, i.e., the quotient M/S is torsion free. Then the construction above applies to the finite index extension $M \supset S \oplus N$, where $N = S^\perp$, giving rise to the kernel $\mathcal{K} \subset \mathcal{S} \oplus \mathcal{N}$. Since both S and N are primitive in M, one has $\mathcal{K} \cap \mathcal{S} = \mathcal{K} \cap \mathcal{N} = 0$; hence, $\mathcal{K}$ is the graph of an anti-isometry κ between certain subgroups $\mathcal{S}' \subset \mathcal{S}$ and $\mathcal{N}' \subset \mathcal{N}$. If M is unimodular, then $\mathcal{S}' = \mathcal{S}$ and $\mathcal{N}' = \mathcal{N}$, i.e., κ is an anti-isometry $\mathcal{S} \to \mathcal{N}$. Similarly, if M is p-unimodular for a certain prime p, then $\mathcal{S}'_p = \mathcal{S}_p$ and $\mathcal{N}'_p = \mathcal{N}_p$, i.e., κ is an anti-isometry $\mathcal{S}_p \to \mathcal{N}_p$. In particular, $\ell(\mathcal{S}) = \ell(\mathcal{N})$ (respectively, $\ell_p(\mathcal{S}) = \ell_p(\mathcal{N})$). Combining these observations with Sect. 3.2(4), we arrive at the following statement.

Lemma 3.1. *Let p be a prime, and let $L \supset S$ be a p-unimodular extension of a nondegenerate lattice S. Denote by $\tilde{S}$ the primitive hull of S in L, and let $\mathcal{K}$ be the kernel of the finite index extension $\tilde{S} \supset S$. Then* $\operatorname{rk} S^\perp \geqslant \ell_p(\mathcal{S}) - 2\ell_p(\mathcal{K})$. □

4 The Alexander Module

Here we discuss (a version of) the Alexander module of a plane curve and its relation to the resolution lattice in the homology of the double covering of the plane ramified at the curve.

4.1 The Reduced Alexander Module

Let π be a group, and let $\kappa\colon \pi \twoheadrightarrow \mathbb{Z}_2$ be an epimorphism. Set $K = \operatorname{Ker}\kappa$ and define the *Alexander module* of π (more precisely, of κ) as the $\mathbb{Z}[\mathbb{Z}_2]$-module $A_\pi = K/[K,K]$, the generator t of $\mathbb{Z}_2$ acting via $x \mapsto [\bar{t}^{-1}\bar{x}\bar{t}] \in A_\pi$, where $\bar{t} \in \pi$ and $\bar{x} \in K$ are some representatives of t and x, respectively. (We simplify the usual definition and consider only the case needed in the sequel. A more general version and further details can be found in A. Libgober [7].)

Let $B \subset \mathbb{P}^1 \times \mathbb{P}^1$ be an irreducible curve of even bidegree $(d,d) = (2k,2k)$, and let $\pi = \pi_1(\mathbb{P}^1 \times \mathbb{P}^1 \smallsetminus B)$. Recall that $\pi/[\pi,\pi] = \mathbb{Z}_{2k}$; hence, there is a unique epimorphism $\kappa\colon \pi \twoheadrightarrow \mathbb{Z}_2$. The resulting Alexander module $A_B = A_\pi$ will be called the *Alexander module* of B. The *reduced Alexander module* $\tilde{A}_B$ is the kernel of the canonical homomorphism $A_B \to \mathbb{Z}_k \subset \pi/[\pi,\pi]$. There is a natural exact sequence

$$0 \longrightarrow \tilde{A}_B \longrightarrow A_B \longrightarrow \mathbb{Z}_k \longrightarrow 0 \tag{2}$$

of $\mathbb{Z}[\mathbb{Z}_2]$-modules (where the $\mathbb{Z}_2$-action on $\mathbb{Z}_k$ is trivial). The following statement is essentially contained in Zariski [10].

Lemma 4.1. *The exact sequence* (2) *splits*: *one has* $A_B = \tilde{A}_B \oplus \operatorname{Ker}(1-t)$, *where* t *is the generator of* $\mathbb{Z}_2$. *Furthermore,* $\tilde{A}_B$ *is a finite group free of* 2*-torsion, and the action of* t *on* $\tilde{A}_B$ *is via the multiplication by* (-1).

Proof. Since A_B is a finitely generated abelian group, to prove that it is finite and free of 2-torsion, it suffices to show that $\operatorname{Hom}_{\mathbb{Z}}(\tilde{A}_B, \mathbb{Z}_2) = 0$. Assume the contrary. Then the $\mathbb{Z}_2$-action in the 2-group $\operatorname{Hom}_{\mathbb{Z}}(\tilde{A}_B, \mathbb{Z}_2)$ has a fixed nonzero element, i.e., there is an equivariant epimorphism $\tilde{A}_B \twoheadrightarrow \mathbb{Z}_2$. Hence, π factors to a group G that is an extension $0 \to \mathbb{Z}_2 \to G \to \mathbb{Z}_{2k} \to 0$. The group G is necessarily abelian, and it is strictly larger than $\mathbb{Z}_{2k} = \pi/[\pi,\pi]$. This is a contradiction.

Since $\tilde{A}_B$ is finite and free of 2-torsion, one can divide by 2, and there is a splitting $\tilde{A}_B = \tilde{A}^+ \oplus \tilde{A}^-$, where $\tilde{A}^\pm = \operatorname{Ker}[(1 \pm t)\colon \tilde{A}_B \to \tilde{A}_B]$. Then π factors to a group G that is a central extension $0 \to \tilde{A}^+ \to G \to \mathbb{Z}_{2k} \to 0$, and as above, one concludes that $\tilde{A}^+ = 0$, i.e., t acts on $\tilde{A}_B$ via (-1).

Pick a representative $a' \in A_B$ of a generator of $\mathbb{Z}_k = A_B/\tilde{A}_B$. Then obviously, $(1-t)a' \in \tilde{A}_B$, and replacing a' with $a' + \frac{1}{2}(1-t)a'$, one obtains a t-invariant representative $a \in \operatorname{Ker}(1-t)$. The multiple $ka \in \tilde{A}_B$ is both invariant and skew-invariant; since $\tilde{A}_B$ is free of 2-torsion, $ka = 0$, and the sequence splits. □

4.2 The Double Covering of $\mathbb{P}^1 \times \mathbb{P}^1$

Let $B \subset \mathbb{P}^1 \times \mathbb{P}^1$ be an irreducible curve of even bidegree $(d,d) = (2k,2k)$ and with simple singularities only. Consider the double covering $X \to \mathbb{P}^1 \times \mathbb{P}^1$ ramified at B and denote by $\tilde{X}$ the minimal resolution of singularities of X. Let $\tilde{B} \subset \tilde{X}$ be the proper transform of B, and let $E \subset \tilde{X}$ be the exceptional divisor contracted by the blowdown $\tilde{X} \to X$.

Recall that the minimal resolution of a simple surface singularity is diffeomorphic to its perturbation; see, e.g., [2]. Hence, $\tilde{X}$ is diffeomorphic to the double covering of $\mathbb{P}^1 \times \mathbb{P}^1$ ramified at a nonsingular curve. In particular, $\pi_1(\tilde{X}) = 0$, and one has

$$b_2(X) = \chi(X) - 2 = 8k^2 - 8k + 6, \quad \sigma(X) = -4k^2. \tag{3}$$

4.3 An Estimate on the Discriminant Group

Set $L = H_2(\tilde{X})$. We regard L as a lattice via the intersection index pairing on $\tilde{X}$. (Since $\tilde{X}$ is simply connected, L is a free abelian group. It is a unimodular lattice by Poincaré duality.) Let $\Sigma \subset L$ be the sublattice spanned by the components of E, and let $\tilde{\Sigma} \subset L$ be the primitive hull of Σ. Recall that Σ is a negative definite lattice. Further, let $h_1, h_2 \subset L$ be the classes of the pullbacks of a pair of generic generatrices of $\mathbb{P}^1 \times \mathbb{P}^1$, so that $h_1^2 = h_2^2 = 0$, $h_1 \cdot h_2 = 2$.

Lemma 4.2. *If a curve B as above is irreducible, then there are natural isomorphisms $\tilde{A}_B = \operatorname{Hom}_{\mathbb{Z}}(\mathcal{K}, \mathbb{Q}/\mathbb{Z}) = \operatorname{Ext}_{\mathbb{Z}}(\mathcal{K}, \mathbb{Z})$, where $\mathcal{K}$ is the kernel of the extension $\tilde{\Sigma} \supset \Sigma$.*

Proof. One has $A_B = H_1(\tilde{X} \smallsetminus (\tilde{B} + E))$ as a group, the $\mathbb{Z}_2$-action being induced by the deck translation of the covering. Hence, by Poincaré–Lefschetz duality, A_B is the cokernel of the inclusion homomorphism $i^* \colon H^2(\tilde{X}) \to H^2(\tilde{B} + E)$.

On the other hand, there is an orthogonal (with respect to the intersection index form in $\tilde{X}$) decomposition $H_2(\tilde{B} + E) = \Sigma \oplus \langle b \rangle$, where $b = k(h_1 + h_2)$ is the class realized by the divisorial pullback of B in $\tilde{X}$. The cokernel of the restriction $i^* \colon H^2(X) \to \langle b \rangle^*$ is a cyclic group $\mathbb{Z}_k$ fixed by the deck translation. Hence, in view of Lemma 4.1,

$$\tilde{A}_B = \operatorname{Coker}[i^* \colon H^2(\tilde{X}) \to H^2(E)] = \operatorname{Coker}[L^* \to \Sigma^*] = \operatorname{discr} \Sigma / \mathcal{K}^\perp.$$

(We use the splitting $L^* \twoheadrightarrow \tilde{\Sigma}^* \to \Sigma^*$, the first map being an epimorphism, since $L/\tilde{\Sigma}$ is torsion free.) Since the discriminant form is nondegenerate (see Sect. 3.2(1)), one has $\operatorname{discr} \Sigma / \mathcal{K}^\perp = \operatorname{Hom}_{\mathbb{Z}}(\mathcal{K}, \mathbb{Q}/\mathbb{Z})$.

Since $\mathcal{K}$ is a finite group, applying the functor $\operatorname{Hom}_{\mathbb{Z}}(\mathcal{K}, \cdot)$ to the short exact sequence $0 \to \mathbb{Z} \to \mathbb{Q} \to \mathbb{Q}/\mathbb{Z} \to 0$, one obtains an isomorphism $\operatorname{Hom}_{\mathbb{Z}}(\mathcal{K}, \mathbb{Q}/\mathbb{Z}) = \operatorname{Ext}_{\mathbb{Z}}(\mathcal{K}, \mathbb{Z})$. □

Corollary 4.3. *In the notation of Lemma 4.2, if B is irreducible and the group $\pi_1(\mathbb{P}^1 \times \mathbb{P}^1 \smallsetminus B)$ is abelian, then $\mathcal{K} = 0$.* □

Corollary 4.4. *In the notation of Lemma 4.2, if B is an irreducible curve of bidegree (d,d), $d = 2k \geqslant 2$, then $\mathcal{K}$ is free of 2-torsion and $\ell(\mathcal{K}) \leqslant d-2$.*

Proof. Due to Lemma 4.2, one can replace $\mathcal{K}$ with $\tilde{A}_B$. Then the statement on the 2-torsion is given by Lemma 4.1, and it suffices to estimate the numbers $\ell_p(\tilde{A}_B) = \ell(\tilde{A}_B \otimes \mathbb{Z}_p)$ for odd primes p.

Due to the Zariski–van Kampen theorem [5] applied to one of the two rulings of $\mathbb{P}^1 \times \mathbb{P}^1$, there is an epimorphism $\pi_1(L \smallsetminus B) = F_{d-1} \twoheadrightarrow \pi_1(\mathbb{P}^1 \times \mathbb{P}^1 \smallsetminus B)$, where L is a generic generatrix of $\mathbb{P}^1 \times \mathbb{P}^1$ and F_{d-1} is the free group on $d-1$ generators. Hence, A_B is a quotient of the Alexander module

$$A_{F_{d-1}} = \mathbb{Z}[\mathbb{Z}_2]/(t-1) \oplus \bigoplus_{d-2} \mathbb{Z}[\mathbb{Z}_2].$$

For an odd prime p, there is a splitting $A_{F_{d-1}} \otimes \mathbb{Z}_p = A_p^+ \oplus A_p^-$ (over the field $\mathbb{Z}_p$) into the eigenspaces of the action of $\mathbb{Z}_2$, and due to Lemma 4.1, the group $\tilde{A}_B \otimes \mathbb{Z}_p$ is a quotient of $A_p^- = \bigoplus_{d-2} \mathbb{Z}_p$. □

Remark 4.5. All statements in this section hold for pseudoholomorphic curves as well; cf. Sect. 2.3. For Corollary 4.4, it suffices to assume that B is a small perturbation of an algebraic curve of bidegree (d,d). Then one still has an epimorphism $F_{d-1} \twoheadrightarrow \pi_1(\mathbb{P}^1 \times \mathbb{P}^1 \smallsetminus B)$, and the proof applies literally.

5 Proof of Theorem 1.2

As explained in Sect. 2, it suffices to prove Theorem 2.4. We consider the cases of d even and d odd separately.

5.1 Preliminary Observations

Let $B \subset \mathbb{P}^1 \times \mathbb{P}^1$ be an irreducible curve of even bidegree (d,d), $d = 2k$. Assume that all singularities of B are simple and let $\tilde{X}$ be the minimal resolution of singularities of the double covering $X \to \mathbb{P}^1 \times \mathbb{P}^1$ ramified at B; cf. Sect. 4.2. As in Sect. 4.3, consider the unimodular lattice $L = H_2(\tilde{X})$.

Let $c\colon \tilde{X} \to \tilde{X}$ be a real structure on $\tilde{X}$, and denote by $L^{\pm}$ the (± 1)-eigenlattices of the induced involution c_* of L. The following statements are well known:

1. $L^{\pm}$ are the orthogonal complements of each other;
2. $L^{\pm}$ are p-unimodular for any odd prime p;
3. One has $\sigma_+(L^+) = \sigma_+(L^-) - 1$.

Since also $\sigma_+(L^+) + \sigma_+(L^-) = \sigma_+(L) = 2k^2 - 4k + 3$, see (3), one arrives at $\sigma_+(L^+) = \sigma_+(L^-) - 1 = (k-1)^2$ and, further, at

$$\operatorname{rk} L^- = (7k^2 - 6k + 5) - \sigma_-(L^+). \tag{4}$$

Remark 5.1. The common proof of Property 5.1(3) uses the Hodge structure. However, there is another (also very well known) proof that also applies to almost complex manifolds. Let $\tilde{X}_{\mathbb{R}} = \operatorname{Fix} c$ be the real part of $\tilde{X}$. Then the normal bundle of $\tilde{X}_{\mathbb{R}}$ in $\tilde{X}$ is i times its tangent bundle; hence, the normal Euler number $\tilde{X}_{\mathbb{R}} \circ \tilde{X}_{\mathbb{R}}$ equals (-1) times the index of any tangent vector field on $\tilde{X}_{\mathbb{R}}$, i.e., $-\chi(\tilde{X}_{\mathbb{R}})$. Now one has $\sigma(L^+) - \sigma(L^-) = \tilde{X}_{\mathbb{R}} \circ \tilde{X}_{\mathbb{R}} = -\chi(\tilde{X}_{\mathbb{R}})$ (by the Hirzebruch G-signature theorem) and $\operatorname{rk} L^+ - \operatorname{rk} L^- = \chi(\tilde{X}_{\mathbb{R}}) - 2$ (by the Lefschetz fixed point theorem). Adding the two equations, one obtains 5.1(3).

5.2 *The Case of $d = 2k$ Even*

Perturbing, if necessary, B in the class of real pseudoholomorphic curves, see Sect. 2.3, one can assume that all singularities of B are c real ordinary cusps and n ordinary nodes, where

$$c = 2d + 2g - 2 \quad \text{and} \quad n = d^2 - 4d - 1 - 3g; \tag{5}$$

see Theorem 2.3 and (1). Let $n = r + 2s$, where r and s are respectively the numbers of real nodes and pairs of conjugate nodes.

5.3 *The Contribution of the Singular Points*

Consider the double covering $\tilde{X}$, see Sect. 4.2, lift the real structure on $\mathbf{E}$ to a real structure c on $\tilde{X}$, and let $L^{\pm} \subset L$ be the corresponding eigenlattices; see Sect. 5.1. In the notation of Sect. 4.3, let $\Sigma^{\pm} = \Sigma \cap L^{\pm}$. Then

- Each real cusp of B contributes a sublattice $\mathbf{A}_2$ to Σ^-;
- Each real node of B contributes a sublattice $\mathbf{A}_1 = [-2]$ to Σ^-;
- Each pair of conjugate nodes contributes $[-4]$ to Σ^- and $[-4]$ to Σ^+.

In addition, the classes h_1, h_2 of two generic generatrices of $\mathbf{E}$ span a hyperbolic plane orthogonal to Σ; see Sect. 4.3. It contributes

- A sublattice $[4] \subset L^-$ spanned by $h_1 + h_2$, and
- A sublattice $[-4] \subset L^+$ spanned by $h_1 - h_2$.

(Recall that any real structure reverses the canonical complex orientation of pseudoholomorphic curves.)

5.4 End of the Proof

All sublattices of L^+ described above are negative definite; hence, their total rank $s+1$ contributes to $\sigma_-(L^+)$. The total rank $2c+r+s+1$ of the sublattices of L^- contributes to the rank of $S^- = \Sigma^- \oplus [4] \subset L^-$. Due to (4), one has

$$2c+n+2+\operatorname{rk} S^\perp \leqslant 7k^2-6k+5, \tag{6}$$

where $S^\perp$ is the orthogonal complement of S^- in L^-. All summands of S^- other than $\mathbf{A}_2$ are 3-unimodular, whereas $\operatorname{discr}\mathbf{A}_2$ is the group $\mathbb{Z}_3$ spanned by an element of square $\frac{1}{3}$ mod $\mathbb{Z}$. Let $\tilde{S}^- \supset S^-$ and $\tilde{\Sigma} \supset \Sigma$ be the primitive hulls, and denote by $\mathcal{K}^-$ and $\mathcal{K}$ the kernels of the corresponding finite index extensions; see Sect. 3.3. Clearly, $\ell_3(\mathcal{K}^-) \leqslant \ell_3(\mathcal{K})$, and due to Corollary 4.4 (see also Remark 4.5), one has $\ell_3(\mathcal{K}) \leqslant d-2$. Then, using Lemma 3.1, one obtains $\operatorname{rk} S^\perp \geqslant c-2(d-2)$, and combining the last inequality with (6), one arrives at

$$3c+n-2(d-2) \leqslant 7k^2-6k+3.$$

It remains to substitute the expressions for c and n given by (5) and solve for g to get

$$g \leqslant k^2-2k+\frac{2}{3}.$$

Since g is an integer, the last inequality implies $g \leqslant G_0(2k)$ as in Theorem 2.4.

5.5 The Case of $d = 2k-1$ Odd

As above, one can assume that B has c real ordinary cusps and $n = r+2s$ ordinary nodes; see (5). Furthermore, one can assume that $c > 0$, since otherwise, $g = 0$ and $d = 1$. Then B has a real cusp, and hence a real smooth point P.

Let L_1, L_2 be the two generatrices of $\mathbf{E}$ passing through P. Choose P generic, so that each L_i, $i = 1,2$, intersects B transversally at d points, and consider the real curve $B' = B + L_1 + L_2$ of even bidegree $(2k,2k)$, applying to it the same double covering arguments as above. In addition to the nodes and cusps of B, the new curve B' has $(d-1)$ pairs of conjugate nodes and a real triple (type $\mathbf{D}_4$) point at P (with one real and two complex conjugate branches). Hence, in addition to the classes listed in Sect. 5.3, there are

- $(d-1)$ copies of $[-4]$ in each Σ^+, Σ^- (from the new conjugate nodes),
- A sublattice $[-4] \subset \Sigma^+$ (from the type $\mathbf{D}_4$ point), and
- A sublattice $\mathbf{A}_3 \subset \Sigma^-$ (from the type $\mathbf{D}_4$ point).

Thus, inequality (6) turns into

$$2c+n+2(d-1)+4+2+\operatorname{rk} S^\perp \leqslant 7k^2-6k+5.$$

We will show that $\operatorname{rk} S^{\perp} \geqslant c$. Then, substituting the expressions for c and n, see (5), and solving the resulting inequality in g, one will obtain $g \leqslant G_0(2k-1)$, as required.

5.6 *An Estimate on* rk$S^{\perp}$

In view of Lemma 3.1, in order to prove that $\operatorname{rk} S^{\perp} \geqslant c$, it suffices to show that $\ell_3(\mathcal{K}) = 0$ (cf. similar arguments in Sect. 5.4).

Perturb B' to a pseudoholomorphic curve B'', keeping the cusps of B' and resolving the other singularities. (It would suffice to resolve the singular points resulting from the intersection $B \cap L_1$.) Then, applying the Zariski–van Kampen theorem [5] to the ruling containing L_1, it is easy to show that the fundamental group $\pi_1(\mathbb{P}^1 \times \mathbb{P}^1 \smallsetminus B'')$ is cyclic.

Indeed, let U be a small tubular neighborhood of L_1 in $\mathbb{P}^1 \times \mathbb{P}^1$, and let $L'' \subset U$ be a generatrix transversal to B''. The epimorphism $\pi_1(L_1'' \smallsetminus B'') \twoheadrightarrow \pi_1(\mathbb{P}^1 \times \mathbb{P}^1 \smallsetminus B'')$ given by the Zariski–van Kampen theorem factors through $\pi_1(U \smallsetminus B'')$, and the latter group is cyclic.

On the other hand, the new double covering $\tilde{X}'' \to \mathbb{P}^1 \times \mathbb{P}^1$ ramified at B'' is diffeomorphic to $\tilde{X}$, and the diffeomorphism can be chosen identical over the union of a collection of Milnor balls about the cusps of B'. Thus, since $\operatorname{discr} \mathbf{A}_1$ and $\operatorname{discr} \mathbf{D}_4$ are 2-torsion groups, the perturbation does not change $\mathcal{K} \otimes \mathbb{Z}_3$, and Corollary 4.3 (see also Remark 4.5) implies that $\mathcal{K} \otimes \mathbb{Z}_3 = 0$. □

Acknowledgments I am grateful to T. Ekedahl and B. Shapiro for the fruitful discussions of the subject. This work was completed during my participation in the special semester on Real and Tropical Algebraic Geometry held at *Centre Interfacultaire Bernoulli, École polytechnique fédérale de Lausanne*. I extend my gratitude to the organizers of the semester and to the administration of *CIB*.

References

1. A. Degtyarev, T. Ekedahl, I. Itenberg, B. Shapiro, M. Shapiro: On total reality of meromorphic functions. Ann. Inst. Fourier (Grenoble) **57**, 2015–2030 (2007)
2. A. H. Durfee: Fifteen characterizations of rational double points and simple critical points. Enseign. Math. (2) **25**, 131–163 (1979)
3. T. Ekedahl, B. Shapiro, M. Shapiro: First steps towards total reality of meromorphic functions. Moscow Math. J. (1) **222**, 95–106 (2006)
4. A. Eremenko, A. Gabrielov: Rational functions with real critical points and the B. and M. Shapiro conjecture in real enumerative geometry. Ann. of Math. (2) **155**, 105–129 (2002)
5. E. R. van Kampen: On the fundamental group of an algebraic curve. Amer. J. Math. **55** 255–260 (1933)
6. E. Mukhin, V. Tarasov, A. Varchenko: The B. and M. Shapiro conjecture in real algebraic geometry and the Bethe ansatz. Ann. of Math. (2) **170**, 863–881 (2009)
7. A. Libgober: Alexander modules of plane algebraic curves. Contemporary Math. **20**, 231–247 (1983)

8. V. V. Nikulin: Integral symmetric bilinear forms and some of their geometric applications,. Izv. Akad. Nauk SSSR Ser. Mat. **43**, 111–177 (1979) (Russian). English transl. in Math. USSR–Izv. **14**, 103–167 (1980)
9. S. Yu. Orevkov: Classification of flexible M-curves of degree 8 up to isotopy. Geom. Funct. Anal. **12**, 723–755 (2002)
10. O. Zariski: On the irregularity of cyclic multiple planes. Ann. of Math. **32**, 485–511 (1931)

On the Number of Components of a Complete Intersection of Real Quadrics

Alex Degtyarev, Ilia Itenberg, and Viatcheslav Kharlamov

To Oleg Viro
viros / viro : *Terme gaulois désignant ce qui est juste, vrai, sincère …*
X. Delamarre, Dictionnaire de la langue gauloise
Errance, Paris, 2003

Abstract Our main results concern complete intersections of three real quadrics. We prove that the maximal number $B_2^0(N)$ of connected components that a regular complete intersection of three real quadrics in $\mathbb{P}^N$ may have differs at most by one from the maximal number of ovals of the submaximal depth $[(N-1)/2]$ of a real plane projective curve of degree $d = N+1$. As a consequence, we obtain a lower bound $\frac{1}{4}N^2 + O(N)$ and an upper bound $\frac{3}{8}N^2 + O(N)$ for $B_2^0(N)$.

Keywords Betti number • Quadric • Complete intersection • Theta characteristic

A. Degtyarev (✉)
Bilkent University, 06800 Ankara, Turkey
e-mail: degt@fen.bilkent.edu.tr

I. Itenberg
Université Pierre et Marie Curie and Institut Universitaire de France, Institut de Mathématiques de Jussieu, 4 place Jussieu, 75005 Paris, France
e-mail: itenberg@math.jussieu.fr

V. Kharlamov
Université de Strasbourg and IRMA, 7 rue René Descartes, 67084 Strasbourg Cedex, France
e-mail: viatcheslav.kharlamov@math.unistra.fr

I. Itenberg et al. (eds.), *Perspectives in Analysis, Geometry, and Topology: On the Occasion of the 60th Birthday of Oleg Viro*, Progress in Mathematics 296,
DOI 10.1007/978-0-8176-8277-4_5,

1 Introduction

1.1 Statement of the Problem and Principal Results

The question of the maximal number of connected components that a real projective variety of a given (multi)degree may have remains one of the most difficult and least understood problems in topology of real algebraic varieties. Besides the trivial case of varieties of dimension zero, essentially the only general situation in which this problem is solved is that of curves: the answer is given by the famous Harnack inequality in the case of plane curves [15], and by a combination of the Castelnuovo–Halphen [7, 14] and Harnack–Klein [19] inequalities in the case of curves in projective spaces of higher dimension; see [16] and [25].

The immediate generalization of the Harnack inequality given by the Smith theory, the *Smith inequality* (see, *e.g.*, [9]), involves all Betti numbers of the real part, and the resulting bound is too rough when applied to the problem of the number of connected components in a straightforward manner (see, *e.g.*, the discussion in Sect. 6.7).

In this paper, we address the problem of the maximal number of connected components in the case of varieties defined by equations of degree two, *i.e.*, complete intersections of quadrics. To be more precise, let us denote by

$$B^0_r(N), \quad 0 \leqslant r \leqslant N-1,$$

the maximal number of connected components that a regular complete intersection of $r+1$ real quadrics in $\mathbb{P}^N_{\mathbb{R}}$ may have. Certainly, as we study regular complete intersections of even degree, the actual number of connected components covers the whole range of values between 0 and $B^0_r(N)$.

In the following three extremal cases, the answer is easy and well known:

- $B^0_0(N) = 1$ for all $N \geqslant 2$ (a single quadric),
- $B^0_1(N) = 2$ for all $N \geqslant 3$ (intersection of two quadrics), and
- $B^0_{N-1}(N) = 2^N$ for all $N \geqslant 1$ (intersection of dimension zero).

To our knowledge, very little was known in the next case $r = 2$ (intersection of three quadrics); even the fact that $B^0_2(N) \to \infty$ as $N \to \infty$ does not seem to have been observed before. Our principal result here is the following theorem, providing a lower bound $\frac{1}{4}N^2 + O(N)$ and an upper bound $\frac{3}{8}N^2 + O(N)$ for $B^0_2(N)$.

Theorem 1.1. *For all* $N \geqslant 4$, *one has*

$$\frac{1}{4}(N-1)(N+5) - 2 < B^0_2(N) \leqslant \frac{3}{2}k(k-1) + 2,$$

where $k = [\frac{1}{2}N] + 1$.

The proof of Theorem 1.1 found in Sect. 4.5 is based on a real version of the Dixon correspondence [10] between nets of quadrics (*i.e.*, linear systems generated by three independent quadrics) and plane curves equipped with a nonvanishing even theta characteristic. Another tool is a spectral sequence due to Agrachev [1], which computes the homology of a complete intersection of quadrics in terms of its spectral variety. The following intermediate result seems to be of independent interest.

Definition 1.2. Define the *Hilbert number* $\operatorname{Hilb}(d)$ as the maximal number of ovals of the submaximal depth $[d/2]-1$ that a nonsingular real plane algebraic curve of degree d may have. (Recall that the depth of an oval of a curve of degree $d = N+1$ does not exceed $[(N+1)/2]$. A brief introduction to topology of nonsingular real plane algebraic curves can be found in Sect. 4.2.)

Theorem 1.3. *For any integer $N \geqslant 4$, one has*

$$\operatorname{Hilb}(N+1) \leqslant B_2^0(N) \leqslant \operatorname{Hilb}(N+1)+1.$$

This theorem is proved in Sect. 4.4. The few known values of $\operatorname{Hilb}(N+1)$ and $B_2^0(N)$ are given by the following table.

N	3	4	5	6	7
$\operatorname{Hilb}(N+1)$	4	6	9	13	17 or 18
$B_2^0(N)$	8	6	10	13 or 14	17, 18, or 19

It is worth mentioning that $B_2^0(4) = \operatorname{Hilb}(5)$, whereas $B_2^0(5) = \operatorname{Hilb}(6)+1$; see Sects. 6.4 and 6.5, respectively. (The case $N = 3$ is not covered by Theorem 1.3.) At present, we do not know the precise relation between the two sequences.

For completeness, we also discuss another extremal case, namely that of curves. Here, the maximal number of components is attained on the M-curves, and the statement should be a special case of the general Viro–Itenberg construction producing maximal complete intersections of any multidegree (see [17] for a simplified version of this construction). The result is the following theorem, which is proved in Sect. 5 by means of a Harnack-like construction.

Theorem 1.4. *For all $N \geqslant 2$, one has $B_{N-2}^0(N) = 2^{N-2}(N-3)+2$.*

1.2 Conventions

Unless indicated explicitly, the coefficients of all homology and cohomology groups are $\mathbb{Z}_2$. For a compact complex curve C, we freely identify $H^1(C;R) = H_1(C;R)$ (for any coefficient ring R) *via* Poincaré duality. We do not distinguish among line bundles, invertible sheaves, and classes of linear equivalence of divisors, switching freely from one to another.

A *quod erat demonstrandum* symbol □ after a statement means that no proof will follow: the statement is obvious and the proof is straightforward, the proof has already been explained, or a reference is given at the beginning of the section.

1.3 Content of the Paper

The bulk of the paper, except Sect. 5, where Theorem 1.4 is proved, is devoted to the proof of Theorems 1.1 and 1.3. In Sect. 2, we collect the necessary material on the theta characteristics, including the real version of the theory. In Sect. 3, we introduce and study the spectral curve of a net and discuss Dixon's correspondence. The aim is to introduce the Spin- and index (semi)orientations of the real part of the spectral curve, the former coming from the theta characteristic, and the latter directly from the topology of the net, and to show that the two semiorientations coincide. In Sect. 4, we introduce the Agrachev spectral sequence that computes the Betti numbers of the common zero locus of a net in terms of its index function. The sequence is used to prove Theorem 1.3, relating the number $B_2^0(N)$ in question and the topology of real plane algebraic curves of degree $N+1$. Then we cite a few known estimates on the number of ovals of a curve (see Corollaries 4.16 and 4.18) and deduce Theorem 1.1. Finally, in Sect. 6, we discuss a few particular cases of nets and address several related questions.

2 Theta Characteristics

For the reader's convenience, we cite a number of known results related to the (real) theta characteristics on algebraic curves. Appropriate references are given at the beginning of each section.

To avoid various "boundary effects," we consider only curves of genus at least 2. For real curves, we assume that the real part is nonempty.

2.1 Complex Curves (see [5, 20])

Recall that a *theta characteristic* on a nonsingular compact complex curve C is a line bundle θ on C such that θ^2 is isomorphic to the canonical bundle K_C. In topological terms, a theta characteristic is merely a Spin-structure on the topological surface C. One associates with a theta characteristic θ the integer $h(\theta) = \dim H^0(C;\theta)$ and its $\mathbb{Z}_2$-residue $\phi(\theta) = h(\theta) \bmod 2$.

Let $\mathfrak{S} \subset \operatorname{Pic}^{g-1} C$ be the set of theta characteristics on C. The map $\phi\colon \mathfrak{S} \to \mathbb{Z}_2$ has the following fundamental properties:

1. ϕ is preserved under deformations;
2. ϕ is a quadratic extension of the intersection index form;
3. $\operatorname{Arf}\phi = 0$ (equivalently, ϕ vanishes at $2^{g-1}(2^g+1)$ points).

To make the meaning of items (2) and (3) precise, notice that $\mathfrak{S}$ is an affine space over $H^1(C)$, so that (2) is equivalent to the identity

$$\phi(a+x+y) - \phi(a+x) - \phi(a+y) + \phi(a) = \langle x, y\rangle,$$

while $\operatorname{Arf}\phi$ is the usual Arf-invariant of ϕ after $\mathfrak{S}$ is identified with $H^1(C)$ by choosing for zero any element $\theta \in \mathfrak{S}$ with $\phi(\theta) = 0$.

A theta characteristic θ is called *even* if $\phi(\theta) = 0$; otherwise, it is called *odd*. An even theta characteristic is called *nonvanishing* (or *nonzero*) if $h(\theta) = 0$.

Recall that there are canonical bijections between the set of Spin-structures on C, the set of quadratic extensions of the intersection index form on $H^1(C)$, and the set of theta characteristics on C. In particular, a theta characteristic $\theta \in \mathfrak{S}$ is uniquely determined by the quadratic function ϕ_θ on $H^1(C)$ given by $\phi_\theta(x) = \phi(\theta + x) - \phi(\theta)$. One has $\operatorname{Arf}\phi_\theta = \phi(\theta)$.

2.2 *The Moduli Space*

The moduli space of pairs (C, θ), where C is a curve of a given genus and θ is a theta characteristic on C, has two connected components, formed by even and odd theta characteristics. If C is restricted to nonsingular plane curves of a given degree d, the result is almost the same. Namely, if d is even, there are still two connected components, while if d is odd, there is an additional component formed by the pairs $(C, \frac{1}{2}(d-3)H)$, where H is the hyperplane section divisor. In topological terms, the extra component consists of the pairs $(C, \mathcal{R})$, where $\mathcal{R}$ is the Rokhlin function (see [26]); it is even if $d = \pm 1 \bmod 8$ and odd if $d = \pm 3 \bmod 8$. The other theta characteristics still form two connected components, distinguished by the parity.

2.3 *Real Curves (see [12, 13])*

Now let C be a real curve, *i.e.*, a complex curve equipped with an antiholomorphic involution $c\colon C \to C$ (a *real structure*). Recall that we always assume that the genus $g = g(C)$ is greater than 1 and that the *real part* $C_{\mathbb{R}} = \operatorname{Fix} c$ is nonempty.

Consider the set

$$\mathfrak{S}_{\mathbb{R}} = \mathfrak{S} \cap \operatorname{Pic}^{g-1}_{\mathbb{R}} C$$

of *real* (*i.e.*, c-invariant) theta characteristics. Under the assumptions above, there are canonical bijections between $\mathfrak{S}_{\mathbb{R}}$, the set of c-invariant Spin-structures on C, and the set of c_*-invariant quadratic extensions of the intersection index form on the group $H^1(C)$.

The set $\mathfrak{S}_{\mathbb{R}}$ of real theta characteristics is a principal homogeneous space over the $\mathbb{Z}_2$-torus $(J_{\mathbb{R}})_2$ of torsion 2 elements in the real part $J_{\mathbb{R}}(C)$ of the Jacobian $J(C)$. In particular, $\operatorname{Card}\mathfrak{S}_{\mathbb{R}} = 2^{g+r}$, where $r = b_0(C_{\mathbb{R}}) - 1$. More precisely, $\mathfrak{S}_{\mathbb{R}}$ admits a free action of the $\mathbb{Z}_2$-torus $(J^0_{\mathbb{R}})_2$, where $J^0_{\mathbb{R}} \subset J_{\mathbb{R}}$ is the component of zero. This action has 2^r orbits, which are distinguished by the restrictions $\phi_\theta\colon (J^0_{\mathbb{R}})_2 \to \mathbb{Z}_2$, which are linear forms. Indeed, the form ϕ_θ depends only on the orbit of an element $\theta \in \mathfrak{S}_{\mathbb{R}}$, and the forms defined by elements θ_1, θ_2 in distinct orbits of the action differ.

Alternatively, one can distinguish the orbits above as follows. Realize an element $\theta \in \mathfrak{S}_{\mathbb{R}}$ by a real divisor D, and for each real component $C_i \subset C_{\mathbb{R}}$, $1 \leqslant i \leqslant r+1$, count the residue $c_i(\theta) = \operatorname{Card}(C_i \cap D) \bmod 2$. The residues $(c_i(\theta)) \in \mathbb{Z}_2^{r+1}$ are subject to relation $\sum c_i(\theta) = g - 1 \mod 2$ and determine the orbit.

In most cases, within each of the above 2^r orbits, the numbers of even and odd theta characteristics coincide. The only exception to this rule is the orbit given by $c_1(\theta) = \cdots = c_r(\theta) = 1$ in the case where C is a dividing curve.

Lemma 2.1. *With one exception, any (real) even theta characteristic on a (real) nonsingular plane curve becomes nonvanishing after a small (real) perturbation of the curve in the plane. The exception is Rokhlin's theta characteristic* $\frac{1}{2}(d-3)H$ *on a curve of degree* $d = \pm 1 \bmod 8$*; see Sect. 2.2.*

Proof. As is well known, the vanishing of a theta characteristic is an analytic condition with respect to the coefficients of the curve. (Essentially, this statement follows from the fact that the Riemann Θ-divisor depends on the coefficients analytically.) Hence, in the space of pairs (C,θ), where C is a nonsingular plane curve of degree d and θ is an even theta characteristic on C, the pairs (C,θ) with nonvanishing θ form a Zariski-open set. Since there does exist a curve of degree d with a nonvanishing even theta characteristic (*e.g.*, any nonsingular spectral curve; see Sect. 3.3), this set is nonempty and hence dense in the (only) component formed by the even theta characteristics other than $\frac{1}{2}(d-3)H$. □

3 Linear Systems of Quadrics

3.1 Preliminaries

Consider an injective linear map $x \mapsto q_x$ from $\mathbb{C}^{r+1}$ to the space $S^2\mathbb{C}^{N+1}$ of homogeneous quadratic polynomials on $\mathbb{C}^{N+1}$. It defines a linear system of quadrics in $\mathbb{P}^N$ of dimension r, *i.e.*, an r-subspace in the projective space $C_2(\mathbb{P}^N)$ of quadrics. Conversely, any linear system of quadrics is defined by a unique, up to obvious equivalence, linear map as above.

Occasionally, we will fix coordinates $(u_0, \dots, u_N)$ in $\mathbb{C}^{N+1}$ and represent q_x by a matrix Q_x, so that $q_x(u) = \langle Q_x u, u \rangle$. Clearly, the map $x \mapsto Q_x$ is also linear.

Define the common zero set

$$V = \left\{ u \in \mathbb{P}^N \;\middle|\; q_x(u) = 0 \text{ for all } x \in \mathbb{C}^{r+1} \right\} \subset \mathbb{P}^N$$

and the *Lagrange hypersurface*

$$L = \left\{ (x,u) \in \mathbb{P}^r \times \mathbb{P}^N \;\middle|\; q_x(u) = 0 \right\} \subset \mathbb{P}^r \times \mathbb{P}^N.$$

(As usual, the vanishing condition $q_x(u) = 0$ does not depend on the choice of the representatives of x and u.) The following statement is straightforward.

Lemma 3.1. *An intersection of quadrics V is regular if and only if the associated Lagrange hypersurface L is nonsingular.* □

3.2 The Spectral Variety

Define the *spectral variety* C of a linear system of quadrics $x \mapsto q_x$ *via*

$$C = \left\{ x \in \mathbb{P}^r \;\middle|\; \det Q_x = 0 \right\} \subset \mathbb{P}^r.$$

Clearly, this definition does not depend on the choice of the matrix representation $x \mapsto Q_x$: the spectral variety is formed by the elements of the linear system that are singular quadrics. More precisely (as a scheme), C is the intersection of the linear system with the discriminant hypersurface $\Delta \subset C_2(\mathbb{P}^N)$.

In what follows, we assume that C is a proper subset of $\mathbb{P}^r$, *i.e.*, we exclude the possibility $C = \mathbb{P}^r$, since in that case, all quadrics in the system have a common singular point, and hence V is not a regular intersection. Under this assumption, $C \subset \mathbb{P}^r$ is a hypersurface of degree $N+1$, possibly not reduced. By the dimension argument, C is necessarily singular whenever $r \geqslant 3$. Furthermore, even in the case $r = 1$ or 2 (*pencils* or *nets*), the spectral variety of a regular intersection may still be singular.

Lemma 3.2. *Let x be an isolated point of the spectral variety C of a pencil. Then x is a simple point of C if and only if the quadric $\{q_x = 0\}$ has a single singular point, and this point is not a base point of the pencil.* □

Corollary 3.3. *For any r, if $x \in C$ is a smooth point, then* corank $q_x = 1$, *i.e., the quadric $\{q_x(u) = 0\}$ has a single singular point.*

Proof. Restrict the system to a generic pencil through the point. □

Lemma 3.4. *If the spectral variety C of a linear system is nonsingular, then the complete intersection V is regular.*

Proof. It is easy to see that the common zero set V of a linear system is a regular complete intersection if and only if none of the members of the system has a singular point in V. Thus, it suffices to observe that a generic pencil through a point $x \in C$ is transversal to C; hence, x is a simple point of its discriminant variety, and the only singular point of $\{q_x = 0\}$ is not in V; see Lemma 3.2. □

3.3 Dixon's Theorem (see [10, 11])

From now on, we confine ourselves to the case of nets, *i.e.*, $r = 2$. As explained in Sect. 3.2, each net gives rise to its spectral curve, which is a curve $C \subset \mathbb{P}^2$ of degree $d = N + 1$; if the net is generic, C is nonsingular.

Assume that the spectral curve C is nonsingular. Then at each point $x \in C$, the kernel $\operatorname{Ker} Q_x \subset \mathbb{R}^{N+1}$ is a 1-subspace; see Lemma 3.3. The correspondence $x \mapsto \operatorname{Ker} Q_x$ defines a line bundle $\mathcal{K}$ on C, or, after a twist, a line bundle $\mathcal{L} = \mathcal{K}(d-1)$. The latter has the following properties: $\mathcal{L}^2 = \mathcal{O}_C(d-1)$ (so that $\deg \mathcal{L} = \frac{1}{2}d(d-1)$) and $H^0(C, \mathcal{L}(-1)) = 0$. Thus, switching to $\theta = \mathcal{L}(-1)$, we obtain a nonvanishing even theta characteristic on C; it is called the *spectral theta characteristic* of the net.

The following theorem is due to Dixon [10].

Theorem 3.5. *Given a nonvanishing even theta characteristic θ on a nonsingular plane curve C of degree $N+1$, there exists a unique, up to projective transformation of $\mathbb{P}^N$, net of quadrics in $\mathbb{P}^N$ such that C is its spectral curve and θ is its spectral theta characteristic.* □

3.4 The Dixon Construction

The original proof by Dixon contains an explicit construction of the net. We outline this construction below. Pick a basis $\phi_{11}, \phi_{12}, \dots, \phi_{1d} \in H^0(C, \mathcal{L})$ and let

$$v_{11} = \phi_{11}^2,\ v_{12} = \phi_{11}\phi_{12},\ \dots,\ v_{1d} = \phi_{11}\phi_{1d} \in H^0(C, \mathcal{L}^2).$$

Since the restriction map $H^0(\mathbb{P}^2; \mathcal{O}_{\mathbb{P}^2}(d-1)) \to H^0(C; \mathcal{O}_C(d-1)) = H^0(C; \mathcal{L}^2)$ is onto, we can regard v_{1i} as homogeneous polynomials of degree $d-1$ in the coordinates x_0, x_1, x_2 in $\mathbb{P}^2$. Let also $U(x_0, x_1, x_2) = 0$ be the equation of C. The curve $\{v_{12} = 0\}$ passes through all points of intersection of C and $\{v_{11} = 0\}$. Hence, there are homogeneous polynomials v_{22}, w_{1122} of degrees $d-1$, $d-2$, respectively, such that

$$v_{12}^2 = v_{11}v_{22} - Uw_{1122}.$$

In the same way, we get polynomials v_{rs}, w_{11rs}, $2 \leqslant s, r \leqslant d$, such that

$$v_{1r}v_{1s} = v_{11}v_{rs} - Uw_{11rs}.$$

Obviously, $v_{rs} = v_{sr}$ and $w_{11rs} = w_{11sr}$. It is shown in [10] that

1. the algebraic complement A_{rs} in the $(d \times d)$ symmetric matrix $[v_{ij}]$ is of the form $U^{d-2}\beta_{rs}$, where β_{rs} are certain linear forms, and
2. the determinant $\det[\beta_{ij}]$ is a constant nonzero multiple of U.

(It is the nonvanishing of the theta characteristic that is used to show that the latter determinant is nonzero.) Thus, C is the spectral curve of the net $Q_x = [\beta_{ij}]$.

It is immediate that the construction works over any field of characteristic zero. Hence, we obtain the following real version of the Dixon theorem.

Theorem 3.6. *Given a nonsingular real plane curve C of degree $N+1 \geqslant 4$ with nonempty real part and a real nonvanishing even theta characteristic θ on C, there exists a unique, up to real projective transformation of $\mathbb{P}^N_{\mathbb{R}}$, real net of quadrics in $\mathbb{P}^N_{\mathbb{R}}$ such that C is its spectral curve and θ is its spectral theta characteristic.* □

3.5 The Spin-*Orientation (cf. [21, 22])*

Let (C, c) be a real curve equipped with a real theta characteristic θ. As above, assume that $C_{\mathbb{R}} \neq \varnothing$. Then the real structure of C lifts to a real structure (*i.e.*, a fiberwise antilinear involution) $c\colon \theta \to \theta$, which is unique up to a phase factor $e^{i\phi}$, $\phi \in \mathbb{R}$. If an isomorphism $\theta^2 = K_C$ is fixed, one can choose a lift compatible with the canonical action of c on K_C; such a lift is unique up to multiplication by i.

Fix a lift $c\colon \theta \to \theta$ as above and pick a c-real meromorphic section ω of θ. Then ω^2 is a real meromorphic 1-form with zeros and poles of even multiplicities. Therefore, it determines an orientation of $C_{\mathbb{R}}$. This orientation does not depend on the choice of ω, and it is reversed when switching from c to ic. Thus it is, in fact, a semiorientation of $C_{\mathbb{R}}$; it is called the Spin-*orientation* defined by θ.

The definition above can be made closer to Dixon's original construction outlined in Sect. 3.4. One can replace ω by a meromorphic section ω' of $\theta(1)$ and treat $(\omega')^2$ as a real meromorphic 1-form with values in $\mathcal{O}_C(2)$; the latter is trivial over $C_{\mathbb{R}}$.

The following, more topological, definition is equivalent to the previous one. Recall that a semiorientation is essentially a rule comparing orientations of pairs of components. Let θ be a c-invariant Spin-structure on C, and let $\phi_\theta\colon H_1(C) \to \mathbb{Z}_2$ be the associated quadratic extension. Pick a point p_i on each real component C_i of $C_{\mathbb{R}}$. For each pair p_i, p_j, $i \neq j$, pick a simple smooth path connecting p_i and p_j in the complement $C \smallsetminus C_{\mathbb{R}}$ and transversal to $C_{\mathbb{R}}$ at the ends, and let γ_{ij} be the loop obtained by combining the path with its c-conjugate. Pick an orientation of C_i and transfer it to C_j by a vector field normal to the path above. The two orientations are considered coherent with respect to the Spin-orientation defined by θ if and only if $\phi_\theta([\gamma_{ij}]) = 0$.

Lemma 3.7. *Assuming that* $C_{\mathbb{R}} \neq \varnothing$, *any semiorientation of* $C_{\mathbb{R}}$ *is the* Spin-*orientation for a suitable real even theta characteristic.* □

Proof. Observe that the c_*-invariant classes $\{[\gamma_{1i}], [C_i]\}$ with $i \geqslant 2$ (see above) form a standard symplectic basis in a certain nondegenerate c_*-invariant subgroup $S \subset H_1(C)$. On the complement $S^\perp$, one can pick any c_*-invariant quadratic extension with Arf-invariant 0. Then the values $\phi_\theta([\gamma_{1i}])$ can be chosen arbitrarily (thus producing any given Spin-orientation), and the values $\phi_\theta([C_i])$, $i \geqslant 2$, can be adjusted (*e.g.*, made all 0) to make the resulting theta characteristic even. □

3.6 Alternating Semiorientations

If $d = \pm 1 \bmod 8$ and θ is the exceptional theta characteristic $\frac{1}{2}(d-3)H$, see Sect. 2.2, then the Spin-orientation defined by θ is given by the residue $\operatorname{res}(p^2\Omega/U)$, where $U = 0$ is the equation of C as above, $\Omega = x_0 \mathrm{d}x_1 \wedge \mathrm{d}x_2 - x_1 \mathrm{d}x_0 \wedge \mathrm{d}x_2 + x_2 \mathrm{d}x_0 \wedge \mathrm{d}x_1$ is a nonvanishing section of $K_{\mathbb{P}^2}(3) \cong \mathcal{O}_{\mathbb{P}^2}$, and p is any real homogeneous polynomial of degree $(d-3)/2$. In affine coordinates, this orientation is given by $p^2 \mathrm{d}x \wedge \mathrm{d}y/\mathrm{d}U$. Such a semiorientation is called *alternating*: it is the only semiorientation of $C_{\mathbb{R}}$ induced by alternating orientations of the components of $\mathbb{P}^2_{\mathbb{R}} \smallsetminus C_{\mathbb{R}}$.

3.7 The Index Function (cf. [2])

Fix a real dimension r linear system $x \mapsto q_x$ of quadrics in $\mathbb{P}^N$. Consider the sphere $S^r = (\mathbb{R}^{r+1} \smallsetminus 0)/\mathbb{R}_+$ and denote by $\tilde{C} \subset S^r$ the pullback of the real part $C_{\mathbb{R}} \subset \mathbb{P}^r_{\mathbb{R}}$ of the spectral variety under the double covering $S^r \to \mathbb{P}^r_{\mathbb{R}}$. Define the *index function*

$$\operatorname{ind}\colon S^r \to \mathbb{Z}$$

by sending a point $x \in S^r$ to the negative index of inertia of the quadratic form q_x. The following statement is obvious. (For item 3.8(3), one should use Corollary 3.3.)

Proposition 3.8. *The index function* ind *has the following properties*:

1. ind *is lower semicontinuous*;
2. ind *is locally constant on* $S^r \smallsetminus \tilde{C}$;
3. ind *jumps by* ± 1 *when crossing* $\tilde{C}$ *transversally at its regular point*;
4. *one has* $\operatorname{ind}(-x) = N + 1 - (\operatorname{ind} x + \operatorname{corank} q_x)$. □

Due to Proposition 3.8(3), ind defines a coorientation of $\tilde{C}$ at all its smooth points. This coorientation is reversed by the antipodal map $a\colon S^r \to S^r$, $x \mapsto -x$; see Proposition 3.8(4). If $r = 2$ and the real part $C_{\mathbb{R}}$ of the spectral curve is nonsingular, this coorientation defines a semiorientation of $C_{\mathbb{R}}$ as follows: pick an orientation of S^2 and use it to convert the coorientation to an orientation $\tilde{o}$ of $\tilde{C}$; since the antipodal map a is orientation reversing, $\tilde{o}$ is preserved by a and hence descends to an orientation o of $C_{\mathbb{R}}$. The latter is defined up to total reversing (due to the initial choice of an orientation of S^2); hence, it is in fact a semiorientation. It is called the *index orientation* of $C_{\mathbb{R}}$.

Conversely, any semiorientation of $C_{\mathbb{R}}$ can be defined as above by a function $\mathrm{ind}\colon S^r \to \mathbb{Z}$ satisfying Proposition 3.8(1)–(4); the latter is unique up to the antipodal map.

Theorem 3.9 (cf. [29]). *Assume that the real part $C_{\mathbb{R}}$ of the spectral curve of a real net of quadrics is nonsingular. Then the index orientation of $C_{\mathbb{R}}$ coincides with its* Spin*-orientation defined by the spectral theta characteristic.*

Proof. The Spin-semiorientation of $C_{\mathbb{R}}$ is given by the residue $\mathrm{res}(v_{11}\Omega/U)$, cf. Sect. 3.6, and it is sufficient to check that the index function is larger on the side of $U = 0$ where $v_{11}/U > 0$. Since the kernel $\sum x_i Q_i$, regarded as a section of the projectivization of the trivial bundle over C, is given by $v = (v_{11}, v_{12}, \dots, v_{1d})$, it remains to observe that over $C_{\mathbb{R}}$, one has $\sum x_i \langle Q_i v, v\rangle = v_{11} \det(\sum x_i Q_i) = v_{11} U$. □

Theorem 3.10. *Let C be a nonsingular real plane curve of degree $d = N+1$ and with nonempty real part, and let o be a semiorientation of $C_{\mathbb{R}}$. Assume that either $d \neq \pm 1 \bmod 8$ or o is not the alternating semiorientation; see Sect. 3.6. Then after a small real perturbation of C, there exists a regular intersection of three real quadrics in $\mathbb{P}^N_{\mathbb{R}}$ that has C as its spectral curve and o as its spectral* Spin*-orientation.*

Remark 3.11. According to Sect. 3.6, the only case not covered by Theorem 3.10 is that in which $d = \pm 1 \bmod 8$ and the index function ind assumes only the two middle values $(d \pm 1)/2$.

Proof (of Theorem 3.10). By Lemma 3.7, there exists a real even theta characteristic θ that has o as its Spin-orientation. Using Lemma 2.1, one can make θ nonvanishing by a small real perturbation of C, and it remains to apply Theorem 3.6. □

4 The Topology of the Zero Locus of a Net

4.1 The Spectral Sequence

Consider a real dimension r linear system $x \mapsto q_x$ of quadrics in $\mathbb{P}^N$; see Sect. 3.1 for the notation. Let $V_{\mathbb{R}} \subset \mathbb{P}^N_{\mathbb{R}}$, $C_{\mathbb{R}} \subset \mathbb{P}^r_{\mathbb{R}}$, and $L_{\mathbb{R}} \subset \mathbb{P}^r_{\mathbb{R}} \times \mathbb{P}^N_{\mathbb{R}}$ be the real parts of the common zero set, spectral variety, and Lagrange hypersurface, respectively.

Consider the sphere $S^r = (\mathbb{R}^{r+1} \smallsetminus 0)/\mathbb{R}_+$ and the lift $\tilde{C} \subset S^r$ of $C_{\mathbb{R}}$, cf. Sect. 3.7, and let $\tilde{L} = \{(x,u) \in S^r \times \mathbb{P}^N_{\mathbb{R}} \,|\, q_x(u) = 0\} \subset S^r \times \mathbb{P}^N_{\mathbb{R}}$ be the lift of $L_{\mathbb{R}}$ and

$$L_+ = \left\{(x,u) \in S^r \times \mathbb{P}^N_{\mathbb{R}} \,\middle|\, q_x(u) > 0\right\} \subset S^r \times \mathbb{P}^N_{\mathbb{R}}$$

its *positive complement.* (Clearly, the conditions $q_x(u) = 0$ and $q_x(u) > 0$ do not depend on the choice of representatives of $x \in S^r$ and $u \in \mathbb{P}^N_{\mathbb{R}}$.)

Lemma 4.1. *The projection $S^r \times \mathbb{P}^N_{\mathbb{R}} \to \mathbb{P}^N_{\mathbb{R}}$ restricts to a homotopy equivalence $L_+ \to \mathbb{P}^N_{\mathbb{R}} \smallsetminus V_{\mathbb{R}}$.*

Proof. Denote by p the restriction of the projection to L_+. The pullback $p^{-1}(u)$ of a point $u \in V_{\mathbb{R}}$ is empty; hence, p sends L_+ to $\mathbb{P}^N_{\mathbb{R}} \smallsetminus V_{\mathbb{R}}$. On the other hand, the restriction $L_+ \to \mathbb{P}^N_{\mathbb{R}} \smallsetminus V_{\mathbb{R}}$ is a locally trivial fibration, and for each point $u \in \mathbb{P}^N_{\mathbb{R}} \smallsetminus V_{\mathbb{R}}$, the fiber $p^{-1}(u)$ is the open hemisphere $\{x \in S^r \,|\, q_x(u) > 0\}$, hence contractible. □

Proposition 4.2. *Let $r = 2$. Then, one has $b^0(L_+) = b^1(L_+) = 1$, and if $V_{\mathbb{R}}$ is nonsingular, also $b^2(L_+) = b^0(V_{\mathbb{R}}) + 1$.*

Proof. Due to Lemma 4.1, one has $H^*(L_+) = H^*(\mathbb{P}^N_{\mathbb{R}} \smallsetminus V_{\mathbb{R}})$, and the statement of the proposition follows from the Poincaré–Lefschetz duality $H^i(\mathbb{P}^N_{\mathbb{R}} \smallsetminus V_{\mathbb{R}}) = H_{N-i}(\mathbb{P}^N_{\mathbb{R}}, V_{\mathbb{R}})$ and the exact sequence of the pair $(\mathbb{P}^N_{\mathbb{R}}, V_{\mathbb{R}})$. For the last statement, one needs in addition to know that the inclusion homomorphism $H_{N-3}(V_{\mathbb{R}}) \to H_{N-3}(\mathbb{P}^N_{\mathbb{R}})$ is trivial, *i.e.*, that every 3-plane P intersects each component of $V_{\mathbb{R}}$ at an even number of points. By restricting the system to a 4-plane containing P, one reduces the problem to the case $N = 4$. In this case, $V \subset \mathbb{P}^N$ is the canonical embedding of a genus 5 curve, *cf.* Sect. 6.4, and the statement is obvious. □

From now on, we assume that the real part $C_{\mathbb{R}}$ of the spectral hypersurface is nonsingular. Consider the ascending filtration

$$\varnothing = \Omega_{-1} \subset \Omega_0 \subset \Omega_1 \subset \cdots \subset \Omega_{N+1} = S^r, \quad \Omega_i = \{x \in S^r \,|\, \operatorname{ind} x \leqslant i\}. \tag{4.1}$$

Due to Proposition 3.8(1), all Ω_i are closed subsets.

Theorem 4.3 (cf. [1]). *There is a spectral sequence*

$$E_2^{pq} = H^p(\Omega_{N-q}) \Rightarrow H^{p+q}(L_+).$$

Proof. The sequence in question is the Leray spectral sequence of the projection $\pi\colon L_+ \to S^r$. Let $\mathcal{Z}_2$ be the constant sheaf on L_+ with the fiber $\mathbb{Z}_2$. Then the sequence is $E_2^{pq} = H^p(S^r, R^q\pi_*\mathcal{Z}_2) \Rightarrow H^{p+q}(L_+)$. Given a point $x \in S^r$, the stalk $(R^q\pi_*\mathcal{Z}_2)|_x$ equals $H^q(\pi^{-1}U_x)$, where $U_x \ni x$ is a small neighborhood of x regular with respect to a triangulation of S^r compatible with the filtration. If $x \notin \tilde{C}$ and $\operatorname{ind} x = i$, then

$$\pi^{-1}U_x \sim \pi^{-1}x = \{u \in \mathbb{P}^N_{\mathbb{R}} \,|\, q_x(u) > 0\} \sim \mathbb{P}^{N-i}_{\mathbb{R}}.$$

(The fiber $\pi^{-1}x$ is a D^i-bundle over $\mathbb{P}_{\mathbb{R}}^{N-i}$; if $i=N+1$, the fiber is empty.) Besides, if $x \in \tilde{C}$ and $\operatorname{ind} x = i$, then $\pi^{-1}U_x \sim \pi^{-1}x'$, where $x' \in (U_x \cap \Omega_i) \smallsetminus \tilde{C}$. Thus, $(R^q\pi_*\mathcal{Z}_2)|_x = \mathbb{Z}_2$ or 0 if x does (respectively does not) belong to Ω_{N-q}, and the statement is immediate. □

Remark 4.4. It is worth mentioning that there are spectral sequences, similar to the one introduced in Theorem 4.3, that compute the cohomology of the double coverings of $\mathbb{P}_{\mathbb{R}}^N \smallsetminus V_{\mathbb{R}}$ and $V_{\mathbb{R}}$ sitting in S^N; see [2].

4.2 Elements of Topology of Real Plane Curves

Let $C \subset \mathbb{P}^2$ be a nonsingular real curve of degree d. Recall that the real part $C_{\mathbb{R}}$ splits into a number of *ovals* (*i.e.*, embedded circles contractible in $\mathbb{P}_{\mathbb{R}}^2$), and if d is odd, one *one-sided component* (*i.e.*, an embedded circle isotopic to $\mathbb{P}_{\mathbb{R}}^1$.) The complement of each oval $\mathfrak{o}$ has two connected components, exactly one of them being contractible; this contractible component is called the *interior* of $\mathfrak{o}$.

On the set of ovals of $C_{\mathbb{R}}$, there is a natural partial order: an oval $\mathfrak{o}$ is said to *contain* another oval $\mathfrak{o}'$, $\mathfrak{o} \prec \mathfrak{o}'$, if $\mathfrak{o}'$ lies in the interior of $\mathfrak{o}$. An oval is called *empty* if it does not contain another oval. The *depth* $\operatorname{dp}\mathfrak{o}$ of an oval $\mathfrak{o}$ is the number of elements in the maximal descending chain starting at $\mathfrak{o}$. (Such a chain is unique.) Every oval $\mathfrak{o}$ of depth >1 has a unique immediate predecessor; it is denoted by $\operatorname{pred}\mathfrak{o}$.

A *nest* of C is a linearly ordered chain of ovals of $C_{\mathbb{R}}$; the *depth* of a nest is the number of its elements. The following statement is a simple and well-known consequence of Bézout's theorem.

Proposition 4.5. *Let C be a nonsingular real plane curve of degree d. Then*

1. *C cannot have a nest of depth greater than $D_{\max} = D_{\max}(d) = [d/2]$;*
2. *if C has a nest of depth $D_{\max}$ (a* maximal nest*), it has no other ovals;*
3. *if C has a nest $\mathfrak{o}_1 \prec \cdots \prec \mathfrak{o}_k$ of depth $k = D_{\max} - 1$ (a* submaximal nest*) but no maximal nest, then all ovals other than $\mathfrak{o}_1, \ldots, \mathfrak{o}_{k-1}$ are empty.* □

Let $\operatorname{pr}\colon S^2 \to \mathbb{P}_{\mathbb{R}}^2$ be the orientation double covering, and let $\tilde{C} = \operatorname{pr}^{-1} C_{\mathbb{R}}$. The pullback of an oval $\mathfrak{o}$ of C consists of two disjoint circles $\mathfrak{o}'$, $\mathfrak{o}''$; such circles are called *ovals* of $\tilde{C}$. The antipodal map $x \mapsto -x$ of S^2 induces an involution on the set of ovals of $\tilde{C}$; we denote it by a bar: $\mathfrak{o} \mapsto \bar{\mathfrak{o}}$. The pullback of the one-sided component of $C_{\mathbb{R}}$ is connected; it is called the *equator*. The *tropical components* are the components of $S^2 \smallsetminus \tilde{C}$ whose image is the (only) component of $\mathbb{P}_{\mathbb{R}}^2 \smallsetminus C_{\mathbb{R}}$ outer to all ovals. The *interior* $\operatorname{int}\mathfrak{o}$ of an oval $\mathfrak{o}$ of $\tilde{C}$ is the component of the complement of $\mathfrak{o}$ that projects to the interior of $\operatorname{pr}\mathfrak{o}$ in $\mathbb{P}_{\mathbb{R}}^2$. As in the case of $C_{\mathbb{R}}$, one can use the notion of interior to define the partial order, depth, nests, *etc*. The projection pr and the antipodal involution induce strictly increasing maps of the sets of ovals.

Now consider an (abstract) index function ind: $S^2 \to \mathbb{Z}$ satisfying Proposition 3.8(1)–(4) (where $N+1=d$ and $\operatorname{corank} q_x = \chi_{\tilde{C}}(x)$ is the characteristic function of $\tilde{C}$) and use it to define the filtration Ω_* as in (4.1). For an oval $\mathfrak{o}$ of $\tilde{C}$, define $i(\mathfrak{o})$ as the value of ind immediately inside $\mathfrak{o}$. (Note that in general, $i(\mathfrak{o})$ is *not* the restriction of ind to $\mathfrak{o}$ as a subset of S^r.) Then in view of Proposition 3.8, one has

$$\frac{1}{2}N \leqslant \operatorname{ind}|_T \leqslant \frac{1}{2}N+1 \quad \text{for each tropical component } T, \tag{4.2}$$

$$i(\mathfrak{o}) \leqslant N+1-(D_{\max}-\operatorname{dp}\mathfrak{o}) \quad \text{for each oval } \mathfrak{o}, \tag{4.3}$$

$$i(\mathfrak{o}) = N+1-(D_{\max}-\operatorname{dp}\mathfrak{o}) \bmod 2 \quad \text{if } N \text{ is odd.} \tag{4.4}$$

Here, as above, $D_{\max} = [(N+1)/2]$. Keeping in mind the applications, we state the restrictions in terms of $N = d-1$. (Certainly, the congruence (4.4) simplifies to $i(\mathfrak{o}) + \operatorname{dp}\mathfrak{o} = D_{\max} \bmod 2$; however, we leave it in a form convenient for further applications.)

Corollary 4.6. *If $N \geqslant 5$, then Ω_{N-2} contains the tropical components.* □

Lemma 4.7. *Assume that $b^0(\Omega_q) > 1$ for some integer $q > \frac{1}{2}N$. Then $\tilde{C}$ has a nest $\mathfrak{o} \prec \mathfrak{o}'$ such that $i(\mathfrak{o}) = q+1$ and $i(\mathfrak{o}') = q$.*

Proof. Due to (4.2), the assumption $q > \frac{1}{2}N$ implies that Ω_q contains the tropical components. Then one can take for $\mathfrak{o}'$ the oval bounding from outside another component of Ω_q, and let $\mathfrak{o} = \operatorname{pred}\mathfrak{o}'$. □

Lemma 4.8. *Let $N \geqslant 7$, and assume that the curve $\tilde{C}$ has a nest $\mathfrak{o}_{k-1} \prec \mathfrak{o}_k$, $k = D_{\max}-1$, with $i(\mathfrak{o}_{k-1}) = N-1$. Then $\Omega_{N-2} \supset S^2 \setminus \operatorname{int}\operatorname{pred}\mathfrak{o}_{k-1}$.*

Proof. Due to (4.3), the nest $\mathfrak{o}_{k-1} \prec \mathfrak{o}_k$ in the statement can be completed to a submaximal nest $\mathfrak{o}_1 \prec \ldots \prec \mathfrak{o}_k$.

First, assume that either $\tilde{C}$ has no maximal nest or the innermost oval of $\tilde{C}$ is inside $\mathfrak{o}_k$. Then $\operatorname{dp}\mathfrak{o}_s = s$ and $\mathfrak{o}_s = \operatorname{pred}\mathfrak{o}_{s+1}$ for all s. Due to (4.3) and Proposition 3.8(3), one has $i(\mathfrak{o}_{k-s}) = N-s$ for $s = 1,\ldots,k-1$. Then, due to Proposition 3.8(4), $i(\bar{\mathfrak{o}}_{k-s}) = s+1$ for $s = 1,\ldots,k-1$ and hence $i(\bar{\mathfrak{o}}_k) \leqslant 3$.

Proposition 4.5 implies that all ovals other than $\mathfrak{o}_{k-s}$, $\bar{\mathfrak{o}}_{k-s}$, $s = 0,\ldots,k-1$, are empty. For such an oval $\mathfrak{o}$, one has $i(\mathfrak{o}) \leqslant [\frac{1}{2}N]+2$ if $\operatorname{dp}\mathfrak{o} = 1$, see (4.2), and $i(\mathfrak{o}) \leqslant 4$, $s+2$, or $N+1-s$ if $\operatorname{pred}\mathfrak{o} = \bar{\mathfrak{o}}_k$, $\bar{\mathfrak{o}}_{k-s}$, or $\mathfrak{o}_{k-s}$, respectively, $s = 1,\ldots,k-1$. From the assumption $N \geqslant 7$, it follows that for any oval $\mathfrak{o}$, one has $i(\mathfrak{o}) \leqslant N-2$ unless $\mathfrak{o} \succcurlyeq \mathfrak{o}_{k-2}$. Together with Corollary 4.6, this observation implies the statement.

The case in which $\tilde{C}$ has another oval $\mathfrak{o}' \prec \mathfrak{o}_i$ for some $i \leqslant k-1$ is treated similarly. In this case, $N = 2k$ is even, see (4.4), and by renumbering the ovals consecutively from k down to 0, one has $i(\mathfrak{o}_{k-s}) = N-s$, $i(\bar{\mathfrak{o}}_{k-s}) = s+1$ for $s = 1,\ldots,k$. □

Note that, in particular, the condition $N \geqslant 7$ in Lemma 4.8 is necessary for $\mathfrak{o}_{k-1}$ to have a predecessor, *i.e.*, for the statement to make sense. The remaining two interesting cases $N = 5$ and 6 are treated in the next lemma.

Lemma 4.9. *Let $N = 5$ or 6, and assume that the curve $\tilde{C}$ has a nest $\mathfrak{o}_1 \prec \mathfrak{o}_2$ with $i(\mathfrak{o}_1) = N - 1$. Then either*

1. *for each pair $\mathfrak{o}$, $\bar{\mathfrak{o}}$ of nonempty antipodal ovals of depth 1, the set Ω_{N-2} contains the interior of exactly one of them, or*
2. *$N = 5$ and $\tilde{C}$ has another nested oval $\mathfrak{o}_3 \succ \mathfrak{o}_2$ with $i(\mathfrak{o}_3) = 2$. (These data determine the index function uniquely.)*

Proof. The proof repeats literally that of Lemma 4.8, with a careful analysis of the inequalities that do not hold for small values of N. □

4.3 The Estimates

In this section, we consider a net of quadrics ($r = 2$) and assume that the spectral curve $\tilde{C} \subset S^2$ is nonsingular. Furthermore, we can assume that the index function takes values between 1 and N, since otherwise, the net would contain an empty quadric and one would have $V = \varnothing$. Thus, one has $\varnothing = \Omega_{-1} = \Omega_0$ and $\Omega_N = \Omega_{N+1} = S^2$.

Set $i_{\max} = \max_{x \in S^2} \operatorname{ind} x$. Thus, we assume that $i_{\max} \leqslant N$.

In addition, we can assume that $N \geqslant 3$, since for $N \leqslant 2$, a regular intersection of three quadrics in $\mathbb{P}^N$ is empty.

The spectral sequence E_r^{pq} given by Theorem 4.3 is concentrated in the strip $0 \leqslant p \leqslant 2$, and all potentially nontrivial differentials are $d_2^{0,q}\colon E_2^{0,q} \to E_2^{2,q-1}$, $q \geqslant 1$. Furthermore, one has

$$E_2^{0,q} = E_2^{2,q} = \mathbb{Z}_2, \quad E_2^{1,q} = 0 \quad \text{for } q = 0, \ldots, N - i_{\max}, \tag{4.5}$$

$$E_2^{0,q} = E_2^{1,q} = E_2^{2,q} = 0 \quad \text{for } q \geqslant i_{\max}, \quad \text{and} \tag{4.6}$$

$$E_2^{2,q} = 0 \quad \text{for } q > N - i_{\max}. \tag{4.7}$$

In particular, it follows that $d_2^{0,q} = 0$ for $q > N + 1 - i_{\max}$.

Corollary 4.10. *If $i_{\max} \leqslant N - 2$, then $b^0(V_{\mathbb{R}}) \leqslant 1$.* □

The assertion of Corollary 4.10 was first observed by Agrachev [1].

Lemma 4.11. *With one exception, $d_2^{0,1} = 0$. The exception is a curve $\tilde{C}$ with a maximal nest $\mathfrak{o}_1 \prec \ldots \prec \mathfrak{o}_k$, $k = D_{\max}$, so that $i(\mathfrak{o}_k) = N - 1$ and $i(\mathfrak{o}_{k-1}) = N$. In this exceptional case, one has $b^0(V_{\mathbb{R}}) = 1$.*

Proof. Since $b^1(L_+) = 1$, see Proposition 4.2, and $E_2^{1,0} = 0$, the differential $d_2^{0,1}$ is nontrivial if and only if $b^0(\Omega_{N-1}) = \dim E_2^{0,1} > 1$. Since $N \geqslant 3$, the exceptional case is covered by Lemma 4.7 and Proposition 4.5. □

Corollary 4.12. *If $i_{\max} = N-1 \geqslant 2$, then one has $b^0(\Omega_{N-2}) - 1 \leqslant b^0(V_{\mathbb{R}}) \leqslant b^0(\Omega_{N-2})$.* □

Lemma 4.13. *Assume that $i_{\max} = N-1 \geqslant 4$ and that $b^0(\Omega_{N-2}) > 1$. Then $\beta \leqslant b^0(V_{\mathbb{R}}) \leqslant \beta+1$, where β is the number of ovals of $C_{\mathbb{R}}$ of depth $D_{\max}-1$.*

Proof. In view of Corollary 4.12, it suffices to show that $b^0(\Omega_{N-2}) = \beta+1$. Due to Lemma 4.7, the curve $\tilde{C}$ has a nest $\mathfrak{o} \prec \mathfrak{o}'$ with $i(\mathfrak{o}) = N-1$, and then the set Ω_{N-2} is described by Lemmas 4.8 and 4.9 and the assumption $i_{\max} = N-1$. In view of Proposition 4.5, each oval of $C_{\mathbb{R}}$ of depth $\operatorname{dp}\mathfrak{o}+1$ is inside $\operatorname{pr}\mathfrak{o}$, thus contributing an extra unit to $b^0(\Omega_{N-2})$. □

Lemma 4.14. *Assume that $i_{\max} = N \geqslant 5$. Then $b^0(V_{\mathbb{R}})$ is equal to the number β of ovals of $C_{\mathbb{R}}$ of depth $D_{\max}-1$.*

Proof. If $(\tilde{C}, \operatorname{ind})$ is the exceptional index function mentioned in Lemma 4.11, then $b^0(V_{\mathbb{R}}) = \beta = 1$, and the statement holds. Otherwise, both differentials $d_2^{0,1}$ and $d_2^{0,2}$ vanish, see (4.7), and, using Proposition 4.2 and (4.5), one concludes that $b^0(V_{\mathbb{R}}) = \dim E_2^{0,2} + \dim E_2^{1,1} = b^0(\Omega_{N-2}) + b^1(\Omega_{N-1})$.

Pick an oval $\mathfrak{o}'$ with $i(\mathfrak{o}') = N$ and let $\mathfrak{o} = \operatorname{pred}\mathfrak{o}'$. The topology of Ω_{N-2} is given by Lemmas 4.8 and 4.9. If $\operatorname{dp}\mathfrak{o} = D_{\max}-1$, then $b^0(V_{\mathbb{R}}) = \beta = 1$. Otherwise ($\operatorname{dp}\mathfrak{o} = D_{\max}-2$), one has $b^0(\Omega_{N-2}) = \beta_- + 1$ and $b^1(\Omega_{N-1}) = \beta_+ - 1$, where $\beta_- \geqslant 0$ and $\beta_+ > 0$ are the numbers of ovals $\mathfrak{o}'' \succ \mathfrak{o}$ with $i(\mathfrak{o}'') = N-2$ and N, respectively; due to Proposition 4.5, one has $\beta_- + \beta_+ = \beta$. □

4.4 Proof of Theorem 1.3

The case $N = 4$ is covered by Theorem 1.4. (Note that $B_2^0(4) = \operatorname{Hilb}(5)$; see Sect. 6.4.) Alternatively, one can treat this case manually, trying various index functions on a curve of degree 5.

Assume that $N \geqslant 5$. The upper bound on $B_2^0(N)$ follows from Corollary 4.10 and Lemmas 4.13 and 4.14. For the lower bound, pick a generic real curve C of degree $d = N+1$ with $\operatorname{Hilb}(d)$ ovals of depth $[d/2]-1$. Select an oval $\mathfrak{o}_i$ in each pair $(\mathfrak{o}_i, \bar{\mathfrak{o}}_i)$ of antipodal outermost ovals of $\tilde{C}$; if d is even, make sure that all selected ovals are in the boundary of the same tropical component. Take for ind the "monotonous" function defined *via* $i(\mathfrak{o}) = N+1-(D_{\max} - \operatorname{dp}\mathfrak{o})$ if $\mathfrak{o} \succcurlyeq \mathfrak{o}_i$ and $i(\mathfrak{o}) = D_{\max} - \operatorname{dp}\mathfrak{o}$ if $\mathfrak{o} \succcurlyeq \bar{\mathfrak{o}}_i$ for some i; see Fig. 1 (where the cases $N = 7$ and $N = 8$ are shown schematically). Due to Theorem 3.10 (see also Remark 3.11), the pair (C, ind) is realized by a net of quadrics, and for this net one has $b^0(V) = \operatorname{Hilb}(d)$; see Lemma 4.14. □

4.5 Proof of Theorem 1.1

In this section, we make an attempt to estimate the Hilbert number $\operatorname{Hilb}(d)$ introduced in Definition 1.2.

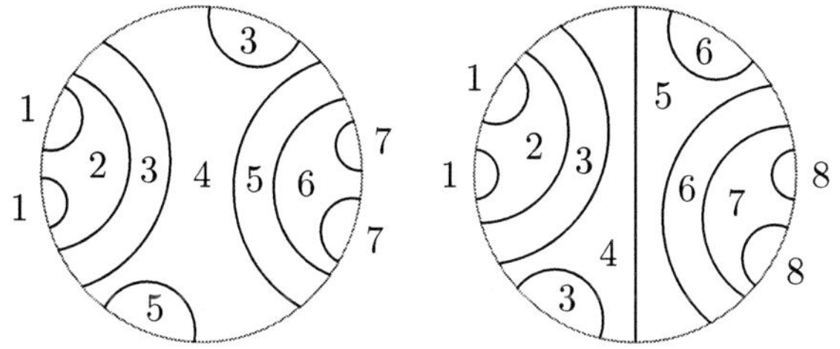

Fig. 1 "Monotonous" index functions ($N = 7$ and $N = 8$)

Let C be a nonsingular real plane algebraic curve of degree d. An oval of $C_{\mathbb{R}}$ is said to be *even* (*odd*) if its depth is odd (respectively even). An oval is called *hyperbolic* if it has more than one immediate successor (in the partial order defined in Sect. 4.2).

The following statement is known as the *generalized Petrovsky inequality.*

Theorem 4.15 (see [4]). *Let C be a nonsingular real plane curve of even degree $d = 2k$. Then*

$$p - n^- \leqslant \frac{3}{2}k(k-1) + 1, \quad n - p^- \leqslant \frac{3}{2}k(k-1),$$

where p, n are the numbers of even/odd ovals of $C_{\mathbb{R}}$, and p^-, n^- are the numbers of even/odd hyperbolic ovals. □

Corollary 4.16. *One has* $\operatorname{Hilb}(d) \leqslant \frac{3}{2}k(k-1) + 1$, *where* $k = [(d+1)/2]$.

Proof. Let $d = 2k$ be even, and let C be a curve of degree d with $m > 1$ ovals of depth $k-1$. All submaximal ovals are situated inside a nest $\mathfrak{o}_1 \prec \ldots \prec \mathfrak{o}_{k-2}$ of depth $D_{\max} - 2 = k - 2$; see Proposition 4.5. Assume that $k = 2l$ is even. Then the submaximal ovals are even, and one has $p \geqslant m + l - 1$, counting as well the even ovals $\mathfrak{o}_1, \mathfrak{o}_3, \ldots, \mathfrak{o}_{2l-1}$ in the nest. On the other hand, $n^- \leqslant l - 1$, since all odd ovals other than $\mathfrak{o}_2, \mathfrak{o}_4, \ldots, \mathfrak{o}_{2l}$ are empty, hence not hyperbolic; see Proposition 4.5 again. Hence, the statement follows from the first inequality in Theorem 4.15. The case of k odd is treated similarly, using the second inequality in Theorem 4.15.

Let $d = 2k - 1$ be odd, and consider a real curve C of degree d with a nest $\mathfrak{o}_1 \prec \ldots \prec \mathfrak{o}_{k-3}$ of depth $D_{\max} - 2 = k - 3$ and $m \geqslant 2$ ovals $\mathfrak{o}', \mathfrak{o}'', \ldots$ of depth $k-2$. Pick a pair of points p' and p'' inside $\mathfrak{o}'$ and $\mathfrak{o}''$, respectively, and consider the line $L = (p_1 p_2)$. From Bézout's theorem, it follows that all points of intersection of L and C are one point on the one-sided component of $C_{\mathbb{R}}$ and a pair of points on each of the ovals $\mathfrak{o}_1, \ldots, \mathfrak{o}_{k-3}, \mathfrak{o}', \mathfrak{o}''$. Furthermore, the pair $\mathfrak{o}'$, $\mathfrak{o}''$ can be chosen so that all other innermost ovals of $C_{\mathbb{R}}$ lie to one side of L in the interior of $\mathfrak{o}_{k-1}$ (which is divided by L into two components).

According to Brusotti's theorem [6], the union $C + L$ can be perturbed to form a nonsingular curve of degree $2k$ with m ovals of depth $k-1$; see Fig. 2 (where the curve and its perturbation are shown schematically in gray and black, respectively). Hence, the statement follows from the case of even degree considered above. □

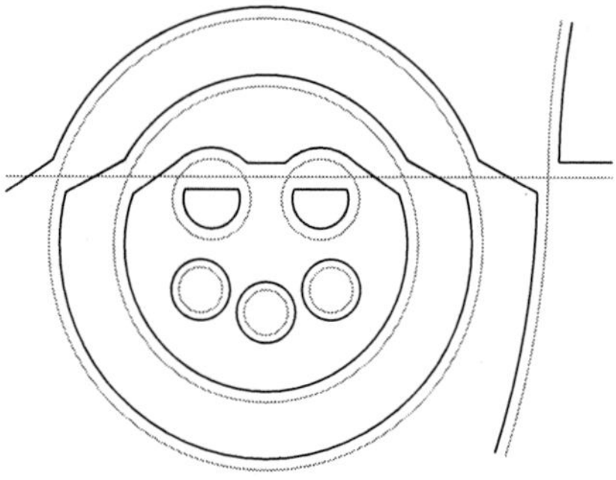

Fig. 2 The perturbation of $C+L$ with a deep nest

Theorem 4.17 (see [16]). *For each integer $d \geqslant 4$, there is a nonsingular curve of degree d in $\mathbb{P}^2_{\mathbb{R}}$ with*

- *Four ovals of depth 1 if $d = 4$,*
- *Six ovals of depth 1 if $d = 5$,*
- *$k(k+1) - 3$ ovals of depth $[d/2] - 1$ if $d = 2k$ is even,*
- *$k(k+2) - 3$ ovals of depth $[d/2] - 1$ if $d = 2k+1$ is odd.* □

For the reader's convenience, we give a brief outline of the original construction due to Hilbert that produces curves as in Theorem 4.17.

For even degrees, one can use an inductive procedure that produces a sequence of curves $C^{(2k)}$, $\deg C^{(2k)} = 2k$. Let $E = \{p_E = 0\}$ be an ellipse in $\mathbb{P}^2_{\mathbb{R}}$. The curve $C^{(2)}$ is defined by a polynomial $p^{(2)}$ of the form

$$p^{(2)} = p_E + \varepsilon^{(2)} l_1^{(2)} l_2^{(2)},$$

where $\varepsilon^{(2)} > 0$ is a real number, $|\varepsilon^{(2)}| \ll 1$, and $l_1^{(2)}$ and $l_2^{(2)}$ are real polynomials of degree 1 such that the pair of lines $\{l_1^{(2)} l_2^{(2)} = 0\}$ intersects E at four distinct real points. The intersection of the exterior of E and the interior of $C^{(2)}$ is formed by two disks $D_1^{(2)}$ and $D_2^{(2)}$.

Inductively, we construct curves $C^{(2k)} = \{p^{(2k)} = 0\}$ with the following properties:

1. $C^{(2k)}$ has an oval $\mathfrak{o}^{(2k)}$ of depth $k-1$ such that $\mathfrak{o}^{(2k)}$ intersects E at $4k$ distinct points, the orders of the intersection points on $\mathfrak{o}^{(2k)}$ and E coincide, and the intersection of the exterior of E and the exterior of $\mathfrak{o}^{(2k)}$ consists of a Möbius strip and $2k-1$ disks $D_1^{(2k)}, \ldots, D_{2k-1}^{(2k)}$ (shaded in Fig. 3);
2. one has

$$p^{(2k)} = p^{(2k-2)} p_E + \varepsilon^{(2k)} l_1^{(2k)} \ldots l_{2k}^{(2k)},$$

where $\varepsilon^{(2k)}$ is a real number, $|\varepsilon^{(2k)}| \ll 1$, and $l_1^{(2k)}, \ldots, l_{2k}^{(2k)}$ are certain polynomials of degree 1 such that the union of lines $\{l_1^{(2k)} \ldots l_{2k}^{(2k)} = 0\}$ intersects E at $4k$ distinct real points, all points belonging to $\partial D_1^{(2k-2)}$ (see Fig. 4).

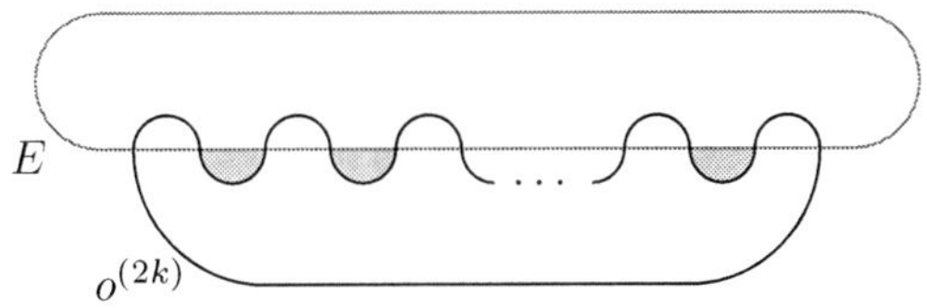

Fig. 3 The oval $\mathfrak{o}^{(2k)}$ and the disks $D_i^{(2k)}$ (*shaded*)

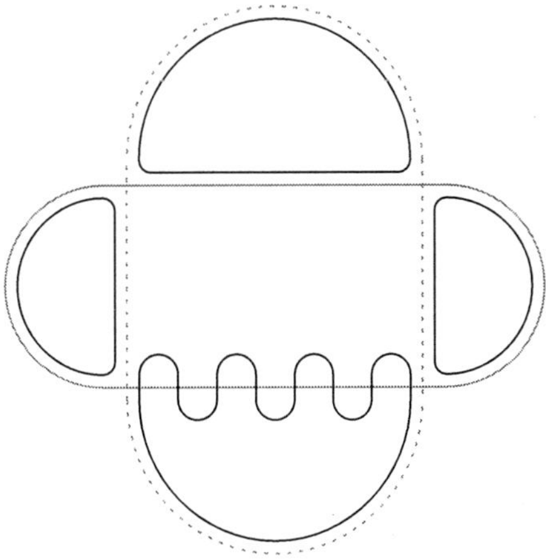

Fig. 4 Hilbert's construction in degree 4

The sign of $\varepsilon^{(2k)}$ is chosen so that $\mathfrak{o}^{(2k-2)} \cup E$ produces $4k-4$ ovals of $C^{(2k)}$.

The above properties imply that each curve $C^{(2k)}$ has the required number of ovals of depth $k-1$ (and the next curve $C^{(2k+2)}$ still satisfies condition 1).

The curves $C^{(2k+1)}$ of odd degree $2k+1$ are constructed similarly, starting from a curve $C^{(3)}$ defined by a polynomial of the form

$$p^{(3)} = l p_E + \varepsilon^{(3)} l_1^{(3)} l_2^{(3)} l_3^{(3)},$$

where $\varepsilon^{(3)} > 0$ is a sufficiently small real number, l is a polynomial of degree 1 defining a line disjoint from E, and $l_1^{(3)}$, $l_2^{(3)}$, and $l_3^{(3)}$ are polynomials of degree 1 such that the union of lines $\{l_1^{(3)} l_2^{(3)} l_3^{(3)} = 0\}$ intersects E at six distinct real points.

Corollary 4.18. *One has* $\mathrm{Hilb}(d) > \frac{1}{4}(d-2)(d+4) - 2.$ □

Remark 4.19. S. Orevkov informed us, see [24], that there are real algebraic curves of degree d with

$$\frac{9}{32}d^2 + O(d)$$

ovals of depth $[\frac{d}{2}] - 1$, and that he expects that this estimate is still not sharp. In the category of real pseudoholomorphic curves, Orevkov achieved as many as $\frac{1}{3}d^2 + O(d)$ ovals of submaximal depth.

Proof of Theorem 1.1

The statement of the theorem follows from Theorem 1.3 and the bounds on $\mathrm{Hilb}(d)$ given by Corollaries 4.16 and 4.18. □

5 Intersections of Quadrics of Dimension One

In this section, we consider the case $N = r+2$, *i.e.*, one-dimensional complete intersections of quadrics.

5.1 Proof of Theorem 1.4: The Upper Bound

Let V be a regular complete intersection of $N-1$ quadrics in $\mathbb{P}^N$. Iterating the adjunction formula, one finds that the genus $g(V)$ of the curve V satisfies the relation

$$2g(V)-2 = 2^{N-1}\big(2(N-1)-(N+1)\big) = 2^{N-1}(N-3);$$

hence, $g(V) = 2^{N-2}(N-3)+1$, and the Harnack inequality gives the upper bound $B^0_{N-2}(N) \leq 2^{N-2}(N-3)+2$. □

5.2 Proof of Theorem 1.4: The Construction

To prove the lower bound $B^0_{N-2}(N) \geqslant 2^{N-2}(N-3)+2$, for each integer $N \geqslant 2$ we construct a homogeneous quadratic polynomial $q^{(N)} \in \mathbb{R}[x_0,\ldots,x_N]$ and a pair $(l_1^{(N)}, l_2^{(N)})$ of linear forms $l_i^{(N)} \in \mathbb{R}[x_0,\ldots,x_N]$, $i=1,2$, with the following properties:

1. the common zero set $V^{(N)} = \{q^{(2)} = \ldots = q^{(N)} = 0\} \subset \mathbb{P}^N$ is a regular complete intersection;
2. the real part $V_{\mathbb{R}}^{(N)}$ has $2^{N-2}(N-3)+2$ connected components;
3. there is a distinguished component $\mathfrak{o}^{(N)} \subset V_{\mathbb{R}}^{(N)}$, which has two disjoint closed arcs $A_1^{(N)}, A_2^{(N)}$ such that the interior of $A_i^{(N)}$, $i=1,2$, contains all 2^{N-1} points of intersection of the hyperplane $L_i^{(N)} = \{l_i^{(N)} = 0\}$ with $V^{(N)}$.

Property (2) gives the desired lower bound.

The construction is by induction. Let

$$l_1^{(2)} = x_2, \quad l_2^{(2)} = x_2 - x_1, \quad \text{and} \quad q^{(2)} = l_1^{(2)} l_2^{(2)} + (x_1-x_0)(x_1-2x_0).$$

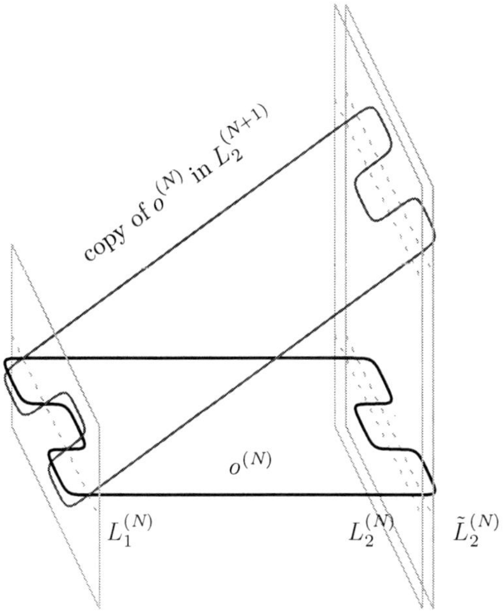

Fig. 5 Construction of a one-dimensional intersection of quadrics

Assume that for all integers $2 \leqslant k \leqslant N$, polynomials $q^{(k)}$, $l_1^{(k)}$, and $l_2^{(k)}$ satisfying conditions (1)–(3) above are constructed. Let $\tilde{l}_2^{(N)} = l_2^{(N)} - \delta^{(N)} l_1^{(N)}$, where $\delta^{(N)} > 0$ is a real number so small that for all $t \in [0, \delta^{(N)}]$, the line $\{l_2^{(N)} - t^{(N)} l_1^{(N)} = 0\}$ intersects $\mathfrak{o}^{(N)}$ at 2^{N-1} distinct real points all of which belong to the arc $A_2^{(N)}$. Put

$$l_1^{(N+1)} = x_{N+1} \quad \text{and} \quad l_2^{(N+1)} = x_{N+1} - l_1^{(N)}.$$

The intersection of the cone $\{q^{(2)} = \cdots = q^{(N)} = 0\} \subset \mathbb{P}_{\mathbb{R}}^{N+1}$ (over $V_{\mathbb{R}}^{(N)}$) and the hyperplane $L_2^{(N+1)} = \{l_2^{(N+1)} = 0\}$ is a copy of $V_{\mathbb{R}}^{(N)}$; see Fig. 5.

Put

$$q^{(N+1)} = l_1^{(N+1)} l_2^{(N+1)} + \varepsilon^{(N+1)} l_2^{(N)} \tilde{l}_2^{(N)},$$

where $\varepsilon^{(N+1)} > 0$ is a sufficiently small real number. One can observe that on the hyperplane $\{l_1^{(N)} = 0\} \subset \mathbb{P}_{\mathbb{R}}^{N+1}$, the polynomial $q^{(N+1)}$ has no zeros outside the subspace $\{x_{N+1} = l_2^{(N)} = 0\}$.

The new curve $V^{(N+1)}$ is a regular complete intersection, and its real part has $2^{N-1}(N-2)+2$ connected components. Indeed, each component $\mathfrak{o} \subset V_{\mathbb{R}}^{(N)}$ other than $\mathfrak{o}^{(N)}$ gives rise to two components of $V_{\mathbb{R}}^{(N+1)}$, whereas $\mathfrak{o}^{(N)}$ gives rise to 2^{N-1} components of $V_{\mathbb{R}}^{(N+1)}$, each component being the perturbation of the union $\mathfrak{a}_j \cup \mathfrak{a}'_j$, where $\mathfrak{a}_j \subset \mathfrak{o}^{(N)}$, $j = 0, \ldots, 2^{N-1}-1$, is the arc bounded by two consecutive (in $\mathfrak{o}^{(N)}$) points of the intersection $L_1^{(N)} \cap \mathfrak{o}^{(N)}$ (and not containing other intersection points), and $\mathfrak{a}'_j$ is the copy of $\mathfrak{a}_j$ in $L_2^{(N+1)}$. All but one of the arcs $\mathfrak{a}_j$ belong to $A_1^{(N)}$ and produce "small" components; the arc $\mathfrak{a}_0$ bounded by the two outermost (from the point of view of $A_1^{(N)}$) intersection points produces the "long" component, which we take for $\mathfrak{o}^{(N+1)}$. Finally, observe that the new component $\mathfrak{o}^{(N+1)}$ has two arcs $A_i^{(N+1)}$, $i = 1, 2$, satisfying condition (3) above: they are the perturbations of the arc $A_2^{(N)} \subset L_1^{(N+1)}$ and its copy in $L_2^{(N+1)}$. □

6 Concluding Remarks

In this section, we consider the first few special cases, $N = 2$, 3, 4, and 5, where, in fact, a complete deformation classification can be given. We also briefly discuss the other Betti numbers and the maximality of common zero sets of nets of quadrics; however, we merely outline directions for further investigation, leaving all details for a subsequent paper.

6.1 *Empty Intersections of Quadrics*

Consider a complete intersection V of $(r+1)$ real quadrics in $\mathbb{P}^N$, and assume that $V_{\mathbb{R}} = \varnothing$. Choosing generators $q_0, q_1, \dots, q_r$ of the linear system, we obtain a map

$$S^N = (\mathbb{R}^{N+1} \smallsetminus 0)/\mathbb{R}_+ \to S^r = (\mathbb{R}^{r+1} \smallsetminus 0)/\mathbb{R}_+, \quad u \mapsto (q_0(u), \dots, q_r(u))/\mathbb{R}_+.$$

Clearly, the homotopy class of this map, which can be regarded as an element of the group $\pi_N(S^r)$ modulo the antipodal involution, is a deformation invariant of the system. Furthermore, the map is even (the images of u and $-u$ coincide); hence, it also induces certain maps $\mathbb{P}^N_{\mathbb{R}} \to S^r$, $S^N \to \mathbb{P}^r_{\mathbb{R}}$, and $\mathbb{P}^N_{\mathbb{R}} \to \mathbb{P}^r_{\mathbb{R}}$, and their homotopy classes are also deformation invariant. Below, among other topics, we consider a few special cases in which these classes distinguish empty regular intersections.

In general, the deformation classifications of linear systems of quadrics, quadratic (rational) maps $S^N \to S^r$ (or $\mathbb{P}^N_{\mathbb{R}} \to \mathbb{P}^r_{\mathbb{R}}$), and spectral hypersurfaces (*e.g.*, spectral curves, even endowed with a theta characteristic) are different problems. We will illustrate this by examples.

6.2 *Three Conics*

We start with the case $r = N = 2$, *i.e.*, a net of conics in $\mathbb{P}^2_{\mathbb{R}}$. The spectral curve is a cubic $C \subset \mathbb{P}^2$, and the regularity condition implies that the common zero set must be empty (even over $\mathbb{C}$). There are two deformation classes of complete intersections of three conics; they can be distinguished by the $\mathbb{Z}_2$-Kronecker invariant, *i.e.*, the mod 2 degree of the associated map $\mathbb{P}^2_{\mathbb{R}} \to S^2$; see Sect. 6.1. If deg $= 0 \bmod 2$, the index function takes all three values 0, 1, and 2; otherwise, the index function takes only the middle value 1. (Alternatively, the Kronecker invariant counts the parity of the number of real solutions of the system $q_a = q_b = 0$, $q_c > 0$ in $\mathbb{P}^2_{\mathbb{R}} = \mathbb{P}^N_{\mathbb{R}}$, where a, b, c represent any triple of noncollinear points in $\mathbb{P}^2_{\mathbb{R}} = \mathbb{P}^r_{\mathbb{R}}$).

The classification of generic nets of conics can be obtained using the results of [8]. Note that in considering generic quadratic maps $\mathbb{P}^2_{\mathbb{R}} \to \mathbb{P}^2_{\mathbb{R}}$ rather than regular complete intersections, there are four deformation classes; they can be distinguished by the topology of $C_{\mathbb{R}}$ and the spectral theta characteristic.

6.3 *Spectral Curves of Degree* 4

Our next special case is an intersection of three quadrics in $\mathbb{P}^3_{\mathbb{R}}$. Here, a regular intersection $V_{\mathbb{R}}$ may consist of 0, 2, 4, 6, or 8 real points, and the spectral curve is a quartic $C \subset \mathbb{P}^2_{\mathbb{R}}$. Assuming C nonsingular and computing the Euler characteristic (*e.g.*, using Theorem 4.3 or the general formula for the Euler characteristic found in [3]), one can see that if $V \neq \varnothing$, then the real part $C_{\mathbb{R}}$ consists of $\frac{1}{2}\operatorname{Card} V_{\mathbb{R}}$ empty ovals. In this case, $\operatorname{Card} V_{\mathbb{R}}$ determines the net up to deformation. If $V_{\mathbb{R}} = \varnothing$, then either $C_{\mathbb{R}} = \varnothing$ or $C_{\mathbb{R}}$ is a nest of depth two. Such nets form two deformation classes, the homotopy class of the associated quadratic map (see Sect. 6.1) being either 0 or $1 \in \pi_3(S^2)/\pm 1$.

6.4 *Canonical Curves of Genus* 5 *in* $\mathbb{P}^4_{\mathbb{R}}$

Regular complete intersections of three quadrics in $\mathbb{P}^4$ are canonical curves of genus 5. Thus, the set of projective classes of such (real) intersections is embedded into the moduli space of (real) curves of genus 5. As is known, see, *e.g.*, [28], the image of this embedding is the complement of the strata formed by the hyperelliptic curves, trigonal curves, and curves with a vanishing theta constant. Since each of the three strata has positive codimension, the known classification of real forms of curves of a given genus (applied to $g = 5$) implies that the maximal number of connected components that a regular complete intersection of three real quadrics in $\mathbb{P}^4_{\mathbb{R}}$ may have is $6 = \operatorname{Hilb}(5)$.

6.5 *K3-surfaces of Degree* 8 *in* $\mathbb{P}^5_{\mathbb{R}}$

A regular complete intersection of three quadrics in the projective space of dimension 5 is a $K3$-surface with a (primitive) polarization of degree 8. Thus, as in the previous case, the set of projective classes of intersections is embedded into the moduli space of $K3$-surfaces with a polarization of degree 8, the complement consisting of a few strata of positive codimension (for details, see [27]). In particular, any generic $K3$-surface with a polarization of degree 8 is indeed a complete intersection of three quadrics. The deformation classification of $K3$-surfaces can be obtained using the results of Nikuin [23]. The case of maximal real $K3$-surfaces is particulary simple: there are three deformation classes, distinguished by the topology of the real part, which can be $S_{10} \sqcup S$, $S_6 \sqcup 5S$, or $S_2 \sqcup 9S$. In particular, the maximal number of connected components of a complete intersection of three real quadrics in $\mathbb{P}^5_{\mathbb{R}}$ is $10 = \mathrm{Hilb}(6)+1$. (There is another shape with ten connected components, the $K3$-surface with real part $S_1 \sqcup 9S$. However, one can easily show that a $K3$-surface of degree 8 cannot have ten spheres).

6.6 Other Betti Numbers

The techniques of this paper can be used to estimate the other Betti numbers as well. For $0 \leqslant i < \frac{1}{2}(N-3)$, we would obtain a bound of the form

$$B_2^i(N) - \mathrm{Hilb}_{i+1}(N+1) = O(1),$$

where $B_2^i(N)$ is the maximal i-th Betti number of a regular complete intersection of three real quadrics in $\mathbb{P}^N_{\mathbb{R}}$, and $\mathrm{Hilb}_{i+1}(N+1)$ is the maximal number of ovals of depth $\geqslant (D_{\max} - i - 1) = [\frac{1}{2}(N-1)] - i$ that a nonsingular real plane curve of degree $d = N+1$ may have. The possible discrepancy is due to a couple of unknown differentials in the spectral sequence and the inclusion homomorphism $H^{N-3-i}(\mathbb{P}^N_{\mathbb{R}}) \to H^{N-3-i}(V_{\mathbb{R}})$.

6.7 The Examples are Asymptotically Maximal

Recall that given a real algebraic variety X, the Smith inequality states that

$$\dim H_*(X_{\mathbb{R}}) \leqslant \dim H_*(X). \tag{6.1}$$

(As usual, all homology groups are with $\mathbb{Z}_2$ coefficients.) If equality holds, X is said to be *maximal*, or an *M-variety*. In particular, if X is a nonsingular plane curve of degree $N+1$, the Smith inequality (6.1) implies that the number of connected components of $X_{\mathbb{R}}$ does not exceed $g+1 = \frac{1}{2}N(N-1)+1$. The Hilbert curves used in Sect. 4 to construct nets with a large number of connected components are known to be maximal.

Using the spectral sequence of Theorem 4.3, one can easily see that under the choice of the index function made in the proof of Theorem 1.3, the dimension $\dim H_*(L_+)$ is $2(g+1)+2$ if N is odd and $2(g+1)+1$ if N is even.

On the other hand, for the common zero set V (as for any projective variety) there is a certain constant l such that the inclusion homomorphism $H_i(V_{\mathbb{R}}) \to H_i(\mathbb{P}^N_{\mathbb{R}})$ is nontrivial for all $i \leqslant l$ and trivial for all $i > l$ (see [19]; in our case, $1 \leqslant l < \frac{1}{2}N$). Hence, by Poincaré–Lefschetz duality and Lemma 4.1, one has

$$\dim H_*(L_+) = \dim H_*(\mathbb{P}^N_{\mathbb{R}}, V_{\mathbb{R}}) = \dim H_*(V_{\mathbb{R}}) + N - 2l - 1.$$

Finally, one can easily find $\dim H_*(V)$: it equals $4(k^2 - 1)$ if $N = 2k$ is even and $4(k^2 - k)$ if $N = 2k - 1$ is odd.

Combining the above computations, one observes that the intersections of quadrics constructed from the Hilbert curves using monotonous index functions are asymptotically maximal in the sense that

$$\dim H_*(V_{\mathbb{R}}) = \dim H_*(V) + O(N) = N^2 + O(N).$$

The latter identity shows that the upper bound for $B^0_2(N)$ provided by the Smith inequality is too rough: this bound is of the form $\frac{1}{2}N^2 + O(N)$, whereas, as is shown in this paper, $B^0_2(N)$ does not exceed $\frac{3}{8}N^2 + O(N)$. When the intersection is of even dimension, one can improve the leading coefficient in the bound by combining the Smith inequality and the generalized Comessatti inequality; however, the resulting estimate is still too far from the sharp bound.

6.8 *The Examples are Not Maximal*

Another interesting consequence of the computation of the previous section is the fact that starting from $N = 6$, the complete intersections of quadrics maximizing the number of components are never truly maximal in the sense of the Smith inequality (6.1): one has

$$N + O(1) \leqslant \dim H_*(V) - \dim H_*(V_{\mathbb{R}}) \leqslant 2N + O(1).$$

Using the spectral sequence of Theorem 4.3, one can easily show that a maximal complete intersection V of three real quadrics in $\mathbb{P}^N_{\mathbb{R}}$ must have index function taking values between $\frac{1}{2}(N-1)$ and $\frac{1}{2}(N+3)$ (cf. (4.5)–(4.7)); the real part $V_{\mathbb{R}}$ has large Betti numbers in two or three middle dimensions, (most) other Betti numbers being equal to 1.

Apparently, it is the Harnack M-curves that are suitable for obtaining nets with maximal common zero locus. However, at present we do not know much about the differentials in the spectral sequence or the constant l introduced in the previous section. It may happen that these data are controlled by an extra flexibility in the choice of the real Spin-structure on the spectral curve: in addition to the semiorientation, one can also choose the values on the components of $C_{\mathbb{R}}$.

Acknowledgments This paper was originally inspired by the following question suggested to us by D. Pasechnik and B. Shapiro: is the number of connected components of an intersection of $(r+1)$ real quadrics in $\mathbb{P}^N_{\mathbb{R}}$ bounded by a constant $C(r)$ independent of N?

The paper was essentially completed during our stay at *Centre Interfacultaire Bernoulli, École polytechnique fédérale de Lausanne,* and the final version was prepared during the stay of the second and third authors at the *Max-Planck-Institut für Mathematik,* Bonn. We are grateful to these institutions for their hospitality and excellent working conditions.

The second and third authors acknowledge the support from grant ANR-05-0053-01 of Agence Nationale de la Recherche (France) and a grant of Université Louis Pasteur, Strasbourg.

References

1. A. A. Agrachev: Homology of intersections of real quadrics. Soviet Math. Dokl. **37**, 493–496 (1988)
2. A. A. Agrachev: Topology of quadratic maps and Hessians of smooth maps. In: Itogi Nauki i Tekhniki, **26**, 85–124 (1988) (Russian). English transl. in J. Soviet Math. **49**, 990–1013 (1990)
3. A. A. Agrachev, R. Gamkrelidze: Computation of the Euler characteristic of intersections of real quadrics. Dokl. Acad. Nauk SSSR **299**, 11–14 (1988)
4. V. I. Arnol′d: On the arrangement of ovals of real plane algebraic curves, involutions on four-dimensional manifolds, and the arithmetic of integer-valued quadratic forms. Funct. Anal. Appl. **5**, 169–176 (1971)
5. M. F. Atiyah: Riemann surfaces and Spin-structures. Ann. Sci. École Norm. Sup. (4) **4**, 47–62 (1971)
6. L. Brusotti: Sulla "piccola variazione" di una curva piana algebrica reale. Rom. Acc. L. Rend. (5) $\mathbf{30}_1$, 375–379 (1921)
7. G. Castelnuovo: Ricerche di geometria sulle curve algebriche. Atti. R. Acad. Sci. Torino **24**, 196–223 (1889)
8. A. Degtyarev: Quadratic transformations $\mathbb{R}p^2 \to \mathbb{R}p^2$. In: Topology of real algebraic varieties and related topics, Amer. Math. Soc. Transl. Ser. 2, **173**, 61–71. Amer. Math. Soc., Providence, RI (1996)
9. A. Degtyarev, V. Kharlamov: Topological properties of real algebraic varieties: Rokhlin's way. Russian Math. Surveys **55**, 735–814 (2000)
10. A. C. Dixon: Note on the Reduction of a Ternary Quartic to Symmetrical Determinant. Proc. Cambridge Philos. Soc. **2**, 350–351 (1900–1902)
11. I. V. Dolgachev: Topics in Classical Algebraic Geometry. Part 1. Dolgachev's homepage (2006)
12. B. A. Dubrovin, S. M. Natanzon: Real two-zone solutions of the sine–Gordon equation. Funct. Anal. Appl. **16**, 21–33 (1982)
13. B. H. Gross, J. Harris: Real algebraic curves. Ann. Sci. École Norm. Sup. (4) **14**, 157–182 (1981)
14. G. Halphen: Mémoire sur la classification des courbes gauches algébriques. J. École Polytechnique **52**, 1–200 (1882)
15. A. Harnack: Über die Vielfaltigkeit der ebenen algebraischen Kurven. Math. Ann. **10**, 189–199 (1876)
16. D. Hilbert: Ueber die reellen Züge algebraischer Curven. Math. Ann. **38**, 115–138 (1891)
17. I. Itenberg, O. Viro: Asymptotically maximal real algebraic hypersurfaces of projective space. Proceedings of Gökova Geometry/Topology conference 2006, International Press, 91–105 (2007)
18. V. Kharlamov: Additional congruences for the Euler characteristic. of even-dimensional real algebraic varieties Funct. Anal. Appl. **9**, 134–141 (1975)

19. Ueber eine neue Art von Riemann'schen Flächen. F. Klein: Math. Ann **10**, 398 – 416 (1876)
20. D. Mumford: Theta-characteristics of an algebraic curve. Ann. Sci. École Norm. Sup. (4) **4**, 181–191 (1971)
21. S. M. Natanzon: Prymians of real curves and their applications to the effectivization of Schrödinger operators. Funct. Anal. Appl. **23**, 33–45 (1989)
22. S. M. Natanzon: Moduli of Riemann surfaces, real algebraic curves, and their superanalogs. Translations of Mathematical Monographs, **225**, American Mathematical Society, Providence, RI (2004)
23. V. V. Nikulin: Integral symmetric bilinear forms and some of their. geometric applications Izv. Akad. Nauk SSSR Ser. Mat. **43**, 111–177 (1979) (Russian). English transl. in Math. USSR–Izv. **14**, 103–167 (1980)
24. S. Yu. Orevkov: Some examples of real algebraic and real pseudoholomorphic curves. This volume
25. D. Pecker: Un théorème de Harnack dans l'espace. Bull. Sci. Math. **118**, 475–484 (1994)
26. V. A. Rokhlin: Proof of Gudkov's conjecture.. Funct. Anal. Appl. **6**, 136–138 (1972)
27. B. Saint-Donat: Projective models of $K3$ surfaces. Amer. J. Math. **96**, 602–639 (1974)
28. A. N. Tyurin: On intersection of quadrics. Russian Math. Surveys **30**, 51–105 (1975)
29. V. Vinnikov: Self-adjoint determinantal representations of real plane curves. Math. Ann. **296**, 453–479 (1983)

Rational SFT, Linearized Legendrian Contact Homology, and Lagrangian Floer Cohomology

Tobias Ekholm

This paper is dedicated to Oleg Viro on the occasion of his 60th birthday

Abstract We relate the version of rational symplectic field theory for exact Lagrangian cobordisms introduced in [6] to linearized Legendrian contact homology. More precisely, if $L \subset X$ is an exact Lagrangian submanifold of an exact symplectic manifold with convex end $\Lambda \subset Y$, where Y is a contact manifold and Λ is a Legendrian submanifold, and if L has empty concave end, then the linearized Legendrian contact cohomology of Λ, linearized with respect to the augmentation induced by L, equals the rational SFT of (X, L). Following ideas of Seidel [15], this equality in combination with a version of Lagrangian Floer cohomology of L leads us to a conjectural exact sequence that in particular implies that if $X = \mathbb{C}^n$, then the linearized Legendrian contact cohomology of $\Lambda \subset S^{2n-1}$ is isomorphic to the singular homology of L. We outline a proof of the conjecture and show how to interpret the duality exact sequence for linearized contact homology of [7] in terms of the resulting isomorphism.

Keywords Floer cohomology • Symplectic field theory • Legendrian contact homology • Cobordism • Holomorphic disk

1 Introduction

Let Y be a contact $(2n-1)$-manifold with contact 1-form λ (i.e., $\lambda \wedge (d\lambda)^{n-1}$ is a volume form on Y). The *Reeb vector field* R_λ of λ is the unique vector field that satisfies $\lambda(R_\lambda) = 1$ and $d\lambda(R_\lambda, \cdot) = 0$. The *symplectization* of Y is the symplectic

T. Ekholm (✉)
Department of mathematics, Uppsala University, Box 480, 751 06 Uppsala, Sweden
e-mail: tobias@math.uu.se

I. Itenberg et al. (eds.), *Perspectives in Analysis, Geometry, and Topology: On the Occasion of the 60th Birthday of Oleg Viro*, Progress in Mathematics 296,
DOI 10.1007/978-0-8176-8277-4_6,

manifold $Y \times \mathbb{R}$ with symplectic form $d(e^t\lambda)$, where t is a coordinate in the $\mathbb{R}$-factor. A *symplectic manifold with cylindrical ends* is a symplectic $2n$-manifold X that contains a compact subset K such that $X - K$ is symplectomorphic to a disjoint union of two half-symplectizations $Y^+ \times \mathbb{R}_+ \cup Y^- \times \mathbb{R}_-$, for some contact $(2n-1)$-manifolds $Y^\pm$, where $\mathbb{R}_+ = [0,\infty)$ and $\mathbb{R}_- = (-\infty,0]$. We call $Y^+ \times \mathbb{R}_+$ and $Y^- \times \mathbb{R}_-$ the *positive and negative ends* of X, respectively, and Y^+ and Y^-, $(+\infty)$- *and* $(-\infty)$-*boundaries* of X, respectively.

The relative counterpart of a symplectic manifold with cylindrical ends is a pair (X,L) of a symplectic $2n$-manifold X with a Lagrangian n-submanifold $L \subset X$ (i.e., the restriction of the symplectic form in X to any tangent space of L vanishes) such that outside a compact subset, (X,L) is symplectomorphic to the disjoint union of $\left(Y^+ \times \mathbb{R}_+, \Lambda^+ \times \mathbb{R}_+\right)$ and $\left(Y^- \times \mathbb{R}_-, \Lambda^- \times \mathbb{R}_-\right)$, where $\Lambda^\pm \subset Y^\pm$ are Legendrian $(n-1)$-submanifolds (i.e., $\Lambda^\pm$ are everywhere tangent to the kernels of the contact forms on $Y^\pm$). If the symplectic manifold X is exact (i.e., if the symplectic form ω on X satisfies $\omega = \mathrm{d}\beta$ for some 1-form β) and if the Lagrangian submanifold L is exact as well (i.e., if the restriction $\beta|L$ satisfies $\beta|L = \mathrm{d}f$ for some function f), then we call the pair (X,L) of exact manifolds an *exact cobordism*. We assume throughout the paper that X is simply connected, that the first Chern class of TX, viewed as a complex bundle using any almost complex structure compatible with the symplectic form on X, is trivial, and that the Maslov class of L is trivial as well. (These assumptions are made in order to have well-defined gradings in contact homology algebras over $\mathbb{Z}_2$. In more general cases, one would work with contact homology algebras with suitable Novikov coefficients in order to have appropriate gradings.)

In [6], a version of rational symplectic field theory (SFT) (see [12] for a general description of SFT) for exact cobordisms with good ends was developed; see Sect. 2. (The additional condition that ends be good allows us to disregard Reeb orbits in the ends when setting up the theory. Standard contact spheres as well as 1-jet spaces with their standard contact structures are good.) It associates to an exact cobordism (X,L), where L has k components, a $\mathbb{Z}$-graded filtered $\mathbb{Z}_2$-vector space $\mathbf{V}(X,L)$, with k filtration levels and with a filtration-preserving differential $d^f : \mathbf{V}(X,L) \to \mathbf{V}(X,L)$. Elements in $\mathbf{V}(X,L)$ are formal sums of admissible formal disks in which the number of summands with $(+)$-action below any given number is finite; see Sect. 2 for definitions of these notions.

The differential increases $(+)$-action, and hence if $\mathbf{V}_{[\alpha]}(X,L)$ denotes $\mathbf{V}(X,L)$ divided out by the subcomplex of all formal sums in which all disks have $(+)$-action larger than α, then the differential induces filtration-preserving differentials $d^f_\alpha : \mathbf{V}_{[\alpha]}(X,L) \to \mathbf{V}_{[\alpha]}(X,L)$ with associated spectral sequences $\left\{E^{p,q}_{r;[\alpha]}(X,L)\right\}_{r=1}^k$. The projection maps $\pi^\alpha_\beta : \mathbf{V}_{[\alpha]}(X,L) \to \mathbf{V}_{[\beta]}(X,L)$, $\alpha > \beta$, give an inverse system of chain maps. The limit $E^*_r(X,L) = \varprojlim_\alpha E^*_{r;[\alpha]}(X,L)$ is invariant under deformations of (X,L) through exact cobordisms with good ends and in particular under deformations of L through exact Lagrangian submanifolds with cylindrical ends; see Theorem 2.1.

In this paper we will use only the simplest version of the theory just described, which is as follows. Let (X,L) be an exact cobordism such that L is connected and without negative end, i.e., $\Lambda^- = \emptyset$. In this case, admissible formal disks have only one positive puncture, and we identify (a quotient of) $\mathbf{V}(X,L)$ with the $\mathbb{Z}_2$-vector space of formal sums of Reeb chords of $\Lambda = \Lambda^+$. Furthermore, our assumptions on $\pi_1(X)$ and vanishing of $c_1(TX)$ and of the Maslov class of L imply that the grading of a formal disk depends only on the Reeb chord at its positive puncture. We let $|c|$ denote the grading of a chord $c \in \mathbf{V}(X,L)$. Rational SFT then provides a differential

$$d^f : \mathbf{V}(X,L) \to \mathbf{V}(X,L)$$

with $|d^f(c)| = |c| + 1$ that increases action in the sense that if $\mathfrak{a}(c)$ denotes the action of the Reeb chord c and if the Reeb chord b appears with nonzero coefficient in ∂c, then $\mathfrak{a}(b) > \mathfrak{a}(c)$. Furthermore, since L is connected, the spectral sequences have only one level, and

$$E_1^*(X,L) = \varprojlim_\alpha \left(\ker d_\alpha^f / \operatorname{im} d_\alpha^f\right).$$

Our first result relates $E_1^*(X,L)$ to linearized Legendrian contact cohomology; see Sect. 3. Legendrian contact homology was introduced in [5, 12]. It was worked out in detail in special cases including 1-jet spaces in [9–11]. From the point of view of Legendrian contact homology, an exact cobordism (X,L) with good ends induces a chain map from the contact homology algebra of (Y^+,Λ^+) to that of (Y^-,Λ^-). In particular, if $\Lambda^- = \emptyset$, then the latter equals the ground field $\mathbb{Z}_2$ with the trivial differential. Such a chain map ϵ is called an *augmentation*, and it gives rise to a linearization of the contact homology algebra of Λ^+. That is, it endows the chain complex $Q(\Lambda)$ generated by Reeb chords of Λ with a differential ∂^ϵ.

The resulting homology is called *ϵ-linearized contact homology* and denoted by $\mathrm{LCH}_*(Y,\Lambda;\epsilon)$. We let $\mathrm{LCH}^*(Y;\Lambda;\epsilon)$ be the homology of the dual complex $Q'(\Lambda) = \mathrm{Hom}(Q(\Lambda);\mathbb{Z}_2)$ and call it the *ϵ-linearized contact cohomology* of Λ.

We say that (X,L) satisfies a *monotonicity condition* if there are constants C_0 and $C_1 > 0$ such that for any Reeb chord c of $\Lambda \subset Y$, $|c| > C_1\mathfrak{a}(c) + C_0$. Note that if Y is a 1-jet space or the sphere endowed with a generic small perturbation of the standard contact form and if Λ is in general position with respect to the Reeb flow, then (X,L) satisfies a monotonicity condition.

Theorem 1.1. *Let (X,L) be an exact cobordism with good ends. Let (Y,Λ) denote the positive end of (X,L) and assume that the $(-\infty)$-boundary of L is empty. Let ϵ denote the augmentation on the contact homology algebra of Λ induced by L. Then the natural map $Q'(\Lambda) \to \mathbf{V}(X,L)$, which takes an element in $Q'(\Lambda)$ thought of as a formal sum of covectors dual to Reeb chords in $Q(\Lambda)$ to the corresponding formal sum of Reeb chords in $\mathbf{V}(X,L)$, is a chain map. Furthermore, if (X,L) satisfies a monotonicity condition, then the corresponding map on homology*

$$\mathrm{LCH}^*(Y,\Lambda;\epsilon) \to E_1^*(X,L),$$

is an isomorphism.

Theorem 1.1 is proved in Sect. 3.2. We point out that when (X,L) satisfies a monotonicity condition, it follows from this result that $\mathrm{LCH}_*(Y,\Lambda;\epsilon)$ depends only on the symplectic topology of (X,L).

We next consider two exact cobordisms (X,L_0) and (X,L_1) with good ends and with the following properties: both L_0 and L_1 have empty $(-\infty)$-boundaries; if Λ_j denotes the $(+\infty)$-boundary of L_j, then $\Lambda_0 \cap \Lambda_1 = \emptyset$; L_0 and L_1 intersect transversely; and the Reeb flow of Λ_0 along a Reeb chord connecting Λ_0 to Λ_1 is transverse to Λ_1 at its endpoint. For such pairs of exact cobordisms we define Lagrangian Floer cohomology $HF^*(X;L_0,L_1)$ as an inverse limit of the cohomologies $HF^*_{[\alpha]}(X;L_0,L_1)$ of cochain complexes $C_{[\alpha]}(X;L_0,L_1)$ generated by Reeb chords between Λ_0 and Λ_1 of action at most α and by points in $L_0 \cap L_1$. This Floer cohomology has a relative $\mathbb{Z}$-grading and is invariant under exact deformations of L_1.

Consider an exact cobordism (X,L) where L has empty $(-\infty)$-boundary and $(+\infty)$-boundary Λ. Let L' be a slight push-off of L, which is an extension of a small push-off Λ' of Λ along the Reeb vector field.

Conjecture 1.2. For any $\alpha > 0$, there is a long exact sequence

$$\begin{array}{lllll} \cdots \xrightarrow{\delta_{\alpha;L,L'}} & E^*_{1;[\alpha]}(X,L) & \longrightarrow & HF^*_{[\alpha]}(X;L,L') & \longrightarrow & H_{n-*}(L) \\ \xrightarrow{\delta_{\alpha;L,L'}} & E^{*+1}_{1;[\alpha]}(X,L) & \longrightarrow & HF^{*+1}_{[\alpha]}(X;L,L') & \longrightarrow & H_{n-*-1}(L) \\ \xrightarrow{\delta_{\alpha;L,L'}} & \cdots, & & & & \end{array} \tag{1}$$

where $H_*(L)$ is the ordinary homology of L with $\mathbb{Z}_2$-coefficients. It follows in particular that if $X = \mathbb{C}^n$ or $X = J^1(\mathbb{R}^{n-1}) \times \mathbb{R}$, then $HF^*(X;L,L') = 0$, and the map $\delta_{L,L'}\colon H_{n-*+1}(L) \to E_1^*(X,L) \approx \mathrm{LCH}^*(Y,\Lambda;\epsilon)$ induced by the maps $\delta_{\alpha;L,L'}$ is an isomorphism.

The author learned about the isomorphism above, between linearized contact homology of a Legendrian submanifold with a Lagrangian filling and the ordinary homology of the filling, from Seidel [15], who explained it using an exact sequence in wrapped Floer homology [1, 14] similar to (1). Borrowing Seidel's argument, we outline in Sect. 4.4 a proof of Conjecture 1.2 in which the Lagrangian Floer cohomology $HF(X;L,L')$ plays the role of wrapped Floer homology.

In [7], a duality exact sequence for linearized contact homology of a Legendrian submanifold $\Lambda \subset Y$, where $Y = P \times \mathbb{R}$ for some exact symplectic manifold P and where the projection of Λ into P is displaceable, was found. In what follows, we restrict attention to the case $Y = J^1(\mathbb{R}^{n-1})$. Then every compact Legendrian submanifold has displaceable projection, and the duality exact sequence is the following, where ϵ denotes any augmentation and where we suppress the ambient manifold $Y = J^1(\mathbb{R}^{n-1})$ from the notation:

$$\cdots \xrightarrow{\rho} H_{k+1}(\Lambda) \xrightarrow{\sigma} \mathrm{LCH}^{(n-1)-k-1}(\Lambda;\epsilon) \xrightarrow{\theta} \mathrm{LCH}_k(\Lambda;\epsilon)$$
$$\xrightarrow{\rho} H_k(\Lambda) \xrightarrow{\sigma} \mathrm{LCH}^{(n-1)-k}(\Lambda;\epsilon) \xrightarrow{\theta} \mathrm{LCH}_{k-1}(\Lambda;\epsilon)\cdots. \tag{2}$$

Here, if $\beta = \rho(\alpha) \in H_k(\Lambda)$, then the Poincaré dual $\gamma \in H_{n-k}(\Lambda)$ of β satisfies $\langle\sigma(\gamma),\alpha\rangle = 1$, where $\langle,\rangle$ is the pairing between the homology and cohomology of $Q(\Lambda)$. Furthermore, the maps ρ and σ are defined through a count of rigid configurations of holomorphic disks with boundary on Λ with a flow line emanating from its boundary, and the map θ is defined through a count of rigid holomorphic disks with boundary on Λ with two positive punctures.

In $J^1(\mathbb{R}^{n-1})$, a generic Legendrian submanifold has finitely many Reeb chords. Furthermore, if L is an exact Lagrangian cobordism in the symplectization $J^1(\mathbb{R}^{n-1}) \times \mathbb{R}$ with empty $(-\infty)$-boundary, then L is displaceable. Hence, both Theorem 1.1 and Conjecture 1.2 give isomorphisms. Combining (1) and (2) leads to the following.

Corollary 1.3. *Let L be an exact Lagrangian cobordism in $J^1(\mathbb{R}^{n-1}) \times \mathbb{R}$ with empty $(-\infty)$-boundary and with $(+\infty)$-boundary Λ and let ϵ denote the augmentation on the contact homology algebra of Λ induced by L. Then the following diagram with exact rows commutes, and all vertical maps are isomorphisms:*

$$\begin{array}{ccccccc}
H_{k+1}(\Lambda) & \longrightarrow & H_{k+1}(L) & \longrightarrow & H_{k+1}(L,\Lambda) & \longrightarrow & H_k(\Lambda) \\
\cdots \quad \downarrow \mathrm{id} & & \downarrow \delta_{L,L'} & & \downarrow H^{-1}\circ\delta'_{L,L'} & & \downarrow \mathrm{id} \quad \cdots \\
H_{k+1}(\Lambda) & \xrightarrow{\sigma} & \mathrm{LCH}^{n-k-2}(\Lambda;\epsilon) & \longrightarrow & \mathrm{LCH}_k(\Lambda;\epsilon) & \xrightarrow{\rho} & H_k(\Lambda)
\end{array}$$

Here the top row is the long exact homology sequence of (L,Λ), the bottom row is the duality exact sequence, the map $\delta_{L,L'}$ is the map in Conjecture 1.2, the map $\delta'_{L,L'}$ is analogous to $\delta_{L,L'}$, and the map H counts disks in the symplectization with boundary on Λ and with two positive punctures; see Sect. 4.5 for details.

The proof of Corollary 1.3 is discussed in Sect. 4.5.

2 A Brief Sketch of Relative SFT of Lagrangian Cobordisms

Although we will use only the simplest version of relative SFT introduced in [6] in this paper, we give a brief introduction to the full theory for two reasons. First, it is reasonable to expect that this theory is related to product structures on linearized contact homology, see [4], in much the same way as the simplest version of the theory appears in Conjecture 1.2. Second, some of the moduli spaces of holomorphic disks that we will make use of are analogous to those needed for more involved versions of the theory.

2.1 Formal and Admissible Disks

In order to describe relative rational SFT, we introduce the following notation. Let (X,L) be an exact cobordism with ends $(Y^{\pm}\times\mathbb{R}_{\pm},\Lambda^{\pm}\times\mathbb{R}_{\pm})$. Write $(\bar{X},\bar{L})$ for a compact part of (X,L) obtained by cutting the infinite parts of the cylindrical ends off at some $|t|=T>0$. We will sometimes think of Reeb chords of $\Lambda^{\pm}$ in the $(\pm\infty)$-boundary as lying in $\partial\bar{X}$ with endpoints on $\partial\bar{L}$. A *formal disk* of (X,L) is a homotopy class of maps of the 2-disk D, with m marked disjoint closed subintervals in ∂D, into $\bar{X}$, where the m marked intervals are required to map in an orientation-preserving (reversing) manner to Reeb chords of $\partial\bar{L}$ in the $(+\infty)$-boundary (in the $(-\infty)$-boundary) and where remaining parts of the boundary ∂D map to $\bar{L}$.

If L^b and L^a are exact Lagrangian cobordisms in X^b and X^a, respectively, such that a component (Y,Λ) of the $(-\infty)$-boundary of (X^a,L^a) agrees with a component of the $(+\infty)$-boundary of (X^b,L^b), then these cobordisms can be joined to an exact cobordism L^{ba} in X^{ba}, where (X^{ba},L^{ba}) is obtained by gluing the positive end (Y,Λ) of (X^b,L^b) to the corresponding negative end of (X^a,L^a). Furthermore, if v^b and v^a are collections of formal disks of (X^b,L^b) and (X^a,L^a), respectively, then we can construct formal disks in L^{ba} in the following way: start with a disk v_1^a from v^a, and let $c_1,\dots,c_{r_1}$ denote the Reeb chords at its negative punctures. Attach positive punctures of disks $v_{1;1}^b,\dots,v_{1;r_1}^b$ in v^b mapping the Reeb chords $c_1,\dots,c_{r_1}$ to the corresponding negative punctures of the disk v_1^a. This gives a disk v_1^{ba} with some positive punctures mapping to chords $c_1,\dots,c_{r_2}$ of Λ. Attach negative punctures of the disk $v_{2;1}^a,\dots,v_{2;r_2}^a$ to v_1^{ba} at $c_1,\dots,c_{r_2}$. This gives a disk v_2^{ba} with some negative punctures mapping to Reeb chords in Λ. Continue this process until there are no punctures mapping to Λ. We call the resulting disk a formal disk in L^{ba} *with factors from v^a and v^b*.

Assume that the set of connected components of L has been subdivided into subsets L_j so that L is a disjoint union $L=L_1\cup\dots\cup L_k$, where each L_j is a collection of connected components of L. We call $L_1,\dots,L_k$ the *pieces* of L. With respect to such a subdivision, Reeb chords fall into two classes: *pure*, with both endpoints on the same piece, and *mixed*, with endpoints on distinct pieces.

A formal disk represented by a map $u\colon D\to\bar{X}$ is *admissible* if for any arc α in D that connects two unmarked segments in ∂D that are mapped to the same piece by u, all marked segments on the boundary of one of the components of $D-\alpha$ map to pure Reeb chords in the $(-\infty)$-boundary.

2.2 Holomorphic Disks

Let (X,L) be an exact cobordism. Fix an almost complex structure J on X that is adjusted to its symplectic form. Let S be a punctured Riemann surface with complex structure j and with boundary ∂S. A J-holomorphic curve with boundary on L is a map $u\colon S\to X$ such that

$$du+J\circ du\circ j=0$$

and such that $u(\partial S) \subset L$. For details on holomorphic curves in this setting we refer to [6, Appendix B] and references therein. Here we summarize the main properties we will use.

By definition, an adjusted almost complex structure J is invariant under $\mathbb{R}$-translations in the ends of X and pairs the Reeb vector field in the $(\pm\infty)$-boundary with the symplectization direction. Consequently, strips that are cylinders over Reeb chords as well as cylinders over Reeb orbits are J-holomorphic. Furthermore, any J holomorphic disk of finite energy is asymptotic to such Reeb chord strips at its boundary punctures and to Reeb orbit cylinders at interior punctures; see [6, Sect. B.1]. We say that a puncture of a J-holomorphic disk is positive (negative) if the disk is asymptotic to a Reeb chord strip (Reeb orbit cylinder) in the positive (negative) end of (X,L). Note that exactness of (X,L) and the fact that the symplectic form is positive on J-complex tangent lines imply that any J-holomorphic curve has at least one positive puncture.

These results on asymptotics imply that any J-holomorphic disk in X with boundary on L determines a formal disk. Let $\mathcal{M}(v)$ denote the moduli space of J-holomorphic disks with associated formal disk equal to v. The formal dimension of $\mathcal{M}(v)$ is determined by the Fredholm index of the linearized $\bar{\partial}_J$-operator along a representative of v; see [6, Sect. 3.1].

A sequence of J-holomorphic disks with boundary on L may converge to a broken disk of two components that intersects at a boundary point. We will refer to this phenomenon as boundary bubbling. However, if all elements in the sequence have only one positive puncture, then boundary bubbling is impossible by exactness: each component in the limit curve must have at least one positive puncture. The reason for using admissible disks to set up relative SFT is the following: In a sequence of holomorphic disks with corresponding formal disks admissible, boundary bubbling is impossible for topological reasons. As a consequence, if v is a formal disk, then the boundary of $\mathcal{M}(v)$ consists of several level J-holomorphic disks and spheres joined at Reeb chords or at Reeb orbits; see [2].

Recall from Sect. 1 that we require the ends of our exact cobordisms (X,L) to be good. The precise formulation of this condition is as follows. If γ^+ (γ^-) is a Reeb orbit in the $(+\infty)$-boundary Y^+ (in the $(-\infty)$-boundary Y^-) of X, then the formal dimension of any moduli space of holomorphic spheres in X (in $Y^- \times \mathbb{R}$) with positive puncture at γ^+ (at γ^-) is ≥ 2. Together with transversality arguments these conditions guarantee that broken curves in the boundary of $\mathcal{M}(v)$, where v is an admissible formal disk, cannot contain any spheres if $\dim(\mathcal{M}(v)) \leq 1$, or if $\dim(\mathcal{M}(v)) = 2$ when (X,L) is a trivial cobordism; see [6, Lemma B.6]. In particular, in the boundary of $\mathcal{M}(v)$, where $\dim(\mathcal{M}(v))$ satisfies these dimensional constraints and where v is admissible, there can be only two level curves, all pieces of which are admissible disks; see [6, Lemma 2.5].

Under our additional assumptions ($\pi_1(X)$ trivial, first Chern class of X and Maslov class of L vanish), the grading of a formal disk depends only on the Reeb chords at its punctures. For later reference, we describe this more precisely in the case that L is connected and its $(-\infty)$-boundary is empty. Let (Y,Λ) denote the $(+\infty)$-boundary of (X,L). If c is a Reeb chord of $\Lambda \subset Y$, then let γ be any path in L

joining its endpoints. Since X is simply connected, $\gamma \cup c$ bounds a disk $\Gamma\colon D \to X$. Fix a trivialization of TX along Γ such that the linearized Reeb flow along c is represented by the identity transformation with respect to this trivialization. Then the tangent space $T_s(\Lambda \times \mathbb{R})$ at the initial point s of c is transported to a subspace $V_s \times \mathbb{R}$ in the tangent space T_eX at the final point e of c where V_s is transverse to $T_e\Lambda$ in the contact hyperplane ξ_e at e. Let R denote a negative rotation along the complex angle taking V_s to $T_e\Lambda$ in ξ_e; see [6, Sect. 3.1]. Then the Lagrangian tangent planes of L along γ capped off with R form a loop Δ_Γ of Lagrangian subspaces in $\mathbb{C}^n$ with respect to the trivialization, and if $\mathcal{M}(c)$ denotes the moduli space of holomorphic disks in X with one positive boundary puncture at which they are asymptotic to the Reeb chord strip of the Reeb chord c, then

$$\dim(\mathcal{M}(c)) = n - 3 + \mu(\Delta_\Gamma) + 1,$$

where μ denotes the Maslov index; see [6, p. 655]. To see that this is independent of Γ, note that the difference of two trivializations along the disks is measured by $c_1(TX)$. To see that it is independent of the path γ, note that the difference in the dimension formula corresponding to two different paths γ and γ' is measured by the Maslov class of L evaluated on the loop $\gamma \cup -\gamma'$. Define

$$|c| = \dim(\mathcal{M}(c)). \tag{3}$$

If a and $b_1,\dots,b_m$ are Reeb chords of Λ and if $\mathcal{M}(a;b_1,\dots,b_m)$ denotes the moduli space of holomorphic disks in $Y \times \mathbb{R}$ with boundary on $\Lambda \times \mathbb{R}$ with positive puncture at the Reeb chord a and negative punctures at the Reeb chords $b_1,\dots,b_k$ in the order given by following the boundary orientation of the disk starting at the positive puncture, then additivity of the index gives

$$\dim(\mathcal{M}(a;b_1,\dots,b_k)) = |a| - \sum_j |b_j|. \tag{4}$$

2.3 Hamiltonian and Potential Vectors and Differentials

Let (X,L) be an exact cobordism and let v be a formal disk of (X,L). Define the $(+)$-action of v as the sum of the actions

$$\mathfrak{a}(c) = \int_c \lambda^+$$

over the Reeb chords c at their positive punctures. Here λ^+ is the contact form in the $(+\infty)$-boundary Y^+ of X. Note that for generic Legendrian $(+\infty)$-boundary, $\Lambda^+ \subset Y^+$, the set of actions of Reeb chords, is a discrete subset of $\mathbb{R}$. Let $\mathbf{V}(X,L)$ denote the $\mathbb{Z}$-graded vector space over $\mathbb{Z}_2$ with elements that are formal sums of admissible formal disks that contain only a finite number of summands below any given $(+)$-action. The grading on $\mathbf{V}(X,L)$ is the following: the degree of a formal disk v is the formal dimension of the moduli space $\mathcal{M}(v)$ of J-holomorphic disks homotopic to the formal disk. We use the natural filtration

$$0 \subset F^k\mathbf{V}(X,L) \subset \cdots \subset F^2\mathbf{V}(X,L) \subset F^1\mathbf{V}(X,L) = \mathbf{V}(X,L)$$

of $\mathbf{V}(X,L)$, where k is the number of pieces of L and where the filtration level is determined by the number of positive punctures. (It is straightforward to check that an admissible formal disk has at most k Reeb chords at the positive end.)

We will define a differential $d^f\colon \mathbf{V}(X,L) \to \mathbf{V}(X,L)$ that respects this filtration using 1-dimensional moduli spaces of holomorphic disks. To this end, fix an almost complex structure J on X that is compatible with the symplectic form and adjusted to $d(e^t\lambda^\pm)$ in the ends, where $\lambda^\pm$ is the contact form in the $(\pm\infty)$-boundary. Assume that J is generic with respect to 0- and 1-dimensional moduli spaces of holomorphic disks; see [6, Lemma B.8]. Since J is invariant under translations in the ends, $\mathbb{R}$ acts on moduli spaces $\mathcal{M}(u)$, where u is a formal disk of $(Y^\pm \times \mathbb{R}, \Lambda^\pm \times \mathbb{R})$. In this case we define the reduced moduli spaces as $\widehat{\mathcal{M}}(u) = \mathcal{M}(u)/\mathbb{R}$. Let $h^\pm \in \mathbf{V}(Y^\pm \times \mathbb{R}, \Lambda^\pm \times \mathbb{R})$ denote the vector of admissible formal disks in $Y^\pm \times \mathbb{R}$ with boundary on $\Lambda^\pm \times \mathbb{R}$ represented by J-holomorphic disks:

$$h^\pm = \sum_{\dim(\widehat{\mathcal{M}}(v))=0} |\widehat{\mathcal{M}}(v)|\, v \in \mathbf{V}(\Lambda^\pm \times \mathbb{R}), \tag{5}$$

where the sum ranges over all formal disks of $\Lambda^\pm \times \mathbb{R}$ and where $|\widehat{\mathcal{M}}|$ denotes the mod-2 number of points in the compact 0-manifold $\widehat{\mathcal{M}}$. We call h^+ and h^- the *Hamiltonian vectors* of the positive and negative ends, respectively. Similarly, let f denote the generating function of rigid disks in the cobordism:

$$f = \sum_{\dim(\mathcal{M}(v))=0} |\mathcal{M}(v)|\, v \in \mathbf{V}(X,L), \tag{6}$$

where the sum ranges over all formal disks of (X,L). We call f the *potential vector of* (X,L).

We view elements w in $\mathbf{V}(Y^\pm \times \mathbb{R}, \Lambda^\pm \times \mathbb{R})$ and $\mathbf{V}(X,L)$ as sets of admissible formal disks, where the set consists of those formal disks that appear with nonzero coefficient in w. Define the differential $d^f\colon \mathbf{V}(X,L) \to \mathbf{V}(X,L)$ as the linear map such that if v is an admissible formal disk (a generator of $\mathbf{V}(X,L)$), then $d^f(v)$ is the sum of all admissible formal disks obtained in the following way:

(i) Attach a positive puncture of v to a negative puncture of an h^+-disk.
(ii) Then attach f-disks at remaining negative punctures of the h^+-disk, or
(iii) Attach a negative puncture of v to a positive puncture of an h^--disk.
(iv) Then attach f-disks at remaining positive punctures of the h^--disk.

The fact that this is a differential is a consequence of the product structure of the boundary of the moduli space mentioned above in the case of 1-dimensional moduli spaces; see [6, Lemma 3.7]. Furthermore, the differential increases the grading by 1 and respects the filtration, since any disk in $h^{\pm}$ or in f has at least one positive puncture.

2.4 The Rational Admissible SFT Spectral Sequence

Fix $\alpha > 0$. If $\mathbf{V}_{[\alpha+]}(X,L) \subset \mathbf{V}(X,L)$ denotes the subspace of formal sums of formal disks with $(+)$-action at least α, then since holomorphic disks have positive symplectic area, it follows that $d^f\big(\mathbf{V}_{[\alpha+]}(X,L)\big) \subset \mathbf{V}_{[\alpha+]}(X,L)$. If $\mathbf{V}_{[\alpha]}(X,L) = \mathbf{V}(X,L)/\mathbf{V}_{[\alpha+]}(X,L)$, then $\mathbf{V}_{[\alpha]}(X,L)$ is isomorphic to the vector space generated by formal disks of $(+)$-action less than α, and there is a short exact sequence of chain complexes

$$0 \longrightarrow \mathbf{V}_{[\alpha+]}(X,L) \longrightarrow \mathbf{V}(X,L) \longrightarrow \mathbf{V}_{[\alpha]}(X,L) \longrightarrow 0.$$

The quotients $\mathbf{V}_{[\alpha]}(X,L)$ form an inverse system

$$\pi^{\alpha}_{\beta} : \mathbf{V}_{[\alpha]}(X,L) \to \mathbf{V}_{[\beta]}(X,L), \quad \alpha > \beta,$$

of graded chain complexes, where π^{β}_{α} are the natural projections. Consequently, the k-level spectral sequences corresponding to the filtrations

$$0 \subset F^k\mathbf{V}_{[\alpha]}(X,L) \subset \cdots \subset F^2\mathbf{V}_{[\alpha]}(X,L) \subset F^1\mathbf{V}_{[\alpha]}(X,L) = \mathbf{V}_{[\alpha]}(X,L),$$

which we denote by

$$\left\{E^{p,q}_{r;[\alpha]}(X,L)\right\}_{r=1}^{k},$$

form an inverse system as well, and we define the rational admissible SFT invariant as

$$\{E^{p,q}_r(X,L)\}_{r=1}^k = \varprojlim_\alpha \left\{E^{p,q}_{r;[\alpha]}(X,L)\right\}_{r=1}^{k}.$$

This is in general not a spectral sequence, but it is so under some finiteness conditions. The following result is a consequence of [6, Theorems 1.1 and 1.2].

Theorem 2.1. *Let (X,L) be an exact cobordism with a subdivision $L = L_1 \cup \cdots \cup L_k$ into pieces. Then $\left\{E^{p,q}_r(X,L)\right\}$ does not depend on the choice of adjusted almost complex structure J, and is invariant under deformations of (X,L) through exact cobordisms with good ends.*

Proof. Any such deformation can be subdivided into a compactly supported deformation and a Legendrian isotopy at infinity. Deformations of the former type are shown in [6, Theorem 1.1.] to induce isomorphisms of $\{E_r^{p,q}(X,L)\}$.

To show that a deformation of the latter type induces an isomorphism, we note that it gives rise to an invertible exact cobordism, see [6, Appendix A], and use the same argument as in the proof of [6, Theorem 1.2.] as follows. Let C_{01} be the exact cobordism of the Legendrian isotopy at infinity and let C_{10} be its inverse cobordism. We use the symbol $A\#B$ to denote the result of joining two cobordisms along a common end. Consider first the cobordism

$$L\#C_{01}\#C_{10}.$$

Since this cobordism can be deformed by a compact deformation to L, we find that the composition of the maps Φ: $\mathbf{V}(X,L\#C_{01}) \to \mathbf{V}(X,L\#C_{01}\#C_{10})$ and Ψ: $\mathbf{V}(X,L) \to \mathbf{V}(X,L\#C_{01})$ is chain homotopic to the identity. Hence Ψ is injective on homology. Consider second the cobordism

$$L\#C_{01}\#C_{10}\#C_{01}.$$

Since this cobordism can be deformed to $L\#C_{01}$, we find similarly that there is a map Θ such that $\Psi \circ \Theta$ is chain homotopic to the identity on $\mathbf{V}(X,L\#C_{01})$; hence Ψ is surjective on homology as well.

2.5 A Simple Version of Rational Admissible SFT

As mentioned in Sect. 1, in the present paper, we will use the rational admissible spectral sequence in the simplest case: for (X,L), where L has only one component. Since there is only one piece, the spectral sequence has only one level, and

$$E_1^{1,q}(X,L) = \varprojlim_\alpha E_{1;[\alpha]}^{1,q}(X,L) = \varprojlim_\alpha \ker(d_\alpha^f)/\operatorname{im}(d_\alpha^f)$$

is the invariant that we will compute. To simplify things further, we will work not with the chain complex $\mathbf{V}(X,L)$ as described above but with the quotient of it obtained by forgetting the homotopy classes of formal disks. We view this quotient, using our assumption that $\pi_1(X)$ is trivial, as the space of formal sums of Reeb chords of the $(+\infty)$-boundary Λ of L. Further, our assumptions $c_1(TX) = 0$ and vanishing Maslov class of L imply that the grading descends to the quotient; see (3). For simplicity, we keep the notation $\mathbf{V}(X,L)$ and $\mathbf{V}_{[\alpha]}(X,L)$ for the corresponding quotients.

3 Legendrian Contact Homology, Augmentation, and Linearization

In this section we will define Legendrian contact homology and its linearization. We work in the following setting: (X,L) is an exact cobordism with good ends, the $(-\infty)$-boundary of L is empty, and the $(+\infty)$-boundary of (X,L) will be denoted by (Y,Λ).

Recall that the assumption on good ends allows us to disregard Reeb orbits. Furthermore, our additional assumptions on (X,L), i.e., $\pi_1(X)$ trivial and first Chern class and Maslov class trivial, allows us to work with coefficients in $\mathbb{Z}_2$ and still retain the grading.

3.1 Legendrian Contact Homology

Assume that $\Lambda \subset Y$ is generic with respect to the Reeb flow on Y. If c is a Reeb chord of Λ, let $|c| \in \mathbb{Z}$ be as in (3).

Definition 3.1. The DGA of (Y,Λ) is the unital noncommutative algebra $\mathcal{A}(Y,\Lambda)$ over $\mathbb{Z}_2$ generated by the Reeb chords of Λ. The grading of a Reeb chord c is $|c|$.

Definition 3.2. The contact homology differential is the map $\partial\colon \mathcal{A}(Y,\Lambda) \to \mathcal{A}(Y,\Lambda)$ that is linear over $\mathbb{Z}_2$ satisfies the Leibniz rule, and is defined as follows on generators:

$$\partial c = \sum_{\dim(\mathcal{M}(c;\overline{b}))=1} |\widehat{\mathcal{M}}(c;\overline{b})|\overline{b},$$

where c is a Reeb chord and $\overline{b} = b_1,\dots,b_k$ is a word of Reeb chords. (For notation, see (4).)

We give a brief explanation of why ∂ in Definition 3.2 is a differential, i.e., why $\partial^2 = 0$. Consider the boundary of the 2-dimensional moduli space $\mathcal{M}$ (which becomes 1-dimensional after the $\mathbb{R}$-action has been divided out) of holomorphic disks with one positive puncture at a. As explained in Sect. 2.2, the boundary of such a moduli space consists of two level curves such that all components except two are Reeb chord strips. Since these configurations are exactly what is counted by $\partial^2 c$ and since they correspond to the boundary points of the compact 1-manifold $\mathcal{M}/\mathbb{R}$, we conclude that $\partial^2 c = 0$.

Definition 3.3. An *augmentation* of $\mathcal{A}(Y,\Lambda)$ is a chain map $\epsilon\colon \mathcal{A}(Y,\Lambda) \to \mathbb{Z}_2$, where $\mathbb{Z}_2$ is equipped with the trivial differential.

Given an augmentation ϵ, define the algebra isomorphism $E_\epsilon \colon \mathcal{A}(Y,\Lambda) \to \mathcal{A}(Y,\Lambda)$ by letting

$$E_\epsilon(c) = c + \epsilon(c),$$

for each generator c. Consider the word-length filtration of $\mathcal{A}(Y,\Lambda)$,

$$\mathcal{A}(Y,\Lambda) = \mathcal{A}_0(Y,\Lambda) \supset \mathcal{A}_1(Y,\Lambda) \supset \mathcal{A}_2(Y,\Lambda) \supset \cdots .$$

The differential $\partial^\epsilon = E_\epsilon \circ \partial \circ E_\epsilon^{-1} \colon \mathcal{A}(Y,\Lambda) \to \mathcal{A}(Y,\Lambda)$ respects this filtration: $\partial^\epsilon(\mathcal{A}_j(Y,\Lambda)) \subset \mathcal{A}_j(Y,\Lambda)$. In particular, we obtain the *ϵ-linearized differential*

$$\partial_1^\epsilon \colon \mathcal{A}_1(Y,\Lambda)/\mathcal{A}_2(Y,\Lambda) \to \mathcal{A}_1(Y,\Lambda)/\mathcal{A}_2(Y,\Lambda). \tag{7}$$

Definition 3.4. The *ϵ-linearized contact homology* is the $\mathbb{Z}_2$-vector space

$$\mathrm{LCH}_*(Y,\Lambda;\epsilon) = \ker(\partial_1^\epsilon)/\operatorname{im}(\partial_1^\epsilon). \tag{8}$$

For simpler notation below, we write

$$Q(Y,\Lambda) = \mathcal{A}_1(Y,\Lambda)/\mathcal{A}_2(Y,\Lambda)$$

and think of $Q(Y,\Lambda)$ as the graded vector space generated by the Reeb chords of Λ. Furthermore, the augmentation will often be clear from the context, and we will drop it from the notation and write the differential as

$$\partial_1 \colon Q(Y,\Lambda) \to Q(Y,\Lambda).$$

Consider an exact cobordism (X,L) with $(+\infty)$-boundary (Y^+,Λ^+) and $(-\infty)$-boundary (Y^-,Λ^-). Define the algebra map $\Phi\colon \mathcal{A}(Y^+,\Lambda^+) \to \mathcal{A}(Y^-,\Lambda^-)$ by mapping generators c of $\mathcal{A}(Y^+,\Lambda^+)$ as follows:

$$\Phi(c) = \sum_{\dim(\mathcal{M}(c;\overline{b}))=0} |\mathcal{M}(c;\overline{b})|\,\overline{b},$$

where $\overline{b} = b_1,\dots,b_k$ is a word of Reeb chords of Λ^-, where $\mathcal{M}(c;\overline{b})$ denotes the moduli space of holomorphic disks in (X,L) with boundary on L, with positive puncture at a and negative punctures at $b_1,\dots,b_k$. An argument completely analogous to the argument above showing that $\partial^2 = 0$, looking at the boundary of 1-dimensional moduli spaces, shows that $\Phi \circ \partial^+ = \partial^- \circ \Phi$, where $\partial^\pm$ is the differential on $\mathcal{A}(Y^\pm,\Lambda^\pm)$, i.e., that Φ is a chain map. Consequently, if $\epsilon^- \colon \mathcal{A}(Y^-,\Lambda^-) \to \mathbb{Z}_2$ is an augmentation, then so is $\epsilon^+ = \epsilon^- \circ \Phi$. In particular if $\Lambda^- = \emptyset$ and $(Y,\Lambda) = (Y^+,\Lambda^+)$, then $\mathcal{A}(Y^-,\Lambda^-) = \mathbb{Z}_2$ with the trivial differential, and $\epsilon = \epsilon^+ = \Phi$ is an augmentation of $\mathcal{A}(Y,\Lambda)$.

3.2 Proof of Theorem 1.1

If $Q_{[\alpha]}(Y,\Lambda)$ denotes the subspace of $Q(Y,\Lambda)$ generated by Reeb chords c of action $\mathfrak{a}(c) < \alpha$, then $Q_{[\alpha]}(Y,\Lambda)$ is a subcomplex of $Q(Y,\Lambda)$. By definition, the map that takes a Reeb chord c viewed as a generator of $\mathbf{V}_{[\alpha]}(X,L)$ to the dual c^* of c in the cochain complex $Q'_{[\alpha]}(Y,\Lambda)$ of $Q_{[\alpha]}(Y,\Lambda)$ is an isomorphism intertwining the respective differentials. To prove the theorem, since the complex is finite-dimensional below the action α, it remains only to show that the monotonicity condition implies $H^r(Q'_{[\alpha]}(Y,\Lambda)) = H^r(Q'(Y,\Lambda))$ for $\alpha > 0$ large enough. This is straightforward: if $|c| = C_1\mathfrak{a}(c) + C_0$, then

$$H^r(Q'_{[\alpha]}(Y,\Lambda)) = H^r(Q'(Y,\Lambda)),$$

for $\alpha > \frac{r+1-C_0}{C_1}$. □

4 Lagrangian Floer Cohomology of Exact Cobordisms

In this section we introduce a Lagrangian Floer cohomology of exact cobordisms. It is a generalization of the two-copy version of the relative SFT of an exact cobordism (X,L), $L = L_0 \cup L_1$, to the case that each piece of L is embedded but $L_0 \cap L_1 \neq \emptyset$. To prove that this theory has the desired properties, we will use a mixture of results from Floer homology of compact Lagrangian submanifolds and the SFT framework explained in Sect. 2. After setting up the theory, we state a conjectural lemma about how moduli spaces of holomorphic disks with boundary on $L \cup L'$, where L' is a small perturbation of L, can be described in terms of holomorphic disks with boundary on L and a version of Morse theory on L. We then show how Conjecture 1.2 and Corollary 1.3 follow from this conjectural description.

Remark 4.1. The Lagrangian Floer cohomology considered here is closely related to the wrapped Floer cohomology introduced in [1, 14]. Indeed, in analogy with results relating the symplectic homology of a Liouville domain to the linearized contact homology of its boundary, see [2], one expects that the Lagrangian Floer cohomology considered here is isomorphic to the wrapped Floer cohomology.

4.1 The Chain Complex

Let X be a simply connected exact symplectic cobordism with $c_1(TX) = 0$ and with good ends. Let L_0 and L_1 be exact Lagrangian cobordisms in X with empty negative ends and with trivial Maslov classes. In other words, (X,L_0) and (X,L_1)

are exact cobordisms with empty $(-\infty)$-boundaries. Let the $(+\infty)$-boundaries of (X, L_0) and (X, L_1) be (Y, Λ_0) and (Y, Λ_1), respectively.

Define

$$C(X;L_0,L_1) = C_\infty(X;L_0,L_1) \oplus C_0(X;L_0,L_1)$$

as follows. The summand $C_\infty(X;L_0,L_1)$ is the $\mathbb{Z}_2$-vector space of formal sums of Reeb chords that start on Λ_0 and end on Λ_1. The summand $C_0(X;L_0,L_1)$ is the $\mathbb{Z}_2$-vector space generated by the transverse intersection points in $L_0 \cap L_1$.

In order to define the grading and a differential on $C(X;L_0,L_1)$, we will consider the following three types of moduli spaces. The first type was considered already in (4); if a is a Reeb chord and $\overline{b} = b_1, \ldots, b_k$ is a word of Reeb chords, we write

$$\mathcal{M}(a;\overline{b})$$

for the moduli space of holomorphic curves in the symplectization $Y \times \mathbb{R}$ with boundary on $\Lambda_0 \times \mathbb{R} \cup \Lambda_1 \times \mathbb{R}$, with positive puncture at a and negative punctures at $b_1, \ldots, b_k$.

The second kind is the standard moduli spaces for Lagrangian Floer homology; if x and y are intersection points of L_0 and L_1, we write

$$\mathcal{M}(x;y)$$

for the moduli space of holomorphic disks in X with two boundary punctures at which the disks are asymptotic to x and y, with boundary on $L_0 \cup L_1$, and that are such that in the orientation on the boundary induced by the complex orientation, the incoming boundary component at x maps to L_0.

Finally, the third kind is a mixture of these; if c is a Reeb chord connecting Λ_0 to Λ_1 and if y is an intersection point of L_0 and L_1, we write $\mathcal{M}(c;y)$ for the moduli space of holomorphic disks in X with two boundary punctures; at one, the disk has a positive puncture at c, and at the other, the disk is asymptotic to y.

If a and c are both Reeb chord generators, then we define the grading difference between a and c to equal the formal dimension $\dim(\mathcal{M}(a;c))$. If g is a Reeb chord or an intersection point generator and if x is an intersection point generator, then we take the grading difference between g and x to equal $\dim(\mathcal{M}(g;x)) + 1$. Our assumptions on the exact cobordism guarantee that this is well defined.

Write $C = C(X;L_0,L_1)$, $C_0 = C_0(X;L_0,L_1)$, and $C_\infty = C_\infty(X;L_0,L_1)$. Define the differential $d\colon C \to C$, using the decomposition $C = C_\infty \oplus C_0$ given by the matrix

$$d = \begin{pmatrix} d_\infty & \rho \\ 0 & d_0 \end{pmatrix},$$

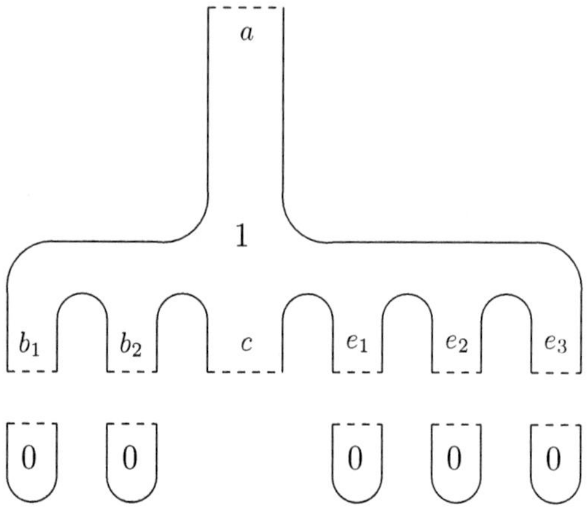

Fig. 1 A disk configuration contributing to the differential of c. Reeb chords are *dashed*. Numbers inside disk components indicate their dimension

where $d_\infty\colon C_\infty \to C_\infty$, $\rho\colon C_0 \to C_\infty$, and $d_0\colon C_0 \to C_0$ are defined as follows. Let c be a Reeb chord from Λ_0 to Λ_1 and let $\theta\colon \mathcal{A}(Y,\Lambda_0) \to \mathbb{Z}_2$ and $\epsilon\colon \mathcal{A}(Y,\Lambda_1) \to \mathbb{Z}_2$ denote the augmentations induced by L_0 and L_1, respectively. Define

$$d_\infty(c) = \sum_{\dim(\mathcal{M}(a;\overline{b}c\overline{e}))=1} |\widehat{\mathcal{M}}(a;\overline{b}c\overline{e})|\epsilon(\overline{b})\theta(\overline{e})\,a, \tag{9}$$

where $\overline{b}$ and $\overline{e}$ are words of Reeb chords from Λ_0 to Λ_0 and from Λ_1 to Λ_1, respectively; see Fig. 1. Let x be an intersection point of L_0 and L_1. Define

$$\rho(x) = \sum_{\dim(\mathcal{M}(c;x))=0} |\mathcal{M}(c;x)|\,c, \tag{10}$$

where c is a Reeb chord from Λ_0 to Λ_1, and

$$d_0(x) = \sum_{\dim(\mathcal{M}(y;x))=0} |\mathcal{M}(y;x)|\,y, \tag{11}$$

where y is an intersection point of L_0 to L_1. See Fig. 2. Then d increases the grading by 1.

Lemma 4.2. *The map d is a differential, i.e., $d^2 = 0$.*

Proof. We first check that $d_\infty^2 = 0$. Let c and a be Reeb chords of index difference 2. Consider two holomorphic disks in $\mathcal{M}(a';\overline{b}_- c\overline{e}_-)$ and $\mathcal{M}(a;\overline{b}_+ a'\overline{e}_+)$ contributing to the coefficient of a in $d^2(c)$. Gluing these two 1-dimensional families at a' and completing with Reeb chord strips at chords in $\overline{b}_+$ and $\overline{e}_+$, we find that the broken disk corresponds to one endpoint of a reduced moduli space $\widehat{\mathcal{M}}(a;\overline{b}c\overline{e})$, where $\overline{b} = \overline{b}_+\overline{b}_-$ and $\overline{e} = \overline{e}_-\overline{e}_+$. Note that there are three possible types of breaking at the boundary of $\widehat{\mathcal{M}}(a;\overline{b}c\overline{e})$:

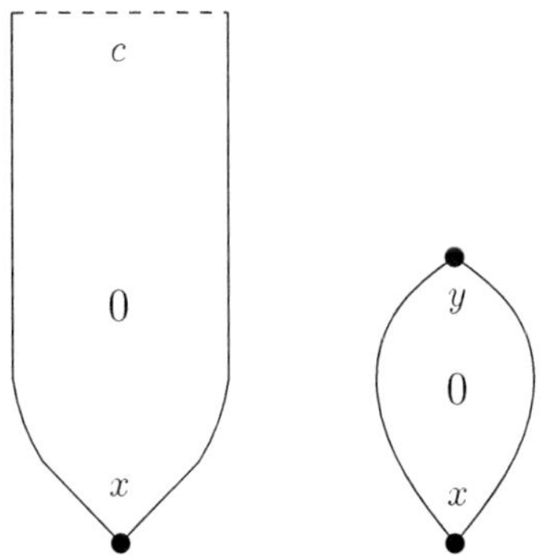

Fig. 2 Disks contributing to the differential of x. Reeb chords are *dashed*, and double points appear as *dots*. Numbers inside disk components indicate their dimension

(a) Breaking at a Reeb chord from Λ_0 to Λ_1,
(b) Breaking at a Reeb chord from Λ_0 to Λ_0,
(c) Breaking at a Reeb chord from Λ_1 to Λ_1.

Summing the formal disks of all these boundary configurations gives 0, since the ends of a compact 1-manifold cancel in pairs. To interpret this algebraically, we define

$$\tilde{d}_\infty\colon \mathcal{A}(Y,\Lambda_0)\otimes C_\infty\otimes\mathcal{A}(Y,\Lambda_1)\to\mathcal{A}(Y,\Lambda_0)\otimes C_\infty\otimes\mathcal{A}(Y,\Lambda_1)$$

as follows on generators:

$$\tilde{d}_\infty(w_0\otimes c\otimes w_1)=\sum_{\dim(\mathcal{M}(a;\overline{b}c\overline{e}))=1}|\widehat{\mathcal{M}}(a;\overline{b}c\overline{e})|\,w_0\overline{b}\otimes a\otimes\overline{e}w_1.$$

Then the cancellation mentioned above implies with $\hat{c}=1\otimes c\otimes 1$ that

$$\tilde{d}_\infty^2(\hat{c})+(\partial_0\otimes 1\otimes 1)(\tilde{d}_\infty(\hat{c}))+(1\otimes 1\otimes\partial_1)(\tilde{d}_\infty(\hat{c}))=0,$$

where $\partial_j\colon\mathcal{A}(Y,\Lambda_j)\to\mathcal{A}(Y,\Lambda_j)$, $j=0,1$, denotes the contact homology differential. Here the first term corresponds to breaking of type (a), the second to type (c), and the third to type (b). By definition,

$$d_\infty(c)=(\epsilon\otimes 1\otimes\theta)(\tilde{d}_\infty(\hat{c})).$$

Consequently,

$$\begin{aligned}d_\infty^2(c)&=(\epsilon\otimes 1\otimes\theta)(\tilde{d}_\infty^2(\hat{c}))\\&=\Big(\epsilon\otimes 1\otimes\theta\Big)\Big((\partial_0\otimes 1\otimes 1)(\tilde{d}_\infty(\hat{c}))+(1\otimes 1\otimes\partial_1)(\tilde{d}_\infty(\hat{c}))\Big)=0,\end{aligned}$$

since $\epsilon\circ\partial_0=0$ and $\theta\circ\partial_1=0$.

The equation $d_0^2 = 0$ follows as in usual Lagrangian Floer homology from the fact that the terms contributing to d_0^2 are in 1-to-1 correspondence with the ends of the 1-dimensional moduli spaces of the form $\mathcal{M}(x;z)$, where x and z are intersection points of grading difference 2.

Finally, to see that $d_\infty \circ \rho + \rho \circ d_0 = 0$, we consider 1-dimensional moduli spaces of the form $\mathcal{M}(c,x)$, where c is a Reeb chord from Λ_0 to Λ_1 and where x is an intersection point. The analysis of breaking, see [2], in combination with standard arguments from Lagrangian Floer theory, shows that there are two possible breakings in the boundary of $\mathcal{M}(c,x)$:

(a) Breaking at an intersection point,
(b) Breaking at Reeb chords.

In case (a), the broken configuration contributes to $\rho \circ d_0$. In case (b), the disk has two levels: the top level is a curve in a moduli space $\mathcal{M}(a;\overline{b}c\overline{e})$, and the second level is a collection of rigid disks in X with boundary on L_0 and L_1 and with positive punctures at Reeb chords in $\overline{b}$ and $\overline{e}$, respectively, and a rigid disk in $\mathcal{M}(c;x)$. Such a configuration contributes to $d_\infty \circ \rho$, and the desired equation follows.

As above, we let $\mathfrak{a}(c)$ denote the action of a Reeb chord c. Define the action $\mathfrak{a}(x) = 0$ for intersection points $x \in L_0 \cap L_1$. Then the differential on $C = C(X;L_0,L_1)$ increases the action. Define $C_{[\alpha+]} \subset C$ as the subcomplex of formal sums in which all summands have action at least α and let $C_{[\alpha]} = C/C_{[\alpha+]}$ denote the corresponding quotient complex. Let $FH^*_{[\alpha]}(X;L_0,L_1)$ denote the cohomology of $C_{[\alpha]}$ and note that the natural projections give an inverse system of cochain maps

$$\pi^\alpha_\beta \colon C_{[\alpha]} \to C_{[\beta]}, \quad \alpha > \beta.$$

Define the Lagrangian Floer cohomology $FH^*(X;L_0,L_1)$ as the inverse limit of the corresponding inverse system of cohomologies

$$FH^*(X;L_0,L_1) = \varprojlim_\alpha FH^*_{[\alpha]}(X;L_0,L_1).$$

4.2 Chain Maps and Invariance

Our proof of the invariance of the Lagrangian Floer cohomology $FH^*(X;L_0,L_1)$ under isotopies of L_1 uses three ingredients: homology isomorphisms induced by compactly supported isotopies, chain maps induced by joining cobordisms, and chain homotopies induced by compactly supported deformations of adjoined cobordisms. Before proving invariance, we consider these three separately.

4.2.1 Compactly Supported Deformations

Let (X,L_0) and (X,L_1) be exact cobordisms as above. We first consider deformations of (X,L_1) that are fixed in the positive end. More precisely, let L_1^t, $0 \le t \le 1$, be a 1-parameter family of exact Lagrangian cobordisms such that the positive end Λ_1^t is fixed at Λ_1 for $0 \le t \le 1$. Our proof of invariance is a generalization of a standard argument in Floer theory.

We first consider changes of the chain complex. Assuming that L_1^t is generic, there is a finite number of birth/death instances $0 < t_1 < \cdots < t_m < 1$ when two double points cancel or are born at a standard Lagrangian tangency moment. (At such a moment t_j, there is exactly one nontransverse intersection point $x \in L_0 \cap L_1^{t_j}$, $\dim(T_xL_0 \cap T_xL_1^{t_j}) = 1$, and if v denotes the deformation vector field of L_1 at x, then v is not symplectically orthogonal to $T_xL_0 \cap T_xL_1^{t_j}$.)

For $0 \le t \le 1$, let $\mathcal{M}^t$ denote a moduli space of the form $\mathcal{M}^t(g;x)$ or $\mathcal{M}^t(c)$, where x is an intersection point, c a Reeb chord, and g either a Reeb chord or an intersection point of holomorphic disks as considered in the definition of the differential, with boundary on L_0 and L_1^t. If L_1^t is chosen generically, then such a moduli space $\mathcal{M}^t$ is empty for all t, provided $\dim(\mathcal{M}^t) < -1$ and there is a finite number of instances $0 < \tau_1 < \cdots < \tau_k < 1$ where there is exactly one disk of formal dimension -1 that is transversely cut out as a 0-dimensional parameterized moduli space; see [6, Lemma B.8]. We call these instances (-1)*-disk instances*. Furthermore, for generic L_1^t, birth/death instances and (-1)-disk instances are distinct.

For $I \subset [0,1]$, consider the parameterized moduli space of disks

$$\mathcal{M}_I = \cup_{t \in I} \mathcal{M}^t,$$

where $\dim(\mathcal{M}^t) = 0$. If I contains neither (-1)-disk instances nor birth/death instances, then $\mathcal{M}_I$ is a 1-manifold with boundary that consists of rigid disks in $\mathcal{M}^t$ and $\mathcal{M}^{t'}$, where $\partial I = \{t,t'\}$, and it follows by the definition of the differential that the chain complexes $C(X,L_0,L_1^t)$ and $C(X;L_0,L_1^{t'})$ are canonically isomorphic. If, on the other hand, I does contain (-1)-disk instances or birth/death instances, then $\mathcal{M}_I$ has additional boundary points corresponding to broken disks at these instances. It is clear that in order to show invariance, it is enough to show that the homology is unchanged over intervals containing only one (-1)-disk instance or birth/death. For simpler notation, we take $I = [-1,1]$ and assume that there is a (-1)-disk instance or a birth/death at $t = 0$.

We start with the case of a (-1)-disk. There are two cases to consider: either the (-1)-disk is mixed (i.e., has boundary components mapping both to L_0 and L_1^0), or it is pure (i.e., all of its boundary maps to L_1^0). Consider first the case of a mixed (-1)-disk. Since L_1^t is fixed at infinity, the (-1)-disk must lie in a moduli space $\mathcal{M}(g;x)$, where g is a Reeb chord or an intersection point and where x is an intersection point. Write $C(-) = C(X;L_0,L_1^{-1})$ and $C(+) = C(X;L_0,L_1^1)$ and let d^- and d^+ denote the corresponding differentials. Note that there is a canonical identification between generators of $C(-)$ and $C(+)$.

Lemma 4.3. *Let $\phi\colon C(-) \to C(+)$ be the linear map defined on generators h as follows:*

$$\phi(h) = \begin{cases} x+g & \text{if } h = x, \\ h & \text{otherwise.} \end{cases}$$

If $\Phi(h) = \phi(h + d^- h)$, then $\Phi\colon C(-) \to C(+)$ is a chain isomorphism.

Proof. We first note that the differentials d^+ and d^- agree on the canonically isomorphic subspaces $C_\infty(-)$ and $C_\infty(+)$. In order to study the remaining part of the differential, we consider parameterized 1-dimensional moduli spaces of the form $\mathcal{M}_I(h;y)$, where y is an intersection point and where h is either an intersection point or a Reeb chord. Note that disks at a fixed generic instance that lie in a parameterized moduli of dimension 1 are exactly those that contribute to the differential. Write $\mathcal{M}_I(g;x)$ for the transversely cut-out 0-manifold that is the parameterized moduli space containing the (-1)-disk.

If all punctures at intersection points are considered mixed, then any disk that contribute to the differential is admissible, see Sect. 2.1, and since the (-1)-disk is mixed, it follows from [6, Lemma B.9] (or from a standard result in Floer theory in case both generators are double points; see [13, Lemma 3.5 and Proposition 4.2]) that the boundary of the compactified 1-manifold $\mathcal{M}_I(h;y)$ satisfies the following:

- If $h \neq g$, $y \neq x$, and $[s,t] \subset I$, then

$$\partial \mathcal{M}_{[s,t]}(h;y) = \mathcal{M}^s(h;y) \cup \mathcal{M}^t(h;y);$$

- If $h = g$, then $y \neq x$ and

$$\partial \mathcal{M}_I(g;y) = \mathcal{M}^{-1}(g;y) \cup \mathcal{M}^1(g;y) \cup (\mathcal{M}_I(g;x) \times \mathcal{M}^0(x;y));$$

- If $y = x$, then $h \neq g$ and

$$\partial \mathcal{M}_I(h;x) = \mathcal{M}^{-1}(h;x) \cup \mathcal{M}^1(h;x) \cup (\widehat{\mathcal{M}}^0(h;g) \times \mathcal{M}_I(g;x)).$$

Here $\widehat{\mathcal{M}}^0(h,g)$ denotes the moduli space divided by the $\mathbb{R}$-action if both h and g are Reeb chords, and the moduli space itself otherwise.

Translating this into algebra, we find that for any generator h the following holds:

$$d^+ h = \begin{cases} d^- h + \phi(d^- h) & \text{if } |h| = |x| - 1, \\ d^- h + x^*(h)\, d^- g & \text{if } |h| = |x|, \\ d^- h & \text{otherwise,} \end{cases}$$

where $x^*\colon C(+) \to \mathbb{Z}_2$ is the map given by $x^*(h) = 0$ if $h \neq x$ and $x^*(h) = 1$ if $h = x$. The lemma follows.

Consider next the case of a pure (-1)-disk. Since L_1^t is fixed at infinity, such a disk must lie in a moduli space $\mathcal{M}(c)$, where c is a Reeb chord of Λ_1. This case is less straightforward than the case of a mixed (-1)-disk considered above. In order to get control of the resulting change in differential, we need to introduce abstract perturbations of the $\bar{\partial}_J$-operator near the moduli space of J-holomorphic disks. We give a short description here and refer to [6, Sect. B.6] for details. Note first that the change in differential is caused by the change that the augmentation induced by L_1^t undergoes as t passes 0. To describe this change, we study parameterized moduli spaces of the form $\mathcal{M}_I(b)$, $I = [-1,1]$ of dimension 1, where b is a Reeb chord of Λ_1. A priori, broken disks in the boundary of such a moduli space consist of a several level disks with one positive puncture at b and several negative punctures at $c_1,\dots,c_m$, where each c_j is capped off with a disk in $\mathcal{M}(c_j)$ and the only requirement is that the sum of dimensions of the components equal 0. In particular, if $c_j = c$ for several indices j, then since the cap at c has dimension -1, the sum of dimensions over the disks in the symplectization must be larger than 1.

This situation is impossible to control algebraically. In order to gain algebraic control, a perturbation that time orders the complex structures at the negative pure Λ_1-punctures of any disk in the symplectization $(Y\times\mathbb{R},\Lambda_1\times\mathbb{R})$ is introduced. This perturbation needs to be extended over the entire moduli space of 1-punctured holomorphic disks (below a fixed $(+)$-action) in the symplectization $(Y\times\mathbb{R},\Lambda_1\times\mathbb{R})$. Such a perturbation is defined energy level by energy level starting from the lowest one. The time ordering of the negative punctures implies that only one (-1)-disk at a time can be attached to any disk in the symplectization. It is important to note that the time ordering itself may introduce new (-1)-disks, but the positive punctures of such introduced disks all lie close to $t=0$. More precisely, the count of (-1)-disks with positive puncture at a Reeb chord b depends on the perturbation used on energy levels below $\mathfrak{a}(b)$, and the positive puncture of any such (-1)-disk lies close to $t=0$ compared to the size of the time-ordering perturbation of negative punctures mapping to b.

Let ϵ^- and ϵ^+ denote the augmentations on $\mathcal{A}(Y,\Lambda)$ induced by (X,L_1^{-1}) and (X,L_1^1), respectively. It is a consequence of [6, Lemma B.15] (which uses the perturbation scheme above) that there is a map K from the set of generators of $\mathcal{A}(Y,\Lambda_1)$ into $\mathbb{Z}_2$ such that if c is a Reeb chord, then $K(c)$ counts (-1)-disks with positive puncture at c, and such that

$$\epsilon^-(c)+\epsilon^+(c) = \Omega_K(\partial c). \tag{12}$$

Here $\partial\colon \mathcal{A}(Y,\Lambda_1)\to\mathcal{A}(Y,\Lambda_1)$ is the contact homology differential, and if $w = b_1\dots b_m$ is a word of Reeb chords, then

$$\Omega_K(w) = \sum_j \epsilon^-(b_1)\dots\epsilon^-(b_{j-1})K(b_j)\epsilon^+(b_{j+1})\dots\epsilon^+(b_m).$$

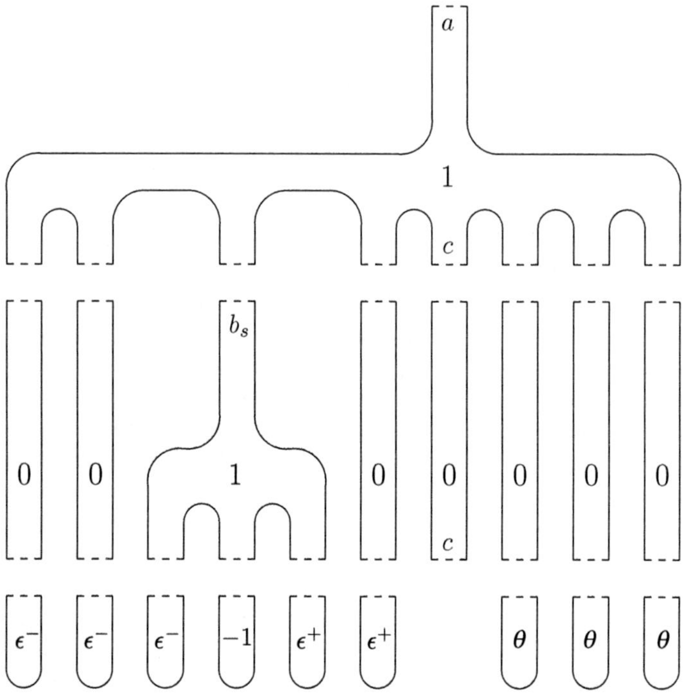

Fig. 3 A disk contributing to $d^+(c)+d^-(c)$: the disk breaks at the pure Reeb chord b_s, and the (-1)-disk is attached to one of the negative ends of the disk with positive puncture at b_s

Define the linear map $\phi\colon C(-)\to C(+)$ as

$$\phi(c) = \sum_{\dim(\mathcal{M}(a;\overline{b}c\overline{e}))=1} |\widehat{\mathcal{M}}(a;\overline{b}c\overline{e})|\Omega_K(\overline{b})\theta(\overline{e})\, a$$

for generators $c\in C_\infty(-)$ and $\phi(x)=0$ for generators $x\in C_0(-)$.

Lemma 4.4. *The map* $\Phi\colon C(-)\to C(+)$,

$$\Phi(c) = c+\phi(c),$$

is a chain isomorphism.

Proof. The map is an isomorphism, since the action of any chord in $\phi(c)$ is larger than that of c. We thus need only show that it is a chain map, or in other words, that

$$d^+ + d^- = d^+\circ\phi + \phi\circ d^-.$$

Consider first the operator on the left-hand side acting on a Reeb chord c. According to (12), broken disk configurations that contribute to $d^+(c)+d^-(c)$ are of the following form; see Fig. 3:

1. The top level is a 1-dimensional disk in $Y \times \mathbb{R}$ with positive puncture at a Reeb chord a connecting Λ_0 to Λ_1, followed by k negative punctures at Reeb chords $b_1, \dots, b_k$ connecting Λ_1 to itself, followed by a negative puncture at c connecting Λ_0 to Λ_1, in turn followed by r negative punctures at Reeb chords $e_1, \dots e_r$ connecting Λ_0 to itself.
2. The middle level consists of Reeb chord strips at all negative punctures except b_s for some $1 \leq s \leq k$. At b_s, a 1-dimensional disk in $Y \times \mathbb{R}$ with boundary on $\Lambda_1 \times \mathbb{R}$ and with negative punctures at $q_1, \dots, q_m$ is attached.
3. The bottom level consists of rigid disks with boundary on L_0 and positive puncture at e_j attached at all punctures e_j, $j = 1, \dots, r$, rigid disks with boundary on L_1^{-1} attached at punctures b_j, $1 \leq j \leq s-1$, and at punctures q_j, $1 \leq j < v$, a (-1)-disk attached at q_v, and rigid disks with boundary on L_1^1 at punctures b_j, $s+1 \leq j \leq k$, and q_j, $v < j \leq m$.

Consider gluing the top and middle levels above in the symplectization. This gives one boundary component of a reduced 1-dimensional moduli space. The other boundary component corresponds to one of three breakings: at a pure Λ_1-chord, at a chord connecting Λ_0 to Λ_1, or at a pure Λ_0 chord. The first type of breaking contributes to $d_-(c) + d_+(c)$ as well; see Fig. 3. The second type contributes to either $d^+ \circ \phi(c)$ or $\phi \circ d^-(c)$ depending on the factor to which the (-1)-end goes – see Figs. 4 and 5, respectively – and finally, the total contribution of the third type of breaking is 0, since $\theta \circ \partial = 0$, where $\partial\colon \mathcal{A}(Y, \Lambda_0) \to \mathcal{A}(Y, \Lambda_0)$ is the contact homology differential and where $\theta\colon \mathcal{A}(Y, \Lambda_0) \to \mathbb{Z}_2$ is the augmentation; see Fig. 6.

Next, consider the operators acting on a double point x. The chain map property in this case follows from an argument similar to the one just given. Additional boundary components of the parameterized moduli space $\mathcal{M}^t(c;x)$ correspond to two level disks with top level a disk as in (1) above and with bottom level as in (3) above with the addition that there is a rigid disk with positive puncture at c and a puncture at x. We conclude that Φ is a chain map.

Finally, consider a birth/death moment involving intersection points x and y. Assume that $x, y \in C(+)$ (birth moment). Then $d^+x = y + v$, where v does not contain any y-term; see [13, Lemma 3.7 and Proposition 5.1]. Define the map $\Phi\colon C(+) \to C(-)$ by

$$\Phi(x) = 0, \quad \Phi(y) = v, \quad \text{and} \quad \Phi(w) = w \text{ for } w \neq x, y.$$

Define the map $\Psi\colon C(-) \to C(+)$ by

$$\Psi(c) = c + y^*(d^+c)x.$$

Lemma 4.5. *The maps Φ and Ψ are chain maps that induce isomorphisms on homology.*

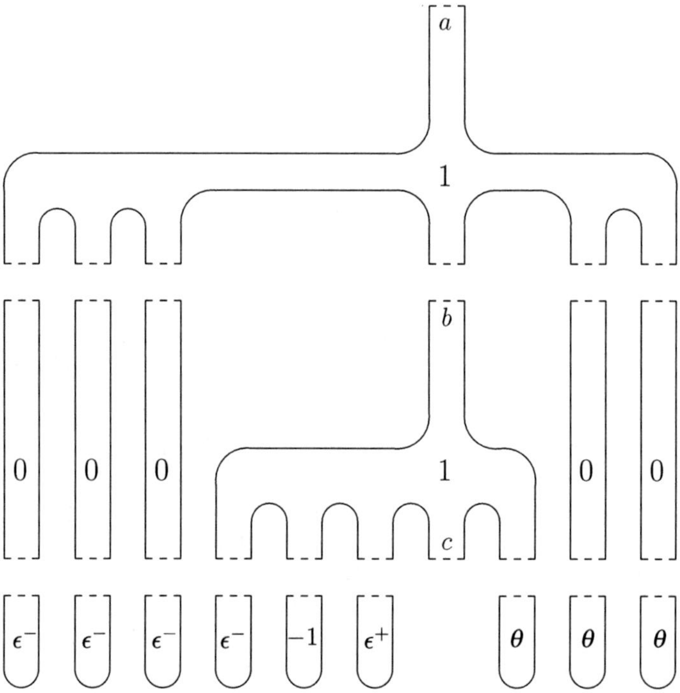

Fig. 4 A disk contributing to $d^+ \circ \phi(c)$: the disk breaks at the mixed chord b, and the (-1)-disk is attached to the disk with positive puncture at b

Proof. A straightforward generalization of the gluing theorem [9, Proposition 2.16] (in the case that only one disk is glued at the degenerate intersection) shows that if g is a generator of $C(-)$, then

$$d^-(c) = d^+(c) + y^*(d^+(c))v.$$

The lemma then follows from a straightforward calculation.

4.2.2 Joining Cobordisms

Let (X, L_0) and (X, L_1) be cobordisms as above. Consider the trivial cobordism $(Y \times \mathbb{R}, \Lambda_0 \times \mathbb{R})$ and some cobordism $(Y \times \mathbb{R}, L_1^a)$, where the $(-\infty)$-boundary of L_1^a equals Λ_1 and its $(+\infty)$-boundary equals Λ_1^a. Then we can join these cobordisms to (X, L_0) and (X, L_1), respectively. This results in a new pair of cobordisms (X, L_0) and $(X, \tilde{L}_1)$. Consider the subdivision

$$C(X; L_0, \tilde{L}_1) = C_\infty(X; L_0, \tilde{L}_1) \oplus C_0(X; L_0, \tilde{L}_1).$$

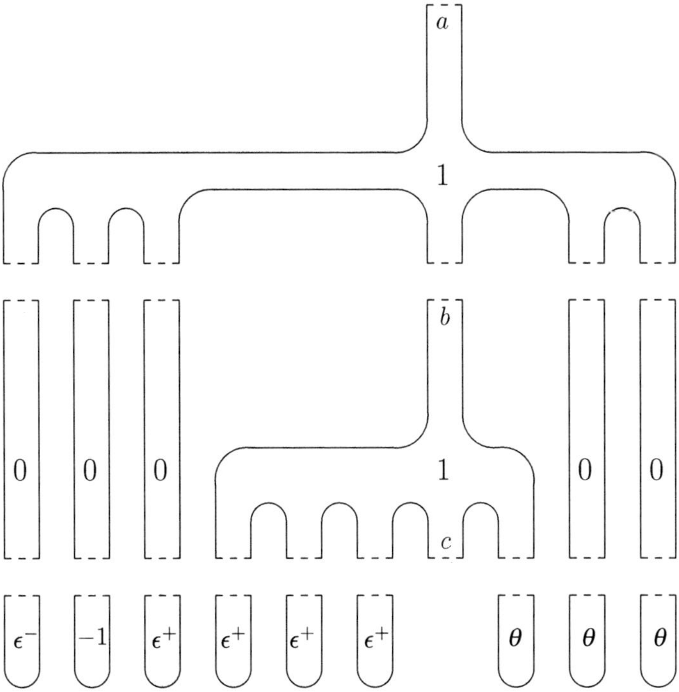

Fig. 5 A disk contributing to $\phi \circ d^-(c)$: the disk breaks at the mixed chord b, and the (-1)-disk is attached to the disk with positive puncture at a

If $z \in L_0 \cap \tilde{L}_1$, then either $z \in L_0 \cap L_1$ or $z \in (\Lambda_0 \times \mathbb{R}) \cap L_1^a$, and we have the further subdivision

$$C_0(X; L_0, \tilde{L}_1) = C_0(X; L_0, L_1) \oplus C_0(Y \times \mathbb{R}; \Lambda_0 \times \mathbb{R}, L_1^a).$$

We define a map

$$\Phi \colon C_\infty(X; L_0, L_1) \oplus C_0(X; L_0, L_1) \to \\ C_\infty(X; L_0, \tilde{L}_1) \oplus C_0(Y \times \mathbb{R}; \Lambda_0 \times \mathbb{R}, L_1^a) \oplus C_0(X; L_0, L_1),$$

with matrix

$$\begin{pmatrix} \phi_\infty & 0 \\ \phi_0 & 0 \\ 0 & \mathrm{id} \end{pmatrix},$$

as follows, using moduli spaces of holomorphic disks in $Y \times \mathbb{R}$ with boundary on $(\Lambda_0 \times \mathbb{R}) \cup L_1^a$. If c is a generator of $C_\infty(X; L_0, L_1)$, then c is a Reeb chord from Λ_0 to Λ_1. Thinking of c as lying in the negative end of $(Y \times \mathbb{R}, \Lambda_0 \times \mathbb{R} \cup L_1^a)$, we define

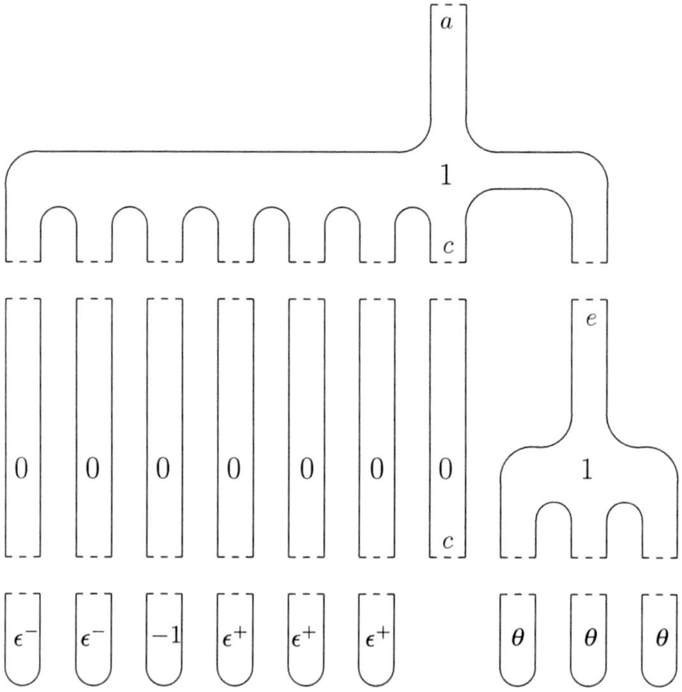

Fig. 6 Disks with total contribution 0: the disk breaks at the pure Λ_0 chord e, and the total contribution vanishes, since $\theta \circ \partial = 0$

$$\phi_0(c) = \sum_{\dim \mathcal{M}(z;\overline{b}c\overline{e})=0} |\mathcal{M}(x;\overline{b}c\overline{e})|\epsilon(\overline{b})\theta(\overline{e})\, z,$$

where the sum ranges over intersection points $z \in C_0(Y \times \mathbb{R}; \Lambda \times \mathbb{R}, L_1^a)$, and

$$\phi_\infty(c) = \sum_{\dim \mathcal{M}(a;\overline{b}c\overline{e})=0} |\mathcal{M}(a;\overline{b}c\overline{e})|\epsilon(\overline{b})\theta(\overline{e})\, a,$$

where the sum ranges over Reeb chords a connecting Λ_0 to Λ_1^a, i.e., over generators of $C_\infty(X; L_0, \tilde{L}_1)$.

Lemma 4.6. *The map Φ is a chain map. That is, if d and $\tilde{d}$ denote the differentials on $C(X, L_0, L_1)$ and $C(X, L_0, \tilde{L}_1)$, respectively, then*

$$\tilde{d} \circ \Phi = \Phi \circ d.$$

Proof. As above, we write the differentials d and $\tilde{d}$ in matrix form with respect to the splittings $C = C_\infty \oplus C_0$:

$$d = \begin{pmatrix} d_\infty & \rho \\ 0 & d_0 \end{pmatrix} \quad \text{and} \quad \tilde{d} = \begin{pmatrix} \tilde{d}_\infty & \tilde{\rho} \\ 0 & \tilde{d}_0 \end{pmatrix}.$$

Furthermore, we decompose $\tilde{d}_0$ as $\tilde{d}_0 = \tilde{d}_0' \oplus \tilde{d}_0''$ with respect to the decomposition

$$C_0(X;L_0,\tilde{L}_1) = C_0(Y \times \mathbb{R};\Lambda_0 \times \mathbb{R}, L_1^a) \oplus C_0(X;L_0,L_1),$$

i.e., d_0' maps into the first summand and d_0'' into the second.

Consider first an intersection point $x \in L_0 \cap L_1$. In this case, we must show that

$$\phi_\infty(\rho x) + \phi_0(\rho x) + d_0 x = \tilde{\rho} x + \tilde{d}_0' x + \tilde{d}_0'' x.$$

Note first that holomorphic disks that contribute to $d_0 x$ also contribute to $\tilde{d}_0'' x$, and hence the last terms on the left- and right-hand sides cancel.

A moduli space contributing to $\tilde{\rho} x$ or $\tilde{d}_0' x$ is of the form $\mathcal{M}(g;x)$, where g is respectively a Reeb chord connecting Λ_0 to Λ_1^a or an intersection point in $(\Lambda_0 \times \mathbb{R}) \cap L_1^a$. Consider stretching along the hypersurface $(Y, \Lambda_0 \cup \Lambda_1)$ where the cobordisms are joined. It is a consequence of [2] that families of disks in $\mathcal{M}(g;x)$ converge to broken disks with one part in $(X, L_0 \cup L_1)$, one in $(Y \times \mathbb{R}; \Lambda_0 \times \mathbb{R} \cup L_1^a)$, and possibly other levels in the symplectizations, in the limit. Since $\dim(\mathcal{M}(g;x)) = 0$, every level in the limit must have dimension 0 by transversality. By admissibility of the disks in $\mathcal{M}(g;x)$ there is exactly one Reeb chord c' connecting Λ_0 to Λ_1 in the hypersurface Y in the broken disk that arises in the limit. By definition, the sum of the two first terms on the left-hand side counts broken disks of this type, and the chain map equation follows in this case.

Consider second a Reeb chord c connecting Λ_0 to Λ_1. In this case, we must show that

$$\phi_\infty(d_\infty c) + \phi_0(d_\infty c) = \tilde{d}_\infty(\phi_\infty c) + \tilde{\rho}(\phi_0 c) + \tilde{d}_0''(\phi_0 c).$$

To show that the chain map equation holds in this case, we first consider 1-dimensional moduli spaces $\mathcal{M}(z;\overline{b}c\overline{e})$ of disks in $Y \times \mathbb{R}$ with boundary on $(\Lambda_0 \times \mathbb{R}) \cup L_1^a$. Here z is an intersection point in $(\Lambda_0 \times \mathbb{R}) \cap L_1^a$, $\overline{b} = b_1, \dots, b_k$, and $\overline{e} = e_1, \dots, e_r$ are words of Reeb chords connecting Λ_1 (i.e., the negative end of L_1^a) to itself and connecting Λ_0 to itself, respectively, and such that $|b_j| = 0$ for all j, and $|e_l| = 0$ for all l. By transversality and admissibility, the boundary points of such a moduli space consists of a broken disks with two components that are either both 0-dimensional and joined at an intersection point, or a 1-dimensional disk in the negative end joined at a Reeb chord to a 0-dimensional disk. The former broken disks contribute to $d_0''(\phi_0 c)$ and the latter to $\phi_0(d_\infty c)$. Thus these two terms cancel.

Second, we consider 1-dimensional moduli spaces $\mathcal{M}(a;\overline{b}c\overline{e})$ of disks in $Y \times \mathbb{R}$ with boundary on $(\Lambda_0 \times \mathbb{R}) \cup L_1^a$. Here a is a Reeb chord connecting Λ_0 to Λ_1^a at the positive end of $Y \times \mathbb{R}$, with $\overline{b} = b_1, \dots, b_k$ and $\overline{e} = e_1, \dots, e_r$ as above. By transversality and admissibility, the boundary of such a moduli space consists of two level broken disks of the following form.

- Broken disks with two 0-dimensional components joined at an intersection point. Such disks contribute to $\tilde{\rho}(\phi_0 c)$.
- Broken disks with one 1-dimensional component in the positive end joined at Reeb chords to 0-dimensional disks. Such disks contribute to $\tilde{d}_\infty(\phi_\infty c)$.
- Broken disks with one 1-dimensional component in the negative end joined at a mixed Reeb chord to a 0-dimensional disk. Such disks contribute to $\phi_\infty(d_\infty c)$.
- Broken disks with one 1-dimensional component in the negative end joined at a pure Reeb chord to a 0-dimensional disk. Contributions from such disks cancel, since $\theta \circ \partial = 0$ and $\epsilon \circ \partial = 0$.

It follows that the first term on the left-hand side cancels with the sum of the two first terms on the right-hand side. The lemma follows.

4.2.3 Joining Maps and Deformations

We next consider generic 1-parameter families of cobordisms as considered in Sect. 4.2.2. More precisely, let (X, L_0) and (X, L_1) be exact cobordisms as usual. Let $(Y \times \mathbb{R}, L_1^a(t))$, $t \in [a,b]$, be a 1-parameter family of exact cobordisms that is constant outside a compact set and such that the $(-\infty)$-boundary of $L_1^a(t)$ equals Λ_1 and its $(+\infty)$-boundary equals Λ_1^a. Adjoining $(Y \times \mathbb{R}, L_1^a(t))$ to (X, L_1), we obtain a 1-parameter family of cobordisms $(X, \tilde{L}_1(t))$, and for generic t in $[a,b]$, where moduli spaces are transversely cut out, corresponding chain maps

$$\Phi_t \colon C(X; L_0, L_1) \to C(X; L_0, \tilde{L}_1(t)).$$

Furthermore, since $(X, \tilde{L}_1(t))$ and $(X, \tilde{L}_1(t'))$ are related by a compact deformation, Lemmas 4.3–4.5 provide chain maps

$$\Psi_{tt'} \colon C(X; L_0, L_1(t)) \to C(X; L_0, \tilde{L}_1(t')),$$

where t and t' are generic, that induce isomorphisms on homology. In fact, unless the interval between t and t' contains (-1)-disk instances or birth/death instances, the map $\Psi_{tt'}$ is the canonical identification map on generators. Here the births/deaths take place in the added cobordism, and the (-1)-disk instances correspond to (-1)-disk instances in the added cobordism. As we shall see below, if the interval between t and t' contains a birth/death or a (-1)-disk instance, then there is a chain homotopy connecting the chain maps $\Psi_{tt'} \circ \Phi_t$ and $\Phi_{t'}$. The proofs of these results are similar to the proofs of results in Sects. 4.2.1 and 4.2.2, and many details from there will not be repeated. For convenient notation below we take $t = -1, t' = 1$ and assume that the critical instance is at $t = 0$. Furthermore, we write $C = C(X, L_0, L_1)$ with differential d, $\tilde{C}_\pm = C(X, L_0, \tilde{L}_1(\pm 1))$ with differential $\tilde{d}_\pm$, $\Phi_\pm = \Phi_{\pm 1}$, and $\Psi = \Psi_{-11}$.

Consider first the case of a pure (-1)-disk in $(Y \times \mathbb{R}, L_1^a(t))$. Applying our perturbation scheme that time orders the negative punctures of disks in the positive end $(Y \times \mathbb{R}, \Lambda_1^a \times \mathbb{R})$, we see that such a disk gives rise to several (-1)-disks in

$(Y \times \mathbb{R}, L_1^a(t))$. The (-1)-disks in the total cobordism $(X, \tilde{L}_1(t))$ are then these (-1)-disks in $(Y \times \mathbb{R}, L_1^a(t))$, capped off with 0-dimensional rigid disks in (X, L_1); see [6, Lemmas 4.3 and 4.4].

Lemma 4.7. *If there is a pure (-1)-disk at $t = 0$, then the following diagram commutes:*

$$\begin{array}{ccc} C & \xrightarrow{\Phi_-} & \tilde{C}_- \\ {\scriptstyle \mathrm{id}}\downarrow & & \downarrow{\scriptstyle \Psi} \\ C & \xrightarrow[\Phi_+]{} & \tilde{C}_+ \end{array}.$$

Proof. Consider the parameterized 1-dimensional moduli space corresponding to a 0-dimensional moduli space contributing to Φ_+. A boundary component at $t = -1$ contributes to $\Psi \circ \Phi_-$. A boundary component in the interior of $[-1, 1]$ is a broken disk consisting of a 1-dimensional disk in an end and a (-1)-disk and 0-disks in the cobordisms. If the 1-dimensional disk lies in the upper end, then the broken disk contributes to $\Psi \circ \Phi_-$, and as usual, the total contribution of disks with a 1-dimensional disk in the lower end is 0, since augmentations are chain maps. The result follows.

Second, consider the case of a mixed (-1)-disk. By admissibility, any disk contributing to the chain maps then contains at most one such (-1)-disk; see [6, Lemma 2.8]. Define the map $K \colon C \to \tilde{C}_+$ as follows:

$$K(x) = 0$$

if x is an intersection point generator, and

$$K(c) = \sum_{\dim(\mathcal{M}_I(g;\overline{b}c\overline{e}))=0} |\mathcal{M}_I(g;\overline{b}c\overline{e})| \epsilon(\overline{b}) \theta(\overline{e}) g$$

if c is a Reeb chord generator, where $\mathcal{M}_I(g;\overline{b}c\overline{e})$ is the parameterized moduli space of disks in $Y \times \mathbb{R}$ with boundary on $\Lambda_0 \times \mathbb{R}$ and $L_1^a(t)$, and where $\overline{b}$ and $\overline{e}$ are Reeb chords of Λ_0 and Λ_1, respectively.

Lemma 4.8. *If there is a mixed (-1)-disk at $t = 0$, then the chain maps in the diagram*

$$\begin{array}{ccc} C & \xrightarrow{\Phi_-} & \tilde{C}_- \\ {\scriptstyle \mathrm{id}}\downarrow & & \downarrow{\scriptstyle \Psi} \\ C & \xrightarrow[\Phi_+]{} & \tilde{C}_+ \end{array}$$

satisfy $\Psi \circ \Phi_- + \Phi_+ = K \circ d + \tilde{d}_+ \circ K$.

Proof. Consider first the left-hand side acting on an intersection point $x \in L_0 \cap L_1$. Both maps Φ_- and Φ_+ are then inclusions, and by definition, $\Psi(x)$ counts (-1)-disks in $\mathcal{M}(g;x)$, where g is a generator. Since there are no (-1)-disks in the lower cobordism, we find that in a moduli space that contributes to $\Psi(x)$, the generator g is either an intersection point or a Reeb chord in the upper cobordism. Consider now the splittings of a (-1)-disk as we stretch over the joining hypersurface: it splits into a (-1)-disk in the upper cobordism and 0-dimensional disks in the lower. By definition, the count of such split disks is $K(d(x))$. Since $K(x) = 0$, the chain map equation follows.

Consider next the left-hand side acting on a Reeb chord generator c. Here $\Phi_\pm(c)$ are given by counts of 0-dimensional disks in the upper cobordism, and Ψ is a count of (-1)-disks emanating at double points. In particular, $\Psi(a) = a$ for any Reeb chord. Consider now a moduli space that contributes to Φ_+. The boundary of the corresponding parameterized moduli space consists of rigid disks over endpoints as well as broken disks with a 1-dimensional disk in either symplectization end and a (-1)-disk in the cobordism. The total count of such disks gives the desired equation after one observes that for the usual reason, splittings at pure chords do not contribute.

Third, consider a birth/death instance.

Lemma 4.9. *If there is a birth/death instance at $t = 0$, then the following diagram commutes:*

$$\begin{array}{ccc} C & \xrightarrow{\Phi_-} & \tilde{C}_- \\ {\scriptstyle \mathrm{id}}\uparrow & & \uparrow{\scriptstyle \Psi}. \\ C & \xrightarrow[\Phi_+]{} & \tilde{C}_+ \end{array}$$

Proof. Assume that the canceling pair of double points is (x,y) with $\tilde{d}_+ x = y + v$ as in Lemma 4.5. As was the case there, disks from x to v can on the one hand be glued to rigid disks ending at y, resulting in rigid disks, and on the other, can be glued to rigid disks ending at x, resulting in nonrigid disks. Commutativity then follows from a straightforward calculation.

4.2.4 Invariance

Let (X, L_0) and (X, L_1) be exact cobordisms as above.

Theorem 4.10. *The Lagrangian Floer homology $FH^*(X; L_0, L_1)$ is invariant under exact deformations of L_1.*

Proof. The proof is similar to the proof of Theorem 2.1: Any deformation considered can be subdivided into a compactly supported deformation and a Legendrian isotopy at infinity. The former type induces isomorphisms on homology by

Lemmas 4.3–4.5. The latter type of deformation gives rise to an invertible exact cobordism, which in turn gives a chain map on homology by Lemma 4.6. Lemmas 4.7–4.9 show that on the homology level these maps are independent of compact deformations of the cobordism added. A word-for-word repetition of the proof of Theorem 2.1 then finishes the proof.

4.3 Holomorphic Disks for $FH^*(X;L,L')$

Let (X,L) be an exact cobordism as considered above. Let L' denote a copy of L. In order to make L and L' transverse, we identify a neighborhood of $L \subset X$ with the cotangent bundle T^*L. Pick a Morse function $\tilde{F}\colon L \to \mathbb{R}$ such that

$$\tilde{F}(y,t) = C + t,$$

where C is a constant, for $(y,t) \in \Lambda \times [T,\infty)$ for some $T > 0$. Cut the function $\tilde{F}$ off outside a small neighborhood of L and let L'' be the image of L under the time 1-flow of the Hamiltonian vector field of $\epsilon\tilde{F}$ for small ϵ. Then L intersects L'' transversely. However, in the end where $\tilde{F} = C + t$, the Hamiltonian just shifts Λ along the Reeb flow, so that there is a Reeb chord from Λ to Λ'' at every point of Λ. In order to perturb our way out from this Morse–Bott situation, we identify a neighborhood of $\Lambda \subset Y$ with $J^1(\Lambda)$, fix a Morse function $f\colon \Lambda \to \mathbb{R}$, and let Λ' denote the graph of the 1-jet extension of f. Then Λ and Λ' are contact isotopic via the contact isotopy generated by the time-dependent contact Hamiltonian $H_t(q,p,z) = \psi(t)f(q)$, where $(q,p,z) \in T^*\Lambda \times \mathbb{R}$ and $\psi(t)$ is a cut-off function. Take f and $\frac{d\psi}{dt}$ very small and adjoin the cobordism $Y \times \mathbb{R}$ with the symplectic form $de^t(\lambda - H_t)$ to (X,L). This gives the desired L' with Λ' as $(+\infty)$-boundary.

We will state a conjectural lemma that gives a description of holomorphic disks with boundary on $L \cup L'$. To this end, we first describe a version of Morse theory on L and then discuss intersection points in $L \cap L'$ and Reeb chords of $\Lambda \cup \Lambda'$.

Consider the gradient equation $\dot{x} = \nabla F(x)$ of the function $F\colon L \to \mathbb{R}$, where

$$F(x) = \tilde{F}(x) + \psi(t)f(y),$$

where we write $x = (y,t) \in \Lambda \times [T,\infty)$. It is easy to see that if a solution of $\dot{x} = \nabla F(x)$ leaves every compact, then it is exponentially asymptotic to a *critical-point solution* of the form $s \mapsto (y_0, s)$, where $\nabla f(y_0) = 0$ in $\Lambda \times [0,\infty)$. Furthermore, every sequence of solutions of $\dot{x} = \nabla F(x)$ has a subsequence that converges to a several-level solution with one level in L and levels in $\Lambda \times \mathbb{R}$ that are solutions to the gradient equation of $f + t$, asymptotic to critical-point solutions. We call solutions that have formal dimension 0 (after dividing out reparameterization) and that are transversely cut out *rigid flow lines*. Consider a moduli space $\mathcal{M}(c)$ of holomorphic disks in X

with boundary on L or a moduli space $\mathcal{M}(c;\overline{b})$ in $Y \times \mathbb{R}$ with boundary on $\Lambda \times \mathbb{R}$. Marking a point on the boundary, there is an evaluation map $\mathrm{ev}\colon \mathcal{M}^*(c) \to L$ or $\mathrm{ev}\colon \mathcal{M}^*(c;\overline{b}) \to \Lambda \times \mathbb{R}$; see [7, Sect. 6]. Consider now a flow line of F or of $f+t$ that hits the image of ev. We call such a configuration a *generalized disk*, and we say that it is rigid if it has formal dimension 0 (after dividing out the $\mathbb{R}$-translation in $Y \times \mathbb{R}$) and if it is transversely cut out.

Note that points in $L \cap L'$ are in 1-to-1 correspondence with critical points of F. We use the terms "intersection point" and "critical point" of F interchangeably. Note also that the Reeb chords connecting Λ to Λ' are of two kinds: *short chords*, which correspond to critical points of f, and *long chords* close to each Reeb chord of Λ. As with intersection points, we will sometimes identify critical points of f with their corresponding short Reeb chords. Also note that to each Reeb chord of Λ there is a unique Reeb chord of Λ'.

The following conjectural lemma is an analogue of [7, Theorem 3.6].

Lemma 4.11 (Conjectural). *Let $\overline{b}'$ denote a word of Reeb chords of Λ' and let $\overline{b}$ denote the corresponding word of Reeb chords of Λ. Let also $\overline{e}$ denote a word of Reeb chords of Λ. If c is a long Reeb chord connecting Λ to Λ', then let $\hat{c}$ denote the corresponding Reeb chord of Λ.*

For sufficiently small shift L' of L, there are the following 1-to-1 correspondences:

1. *If a and c are long Reeb chords, then rigid disks in $\mathcal{M}(a;\overline{b}'c\overline{e})$ correspond to rigid disks in $\mathcal{M}(\hat{a};\overline{b}\hat{c}\,\overline{e})$.*
2. *If a is a long Reeb chord and c is a short Reeb chord, then rigid disks in $\mathcal{M}(a;\overline{b}'c\overline{e})$ correspond to rigid generalized disks with disk component in $\mathcal{M}(\hat{a};\overline{b}\,\overline{e})$ and with flow line asymptotic to the critical-point solution of c at $-\infty$ and ending at a boundary point between the last $\overline{b}$-chord and the first $\overline{e}$-chord.*
3. *If a and c are short Reeb chords, then rigid disks in $\mathcal{M}(a;c)$ correspond to rigid flow lines asymptotic to the critical-point solutions of c and of a at $-\infty$ and $+\infty$, respectively.*
4. *If x is an intersection point and a is a long Reeb chord, then rigid disks in $\mathcal{M}(a;x)$ correspond to rigid generalized disks with disk component in $\mathcal{M}(\hat{a})$ and with gradient line starting at x and ending at the boundary of the disk.*
5. *If x is an intersection point and c is a short Reeb chord, then rigid disks in $\mathcal{M}(c;x)$ correspond to rigid flow lines starting at x and asymptotic to the critical-point solution of c at $+\infty$.*

Here items (1) to (3) follow from [7, Theorem 3.6] in combination with [8, Sect. 2.7] in the special case $Y = P \times \mathbb{R}$, where P is an exact symplectic manifold. Proofs of (4) and (5) would require an analysis analogous to that of [7, Sect. 6] carried out for a symplectization, taking into account the interpolation region used in the construction of L'.

4.4 Outline of Proof of Conjecture 1.2

Choose a grading on $C(X;L,L')$ such that if c is a long Reeb chord connecting Λ to Λ', then the degree $|c|$ satisfies $|c| = |\hat{c}|$, where $\hat{c}$ is the corresponding Reeb chord of Λ; see (3). Let $C = C_{[\alpha]}(X;L,L')$ and consider the decomposition

$$C = C_+ \oplus C_0,$$

where C_+ is generated by long Reeb chords and where C_0 is generated by short Reeb chords and double points. Then C_+ is a subcomplex, and we have the exact sequence

$$0 \longrightarrow C_+ \longrightarrow C \longrightarrow \widehat{C} \longrightarrow 0,$$

where $\widehat{C} = C/C_+$. If L and L' are sufficiently close, then the augmentations ϵ and θ agree, and Lemma 4.11(1) implies that the differential on C_+ is identical to that on $\mathbf{V}_{[\alpha]}(X,L)$. Furthermore, Lemma 4.11(3),(4) implies that the differential on $\widehat{C}$ is that of the Morse complex of L, and the existence of the exact sequence follows.

Consider next the isomorphism statement. Since $\mathbb{C}^n$ and $J^1(\mathbb{R}^{n-1}) \times \mathbb{R}$ satisfy monotonicity conditions, the homology of C in a fixed degree can be computed using a fixed sufficiently large energy level. Furthermore, in $\mathbb{C}^n$ or $J^1(\mathbb{R}^{n-1})$, L is displaceable, i.e., L' can be moved by Hamiltonian isotopy in such a way that $L \cap L' = \emptyset$ and so that there are no Reeb chords connecting Λ and Λ'. Hence by the invariance of Lagrangian Floer cohomology proved in Theorem 4.10, the total complex C is acyclic. The theorem follows. □

4.5 Outline of Proof of Corollary 1.3

In order to discuss Corollary 1.3, we first describe the duality exact sequence (2) in more detail.

4.5.1 Properties of the Duality Exact Sequence

We recall how the exact sequence (2) was constructed. Let Λ'' be a copy of Λ shifted a large distance (compared to the length of any Reeb chord of Λ) away from Λ in the Reeb direction and then perturbed slightly by a Morse function f. The part of the linearized contact homology complex of $\Lambda \cup \Lambda''$ generated by mixed Reeb chords was split as a direct sum $Q \oplus C \oplus P$. Here Q is generated by the mixed Reeb chords near Reeb chords of Λ that connect the lower sheet of Λ to the upper sheet of Λ'', C is generated by the Reeb chords that correspond to critical points of f, and P is generated by the Reeb chords near Reeb chords of Λ that connect the upper sheet of Λ to the lower sheet of Λ''. The linearized contact homology differential ∂ on $Q \oplus C \oplus P$ has the form

$$\partial = \begin{pmatrix} \partial_q & 0 & 0 \\ \rho & \partial_c & 0 \\ \eta & \sigma & \partial_p \end{pmatrix}.$$

Using the analogue of Lemma 4.11, the subcomplex (P, ∂_p) can be shown to be isomorphic to the dual complex of the complex (Q, ∂_q) using the natural pairing that pairs Reeb chords in Q and P that are close to the same Reeb chord of Λ. Furthermore, the complex (Q, ∂_q) is canonically isomorphic to the linearized contact homology complex $(Q(\Lambda), \partial_1)$.

The next step is the observation that since Λ is displaceable (in the sense above), the complex $C \oplus Q \oplus P$ is acyclic. Using the coarser decomposition $(Q \oplus C) \oplus P$ and writing

$$\partial = \begin{pmatrix} \partial_{qc} & 0 \\ H & \partial_p \end{pmatrix},$$

the chain map induced by

$$H = \begin{pmatrix} \eta & \sigma \end{pmatrix}$$

induces an isomorphism on homology between $Q \oplus C$ and P. Here the map $\sigma \colon C \to P$ counts generalized trees, whereas the map $\eta \colon Q \to P$ counts disks with boundary on Λ and with two positive punctures disks. (To make sense of the latter count and have transversely cut-out moduli spaces, actual disks counted have boundary on Λ and on a nearby copy Λ'.) The exact sequence (2) is then constructed from the long exact sequence of the short exact sequence for the complex $C \oplus Q$.

Below we will also make use of the other splitting of the acyclic complex $Q \oplus C \oplus P$ as $Q \oplus (C \oplus P)$ with differential

$$\partial = \begin{pmatrix} \partial_q & 0 \\ H' & \partial_{cp} \end{pmatrix},$$

where

$$H' = \begin{pmatrix} \rho \\ \eta \end{pmatrix}$$

induces an isomorphism on homology: our proof of Corollary 1.3 relates the maps H and H' to isomorphisms coming from Lagrangian Floer homology.

Let C_* denote the Morse complex for $f\colon \Lambda \to \mathbb{R}$ and let $\bar{C}^*$ denote the Morse complex for $-f$. Then $C_* = \bar{C}^{n-1-*}$, and we have the following diagram of chain maps:

$$\begin{array}{ccccccc}
\longrightarrow & \bar{C}^{n-k-2} & \longrightarrow & P^{n-k-1} & \longrightarrow & P^{n-k-1} \oplus \bar{C}^{n-k-1} \\
 & \Big\| & & H\Big\uparrow & & \Big\uparrow H' \\
\longrightarrow & C_{k+1} & \longrightarrow & Q_{k+1} \oplus C_{k+1} & \longrightarrow & Q_{k+1} \\
\longrightarrow & \bar{C}^{n-k-1} & & & & \\
 & \Big\| & & \cdots & & \\
\longrightarrow & C_k & & & &
\end{array} \tag{13}$$

Lemma 4.12. *The diagram* (13) *commutes after passing to homology.*

Proof. The first and last squares commute already on the chain level by definition of the differential. Commutativity of the middle square can be seen as follows. Starting at the lower left corner with an element from Q_{k+1}, it is clear that the results of going up then right, and right then up have common component in the P^{n-k-1}-summand. Using the commutativity already established, we see that starting with an element in C_{k+1} and going up, then right is the same thing as first pulling that element back to the left and then going up and two steps to the right. This vanishes in homology by exactness and hence gives the same result as going right then up. Finally, going right then up and projecting to the $\bar{C}^{n-k-1}$-component is the same as going right twice and then up. This vanishes in homology by exactness and gives the same result as going the other way.

4.5.2 Proof of Corollary 1.3

We write down the diagram on the chain level. Let (C_*, ∂_C) denote the Morse complex of the function $-f\colon \Lambda \to \mathbb{R}$ and let (I_*, ∂_I) denote the Morse complex of the function $-F\colon L \to \mathbb{R}$ that computes the relative homology of $(\bar{L}, \partial\bar{L})$. As above, we write $\overline{C}^*$ and $\overline{I}^*$ for the corresponding cochain complexes of f and F, respectively.

Then the complex that computes the homology for L is $(I \oplus C, \partial_{IC})$, where

$$\partial_{IC} = \begin{pmatrix} \partial_I & 0 \\ \mu & \partial_C \end{pmatrix},$$

where μ counts flow lines of F that start at a critical point of F and are asymptotic to a critical-point solution at $+\infty$. We then have the following diagram:

$$
\begin{array}{ccccccc}
C_{k+1} & \longrightarrow & I_{k+1} \oplus C_{k+1} & \longrightarrow & I_{k+1} & \longrightarrow & C_k \\
\| & & \| & & \| & & \| \\
\bar{C}^{n-k-1} & \longrightarrow & \bar{I}^{n-k-1} \oplus \bar{C}^{n-k-1} & \longrightarrow & \bar{I}_{n-k-1} & \longrightarrow & \bar{C}_{n-k} \\
\| & & \delta_{L,L'} \downarrow & & \downarrow \delta'_{L,L'} & & \| \\
\bar{C}^{n-k-1} & \longrightarrow & P^{n-k} & \longrightarrow & P^{n-k} \oplus \bar{C}^{n-k} & \longrightarrow & \bar{C}^{n-k},
\end{array}
\tag{14}
$$

where $\delta'_{L,L'}$ is the chain map inducing an isomorphism on homology from the splitting $C(X;L,L') = (P \oplus C) \oplus I$ instead of the splitting $P \oplus (C \oplus I)$ used in the proof sketch of Conjecture 1.2. Note that the bottom row of (14) is the same as the top row of (13). Joining the diagrams along this row, we see that Corollary 1.3 follows once we show that (14) commutes on the homology level. For the left and right squares, this is true already on the chain level by definition of the differential. The fact that the middle square commutes for element in $\bar{I}^{n-k-1}$ after projection to P^{n-k-1} is immediate from the definition. The same argument, using commutativity of exterior squares, as in the proof of Lemma 4.12 then gives commutativity on the homology level. □

Acknowledgement The author acknowledges support from the Göran Gustafsson Foundation for Research in Natural Sciences and Medicine.

References

1. M. Abouzaid, P. Seidel, *An open string analogue of Viterbo functoriality*, Geom. Topol. **14** (2010), no. 2, 627–718.
2. F. Bourgeois, Y. Eliashberg, H. Hofer, K. Wysocki, E. Zehnder, *Compactness results in symplectic field theory*, Geom. Topol. **7** (2003), 799–888.
3. F. Bourgeois, A. Oancea, *An exact sequence for contact- and symplectic homology*, Invent. Math. **175** (2009), 611–680.
4. G. Civan, P. Koprowski, J. Etnyre, J. Sabloff, A. Walker, *Product structures for Legendrian contact homology*, Math. Proc. Cambridge Philos. Soc. **150** (2011), no. 2, 291–311.
5. Y. Chekanov, *Differential algebra of Legendrian links*, Invent. Math. **150** (2002), 441–483.
6. T. Ekholm, *Rational symplectic field theory over $\mathbb{Z}_2$ for exact Lagrangian cobordisms* J. Eur. Math. Soc. (JEMS) **10** (2008), no. 3, 641–704.
7. T. Ekholm, J. Etnyre, J. Sabloff, *A duality exact sequence for Legendrian contact homology*, preprint, Duke Math. J. **150** (2009), no. 1, 1–75.
8. T. Ekholm, J. Etnyre, M. Sullivan, *Non-isotopic Legendrian submanifolds in $\mathbb{R}^{2n+1}$*, J. Differential Geom. **71** (2005), no. 1, 85–128.
9. T. Ekholm, J. Etnyre, M. Sullivan, *The contact homology of Legendrian submanifolds in $\mathbb{R}^{2n+1}$*, J. Differential Geom. **71** (2005), no. 2, 177–305.
10. T. Ekholm, J. Etnyre, M. Sullivan, *Orientations in Legendrian contact homology and exact Lagrangian immersions*, Internat. J. Math. **16** (2005), no. 5, 453–532.
11. T. Ekholm, J. Etnyre, M. Sullivan, *Legendrian contact homology in $P \times \mathbb{R}$*, Trans. Amer. Math. Soc. **359** (2007), no. 7, 3301–3335.

12. Y. Eliashberg, A. Givental, H. Hofer, *Introduction to symplectic field theory*, GAFA 2000 (Tel Aviv, 1999). Geom. Funct. Anal. 2000, Special Volume, Part II, 560–673.
13. A. Floer *Morse theory for Lagrangian intersections*, J. Differential Geom. **28** (1988), no. 3, 513–547.
14. K. Fukaya, P. Seidel and I. Smith, *The symplectic geometry of cotangent bundles from a categorical viewpoint*, in Homological Mirror Symmetry, Springer Lecture Notes in Physics 757, (2008) (Kapustin, Kreuzer and Schlesinger, eds).
15. P. Seidel, private communication.

Topology of Spaces of S-Immersions

Yakov Eliashberg and Nikolai Mishachev

To Oleg Viro of his 60th birthday

Abstract We use the wrinkling theorem proven in [EM97] to fully describe the homotopy type of the space of S-immersions, i.e., equidimensional folded maps with prescribed folds.

Keywords S-immersion • h-principle • Wrinkling • Zigzag • Folded mapping

1 Equidimensional Folded Maps

Let V and W be two manifolds of dimension q. The manifold V will always be assumed *closed* and *connected*. A map $f : V \to W$ is called *folded* if it has only fold-type singularities. We will discuss in this paper an h-principle for folded maps $f : V \to W$ with a *prescribed fold* $\Sigma^{10}(f) \subset V$. Folded maps with the fold $S = \Sigma^{1,0}(f)$ are also called *S-immersions*; see [El70]. Throughout this paper we assume that $S \neq \varnothing$. We provide in this paper an essentially complete description of the homotopy type of the spaces of S-immersions. More precisely, we prove that in most cases, one has a result of h-principle type, while sometimes the space of S-immersions may have anumber of additional components of a different nature. The topology of these components is also fully described.

Y. Eliashberg (✉)
Department of Mathematics, Stanford University, 450 Serra Mall, Bldg. 380,
Stanford, CA 94305-2125, USA
e-mail: eliash@math.stanford.edu

N. Mishachev
Lipetsk Technical University, Moskovskaja ul. 30, Lipetsk 398055, Russia
e-mail: mishachev@lipetsk.ru

I. Itenberg et al. (eds.), *Perspectives in Analysis, Geometry, and Topology: On the Occasion of the 60th Birthday of Oleg Viro*, Progress in Mathematics 296,
DOI 10.1007/978-0-8176-8277-4_7,

Remarks. 1. The results proven in this paper were formulated in [El70, El72], but the proof of the injectivity part of the h-principle claim has never been published previously.
2. The approach described in this paper to the problem of construction of mappings with prescribed singularities can be generalized to the case of maps $V^n \to W^q$ with $n > q$. However, in contrast to the case $n = q$, the results that can be proven using the current techniques are essentially equivalent to the results of [El72].
3. The subject of this paper is the geometry of singularities in the *source* manifold. The geometry of singularities in the *image* is much more subtle; see [Gr07].

1.1 S-Immersions

We refer the reader to Sect. 4 below for the definition and basic properties of folds and wrinkles, and for the formulation of the wrinkling theorem from [EM97].

Let $f : V \to W$ be an S-immersion, i.e., a map with only a fold-type singularity $S = \Sigma^{1,0}(f)$. The fold $S \subset V$ has a neighborhood U that admits an involution $\alpha_{\mathrm{loc}} : U \to U$ such that $f \circ \alpha_{\mathrm{loc}} = f$. In particular, if S divides V into two submanifolds $V_\pm$ with common boundary $\partial V_\pm = S$, then an S-immersion is just a pair of immersions $f_\pm : V_\pm \to W$ such that $f_+ = f_- \circ \alpha_{\mathrm{loc}}$ near S. Denote by $T_S V$ an n-dimensional tangent bundle over V that is obtained from TV by regluing TV along S with $\mathrm{d}\alpha_{\mathrm{loc}}$. For example, if $V = S^q$ and S is the equator $S^{q-1} \subset S^q$, then $T_S V = S^q \times \mathbb{R}^q$. We will call $T_S V$ the *tangent bundle of V folded along S*. The differential $\mathrm{d}f : TV \to TW$ of any S-immersion $f : V \to W$ has a canonical (bijective) regularization $d_S f : T_S V \to TW$, the *folded differential* of f.

Let us denote by $\mathfrak{M}(V,W,S)$ the space of S-immersions $V \to W$. We also consider the space $\mathfrak{m}(V,W,S)$, a formal analogue of $\mathfrak{M}(V,W,S)$, which consists of bijective homomorphisms $T_S V \to TW$. The folded differential induces a natural inclusion $d : \mathfrak{M}(V,W,S) \to \mathfrak{m}(V,W,S)$. Our goal is to study the homotopic properties of the map d. We prove that in most cases, the map d is a (weak) homotopy equivalence. However, there are some exceptional cases that arise when the map d is a homotopy equivalence on some of the components of $\mathfrak{M}(V,W,S)$, while the structure of the remaining components can also be completely understood.

1.2 Taut–Soft Dichotomy for S-Immersions

A map $f \in \mathfrak{M}(V,W,S)$ is called *taut* if there exists an involution $\alpha : V \to V$ such that $\mathrm{Fix}\,\alpha = S$ and $f \circ \alpha = f$. We denote by $\mathfrak{M}_{\mathrm{taut}}(V,W,S)$ the subspace of $\mathfrak{M}(V,W,S)$ that consists of taut maps. Non-taut maps are called *soft*, and we define $\mathfrak{M}_{\mathrm{soft}}(V,W,S) := \mathfrak{M}(V,W,S) \setminus \mathfrak{M}_{\mathrm{taut}}(V,W,S)$. A map $f \in \mathfrak{M}_{\mathrm{taut}}(V,W,S)$ uniquely determines the corresponding involution α, and thus we have:

1.2.1. (Topological Structure of the Space of Taut S-Immersions) *The space* $\mathfrak{M}_{\text{taut}}(V,W,S)$ *is the space of pairs* (α,h), *where* $\alpha : V \to V$ *is an involution with* $\operatorname{Fix}\alpha = S$ *and* h *is an immersion of the quotient manifold* V/α *with boundary* S *to* W.

Of course, for most pairs (V,S), the space $\mathfrak{I}(V,S)$ of such involutions is empty, and hence in such cases the space $\mathfrak{M}_{\text{taut}}(V,W,S)$ is empty as well.

The topology of thc space $\mathfrak{M}_{\text{taut}}(V,W,S)$ is especially simple if S divides V into two submanifolds $V_\pm$ with common boundary $\partial V_\pm = S$, e.g., when both manifolds V and S are orientable. Clearly, we have the following result:

1.2.2. (Topological Structure of the Space of Taut S-Immersions: The Orientable Case) *If* S *divides* V *into two submanifolds* $V_\pm$ *with common boundary* $\partial V_\pm = S$ *such that there exists a diffeomorphism* $V_+ \to V_-$ *fixed along the boundary, then the space* $\mathfrak{M}_{\text{taut}}(V,W,S)$ *is homeomorphic to the product*

$$\operatorname{Diff}_S(V_+) \times \operatorname{Imm}(V_+,W),$$

where $\operatorname{Diff}_S(V_+)$ *is the group of diffeomorphisms* $V_+ \to V_+$ *fixed at the boundary together with their* ∞*-jet, and* $\operatorname{Imm}(V_+,W)$ *is the space of immersions* $V_+ \to W$.

Note that according to Hirsch's theorem [Hi], the space $\operatorname{Imm}(V_+,W)$ is homotopy equivalent to the space $Iso(V,W)$ of fiberwise isomorphic bundle maps $TV_+ \to TW$. For instance, when $V = S^q$, $W = \mathbb{R}^q$, and S is the equator $S^{q-1} \subset S^q$, then we get

$$\mathfrak{M}_{\text{taut}}(S^q,\mathbb{R}^q,S^{q-1}) \overset{\text{h.e.}}{\simeq} \operatorname{Diff}_{\partial D^q} D^q \times O(q).$$

In particular, when $q = 2$, the space of taut S^1-immersions $S^2 \to \mathbb{R}^2$ that preserve orientation on S^2_+ is homotopy equivalent to S^1.

1.2.3. (Subspaces of Taut and Soft S-Immersions are Open–Closed) *Let* $S \subset V$ *be a closed* $(q-1)$*-dimensional submanifold of* V. *Then the subspaces* $\mathfrak{M}_{\text{taut}}(V,W,S) \subset \mathfrak{M}(V,W,S)$ *and* $\mathfrak{M}_{\text{soft}}(V,W,S)$ *are open and closed, i.e., they consist of whole connected components of* $\mathfrak{M}(V,W,S)$.

Proof. Given any $f \in \mathfrak{M}(V,W,S)$ there exists $\varepsilon = \varepsilon(f) > 0$ such that the local involution $\alpha^f_{\text{loc}} : \mathcal{O}p\,S \to \mathcal{O}p\,S$ with $f \circ \alpha^f_{\text{loc}} = f$ is defined on an ε-tubular neighborhood $U_\varepsilon \supset S$. If f is taut, then α^f_{loc} extends as a global involution $\alpha^f : V \to V$ such that $f \circ \alpha^f = f$. Hence, for any $x \in V \setminus U_{\varepsilon/2}$, the distance $d(x,\alpha^f(x))$ is greater than or equal to ε. This implies that there exists a $\delta(\varepsilon) > 0$ such that if $\|f' - f\|_{C^2} < \delta(\varepsilon)$, then $\varepsilon(f') > \frac{\varepsilon}{2}$. Therefore, the local involution $\alpha^{f'}_{\text{loc}}$ is defined on $U_{\varepsilon(f)/2}$. We extend it to $V \setminus U_{\varepsilon(f)/2}$ as a global involution $\alpha^{f'}$ by defining $\alpha^{f'}(x)$ as the unique point from $(f')^{-1}(x)$ whose distance from $\alpha^f(x)$ is less than $\frac{\varepsilon(f)}{2}$. Hence, $\mathfrak{M}_{\text{taut}}(V,W,S)$ is open. On the other hand, using the same argument together with the implicit

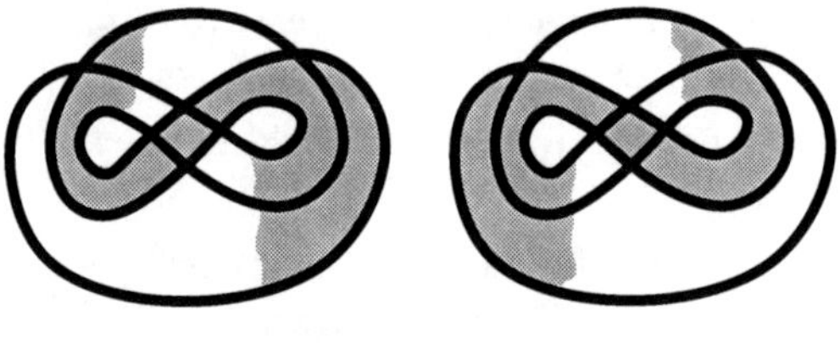

Fig. 1 $f \in \mathfrak{M}_{\text{soft}}(S^2, \mathbb{R}^2, S^1)$ as a pair $f_1, f_2 : D^2 \to \mathbb{R}^2$

function theorem, we conclude that given a sequence $f_n \in \mathfrak{M}_{\text{taut}}(V, W, S)$ such that $f_n \overset{C^2}{\to} f$, then $\alpha^{f_n} \overset{C^1}{\to} \alpha^f$, and hence $f \in \mathfrak{M}_{\text{taut}}(V, W, S)$ and $\mathfrak{M}_{\text{taut}}(V, W, S)$ is closed. □

The following theorem completes the description of the homotopy type of spaces of S-immersions.

1.2.4. (Homotopic Structure of the Space of Soft S-Immersions) *Let $S \subset V$ be a closed nonempty $(q-1)$-dimensional submanifold of V. Suppose that $q \geq 3$, or $q = 2$ but W is open. Then the inclusion*

$$d : \mathfrak{M}_{\text{soft}}(V, W, S) \to \mathfrak{m}(V, W, S)$$

is a (weak) homotopy equivalence. In particular, when the space $\mathfrak{I}(V, S)$ is empty (and hence $\mathfrak{M}_{\text{taut}}(V, W, S)$ is empty as well), then $d : \mathfrak{M}(V, W, S) \to \mathfrak{m}(V, W, S)$ is a (weak) homotopy equivalence.

For instance, when $V = S^q, W = \mathbb{R}^q$ and S is the equator $S^{q-1} \subset S^q$, then we get

$$\mathfrak{M}_{\text{soft}}(S^q, \mathbb{R}^q, S^{q-1}) \overset{\text{h.e.}}{\simeq} O(q) \times \Omega_q(O(q)),$$

where $\Omega_q(O(q))$ is the based q-loops space of the orthogonal group $O(q)$. In particular, using also Proposition 1.2.2, we conclude that the space of *all* S^1-immersions $S^2 \to \mathbb{R}^2$ that preserve orientation on S^2_+ consists of two components homotopy equivalent to S^1. The standard projection represents the taut component, while an example of a soft S^1-immersion $f : S^2 \to \mathbb{R}^2$ (as a pair of maps $f_1, f_2 : D^2 \to \mathbb{R}^2$, $f_1|_{S^1} = f_2|_{S^1}$) is presented in Fig. 1.

Remarks. 1. Theorem 1.2.4 was formulated in [El72], but only the epimorphism part of the statement was proven there.
2. Theorem 1.2.4 trivially holds for $q = 1$ and any W.
3. In the case $q = 2$ and W is a closed surface, we can prove only that the map d induces an epimorphism on homotopy groups.

2 Zigzags

2.1 Zigzags and Soft S-Immersions

A *model zigzag* is any smooth function $\mathcal{Z} : [a,b] \to \mathbb{R}$ such that

- $\mathcal{Z}$ is increasing near a and b;
- $\mathcal{Z}$ has exactly two interior nondegenerate critical points $m, M \in (a,b)$, $m < M$;
- $\mathcal{Z}(b) > \mathcal{Z}(a)$.

If, in addition, $\mathcal{Z}(b) > \mathcal{Z}(M)$ and $\mathcal{Z}(a) < \mathcal{Z}(m)$, then the model zigzag is called *long*.

Example. The function $\mathcal{Z}(z) = z^3 - 3z$ is a model zigzag on any interval $[a,b]$ such that $a < -1$ and $b > 1$ and $\mathcal{Z}(b) > \mathcal{Z}(a)$. If $a < -2$ and $b > 2$, then $\mathcal{Z}$ is long.

An embedding $h : [a,b] \to V$ is called a *zigzag* (respectively *long zigzag*) of a map $f \in \mathfrak{M}(V,W,S)$ if

- It is transversal to S, and
- The composition $f \circ h$ can be presented as $g \circ \mathcal{Z}$, where $\mathcal{Z} : [a,b] \to \mathbb{R}$ is a model zigzag (respectively long model zigzag) and $g : [A,B] \to W$ is an immersion defined on an interval $[A,B]$ such that $(A,B) \supset \mathcal{Z}([a,b])$ (Fig. 2).

The image $h([a,b])$ of the embedding h will also be called a (long) zigzag. The points $h(a), h(b) \in V$ are called the *endpoints* of the zigzag h. Note that:

2.1.1. (Extension of Long Zigzags) *Let $h : [a,b] \to V$ be a* long *zigzag. Then any extension $h' : [a',b'] \to V$ of the embedding h that does not have additional intersection points with S is a long zigzag.*

The implicit function theorem implies the following:

2.1.2. (Local Flexibility of Zigzags) *Let $f_t \in \mathfrak{M}(V,W,S)$ defined for $t \in \mathcal{O}p\,0 \subset \mathbb{R}$. Suppose that f_0 admits a zigzag $h_0 : [a,b] \to V$ such that the composition $f_0 \circ h_0$ can be factored as $g_0 \circ \mathcal{Z}$ for an immersion $g_0 : [A,B] \to V$. Suppose that g_0 is included in a family of immersions $g_t : [A,B] \to W$, $t \in \mathcal{O}p\,0$. Then there exists a family of zigzags $h_t : [a,b] \to V$ defined for $t \in \mathcal{O}p\,0$ such that $f_t \circ h_t = g_t \circ \mathcal{Z}$.*

2.1.3. (Softness Criterion) *A map $f \in \mathfrak{M}(V,W,S)$ is soft if and only if it admits a zigzag.*

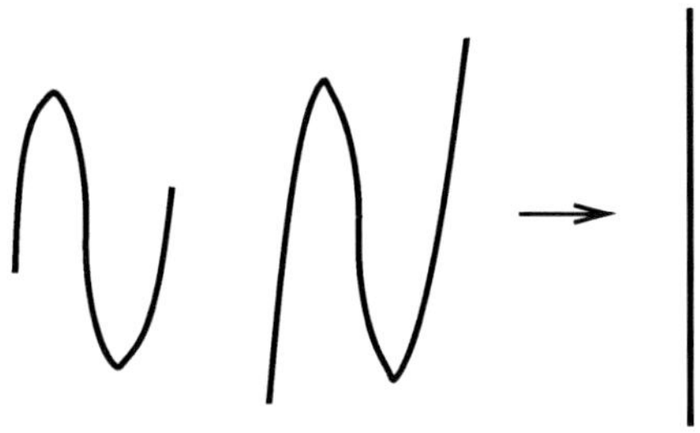

Fig. 2 Zigzag and long zigzag

Proof. Existence of any zigzag is incompatible with the existence of an involution, and hence a map admitting a zigzag is soft. On the other hand, if $f \in \mathfrak{M}(V,W,S)$ does not admit a zigzag, we can define an involution $\alpha = \alpha^f : V \to V$, $f \circ \alpha = f$, as follows. Given $v \in V$, take any arc $C \subset V$ connecting v in $V \setminus S$ with a point $s \in S$. Then the absence of zigzags guarantees that $f^{-1}(C)$ contains a unique candidate v' for $\alpha(v)$. Similarly, if for another path we had another candidate v'', this would create a zigzag. Hence, the involution $\alpha : V \to V$ with $f \circ \alpha = f$ is correctly defined, and therefore f is taut. □

2.2 *Zigzags Adjacent to a Chamber*

The components of $V \setminus S$ are called *chambers*. We say that a zigzag h is *adjacent to a chamber* C if this chamber contains one of the endpoints of the zigzag.

A family of maps $f_s \in \mathfrak{M}_{\text{soft}}(V,W,S)$, $s \in K$, parameterized by a connected compact set K can be viewed as a fibered map $\widetilde{f} : K \times V \to K \times W$. We will call such maps *fibered (over K) S*-immersions. A fibered S-immersion is called *soft* if it is fiberwise soft. According to 1.2.3, $\widetilde{f}$ is soft if it is soft over a point $s \in K$. We denote the space of soft S-immersions fibered over K by $\mathfrak{M}^K_{\text{soft}}(V,W,S)$. A family Z_s of zigzags for f_s, $s \in K$, will be referred to as a *fibered* over K zigzag Z. A fibered zigzag is called *special* if the projections $f_s(Z^s)$ are independent of $s \in K$. We say that a fibered soft map $\widetilde{f} : K \times V \to K \times W$ admits a *set of fibered (respectively special fibered) zigzags subordinated to a covering* $K = \bigcup_1^N U_j$ if there exist fibered (respectively special fibered) *disjoint* zigzags $\widetilde{Z}_1, \dots, \widetilde{Z}_N$ for the fibered maps $\widetilde{f}_j = \widetilde{f}|_{U_j \times V} : U_j \times V \to U_j \times W, j = 1, \dots, N$.

2.2.1. (Set of Zigzags Subordinated to an Inscribed Covering) *Let $\widetilde{Z}_1, \dots, \widetilde{Z}_N$ be a set of (special) fibered zigzags for a fibered map $\widetilde{f}$ subordinated to a covering $K = \bigcup_1^N U_j$. Let $K = \bigcup_1^{N'} U'_j$ be another covering inscribed in the first one, i.e., for every U'_i, $i = 1, \dots, N'$, there is U_j, $j = 1, \dots, N$, such that $U'_i \subset U_j$. Then there exists a set of fibered (special) zigzags $\widetilde{Z}'_1, \dots, \widetilde{Z}'_{N'}$ subordinated to the covering $K = \bigcup_1^{N'} U'_j$ such that for each $s \in U'_j$, the zigzag Z'^s_j is C^∞-close to the zigzag Z^s_i for some $i = 1, \dots, N$.*

Proof. Lemma 2.1.2 implies that a neighborhood of any (special) fibered zigzag is foliated by (special) fibered zigzags, and hence near any (special) fibered zigzag $\widetilde{Z}$ one can always find arbitrarily many disjoint copies of $\widetilde{Z}$. □

We would like to prove that a fibered soft map admits a set of fibered zigzags adjacent to any of its chambers C. Below, we explain two methods for proving this. However, each of the methods requires some additional assumptions. The first one

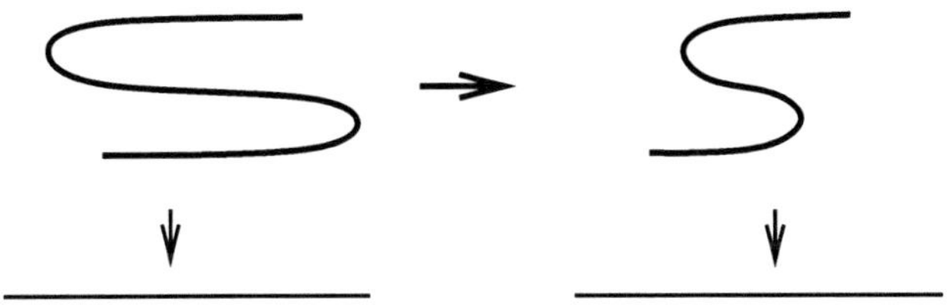

Fig. 3 Making zigzags long

works for any W, but only if $q \geq 3$, while the second works for $q \geq 2$, but only if the target manifold is open. We do not know whether the statement still holds if W is a closed surface.[1]

2.2.2. (Fibered Zigzags Adjacent to a Chamber) *Let $\widetilde{f}$ be a fibered over K soft S-immersion. Suppose that $q \geq 3$, or $q = 2$ but W is open. Then given any chamber $C \subset V \setminus S$, there is a homotopy $\widetilde{f}_t \in \mathfrak{M}^K_{\text{soft}}(V,W,S)$, $t \in [0,1]$, such that $\widetilde{f}_0 = \widetilde{f}$ and $\widetilde{f}_1$ admits a set of special fibered zigzags adjacent to the chamber C.*

CASE $q \geq 3$. We begin with two lemmas.

2.2.3. (Making Zigzags Long) *Let $\widetilde{f} \in \mathfrak{M}^K(V,W,S)$. Suppose that $q \geq 3$. Then there exist a homotopy $\widetilde{f}_t \in \mathfrak{M}^K(V,W,S)$, $t \in [0,1]$, and a covering $K = \bigcup_1^N U_j$ such that $\widetilde{f}_0 = \widetilde{f}$ and $\widetilde{f}_1$ admits a set of special fibered long zigzags subordinated to a covering inscribed in the covering $K = \bigcup_1^N U_j$.*

Proof. Choose a zigzag $h_s : [a,b] \to V$ for each f_s, $s \in K$, in the family $\widetilde{f}$. Denote by g_s the corresponding immersion $[A,B] \to W$. According to Lemma 2.1.2, there is a neighborhood $U_s \ni s$ in K such that for each $s' \in U_s$ there exists a zigzag $h'_{s'} : [a,b] \to V$ such that it factors through the same immersion g_s. Due to compactness of K, we can choose a finite subcovering $U_j = U_{s_j}$, $j = 1,\dots,N$. Lemma 2.1.2 further implies that if we C^∞-small perturb the immersions $g_j := g_{s_j}$, then there still exist for each $j = 1,\dots,N$ and $s \in U_j$ zigzags $Z_{s,j} \subset V$ that factor through them. Hence, if $q \geq 3$, a general position argument allows us to assume that the images of the immersions g_j, and hence of the fibered zigzags $\widetilde{Z}_j$, do not intersect. Now for each $j = 1,\dots,N$, there exists a deformation $f_{s,t}$, $s \in U_j$, $t \in [0,1]$, of the family f_s supported for each s in an arbitrarily small neighborhood of Z_s that makes the zigzags long; see Fig. 3. But the images of constructed zigzags do not intersect, and hence if the neighborhoods of zigzags are chosen sufficiently small, then the above deformation can be done simultaneously over all U_j, $j = 1,\dots,N$. □

[1]The second approach also works for $q = 1$ and any W. The case $W = T^2$ can also be treated by a slight modification of this method.

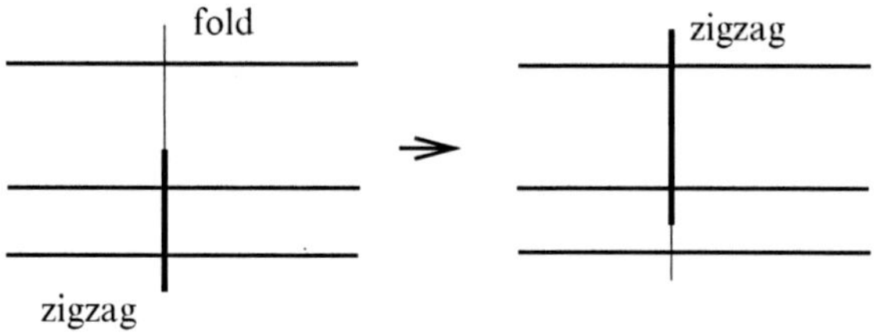

Fig. 4 Penetration through a wall

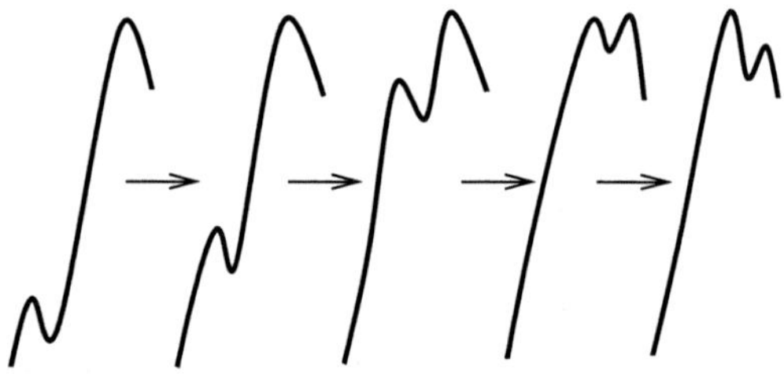

Fig. 5 Homotopy f_t on $h([a,c'])$

2.2.4. (Penetration Through Walls) *Let C be one of the chambers for $f \in \mathfrak{M}(V,W,S)$. Let $h : [a,c] \to V$ be an embedding such that*

- *There exists $b \in (a,c)$ such that $h|_{[a,b]}$ is a long zigzag for f;*
- *h is transversal to S and $h(c) \in C$.*

Then there exists a deformation $f_t \in \mathfrak{M}(V,W,S)$, $t \in [0,1]$, with $f_0 = f$ that is supported in a neighborhood of the image $h([a,c])$ and such that for some $b' \in (a,c)$, the embedding $h|_{[b',c]}$ is a zigzag for f_1 adjacent to C.

Proof. Suppose that the embedding $h|_{[b,c]}$ intersects the wall S in a sequence of points $h(p_1),\dots,h(p_k)$. We can consequently push the original zigzag through these points. Let $k = 1$. By a small perturbation of f near $h([a,c])$ we can make h invariant with respect to the local involution α_{loc} on $\mathcal{O}p\, p$. Then there exists $c' \in (p,c)$ such that $f \circ h|_{[a,c']}$ can be factored as $[a,c'] \to \mathbb{R} \to W$. Then the deformation shown in Figs. 4 and 5 allows us to move the zigzag from $[a,b]$ to $[b',c']$, where $c > p_1$, and thus the new long zigzag ends in the next chamber through which the embedding h travels. For $k > 1$ we apply inductively the same procedure. □

Proof of Lemma 2.2.2 for $q \geq 3$. We begin with a set of fibered long zigzags $\widetilde{Z}_j$ subordinated to a covering $K = \bigcup_1^N U_j$. Given a point $s \in U_j$, for each $j = 1,\dots,N$ we denote by Z_j^s the zigzag over $s \in U_j$, and by $h_j^s : [a,b] \to V$ its parameterization.

Fix $j = 1,\dots,N$ and denote by C_0 one of the chambers to which the zigzag $\widetilde{Z}_j$ is adjacent, say $h_j^s(b) \in C_0$ for $s \in U_j$. Passing, if necessary, to a set of zigzags subordinated to a finer covering of K, we can extend the family of embeddings h_j^s, $s \in U_j$, to a family of embeddings $[a,c] \to V$, $c > b$, still denoted by h_j^s, such that

- For each $s \in U_j$ the embedding h_j^s is transversal to S and $H_j^s(c) \in C$;
- The image $f_j^s([a,c]) \subset W$ is independent of s and disjoint from images of all other zigzags.

Using Lemma 2.2.4, we can construct a deformation of the fibered map $\widetilde{f}|_{U_j}$ supported in $\mathcal{O}p\,\widetilde{h}_j([a,c])$ that creates a fibered zigzag adjacent to the chamber C. Note that for different $i = 1,\ldots,k$, and $j = 1,\ldots,N$, all these deformations are supported in nonintersecting neighborhoods, and hence can be done simultaneously. □

CASE OF AN OPEN W AND $q \geq 2$. Let us consider a function $\phi : W \to \mathbb{R}$ without critical points. Let $\mathcal{F}$ be a 1-dimensional foliation by the gradient trajectories of ϕ for some Riemannian metric.

2.2.5. (The Case of K a Point) *Any $f \in \mathfrak{M}(V,W,S)$ admits a zigzag adjacent to any of its chambers that projects to one of the leaves of the foliation $\mathcal{F}$.*

Proof. Let us call a leaf L of $\mathcal{F}$ *regular* for f if the map $f|_S : S \to W$ is transversal to it. Sard's theorem implies that a generic leaf of $\mathcal{F}$ is regular. Let C be one of the chambers. Given a leaf L that is regular for f, we denote by C_L the union of those connected components of the closed 1-dimensional manifold $f^{-1}(L) \subset V$ that intersect C. We claim that for some regular leaf L, there exists a nonempty component Σ of C_L such that $f|_S : S \to L$ has more than two critical points. Indeed, otherwise we could reconstruct an involution $\alpha : V \to V$ such that $f \circ \alpha = f$, which would imply that f is taut. The intersection of C with the circle Σ consists of one or several arcs. Let A be one of these arcs. Its endpoints, p_1 and p_2, belong to the fold S, and are critical points of $f|_\Sigma : \Sigma \to L$. Let us assume that p_1 is a local minimum and p_2 is a local maximum. Recall that the leaf L is oriented by the gradient vector field of the function ϕ. We orient the arc A from p_1 to p_2 and orient the circle Σ accordingly. Let p_3 be the next critical point of $f|_\Sigma$. Choose a point $q_1 \in A$ close to p_1 and a point q_2 close to p_3 and *after* it in terms of the orientation. If $f(p_3) > f(p_1)$, then the arc $Z = [q_1,q_2]$ is a zigzag adjacent to C beginning at q_1 and ending at q_2. If $f(p_3) < f(p_1)$, then the same arc $Z = [q_1,q_2]$ is again is a zigzag adjacent to C, but beginning at q_2 and ending at q_1. □

Proof of Lemma 2.2.2 for an open W and $q \geq 2$. Lemma 2.2.5 implies that over any point $s \in K$, the map f_s admits a zigzag adjacent to the chamber C. Moreover, we can assume that all these zigzags lie over different leaves of $\mathcal{F}$. These zigzags extend to a neighborhood $\mathcal{O}p\,s \in K$ as fibered zigzags over this neighborhood that for each $s' \in \mathcal{O}p\,s$ project to the same leaves of $\mathcal{F}$. Hence, we can choose a finite covering $K = \bigcup_1^N U_j$ by neighborhoods U_j over which there exist ample sets of special fibered zigzags. Clearly we can arrange that zigzags over different U_j project to different leaves of $\mathcal{F}$, and thus their images do not intersect. □

3 Proof of the Main Theorem

3.1 Wrinkled S-Immersions

A map $f : V \to W$ is called a *wrinkled S-immersion* if it has S as its fold singularity and in the complement of S it is a wrinkled map; see Sect. 4.2. Let C be one of the chambers (i.e., a connected component of $V \setminus S$). We denote by $\mathfrak{M}_w(V,W,S,C)$ the space of wrinkled S-immersions that have all wrinkles in the chamber C. We will call C the *designated* chamber. The regularized differential construction (see Sect. 4.2) provides a map $d_R : \mathfrak{M}_w(V,W,S,C) \to \mathfrak{m}(V,W,S)$, and the wrinkling Theorem 4.4.2 implies the following result:

3.1.1. (From Formal to Wrinkled S-Immersions) *The map d_R is a (weak) homotopy equivalence.*

Proof. Let $C_1,\dots,C_l$ be a sequence of all chambers in $V \setminus S$ such that $C_l = C$ and each C_i, $i < l$, has a common wall with C_j, $j > i$. We apply Hirsch's h-principle (see [Hi]) for equidimensional immersions of open (i.e., nonclosed!) manifolds to $\overline{C}_1$ and then make a fold on $\partial\overline{C}_1$. Next, we apply the relative version the same h-principle to the pair $(C_2,\partial C_1)$ and so on. At the last step, we apply the wrinkling Theorem 4.2 to the pair $(\overline{C},\partial\overline{C})$. □

A slightly stronger statement can be formulated in the language of fibered maps. A fibered over K map is called a *fibered wrinkled S-immersion* if it has S as its fiberwise fold singularity, and in the complement of $K \times S \subset K \times V$ it is a fibered wrinkled map. One can also talk about fibered over K formal S-immersions, i.e., parameterized by K families $F_s \in \mathfrak{m}(V,W,S)$. Then a regularized differential of a fibered S-immersion is a fibered formal S-immersion. The following proposition is a slight improvement of the above homotopy equivalence claim, and it also follows from Theorem 4.4.2.

3.1.2. (From Formal to Wrinkled S-Immersions; A Fibered Version) *Given any fibered over K formal S-immersion $\widetilde{F} \in \mathfrak{m}^K(V,W,S)$, there exists a fibered over K wrinkled map $\widetilde{g} \in \mathfrak{M}^K_w(V,W,S,C)$ whose regularized fibered differential is homotopic to $\widetilde{F}$. Moreover, if over a closed $L \subset K$ we have $\widetilde{F} = \mathrm{d}\widetilde{f}$, where $\widetilde{f}$ is a genuine fibered S-immersion, then the map $\widetilde{g}$ can be chosen equal to $\widetilde{f}$ over L, and the homotopy can be made fixed over L.*

Our next goal is to supply a wrinkled S-immersion with zigzags adjacent to the designated chamber.

3.1.3. (Zigzags for Fibered Wrinkled S-Immersions) *Let $\widetilde{f}$ be a fibered over K wrinkled map that has all wrinkles in the chamber C. Suppose that over a closed subset $L \subset K$, the fibered map $\widetilde{f}$ consists of genuine (i.e., nonwrinkled) S-immersions. Let $\bigcup_1^N U_j \supset L$ be a covering of L by contractible and open in K sets, and let $\widetilde{Z}_1,\dots,\widetilde{Z}_N$ be a set of fibered zigzags adjacent to C subordinated to the covering $\bigcup_1^N U_j$. Then there is a homotopy $\widetilde{f}_t \in \mathfrak{M}^K_w(V,W,S,C)$, $t \in [0,1]$, $\widetilde{f}_0 = \widetilde{f}$, such that*

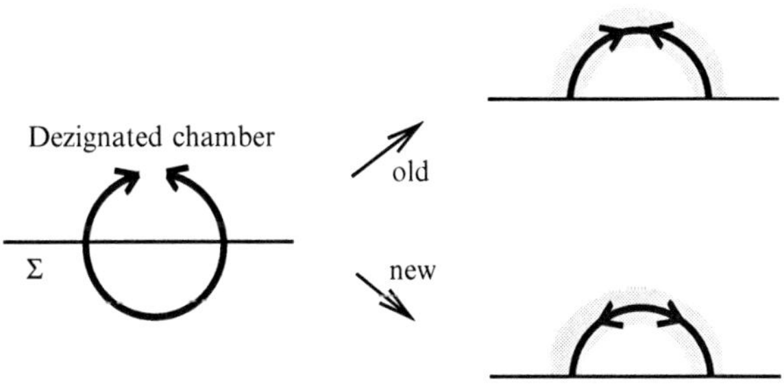

Fig. 6 Local birth of a zigzag

- $\widetilde{f}_t$ *is fixed over* L *in a neighborhood of the zigzags* $\widetilde{Z}_1,\dots,\widetilde{Z}_N$;
- *There exists a zigzag* $\widetilde{Z}$, *fibered over a domain* U, $K\setminus\bigcup_1^N U_j\subset U\subset K\setminus \mathcal{O}p\,L$, *for* $\widetilde{f}_1$, *adjacent to* C *and disjoint from* $\widetilde{Z}_1,\dots,\widetilde{Z}_N$.

Let us define $\Sigma:=\partial C\subset S$. The following two lemmas will be needed in the proof of Lemma 3.1.3.

3.1.4. *Let* $U_1,\dots,U_N$ *be as in Lemma 3.1.3 and* $\sigma_j: U_j\to U_j\times\Sigma$, $j=1,\dots,N$, *disjoint sections. Then there exists a section* $\sigma: K\to K\times\Sigma$ *disjoint from the sections* $\sigma_1,\dots,\sigma_N$ *and homotopic to a constant section.*

Proof. Arguing by induction over $j=1,\dots,N$, we construct a fiberwise isotopy $\widetilde{g}_t: K\times V\to V$, $t\in[0,1]$, that makes sections $\sigma_j: U_j\to U_j\times\Sigma$ constant, i.e., $\widetilde{g}_1\circ\sigma_j(s)=(s,c_j)$, $c_j\in\Sigma$, $j=1,\dots,N$. Take a point $c\in\Sigma$ different from $c_1,\dots,c_N$ and define $\sigma'(s):=(s,c)$ for any $s\in K$. Then the section $\sigma=\widetilde{g}_1^{-1}\circ\sigma': K\to K\to V$ has the required properties. □

3.1.5. (Local Birth of a Zigzag Near σ) *Let* $M\subset K$ *be a closed subset and* $\sigma: M\to M\times\Sigma\subset K\times V$ *a section. Then there exists a deformation* $\widetilde{f}_t\in\mathfrak{M}^K_w(V,W,S,C)$ *of* $\widetilde{f}_0=\widetilde{f}$ *that is supported in* $\mathcal{O}p\,\sigma(M)\subset K\times V$ *such that* $\widetilde{f}_1$ *admits a fibered over* $\mathcal{O}p\,M$ *zigzag adjacent to* C *and supported in* $\mathcal{O}p\,\sigma(M)\subset K\times V$.

Proof. First we give a sketch of the construction. Take a fibered embedding $\widetilde{h}: M\times I\to M\times\mathcal{O}p\,\Sigma$ such that for all $s\in M$, the image $h(s\times I)$ lies on a small α_{loc}-invariant circle near $\sigma(s)\in\Sigma$; see Fig. 6. There exists a homotopy $\widetilde{g}_t$ of the map $\widetilde{g}_0=\widetilde{f}|_{\mathcal{O}p\,h(M\times I)}$ such that $\widetilde{g}_1$ has a fibered zigzag over M; see Fig. 6. This local homotopy can be extended as a homotopy $\widetilde{f}_t\in\mathfrak{M}^K_w(V,W,S,C)$ of the whole fibered map $\widetilde{f}$.

Let us give now a more detailed description. Let $S^1\subset\mathbb{C}$ be the unit circle, and $\exp:\mathbb{R}\to S^1$ a covering map $u\mapsto e^{iu}$, $u\in\mathbb{R}$. Choose a neighborhood $\Omega\supset\sigma(M)$. Let $\widetilde{h}: M\times S^1\to\Omega\subset M\times V$ be a fibered embedding such that for all $s\in M$, the image $h^s(S^1)$ is a small α_{loc}-invariant circle near $\sigma(s)\in\Sigma$; see Fig. 6. We can assume

$h^s(\exp u) \in \overline{C}$ for $u \in [\pi/2, 3\pi/2]$ and $h^s(\exp u) \notin \overline{C}$ for $u \in (-\pi/2, \pi/2)$. Consider also a fibered embedding

$$\widetilde{\varphi} : \mathcal{O}p M \times B \to \mathcal{O}p M \times C \subset \Omega,$$

where B is an open n-ball, such that $\varphi^s(B)$ is a small ball centered at the point $h^s(-1) \in C$ and such that $\varphi^s(D) \cap h^s(S^1) = h^s(\exp((\pi - \varepsilon, \pi + \varepsilon))$, $s \in M$. There exists a compactly supported fibered regular homotopy (see Fig. 6)

$$\widetilde{\psi}_t : \Omega \setminus \widetilde{\varphi}(\mathcal{O}p M \times B) \to \Omega, \ t \in [0,1],$$

such that

$$\widetilde{\psi}_t(\widetilde{h}(s, \exp u)) = \widetilde{h}(s, \exp(1 + t/2)u), \ s \in M, \ u \in [-\pi + \varepsilon, \pi - \varepsilon].$$

Notice that ψ_1^s maps an embedded arc $h^s([-\pi + \varepsilon, \pi - \varepsilon])$ onto an overlapping arc $h^s([-\frac{3}{2}(-\pi + \varepsilon), \frac{3}{2}(\pi - \varepsilon))$. Using Theorem 4.4.2, we can extend the regular homotopy $\widetilde{\psi}_t$ to a compactly supported fibered wrinkled homotopy $\Omega \to \Omega$, and then further extend it to the rest of $K \times V$ as the identity map. We will use the same notation $\widetilde{\psi}_t$ for this extension. Finally, the wrinkled homotopy $\widetilde{f}_t = \widetilde{f}_0 \circ \widetilde{\psi}_t : K \times V \to K \times V$ connects $\widetilde{f}_0$ with a wrinkled map $\widetilde{f}_1$ that has the embedding $\widetilde{\psi}_1 \circ \widetilde{h}|_{M \times [-\pi - \varepsilon, \pi + \varepsilon])}$ as its fibered over M zigzag. □

Proof of Proposition 3.1.3. Note that the zigzags $\widetilde{Z}_j$ over $U_j \subset K$, $j = 1, \dots, N$, intersect the closure $\overline{C}$ of the chamber C along intervals with one end on Σ and the second inside C. Taking the endpoints in Σ, we get sections $\sigma_j : U_j \to U_j \times \Sigma$, $j = 1, \dots, N$. Let us apply Lemma 3.1.4 and construct a section $\sigma : K \to K \times \Sigma$, disjoint from the sections $\sigma_1, \dots, \sigma_N$. Let $M \subset K \setminus L$ be a closed set such that $K \setminus M \subset \bigcup_1^N U_j$. There exists a neighborhood $\Omega \supset \sigma(M) \subset K \times V$ that does not intersect the fibered zigzags $\widetilde{Z}_j$, $j = 1, \dots, N$. Let $U \supset M$ be an open neighborhood whose closure is contained in $K \setminus L$ and such that $\sigma(U) \subset \Omega$. We conclude the proof of Lemma 3.1.3 by applying Lemma 3.1.5. □

3.2 Engulfing Wrinkles by Zigzags

3.2.1. (Getting rid of Wrinkles) *Suppose that $\widetilde{f} \in \mathfrak{M}_w^K(V, W, S, C)$ admits a set of fibered zigzags adjacent to the chamber C. Let $\widetilde{f}$ be a genuine fibered S-immersion over a closed $L \subset K$. Then $\widetilde{f}$ is homotopic in $\mathfrak{M}_w^K(V, W, S, C)$ to a genuine fibered S-immersion via a homotopy fixed over L.*

Proof. Let $\widetilde{Z}_1, \dots, \widetilde{Z}_N$ be fibered zigzags adjacent to C and subordinated to a covering $\bigcup_1^N U_j = K$. First of all, we can apply the enhanced wrinkling theorem (see the remark after Theorem 4.4.2) to $\widetilde{f}|_{K \times C}$ and get a modified $\widetilde{f}$ such that each fibered wrinkle is supported over one of the elements of the covering. Using Lemma 2.2.1, we can assume that wrinkles and zigzags are in 1-to-1 correspondence with the elements of the covering. We will eliminate inductively all the wrinkles by a procedure that we call *engulfing of wrinkles by zigzags*. We will discuss this construction only in the nonparametric case, i.e., when K is a point. The case of a general K differs only in the notation.

Let $w = w(q)$ be the standard wrinkle with membrane D^q (see 4.2). Let us recall that w is a fibered over $\mathbb{R}^{q-1}$ map $\mathbb{R}^q \to \mathbb{R}^q$ defined by the formula

$$(y,z) \mapsto \left(y, z^3 + 3(|y|^2 - 1)z\right),$$

where $y \in \mathbb{R}^{q-1}$, $z \in \mathbb{R}^1$, and $|y|^2 = \Sigma_1^{q-1} y_i^2$. Note that for a sufficiently small $\varepsilon > 0$ we have $w(D^q_{1+\varepsilon}) \subset \{|y| \leq 1+\varepsilon, |z| \in [-3,3]\} = D^{q-1}_{1+\varepsilon} \times [-3,3]$.

We will need the following lemma.

3.2.2. (Standard Model for Engulfing) *Let us consider a fibered over $D^{q-1}_{1+\varepsilon}$ map $\Gamma : D^{q-1}_{1+\varepsilon} \times [a,c] \to D^{q-1}_{1+\varepsilon} \times [a,c]$ such that*

- *for some $b \in (a,c)$, the restriction*

$$\Gamma|_{D^{q-1}_{1+\varepsilon} \times [a,b]} : D^{q-1}_{1+\varepsilon} \times [a,b] \to D^{q-1}_{1+\varepsilon} \times [a,c]$$

  is a fibered over $D^{q-1}_{1+\varepsilon}$ zigzag;
- *the restriction*

$$\Gamma|_{D^{q-1}_{1+\varepsilon} \times [b,c]} : D^{q-1}_{1+\varepsilon} \times [b,c] \to D^{q-1}_{1+\varepsilon} \times [a,c]$$

  is a fibered over $D^{q-1}_{1+\varepsilon}$ wrinkled map with a unique wrinkle whose cusp locus projects to the sphere ∂D^{q-1}. In other words, there exist fibered over $D^{q-1}_{1+\varepsilon}$ embeddings $\alpha : D^q_{1+\varepsilon} \to D^{q-1}_{1+\varepsilon} \times [b,c]$ and $\beta : D^{q-1}_{1+\varepsilon} \times [-3,3] \to D^{q-1}_{1+\varepsilon} \times [a,c]$ such that $\Gamma \circ \alpha = \beta \circ w$.

Then there exists a fibered over $D^{q-1}_{1+\varepsilon}$ homotopy

$$\Gamma_t : D^{q-1}_{1+\varepsilon} \times [a,c] \to D^{q-1}_{1+\varepsilon} \times [a,c], t \in [0,1],$$

that begins with $\Gamma_0 = \Gamma$, is fixed near $D^{q-1}_{1+\varepsilon} \times [a,c]$, and satisfies the following conditions:

- *$\Gamma_t|_{D^{q-1}_{1+\varepsilon} \times [a,b]}$ is the homotopy in the space of fibered folded maps;*
- *$\Gamma_t|_{D^{q-1}_{1+\varepsilon} \times [b,c]}$ is a fibered wrinkled homotopy that eliminates the wrinkle of the map $\Gamma_0 = \Gamma$, i.e., the map $\Gamma_1|_{D^{q-1}_{1+\varepsilon} \times [b,c]}$ is nonsingular.*

We will call the homotopy Γ_t *engulfing of the wrinkle w by a zigzag.*

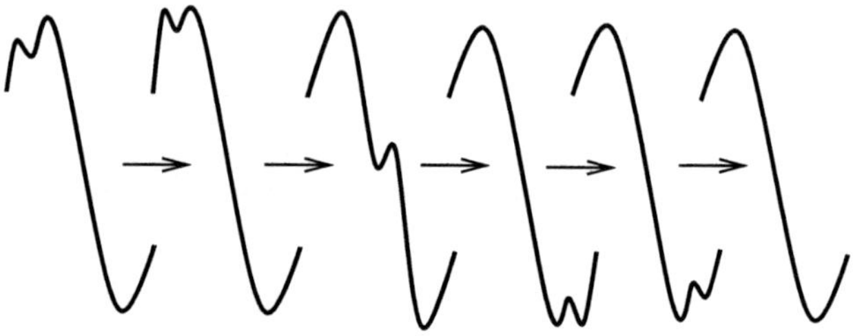

Fig. 7 Homotopy Γ_t on $y \times [a,c]$, $y \in \text{Int}\, D^q$

Fig. 8 Homotopy Γ_t, $q = 2$

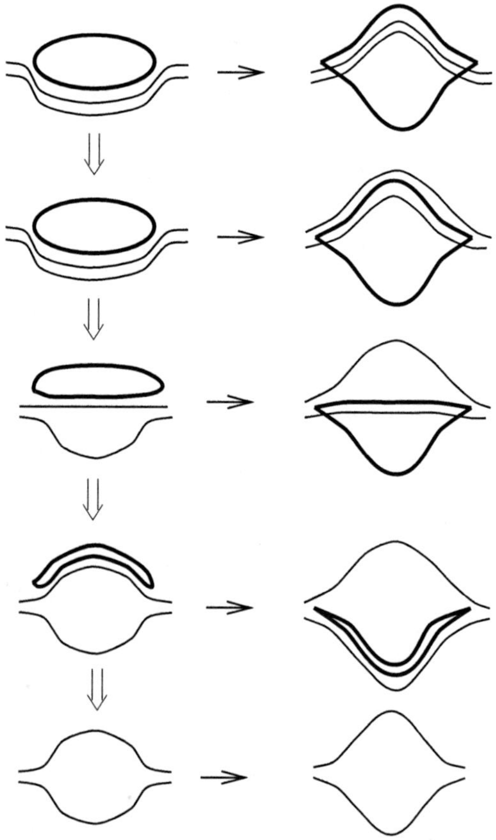

Proof. For $y \in \text{Int}\, D^{q-1}$, the homotopy Γ_t is shown in Fig. 7. This deformation can be done smoothly depending on the parameter $y \in D^{q-1}_{1+\varepsilon}$ and can be made to die out on $[1, 1+\varepsilon]$. See also Fig. 8, where the deformation Γ_t is shown for $q = 2$. □

We continue now the proof of Lemma 3.2.1 by reducing it to the standard engulfing model 3.2.2. Let us recall that there is 1-to-1 correspondence between the wrinkles and zigzags. Let $h : [a,b] \to V$ be a zigzag of a map f adjacent

Fig. 9 Embedding $\widehat{H}$

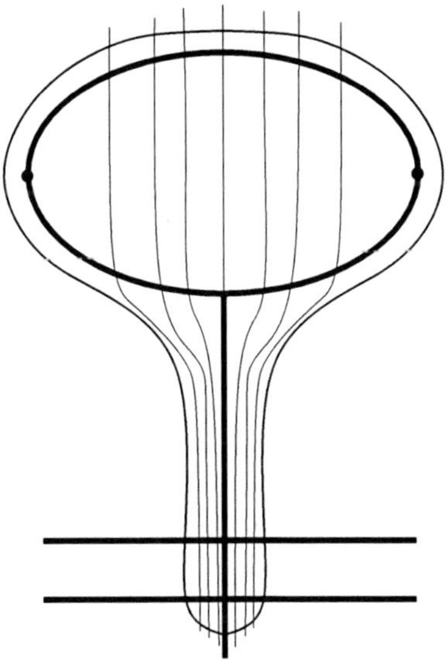

to the designated chamber C. The embedding h extends to an embedding $H : D^{q-1}_{1+\varepsilon} \times [a,b] \to V$ onto a small neighborhood of the zigzag $Z = h([a,b])$ such that $H|_{y\times[a,b]}$ is a zigzag for each $y \in D^{q-1}$. Take the corresponding wrinkle with membrane $D \subset C$. By definition, this means that for a sufficiently small $\varepsilon > 0$, there exists an embedding $\alpha : D^q_{1+\varepsilon} \to V$ such that $\alpha(D^q) = D$ and $f \circ \alpha = g \circ w(q)$ for an embedding $g : D^{q-1}_{1+\varepsilon} \times [-3,3] \to W$.

As is clear from Fig. 9, there exists an extension of the embedding H to an embedding $\widehat{H} : D^{q-1}_{1+\varepsilon} \times [a,c] \to V$, $c > b$, such that

- $\widehat{H}(D^{q-1}_{1+\varepsilon} \times [b,c]) \supset D$;
- The map $f \circ \widehat{H} : D^{q-1}_{1+\varepsilon} \times [a,c]$ can be written as $g \circ \Gamma$, where Γ is a fibered over $D^{q-1}_{1+\varepsilon}$ map $D^{q-1}_{1+\varepsilon} \times [a,c] \to D^{q-1}_{1+\varepsilon} \times [a,c]$ and $g : D^{q-1}_{1+\varepsilon} \times [a,c] \to W$ is an immersion;
- The restriction of Γ to $D^{q-1}_{1+\varepsilon} \times [b,c]$ is a fibered over $D^{q-1}_{1+\varepsilon}$ wrinkled map with the unique wrinkle whose cusp locus projects to the sphere $\partial D^{q-1} \subset D^{q-1}_{1+\varepsilon}$.

Thus we are in a position to apply Lemma 3.2.2. Let

$$\Gamma_t : [a,c] \times D^{q-1}_{1+\varepsilon} \to [a,c] \times D^{q-1}_{1+\varepsilon},\ t \in [0,1],$$

be the engulfing homotopy of $\Gamma = \Gamma_0$ constructed in Lemma 3.2.2, which eliminates the wrinkle. Note that the homotopy Γ_t is fixed near $\partial\left([a,c] \times D^{q-1}_{1+\varepsilon}\right)$. This enables us to define the required wrinkled homotopy $f_t : V \to W$, $t \in [0,1]$, which eliminates the wrinkle w by setting it equal to f on $V \setminus \widehat{H}([a,c] \times D^{q-1}_{1+\varepsilon})$, and to $g \circ \Gamma_t \circ \widehat{H}^{-1}$ on $\widehat{H}([a,c] \times D^{q-1}_{1+\varepsilon})$. $\square$

3.3 Proof of Theorem 1.2.4

Let us prove that the map d induces an injective homomorphism on π_{k-1}, i.e., that $\pi_k(\mathfrak{m}(V,W,S),\mathfrak{M}_{\text{soft}}(V,W,S)) = 0$. Define $K := D^k$, $L := \partial D^k = S^{k-1}$. We need to show that if $\widetilde{F} \in \mathfrak{m}^K(V,W,S)$ is a fibered formal S-immersion that is equal to $d\widetilde{f}$ over L for $\widetilde{f} \in \mathfrak{M}^L_{\text{soft}}(V,W,S)$, then there exists a homotopy $\widetilde{F}_t$ of $\widetilde{F}$ in $(\mathfrak{m}^K(V,W,S),\mathfrak{M}^L_{\text{soft}}(V,W,S)$ such that $\widetilde{F}_1 = d\widetilde{f}_1$ for $\widetilde{f}_1 \in \mathfrak{M}^K_{\text{soft}}(V,W,S)$.

The construction of the required homotopy can be done in four steps:

- Using Lemma 2.2.2, we first deform $\widetilde{F}$ in $(\mathfrak{m}^K(V,W,S),\mathfrak{M}^L(V,W,S))$ so that the new $\widetilde{F}$ admits over L a set of special fibered zigzags adjacent to a designated chamber C.
- Using Lemma 3.1.1, we further deform $\widetilde{F}$ into a differential of a fibered wrinkled S-immersion $\widetilde{f}$ such that all the wrinkles belong to the chamber C.
- Using Lemma 3.1.3, we can further deform $\widetilde{f}$ in such a way that the new $\widetilde{f}$ admits a set of zigzags adjacent to the chamber C. Furthermore, we can choose this set of zigzags in such a way that it includes the set of zigzags over L.
- Using Lemma 3.2.1, one can engulf all the fibered wrinkles.

The proof of the subjectivity claim is similar but does not use the first step.

4 Appendix: Folds, Cusps, and Wrinkles

We recall here, for the reader's convenience, some definitions and results from [EM97]. We consider here only the equidimensional case $n = q$, and this allows us to simplify definitions, notation, and the like from what appears in [EM97].

4.1 Folds and Cusps

Let V and W be smooth manifolds of the same dimension q. For a smooth map $f : V \to W$ we will denote by $\Sigma(f)$ the set of its singular points, i.e.,

$$\Sigma(f) = \{p \in V,\ \text{rank}\, d_p f < q\}.$$

A point $p \in \Sigma(f)$ is called a *fold*-type singularity or a *fold* of index s if near the point p, the map f is equivalent to the map $\mathbb{R}^{q-1} \times \mathbb{R}^1 \to \mathbb{R}^{q-1} \times \mathbb{R}^1$ given by the formula $(y,x) \to (y,\ x^2)$, where $y = (y_1,\dots,y_{q-1}) \in \mathbb{R}^{q-1}$.

Let $q > 1$. A point $p \in \Sigma(f)$ is called a *cusp* of index $s + \frac{1}{2}$ if near the point p, the map f is equivalent to the map $\mathbb{R}^{q-1} \times \mathbb{R}^1 \to \mathbb{R}^{q-1} \times \mathbb{R}^1$ given by the formula $(y,z) \to (y, z^3 + 3y_1 z)$, where $z \in \mathbb{R}^1$, $y = (y_1,\dots,y_{q-1}) \in \mathbb{R}^{q-1}$.

For $q \geq 1$, a point $p \in \Sigma(f)$ is called an *embryo* of index $s + \frac{1}{2}$ if f is equivalent near p to the map $\mathbb{R}^{q-1} \times \mathbb{R}^1 \to \mathbb{R}^{q-1} \times \mathbb{R}^1$ given by the formula $(y,z) \to (y, z^3 + 3|y|^2 z)$, where $y \in \mathbb{R}^{q-1}$, $z \in \mathbb{R}^1$, $|y|^2 = \Sigma_1^{q-1} y_i^2$. The set of all folds of f is denoted by $\Sigma^{10}(f)$, the set of cusps by $\Sigma^{11}(f)$, and the closure $\overline{\Sigma^{10}(f)}$ by $\Sigma^1(f)$.

Observe that folds and cusps are stable singularities for individual maps, while embryos are stable singularities only for 1-parametric families of mappings. For a generic perturbation of an individual map, embryos either disappear or give birth to wrinkles, which we consider in the next section.

4.2 *Wrinkles and Wrinkled Maps*

Consider the map $w(q) : \mathbb{R}^{q-1} \times \mathbb{R}^1 \to \mathbb{R}^{q-1} \times \mathbb{R}^1$ given by the formula

$$(y,z) \mapsto \left(y, z^3 + 3(|y|^2 - 1)z\right),$$

where $y \in \mathbb{R}^{q-1}$, $z \in \mathbb{R}^1$ and $|y|^2 = \Sigma_1^{q-1} y_i^2$.

The singularity $\Sigma^1(w(q))$ is the $(q-1)$-dimensional sphere $S^{q-1} \subset \mathbb{R}^q$. Its equator $\{z = 0, |y| = 1\} \subset \Sigma^1(w(q))$ consists of cusp points, while the upper and lower hemispheres consist of folds points (see Fig. 10). We will call the q-dimensional bounded by $\Sigma^1(w)$ disk $D^q = \{z^2 + |y|^2 \leq 1\}$ the *membrane* of the wrinkle.

A map $f : U \to W$ defined on an open subset $U \subset V$ is called a *wrinkle* if it is equivalent to the restriction of $w(q)$ to an open neighborhood U of the disk $D^q \subset \mathbb{R}^q$. Sometimes, the term "wrinkle" can be also used also for the singularity $\Sigma(f)$ of the wrinkle f.

Observe that for $q = 1$, the wrinkle is a function with two nondegenerate critical points of indices 0 and 1 given in a neighborhood of a gradient trajectory that connects the two points.

Restrictions of the map $w(q)$ to subspaces $y_1 = t$, viewed as maps $\mathbb{R}^{q-1} \to \mathbb{R}^{q-1}$ are nonsingular maps for $|t| > 1$, equivalent to $w(q-1)$ for $|t| < 1$ and to embryos for $t = \pm 1$.

Although the differential $dw(q) : T(\mathbb{R}^q) \to T(\mathbb{R}^q)$ degenerates at points of $\Sigma(w)$, it can be canonically *regularized*. Namely, we can change the element $3(z^2 + |y|^2 - 1)$ in the Jacobi matrix of $w(q)$ to a function γ that coincides with $3(z^2 + |y|^2 - 1)$ outside an arbitrarily small neighborhood U of the disk D^q and does not vanish on U. The new bundle map $\mathcal{R}(dw) : T(\mathbb{R}^q) \to T(\mathbb{R}^q)$ provides a homotopically canonical extension of the map $dw : T(\mathbb{R}^q \setminus U) \to T(\mathbb{R}^q)$ to an epimorphism (fiberwise surjective bundle map) $T(\mathbb{R}^q) \to T(\mathbb{R}^q)$. We call $\mathcal{R}(dw)$ the *regularized differential* of the map $w(q)$.

A map $f : V \to W$ is called *wrinkled* if there exist disjoint open subsets $U_1, \ldots, U_l \subset V$ such that $f|_{V \setminus U}$, $U = \bigcup_1^l U_i$, is an immersion (i.e., has rank q) and for each $i = 1, \ldots, l$, the restriction $f|_{U_i}$ is a wrinkle. Observe that the sets U_i, $i = 1, \ldots, l$, are included in the structure of a wrinkled map.

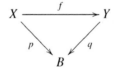

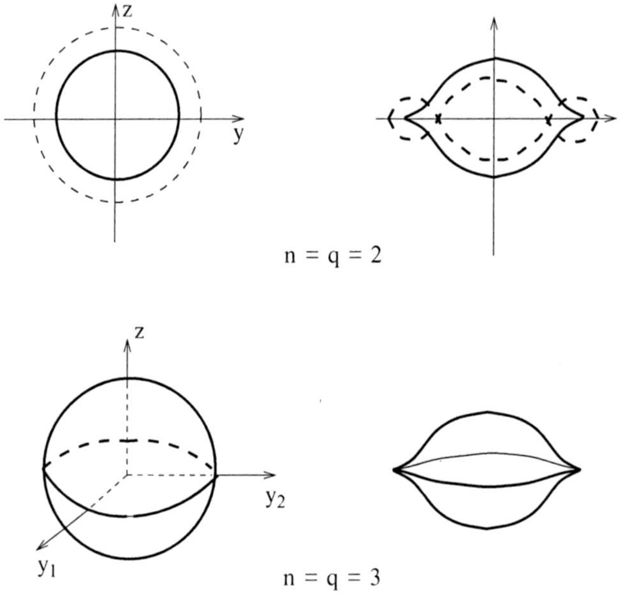

Fig. 10 Wrinkles. Pictures in the source and in the image

The singular locus $\Sigma(f)$ of a wrinkled map f is a union of $(q-1)$-dimensional wrinkles $S_i = \Sigma^1(f|_{U_i}) \subset U_i$. Each S_i has a $(q-2)$-dimensional equator $T_i \subset S_i$ of cusps that divides S_i into two hemispheres of folds of two neighboring indices. The differential $\mathrm{d}f : T(V) \to T(W)$ can be regularized to obtain an epimorphism $\mathcal{R}(\mathrm{d}f) : T(V) \to T(W)$. To get $\mathcal{R}(\mathrm{d}f)$, we regularize $\mathrm{d}f|_{U_i}$ for each wrinkle $f|_{U_i}$.

4.3 Fibered Wrinkles

All the notions from 4.2 can be extended to the parametric case.

A *fibered* (over B) *map* is a commutative diagram

$$\begin{array}{ccc} X & \xrightarrow{\quad f \quad} & Y \\ & {\scriptstyle p}\searrow \quad \swarrow{\scriptstyle q} & \\ & B & \end{array}$$

where p and q are submersions. A fibered map can be also denoted simply by $f : X \to Y$ if B, p, and q are implied from the context.

For a fibered map

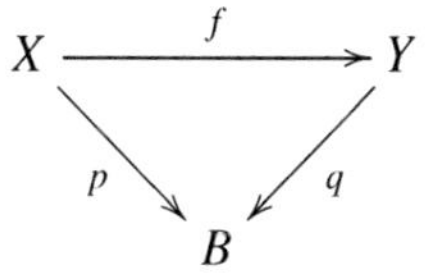

we denote by $T_B X$ and $T_B Y$ the subbundles $\operatorname{Ker} p \subset TX$ and $\operatorname{Ker} q \subset TY$. They are tangent to foliations of X and Y formed by preimages

$$p^{-1}(b) \subset X\,,\; q^{-1}(b) \subset Y\,,\; b \in B\,.$$

The fibered homotopies, fibered differentials, fibered submersions, and so on are naturally defined in the category of fibered maps (see [EM97]). For example, the *fibered differential* of $f : X \to Y$ is the restriction

$$d_B f = \mathrm{d}f|_{T_B X} : T_B X \to T_B Y\,.$$

Notice that $d_B f$ itself is a map fibered over B.

Two fibered maps $f : X \to Y$ over B and $f : X' \to Y'$ over B' are called *equivalent* if there exist open subsets $A \subset B$, $A' \subset B'$, $Z \subset Y$, $Z' \subset Y'$ with $f(X) \subset Z$, $p(X) \subset A$, $f'(X') \subset Z'$, $p'(X') \subset A'$ and diffeomorphisms $\varphi : X \to X'$, $\psi : Z \to Z'$, $s : A \to A'$ such that they form the following commutative diagram:

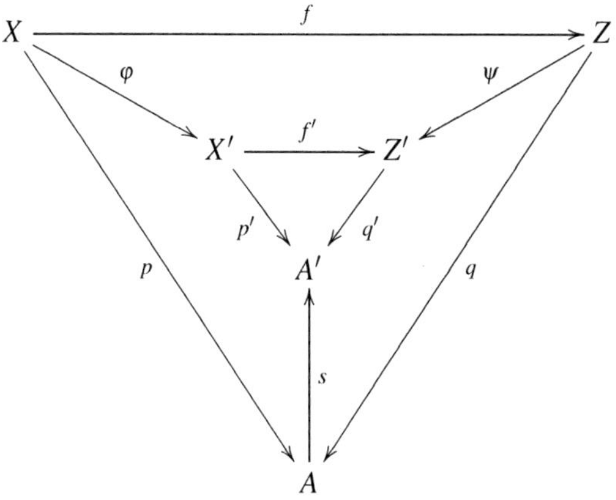

For any integer $k > 0$, the map $w(k+q)$ can be considered a fibered map $w_k(k+q)$ over $\mathbb{R}^k \times \mathbf{0} \subset \mathbb{R}^{k+q}$. A fibered map equivalent to the restriction of $w_k(k+q)$ to an open neighborhood $U^{k+q} \supset D^{k+q}$ is called a *fibered wrinkle*. The regularized differential $\mathcal{R}(\mathrm{d}w_k(k+q))$ is a fibered (over $\mathbb{R}^k$) epimorphism

$$\mathbb{R}^k \times T(\mathbb{R}^{q-1} \times \mathbb{R}^1) \xrightarrow{\mathcal{R}(\mathrm{d}w_k(q))} \mathbb{R}^k \times T(\mathbb{R}^{q-1} \times \mathbb{R}^1)$$

A fibered map $f : V \to W$ is called a *fibered wrinkled map* if there exist disjoint open sets $U_1, U_2, \dots, U_l \subset V$ such that $f|_{V \setminus U}$, $U = \bigcup_1^l U_i$, is a fibered submersion and for each $i = 1, \dots, l$, the restriction $f|_{U_i}$ is a fibered wrinkle. The restrictions of a fibered wrinkled map to a fiber may have, in addition to wrinkles, embryo singularities.

Similarly to the nonparametric case, one can define the regularized differential of a fibered over B wrinkled map $F : V \to W$, which is a fibered epimorphism $\mathcal{R}(d_B f) : T_B V \to T_B W$.

4.4 The Wrinkling Theorem

The following theorem 4.4.1 and its parametric version 4.4.2 are adaptations of the results of our paper [EM97] to the simplest case $n = q$. In fact, in this case the results below can also be deduced from a theorem of V. Poénaru; see [Po].

4.4.1. (Wrinkled Mappings) *Let $F : T(V) \to T(W)$ be an epimorphism that covers a map $f : V \to W$. Suppose that f is an immersion on a neighborhood of a closed subset $K \subset V$, and F coincides with $\mathrm{d}f$ over that neighborhood. Then there exists a wrinkled map $g : V \to W$ that coincides with f near K and such that $\mathcal{R}(\mathrm{dg})$ and F are homotopic relations. $T(V)|_K$. Moreover, the map g can be chosen arbitrarily C^0-close to f and with arbitrarily small wrinkles.*

4.4.2. (Fibered Wrinkled Mappings) *Let $f : V \to W$ be a fibered over B map covered by a fibered epimorphism $F : T_B(V) \to T_B(W)$. Suppose that f is a fibered immersion on a neighborhood of a closed subset $K \subset V$, and F coincides with $\mathrm{d}f$ near a closed subset $K \subset V$. Then there exists a fibered wrinkled map $g : V \to W$ that extends f from a neighborhood of K and such that the fibered epimorphisms $\mathcal{R}(\mathrm{dg})$ and F are homotopic rel. $T_B(V)|_K$. Moreover, the map g can be chosen arbitrarily C^0-close to f and with arbitrarily small fibered wrinkles.*

Remark. The proof (see [EM97]) also gives the following useful enhancement for 4.4.2: given an open covering $\{U_i\}_{i=1,\dots,N}$ of B, one can always choose g such that for each fibered wrinkle there exists U_i such that $p(D^{k+q}) \subset U_i$, where D^{k+q} is the membrane of the wrinkle.

Acknowledgement Partially supported by NSF grants DMS-0707103 and DMS 0244663.

References

[El70] Y. Eliashberg, *On singularities of folding type*, Izv. Akad. Nauk SSSR, Ser. mat., **34**(1970), 1111–1127.

[El72] Y. Eliashberg, *Surgery of singularities of smooth maps*, Izv. Akad. Nauk SSSR Ser. Mat., **36**(1972), 1321–1347.

[EM97] Y. Eliashberg and N. Mishachev, *Wrinkling of smooth mappings and its applications I*, Invent. Math., **130**(1997), 345 369.

[Gr07] M. Gromov, *Singularities, Expanders and Topology of Maps. Part 1: Homology versus Volume in the Spaces of Cycles.* Preprint, Institut des Hautes Etudes Scientifiques, 2007.

[Hi] M.W. Hirsch, *On imbedding differentiable manifolds in Euclidean space*, Ann. of Math. **73**(1961), 566–571.

[Po] V. Poénaru, *Homotopy theory and differentiable singularities*, Manifolds-Amsterdam 1970 (Proc. Nuffic Summer School), Springer Lecture Notes in Math., **197**(1971), 106–132.

On Continuity of Quasimorphisms for Symplectic Maps

Michael Entov, Leonid Polterovich, Pierre Py, with an Appendix by Michael Khanevsky

Dedicated to Oleg Viro on the occasion of his 60th birthday

Abstract We discuss C^0-continuous homogeneous quasimorphisms on the identity component of the group of compactly supported symplectomorphisms of a symplectic manifold. Such quasimorphisms extend to the C^0-closure of this group inside the homeomorphism group. We show that for standard symplectic balls of any dimension, as well as for compact oriented surfaces other than the sphere, the space of such quasimorphisms is infinite-dimensional. In the case of surfaces, we give a user-friendly topological characterization of such quasimorphisms. We also present an application to Hofer's geometry on the group of Hamiltonian diffeomorphisms of the ball.

Keywords Symplectomorphism • Quasimorphism • Calabi homomorphism • Hofer metric

M. Entov (✉)
Department of Mathematics, Technion – Israel Institute of Technology, Israel
e-mail: entov@math.technion.ac.il

L. Polterovich
School of Mathematical Sciences, Tel Aviv University, Israel

Department of Mathematics, University of Chicago, USA
e-mail: polterov@post.tau.ac.il

P. Py
Department of Mathematics, University of Chicago, USA
e-mail: pierre.py@math.uchicago.edu

M. Khanevsky
School of Mathematical Sciences, Tel Aviv University, Israel
e-mail: khanev@post.tau.ac.il

I. Itenberg et al. (eds.), *Perspectives in Analysis, Geometry, and Topology: On the Occasion of the 60th Birthday of Oleg Viro*, Progress in Mathematics 296,
DOI 10.1007/978-0-8176-8277-4_8,

1 Introduction and Main Results

1.1 Quasimorphisms on Groups of Symplectic Maps

Let (Σ,ω) be a compact connected symplectic manifold (possibly with nonempty boundary $\partial\Sigma$). Denote by $\mathcal{D}(\Sigma,\omega)$ the identity component of the group of symplectic C^∞-diffeomorphisms of Σ whose supports lie in the interior of Σ. Write[1] $\mathcal{H}(\Sigma,\omega)$ for the C^0-closure of $\mathcal{D}(\Sigma,\omega)$ in the group of homeomorphisms of Σ supported in the interior of Σ. We always equip Σ with a distance d induced by a Riemannian metric on Σ, and view the C^0-topology on the group of homeomorphisms of Σ as the topology defined by the metric $dist(\phi,\psi)=\max_{x\in\Sigma} d(x,\psi^{-1}\phi(x))$.

The study of the algebraic structure of the groups $\mathcal{D}(\Sigma,\omega)$ was pioneered by Banyaga; see [2, 4]. For instance, when Σ is closed, he calculated the commutator subgroup of $\mathcal{D}(\Sigma,\omega)$ and showed that it is simple. However, the algebraic structure of the groups $\mathcal{H}(\Sigma,\omega)$ is much less understood. Even for the standard two-dimensional disk $\mathbb{D}^2$, it is still unknown whether $\mathcal{H}(\mathbb{D}^2)$ coincides with its commutator subgroup (see [10] for a comprehensive discussion). In the present paper, we focus on a particular algebraic feature of the groups $\mathcal{H}(\Sigma,\omega)$: homogeneous quasimorphisms.

Recall that a homogeneous quasimorphism on a group Γ is a map $\mu:\Gamma\to\mathbf{R}$ that satisfies the following two properties:

1. There exists a constant $C(\mu)\geq 0$ such that $|\mu(xy)-\mu(x)-\mu(y)|\leq C(\mu)$ for any x, y in Γ.
2. $\mu(x^n)=n\mu(x)$ for all $x\in\Gamma$ and $n\in\mathbf{Z}$.

Let us recall two well-known properties of homogeneous quasimorphisms that will be useful in the sequel: they are invariant under conjugation, and their restrictions to abelian subgroups are homomorphisms.

The space of all homogeneous quasimorphisms is an important algebraic invariant of the group. Quasimorphisms naturally appear in the theory of bounded cohomology and are crucial in the study of the commutator length [6]. We refer to [6, 14, 23] or [28] for a more detailed introduction to the theory of quasimorphisms.

Recently, several authors discovered that certain groups of diffeomorphisms preserving a volume or a symplectic form carry homogeneous quasimorphisms; see [5, 7, 17–19, 22, 41, 44, 45]. However, in many cases explicit constructions of nontrivial quasimorphisms on $\mathcal{D}(\Sigma,\omega)$ require a certain type of smoothness in an essential manner. Nevertheless, as we shall show below, some homogeneous quasimorphisms can be extended from $\mathcal{D}(\Sigma,\omega)$ to $\mathcal{H}(\Sigma,\omega)$.

Our first result deals with the case of the Euclidean unit ball $\mathbb{D}^{2n}$ in the standard symplectic linear space.

[1] We abbreviate $\mathcal{D}(\Sigma)$ and $\mathcal{H}(\Sigma)$ whenever the symplectic form ω is clear from the context.

Theorem 1. *The space of homogeneous quasimorphisms on $\mathcal{H}(\mathbb{D}^{2n})$ is infinite-dimensional.*

The proof is given in Sect. 2. Next, we focus on the case of a compact connected surface Σ equipped with an area form. Note that in this case, $\mathcal{H}(\Sigma)$ coincides with the identity component of the group of all area-preserving homeomorphisms supported in the interior of Σ; see [40] or [48].

Theorem 2. *Let Σ be a compact connected oriented surface other than the sphere $\mathbb{S}^2$, equipped with an area form. The space of homogeneous quasimorphisms on $\mathcal{H}(\Sigma)$ is infinite-dimensional.*

The proof is given in Sect. 4. This result is new, for instance, in the case of the 2-torus. The case of the sphere is still out of reach; see Sect. 5.2 for a discussion. Interestingly enough, for balls of any dimension and for the two-dimensional annulus, all our examples of homogeneous quasimorphisms on $\mathcal{H}$ are based on Floer theory. When Σ is of genus greater than one, the group $\mathcal{H}(\Sigma)$ carries plenty of homogeneous quasimorphisms, and the statement of Theorem 2 readily follows from the work of Gambaudo and Ghys [22].

As an immediate application, Theorems 1 and 2 yield that if Σ is a ball or a compact oriented surface other than the sphere, then the stable commutator length is unbounded on the commutator subgroup of $\mathcal{H}(\Sigma)$. This is a standard consequence of Bavard's theory [6].

1.2 Detecting Continuity

A key ingredient of our approach is the following proposition, due to Shtern [47]. It is a simple (nonlinear) analogue of the fact that linear forms on a topological vector space are continuous if and only if they are bounded in a neighborhood of the origin.

Proposition 1 ([47]). *Let Γ be a topological group and $\mu : \Gamma \to \mathbf{R}$ a homogeneous quasimorphism. Then μ is continuous if and only if it is bounded on a neighborhood of the identity.*

Proof. We only prove the "if" part. Assume that $|\mu|$ is bounded by $K > 0$ on an open neighborhood $\mathcal{U}$ of the identity. Let $g \in \Gamma$. For each $p \in \mathbf{N}$, define

$$\mathcal{V}_p(g) := \{h \in \Gamma \mid h^p \in g^p\mathcal{U}\}.$$

It is easy to see that $\mathcal{V}_p(g)$ is an open neighborhood of g. Pick any $h \in \mathcal{V}_p(g)$. Then $h^p = g^p f$ for some $f \in \mathcal{U}$. Therefore

$$|\mu(h^p) - \mu(g^p) - \mu(f)| \leq C(\mu),$$

and hence

$$|\mu(h)-\mu(g)| \leq \frac{C(\mu)+K}{p},$$

which immediately yields the continuity of μ at g. □

Let us discuss in greater detail the extension problem for quasimorphisms. The next proposition shows that C^0-continuous homogeneous quasimorphisms on $\mathcal{D}(\Sigma)$ extend to $\mathcal{H}(\Sigma)$.

Proposition 2. *Let Λ be a topological group and let $\Gamma \subset \Lambda$ be a dense subgroup. Any continuous homogeneous quasimorphism on Γ extends to a continuous homogeneous quasimorphism on Λ.*

Proof. Since μ is continuous, it is bounded by a constant $C > 0$ on an open neighborhood $\mathcal{U}$ of the identity in Γ. Since $\mathcal{U}$ is open in Γ, there exists $\mathcal{U}'$, open in Λ, such that $\mathcal{U} = \mathcal{U}' \cap \Gamma$. We fix an open neighborhood $\mathcal{O}$ of the identity in Λ such that $\mathcal{O}^2 \subset \mathcal{U}'$ and $\mathcal{O} = \mathcal{O}^{-1}$. Given $g \in \Lambda$ and $p \in \mathbf{N}$, we define as before

$$\mathcal{V}_p(g) := \left\{h \in \Lambda \mid h^p \in g^p \mathcal{O}\right\}.$$

Pick a sequence $\{h_k\}$ in Γ such that each h_k lies in $\mathcal{V}_1(g) \cap \ldots \cap \mathcal{V}_k(g)$. For $k \geq p$, we can write $h_k^p = g^p g_{k,p}$ ($g_{k,p} \in \mathcal{O}$). If $k_1, k_2 \geq p$, we can write

$$h_{k_1}^p = h_{k_2}^p g_{k_2,p}^{-1} g_{k_1,p}, \quad g_{k_2,p}^{-1} g_{k_1,p} \in \mathcal{U}.$$

Hence, we have the inequality

$$\left|\mu(h_{k_1}) - \mu(h_{k_2})\right| \leq \frac{C + C(\mu)}{p} \quad (k_1, k_2 \geq p),$$

and $\{\mu(h_p)\}$ is a Cauchy sequence in $\mathbf{R}$. Denote its limit by $\mu'(g)$. One can check easily that the definition is correct and that for any sequence $g_i \in \Gamma$ converging to $g \in \Lambda$, one has $\mu(g_i) \to \mu'(g)$. This readily yields that the resulting function $\mu' : \Lambda \to \mathbf{R}$ is a homogeneous quasimorphism extending μ. Its continuity follows from Proposition 1. □

In view of this proposition, all we need for the proof of Theorems 1 and 2 is to exhibit nontrivial homogeneous quasimorphisms on $\mathcal{D}(\Sigma)$ that are continuous in the C^0-topology. This leads us to the problem of continuity of homogeneous quasimorphisms, which is highlighted in the title of the present paper.

Remark 1. Note that all the concrete quasimorphisms that we know on groups of diffeomorphisms are continuous in the C^1-topology.

1.3 The Calabi Homomorphism and Continuity on Surfaces

It is a classical fact that the Calabi homomorphism is not continuous in the C^0-topology; see [21]. We will discuss the example of the unit ball in $\mathbf{R}^{2n}$ and then explain why the reason for the discontinuity of the Calabi homomorphism is, in a sense, universal.

First, let us recall the definition of the group of Hamiltonian diffeomorphisms of a symplectic manifold (Σ, ω). Given a smooth function $F : \Sigma \times S^1 \to \mathbf{R}$ supported in Interior$(\Sigma) \times S^1$, consider the time-dependent vector field sgradF_t given by $i_{\mathrm{sgrad}F_t}\omega = -\mathrm{d}F_t$, where $F_t(x)$ stands for $F(x,t)$. The flow f_t of this vector field is called the *Hamiltonian flow generated by the Hamiltonian function* F, and its time-one map f_1 is called the *Hamiltonian diffeomorphism generated by* F. Hamiltonian diffeomorphisms form a normal subgroup of $\mathcal{D}(\Sigma, \omega)$, denoted by $\mathrm{Ham}(\Sigma, \omega)$ or just by $\mathrm{Ham}(\Sigma)$. The quotient $\mathcal{D}(\Sigma)/\mathrm{Ham}(\Sigma)$ is isomorphic to a quotient of the group $H^1_{\mathrm{comp}}(\Sigma, \mathbf{R})$. In particular, $\mathcal{D}(\Sigma) = \mathrm{Ham}(\Sigma)$ for $\Sigma = \mathbb{D}^{2n}$ or for $\Sigma = \mathbb{S}^2$. We refer to [38] for the details.

Example 1. Let $\Sigma = \mathbb{D}^{2n}$ be the closed unit ball in $\mathbf{R}^{2n}$ equipped with the symplectic form $\omega = \mathrm{d}p \wedge \mathrm{d}q$. Take any diffeomorphism $f \in \mathrm{Ham}(\mathbb{D}^{2n})$ and choose a Hamiltonian F generating f. The value

$$\mathrm{Cal}(f) := \int_0^1 \int_{\mathbb{D}^{2n}} F(p,q,t)\,\mathrm{d}p\,\mathrm{d}q\,\mathrm{d}t$$

depends only on f and defines the *Calabi homomorphism* $\mathrm{Cal} : \mathcal{D}(\mathbb{D}^{2n}) \to \mathbf{R}$ [13].

Take a sequence of time-independent Hamiltonians F_i supported in balls of radii $\frac{1}{i}$ such that $\int_{\mathbb{D}^{2n}} F_i \,\mathrm{d}p\,\mathrm{d}q = 1$. The corresponding Hamiltonian diffeomorphisms f_i C^0-converge to the identity and satisfy $\mathrm{Cal}(f_i) = 1$. We conclude that the Calabi homomorphism is discontinuous in the C^0-topology.

In the remainder of this section, let us return to the case in which Σ is a compact connected surface equipped with an area form. Our next result shows, roughly speaking, that for a quasimorphism μ on $\mathrm{Ham}(\Sigma)$, its nonvanishing on a sequence of Hamiltonian diffeomorphisms f_i supported in a collection of shrinking balls is the only possible reason for discontinuity. The next remark is crucial for understanding this phenomenon. Observe that support$(f^N) \subset$ support(f) for any diffeomorphism f. Thus in the statement above, nonvanishing yields unboundedness: if $\mu(f_i) \neq 0$ for all i, then the sequence $\mu(f_i^{N_i}) = N_i \mu(f_i)$ is unbounded for an appropriate choice of N_i.

Theorem 3. *Let* $\mu : \mathrm{Ham}(\Sigma) \to \mathbf{R}$ *be a homogeneous quasimorphism. Then* μ *is continuous in the* C^0*-topology if and only if there exists* $a > 0$ *such that the following property holds: For any disk* $D \subset \Sigma$ *of area less than* a*, the restriction of* μ *to the group* $\mathrm{Ham}(D)$ *vanishes.*

Here, by a disk in Σ we mean the image of a smooth embedding $\mathbb{D}^2 \hookrightarrow \Sigma$. We view it as a surface with boundary equipped with the area form that is the restriction of the area form on Σ. The "only if" part of the theorem is elementary. It extends to certain four-dimensional symplectic manifolds (see Remark 2 below). The proof of the "if" part is more involved, and no extension to higher dimensions is available to us so far (see the discussion in Sect. 5.3 below).

Corollary 1. *Let* $\mu : \mathcal{D}(\Sigma) \to \mathbf{R}$ *be a homogeneous quasimorphism. Suppose that the following hold:*

(i) There exists $a > 0$ *such that for any disk* $D \subset \Sigma$ *of area less than* a*, the restriction of* μ *to the group* $\mathrm{Ham}(D)$ *vanishes.*
(ii) The restriction of μ *to each one-parameter subgroup of* $\mathcal{D}(\Sigma)$ *is linear.*

Then μ *is continuous in the* C^0*-topology.*

Note that assumption (ii) is indeed necessary, provided one believes in the axiom of choice. Indeed, assuming that Σ is not $\mathbb{D}^2$, $\mathbb{S}^2$, or $\mathbb{T}^2$, the quotient $\mathcal{D}(\Sigma)/\mathrm{Ham}(\Sigma)$ is isomorphic to the additive group of the vector space $V := H^1_{\mathrm{comp}}(\Sigma, \mathbf{R}) \neq \{0\}$. Define a quasimorphism $\mu : \mathcal{D}(\Sigma) \to \mathbf{R}$ as the composition of the projection $\mathcal{D}(\Sigma) \to V$ with a *discontinuous* homomorphism $V \to \mathbf{R}$. The homomorphism μ satisfies (i), since it vanishes on $\mathrm{Ham}(\Sigma)$, and it is obviously discontinuous.

The criteria of continuity stated in Theorem 3 and Corollary 1 are proved in Sect. 3. They will be used in Sect. 4 in order to verify C^0-continuity of a certain family of quasimorphisms on $\mathcal{D}(\mathbb{T}^2)$ introduced in [22] and explored in [46], which will enable us to complete the proof of Theorem 2.

1.4 An Application to Hofer's Geometry

Here we concentrate on the case of the unit ball $\mathbb{D}^{2n} \subset \mathbf{R}^{2n}$. For a diffeomorphism $f \in \mathrm{Ham}(\mathbb{D}^{2n})$, define its Hofer norm [26] as

$$\|f\|_H := \inf \int_0^1 \left(\max_{z \in \mathbb{D}^{2n}} F(z,t) - \min_{z \in \mathbb{D}^{2n}} F(z,t) \right) \mathrm{d}t,$$

where the infimum is taken over all Hamiltonian functions F generating f. Hofer's famous result states that $d_{\mathrm{H}}(f,g) := \|fg^{-1}\|_{\mathrm{H}}$ is a nondegenerate bi-invariant metric on $\mathrm{Ham}(\mathbb{D}^{2n})$. It is called *Hofer's metric* (see also [31, 42] for Hofer's metric on general symplectic manifolds). It turns out that the quasimorphisms that we construct in the proof of Theorem 1 are Lipschitz with respect to Hofer's metric. Hence, our proof of Theorem 1 yields the following result:

Proposition 3. *The space of homogeneous quasimorphisms on the group* $\mathrm{Ham}(\mathbb{D}^{2n})$ *that are both continuous for the* C^0*-topology and Lipschitz for Hofer's metric is infinite-dimensional.*

The relation between Hofer's metric and the C^0-metric on $\mathrm{Ham}(\Sigma)$ is subtle. First of all, the C^0-metric is never continuous with respect to Hofer's metric. Furthermore, arguing as in Example 1, one can show that Hofer's metric on $\mathrm{Ham}(\mathbb{D}^{2n})$ is not continuous in the C^0-topology. However, for $\mathbf{R}^{2n}$ equipped with the standard symplectic form $\mathrm{d}p \wedge \mathrm{d}q$ (informally speaking, this corresponds to the case of a ball of infinite radius), Hofer's metric is continuous for the C^0-Whitney topology [27].

An attempt to understand the relationship between Hofer's metric and the C^0-metric led Le Roux [34] to the following problem. Let $\mathscr{E}_C \subset \mathrm{Ham}(\mathbb{D}^{2n})$ be the complement of the closed ball (in Hofer's metric) of radius C centered at the identity:

$$\mathscr{E}_C := \{f \in \mathrm{Ham}(\mathbb{D}^{2n}), d_H(f, \mathbb{1}) > C\}.$$

Le Roux asked the following: Is it true that $\mathscr{E}_C$ has nonempty interior in the C^0-topology for any $C > 0$?

The energy-capacity inequality [26] states that if $f \in \mathrm{Ham}(\mathbb{D}^{2n})$ displaces $\phi(\mathbb{D}^{2n}(r))$, where ϕ is any symplectic embedding of the Euclidean ball of radius r, then Hofer's norm of f is at least πr^2. (We say that f *displaces a set* U if $f(U) \cap \bar{U} = \emptyset$.) By Gromov's packing inequality [25], this could happen only when $r^2 \leq 1/2$. Since any Hamiltonian diffeomorphism that is C^0-close to f also displaces a slightly smaller ball $\phi(D^{2n}(r'))$ $(r' < r)$, we get that $\mathscr{E}_C$ indeed has nonempty interior in the C^0-sense for $C < \pi/2$. Using our quasimorphisms, we get an affirmative answer to Le Roux's question even for large values of C.

Corollary 2. *For any $C > 0$, the set $\mathscr{E}_C$ has nonempty interior in the C^0-topology.*

Proof. The statement follows simply from the existence of a nontrivial homogeneous quasimorphism $\mu : \mathrm{Ham}(\mathbb{D}^{2n}) \to \mathbf{R}$ that is both continuous in the C^0-topology and Lipschitz with respect to Hofer's metric. Indeed, choose a diffeomorphism f such that

$$\frac{|\mu(f)|}{\mathrm{Lip}(\mu)} \geq C + 1,$$

where $\mathrm{Lip}(\mu)$ is the Lipschitz constant of μ with respect to Hofer's metric. There is a neighborhood O of f in $\mathrm{Ham}(\mathbb{D}^{2n})$ in the C^0-topology on which $|\mu| > C \cdot \mathrm{Lip}(\mu)$. We get that $\|g\|_H > C$ for $g \in O$, and hence $O \subset \mathscr{E}_C$. This proves the corollary. □

Note that Le Roux's question makes sense on any symplectic manifold. For certain closed symplectic manifolds with infinite fundamental group one can easily get a positive answer using the energy-capacity inequality in the universal cover (as in [32, 33]). However, for closed simply connected manifolds (and already for the case of the 2-sphere), the question is wide open.

2 Quasimorphisms for the Ball

In this section we prove Theorem 1. Denote by $\mathbb{D}^{2n}(r)$ the Euclidean ball $\{|p|^2+|q|^2 \leq r^2\}$, so that $\mathbb{D}^{2n} = \mathbb{D}^{2n}(1)$. We say that a set U in a symplectic manifold (Σ, ω) is *displaceable* if there exists $\phi \in \mathrm{Ham}(\Sigma)$ that displaces it: $\phi(U) \cap \bar{U} = \emptyset$. A quasimorphism $\mu : \mathrm{Ham}(\Sigma) \to \mathbf{R}$ will be called *Calabi* if for any displaceable domain $U \subset M$ such that $\omega|_U$ is exact, one has $\mu|_{\mathrm{Ham}(U)} = \mathrm{Cal}|_{\mathrm{Ham}(U)}$.

We will use the following result, established in [18]: there exists $a > 0$ such that the group $\mathrm{Ham}(\mathbb{D}^{2n}(1+a))$ admits an infinite-dimensional space of quasimorphisms that are Lipschitz in Hofer's metric, vanish on $\mathrm{Ham}(U)$ for every displaceable domain $U \subset \mathbb{D}^{2n}(1+a)$, and do not vanish on $\mathrm{Ham}(\mathbb{D}^{2n})$. These quasimorphisms are obtained by subtracting the appropriate multiple of the Calabi homomorphism from the Calabi quasimorphisms constructed in [9]. We claim that the restriction of each such quasimorphism, say η, to $\mathrm{Ham}(\mathbb{D}^{2n})$ is continuous in the C^0-topology. By Proposition 2, this yields the desired result. By Proposition 1, it suffices to show that for some $\epsilon > 0$ the quasimorphism η is bounded on all $f \in \mathrm{Ham}(\mathbb{D}^{2n})$ such that

$$|f(x) - x| < \epsilon \;\; \forall x \in \mathbb{D}^{2n}. \tag{1}$$

For $c > 0$ define the strip

$$\Pi(c) := \{(p,q) \in \mathbf{R}^{2n} \; : \; |q_n| < c\}.$$

Choose $\epsilon > 0$ so small that $\Pi(2\epsilon) \cap \mathbb{D}^{2n}$ is displaceable in $\mathbb{D}^{2n}(1+a)$. Put $D_\pm := \mathbb{D}^{2n} \cap \{\pm q_n > 0\}$. Observe that $D_\pm$ are displaceable in $\mathbb{D}^{2n}(1+a)$ by a Hamiltonian diffeomorphism that can be represented outside a neighborhood of the boundary as a small vertical shift along the q_n-axis (in the case of D_+, we take the shift that moves it up, and in the case of D_-, the shift that moves it down) composed with a $180°$ rotation in the (p_n, q_n)-plane. The desired boundedness result immediately follows from the following fragmentation-type lemma:

Lemma 1. *Assume that* $f \in \mathrm{Ham}(\mathbb{D}^{2n})$ *satisfies* (1). *Then* f *can be decomposed as* $\theta\phi_+\phi_-$, *where* $\theta \in \mathrm{Ham}(\Pi(2\epsilon) \cap \mathbb{D}^{2n})$ *and* $\phi_\pm \in \mathrm{Ham}(D_\pm)$.

Indeed, η vanishes on $\mathrm{Ham}(U)$ for every displaceable domain $U \subset \mathbb{D}^{2n}(1+a)$. Since $\Pi(2\epsilon) \cap \mathbb{D}^{2n}$ and $D_\pm$ are displaceable, $\eta(\theta) = \eta(\phi_\pm) = 0$. Thus $|\eta(f)| \leq 2C(\eta)$ for every $f \in \mathrm{Ham}(\mathbb{D}^{2n})$ lying in the ϵ-neighborhood of the identity with respect to the C^0-distance, and the theorem follows. It remains to prove the lemma.

Proof of Lemma 1: Denote by S the hyperplane $\{q_n = 0\}$. For $c > 0$ write R_c for the dilation $z \to cz$ of $\mathbf{R}^{2n}$. We assume that all compactly supported diffeomorphisms of $\mathbb{D}^{2n}$ are extended to the whole $\mathbf{R}^{2n}$ by the identity.

Take $f \in \mathrm{Ham}(\mathbb{D}^{2n})$ satisfying (1). Let $\{f_t\}_{0 \leq t \leq 1}$ be a Hamiltonian isotopy supported in $\mathbb{D}^{2n}$ such that $f_t = \mathbb{1}$ for $t \in [0, \delta)$ and $f_t = f$ for $t \in (1-\delta, 1]$ for some $\delta > 0$. Take a smooth function $c : [0,1] \to [1,+\infty)$ that equals 1 near 0 and 1 and satisfies $c(t) > (2\epsilon)^{-1}$ on $[\delta, 1-\delta]$. Consider the Hamiltonian isotopy $h_t = R_{1/c(t)} f_t R_{c(t)}$ of $\mathbf{R}^{2n}$. Note that $h_0 = \mathbb{1}$ and $h_1 = f$. Since $c(t) \geq 1$, we have $h_t z = z$ for $z \notin \mathbb{D}^{2n}$, and h_t is supported in $\mathbb{D}^{2n}$.

We claim that $h_t(S) \subset \Pi(2\epsilon)$. Observe that $R_{c(t)} S = S$. Take any $z \in S$. If $R_{c(t)} z \notin \mathbb{D}^{2n}$, we have that $h_t z = z$. Assume now that $R_{c(t)} z \in \mathbb{D}^{2n}$. Consider the following cases:

- If $t \in (1-\delta, 1]$, then $f_t R_{c(t)}(S) = f(S)$. Thus $f_t R_{c(t)} z \in f(S \cap \mathbb{D}^{2n}) \subset \Pi(2\epsilon)$, where the latter inclusion follows from (1). Therefore $h_t z \in \Pi(2\epsilon)$ since $c(t) \geq 1$.
- If $t \in [\delta, 1-\delta]$, then $h_t z \in \mathbb{D}^{2n}(2\epsilon) \subset \Pi(2\epsilon)$ by our choice of the function $c(t)$.
- If $t \in [0, \delta)$, then $h_t S = S \subset \Pi(2\epsilon)$.

This completes the proof of the claim.

By continuity of h_t, there exists $\kappa > 0$ such that $h_t(\Pi(\kappa)) \subset \Pi(2\epsilon)$ for all t. Cutting off the Hamiltonian of h_t near $h_t(\Pi(\kappa))$, we get a Hamiltonian flow θ_t supported in $\Pi(2\epsilon)$ that coincides with h_t on $\Pi(\kappa)$. Thus, $\theta_t^{-1} h_t$ is the identity on $\Pi(\kappa)$ for all t. It follows that $\theta_t^{-1} h_t$ decomposes into the product of two commuting Hamiltonian flows ϕ_-^t and ϕ_+^t supported in D_- and D_+ respectively. Therefore $f = \theta_1 \phi_-^1 \phi_+^1$ is the desired decomposition. □

3 Proof of the Criterion of Continuity on Surfaces

3.1 A C^0-Small Fragmentation Theorem on Surfaces

Before stating our next result, we recall the notion of *fragmentation* of a diffeomorphism. This is a classical technique in the study of groups of diffeomorphisms; see, e.g., [2, 4, 10]. Given a Hamiltonian diffeomorphism f of a connected symplectic manifold Σ and an open cover $\{U_\alpha\}$ of Σ, one can always write f as a product of Hamiltonian diffeomorphisms each of which is supported in one of the open sets U_α. It is known that the number of factors in such a decomposition is uniform in a C^1-neighborhood of the identity; see [2, 4, 10]. To prove our continuity theorem, we actually need to prove a similar result on surfaces in which one considers diffeomorphisms endowed with the C^0-topology. Such a result appears in [35] in the case when the surface is the unit disk. Observe also that the corresponding fragmentation result is known for volume-preserving homeomorphisms [20].

Theorem 4. *Let Σ be a compact connected surface (possibly with boundary), equipped with an area form. Then for every $a > 0$, there exist a neighborhood $\mathfrak{U}$ of the identity in the group* $\mathrm{Ham}(\Sigma)$ *endowed with the C^0-topology and an integer $N > 0$ such that every diffeomorphism $g \in \mathfrak{U}$ can be written as a product of at most N Hamiltonian diffeomorphisms supported in disks of area less than a.*

This result might be well known to experts and probably can be deduced from the corresponding result for homeomorphisms. However, since the proof is more difficult for homeomorphisms, and in order to keep this paper self-contained, we are going to give a direct proof of Theorem 4 in Sect. 6. Note that this last section is the most technical part of the text. Given the fragmentation result above, one obtains easily a proof of Theorem 3, as we will show now.

3.2 Proof of Theorem 3 and Corollary 1

1. We begin by proving that the condition appearing in the statement of the theorem is necessary for the quasimorphism μ to be continuous. Assume that μ is continuous for the C^0-topology. Then it is bounded on some C^0-neighborhood $\mathcal{U}$ of the identity in $\mathrm{Ham}(\Sigma)$. Choose now a disk D_0 in Σ. If D_0 has a sufficiently small diameter, then $\mathrm{Ham}(D_0) \subset \mathcal{U}$. But since $\mathrm{Ham}(D_0)$ is a subgroup and μ is homogeneous, μ must vanish on $\mathrm{Ham}(D_0)$.

   Now let $a = \mathrm{area}(D_0)$. If D is any disk of area less than a, the group $\mathrm{Ham}(D)$ is conjugate in $\mathrm{Ham}(\Sigma)$ to a subgroup of $\mathrm{Ham}(D_0)$, because for any two disks of the same area in Σ there exists a Hamiltonian diffeomorphism mapping one of the disks onto another; see, e.g., [1, Proposition A.1] for a proof (which, in fact, works for all Σ, though the claim there is stated only for closed surfaces). Hence, μ vanishes on $\mathrm{Ham}(D)$ as required.

Remark 2. This proof extends verbatim to higher-dimensional symplectic manifolds (Σ, ω) that admit a positive constant a_0 with the following property: for every $a < a_0$, all symplectically embedded balls of volume a in the interior of Σ are Hamiltonian isotopic. Here a symplectically embedded ball of volume a is the image of the standard Euclidean ball of volume a in $(\mathbf{R}^{2n}, \mathrm{d}p \wedge \mathrm{d}q)$ under a symplectic embedding. This property holds, for instance, for blowups of rational and ruled symplectic four-manifolds; see [8, 30, 36, 37].

2. We now prove the reverse implication. Assume that a homogeneous quasimorphism μ vanishes on all Hamiltonian diffeomorphisms supported in disks of area less than a. Take the C^0-neighborhood $\mathcal{U}$ of the identity and the integer N from Theorem 4. Then μ is bounded by $(N-1)C(\mu)$ on $\mathcal{U}$, and hence is continuous by Proposition 1. □

   We now prove Corollary 1. Choose compactly supported symplectic vector fields $v_1, \ldots, v_k$ on Σ such that the cohomology classes of the 1-forms $i_{v_j}\omega$ generate $H^1_{\mathrm{comp}}(\Sigma, \mathbf{R})$. Denote by h_i^t the flow of v_i. Let $\mathcal{V}$ be the image of the following map:

$$(-\epsilon, \epsilon)^k \to \mathcal{D}$$
$$(t_1, \ldots, t_k) \mapsto \prod_{i=1}^{k} h_i^{t_i}.$$

Using assumption (i) and applying Theorem 3, we get that the quasimorphism μ is bounded on a C^0-neighborhood, say $\mathcal{U}$, of the identity in $\mathrm{Ham}(\Sigma)$. Thus by (ii) and the definition of a quasimorphism, μ is bounded on $\mathcal{U} \cdot \mathcal{V}$. But the latter set is a C^0-neighborhood of the identity in $\mathcal{D}$. Thus μ is continuous on $\mathcal{D}$ by Proposition 1. □

4 Examples of Continuous Quasimorphisms

In this section we prove Theorem 2 case by case. The case of the disk has already been explained in Sect. 2. This construction generalizes verbatim to all closed surfaces of genus 0 with nonempty boundary, which proves Theorem 2 in this case.

When Σ is a closed surface of genus greater than one, Gambaudo and Ghys constructed in [22] an infinite-dimensional space of homogeneous quasimorphisms on the group $\mathcal{D}(\Sigma)$ satisfying the hypothesis of Theorem 3. These quasimorphisms are defined using 1-forms on the surface and can be thought of as some "quasifluxes." We refer to [22, Sect. 6.1] or to [23, Sect. 2.5] for a detailed description. The fact that these quasimorphisms extend continuously to the identity component of the group of area-preserving homeomorphisms of Σ can be checked easily without appealing to Theorem 3. This was already observed in [23].

In order to settle the case of surfaces of genus one, we shall apply the criterion given by Theorem 3. The quasimorphisms that we will use were constructed by Gambaudo and Ghys in [22]; see also [46]. We recall briefly this construction now.

The fundamental group $\pi_1(\mathbb{T}^2 \setminus \{0\})$ of the once-punctured torus is a free group on two generators, a and b, represented by a parallel and a meridian in $\mathbb{T}^2 \setminus \{0\}$. Let $\mu : \pi_1(\mathbb{T}^2 \setminus \{0\}) \to \mathbf{R}$ be a homogeneous quasimorphism. It is known that there are plenty of such quasimorphisms (see [11], for instance). We will associate to μ a homogeneous quasimorphism $\widetilde{\mu}$ on the group $\mathcal{D}(\mathbb{T}^2)$.

We fix a base point $x_* \in \mathbb{T}^2 \setminus \{0\}$. For all $v \in \mathbb{T}^2 \setminus \{0\}$ we choose a path $\alpha_v(t)$, $t \in [0,1]$, in $\mathbb{T}^2 \setminus \{0\}$ from x_* to v. We assume that the lengths of the paths α_v are uniformly bounded with respect to a Riemannian metric defined on the compact surface obtained by blowing up the origin on $\mathbb{T}^2$. Consider an element $f \in \mathcal{D}(\mathbb{T}^2)$ and fix an isotopy (f_t) from the identity to f. If x and y are distinct points in the torus, we can consider the curve

$$f_t(x) - f_t(y)$$

in $\mathbb{T}^2 \setminus \{0\}$. Its homotopy class depends only on f. We close it to form a loop:

$$\alpha(f,x,y) := \alpha_{x-y} * (f_t(x) - f_t(y)) * \overline{\alpha_{f(x)-f(y)}},$$

where $\overline{\alpha_{f(x)-f(y)}}(t) := \alpha_{f(x)-f(y)}(1-t)$. We have the cocycle relation

$$\alpha(fg,x,y) = \alpha(g,x,y) * \alpha(f,g(x),g(y)).$$

Define a function u_f on $\mathbb{T}^2 \times \mathbb{T}^2 \setminus \Delta$ (where Δ is the diagonal) by $u_f(x,y) = \mu(\alpha(f,x,y))$. From the previous relation and the fact that μ is a quasimorphism, we deduce the relation

$$\left|u_{fg}(x,y) - u_g(x,y) - u_f(g(x),g(y))\right| \leq C(\mu),\ \forall f,g \in \mathcal{D}(\mathbb{T}^2).$$

Moreover, it is not difficult to see that the function u_f is measurable and bounded on $\mathbb{T}^2 \times \mathbb{T}^2 \setminus \Delta$. Hence, the map

$$f \mapsto \int_{\mathbb{T}^2 \times \mathbb{T}^2} u_f(x,y)\mathrm{d}x\mathrm{d}y$$

is a quasimorphism. We denote by $\widetilde{\mu}$ the associated homogeneous quasimorphism

$$\widetilde{\mu}(f) = \lim_{p \to \infty} \frac{1}{p} \int_{\mathbb{T}^2 \times \mathbb{T}^2} u_{f^p}(x,y)\mathrm{d}x\mathrm{d}y.$$

One easily checks that $\widetilde{\mu}$ is linear on any 1-parameter subgroup. The following proposition was established in [46]:

Proposition 4. *Let $f \in \mathrm{Ham}(\mathbb{T}^2)$ be a diffeomorphism supported in a disk D. Then for any homogeneous quasimorphism $\mu : \pi_1(\mathbb{T}^2 \setminus \{0\}) \to \mathbf{R}$, one has*

$$\widetilde{\mu}(f) = 2\mu([a,b]) \cdot \mathrm{Cal}(f),$$

where $\mathrm{Cal} : \mathrm{Ham}(D) \to \mathbf{R}$ *is the Calabi homomorphism.*

By Corollary 1, we get that the quasimorphisms $\widetilde{\mu}$, where μ runs over the set of homogeneous quasimorphisms on $\pi_1(\mathbb{T}^2 \setminus \{0\})$ that take the value 0 on the element $[a,b]$, are all continuous in the C^0-topology. According to [22], this family spans an infinite-dimensional vector space. To complete the proof of Theorem 2 for surfaces of genus 1, we have only to check that the diffeomorphisms that were constructed in [22] in order to establish the existence of an arbitrary number of linearly independent quasimorphisms $\widetilde{\mu}$ can be chosen to be supported in any given subsurface of genus one. But this follows easily from the construction in [22, Sect. 6.2].

5 Discussion and Open Questions

5.1 Is $\mathcal{H}(\mathbb{D}^2)$ Simple? (Le Roux's Work)

Although the algebraic structure of groups of volume-preserving homeomorphisms in dimension greater than 2 is well understood [20], the case of area-preserving homeomorphisms of surfaces is still mysterious. In particular, it is unknown whether the group $\mathcal{H}(\mathbb{D}^2)$ is simple. Some normal subgroups of $\mathcal{H}(\mathbb{D}^2)$ were constructed by

Ghys, Oh, and more recently by Le Roux; see [10] for a survey. However, it is unknown whether any of these normal subgroups is a proper subgroup of $\mathcal{H}(\mathbb{D}^2)$. In [35], Le Roux established that the simplicity of the group $\mathcal{H}(\mathbb{D}^2)$ is equivalent to a certain fragmentation property. Namely, he established the following result (in the following, we assume that the total area of the disk is 1):

The group $\mathcal{H}(\mathbb{D}^2)$ is simple if and only if there exist numbers $\rho' < \rho$ in $(0,1]$ and an integer N such that any homeomorphism $g \in \mathcal{H}(\mathbb{D}^2)$ whose support is contained in a disk of area at most ρ can be written as a product of at most N homeomorphisms whose supports are contained in disks of area at most ρ'.

By a result of Fathi [20], see also [35], g can always be represented as such a product with some, a priori unknown, number of factors.

Remark 3. One can show that the property above depends only on ρ and not of the choice of ρ' smaller than ρ [35].

In the sequel we will denote by G_ε the set of homeomorphisms in $\mathcal{H}(\mathbb{D}^2)$ whose support is contained in an open disk of area at most ε. For an element g$\in \mathcal{H}(\mathbb{D}^2)$ we define (following [12, 35]) $|g|_\varepsilon$ as the minimal integer n such that g can be written as a product of n homeomorphisms of G_ε. Any homogeneous quasimorphism ϕ on $\mathcal{H}(\mathbb{D}^2)$ that vanishes on G_ε gives the following lower bound on $|\cdot|_\varepsilon$:

$$|g|_\varepsilon \geq \frac{|\phi(g)|}{C(\phi)} \quad (g \in \mathcal{H}(\mathbb{D}^2)).$$

In particular, if ϕ vanishes on G_ε but not on $G_{\varepsilon'}$ for some $\varepsilon' > \varepsilon$, then the norm $|\cdot|_\varepsilon$ is unbounded on $G_{\varepsilon'}$.

If $\phi : \mathcal{H}(\mathbb{D}^2) \to \mathbf{R}$ is a homogeneous quasimorphism that is continuous in the C^0-topology, we can define $a(\phi)$ to be the supremum of the positive numbers a satisfying the following property: ϕ vanishes on $\mathrm{Ham}(D)$ for any disk D of area less than or equal to a (for a homogeneous quasimorphism that is not continuous in the C^0-topology, one can define $a(\phi) = 0$). One can think of $a(\phi)$ as the *scale* at which one can detect the nontriviality of ϕ. According to the discussion above, the existence of a nontrivial quasimorphism with $a(\phi) > 0$ implies that the norm $|\cdot|_{a(\phi)}$ is unbounded on the set G_ρ (for any $\rho > a(\phi)$).

According to Le Roux's result, the existence of a sequence of continuous (for the C^0-topology) homogeneous quasimorphisms ϕ_n on $\mathcal{H}(\mathbb{D}^2)$ with $a(\phi_n) \to 0$ would imply that the group $\mathcal{H}(\mathbb{D}^2)$ is not simple. However, for all the examples of quasimorphisms on $\mathcal{H}(\mathbb{D}^2)$ that we know (coming from the continuous quasimorphisms described in Sect. 2), one has $a(\phi) \geq \frac{1}{2}$.

5.2 *Quasimorphisms on* $\mathbb{S}^2$

Consider the sphere $\mathbb{S}^2$ equipped with an area form of total area 1.

Question 1. (i) Does there exist a nonvanishing C^0-continuous homogeneous quasimorphism on $\mathrm{Ham}(\mathbb{S}^2)$?
(ii) If so, can it be made Lipschitz with respect to Hofer's metric?

If the answer to the first question is negative, this would imply that the Calabi quasimorphism constructed in [18] is unique. Indeed, the difference of two Calabi quasimorphisms is continuous in the C^0-topology according to Theorem 3. Note that for surfaces of positive genus, the examples of C^0-continuous quasimorphisms that we gave are related to the existence of many Calabi quasimorphisms [45, 46].

In turn, an affirmative answer to Question 1(ii) would yield the solution of the following problem posed by Misha Kapovich and the second author in 2006. It is known [43] that $\mathrm{Ham}(\mathbb{S}^2)$ carries a one-parameter subgroup, say $L := \{f_t\}_{t\in\mathbf{R}}$, that is a quasigeodesic in the following sense: $\|f_t\|_{\mathrm{H}} \geq c|t|$ for some $c > 0$ and all t. Given such a subgroup, put

$$A(L) := \sup_{\phi\in\mathrm{Ham}(\mathbb{S}^2)} d_{\mathrm{H}}(\phi, L).$$

Question 2. Is $A(L)$ finite or infinite?

The finiteness of $A(L)$ does not depend on the specific quasigeodesic one-parameter subgroup L. Intuitively, the finiteness of $A(L)$ would yield that the whole group $\mathrm{Ham}(\mathbb{S}^2)$ lies in a tube of finite radius around L.

We claim that if $\mathrm{Ham}(\mathbb{S}^2)$ admits a nonvanishing C^0-continuous homogeneous quasimorphism that is Lipschitz in Hofer's metric, then $A(L) = \infty$. Indeed, such a quasimorphism would be independent of the Calabi quasimorphism constructed in [18]. But the existence of two independent homogeneous quasimorphisms on $\mathrm{Ham}(\mathbb{S}^2)$ that are Lipschitz with respect to Hofer's metric implies that $A(L) = \infty$: otherwise, the finiteness of $A(L)$ would imply that Lipschitz homogeneous quasimorphisms are determined by their restriction to L.

5.3 *Quasimorphisms in Higher Dimensions*

Consider the following general question: given a homogeneous quasimorphism on $\mathrm{Ham}(\Sigma^{2n}, \omega)$, is it continuous in the C^0-topology?

The answer is positive, for instance, for quasimorphisms coming from the fundamental group $\pi_1(M)$ [22, 44]. It would be interesting to explore, for instance, the C^0-continuity of a quasimorphism μ given by the difference of a Calabi quasimorphism and the Calabi homomorphism [9, 18] (or more generally, by the difference of two distinct Calabi quasimorphisms). In order to prove the C^0-continuity of μ, one should establish a C^0-small fragmentation lemma with a controlled number of factors in the spirit of Lemma 1 for $\mathbb{D}^{2n}$ or Theorem 4 for surfaces. It is likely that the argument that we used for $\mathbb{D}^{2n}$ could go through without great complications for certain Liouville symplectic manifolds, that is, compact exact symplectic manifolds that admit a conformally symplectic vector field transversal to the boundary, such as the open unit cotangent bundle of the sphere.

Our result for $\mathbb{D}^{2n}$ should also allow the construction of continuous quasimorphisms for groups of Hamiltonian diffeomorphisms of certain symplectic manifolds symplectomorphic to "sufficiently large" open subsets of $\mathbb{D}^{2n}$ (for instance, the open unit cotangent bundle of a torus).

The C^0-small fragmentation problem on general higher-dimensional manifolds looks very difficult. Consider, for instance, the following toy case: find a fragmentation with a controlled number of factors for a C^0-small Hamiltonian diffeomorphism supported in a sufficiently small ball $D \subset \Sigma$. A crucial difference from the situation described in Sect. 2 is that we have no information about the Hamiltonian isotopy $\{f_t\}$ joining f with the identity: it can "travel" far away from D. In particular, when $\dim \Sigma \geq 6$, we do not know whether f lies in $\mathrm{Ham}(D)$. When $\dim \Sigma = 4$, the fact that $f \in \mathrm{Ham}(D)$ (and hence the fragmentation in our toy example) follows from a deep theorem by Gromov based on pseudoholomorphic curves techniques [25]. It would be interesting to apply powerful methods of four-dimensional symplectic topology to the C^0-small fragmentation problem.

6 Proof of the Fragmentation Theorem

In this section we prove Theorem 4. First, we need to recall a few classical results.

6.1 *Preliminaries*

In the course of the proof we will repeatedly use the following result:

Proposition 5. *Let Σ be a compact connected oriented surface, possibly with nonempty boundary $\partial\Sigma$, and let ω_1, ω_2 be two area forms on Σ. Assume that $\int_\Sigma \omega_1 = \int_\Sigma \omega_2$. If $\partial\Sigma \neq \emptyset$, we also assume that the forms ω_1 and ω_2 coincide on $\partial\Sigma$.*

Then there exists a diffeomorphism $f : \Sigma \to \Sigma$, isotopic to the identity, such that $f^*\omega_2 = \omega_1$. Moreover, f can be chosen to satisfy the following properties:

(i) If $\partial\Sigma \neq \emptyset$, then f is the identity on $\partial\Sigma$, and if ω_1 and ω_2 coincide near $\partial\Sigma$, then f is the identity near $\partial\Sigma$.

(ii) If Σ is partitioned into polygons (with piecewise smooth boundaries) such that $\omega_2 - \omega_1$ is zero on the 1-skeleton Γ of the partition and the integrals of ω_1 and ω_2 over each polygon are equal, then f can be chosen to be the identity on Γ.

(iii) The diffeomorphism f can be chosen arbitrarily C^0-close to $\mathbb{1}$, provided ω_1 and ω_2 are sufficiently C^0-close to each other (i.e., $\omega_2 = \chi\omega_1$ for a function χ sufficiently C^0-close to 1).

The existence of f in the case of a closed surface follows from a well-known theorem of Moser [39] (see also [24]). The method of the proof ("Moser's method") can be outlined as follows. Set $\omega_t := \omega_1 + t(\omega_2 - \omega_1)$ and note that the form $\omega_2 - \omega_1$ is exact. Choose a 1-form σ such that $d\sigma = \omega_2 - \omega_1$ and define f as the time-1 flow of the vector field ω_t-dual to σ. In order to prove (i) and (ii), one has to choose a primitive σ for $\omega_2 - \omega_1$ that vanishes near $\partial\Sigma$ or, respectively, on Γ; the construction of such a σ can be easily extracted from [3]. Property (iii) is essentially contained in [39]; it follows easily from the above construction of f, provided we can construct a C^0-small primitive σ for a C^0-small exact 2-form $\omega_2 - \omega_1$, but by [39, Lemma 1], it suffices to do so on a rectangle, and in this case σ can be constructed explicitly.

In fact, a stronger result than (iii) is true. It is known, see [40, 48], that f can be chosen C^0-close to the identity as soon as the two area forms (considered as measures) are close in the weak-$*$ topology. Note that if one of the two forms is the image of the other by a diffeomorphism C^0-close to the identity, the two forms are close in the weak-$*$ topology. However, to keep this text self-contained, we are not going to use this fact, but will prove again directly the particular cases we need.

We equip the surface Σ with a fixed Riemannian metric and denote by d the corresponding distance. For any map $f : X \to \Sigma$ (where X is a closed subset of Σ) we denote by $\|f\| := \max_x d(x, f(x))$ its C^0-norm. Accordingly, the C^0-norm of a smooth function u defined on a closed subset of Σ will be denoted by $\|u\|$.

The following lemmas are the main tools for the proof.

Lemma 2 (Area-preserving extension lemma for disks). *Let $D_1 \subset D_2 \subset D \subset \mathbf{R}^2$ be closed disks such that $D_1 \subset$ Interior $(D_2) \subset D_2 \subset$ Interior (D). Let $\phi : D_2 \to D$ be a smooth area-preserving embedding (we assume that D is equipped with some area form). Then there exists $\psi \in \mathrm{Ham}(D)$ such that*

$$\psi|_{D_1} = \phi \quad \text{and} \quad \|\psi\| \to 0 \text{ as } \|\phi\| \to 0.$$

Lemma 3 (Area-preserving extension lemma for rectangles). *Let $\Pi = [0, R] \times [-c, c]$ be a rectangle and let $\Pi_1 \subset \Pi_2 \subset \Pi$ be two smaller rectangles of the form $\Pi_i = [0, R] \times [-c_i, c_i]$ $(i = 1, 2)$, $0 < c_1 < c_2 < c$. Let $\phi : \Pi_2 \to \Pi$ be an area-preserving embedding (we assume that Π is equipped with some area form) such that:*

- *ϕ is the identity near $0 \times [-c_2, c_2]$ and $R \times [-c_2, c_2]$.*
- *The area in Π bounded by the curve $[0, R] \times y$ and its image under ϕ is zero for some (and hence for all) $y \in [-c_2, c_2]$.*

Then there exists $\psi \in \mathrm{Ham}(\Pi)$ such that

$$\psi|_{\Pi_1} = \phi \quad \text{and} \quad \|\psi\| \to 0 \text{ as } \|\phi\| \to 0.$$

The lemmas will be proved in Sect. 6.3. Let us mention that we implicitly assume in these lemmas that ϕ is close to the inclusion, i.e., that $\|\phi\|$ is small enough. Note that if one is interested only in the existence of ψ, without any control on its norm $\|\psi\|$, these results are standard.

6.2 Construction of the Fragmentation

We are now ready to prove the fragmentation theorem. In the case that Σ is the closed unit disk $\mathbb{D}^2$ in $\mathbf{R}^2$, the theorem has been proved by Le Roux [35, Proposition 4.2]. In general, our proof relies on the case of the disk.

For any $b > 0$ we fix a neighborhood $\mathscr{U}_0(b)$ of the identity in $\mathrm{Ham}(\mathbb{D}^2)$ and an integer $N_0(b)$ such that every element of $\mathscr{U}_0(b)$ is a product of at most $N_0(b)$ diffeomorphisms supported in disks of area at most b. We will prove the following assertion.

For any surface Σ there exist an integer N_1 and disks $(D_j)_{1\leq j\leq N_1}$ in Σ such that for any $\epsilon > 0$ there exists a neighborhood $\mathscr{V}(\epsilon)$ of the identity in $\mathrm{Ham}(\Sigma)$ *with the property that every diffeomorphism $f \in \mathscr{V}(\epsilon)$ can be written as a product $f = g_1 \cdots g_{N_1}$, where each g_i belongs to* $\mathrm{Ham}(D_j)$ *for one of the disks D_j and is ϵ-close to the identity. $(*)$*

Note that there is no restriction in $(*)$ on the areas of the disks D_j. Let us explain how to conclude the proof of Theorem 4 from this assertion. Fix $a > 0$. We can choose, for each i between 1 and N_1, a conformally symplectic diffeomorphism $\psi_i : \mathbb{D}^2 \to D_i$ such that the pullback of the area form on Σ by ψ_i equals the standard area form on the disk $\mathbb{D}^2$ times some constant $\lambda_i > 0$. Here we are using Proposition 5. If ϵ is sufficiently small, $\psi_i^{-1} g_i \psi_i$ is in $\mathscr{U}_0(\frac{a}{\lambda_i})$ for each i, and we can apply the result for the disk to it. This concludes the proof.

Remark 4. It is important that the disks D_i as well as the maps ψ_i are chosen in advance, since we need the neighborhoods $\psi_i \mathscr{U}_0(\frac{a}{\lambda_i})\psi_i^{-1}$ to be known in advance. They determine the neighborhood $\mathscr{V}(\epsilon)$.

We now prove $(*)$. The arguments we use are inspired by the work of Fathi [20]. Fix $\epsilon > 0$. We distinguish two cases: (1) Σ has a boundary, and (2) Σ is closed.

First case. Any compact connected surface with nonempty boundary can be obtained by gluing finitely many 1-handles to a disk. We prove the statement $(*)$ by induction on the number of 1-handles. We already know that $(*)$ is true for a disk (just take $N_1 = 1$ and let D_1 be the whole disk). Assume now that $(*)$ holds for any compact surface with boundary obtained by gluing l 1-handles to the disk. Let Σ be a compact surface obtained by gluing a 1-handle to a compact surface Σ_0, where Σ_0 is obtained from the disk by gluing l 1-handles.

Choose a diffeomorphism (singular at the corners) $\varphi : [-1,1]^2 \to \overline{\Sigma - \Sigma_0}$ sending $[-1,1] \times \{-1,1\}$ into the boundary of Σ_0. Let $\Pi_r = \varphi([-1,1] \times [-r,r])$. Let $\mathscr{V}_1(\epsilon)$

be the neighborhood of the identity in $\mathrm{Ham}(\Sigma_1)$ given by $(*)$ applied to the surface $\Sigma_1 := \Sigma_0 \cup \varphi([-1,1] \times \{s, |s| \geq \frac{1}{4}\})$, and let N_1 be the corresponding integer.

Let $f \in \mathrm{Ham}(\Sigma)$ close to the identity. We apply Lemma 3 to the chain of rectangles $\Pi_{\frac{1}{2}} \subset \Pi_{\frac{3}{4}} \subset \Pi_{\frac{7}{8}}$ and to the restriction of f to $\Pi_{\frac{3}{4}}$ (the hypothesis on the curve $[-1,1] \times \{y\}$ is met because f is Hamiltonian). Here again we are appealing to Proposition 5 to ensure that the pullback of the area form of Σ by φ can be identified with a fixed area form on $\Pi_{\frac{7}{8}}$. We obtain a diffeomorphism ψ supported in $\Pi_{\frac{7}{8}}$ and C^0-close to the identity that coincides with f on $\Pi_{\frac{1}{2}}$. Hence, we can write

$$f = \psi h,$$

where h is supported in Σ_1. Since $f \in \mathrm{Ham}(\Sigma)$ and $\psi \in \mathrm{Ham}(\Pi_{\frac{7}{8}})$, we get that h is Hamiltonian in Σ. Since $H^1_{\mathrm{comp}}(\Sigma_1, \mathbf{R})$ embeds in $H^1_{\mathrm{comp}}(\Sigma, \mathbf{R})$, it means that h actually belongs to $\mathrm{Ham}(\Sigma_1)$.

Define a neighborhood $\mathscr{V}(\epsilon)$ of the identity in $\mathrm{Ham}(\Sigma)$ by the following condition: $f \in \mathscr{V}(\epsilon)$ if first, $\|\psi\| < \epsilon$ (recall that when f converges to the identity, so does ψ) and second, $h \in \mathscr{V}_1(\epsilon)$. Hence, if $f \in \mathscr{V}(\epsilon)$, we can write it as a product of $N_1 + 1$ diffeomorphisms g_i, where each g_i is ϵ-close to the identity and belongs to $\mathrm{Ham}(D_j)$ for some disk $D_j \subset \Sigma$. This proves the claim $(*)$ for Σ in the first case.

Second case. The surface Σ is closed – we view it as a result of gluing a disk to a surface Σ_0 with one boundary component. Choose a diffeomorphism $\varphi : \mathbb{D}^2 \to \overline{\Sigma - \Sigma_0}$ sending the boundary of $\mathbb{D}^2$ into the boundary of Σ_0. Once again, by appealing to Proposition 5, we can assume that the pullback by φ of the area form of Σ is a given area form on $\mathbb{D}^2$. Denote by D_r the image by φ of the disk of radius $r \in [0,1]$ in $\mathbb{D}^2$. Let $\mathscr{V}_1(\epsilon)$ be the neighborhood of the identity given by $(*)$ applied to the surface $\Sigma_1 := \Sigma_0 \cup \varphi(\{z \in \mathbb{D}^2, |z| \geq \frac{1}{4}\})$ and let N_1 be the corresponding integer – recall that in the first case above, we have already proved $(*)$ for Σ_1, which is a surface with boundary.

Let $f \in \mathrm{Ham}(\Sigma)$ close to the identity. We apply Lemma 2 to the chain of disks $D_{\frac{1}{2}} \subset D_{\frac{3}{4}} \subset D_1$ and to the restriction of f to $D_{\frac{3}{4}}$. We obtain a diffeomorphism ψ supported in D_1 and close to the identity that coincides with f on $D_{\frac{1}{2}}$. Hence, we can write

$$f = \psi h,$$

where h is supported in Σ_1. Since $f \in \mathrm{Ham}(\Sigma)$ and $\psi \in \mathrm{Ham}(D_1)$, we get that h is Hamiltonian in Σ. Since Σ_1 has one boundary component, $H^1_{\mathrm{comp}}(\Sigma_1, \mathbf{R})$ embeds in $H^1_{\mathrm{comp}}(\Sigma, \mathbf{R})$, so h actually belongs to $\mathrm{Ham}(\Sigma_1)$. One concludes the proof as in the first case.

This finishes the proof of Theorem 4 (modulo the proofs of the extension lemmas).

6.3 Extension Lemmas

The area-preserving extension lemmas for disks and rectangles will be consequences of the following lemma.

Lemma 4 (Area-preserving extension lemma for annuli). *Let $\mathbb{A} = S^1 \times [-3,3]$ be a closed annulus and let $\mathbb{A}_1 = S^1 \times [-1,1], \mathbb{A}_2 = S^1 \times [-2,2]$ be smaller annuli inside $\mathbb{A}$. Let ϕ be an area-preserving embedding of a fixed open neighborhood of $\mathbb{A}_1$ into $\mathbb{A}_2$ (we assume that $\mathbb{A}$ is equipped with some area form ω) such that for some $y \in [-1,1]$ (and hence for all of them), the curves $S^1 \times y$ and $\phi(S^1 \times y)$ are homotopic in $\mathbb{A}$ and*

$$\text{the area in } \mathbb{A} \text{ bounded by } S^1 \times y \text{ and } \phi(S^1 \times y) \text{ is } 0. \tag{2}$$

Then there exists $\psi \in \mathrm{Ham}(\mathbb{A})$ such that $\psi|_{\mathbb{A}_1} = \phi$ and $\|\psi\| \to 0$ as $\|\phi\| \to 0$.

Moreover, if for some arc $I \subset S^1$ we have that $\phi = \mathbb{1}$ outside a quadrilateral $I \times [-1,1]$ and $\phi(I \times [-1,1]) \subset I \times [-2,2]$, then ψ can be chosen to be the identity outside $I \times [-3,3]$.

Once again, we assume in this lemma that $\|\phi\|$ is small enough. Let us show how this lemma implies the area-preserving extension lemmas for disks and rectangles.

Proof of Lemma 2. Up to replacing D_2 by a slightly smaller disk, we can assume that ϕ is defined in a neighborhood of D_2. Identify some small neighborhood of ∂D_2 with $\mathbb{A} = S^1 \times [-3,3]$ so that ∂D_2 is identified with $S^1 \times 0 \subset \mathbb{A}_1 \subset \mathbb{A}_2 \subset \mathbb{A}$ and $\phi(\mathbb{A}_1) \subset \text{Interior } (\mathbb{A}_2) \subset \mathbb{A} \subset \text{Interior } (D) \setminus \phi(D_1)$.

Apply Lemma 4 and find $h \in \mathrm{Ham}(\mathbb{A})$, $\|h\| \to 0$ as $\epsilon \to 0$, so that $h|_{\mathbb{A}_1} = \phi$. Set $\phi_1 := h^{-1} \circ \phi \in \mathrm{Ham}(D)$. Note that $\phi_1|_{D_1} = \phi$ and ϕ_1 is the identity on $\mathbb{A}_1$. Therefore we can extend $\phi_1|_{D_2 \cup \mathbb{A}_1}$ to D by the identity and obtain the required ψ. □

Proof of Lemma 3. Identify the rectangles $\Pi_1 \subset \Pi_2 \subset \Pi$, by a diffeomorphism, with quadrilaterals $I \times [-1,1] \subset I \times [-2,2] \subset I \times [-3,3]$ in the annulus $\mathbb{A} = S^1 \times [-3,3]$ for some suitable arc $I \subset S^1$ and apply Lemma 4. □

In order to prove Lemma 4, we first need to prove a version of the lemma concerning smooth (not necessarily area-preserving) embeddings.

Lemma 5 (Smooth extension lemma). *Let $\mathbb{A}_1 \subset \mathbb{A}_2 \subset \mathbb{A}$ be as in Lemma 4. Let ϕ be a smooth embedding of a fixed open neighborhood of $\mathbb{A}_1$ into $\mathbb{A}_2$, isotopic to the identity, such that $\|\phi\| \leq \epsilon$ for some $\epsilon > 0$. Then there exists $\psi \in \mathrm{Diff}_{0,c}(\mathbb{A})$ such that ψ is supported in $\mathbb{A}_2$, $\psi|_{\mathbb{A}_1} = \phi$, and $\|\psi\| \leq C\epsilon$, for some $C > 0$, independent of ϕ.*

Moreover, if $\phi = \mathbb{1}$ outside a quadrilateral $I \times [-1,1]$ and $\phi(I \times [-1,1]) \subset I \times [-2,2]$ for some arc $I \subset S^1$, then ψ can be chosen to be the identity outside $I \times [-3,3]$.

Lemma 5 will be proved in Sect. 6.4.

Proof of Lemma 4. As one can easily check using Proposition 5, we can assume without loss of generality that the area form on $\mathbb{A} = S^1 \times [-3,3]$ is $\omega = dx \wedge dy$, where x is the angular coordinate along S^1 and y is the coordinate along $[-3,3]$. All norms and distances are measured with the Euclidean metric on $\mathbb{A}$. Define $\mathbb{A}_+ := S^1 \times [1,2]$, $\mathbb{A}_- := S^1 \times [-2,-1]$.

Assume $\|\phi\| < \epsilon$. By Lemma 5, there exists $f \in \mathrm{Diff}_{0,c}(\mathbb{A}_2)$ such that $\|f\| \le C\epsilon$, and $f = \phi$ on a neighborhood of $\mathbb{A}_1$. Define $\Omega := f^*\omega$. By (2),

$$\int_{\mathbb{A}_+} \Omega = \int_{\mathbb{A}_+} \omega, \int_{\mathbb{A}_-} \Omega = \int_{\mathbb{A}_-} \omega. \tag{3}$$

Note that Ω coincides with ω on a neighborhood of $\partial\mathbb{A}_+$ and $\partial\mathbb{A}_-$. Let us find $h \in \mathrm{Diff}_{0,c}(\mathbb{A}_2)$ such that

- $h|_{\mathbb{A}_1} = \mathbb{1}$,
- $h^*\Omega = \omega$,
- $\|h\| \to 0$ as $\epsilon \to 0$.

Given such an h, we extend fh by the identity to the whole of $\mathbb{A}$. The resulting diffeomorphism of $\mathbb{A}$ is C^0-small (if ϵ is sufficiently small), preserves ω, and belongs to $\mathrm{Diff}_{0,c}(\mathbb{A})$, hence (see, e.g., [50]) also to $\mathcal{D}(\mathbb{A})$. It may not be Hamiltonian, but one can easily make it Hamiltonian by a C^0-small adjustment on $\mathbb{A} \setminus \mathbb{A}_2$. The resulting diffeomorphism $\psi \in \mathrm{Ham}(\mathbb{A})$ will have all the required properties.

Preparations for the construction of *h*. Since on $\mathbb{A}_1$ the map h is required to be the identity, we need to construct it on $\mathbb{A}_+$ and $\mathbb{A}_-$. We will construct $h_+ := h|_{\mathbb{A}_+}$, the case of $\mathbb{A}_-$ being similar. By a rectangle or a square in $\mathbb{A}$ we mean the product of a connected arc in S^1 and an interval in $[-3,3]$.

Let us divide $\mathbb{A}_+ = S^1 \times [1,2]$ into closed squares $K_1, \ldots, K_N$, with a side of size $r = \epsilon^{1/4} > 3\epsilon$ (we assume that ϵ is sufficiently small). Denote by V the set of vertices that are not on the boundary and by E the set of edges that are not on the boundary. Finally, denote by Γ the 1-skeleton of the partition (i.e., the union of all the edges).

For each $v \in V$ denote by $B_v(\delta)$ the open ball in $\mathbb{A}_+$ of radius $\delta > 0$ with center at v. Fix a small positive $\delta_0 < r$ such that for $0 < \delta < \delta_0$, the balls $B_v(\delta)$, $v \in V$, are disjoint and each $B_v(\delta)$ intersects only the edges adjacent to v. Given such a δ, consider for each edge $e \in E$ a small open rectangle $U_e(\delta)$ covering $e \setminus \big(e \cap \cup_{v\in V} B_v(\delta)\big)$ such that

- $U_e(\delta) \cap B_v(\delta) \neq \emptyset$ if and only if v is adjacent to e.
- $U_e(\delta)$ does not intersect any other edge apart from e.
- All the rectangles $U_e(\delta)$, $e \in E$, are mutually disjoint.

Define a neighborhood $U(\delta)$ of Γ by

$$U(\delta) = (\cup_{v\in V} B_v(\delta)) \cup (\cup_{e\in E} U_e(\delta)).$$

For each $\varepsilon_1 > \varepsilon_2 > 0$ we pick a cut-off function $\chi_{\varepsilon_1,\varepsilon_2} : \mathbf{R} \to [0,1]$ that is equal to 1 on a neighborhood of $(-\varepsilon_2, \varepsilon_2)$ and vanishes outside $(-\varepsilon_1, \varepsilon_1)$. Finally, by $C_1, C_2, \ldots$ we will denote positive constants independent of ϵ. The construction of h_+ will proceed in several steps.

Adjusting Ω on Γ. We are going to adjust the form Ω by a diffeomorphism supported inside $U(\delta)$ to make it equal to ω on Γ. One can first construct $h_1 \in \mathrm{Diff}_{0,c}(\mathbb{A}_+)$ supported in $\cup_{v\in V} B_v(2\delta)$ such that $h_1^*\Omega = \omega$ on $\cup_{v\in V} B_v(\delta)$ for some $\delta < \delta_0$ (simply using Darboux charts for Ω and ω). Note that $\|h_1\| < 2\delta$. Write $\Omega' := h_1^*\Omega$. For each $e \in E$ we will construct a diffeomorphism h_e supported in $U_e(\delta)$ so that $h_e^*\Omega' = \omega$ on $l := U_e(\delta) \cap e$ (and thus on the whole e, since Ω' already equals ω on each $B_v(\delta)$).

Without loss of generality, let us assume that e does not lie on $\partial\mathbb{A}_+$ (since Ω' already coincides with ω there) and that $U_e(\delta)$ is of the form $(a,b) \times (-\delta, \delta)$. Write the restriction of Ω' on $l = (a,b) \times 0$ as $\beta(x)\mathrm{d}x \wedge \mathrm{d}y$, $\beta(x) > 0$.

Consider a cut-off function $\chi = \chi_{\delta,\delta/2} : \mathbf{R} \to [0,1]$ and define a vector field $\mathbf{w}(x,y)$ on $U_e(\delta)$ by

$$\mathbf{w}(x,y) = \chi(y)\log(\beta(x))y\frac{\partial}{\partial y}.$$

Note that $\mathbf{w}$ is zero on l and has compact support in $U_e(\delta)$ (the endpoints of l lie in the balls $B_v(\delta)$ on which $\Omega = \omega$ and thus $\beta = 1$ near these endpoints). Let φ_t be the flow of $\mathbf{w}$. A simple calculation shows that

$$\frac{\mathrm{d}}{\mathrm{d}t}\varphi_t^*\omega = \varphi_t^* L_{\mathbf{w}}\omega = \log(\beta(x))\mathrm{e}^{t\log(\beta(x))}\mathrm{d}x \wedge \mathrm{d}y$$

at the point $\varphi_t((x,0)) = (x,0)$. Therefore $\varphi_1^*\omega = \Omega'$ on l. Thus setting $h_e := \varphi_1^{-1}$, we get that $h_e^*\Omega' = \omega$ on l and that $\|h_e\| \le 2\delta$, because h_e preserves the fibers $x \times (-\delta, \delta)$. Set

$$h_2 := \prod_{e\in E} h_e.$$

Since the rectangles $U_e(\delta)$ are disjoint, h_2 is supported in $U(\delta)$ and satisfies the conditions

- $h_2^*\Omega' = \omega$ on Γ.
- $\|h_2\| \le 2\delta$.

The diffeomorphism $h_3 := h_1 h_2 \in \mathrm{Diff}_{0,c}(\mathbb{A}_+)$ satisfies $\|h_3\| \le 4\delta$ and

$$h_3^*\Omega = h_2^*\Omega' = \omega \text{ on } \Gamma.$$

Consider the area form $\Omega'' := h_3^*\Omega$. It coincides with ω on the 1-skeleton Γ and near $\partial\mathbb{A}_+$. Moreover, $\int_{\mathbb{A}_+}\Omega'' = \int_{\mathbb{A}_+}\Omega'$, and hence by (3),

$$\int_{\mathbb{A}_+}\Omega'' = \int_{\mathbb{A}_+}\omega. \tag{4}$$

Adjusting the areas of the squares. In this paragraph we construct a C^0-perturbation $\rho\omega$ of ω that has the same integral as Ω'' on each square K_i.

Making δ sufficiently small, we can assume that $\|h_3\| < \epsilon$. Recall that $r = \epsilon^{1/4} > 3\epsilon$. Therefore the image of one of the squares K_i by h_3 contains a square of area $(r-\epsilon)^2$ and is contained in a square of area $(r+\epsilon)^2$. Hence,

$$\frac{(r-2\epsilon)^2}{r^2} \leq \frac{\int_{K_i}\Omega''}{\int_{K_i}\omega} \leq \frac{(r+2\epsilon)^2}{r^2}.$$

Since $\epsilon/r = \epsilon^{3/4} \to 0$ as $\epsilon \to 0$, we get that if ϵ is sufficiently small, there exists $C_1 > 0$ such that

$$1 - C_1\frac{\epsilon}{r} \leq \frac{\int_{K_i}\Omega''}{\int_{K_i}\omega} \leq 1 + C_1\frac{\epsilon}{r}. \tag{5}$$

Now set $s_i := \int_{K_i}\Omega''$ and $t_i = s_i/r^2 - 1$. By (5),

$$|t_i| \leq C_1\frac{\epsilon}{r} = C_1.\epsilon^{3/4}. \tag{6}$$

For each i we can choose a nonnegative function $\bar{\rho}_i$ supported in the interior of K_i such that $\int_{K_i}\bar{\rho}_i\omega = r^2$ and

$$\|\bar{\rho}_i\|_{C^0} \leq C_2\epsilon^{-1/2} \tag{7}$$

for some constant $C_2 > 0$ independent of i. Define a function ϱ on $\mathbb{A}$ by

$$\varrho := 1 + \sum_{i=1}^{N} t_i\bar{\rho}_i.$$

By (6) and (7), the function ϱ is positive, and the form $\varrho\omega$ converges to ω (in the C^0-sense) as ϵ goes to 0. Moreover, ϱ is equal to 1 on Γ, and the two area forms $\varrho\omega$ and Ω'' have the same integral on each K_i. By (4), one has

$$\int_{\mathbb{A}_+}\varrho\omega = \int_{\mathbb{A}_+}\Omega'' = \int_{\mathbb{A}_+}\omega. \tag{8}$$

Finishing the construction of h_+: Moser's argument. Let us apply Proposition 5, part (ii), to the forms Ω'' and $\varrho\omega$ on $\mathbb{A}_+$. These forms have the same integral over each K_i and coincide on Γ and near the boundary of $\mathbb{A}_+$; therefore, there exists a diffeomorphism $h_4 \in \mathrm{Diff}_{0,c}(\mathbb{A}_+)$ that is the identity on Γ and satisfies $h_4^*\Omega'' = \varrho\omega$. Since h_4 is the identity on Γ and maps each K_i into itself, its C^0-norm is bounded by the diameter of K_i, hence goes to 0 with ϵ.

Finally, apply Proposition 5 to the forms ω and $\varrho\omega$ on $\mathbb{A}_+$: By (8), their integrals over $\mathbb{A}_+$ are the same; they coincide on $\partial\mathbb{A}_+$ and are C^0-close. Therefore, there exists $h_5 \in \mathrm{Diff}_{0,c}(\mathbb{A}_+)$ such that $h_5^*(\varrho\omega) = \omega$ and

$$\|h_5\| \to 0 \text{ as } \epsilon \to 0. \tag{9}$$

Then $h_+ := h_3 h_4 h_5$ is the required diffeomorphism. This finishes the construction of h.

Final observation. Note that if $\phi = \mathbb{1}$ outside a quadrilateral $I \times [-1,1]$ for some arc $I \subset S^1$, then f can be chosen to have the same property. In such a case we need to construct $h_+ \in \mathrm{Diff}_{0,c}(\mathbb{A}_+)$ supported in $I \times [-3,3]$.

Let J be the complement of the interval I in the circle. The partition of $\mathbb{A}_+$ into squares can be chosen so that it extends a partition of $J \times [1,2] \subset \mathbb{A}_+$ into squares of the same size. Going over each step of the construction of h_+ above, we see that since $\Omega = \omega$ on $J \times [1,2]$, each of the maps h_1, h_2, h_3, h_4, h_5 can be chosen to be the identity on each of the squares in $J \times [1,2]$, hence on the whole $J \times [1,2]$. Therefore, h_+, hence h, hence $\psi = fh$, is the identity on $J \times [1,2]$. Moreover, ψ is automatically Hamiltonian in this case. □

6.4 *Proof of the Smooth Extension Lemma*

As in the proof of Lemma 4, we assume that the Riemannian metric on $\mathbb{A} = S^1 \times [-3,3]$ used for the measurements is the Euclidean product metric. We can also assume that the neighborhood of $\mathbb{A}_1$ on which ϕ is defined is, in fact, an open neighborhood of $\mathbb{A}' := S^1 \times [-1.5, 1.5]$ and that $\epsilon \ll 0.5$.

Proof of Lemma 5. Applying Lemma 6 (see the appendix by M. Khanevsky below) to the two curves $S^1 \times \{\pm 1.5\}$ and their images under ϕ, we can find $\psi_1 \in \mathrm{Diff}_{0,c}(\mathbb{A})$, supported in $S^1 \times (-2,-1) \cup S^1 \times (1,2)$, such that ψ_1 coincides with ϕ^{-1} on the curves $\phi(S^1 \times \{\pm 1.5\})$. Moreover, it satisfies $\|\psi_1\| < C'\epsilon$. Define $\psi_2 := \psi_1 \phi$. This map is defined on an open neighborhood of $\mathbb{A}' = S^1 \times [-1.5, 1.5]$ and has the following properties:

- The restriction of ψ_2 to $\mathbb{A}'$ is a diffeomorphism of $\mathbb{A}'$. It is the identity on $\partial\mathbb{A}'$ and coincides with ϕ on $\mathbb{A}_1 = S^1 \times [-1,1] \subset \mathbb{A}'$.
- $\|\psi_2\| < C''\epsilon$, where $C'' := C' + 1$.

We are going to modify ψ_2 (by a C^0-small perturbation) to make it the identity not only on $\partial\mathbb{A}'$ but on an open neighborhood of $\partial\mathbb{A}'$. Then we will extend it by the identity to a diffeomorphism of $\mathbb{A}$ with the required properties.

Since ψ_2 is the identity on $\partial\mathbb{A}'$, by perturbing it slightly near $\partial\mathbb{A}'$ (in the C^0-norm) we can assume that in addition to the properties listed above, near $\partial\mathbb{A}'$ the map ψ_2 preserves the foliation of $\mathbb{A}$ by the circles $S^1 \times y$. Let us explain briefly why (we describe how to perturb ψ_2 near the curve $y = 1.5$; the argument is the same near the other boundary component of $\mathbb{A}'$).

Fix $\alpha > 0$. Since $\psi_2(x, 1.5) = (x, 1.5)$, there exists $\delta > 0$ such that for $|y - 1.5| < \delta$, the curve $\psi_2(S^1 \times \{y\})$ is the graph of a function F_y (depending smoothly on y):

$$\psi_2(S^1 \times \{y\}) = \mathrm{graph}(F_y).$$

Note that $\frac{\partial F_y}{\partial y} > 0$. Choosing δ sufficiently small, we can assume that

$$\sup_{x \in S^1, |y-1.5| \leq \delta} |F_y(x) - 1.5| \leq \alpha \quad \text{and} \quad \delta < \alpha.$$

We can now extend the family of functions $(F_y)_{|y-1.5| \leq \delta}$ to a family of functions $(F_y)_{|y-1.5| \leq \alpha}$ such that $F_y(x) = y$ when $|y - 1.5|$ is close to α and such that we still have the conditions $\frac{\partial F_y}{\partial y} > 0$ and $|F_y(x) - 1.5| \leq \alpha$. By the implicit function theorem, we can now write

$$y = F_{c(x,y)}(x) \quad (x \in S^1, |y - 1.5| \leq \alpha),$$

with $\frac{\partial c}{\partial y} > 0$. Note that $c(x, y) = y$ when $|y - 1.5|$ is close to α. By composing ψ_2 with the C^0-small diffeomorphism h defined by $h(x, y) = (x, c(x, y))$, we obtain the desired perturbation.

The previous perturbation having been performed, we can now assume that for some sufficiently small $r > 0$ the restriction of ψ_2 to $S^1 \times [-1.5, -1.5 + r] \cup S^1 \times [1.5 - r, 1.5]$ has the form

$$\psi_2 : (x, y) \mapsto (x + u(x, y), y),$$

for some smooth function u such that $\|u\| < C''\epsilon$. Choose a cut-off function $\chi = \chi_{1.5, 1.5-r} : \mathbf{R} \to [0, 1]$ and define a map ψ_3 on $\mathbb{A}'$ as follows:

$$\psi_3 := \psi_2 \text{ on } S^1 \times [-1.5 + r, 1.5 - r],$$
$$\psi_3(x, y) := (x + \chi(y) u(x, y), y), \text{ when } |y| \geq 1.5 - r.$$

We now consider the diffeomorphism ψ that equals ψ_3 on $\mathbb{A}'$ and the identity outside $\mathbb{A}'$. It coincides with ϕ on $\mathbb{A}_1$ and satisfies $\|\psi\| < C''\epsilon$. Note that if ϵ is sufficiently small, ψ automatically belongs to the identity component $\mathrm{Diff}_{0,c}(\mathbb{A})$ (this can be easily deduced, for instance, from [15, 16] or [49]). This finishes the construction of ψ in the general case.

Let us now consider the case that $\phi = \mathbb{1}$ outside a quadrilateral $I \times [-1, 1]$ and $\phi(I \times [-1, 1]) \subset I \times [-2, 2]$ for some arc $I \subset S^1$. Then, by Lemma 6, we can assume that ψ_1 is supported in $I \times [-3, 3]$. Then ψ_2 is the identity outside $I \times [-1.5, 1.5]$. When we perturb ψ_2 near $\partial \mathbb{A}'$ to make it preserve the foliation by circles, we can choose the perturbation to be supported in $I \times [-1.5, 1.5]$. Thus $u(x,y)$ would be 0 outside $I \times [-1.5, 1.5]$. This yields that ψ_3, and consequently ψ, is the identity outside $I \times [-3, 3]$. □

7 Appendix by Michael Khanevsky: An Extension Lemma for Curves

For a diffeomorphism ϕ of a compact surface with a Riemannian distance d we write $\|\phi\| = \max d(x, \phi(x))$. The purpose of this appendix is to prove the following extension lemma, which was used in Sect. 6.4 above.

Lemma 6. *Let $A := S^1 \times [-1, 1]$ be an annulus equipped with the Euclidean product metric. Set $L = S^1 \times 0$. Assume that ϕ is a smooth embedding of an open neighborhood of L in A, so that L is homotopic to $\phi(L)$ and $\|\phi\| \le \epsilon$ for some $\epsilon \ll 1$.*

Then there exists a diffeomorphism $\psi \in \mathrm{Diff}_{0,c}(A)$ such that $\psi = \phi$ on L and $\|\psi\| < C'\epsilon$ for some $C' > 0$ independent of ϕ.

Moreover, if $\phi = \mathbb{1}$ outside some arc $I \subset L$ and $\phi(I) \subset I \times [-1, 1]$, then ψ can be made the identity outside $I \times [-1, 1]$.

Proof. We view the coordinate x on A along S^1 as a horizontal one, and the coordinate y along $[-1, 1]$ as a vertical one. If $a, b \in L$ are not antipodal, we denote by $[a, b]$ the shortest closed arc in L between a and b. The proof consists of a few steps. By $C_1, C_2, \ldots$ we will denote some universal positive constants.

Step 1. Shift the curve $\phi(L)$ by 3ϵ upward by a diffeomorphism $\psi_1 \in \mathrm{Diff}_{0,c}(A)$ with $\|\psi_1\| \le C_1\epsilon$, so that $K := \psi_1(\phi(L))$ lies strictly above L (see Fig. 1).

Step 2. Let $x_1, \ldots, x_N$ be points on L chosen in a cyclic order so that the distance between any two consecutive points x_i and x_{i+1} is at most ϵ (here and below, $i+1$ is taken to be 1, if $i = N$).

For each $i = 1, \ldots, N$, consider a vertical ray originating at x_i and assume, without loss of generality, that it is transversal to K and that K is parallel to L near its intersection points with the ray. Among the intersection points of the ray with K choose the closest one to L and denote it by y_i. Denote by r_i the closed vertical interval between x_i and y_i. Choose small disjoint open rectangles U_i, of width at most $\epsilon/3$ and of height at most 4ϵ around each of the intervals r_i.

For each $i = 1, \ldots, N$, it is easy to construct a diffeomorphism $\psi_{2,i}$ supported in U_i that moves a connected arc of $K \cap U_i$ containing y_i by a parallel shift downward

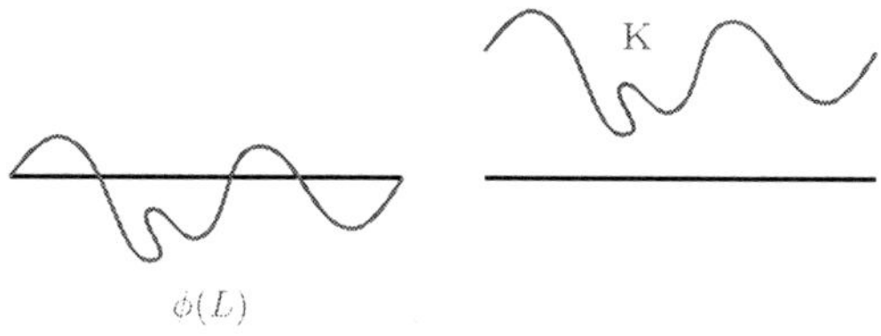

Fig. 1 Shifting L

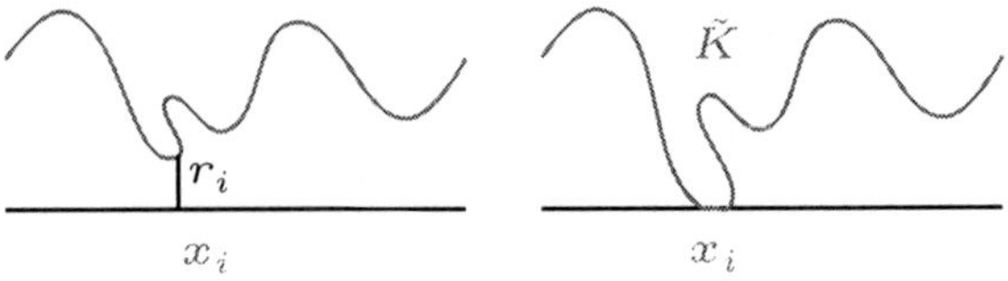

Fig. 2 $\tilde{K}$ coincides with L near x_i

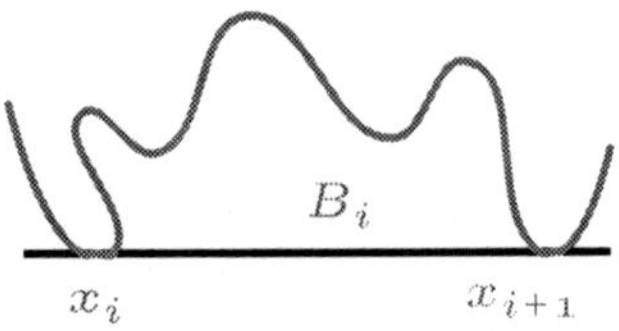

Fig. 3 The open set B_i

into an arc of L containing x_i so that $\psi_{2,i}(K)$ lies completely in $\{y \geq 0\}$. Set $\psi_2 := \prod_{i=1}^{N} \psi_{2,i}$. Clearly, $\|\psi_{2,i}\| \leq C_2\epsilon$ for each i, and therefore, since the supports of all the diffeomorphisms $\psi_{2,i}$ are disjoint, $\|\psi_2\| \leq C_2\epsilon$ as well. Set (see Fig. 2)

$$\tilde{\psi} := \psi_2\psi_1 \in \mathrm{Diff}_{0,c}(A), \quad \tilde{K} := \tilde{\psi}(\phi(L)).$$

Note that $\|\tilde{\psi}\| \leq C_3\epsilon$.

Step 3. Note that the points x_i, $i = 1,\ldots,N$, lie on $\tilde{K}$ and that

$$\tilde{K} \subset \{y \geq 0\}.$$

An easy topological argument shows that in such a case, since the points x_i lie on L in cyclic order, they also lie in the same cyclic order on $\tilde{K}$.

For each i there are two arcs in $\tilde{K}$ connecting x_i and x_{i+1}. Denote by K_i the one homotopic with fixed endpoints to the arc $[x_i, x_{i+1}] \subset L$. Since the points x_i lie on $\tilde{K}$ in the same cyclic order as on L, we see that $K_1,\ldots,K_N$ are precisely the closures of the N open arcs in $\tilde{K}$ obtained by removing the points $x_1,\ldots,x_N$ from $\tilde{K}$.

Let B_i be the open set bounded by K_i and $[x_i, x_{i+1}]$ (see Fig. 3). The B_i are disjoint and have diameter at most $C_4\epsilon$. Let B_i' be disjoint open neighborhoods of the B_i of diameter at most $C_5\epsilon$. Now for each i, the two arcs K_i and $[x_i, x_{i+1}]$ are homotopic in B_i', hence isotopic. Thus, one can find a diffeomorphism $\psi_{3,i} \in \mathrm{Diff}_{0,c}(B_i')$ such that $\psi_{3,i}(K_i) = [x_i, x_{i+1}]$. Since $\psi_{3,i}$ is supported in B_i', we have $\|\psi_{3,i}\| \leq C_5\epsilon$. Set $\psi_3 := \prod_{i=1}^{N} \psi_{3,i}$. Since the supports of all $\psi_{3,i}$ are disjoint, we get $\|\psi_3\| \leq C_5\epsilon$.

Step 4. Define $\psi_4 := \psi_3\tilde{\psi} = \psi_3\psi_2\psi_1$. Clearly, $\psi_4 \in \mathrm{Diff}_{0,c}(A)$ and $\|\psi_4\| \leq C_6\epsilon$. Recall that for each i we have $\psi_3(K_i) = [x_i, x_{i+1}]$ and that each K_i is the shortest arc between x_i and x_{i+1} in $\tilde{K} = \psi_2\psi_1(L)$. Thus ψ_4 maps K into L. The diffeomorphism ψ_4^{-1} satisfies $\psi_4^{-1}(L) = \phi(L)$. We now obtain easily the required ψ by a C^0-small perturbation of ψ_4^{-1}. □

Acknowledgments This text started as an attempt to understand a remark of Dieter Kotschick. We thank him for stimulating discussions and in particular for communicating to us the idea of getting the continuity from the C^0-fragmentation, which appeared in a preliminary version of [29]. The authors would like to thank warmly Frédéric Le Roux for his comments on this work and for the thrilling discussions we had during the preparation of this article, Felix Schlenk for critical remarks on the first draft of this paper, as well as Dusa McDuff for a useful discussion. The third author would like to thank Tel-Aviv University for its hospitality during the spring of 2008, when this work began. The second author expresses his deep gratitude to Oleg Viro for generous help and support at the beginning of his research in topology.

Finally, the authors would like to thank warmly the anonymous referee for his careful reading and for finding several inaccuracies in the first version of the text.

Michael Entov was partially supported by the Israel Science Foundation grant # 881/06. Leonid Polterovich was partially supported by the Israel Science Foundation grant # 509/07. Pierre Py was partially supported by the NSF (grant DMS-0905911).

References

1. M. Akveld and D. Salamon, *Loops of Lagrangian submanifolds and pseudoholomorphic discs*, Geom. and Funct. Analysis **11**, No. 4, (2001), 609–650.
2. A. Banyaga, *Sur la structure du groupe des difféomorphismes qui préservent une forme symplectique*, Comm. Math. Helv. **53** (1978), 174–227.
3. A. Banyaga, *Formes-volume sur les variétés à bord*, Enseignement Math. (2) **20** (1974), 127–131.
4. A. Banyaga, *The structure of classical diffeomorphism groups*, Mathematics and its applications **400**, Kluwer Academic Publishers Group, Dordrecht, 1997.
5. J. Barge and É. Ghys, *Cocycles d'Euler et de Maslov*, Math. Ann. **294** (1992), 235–265.
6. C. Bavard, *Longueur stable des commutateurs*, Enseign. Math. (2) **37**, No. 1-2 (1991), 109–150.
7. G. Ben Simon, *The Nonlinear Maslov index and the Calabi homomorphism*, Commun. Contemp. Math. **9**, No. 6 (2007), 769–780.
8. P. Biran, *Connectedness of spaces of symplectic embeddings*, Internat. Math. Res. Notices **10** (1996), 487–491.
9. P. Biran, M. Entov, and L. Polterovich, *Calabi quasimorphisms for the symplectic ball*, Commun. Contemp. Math. **6**, No. 5 (2004), 793–802.
10. A. Bounemoura, *Simplicité des groupes de transformations de surfaces*, Ensaios Matemáticos **14** (2008), 1–143.
11. R. Brooks, *Some remarks on bounded cohomology*, in *Riemann Surfaces and Related Topics: Proceedings of the 1978 Stony Brook Conference (State Univ. New York, Stony Brook, N.Y., 1978), 53–63*, Ann. of Math. Stud. **97**, Princeton University Press, Princeton, 1981.
12. D. Burago, S. Ivanov, and L. Polterovich, *Conjugation-invariant norms on groups of geometric origin*, in *Groups of Diffeomorphisms: In Honor of Shigeyuki Morita on the Occasion of His 60th Birthday*, Advanced Studies in Pure Mathematics **52**, Math. Society of Japan, Tokyo, 2008.
13. E. Calabi, *On the group of automorphisms of a symplectic manifold*, in *Problems in analysis*, 1–26, Princeton Univ. Press, Princeton, 1970.
14. D. Calegari, *scl*, MSJ Memoirs **20**, Mathematical Society of Japan, Tokyo (2009).
15. C.J. Earle and J. Eells, *The diffeomorphism group of a compact Riemann surface,* Bull. Amer. Math. Soc. **73** (1967), 557–559.
16. C.J. Earle and J. Eells, *A fibre bundle description of Teichmüller theory*, J. Differential Geometry **3** (1969), 19–43.

17. M. Entov, *Commutator length of symplectomorphisms*, Comment. Math. Helv. **79**, No. 1 (2004), 58–104.
18. M. Entov and L. Polterovich, *Calabi quasimorphism and quantum homology*, Int. Math. Res. Not. **30** (2003), 1635–1676.
19. M. Entov and L. Polterovich, *Symplectic quasi-states and semi-simplicity of quantum homology*, in *Toric Topology, 47–70*, Contemporary Mathematics **460**, AMS, Providence, 2008.
20. A. Fathi, *Structure of the group of homeomorphisms preserving a good measure on a compact manifold*, Ann. Sci. École Norm. Sup. (4) **13**, No. 1 (1980), 45–93.
21. J.-M. Gambaudo and É. Ghys, *Enlacements asymptotiques*, Topology **36**, No. 6 (1997), 1355–1379.
22. J.-M. Gambaudo and É. Ghys, *Commutators and diffeomorphisms of surfaces*, Ergodic Theory Dynam. Systems **24**, No. 5 (2004), 1591–1617.
23. É. Ghys, *Knots and dynamics*, in *International Congress of Mathematicians, Vol. I*, 247–277, Eur. Math. Soc., Zürich, 2007.
24. R. E. Greene and K. Shiohama, *Diffeomorphisms and volume-preserving embeddings of noncompact manifolds*, Trans. Amer. Math. Soc. **255** (1979), 403–414.
25. M. Gromov, *Pseudoholomorphic curves in symplectic manifolds,* Invent. Math. **82** (1985), 307–347.
26. H. Hofer, *On the topological properties of symplectic maps*, Proc. of the Royal Soc. of Edinburgh **115A**, No. 1-2 (1990), 25–28.
27. H. Hofer, *Estimates for the energy of a symplectic map*, Comment. Math. Helv. **68**, No. 1 (1993), 48–72.
28. D. Kotschick, *What is... a quasi-morphism?* Notices Amer. Math. Soc. **51**, No. 2 (2004), 208–209.
29. D. Kotschick, *Stable length in stable groups*, in *Groups of Diffeomorphisms: In Honor of Shigeyuki Morita on the Occasion of His 60th Birthday*, Advanced Studies in Pure Mathematics **52**, Math. Society of Japan, Tokyo, 2008.
30. F. Lalonde, *Isotopy of symplectic balls, Gromov's radius and the structure of ruled symplectic 4-manifolds*, Math. Ann. **300** (1994), 273–296.
31. F. Lalonde and D. McDuff, *The geometry of symplectic energy*, Ann. of Math. **141**, No. 2 (1995), 349–371.
32. F. Lalonde and D. McDuff, *Hofer's L^∞-geometry: energy and stability of Hamiltonian flows I*, Invent. Math. **122**, No. 1 (1995), 1–33.
33. F. Lalonde and L. Polterovich, *Symplectic diffeomorphisms as isometries of Hofer's norm*, Topology **36**, No. 3 (1997), 711–727.
34. F. Le Roux, *Six questions, a proposition and two pictures on Hofer distance for Hamiltonian diffeomorphisms on surfaces*, in *Symplectic topology and measure preserving dynamical systems*, 33–40, Contemp. Math., 512, Amer. Math. Soc., Providence, RI, 2010.
35. F. Le Roux, *Simplicity of* $\mathrm{Homeo}(\mathbb{D}^2, \partial\mathbb{D}^2, \mathrm{Area})$ *and fragmentation of symplectic diffeomorphisms*, J. Symplectic Geom. **8** (2010), no. 1, 73–93.
36. D. McDuff, *Remarks on the uniqueness of symplectic blowing up,* in *Symplectic geometry*, D.Salamon ed., 157–167, London Math. Soc. Lecture Note Ser., **192**, Cambridge Univ. Press, Cambridge, 1993.
37. D. McDuff, *From symplectic deformation to isotopy*, in *Topics in symplectic 4-manifolds (Irvine, CA, 1996)*, 85–99, First Int. Press Lect. Ser., I, Int. Press, Cambridge, MA, 1998.
38. D. McDuff and D. Salamon, *Introduction to symplectic topology*, 2nd edition, Oxford University Press, Oxford, 1998.
39. J. Moser, *On the volume elements on a manifold*, Trans. Amer. Math. Soc. **120** (1965), 288–294.
40. Y.-G. Oh, *C^0-coerciveness of Moser's problem and smoothing area preserving homeomorphisms*, preprint, 2006, arXiv:math 0601183.
41. Y. Ostrover, *Calabi quasi-morphisms for some non-monotone symplectic manifolds*, Algebr. Geom. Topol. **6** (2006), 405–434.

42. L. Polterovich, *Symplectic displacement energy for Lagrangian submanifolds*, Ergodic Th. and Dynam. Syst. **13**, No. 2 (1993), 357–367.
43. L. Polterovich, *The geometry of the group of symplectic diffeomorphisms*, Lectures in Mathematics, ETH Zürich, Birkhäuser, Basel, 2001.
44. L. Polterovich, *Floer homology, dynamics and groups*, in *Morse theoretic methods in nonlinear analysis and in symplectic topology*, 417–438, NATO Sci. Ser. II Math. Phys. Chem. **217**, Springer, Dordrecht, 2006.
45. P. Py, *Quasi-morphismes et invariant de Calabi*, Ann. Sci. École Norm. Sup. (4) **39**, No. 1 (2006), 177–195.
46. P. Py, *Quasi-morphismes de Calabi et graphe de Reeb sur le tore*, C. R. Math. Acad. Sci. Paris **343**, No. 5 (2006), 323–328.
47. A. Shtern, *Remarks on pseudocharacters and the real continuous bounded cohomology of connected locally compact groups*, Ann. Global Anal. Geom. **20**, No. 3 (2001), 199–221.
48. J.-C. Sikorav, *Approximation of a volume-preserving homeomorphism by a volume-preserving diffeomorphism*, preprint, 2007, available at http://www.umpa.ens-lyon.fr/~symplexe/publications.php.
49. S. Smale, *Diffeomorphisms of the 2-sphere*, Proc. Amer. Math. Soc. **10** (1959), 621–626.
50. T. Tsuboi, *The Calabi invariant and the Euler class*, Trans. Amer. Math. Soc. **352**, No. 2, (2000), 515–524.

On Symplectic Caps

David T. Gay and András I. Stipsicz

We dedicate this paper to Oleg Viro on the occasion of his 60th birthday

Abstract An important class of contact 3-manifolds comprises those that arise as links of rational surface singularities with reduced fundamental cycle. We explicitly describe symplectic caps (concave fillings) of such contact 3-manifolds. As an application, we present a new obstruction for such singularities to admit rational homology disk smoothings.

Keywords Three-manifold • Symplectic cap • Handle attachment • Link • Open book

1 Introduction

Our understanding of topological properties of (weak) symplectic fillings of certain contact 3-manifolds has improved dramatically in the recent past. These developments have rested on recent results in symplectic topology, most notably on McDuff's characterization of (closed) rational symplectic 4-manifolds [14]. In order to apply results of McDuff, however, *symplectic caps* were needed to close up the fillings at hand. General results of Eliashberg and Etnyre [5,6] showed that such caps do exist in general, but these results can be used powerfully only when a detailed description of the cap is also available. This was the case, for example, for lens spaces with their standard contact structures [12] and for certain 3-manifolds that can be given as links of isolated surface singularities [2, 3, 16].

D.T. Gay (✉) • A.I. Stipsicz
Department of Mathematics, University of Georgia, Athens, Georgia 30602, USA

Rényi Institute of Mathematics, Reáltanoda utca 13–15, Budapest, Hungary
e-mail: d.gay@euclidlab.org; stipsicz@math-inst.hu

I. Itenberg et al. (eds.), *Perspectives in Analysis, Geometry, and Topology: On the Occasion of the 60th Birthday of Oleg Viro*, Progress in Mathematics 296,
DOI 10.1007/978-0-8176-8277-4_9,

In the following we will show an explicit construction of symplectic caps for contact 3-manifolds that can be given as links (with their Milnor fillable structures) of rational singularities with reduced fundamental cycle. In topological terms, this means that the 3-manifold can be given as a plumbing of spheres along a negative definite tree, with the additional assumption that the absolute value of the framing at each vertex is at least the valency of the vertex. The construction of the cap in this case relies on a symplectic handle attachment along a component of the binding of a compatible open-book decomposition. In the terminology of open-book decompositions, our construction coincides with the cap-off procedure initiated and further studied by Baldwin [1].

The success of the rational blowdown procedure (initiated by Fintushel and Stern [8] and then extended by J. Park [17]) led to the search for isolated surface singularities that admit rational homology disk smoothings. Strong restrictions on the combinatorics of the resolution graph of such a singularity were found in [20], and by identifying Neumann's $\overline{\mu}$-invariant with a Heegaard Floer-theoretic invariant of the underlying 3-manifold, further obstructions to the existence of such a smoothing were given in [19]. More recently, the question was answered in [3] for all singularities with star-shaped resolution graphs (in particular, for weighted homogeneous singularities), but the general problem remained open. Motivated by our construction of a symplectic cap for special types of Milnor fillable contact 3-manifolds, we show examples of surface singularities that pass all tests provided by [19, 20] but still do not admit rational homology disk smoothings.

The paper is organized as follows. In Sect. 2, we describe the symplectic handle attachment that caps off a boundary component of a compatible open-book decomposition. Section 3 is devoted to a detailed description of the topology of the symplectic cap, and also an example is worked out. In Sect. 4, we show that certain singularities do not admit rational homology disk smoothings.

2 Symplectic Handle Attachments

Throughout this section, suppose that (Y,ξ) is a strongly convex boundary component of a symplectic 4-manifold (X,ω); that ξ is supported by an open-book decomposition with oriented page Σ, oriented binding $B = \partial\Sigma$, and monodromy h; and that L is a sublink of B. For each component K of L, let pf(K) denote the page-framing of K, the framing induced by the page Σ. Note that if Y_L is the result of performing surgery on Y along each component K of L with framing pf(K), and if $L \neq B$, then the open book on Y induces a natural open book on Y_L with page Σ_L equal to $\Sigma \cup_L (^{|L|}D^2)$, the result of capping off each K with a disk, and with monodromy equal to h extended by the identity on the D^2 caps. (In [1] this construction has been examined from the Heegaard Floer-theoretic point of view.)

If instead, for $|L| = 1$ and $L = K$, Y_K is obtained by surgery along K with framing $\mathrm{pf(K)} \pm 1$, then the open book on Y induces a natural open book on Y_K with page $\Sigma_K = \Sigma$ and with monodromy $h_K = h \circ \tau_K^{\mp 1}$, where τ_K is a right-handed Dehn twist along a circle in the interior of Σ parallel to K. In fact, if $K \neq B$, then surgery with framing $\mathrm{pf(K)} - 1$ coincides with Legendrian surgery along a Legendrian realization of K on the page; hence the 4-dimensional cobordism resulting from the construction supports a symplectic structure. In the following two theorems we extend the existence of such a symplectic structure to the cases in which the surgery coefficients are $\mathrm{pf(K)}$ and $\mathrm{pf(K)} + 1$.

In the first case, in which the surgery coefficient is $\mathrm{pf(K)}$, we have a rather technical extra condition in terms of the existence of a closed 1-form with certain behavior near K. Later, we will state one case in which this condition is always satisfied, but for the moment we leave it technical because the theorem is most general that way. When we discuss the behavior of anything near a component K of B, we always use oriented coordinates (r, μ, λ) near K such that $\mu, \lambda \in S^1$ are the meridional and longitudinal coordinates, respectively, chosen to represent the page framing. In other words, $\mu^{-1}(\theta)$, for any $\theta \in S^1$, is the intersection of a page with this coordinate neighborhood, and the closure of $\lambda^{-1}(\theta)$ is a meridional disk. Also, we assume that ∂_λ points in the direction of the orientation of K, oriented as the boundary of the page.

Theorem 2.1. *Suppose that L is a sublink of B, not equal to B, and that $X_L \supset X$ is the result of attaching a 2-handle to X along each component K of L with framing* $\mathrm{pf(K)}$. *Suppose furthermore that there exists a closed 1-form α_0 defined on $Y \setminus L$ that near each component K of L, has the form $m_K d\mu + l_K d\lambda$ for some constants m_K and l_K, with $l_K > 0$. (The coordinates (r, μ, λ) near K are as described in the preceding paragraph.) Then ω extends to a symplectic form ω_L on X_L, and the new boundary Y_L is ω_L-convex. The new contact structure ξ_L is supported by the natural open book on Y_L described above.*

Proof. Let $\pi\colon Y \setminus B \to S^1$ be the fibration associated with our given open book on Y, and let $\pi_L\colon Y_L \setminus (B \setminus L) \to S^1$ be the fibration for the induced open book on Y_L. Let Z be $[-1, 0] \times Y$ together with the 2-handles attached along $\{0\} \times L \subset \{0\} \times Y$, and identify Y with $\{0\} \times Y$. Thus Z is a cobordism from $\{-1\} \times Y$ to Y_L, and $Y \cap Y_L$ is nonempty and is in fact the complement of a neighborhood of L in Y. We will show that there is a symplectic structure η on Z that on $[-1, 0] \times Y$, is equal to the symplectization of a certain contact form α on Y supported by (B, π) and such that Y_L is η-convex, with induced contact structure ξ_L supported by the natural open book $(B \setminus L, \pi_L)$ on Y_L described above. This proves the theorem.

As mentioned above, for each component K of L we use coordinates (r, μ, λ) on a neighborhood $\nu \cong D^2 \times S^1$ of K, with (r, μ) being polar coordinates on the D^2-factor and λ being the S^1-coordinate, in such a way that $\mu = \pi|_\nu$. Thus the pages are the level sets for μ. We will also add now the convention that r is always parameterized so as to take values in $[0, 1 + \epsilon]$ for some small positive ϵ.

Let ν' be the corresponding neighborhood in Y_L of the belt-sphere for the 2-handle H_K that is attached along K, with corresponding coordinates (r', μ', λ'),

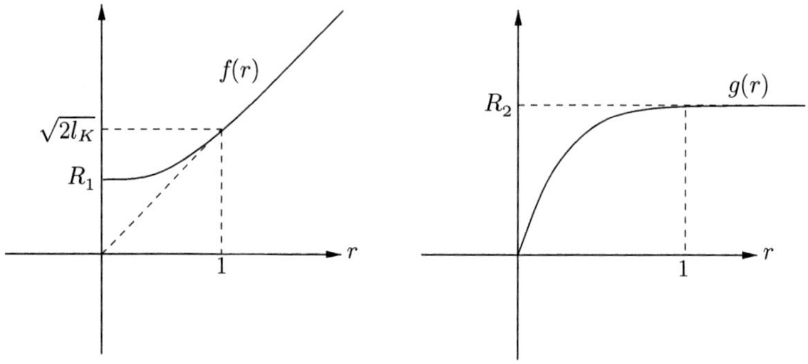

Fig. 1 Graphs of the functions f and g

with the natural diffeomorphism $\nu \setminus \{r=0\} \to \nu' \setminus \{r'=0\}$ given by $r'=r$, $\mu' = -\lambda$, and $\lambda' = \mu$. Note that $\pi_L|_{\nu'} = \lambda'$, which is defined on all of ν'.

There are, of course, many different contact structures supported by the given open book on Y, but they are all isotopic, and up to isotopy, we can always assume that ξ has the following behavior in each neighborhood ν of each component K of L:

1. ξ is (μ,λ)-invariant. That is, there exist functions $F(r)$ and $G(r)$ such that ξ is spanned by ∂_r and $F(r)\partial_\mu + G(r)\partial_\lambda$. We necessarily have $G(0)=0$, and we will adopt the convention that $F(0)>0$, so that $G'(0)<0$ and thus $G(r)<0$ for r close to 0.
2. As r ranges from 0 to 1, ξ makes a full quarter-turn in the (μ,λ)-plane. In other words, the vector $(F(r),G(r)) \in \mathbb{R}^2$ goes from $F(0)>0, G(0)=0$ to $F(1)=0$, $G(1)<0$, with $F(r)>0$ and $G(r)<0$ for all $r \in (0,1)$. (We can make this assumption precisely because $L \neq B$. One way to see this is to think of the construction of a contact structure supported by a given open book as beginning with a Weinstein structure on the page. This Weinstein structure comes from a handle decomposition of the page, and if we choose a handle decomposition starting with collar neighborhoods of the components of L and then adding 1-handles, we will get the desired behavior.)

So now we assume that ξ has the form above.

Next we claim that we can find a contact form α for this ξ satisfying certain special properties. To understand the local properties of α near each K, consider Fig. 1.

This figure shows graphs of two functions f and g, specified by constants R_1, l_K, and R_2. The properties of f and g are as follows:

1. The function f is monotone increasing with $f'(0)=0$ and $f'(r)>0$ for $r>0$.
2. $f(0)=R_1$ and $f(r)=\sqrt{2l_K}r$ for $r \geq 1$. (Hence $\sqrt{2l_K} > R_1$.)
3. $g(0)=0$.
4. The function g is monotone increasing with $g'(r)>0$ on $[0,1)$.
5. $g(r)=R_2$ for $r \geq 1$.

The claim, then, is that there exists a contact form α for ξ such that the following conditions hold:

1. The 1-form $\alpha - \alpha_0$ is a positive contact form on the complement of the neighborhoods of radius $r \leq 1$ of each component K of L, and it also satisfies the support condition for the given open book outside these neighborhoods.
2. For each component K of L there are constants R_1 and R_2 and associated functions f and g, as in Fig. 1 (with the constant l_K coming from $\alpha_0 = m_K \mathrm{d}\mu + l_K \mathrm{d}\lambda$), with $\frac{1}{2}R_2^2 > m_K$ such that in the neighborhood ν of K, α has the form
$$\alpha = \frac{1}{2}g(r)^2 \mathrm{d}\mu + \left(l_K - \frac{1}{2}f(r)^2\right)\mathrm{d}\lambda.$$
(We might need to reparameterize the coordinate r, but only via a reparameterization fixing 0 and 1.)

The condition $\frac{1}{2}R_2^2 > m_K$ is necessary to guarantee that $\alpha - \alpha_0$ is positive contact when $r \geq 1$; this condition will also be used later.

To verify this claim, first choose any contact form α' for ξ satisfying the support condition for the given open book. Now note that for any suitably large constant $k > 0$, $k\alpha' - \alpha_0$ is a positive contact form satisfying the support condition. We know that in ν, $k\alpha' = -G(r)\mathrm{d}\mu + F(r)\mathrm{d}\lambda$ for functions $F(r), G(r)$ such that the vector $(F(r), G(r))$ makes one quarter-turn through the fourth quadrant as r goes from 0 to 1. Because k is large, we may assume that $G(1) < -m_K$. We can then scale $k\alpha'$ by a positive function $\phi(r)$ supported inside $r \leq 1 + \epsilon$ so as to arrange that the pair of functions $(\tilde{F}(r) = \phi(r)F(r), \tilde{G}(r) = \phi(r)G(r))$ has the appropriate shape, and then we let $\frac{1}{2}g(r)^2 = -\tilde{G}(r)$ and $l_K - \frac{1}{2}f(r)^2 = \tilde{F}(r)$. Then we have $\alpha = \phi(r)k\alpha'$.

Now embed ν and ν' in $\mathbb{R}^4$ as follows, using polar coordinates $(r_1, \theta_1, r_2, \theta_2)$ on $\mathbb{R}^4$. The embedding of ν is given by $(r_1 = f(r), \theta_1 = -\lambda, r_2 = g(r), \theta_2 = \mu)$. The embedding of ν' is given by $(r_1 = \sqrt{2l_K}r', \theta_1 = \mu', r_2 = R_2, \theta_2 = \lambda')$. This is illustrated in Fig. 2, which also shows that the region between ν and ν' is precisely our 2-handle H attached along K with framing pf(K). The overlap $\nu \cap \nu'$ is the set $\{r_1 \geq \sqrt{2l_K}, r_2 = R_2\}$, which in ν-coordinates is $\{r \geq 1\}$ and in ν'-coordinates is $\{r' \geq 1\}$.

Consider the standard symplectic form $\omega_0 = r_1 \mathrm{d}r_1 \mathrm{d}\theta_1 + r_2 \mathrm{d}r_2 \mathrm{d}\theta_2$ on $\mathbb{R}^4$. Note that $\omega_0|_\nu = gg'\mathrm{d}r\mathrm{d}\mu - ff'\mathrm{d}r\mathrm{d}\lambda = \mathrm{d}\alpha$, so that H equipped with this symplectic form can be glued symplectically to $[-1,0] \times Y$ with the symplectization of α. Next note that $\omega_0|_{\nu'} = 2l_K r'\mathrm{d}r'\mathrm{d}\mu' = \mathrm{d}\alpha'$, where $\alpha' = \frac{1}{2}(\sqrt{2l_K}r')^2\mathrm{d}\mu' + (\frac{1}{2}R_2^2 - m_K)\mathrm{d}\lambda'$. (Here we see that $\frac{1}{2}R_2^2 > m_K$ is necessary for α' to be a positive contact form and to be supported by the open book inside this neighborhood ν'.) On the overlap $\nu \cap \nu' \subset \mathbb{R}^4$, using the coordinates (r, μ, λ) from ν, we see that $\alpha' = \left(\frac{1}{2}R_2^2 - m_K\right)\mathrm{d}\mu + \big(-\frac{1}{2}\left(\sqrt{2l_K}r\right)^2\big)\mathrm{d}\lambda = \alpha - \alpha_0$. Thus we see that α' extends to the rest of Y_L as $\alpha - \alpha_0$, concluding the proof of the theorem. □

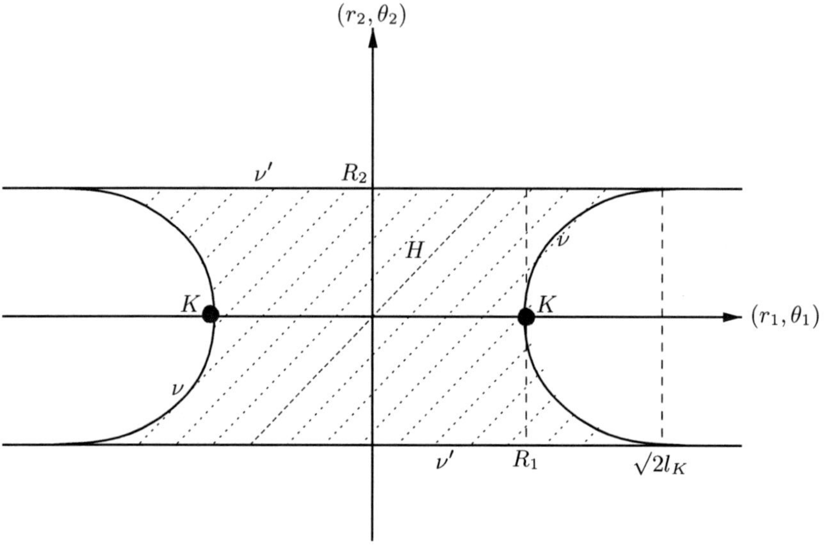

Fig. 2 Embeddings of ν, ν', and H into $\mathbb{R}^4$

In fact, 2-handles can be attached with framing $\mathrm{pf(K)}+1$ to boundary components of a compatible open book, and the symplectic structure will still extend. In this case, however, the convex boundary will become concave. More precisely, we have the following theorem.

Theorem 2.2. *Suppose that $K = B$ and that $X_K \supset X$ is the result of attaching a 2-handle H to X along K with framing* $\mathrm{pf(K)}+1$. *Then ω extends to a symplectic form ω_K on X_K, and the new boundary Y_K is ω_K-concave. The new (negative) contact structure ξ_K is supported by the natural open book on Y_K described above.*

Proof. This is [9, Theorem 1.2]. However in that paper, which predates Giroux's work on open-book decompositions, the terminology is slightly different. [9, Definition 2.4] defines what it means for a transverse link L in a contact 3-manifold (M,ξ) to be "nicely fibered." It is easy to see that if L is the binding of an open book supporting ξ, then L is nicely fibered. (The notion of "nicely fibered" is more general because, in open-book language, it allows for "pages" whose boundaries multiply cover the binding.) Theorem 1.2 in [9] then says that if we attach 2-handles to all the components of a nicely fibered link in the strongly convex boundary of a symplectic 4-manifold, with framings that are more positive than the framings coming from the fibration, then the symplectic form extends across the 2-handles to make the new boundary strongly concave. In our case, we have a single component, and we are attaching with framing exactly one more than the framing coming from the fibration. Finally, [9, Addendum 5.1] characterizes the negative contact structure induced on the new boundary as follows: There exists a constant k such that $\alpha_K = kd\pi - \alpha$ on the complement of the surgery knots.

(Here we are identifying $Y \setminus K$ with the complement in Y_K of the belt sphere for H in the obvious way.) The constant k is simply the appropriate constant such that α_K extends to all of Y_K. Then $\mathrm{d}\pi \wedge \mathrm{d}\alpha_K = -\mathrm{d}\pi \wedge \mathrm{d}\alpha$, which is positive on $-Y_K$. Since $\mathrm{d}\alpha_K = -\mathrm{d}\alpha$, and the Reeb vector field for α is tangent to the level sets for the radial function r on a neighborhood of K (see [9, Definition 2.4]), the Reeb vector field for α_K is necessarily tangent to the new binding of Y_K, and it is not hard to check that it points in the correct direction, so that α_K is supported by the natural open book on Y_K. □

We have the following application. (For a similar result, see [22, Theorem 4']].)

Corollary 2.3. *If the open book on Y is planar (i.e., $\mathrm{genus}(\Sigma) = 0$), then (X, ω) embeds in a closed symplectic 4-manifold (Z, η) that contains a symplectic $(+1)$-sphere disjoint from X.*

As preparation we need the following lemma:

Lemma 2.4. *Let B be the (disconnected) binding of a planar open book on Y, and let $L \subset B$ be the complement of a single component of B. Then there exists a 1-form α_0 on $Y \setminus L$ such that near each component K of L, α_0 has the form $\alpha_0 = m_K \mathrm{d}\mu + l_K \mathrm{d}\lambda$, for $l_K > 0$. (The coordinates near K are as in Theorem 2.1 and are determined by the open book.)*

Proof. Let Y_L be the result of page-framed surgery on L, with the corresponding oriented link $L' \subset Y_L$ (the cores of the surgeries). Note that $Y_L \cong S^3$, because the induced open book on Y_L has disk pages. Thus L' is an oriented link in S^3, and there exists a map $\sigma \colon S^3 \setminus L' \to S^1$ with the closure of each $\sigma^{-1}(\theta)$, for each regular value θ, an oriented Seifert surface for L'. Pull σ back to $Y \setminus L = Y_L \setminus L'$ and let $\alpha_0 = d\sigma$. □

Proof (of Corollary 2.3). Let the components of B be $K_1, \ldots, K_n$. Attach 2-handles to $K_1, \ldots, K_{n-1}$ with framings $\mathrm{pf}(\mathrm{K_i})$, as in Theorem 2.1. This gives $(X', \omega') \supset (X, \omega)$ with ω'-convex boundary (Y', ξ'). Now attach a 2-handle to K_n with framing $\mathrm{pf}(\mathrm{K_n}) + 1$ as in Theorem 2.2; the resulting concave end is S^3 with its negative contact structure supported by the standard disk open book, i.e., the contact structure is the standard negative tight contact structure. Thus we can fill in the concave end with the standard symplectic structure on B^4. Alternatively, we can note that on Y', the positive contact structure ξ' is supported by an open book with page diffeomorphic to a disk. In other words, Y' is diffeomorphic to S^3, and ξ' is the standard positive tight contact structure on S^3. Thus we can remove a standard (B^4, ω_0) from $\mathbb{CP}^2$ with its standard Kähler form, and replace (B^4, ω_0) with (X', ω') to get (Z, η). Since there is a symplectic $(+1)$-sphere in $\mathbb{CP}^2$ disjoint from B^4, we end up with a symplectic $(+1)$-sphere in (Z, η) disjoint from X', and hence disjoint from X. □

By [14], the symplectic 4-manifold Z found in the proof of Corollary 2.3 is diffeomorphic to a blowup of $\mathbb{CP}^2$. Let Z' be the result of anti–blowing down the symplectic $(+1)$-sphere in Z (i.e., Z' is the union of the 4-manifold X' in the proof of the corollary above with B^4). Then Z' (still containing X) is diffeomorphic to the connected sum of a number of copies of $\overline{\mathbb{CP}^2}$. Let D be the closure of $Z' \setminus X$ in Z'; we will call this the *dual configuration* (or *compactification*) for X. Thus we get embeddings of the intersection forms $H_2(X;\mathbb{Z})$ and $H_2(D;\mathbb{Z})$ into a negative definite diagonal lattice, and therefore both $H_2(X;\mathbb{Z})$ and $H_2(D;\mathbb{Z})$ are negative definite.

Remark 2.5. A very similar compactification has been found by Némethi and Popescu-Pampu in [15], using rather different methods.

3 Examples: Rational Surface Singularities with Reduced Fundamental Cycle

Suppose that Γ is a plumbing tree of spheres that is negative definite, and at each vertex the absolute value of the framing is at least the number of edges emanating from the vertex. Every negative definite plumbing graph Γ gives rise to a (not necessarily unique) surface singularity, and the further assumptions on Γ ensure that the singularity has reduced fundamental cycle. According to Laufer's algorithm, for example, this property implies that the singularity is rational; cf. [19, Sect. 3]. The Milnor fillable contact structure on such a 3-manifold is known to be compatible with a planar open-book decomposition [7, 18]. A fairly explicit description of such an open-book decomposition can be given by a construction resting on results of [10]. By [10, Proposition 5.3], the Milnor fillable contact structure is compatible with an open-book decomposition resting on a toric construction (cf. [10, Sect. 4]), and therefore by [10, Proposition 4.2], a compatible planar open book can be explicitly given as follows.

View the tree Γ as a planar graph in $\mathbb{R}^2$ and consider the boundary sphere of an ϵ neighborhood of it in $\mathbb{R}^3$. Suppose that v is a vertex of Γ with framing e_v and valency d_v. Then near v, drill $-e_v - d_v \geq 0$ holes on the sphere. The resulting planar surface will be the page of the open-book decomposition. Consider a parallel circle to each boundary component and further curves near each edge, as shown by the example of Fig. 3. The monodromy of the open-book decomposition is simply the product of the right-handed Dehn twists defined by all these curves on the planar surface.

Consider now the Kirby diagram for Y based on the open-book decomposition as follows: Regard the planar page as a multipunctured disk. (This step involves

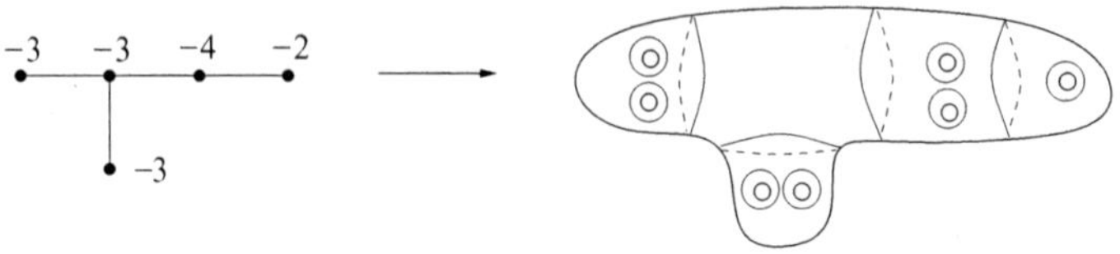

Fig. 3 Light circles on the punctured sphere define the monodromy of the open book

a choice of an "outer circle.") Every hole on the disk defines a 0-framed unknot linking the boundary of the hole, while the light circles defining the monodromy through right-handed Dehn twists give rise to (-1)-framed unknots. In fact, the 0-framed unknots can be turned into dotted circles and then viewed as 4-dimensional 1-handles (for these notions of Kirby calculus, see [11]). These will build up a Lefschetz fibration with fiber diffeomorphic to the page of the open book, and the addition of the (-1)-framed circles corresponds to the vanishing cycles of the Lefschetz fibration, giving the right monodromy.

Having this Kirby diagram for Y, a relative handlebody diagram for the dual configuration D (built on $-Y$) can be easily deduced by performing 0-surgery along all the boundary circles except the outer one. This operation corresponds to capping off all but the last boundary component of the open book defining the Milnor fillable structure on Y. Since after all the capping off we get an open book with a disk as a page, the 4-manifold D is a cobordism from $-Y$ to S^3.

It is usually more convenient to have an absolute handlebody than a relative one, and since the other boundary component of D is S^3, by turning D upside down we can easily derive a handlebody description first for $-D$ and then, after the reversal of the orientation, for D. After appropriate handleslides, in fact, the diagram for D can be given by a simple algorithm. Since we dualize only 2-handles, D can be given by attaching 2-handles to D^4. The framed link can be given by a braid, which is derived from the plumbing tree by the following inductive procedure. To start, we choose a vertex v where the strict inequality $-e_v - d_v > 0$ holds. (Such a vertex always exists; for example, we can take a leaf.) We will choose the outer circle to be the boundary of one of the holes near v. Now associate to every inner boundary component a string and to every light circle a box symbolizing a full negative twist of the strings passing through the box, which in our case consists of those strings that correspond to the boundary components encircled by the light circle. The framing on a string is given by the negative of the "distance" of the boundary component from the outer circle: this distance is simply the number of light circles we have to cross when traveling from the boundary component to the outer circle. Another (obviously equivalent) way of describing the same braid purely in terms of the graph Γ goes as follows: Choose again a vertex v with $-e_v - d_v > 0$, and consider $-e_u - d_u$ strings for each vertex u, except for v, for which we take only $-e_v - d_v - 1$ strings. Introduce a full negative twist on the resulting trivial braid (corresponding to the light circle parallel to the outer circle), and then introduce a further full negative twist for every edge e in the graph, where the strings affected by the negative twist can be characterized by the property that they correspond to vertices that are in a component of $\Gamma - \{e\}$ not containing the distinguished vertex v. Finally, equip every string corresponding to a vertex u with $r_{uv} - 2$, where r_{uv} is the negative of the minimal number of edges we traverse when passing from u to v.

We will demonstrate this procedure through an explicit family of examples. (For a similar result see [21, Theorem 3].) To this end, suppose that the graph Γ_n is given by Fig. 4. It is easy to see that the graphs in the family for $n \geq 1$ are all negative definite, and for $n \geq 2$, they define a rational singularity with reduced fundamental cycle. Assume that $n \geq 3$ and choose a boundary circle near the $(-n-1)$-framed

Fig. 4 An interesting family of plumbing graphs

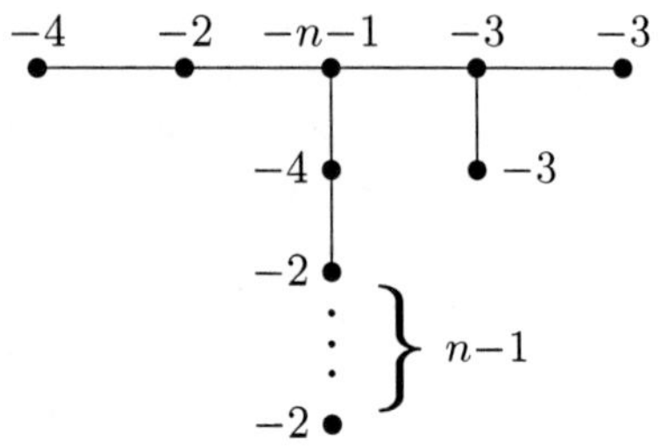

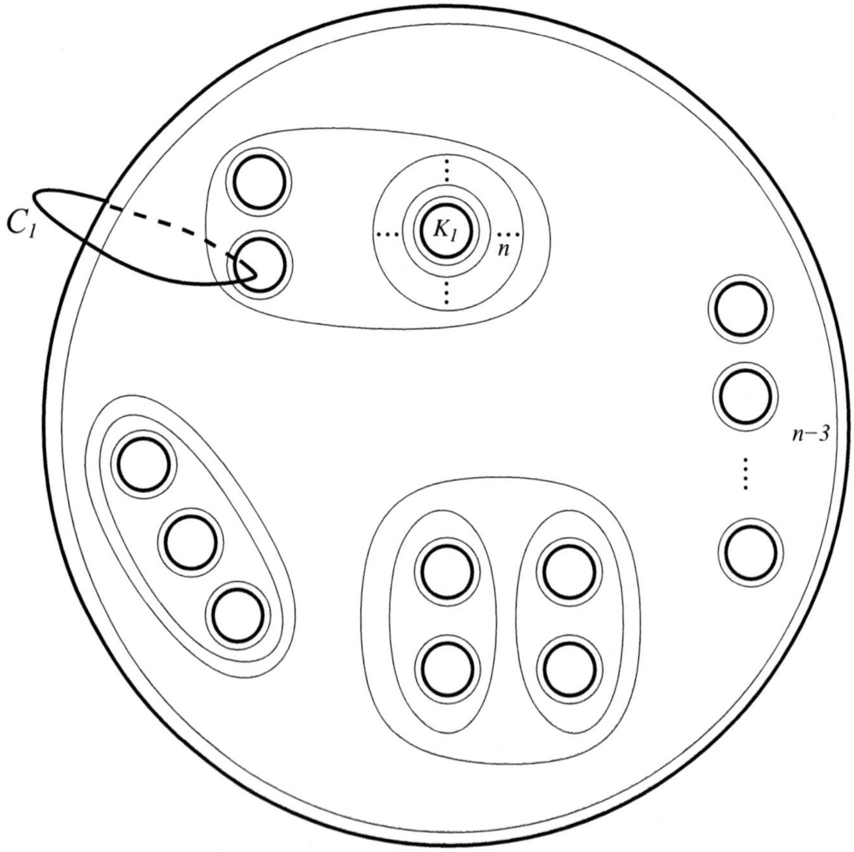

Fig. 5 The light circles on the disk define the monodromy of the open book. There are n concentric light circles around the boundary component labeled by K, and there are $n-3$ boundary circles on the right-hand side of the disk. For each of the interior boundary components there should be a corresponding unknot C_i linking it and the exterior boundary component; here we have drawn only C_1

vertex to be the outer circle. The page of the planar open book, together with the light circles (giving rise to the monodromy through right-handed Dehn twists), is pictured by Fig. 5 (with the circle C_1 disregarded for a moment).

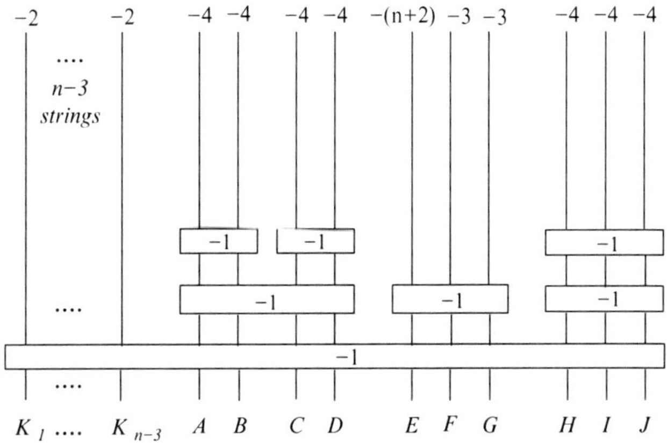

Fig. 6 Boxes in the diagram mean full negative twists

The 0-framed unknots originating from the 1-handles of the Lefschetz fibration become unknots each of which links one of the interior boundary components of the punctured disk once and the exterior boundary once. In the diagram, the unknot labeled C_1 is one of these unknots; we have not drawn the rest because they would only complicate the picture needlessly, but it is important to keep in mind that there is one such unknot for each interior boundary. Putting (-1)-framings on all light circles we get a convenient description of Y. Now add framing 0 to all boundary components except the outer one. The result is a cobordism D from $-Y$ to S^3. Mark all these circles (for example, use the convention of [11] by replacing all framing a with $\langle a \rangle$) and turn D upside down: add 0-framed meridians to the circles corresponding to the boundary components of the open book (these are the curves along which we "capped off" the open book). Now sliding and blowing down marked curves only, we end up with the diagram of $-D$, and by reversing all crossings and multiplying all framings by (-1), eventually we get a Kirby diagram for D, as shown in Fig. 6. (Every box in the diagram means a full negative twist.)

4 The Nonexistence of Rational Homology Disk Smoothings

Next we will demonstrate how the explicit topological description of the dual D can be applied to study smoothings of surface singularities. We start with a simple observation providing an obstruction for a 3-manifold to bound a rational homology disk, i.e., a 4-manifold V with $H_*(V;\mathbb{Q}) = H_*(D^4;\mathbb{Q})$.

Theorem 4.1. *Suppose that the rational homology 3-sphere $-Y$ is the boundary of a compact 4-manifold D with the property that* $\mathrm{rkH_2(D;\mathbb{Z})} = \mathrm{n}$ *and that the* *intersection form $(H_2(D;\mathbb{Z}), Q_D)$ does not embed into the negative definite diagonal lattice $n\langle -1\rangle$ of the same rank. Then Y cannot bound a rational homology disk.*

Proof. Suppose that such a rational homology disk V exists; then $Z = V \cup_Y D$ is a closed, negative definite 4-manifold. By Donaldson's theorem [4], the intersection form of Z is diagonalizable over $\mathbb{Z}$, and by our assumption on V, we get that $\mathrm{rkH}_2(\mathrm{Z};\mathbb{Z}) = \mathrm{rkH}_2(\mathrm{D};\mathbb{Z}) = \mathrm{n}$. Since $H_2(D;\mathbb{Z}) \subset H_2(Z;\mathbb{Z})$ does not embed into $n\langle -1\rangle$, we get a contradiction, implying the result. □

Consider now the plumbing graph Γ_n of Fig. 4, and denote the corresponding 3-manifold by Y_n.

Proposition 4.2. *The 3-manifold Y_n does not bound a rational homology disk 4-manifold for $n \geq 7$.*

Remark 4.3. Notice that elements of this family pass all the tests provided by [20], since these graphs are elements of the family $\mathcal{A}$ of [20]: change the framing of the single (-4)-framed vertex with valency two to (-1) and blow down the graph until it becomes the defining graph of the family $\mathcal{A}$. Also, using the algorithm described, e.g., in [19], it is easy to see that $\det \Gamma_n \equiv n \bmod 2$; hence for odd n, the 3-manifold Y_n admits a unique spin structure. The corresponding Wu class can be given by the (-3)-framed vertex of valency three, the unique (-4)-framed vertex on the long chain, and then every second (-2)-framed vertex. A simple count then shows that for n odd, we have $\overline{\mu}(Y_n) = 0$, and hence the result of [19] provides no obstruction to a rational homology disk smoothing. (For the terminology used in the above argument, see [19].)

Proposition 4.4. *The lattice determined by the intersection form of the dual D_n given by Fig. 6, for $n \geq 7$, does not embed into a negative definite diagonal lattice of the same rank.*

Proof. The labels on the components of the braid in Fig. 6 will be used to represent the corresponding basis elements for the lattice determined by the intersection form of D_n. The rank is $n+7$. Let $E = \{e_1, \ldots, e_{n+7}\}$ be the standard basis for the negative definite diagonal lattice of rank $n+7$, so $e_i \cdot e_j = -\delta_{ij}$. Suppose that the lattice for D_n does embed into the definite diagonal lattice. Then without loss of generality, since $K_i \cdot K_i = -2$ and $K_i \cdot K_j = -1$ otherwise, we may assume that $K_i = e_1 + e_{10+i}$. Furthermore, without loss of generality we may assume that every other one of the basis elements $A, B, \ldots, J$ is of the form $e_1 + x$ where x is an expression in $e_2, \ldots, e_{10}$. Thus each basis element whose square is -3 (i.e., F and G) must be of the form $e_1 \pm u \pm v$, where u and v are distinct elements of the set $\{e_2, \ldots, e_{10}\}$. Each element whose square is -4 (i.e., A, B, C, D, H, I, and J) must be of the form $e_1 \pm q \pm r \pm s$, where q, r, and s are distinct elements of the set $\{e_2, \ldots, e_{10}\}$.

Now we can assume that $F = e_1 + e_2 + e_3$ and $G = e_1 + e_2 + e_4$ (noting that $F \cdot G = -2$). Then we note that none of the expressions for A, B, C, D, H, I, J can contain e_2, e_3, or e_4 for the following reason: For each of $X = A, B, C, D, H, I, J$ there is another basis element Y from this set such that $X \cdot Y = -3$, while $X \cdot X = Y \cdot Y = -4$. Thus if we write $X = e_1 + \alpha a + \beta b + \gamma c$ with $a, b, c \in E$ and $\alpha, \beta, \gamma \in \{-1, 1\}$, then Y must be $Y = e_1 + \alpha a + \beta b + \delta d$, with $d \in E$ and $\delta \in \{-1,1\}$, where a, b, c, and d are distinct elements from the set $\{e_2,\ldots,e_{10}\}$. Now noting that $X \cdot F = X \cdot G = Y \cdot F = Y \cdot G = -1$, we see that if $a = e_2$, then b,c,d must be in $\{e_3,e_4\}$, which cannot happen, because b, c, and d must be distinct. Similarly, b cannot be e_2. If $a = e_3$, then b or c must be e_2, but we have just seen that it cannot be b, so $c = e_2$. But the same argument also shows that $d = e_2$, but $c \neq d$. Similarly, we can rule out $a = e_4$ and also $b = e_3$ and $b = e_4$. But if one of c and d is in the set $\{e_2,e_3,e_4\}$, then one of a and b must also be, so finally we see that none of them can be.

Thus we can now take $H = e_1 + e_5 + e_6 + e_7$. There are then two possibilities for I and J (up to relabeling the members of the sets $\{e_8,e_9,e_{10}\}$ and $\{e_5,e_6,e_7\}$).

Case I: $I = e_1 + e_5 + e_6 + e_8$ and $J = e_1 + e_5 + e_6 + e_9$. In this case, we can see that A,B,C,D cannot contain e_7, e_8, or e_9. So then the only remaining possibilities are all equivalent (after changing signs of basis elements in E) to $A = e_1 + e_5 - e_6 + e_{10}$, but then we cannot find any candidates for B that give $A \cdot B = -3$. This rules out Case I.

Case II: $I = e_1 + e_5 + e_7 + e_8$ and $J = e_1 + e_5 + e_6 + e_8$. To rule out this case, write $A = e_1 + \alpha a + \beta b + \gamma c$, $a,b,c \in \{e_5,e_6,e_7,e_8,e_9,e_{10}\}$, and $\alpha,\beta,\gamma \in \{-1,1\}$. In order to have $A \cdot H = -1$, either none or two of a,b,c must be in the set $\{e_5,e_6,e_7\}$, but not one or three of them. Similarly, using $A \cdot I = -1$, either none or two must be in $\{e_5,e_7,e_8\}$, and using $A \cdot J = -1$, either none or two must be in $\{e_5,e_6,e_8\}$. If it is none in one of these cases, it must be none for all three, but that leaves only e_9 and e_{10} for a, b, and c, an impossibility. Thus it is two in each case. We cannot have one of them e_5, because then we could not have exactly two from all three sets. So we must have $a = e_6$, $b = e_7$, $c = e_8$. But exactly the same argument holds for B, and we can never get $A \cdot B = -3$. Thus Case II is ruled out, concluding the proof of the proposition. □

Proof (of Proposition 4.2). Combine Theorem 4.1 and Proposition 4.4. □

Corollary 4.5. *Suppose that $(S_\Gamma,0)$ is an isolated surface singularity with resolution graph given by Fig. 4. If $n \geq 7$, then $(S_\Gamma,0)$ admits no rational homology disk smoothing, i.e., it has no smoothing V with $H_*(V;\mathbb{Z}) = H_*(D^4;\mathbb{Z})$.* □

Acknowledgments The authors would like to acknowledge support from the Hungarian–South African Bilateral Project NRF 62124 (ZA-15/2006), and we thank András Némethi for many useful conversations. The second author was also supported by OTKA T67928. The authors would also like to thank Chris Wendl for pointing out an important mistake in the first version of this paper.

References

1. J. Baldwin, *Capping off open books and the Ozsváth–Szabó contact invariant*, arXiv:0901.3797.
2. M. Bhupal and K. Ono, *Symplectic fillings of links of quotient surface singularities*, arXiv:0808.3794.

3. M. Bhupal and A. Stipsicz, *Weighted homogeneous singularities and rational homology disk smoothings*, J. Amer. Math.
4. S. Donaldson, *Irrationality and the h-cobordism conjecture*, J. Differential Geom. **26** (1987), 141–168.
5. Y. Eliashberg, *A few remarks about symplectic filling*, Geom. Topol. **8** (2004), 277–293.
6. J. Etnyre, *On symplectic fillings*, Algebr. Geom. Topol. **4** (2004), 73–80.
7. J. Etnyre and B. Ozbagci, *Open books and plumbings*, Int. Math. Res. Not. **2006** Art. ID 72710, 17pp.
8. R. Fintushel and R. Stern, *Rational blowdowns of smooth 4-manifolds*, J. Diff. Geom. **46** (1997), 181–235.
9. D. Gay, *Symplectic 2–handles and transverse links*, Trans. Amer. Math. Soc. **354** (2002), 1027–1047.
10. D. Gay and A. Stipsicz, *Symplectic surgeries and normal surface singularities*, Algebr. Geom. Topol. **9** (2009), no. 4, 2203–2223.
11. R. Gompf and A. Stipsicz, *4-manifolds and Kirby calculus*, AMS Grad. Studies Math. **20**, 1999.
12. P. Lisca, *On symplectic fillings of lens spaces*, Trans. Amer. Math. Soc. **360** (2008), 765–799.
13. P. Lisca, *Lens spaces, rational balls and the ribbon conjecture*, Geom. Topol. **11** (2007), 429–472.
14. D. McDuff, *The structure of rational and ruled symplectic 4-manifolds*, J. Amer. Math. Soc. **3** (1990), 679–712; erratum, **5** (1992), 987–988.
15. A. Némethi and P. Popescu-Pampu, *On the Milnor fibers of sandwiched singularities*, Int. Math. Res. Not. IMRN 2010, no. 6, 1041–1061.
16. H. Ohta and K. Ono, *Simple singularities and symplectic fillings*, J. Differential Geom. **69** (2005), no. 1, 1–42.
17. J. Park, *Seiberg–Witten invariants of generalized rational blow-downs*, Bull. Austral. Math. Soc. **56** (1997), 363–384.
18. S. Schönenberger, *Determining symplectic fillings from planar open books*, J. Symplectic Geom. **5** (2007), 19–41.
19. A. Stipsicz, *On the $\overline{\mu}$–invariant of rational surface singularities*, Proc. Amer. Math. Soc. **136** (2008), 3815–3823.
20. A. Stipsicz, Z. Szabó, and J. Wahl, *Rational blowdowns and smoothings of surface singularities*, J. Topol. **1** (2008), 477–517.
21. J. Wahl, *Construction of QHD smoothings of valency 4 surface singularities*, Geom. Topol. **15** (2011) 1125–1156.
22. C. Wendl, *Non-exact symplectic cobordisms between contact 3-manifolds*, arXiv:1008.2465

Cauchy–Pompeiu-Type Formulas for $\bar{\partial}$ on Affine Algebraic Riemann Surfaces and Some Applications

Gennadi M. Henkin

Dedicated to Oleg Viro on the occasion of his 60th birthday

Abstract We present explicit solution formulas $f = \hat{R}\varphi$ and $u = R_\lambda f$ for the equations $\bar{\partial} f = \varphi$ and $(\partial + \lambda \mathrm{d}z_1)u = f - \mathcal{H}_\lambda f$ on an affine algebraic curve $V \subset \mathbb{C}^2$. Here $\mathcal{H}_\lambda f$ denotes the projection of $f \in \tilde{W}^{1,\tilde{p}}_{1,0}(V)$ to the subspace of pseudoholomorphic (1,0)-forms on V: $\bar{\partial}\mathcal{H}_\lambda f = \bar{\lambda}\mathrm{d}\bar{z}_1 \wedge \mathcal{H}_\lambda f$. These formulas can be interpreted as explicit versions and refinements of the Hodge–Riemann decomposition on Riemann surfaces. The main application consists in the construction of the Faddeev–Green function for $\bar{\partial}(\partial + \lambda \mathrm{d}z_1)$ on V as the kernel of the operator $R_\lambda \circ \hat{R}$. This Faddeev–Green function is the main tool for the solution of the inverse conductivity problem on bordered Riemann surfaces $X \subset V$, that is, for the reconstruction of the conductivity function σ in the equation $d(\sigma d^c U) = 0$ from the Dirichlet-to-Neumann mapping $U\big|_{bX} \mapsto \sigma d^c U\big|_{bX}$. The case $V = \mathbb{C}$ was treated by R. Novikov [N1]. In Sect. 4 we give a correction to the paper [HM], in which the case of a general algebraic curve V was first considered.

Keywords Riemann surface • Inverse conductivity problem • $\bar{\partial}$-method • Homotopy formulas

G.M. Henkin (✉)
Université Pierre et Marie Curie, Mathématiques, case 247, 4 place Jussive, 75252 Paris, FRANCE
e-mail: henkin@math.jussieu.fr

I. Itenberg et al. (eds.), *Perspectives in Analysis, Geometry, and Topology: On the Occasion of the 60th Birthday of Oleg Viro*, Progress in Mathematics 296,
DOI 10.1007/978-0-8176-8277-4_10,

Introduction

This paper is motivated by a problem from two-dimensional electrical impedance tomography, namely the question of how to reconstruct the conductivity function σ on a bordered Riemann surface X from the knowledge of the Dirichlet-to-Neumann mapping $u\big|_{bX} \to \sigma d^c U\big|_{bX}$ for solutions U of the Dirichlet problem

$$d(\sigma d^c U)\big|_X = 0, \ U\big|_{bX} = u, \quad \text{where } d = \partial + \bar{\partial}, \ d^c = i(\bar{\partial} - \partial).$$

For the case $X = \Omega \subset \mathbb{R}^2 \simeq \mathbb{C}$, $z = x_1 + ix_2$, the exact reconstruction scheme was given first by R. Novikov [N1] under a certain restriction on the conductivity function σ. This restriction was eliminated later by A. Nachman [Na].

The scheme consists in the following.

Let $\sigma(x) > 0$ for $x \in \bar{\Omega}$ and $\sigma \in C^{(2)}(\bar{\Omega})$. Put $\sigma(x) = 1$ for $x \in \mathbb{R}^2 \backslash \bar{\Omega}$. The substitution $\psi = \sqrt{\sigma} U$ transforms the equation $d(\sigma d^c U) = 0$ into the equation $dd^c \psi = \frac{dd^c \sqrt{\sigma}}{\sqrt{\sigma}} \psi$ on $\mathbb{R}^2$. From L. Faddeev's [F1] result (with additional arguments [BC2] and [Na]), it follows that for each $\lambda \in \mathbb{C}$, there exists a unique solution $\psi(z, \lambda)$ of the above equation, with asymptotics

$$\psi(z,\lambda) \cdot \mathrm{e}^{-\lambda z} \stackrel{\text{def}}{=} \mu(z,\lambda) = 1 + o(1), \quad z \to \infty.$$

Such a solution can be found from the integral equation

$$\mu(z,\lambda) = 1 + \frac{i}{2} \int\limits_{\xi \in \Omega} g(z - \xi, \lambda) \frac{\mu(\xi,\lambda) dd^c \sqrt{\sigma}}{\sqrt{\sigma}},$$

where the function

$$g(z,\lambda) = \frac{-1}{2i(2\pi)^2} \int\limits_{w \in \mathbb{C}} \frac{\mathrm{e}^{i(w\bar{z} + \bar{w}z)} dw \wedge d\bar{w}}{w(\bar{w} - i\lambda)}, \quad z \in \mathbb{C}, \ \lambda \in \mathbb{C},$$

is called the Faddeev–Green function for the operator $\mu \mapsto \bar{\partial}(\partial + \lambda dz)\mu$.

From the work of R. Novikov [N1] it follows that the function $\psi\big|_{b\Omega}$ can be found through the Dirichlet-to-Neumann mapping by the integral equation

$$\psi(z,\lambda)\big|_{b\Omega} = \mathrm{e}^{\lambda z} + \int\limits_{\xi \in b\Omega} \mathrm{e}^{\lambda(z-\xi)} g(z-\xi,\lambda)(\hat{\Phi}\psi(\xi,\lambda) - \hat{\Phi}_0\psi(\xi,\lambda)),$$

where

$$\hat{\Phi}\psi = \bar{\partial}\psi\big|_{b\Omega}, \ \hat{\Phi}_0\psi = \bar{\partial}\psi_0\big|_{b\Omega}, \ \psi_0\big|_{b\Omega} = \psi, \ \bar{\partial}\partial\psi_0\big|_{\Omega} = 0.$$

By results of R. Beals, R. Coifmann [BC1], P. Grinevich, S. Novikov [GN], and R. Novikov [N2], it then follows that $\psi(z,\lambda)$ satisfies a $\bar{\partial}$-equation of Bers–Vekua type with respect to $\lambda \in \mathbb{C}$:

$$\frac{\partial \psi}{\partial \bar{\lambda}} = b(\lambda)\bar{\psi},$$

where $\lambda \mapsto b(\lambda) \in L^{2+\varepsilon}(\mathbb{C}) \cap L^{2-\varepsilon}(\mathbb{C})$, and $\psi(z,\lambda)e^{-\lambda z} \to 1$ as $\lambda \to \infty$, for all $z \in \mathbb{C}$.

This $\bar{\partial}$-equation combined with R. Novikov's integral equation permits us to find, starting from the Dirichlet-to-Neumann mapping, first the boundary values $\psi\big|_{b\Omega}$, second the "$\bar{\partial}$-scattering data" $b(\lambda)$, and third $\psi\big|_{\Omega}$.

Summarizing, the conductivity function $\sigma\big|_{\Omega}$ is thus retrieved from the given Dirichlet-to-Neumann data by means of the scheme

$$\text{DN data} \to \psi\big|_{b\Omega} \to \bar{\partial}\text{−scattering data} \to \psi\big|_{\Omega} \to \frac{dd^c\sqrt{\sigma}}{\sqrt{\sigma}}\big|_{\Omega}.$$

Main result

We suppose that instead of $\mathbb{C}$, we have a smooth algebraic Riemann surface V in $\mathbb{C}^2$, given by an equation $V = \{z \in \mathbb{C}^2; P(z) = 0\}$, where P is a holomorphic polynomial of degree $d \geq 1$. Put $z_1 = w_1/w_0$, $z_2 = w_2/w_0$, and suppose that the projective compactification $\tilde{V}$ of V in $\mathbb{C}P^2 \supset \mathbb{C}^2$ with coordinates $w = (w_0 : w_1 : w_2)$ intersects $CP^1_\infty = \{z \in \mathbb{C}P^2; w_0 = 0\}$ transversally in d points. In order to extend the Novikov reconstruction scheme to the case of a Riemann surface $V \subset \mathbb{C}^2$, we need first to find an appropriate Faddeev-type Green function for $\bar{\partial}(\partial + \lambda dz_1)$ on V. One can check that for the case $V = \mathbb{C}$, the Faddeev–Green function $g(z,\lambda)$ is a composition of Cauchy–Green Pompeiu kernels for the operators $f \mapsto \varphi = \bar{\partial} f$ and $u \mapsto f = (\partial + \lambda dz)u$, where u, f, and φ are respectively a function, a $(1,0)$-form, and a $(1,1)$-form on $\mathbb{C}$. More precisely, one has the formula

$$g(z,\lambda) = \frac{-1}{i(2\pi)^2} \int\limits_{w \in \mathbb{C}} \frac{e^{\lambda w - \bar{\lambda}\bar{w}} dw \wedge d\bar{w}}{(w+z) \cdot \bar{w}}.$$

The main purpose of this paper is to construct an analogue of the Faddeev–Green function on the Riemann surface V. To do this, we need to find explicit formulas $f = \hat{R}\varphi$ and $u = R_\lambda f$ (with appropriate estimates) for solutions of the two equations $\bar{\partial} f = \varphi$ and $(\partial + \lambda dz_1)u = f - \mathcal{H}_\lambda f$ on V. Here we consider V equipped with the Euclidean volume form $dd^c|z|^2$, and we require $\varphi \in L^1_{1,1}(V)$, $f \in \tilde{W}^{1,\tilde{p}}_{1,0}(V)$, and $u \in L^\infty(V)$, $\tilde{p} > 2$, with $\mathcal{H}_\lambda f$ the projection of f on the subspace of pseudoholomorphic (1,0)-forms on $\tilde{V}$: $\bar{\partial}\mathcal{H}_\lambda f = \bar{\lambda} d\bar{z}_1 \wedge \mathcal{H}_\lambda f$.

The new formulas obtained in this paper for the solution of $\bar{\partial} f = \varphi$ and $(\partial + \lambda dz_1)u = f$ on V can be interpreted as explicit and more precise versions of the classical Hodge–Riemann decomposition results on Riemann surfaces. We will define the Faddeev-type Green function for $\bar{\partial}(\partial + \lambda dz_1)$ on V as the kernel $g_\lambda(z,\xi)$ of the integral operator $R_\lambda \circ \hat{R}$.

Further results

Let $\sigma \in C^{(2)}(V)$, with $\sigma > 0$ on V, and $\sigma \equiv$ const on a neighborhood of $\tilde{V} \backslash V$. Let $a_1, \dots, a_g$ be generic points in this neighborhood, with g the genus of $\tilde{V}$. Using the Faddeev-type Green function constructed here, we have in [HM] obtained natural analogues of all steps of the Novikov reconstruction scheme in the case of a Riemann surface V. In particular, under a smallness assumption on $d \log \sqrt{\sigma}$, the existence (and uniqueness) of the solution $\mu(z,\lambda)$ of the Faddeev-type integral equation

$$\mu(z,\lambda) = 1 + \frac{i}{2} \int\limits_{\xi \in V} g_\lambda(z,\xi) \frac{\mu(\xi,\lambda) dd^c \sqrt{\sigma}}{\sqrt{\sigma}} + i \sum_{l=1}^{g} c_l g_\lambda(z,a_l), \quad z \in V, \ \lambda \in \mathbb{C},$$

holds for any a priori fixed constants $c_1, \dots, c_g$. However (and this was overlooked in [HM]), there exists only one choice of constants $c_l = c_l(\lambda, \sigma)$ for which the integral equation above is equivalent to the differential equation

$$\bar{\partial}(\partial + \lambda dz_1)\mu = \frac{i}{2} \left(\frac{dd^c \sqrt{\sigma}}{\sqrt{\sigma}} \mu \right) + i \sum_{l=1}^{g} c_l \delta(z,a_l),$$

where $\delta(z,a_l)$ are Dirac measures concentrated in the points a_l (see also Sect. 4 below).

1 A Cauchy–Pompeiu-Type Formula on an Affine Algebraic Riemann Surface

By $L_{p,q}(V)$ we denote the space of (p,q)-forms on V whose coefficients are distributions of measure type on V. By $L^s_{p,q}(V)$ we denote the space of (p,q)-forms on V with coefficients absolutely integrable in degree $s \geq 1$ with respect to the Euclidean volume form on V. If $V = \mathbb{C}$ and f is a function from $L^1(\mathbb{C})$ such that $\bar{\partial} f \in L_{0,1}(\mathbb{C})$, then the generalized Cauchy formula has the following form:

$$f(z) = -\frac{1}{2\pi i} \int\limits_{\xi \in \mathbb{C}} \frac{\bar{\partial} f(\xi) \wedge d\xi}{\xi - z}, \quad z \in \mathbb{C}.$$

This formula becomes the classical Cauchy formula when $f = 0$ on $\mathbb{C} \backslash \Omega$ and $f \in \mathcal{O}(\Omega)$, where Ω is some bounded domain with rectifiable boundary in $\mathbb{C}$.

The generalized Cauchy formula was discovered by Pompeiu [P1] in connection with his solution of the Painlevé problem. Pompeiu proved the existence of a nonzero function $f \in \mathcal{O}(\mathbb{C}\backslash E) \cap C(\mathbb{C}) \cap L^1(\mathbb{C})$ for any totally disconnected compact set E of positive Lebesgue measure. The Cauchy–Pompeiu formula has a large number of fundamental applications: in the theory of distributions (L. Schwartz), in approximation problems (E. Bishop, S. Mergelyan, A. Vitushkin), in the solution of the corona problem (L. Carleson), in the theory of pseudoanalytic functions (L. Bers, I. Vekua), and in inverse scattering and integrable equations (R. Beals, R. Coifman, M. Ablowitz, D. Bar Yaacov, A. Fokas).

Motivated by applications to electrical impedance tomography, we develop in this paper the Cauchy–Pompeiu-type formulas on affine algebraic Riemann surfaces $V \subset \mathbb{C}^2$ and give some applications.

Let $\tilde{V}$ be a smooth algebraic curve in $\mathbb{C}P^2$ given by the equation

$$\tilde{V} = \{w \in \mathbb{C}P^2;\ \tilde{P}(w) = 0\},$$

with $\tilde{P}$ a homogeneous holomorphic polynomial in the homogeneous coordinates $w = (w_0 : w_1 : w_2) \in \mathbb{C}P^2$. Without loss of generality, we may suppose the following:

(i) $\tilde{V}$ intersects $\mathbb{C}P^1_\infty = \{w \in \mathbb{C}P^2;\ w_0 = 0\}$ transversally, $\tilde{V} \cap \mathbb{C}P^1_\infty = \{a_1, \dots, a_d\}$, $d = \deg \tilde{P}$;
(ii) $V = \tilde{V} \backslash \mathbb{C}P^1_\infty$ is a connected curve in $\mathbb{C}^2$ with an equation of the form $V = \{z \in \mathbb{C}^2;\ P(z) = 0\}$, where $P(z) = \tilde{P}(1, z_1, z_2)$ such that

$$\left| \frac{\partial P / \partial z_1}{\partial P / \partial z_2} \right| \leq \text{const}(V), \quad \text{if } |z_1| \geq r_0 = \text{const}(V);$$

(iii) For any $z^* \in V$ such that $\frac{\partial P}{\partial z_2}(z^*) = 0$, we have $\frac{\partial^2 P}{\partial z_2^2}(z^*) \neq 0$.

By the Hurwitz–Riemann theorem, the number of such ramification points is equal to $d(d-1)$. Let us equip V with the Euclidean volume form $dd^c|z|^2$.

Notation

Let $\tilde{W}^{1,\tilde{p}}(V) = \{F \in L^\infty(V); \bar{\partial} F \in L^{\tilde{p}}_{0,1}(V)\}$, $\tilde{p} > 2$. Let us denote by $H^p_{0,1}(V)$ the subspace in $L^p_{0,1}(V)$, $1 < p < 2$, consisting of antiholomorphic forms. For all $p \in (1,2)$, the space $H^p_{0,1}(V)$ coincides with the space of antiholomorphic forms on V admitting an antiholomorphic extension to the compactification $\tilde{V} \subset \mathbb{C}P^2$. Hence, by the Riemann–Clebsch theorem, one has $\dim_{\mathbb{C}} H^p_{0,1}(V) = (d-1)(d-2)/2$ for all $p \in (1,2)$.

Proposition 1. *Let $\{V_j\}$ be the connected components of $\{z \in V;\ |z_1| > r_0\}$. Then for all $j \in \{1, \dots, d\}$ there exist operators $R_1:\ L^p_{0,1}(V) \to L^{\tilde{p}}(V)$, $R_0:\ L^p_{0,1}(V) \to \tilde{W}^{1,\tilde{p}}(V)$, and $\mathcal{H}:\ L^p_{0,1}(V) \to H^p_{0,1}(V)$, $1 < p < 2$, $1/\tilde{p} = 1/p - 1/2$, such that for* *all $\Phi \in L^p_{0,1}(V)$, one has the decomposition*

$$\Phi = \bar\partial R\Phi + \mathcal{H}\Phi, \quad \text{where } R = R_1 + R_0, \tag{1.1}$$

$$R_1\Phi = \frac{1}{2\pi i}\int\limits_{\xi\in V} \Phi(\xi)\frac{d\xi_1}{\frac{\partial P}{\partial \xi_2}} \det\left[\frac{\partial P}{\partial \xi}(\xi), \frac{\bar\xi - \bar z}{|\xi - z|^2}\right], \tag{1.2}$$

$$\mathcal{H}\Phi = \sum_{j=1}^{g}\left(\int\limits_V \Phi\wedge\omega_j\right)\bar\omega_j,$$

with $\{\omega_j\}$ an orthonormal basis for the holomorphic (1,0)-forms on $\tilde V$, i.e.,

$$\int\limits_V \omega_j \wedge \bar\omega_k = \delta_{jk}, \quad j,k = 1,2,\dots,g,$$

and

$$\lim_{\substack{z\in V_j \\ z\to\infty}} R\Phi(z) = 0.$$

Remark 1. If $p \in [1,2)$ and $q \in (2,\infty]$, the condition $\Phi \in L^p_{0,1}(V)\cap L^q_{0,1}(V)$ implies that $R\Phi \in C(\tilde V)$.

Remark 2. For the case $V = \mathbb{C} = \{z \in \mathbb{C}^2;\ z_2 = 0\}$, Proposition 1 and Remark 1 reduce to the classical results of Pompeiu [P1], [P2] and of Vekua [V].

Remark 3. Based on the technique of [HP], one can construct an explicit formula not only for the main part R_1 of the R-operator, but for the whole operator R.

Proof of Proposition 1: Let $Q(\xi,z) = \{Q_1(\xi,z), Q_2(\xi,z)\}$ be a pair of holomorphic polynomials in the variables $\xi = (\xi_1,\xi_2)$ and $z = (z_1,z_2)$ such that

$$Q(\xi,\xi) = \frac{\partial P}{\partial \xi}(\xi) \ \text{ and}$$
$$P(\xi) - P(z) = Q_1(\xi,z)(\xi_1 - z_1) + Q_2(\xi,z)(\xi_2 - z_2) \overset{\text{def}}{=} \langle Q(\xi,z), \xi - z\rangle. \tag{1.3}$$

The conditions (i) and (ii) imply that for ε_0 small enough, there exists a holomorphic retraction $z \to r(z)$ of the domain $\mathcal{U}_{\varepsilon_0} = \{z \in \mathbb{C}^2 : |P(z)| < \varepsilon_0\}$ onto the curve V.

Put $\mathcal{U}_{\varepsilon,r} = \{z \in \mathbb{C}^2;\ |P(z)| < \varepsilon,\ |z_1| < r\}$, where $0 < \varepsilon \le \varepsilon_0$ and $r \ge r_0$. Put also $V^c = \{z \in \mathbb{C}^2;\ P(z) = c\}$, where $c \in \mathbb{C}$, $|c| \le \varepsilon_0$, and $\tilde\Phi(z) = \Phi(r(z))$, $z \in \mathcal{U}_{\varepsilon_0}$. The condition $\Phi \in L^p_{0,1}(V)$ and properties of the retraction $z \to r(z)$ together imply that $\bar\partial\tilde\Phi = 0$ on $\mathcal{U}_{\varepsilon_0}$ and

$$\|\tilde\Phi\|_{L^p(V^c)} \le \text{const}(V)\cdot\|\Phi\|_{L^p(V)}, \tag{1.4}$$

uniformly with respect to c, for $|c| \leq \varepsilon_0$. By results from [H] and [Po] we can choose the following explicit solution $\tilde{F}_{\varepsilon,r}$ on $\mathfrak{U}_{\varepsilon,r}$ of the $\bar{\partial}$-equation $\bar{\partial}\tilde{F}_{\varepsilon,r} = \tilde{\Phi}\big|_{\mathfrak{U}_{\varepsilon,r}}$:

$$\tilde{F}_{\varepsilon,r}(z) = \left(\frac{1}{2\pi i}\right)\left\{ \int\limits_{\xi\in\mathfrak{U}_{\varepsilon,r}} \tilde{\Phi}\wedge\det\left[\frac{\bar{\xi}-\bar{z}}{|\xi-z|^2}, \bar{\partial}_\xi\frac{\bar{\xi}-\bar{z}}{|\xi-z|^2}\right]\wedge d\xi_1\wedge d\xi_2 \right.$$
$$+ \int\limits_{\xi\in b\mathfrak{U}_{\varepsilon,r}\colon |\xi_1|=r} \tilde{\Phi}\wedge\left[-\frac{(\bar{\xi}_2-\bar{z}_2)}{(\xi_1-z_1)|\xi-z|^2}\right]\wedge d\xi_1\wedge d\xi_2$$
$$\left. + \int\limits_{\xi\in b\mathfrak{U}_{\varepsilon,r}\colon |P(\xi)|=\varepsilon} \tilde{\Phi}\wedge\det\left[\frac{\bar{\xi}-\bar{z}}{|\xi-z|^2}, \frac{Q}{P(\xi)-P(z)}\right]\wedge d\xi_1\wedge d\xi_2\right\},\ z\in\mathfrak{U}_{\varepsilon,R}. \tag{1.5}$$

Property (1.4) implies that for any $z\in V$, we have

$$\int\limits_{\xi\in\mathfrak{U}_{\varepsilon,r}} \tilde{\Phi}\wedge\det\left[\frac{\bar{\xi}-\bar{z}}{|\xi-z|^2}, \bar{\partial}_\xi\frac{\bar{\xi}-\bar{z}}{|\xi-z|^2}\right]\wedge d\xi_1\wedge d\xi_2 \to 0,\ \varepsilon\to 0 \text{ and}$$
$$\int\limits_{\xi\in b\mathfrak{U}_{\varepsilon,r}\colon |\xi_1|=r} \tilde{\Phi}\wedge\left[-\frac{(\bar{\xi}_2-\bar{z}_2)}{(\xi_1-z_1)|\xi-z|^2}\right]\wedge d\xi_1\wedge d\xi_2 \to 0,\ \varepsilon\to 0,\ r\to\infty.$$

Hence for all $z\in V$ there exists $\lim\limits_{\substack{\varepsilon\to 0\\ r\to\infty}} \tilde{F}_{\varepsilon,r} = \tilde{F}(z)$, where

$$\tilde{F}(z) = -\frac{1}{2\pi i}\int\limits_{\xi\in V} \frac{\Phi d\xi_1}{\frac{\partial P}{\partial\xi_2}(\xi)}\wedge\det\left[\frac{\bar{\xi}-\bar{z}}{|\xi-z|^2}, Q(\xi,z)\right]. \tag{1.6}$$

From (1.5) and (1.6) it follows that

$$\bar{\partial}_z\tilde{F}\big|_V = \Phi(z). \tag{1.7}$$

Now put $F_1 = R_1\Phi$. Using (1.2), (1.3), (1.6), and (1.7), we obtain

$$\bar{\partial}_z F_1(z)\big|_V = \frac{1}{2\pi i}\int\limits_{\xi\in V}\Phi(\xi)\wedge\frac{d\xi_1}{\frac{\partial P}{\partial\xi_2}}\wedge\det\left[\frac{\partial P}{\partial\xi}(\xi), \bar{\partial}_z\frac{\bar{\xi}-\bar{z}}{|\xi-z|^2}\right]$$
$$= \Phi + \frac{1}{2\pi i}\int\limits_{\xi\in V}\Phi(\xi)\wedge\frac{d\xi_1}{\frac{\partial P}{\partial\xi_2}}\wedge\frac{1}{|\xi-z|^4}\det\begin{vmatrix}\frac{\partial P}{\partial\xi_1}(\xi) & \xi_2-z_2\\ \frac{\partial P}{\partial\xi_2}(\xi) & -(\xi_1-z_1)\end{vmatrix}\det\begin{vmatrix}\bar{\xi}_1-\bar{z}_1 & d\bar{z}_1\\ \bar{\xi}_2-\bar{z}_2 & d\bar{z}_2\end{vmatrix}$$
$$= \Phi + K\Phi,$$

where

$$K\Phi = \frac{1}{2\pi i} \int\limits_{\xi \in V} \frac{\Phi(\xi) \wedge d\xi_1}{|\xi - z|^4} \wedge \frac{\langle \frac{\partial P}{\partial \xi}(\xi), \xi - z\rangle \cdot \langle \frac{\partial \bar{P}}{\partial \bar{\xi}}(z), \bar{\xi} - \bar{z}\rangle}{\frac{\partial P}{\partial \xi_2}(\xi) \cdot \frac{\partial \bar{P}}{\partial \bar{\xi}_2}(z)} d\bar{z}_1. \tag{1.8}$$

The estimate $R_1\Phi = F_1 \in L^{\tilde{p}}(V)$ follows from the property $\Phi \in L^p(V)$ and the following estimate of the kernel for the operator R_1:

$$\left| \left(\frac{\partial P}{\partial \xi_2}(\xi) \right)^{-1} \det \left[\frac{\partial P}{\partial \xi}(\xi), \frac{\bar{\xi} - \bar{z}}{|\xi - z|^2} \right] d\xi_1 \right| = O\left(\frac{1}{|\xi - z|} \right) (|d\xi_1| + |d\xi_2|),$$

where $\xi, z \in V$.

For the kernels of the operators $\Phi \mapsto K\Phi$ and $\Phi \mapsto \partial_z K\Phi$ we have the corresponding estimates

$$\left| \frac{\langle \frac{\partial P}{\partial \xi}(\xi), \xi - z\rangle \cdot \langle \frac{\partial \bar{P}}{\partial \bar{\xi}}(z), \bar{\xi} - \bar{z}\rangle \, d\xi_1 \wedge d\bar{z}_1}{\frac{\partial P}{\partial \xi_2}(\xi) \cdot |\xi - z|^4 \cdot \frac{\partial \bar{P}}{\partial \bar{\xi}_2}(z)} \right|$$
$$= \begin{cases} O\left(\frac{1}{1+|z|^2}\right) |(d\xi_1 + d\xi_2) \wedge (d\bar{z}_1 + d\bar{z}_2)| & \text{if } |\xi - z| \le 1, \\ O\left(\frac{1}{|\xi - z|^2}\right) |(d\xi_1 + d\xi_2) \wedge (d\bar{z}_1 + d\bar{z}_2)| & \text{if } |\xi - z| \ge 1. \end{cases} \tag{1.9}$$

$$\left| \partial_z \frac{\langle \frac{\partial P}{\partial \xi}(\xi), \xi - z\rangle \cdot \langle \frac{\partial \bar{P}}{\partial \bar{\xi}}(z), \bar{\xi} - \bar{z}\rangle \, d\xi_1 \wedge d\bar{z}_1}{\frac{\partial P}{\partial \xi_2}(\xi) \cdot |\xi - z|^4 \cdot \frac{\partial \bar{P}}{\partial \bar{\xi}_2}(z)} \right|$$
$$= \begin{cases} O\left(\frac{1}{(1+|z|^2)|\xi - z|}\right) |(d\xi_1 + d\xi_2) \wedge (d\bar{z}_1 + d\bar{z}_2) \wedge (dz_1 + dz_2)| & \text{if } |\xi - z| < 1, \\ O\left(\frac{1}{|\xi - z|^3}\right) |(d\xi_1 + d\xi_2) \wedge (dz_1 + dz_2) \wedge (d\bar{z}_1 + d\bar{z}_2)| & \text{if } |\xi - z| \ge 1, \\ & \xi, z \in V \end{cases} \tag{1.10}$$

These estimates imply that for all $\tilde{p} > 2$ and $p > 1$, one has

$$\Phi_0 \stackrel{\text{def}}{=} K\Phi \in W^{1,\tilde{p}}_{0,1}(V) \cap L^p_{0,1}(V). \tag{1.11}$$

From estimates (1.9)–(1.11), it follows that the (0,1)-form $\Phi_0 = K\Phi$ on V can be considered also as a (0,1)-form on the compactification $\tilde{V}$ of V in $\mathbb{C}P^2$ belonging to the spaces $W^{1,p}_{0,1}(\tilde{V})$ for all $p < 2$, where $\tilde{V}$ is equipped with the projective volume form $dd^c \ln(1 + |z|^2)$.

From the Hodge–Riemann decomposition theorem [Ho], [W], we have

$$\Phi_0 = \bar{\partial}(\bar{\partial}^* G\Phi_0) + \mathcal{H}\Phi_0, \tag{1.12}$$

where $\mathcal{H}\Phi_0 \in H_{0,1}(\tilde{V})$, and G is the Hodge–Green operator for the Laplacian $\bar{\partial}\bar{\partial}^* + \bar{\partial}^*\bar{\partial}$ on $\tilde{V}$ with the properties

$$G(H_{0,1}(\tilde{V})) = 0, \;\; \bar{\partial}G = G\bar{\partial}, \;\; \bar{\partial}^*G = G\bar{\partial}^*.$$

The decomposition (1.12) implies that

$$\bar{\partial}^* G\Phi_0 \in W^{2,p}(\tilde{V}), \; p \in (1,2) \text{ and } \mathcal{H}\Phi_0 \in H_{0,1}(\tilde{V}),$$

and this in turn implies that $\bar{\partial}^* G\Phi_0 \in C(\tilde{V})$. Returning to the affine curve V with the Euclidean volume form, we obtain that

$$\tilde{R}_0\Phi \overset{\text{def}}{=} \bar{\partial}^* GK\Phi\big|_V \in \tilde{W}^{1,\tilde{p}}(V), \; \forall \tilde{p} > 2, \;\; \text{where}$$

$$\tilde{W}^{1,\tilde{p}}(V) \overset{\text{def}}{=} \{F \in L^\infty(V);\; \bar{\partial}F \in L^{\tilde{p}}_{0,1}(V)\},$$

$$\text{and } \mathcal{H}\Phi \overset{\text{def}}{=} \mathcal{H}K\Phi\big|_V \in H^p_{0,1}(V), \; p > 1. \tag{1.13}$$

Now put $\tilde{R} = R_1 + \tilde{R}_0$. Then for all $\Phi \in L^p_{0,1}(V)$, we have $\tilde{R}_0\Phi \in \tilde{W}^{1,\tilde{p}}(V)$ and $\tilde{R}\Phi \in L^\infty(V) \cup L^{\tilde{p}}(V)$.

By Corollary 1 below, which is based only on (1.13), it follows that for any form $\Phi \in L^p_{0,1}(V)$, one has a limit

$$\lim_{\substack{z\to\infty \\ z\in V_j}} \tilde{R}\Phi(z) \overset{\text{def}}{=} \tilde{R}\Phi(\infty_j).$$

Put $R_0\Phi = \tilde{R}_0\Phi - \tilde{R}\Phi(\infty_j)$ and $R\Phi = \tilde{R}\Phi - \tilde{R}\Phi(\infty_j)$. We then have property (1.1) for $R = R_1 + R_0$. This concludes the proof of Proposition 1.

Corollary 1. *Let* $F \in L^\infty(V)$ *and* $\bar{\partial}F \in L^p_{0,1}(V)$, $1 < p < 2$. *Then for all* $j \in \{1,\dots,d\}$, *there exists a limit* $\lim_{\substack{z\to\infty \\ z\in V_j}} F(z) \overset{\text{def}}{=} F(\infty_j)$ *such that* $(F - F(\infty_j))\big|_{V_j} \in L^{\tilde{p}}$.

Proof. Put $\bar{\partial}F = \Phi$. Then by (1.13) we have $\tilde{R}\Phi \in L^\infty(V) \cup L^{\tilde{p}}(V)$ and $\bar{\partial}(F - \tilde{R}\Phi) = \mathcal{H}\Phi$. Then the function $h = F - \tilde{R}\Phi$ is harmonic on V. The estimates $F \in L^\infty(V)$ and $\tilde{R}\Phi \in L^{\tilde{p}}(V) \cup L^\infty(V)$ imply by the Riemann extension theorem that h can be extended to a harmonic function $\tilde{h}$ on $\tilde{V}$. Hence, one has $h = F - \tilde{R}\Phi \equiv \text{const} = c$. This implies that there exists $\lim_{\substack{z\to\infty \\ z\in V_j}} F(z) = c_j \overset{\text{def}}{=} F(\infty_j)$. Corollary 1 is proved.

Corollary 1 admits the following useful reformulation.

Corollary 2. *In the notation of Proposition 1, for any bounded function ψ on V such that $\bar{\partial}\psi \in L^p(V)$, $1 < p < 2$, the following formula is valid:*

$$\psi(z) = \psi(\infty_j) + R_0\bar{\partial}\psi + \frac{1}{2\pi i}\int\limits_{\xi\in V} \bar{\partial}_\xi \left(\frac{\det\left[\frac{\partial P}{\partial \xi}(\xi), \bar{\xi} - \bar{z}\right] d\xi_1}{\frac{\partial P}{\partial \xi_2}(\xi)\cdot|\xi - z|^2}\right) \wedge \psi,$$

where $R_0\bar{\partial}\psi \in \tilde{W}^{1,\tilde{p}}(V)$, $1/\tilde{p} = 1/2 - 1/p$ and $R_0\bar{\partial}\psi(z) \to 0$, for $V_j \ni z \to \infty$.

2 Kernels and Estimates for $\bar{\partial}f = \varphi$ with $\varphi \in L^1_{1,1}(V)$

Let φ be a $(1,1)$-form of class $L^\infty_{1,1}(V)$ with support in $V_0 = \{z \in V;\ |z_1| \le r_0\}$, where r_0 satisfies condition (ii) of Sect. 1.

If $V = \mathbb{C}$, then by classical results from [P1] and [V], the Cauchy–Pompeiu operator

$$\varphi \mapsto \frac{dz}{2\pi i}\int\limits_{\xi\in V_0} \frac{\varphi(\xi)}{\xi - z} \stackrel{\text{def}}{=} \hat{R}\varphi$$

determines a solution $f = \hat{R}\varphi$ for the equation $\bar{\partial}f = \varphi$ on $\mathbb{C}$ with the property

$$f \in W^{1,\tilde{p}}_{1,0}(\mathbb{C}) \cap \mathcal{O}_{1,0}(\mathbb{C}\backslash V_0) \quad \text{for all} \quad \tilde{p} > 2.$$

In this section we derive an analogous result for the case of an affine algebraic Riemann surface $V \subset \mathbb{C}^2$.

Let $V\backslash V_0 = \cup_{j=1}^d V_j$, where $\{V_j\}$ are the connected components of $V\backslash V_0$.

Lemma 1. *Let $\Phi = dz_1 \rfloor \varphi$ and $f = F dz_1 = (R\Phi)dz_1$, where R is the operator from Proposition 1. Then*

(i) $\Phi \in L^p_{0,1}(V_0)$, $p \in [1,2)$, $\Phi = 0$ *on* $V\backslash V_0$*;*

(ii) $F\big|_{V_0} \in W^{1,p}(V_0)$ *for all* $p \in (1,2)$, $f \in \tilde{W}^{1,\tilde{p}}_{1,0}(V)$ *for all* $\tilde{p} \in (2,\infty)$, $\bar{\partial}F = \Phi - \mathcal{H}\Phi$, *where $\mathcal{H}$ is the operator from Proposition 1,* $\bar{\partial}f = \varphi - (\mathcal{H}\Phi)\wedge dz_1$, *and* $\|F\|_{L^\infty(V\backslash V_0)} + \|F\|_{L^{\tilde{p}}(V_0)} \le \text{const}(V,p)\|\Phi\|_{L^p(V)}$, $1/\tilde{p} = 1/p - 1/2$*;*

(iii) If, in addition, $\varphi \in W^{1,\infty}(V)$, *then* $f \in \tilde{W}^{2,\tilde{p}}(V)$.

Proof. (i) The property $\Phi\big|_{V\backslash V_0} = 0$ follows from $\varphi\big|_{V\backslash V_0} = 0$.

Put $V_0^\pm = \{z \in V_0;\ \pm\left|\frac{\partial P}{\partial z_2}\right| \ge \pm\left|\frac{\partial P}{\partial z_1}\right|\}$. The definition $\Phi = dz\rfloor\varphi$ implies that

$$\begin{aligned} \Phi\big|_{V_j^+} &= \Phi^+ d\bar{z}_1, \text{ where } \Phi^+ \in L^\infty(V_0^+) \text{ and} \\ \Phi\big|_{V_j^-} &= \Phi^- d\bar{z}_2/(\partial z_1/\partial z_2), \text{ where } \Phi^- \in L^\infty(V_0^-). \end{aligned} \tag{2.1}$$

The properties (2.1) imply that $\Phi \in L^p_{0,1}(V_0)$ for all $p \in (1,2)$.

(ii) The equalities $\bar{\partial}F = \Phi - \mathcal{H}\Phi$ and $\bar{\partial}f = \varphi - (\mathcal{H}\Phi) \wedge \mathrm{d}z_1$ follow from Proposition 1 together with the definitions $\Phi = \mathrm{d}z_1 \rfloor \varphi$ and $f = F\mathrm{d}z_1$. The inclusions $F \in L^\infty(V \setminus V_0)$ and $F\big|_{V_0} \in W^{1,p}(V_0)$ follow from the formula $F = R\Phi$ and Proposition 1. The inclusion $f \in \tilde{W}^{1,\tilde{p}}_{1,0}(V)$ follows from the equalities $\bar{\partial}f = \varphi - (\mathcal{H}\Phi) \wedge \mathrm{d}z_1$, $f = F\mathrm{d}z_1$, and Proposition 1.

(iii) If, in addition, $\varphi \in W^{1,\infty}_{1,1}(V)$, then $f \in \tilde{W}^{2,\tilde{p}}_{1,0}(V)$. This follows from the equalities above with $\varphi \in W^{1,\infty}_{1,1}(V)$ and supp, $\varphi \subset V_0$.

Lemma 2. *For each (0,1)-form $g \in H^p_{0,1}(V)$ there exists a (1,0)-form $h \in L^p_{1,0}(\tilde{V})$ ($1 \leq p < 2$), unique up to adding holomorphic (1,0)-forms on $\tilde{V}$, such that*

$$\bar{\partial}h\big|_{\tilde{V}} = g\mathrm{d}z_1. \tag{2.2}$$

Proof. For any $g \in H^p_{0,1}(V)$ the $(1,1)$-form $g \wedge \mathrm{d}z_1$ determines a current G on $\tilde{Y}$ by the equality

$$\langle G, \chi \rangle \stackrel{\text{def}}{=} \lim_{R \to \infty} \sum_{j=1}^{d} \left[\int_{V_j} (\chi - \chi_j(\infty)) g\mathrm{d}z_1 + \chi_j(\infty) \int_{\{z \in V_j;\ |z_1| < r\}} g \wedge \mathrm{d}z_1 \right],$$

where $\chi \in C^{(\varepsilon)}(\tilde{V})$, $\varepsilon > 0$, and $\chi_j(\infty) = \lim_{\substack{z \in V_j \\ z \to \infty}} \chi(z)$.

By Serre duality [S], the current G is $\bar{\partial}$-exact on $\tilde{V}$ if and only if

$$\langle G, 1 \rangle = \lim_{R \to \infty} \int_{\{z \in V;\ |z_1| \leq r\}} g \wedge \mathrm{d}z_1 = 0. \tag{2.3}$$

Let us check (2.3). We have

$$\int_{\{z \in V:\ |z_1| \leq r\}} g \wedge \mathrm{d}z_1 = - \int_{\{z \in V:\ |z_1| = r\}} z_1 \wedge g.$$

Putting $w_1 = 1/z_1$ into the right-hand side of this equality, we obtain

$$\int_{\{z \in V:\ |z_1| \leq r\}} g \wedge \mathrm{d}z_1 = - \sum_{j=1}^{d} \int_{|w_1| = 1/r} g_j(\bar{w}_1) \frac{\mathrm{d}\bar{w}_1}{w_1} = 0.$$

Here the last equality follows from the properties

$$g_j(\bar{w}_1)\mathrm{d}\bar{w}_1 = g\big|_{V_j \cap \{|w_1| \leq 1/r\}} \text{ and } \bar{g}_j \in \mathcal{O}(D(0, 1/r)).$$

Hence by (2.3), there exists $h \in L^1_{1,0}(\tilde{V})$ such that equality (2.2) is valid in the sense of currents. Moreover, any solution of (2.2) automatically belongs to $L^p_{1,0}(\tilde{V})$, $1 < p < 2$. Such a solution h of (2.2) is unique up to holomorphic (1,0)-forms on $\tilde{V}$ because the conditions $h \in L^p_{1,0}(\tilde{V})$ and $\bar{\partial} h = 0$ on V imply that h extends as a holomorphic (1,0)-form on $\tilde{V}$.

Notation: Let $\mathcal{H}^{\perp}: H^p_{0,1}(V) \to L^p_{1,0}(\tilde{V})$ $(1 < p < 2)$ be the operator defined by the formula $g \mapsto \mathcal{H}^{\perp} g$, where $\mathcal{H}^{\perp} g$ is the unique solution h of (2.2) in $L^p_{1,0}(\tilde{V})$ with the property

$$\int_V h \wedge \tilde{g} = 0 \quad \text{for all} \quad \tilde{g} \in H^p_{0,1}(V).$$

Lemma 2 guarantees the existence and uniqueness of $H^{\perp} g \in L^p_{1,0}(\tilde{V})$ for any $g \in H^p_{0,1}(V)$.

Proposition 2. *Let R be the operator defined by formula (1.1), and $\mathcal{H}$ the operator defined by formula (1.13). For any $(1,1)$-form $\varphi \in L^1_{1,1}(V) \cap L^{\infty}_{1,1}(V)$ with support in V_0, put*

$$\hat{R}\varphi = R^1\varphi + R^0\varphi, \tag{2.4}$$

where

$$R^1\varphi = (R(\mathrm{d}z_1 \rfloor \varphi))\mathrm{d}z_1, \quad R^0\varphi = \mathcal{H}^{\perp} \circ \mathcal{H}(\mathrm{d}z_1 \rfloor \varphi).$$

Then

$$\bar{\partial}\hat{R}\varphi = \varphi, \tag{2.5}$$

$$f = F\mathrm{d}z_1 = \hat{R}\varphi \in \tilde{W}^{1,\tilde{p}}_{1,0}(V) \text{ for all } \tilde{p} \in (2,\infty),$$

$$F\big|_{V_0} \in W^{1,p}(V_0) \text{ for all } p \in (1,2) \tag{2.6}$$

$$\text{and } f\big|_{V_l} = \sum_{k=1}^{\infty} \frac{c_k^{(l)}}{z_1^k}\mathrm{d}z_1 + b_l \mathrm{d}z_1, \text{ if } |z_1| \geq r_0.$$

Here $l = 1, \ldots, d$, and $b_l = 0$ for $l = j$.

Proof. The properties (2.5) and $f = \hat{R}\varphi \in \tilde{W}^{1,\tilde{p}}_{1,0}(V)$ follow from Proposition 1 and Lemmas 1 and 2. The properties (2.5) and $\varphi\big|_{V \setminus V_0} = 0$ imply analyticity of f on $V \setminus V_0$. The series expansion (2.6) follows from the analyticity of $f\big|_{V \setminus V_0}$ and the inclusion $f\big|_{V \setminus V_0} \in L^{\infty}_{1,0}(V \setminus V_0)$.

Supplement. Let $\tilde{V}_0 = \{z \in V : |z_1| \leq \tilde{r}_0\}$, where $\tilde{r}_0 > r_0$. If $\operatorname{supp}\varphi \subseteq V_0$ and

$$\left(\varphi - \sum_{l=1}^{g} c_l \delta(z, a_l)\right) \in L^{\infty}_{1,1}(V), \text{ where } a_l \in V_{j(l)} \cap \tilde{V}_0,$$

then

$$\left(\hat{R}\varphi - \sum_{l=1}^{g} c_l \hat{R}(\delta(z, a_l))\right) \in W_{1,0}^{1,\tilde{p}}(V).$$

3 Kernels and Estimates for $(\partial + \lambda \mathrm{d}z_1)u = f$, with $f \in \tilde{W}_{1,0}^{1,\tilde{p}}(V)$

If $V = \mathbb{C}$, then the equation $\partial u + \lambda u \mathrm{d}z_1 = f$ was also introduced by Pompeiu [P2]. One can check that this equation can be solved by the explicit formula

$$\mathrm{e}^{\lambda z - \bar{\lambda}\bar{z}} u(z) = \frac{1}{2\pi i} \int\limits_{\xi \in \mathbb{C}} \frac{\mathrm{e}^{\lambda\xi - \bar{\lambda}\bar{\xi}} f(\xi) \mathrm{d}\bar{\xi}}{\bar{\xi} - \bar{z}} \stackrel{\text{def}}{=} \lim_{r \to \infty} \frac{1}{2\pi i} \int\limits_{\{\xi \in \mathbb{C}:\, |\xi| < r\}} s \frac{\mathrm{e}^{\lambda\xi - \bar{\lambda}\bar{\xi}} f(\xi) \mathrm{d}\bar{\xi}}{\bar{\xi} - \bar{z}}.$$

For a Riemann surface $V = \{z \in \mathbb{C}^2 : P(z) = 0\}$ we will obtain the following generalization of this formula.

Proposition 3. *Let* $f = F\mathrm{d}z_1$ *be a* $(1,0)$*-form as in Proposition 2, i.e.,* $F\big|_{V_0} \in W^{1,p}(V_0)$ *for all* $p \in (1,2)$, $f \in \tilde{W}_{1,0}^{1,\tilde{p}}(V)$ *for all* $\tilde{p} \in (2, \infty)$, *and* $\operatorname{supp} \bar{\partial} f \subset V_0$. *Let* $e_\lambda(\xi) = \mathrm{e}^{\lambda\xi_1 - \bar{\lambda}\bar{\xi}_1}$. *Put*

$$\overline{R_1(\bar{e}_\lambda \bar{f})} \stackrel{\text{def}}{=} -\frac{1}{2\pi i} \lim_{r \to \infty} \int\limits_{\{\xi \in V:\, |\xi| < r\}} e_\lambda(\xi) f(\xi) \frac{\mathrm{d}\bar{\xi}_1}{\partial \bar{P} / \partial \bar{\xi}_2} \det\left[\frac{\partial \bar{P}}{\partial \bar{\xi}}(\xi), \frac{\xi - z}{|\xi - z|^2}\right].$$

Put also $\mathcal{H} f \stackrel{\text{def}}{=} \overline{\mathcal{H} \bar{f}}$, *where* $\mathcal{H}$ *is the operator from Proposition 1. Finally, let*

$$u = R_\lambda f = R_\lambda^1 f + R_\lambda^0 f, \tag{3.1}$$

where $R_\lambda^1 f + R_\lambda^0 f = e_{-\lambda}(z) \cdot \overline{R_1(\bar{e}_\lambda \bar{f})} + e_{-\lambda}(z) \cdot \overline{R_0(\bar{e}_\lambda \bar{f})}$, *with* R_1 *and* R_0 *being the operators from Proposition 1.*

Then for all $\lambda \neq 0$ *one has the following:*

(i) $(\partial + \lambda \mathrm{d}z_1) R_\lambda f = f - \mathcal{H}_\lambda(f)$, *where* $\mathcal{H}_\lambda(f) = e_{-\lambda}(z) \mathcal{H}(e_\lambda f)$.
(ii)

$$\|u - u(\infty_l)\|_{L^\infty(V)} \leq$$

$$\mathrm{const}(V, p) \cdot \min\left(\frac{1}{\sqrt{|\lambda|}}, \frac{1}{|\lambda|}\right) \left(\|F\|_{L^{\tilde{p}}(V_0)} + \|F\|_{L^\infty(V \setminus V_0)} + \|\partial F\|_{L^p_{1,0}(V)}\right),$$

$$\|\partial u\|_{L^{\tilde{p}}_{1,0}(V)} \leq \mathrm{const}(V, p) \cdot \|\partial F\|_{L^p_{1,0}(V)}, \text{ where } 1/\tilde{p} = 1/p - 1/2,\ l = 1, \ldots, d.$$

(iii) In addition, if $|\lambda| \le 1$, then

$$\|(1+|z_1|)(u-u(\infty_l))\|_{L^\infty(V_l)} \le \frac{\mathrm{const}(V,\tilde{p})}{\sqrt{|\lambda|}}(\|F\|_{L^{\tilde{p}}(V_0)} + \|F\|_{L^\infty(V\setminus V_0)}),$$

$$\|(1+|z_1|)\partial u\|_{L^\infty_{1,0}(V)} \le \mathrm{const}(V,\tilde{p})(\sqrt{|\lambda|}+1)(\|F\|_{L^{\tilde{p}}(V_0)} + \|F\|_{L^\infty(V\setminus V_0)}), \forall \tilde{p} > 2.$$

Supplement. Put

$$L^{2\pm\varepsilon}(V) = \{u; u\big|_{\tilde{V}_0} \in L^{2-\varepsilon}(\tilde{V}_0),\ \ u\big|_{V\setminus\tilde{V}_0} \in L^{2+\varepsilon}(V\setminus\tilde{V}_0)\}.$$

If $f = f_0 - f_1$, where $f_0 \in \tilde{W}^{1,\tilde{p}}_{1,0}(V)$, $\operatorname{supp}\bar{\partial} f_0 \subset V_0$, and $f_1 = \sum_{l=1}^{g} c_l \hat{R}(\delta(z,a_l))$, $a_l \in V_{j(l)} \cap \tilde{V}_0$, then instead of (i)–(iii) we have (i) and the following conclusion:

(ii′) We have

$$\|R_\lambda f - R_\lambda f(\infty_l)\|_{L^{2+\varepsilon}(V_l)} \le \frac{\mathrm{const}(V,\tilde{p})}{\varepsilon} \min\left(|\lambda|^{-1/2}, |\lambda|^{-1}\right)\left(\|f_0\|_{\tilde{W}^{1,\tilde{p}}_{1,0}} + \sum_{l=1}^{g}|c_l|\right),$$

$$\|\partial R_\lambda f\|_{L^{2\pm\varepsilon}_{1,0}(V)} \le \frac{\mathrm{const}(V,\tilde{p})}{\varepsilon}\left(\|f_0\|_{\tilde{W}^{1,\tilde{p}}_{1,0}} + \sum_{l=1}^{g}|c_l|\right),$$

$$\|\mathcal{H}_\lambda(f)\|_{L^\infty_{1,0}(V)} \le \frac{\mathrm{const}(V,\tilde{p})}{(1+|\lambda|)}\left(\|f_0\|_{\tilde{W}^{1,\tilde{p}}_{1,0}} + \sum_{l=1}^{g}|c_l|\right), \text{ where } \tilde{p} > 2, 0 < \varepsilon < 1/2.$$

Lemma 3. *Put*

$$J(z) = \int_{\{\xi\in\mathbb{C}:\ |\xi|<\rho\}} \frac{\psi(\xi)d\xi\wedge d\bar{\xi}}{|\xi|\cdot|\xi-z|},\ \ z\in\mathbb{C},$$

where $\psi \in L^p(V_0)$, $p > 1$. Then for any $\varepsilon > 0$ and any $\tilde{p} > 2$, one has the estimate

$$\|J(z)\|_{L^{\tilde{p}}(\mathbb{C})} \le \frac{1}{\varepsilon} O(\rho^{(2-2\varepsilon\tilde{p})/\tilde{p}}) \cdot \|\psi\|_{L^{(1+\varepsilon)/\varepsilon}(V_0)}.$$

Proof. Using the notation $\|\psi\|_\varepsilon = \|\psi\|_{L^{(1+\varepsilon)/\varepsilon}(V_0)}$, we obtain from the expression for $J(z)$ the following estimates:

$$|J(z)| \le \left(\int_{|\xi|\le\rho} \frac{|d\xi\wedge d\bar{\xi}|}{|\xi|^{1+\varepsilon}|\xi-z|^{1+\varepsilon}}\right)^{1/(1+\varepsilon)} \cdot \|\psi\|_\varepsilon$$

$$\le O\left(\int_{r=0}^{\rho}\frac{\mathrm{d}r}{r^{\varepsilon}}\int_{0}^{1}\frac{\mathrm{d}\varphi}{(|r-|z||+|z|\varphi)^{1+\varepsilon}}\right)^{1/(1+\varepsilon)}\cdot\|\psi\|_{\varepsilon}$$

$$\le O\left(\int_{r=0}^{\rho}\frac{\mathrm{d}r}{r^{\varepsilon}}\frac{1}{|z|^{1+\varepsilon}}\int_{0}^{1}\frac{\mathrm{d}\varphi}{(|\frac{r}{|z|}-1|+\varphi)^{1+\varepsilon}}\right)^{1/(1+\varepsilon)}\cdot\|\psi\|_{\varepsilon}$$

$$\le\frac{1}{\varepsilon}O\left(\int_{0}^{\rho}\frac{\mathrm{d}r}{r^{\varepsilon}}\frac{1}{|z|}\left(\frac{1}{|r-|z||^{\varepsilon}}-\frac{1}{(|r-|z||+|z|)^{\varepsilon}}\right)\right)^{1/(1+\varepsilon)}\cdot\|\psi\|_{\varepsilon}.$$

From the last estimate we deduce

$$|J(z)|\le\frac{1}{\varepsilon}O\left(\frac{1}{|z|}\left(\int_{0}^{|z|}\frac{\mathrm{d}r}{r^{\varepsilon}|z|^{\varepsilon}}+\int_{|z|}^{\rho}\frac{\varepsilon|z|}{r^{\varepsilon}r^{\varepsilon}r}\right)^{1/(1+\varepsilon)}\|\psi\|_{\varepsilon},\ \text{if}\ |z|\le\rho,$$

$$\text{and}\ |J(z)|\le\frac{1}{\varepsilon}O\left(\frac{1}{|z|}\int_{0}^{\rho}\frac{\mathrm{d}r}{r^{\varepsilon}|z|^{\varepsilon}}\right)\Bigg)^{1/(1+\varepsilon)}\|\psi\|_{\varepsilon},\ \text{if}\ |z|\ge\rho.$$

These equalities imply

$$|J(z)|\le\frac{1}{\varepsilon}O\left(\left(\frac{1}{|z|}\right)^{2\varepsilon/(1+\varepsilon)}\right)\|\psi\|_{\varepsilon},\ \text{ if }\ |z|\le\rho,$$

$$|J(z)|\le\frac{1}{\varepsilon}O\left(\frac{1}{|z|}\rho^{(1-\varepsilon)/(1+\varepsilon)}\right)\|\psi\|_{\varepsilon},\ \text{ if }\ |z|\ge\rho.$$

Putting $|z|=t$, we obtain finally that

$$\|J\|_{L^{\tilde p}(\mathbb{C})}\le\frac{1}{\varepsilon}O\left(\int_{0}^{\rho}\frac{\mathrm{d}t}{t^{2\varepsilon\tilde p/(1+\varepsilon)-1}}+\rho^{\frac{1-\varepsilon}{1+\varepsilon}\tilde p}\int_{\rho}^{\infty}\frac{\mathrm{d}t}{t^{\tilde p-1}}\right)^{1/\tilde p}\|\psi\|_{\varepsilon}\le\frac{1}{\varepsilon}O\left(\rho^{\frac{2-\varepsilon\tilde p}{\tilde p}}\right)\|\psi\|_{\varepsilon}.$$

Lemma 3 is proved.

Proof of Proposition 3:

(i) We have

$$(\partial+\lambda\mathrm{d}z_1)R_\lambda f=(\partial+\lambda\mathrm{d}z_1)e_{-\lambda}(z)\cdot\overline{R(\bar e_\lambda\bar f)}$$
$$=\partial(e_{-\lambda}(z))\cdot\overline{R(\bar e_\lambda\bar f)}+e_{-\lambda}(z)\partial(\overline{R(\bar e_\lambda\bar f)})+\lambda\mathrm{d}z_1e_{-\lambda}(z)\cdot\overline{R(\bar e_\lambda\bar f)}$$

$$= (-\lambda dz_1 + \lambda dz_1)e_{-\lambda}(z) \cdot \overline{R(\bar{e}_\lambda \bar{f})}$$
$$+e_{-\lambda}(z) \cdot (e_\lambda(z) f - \overline{\mathcal{H}\bar{e}_\lambda \bar{f}}) = f - e_{-\lambda}\mathcal{H}(e_\lambda f) \stackrel{\text{def}}{=} f - \mathcal{H}_\lambda f,$$

where we have used the equality (1.1) from Proposition 1.

(iii) Let $r \geq r_0$. Let the functions $\chi_\pm \in C^{(1)}(V)$ be such that $\chi_+ + \chi_- \equiv 1$ on V, $\operatorname{supp}\chi_+ \subset \{\xi \in V : |\xi_1| < 2r\}$, $\operatorname{supp}\chi_- \subset \{\xi \in V : |\xi_1| \geq r\}$, and $|d\chi_\pm| = O(1/r)$. We then have $u = u_+ + u_-$, where

$$u_\pm(z) = R_\lambda(\chi_\pm f). \tag{3.1$_\pm$}$$

Using the properties $f \in L^\infty(V)$ and $|e_\lambda| \equiv 1$, in combination with the equality $\partial u_+ = \chi_+ F dz_1 - \lambda u_+ dz_1 - \mathcal{H}_\lambda(\chi_+ f)$, we obtain for u_+ and $\frac{\partial u_\pm}{\partial z_1}$ the estimates

$$\|(1+|z|)(u_+(z) - u_+(\infty_l))\|_{L^\infty(V_l)} = O(r)\|f\|_{L^\infty_{1,0}(V)},\ l = 1,\ldots,d,$$
$$\|(1+|z|)\partial u_+(z)\|_{L^\infty_{1,0}(V)} = O(\lambda r + 1)\|f\|_{L^\infty_{1,0}(V)}. \tag{3.2}$$

In order to estimate u_-, we transform the expression $(3.1)_-$ using the series expansion (2.6) for $f\big|_{V_j}$, and we integrate by parts. We thus obtain

$$u_-(z) = R_\lambda \chi_- f = R^1_\lambda \chi_- f + R^0_\lambda \chi_- f$$
$$= -\frac{e_{-\lambda}(z)}{2\pi i}\frac{1}{\lambda}\int\limits_{\xi\in V} \frac{e^{\lambda\xi_1 - \bar{\lambda}\bar{\xi}_1}(d\chi_-)F \wedge d\bar{\xi}_1 \det\left[\frac{\partial \bar{P}}{\partial \bar{\xi}}(\xi), \xi - z\right]}{\frac{\partial \bar{P}}{\partial \bar{\xi}_2}(\xi)\cdot|\xi - z|^2}$$
$$+ \frac{e_{-\lambda}(z)}{2\pi i}\frac{1}{\lambda}\sum_j \int\limits_{\xi\in V_j} e^{\lambda\xi_1 - \bar{\lambda}\bar{\xi}_1}\chi_-\left(\sum_{k=1}^{\infty} k\frac{c_k^{(j)}}{\xi_1^{k+1}}\right)\frac{d\xi_1 \wedge d\bar{\xi}_1 \det\left[\frac{\partial \bar{P}}{\partial \bar{\xi}}(\xi), \xi - z\right]}{\frac{\partial \bar{P}}{\partial \bar{\xi}_2}(\xi)\cdot|\xi - z|^2}$$
$$- \frac{e_{-\lambda}(z)}{2\pi i}\frac{1}{\lambda}\int\limits_{\xi\in V} e^{\lambda\xi_1 - \bar{\lambda}\bar{\xi}_1}\chi_- F \partial_\xi\left(\frac{\det\left[\frac{\partial \bar{P}}{\partial \bar{\xi}}(\xi), \xi - z\right]d\bar{\xi}_1}{\frac{\partial \bar{P}}{\partial \bar{\xi}_2}(\xi)\cdot|\xi - z|^2}\right)$$
$$+ e_{-\lambda}(z)\bar{R}_0(e_\lambda \chi_- f), \tag{3.3}$$

where the operator $R_0 = \bar{\partial}^* GK$ is defined by (1.13). Using Corollary 2 we have, in addition,

$$-\frac{e_{-\lambda}(z)}{2\pi i}\frac{1}{\lambda}\int\limits_{\xi\in V} \mathrm{e}^{\lambda\xi_1-\bar\lambda\bar\xi_1}\chi_- F\partial_\xi\left(\frac{\det\left[\frac{\partial\bar P}{\partial\bar\xi}(\xi),\xi-z\right]\mathrm{d}\bar\xi_1}{\frac{\partial\bar P}{\partial\bar\xi_2}(\xi)\cdot|\xi-z|^2}\right)$$

$$=\frac{e_{-\lambda}(z)}{2\pi i}\frac{1}{\lambda}e_\lambda(z)\chi_-(z)F(z)-\frac{e_{-\lambda}(z)}{2\pi i}\frac{1}{\lambda}\bar R_0(\partial(e_\lambda\chi_- F))$$

$$=\frac{1}{2\pi i}\frac{1}{\lambda}\chi_-(z)F(z)-\frac{e_{-\lambda}(z)}{2\pi i}\frac{1}{\lambda}\bar R_0(\partial(e_\lambda\chi_- F)).$$

Putting the last equality into (3.3) and making use of the properties $|e_\lambda|\equiv 1$, $|d\chi_-|=O(1/r)$, $\partial u_-=\chi_- F\mathrm{d}z_1-\lambda u_-\mathrm{d}z_1-\mathcal{H}_\lambda(\chi_- f)$, and the property of R_0, we obtain from Proposition 1

$$\|(1+|z_1|)(u_--u_-(\infty_l))\|_{L^\infty(V_l)}$$

$$=O\left(\frac{1}{|\lambda|r}\right)(\|F\|_{L^{\tilde p}(V_0)}+\|F\|_{L^\infty(V\setminus V_0)})+\|(1+|z_1|)\bar R_0(e_\lambda\chi_- f)\|_{L^\infty(V)}$$

$$+\frac{1}{2\pi|\lambda|}\|(1+|z_1|)\bar R_0\partial(e_\lambda\chi_- F)\|_{L^\infty(V)}\le O\left(\frac{1}{|\lambda|r}\right)\|F\|_{L^{\tilde p}(V)},\ l=1,\ldots,d$$

$$\text{and}\quad \|(1+|z_1|)\frac{\partial u_-}{\partial z_1}\|_{L^\infty(V)}=O(1/r+1)(\|F\|_{L^{\tilde p}(V_0)}+\|F\|_{L^\infty(V\setminus V_0)}).\tag{3.4}$$

The estimates (3.2) and (3.4) imply

$$\|(1+|z_1|)(u-u(\infty_l))\|_{L^\infty(V_l)}$$

$$=O\left(r+\frac{1}{|\lambda|r}\right)(\|F\|_{L^{\tilde p}(V_0)}+\|F\|_{L^\infty(V\setminus V_0)}),$$

$$\text{and}\quad \|(1+|z_1|)\partial u\|_{L^\infty_{1,0}(V)}$$

$$=O(|\lambda|r+1/r+1)(\|F\|_{L^{\tilde p}(V_0)}+\|F\|_{L^\infty(V\setminus V_0)}),\ \forall\tilde p>2.\tag{3.5}$$

Putting in (3.5) $r=r_0/\sqrt{|\lambda|}$, we obtain (iii).

(ii) For proving (ii), let us put $r=\tilde r_0$ and transform $(3.1)_+$ for u_+ in the following way:

$$u_+(z)=R_\lambda\chi_+f=-\frac{e_{-\lambda}(z)}{2\pi i}\frac{1}{\lambda}\int\limits_{|\xi_1|\le r}\frac{\mathrm{e}^{\lambda\xi_1-\bar\lambda\bar\xi_1}\mathrm{d}\chi_+F\wedge\mathrm{d}\bar\xi_1\det\left[\frac{\partial\bar P}{\partial\bar\xi}(\xi),\xi-z\right]}{\frac{\partial\bar P}{\partial\bar\xi_2}(\xi)\cdot|\xi-z|^2}$$

$$
\begin{aligned}
&-\frac{e_{-\lambda}(z)}{2\pi i}\frac{1}{\lambda}\int\limits_{|\xi_1|\le r}\frac{\mathrm{e}^{\lambda\xi_1-\bar\lambda\bar\xi_1}\chi_+\partial F\wedge \mathrm{d}\bar\xi_1\det\left[\frac{\partial\bar P}{\partial\bar\xi}(\xi),\xi-z\right]}{\frac{\partial\bar P}{\partial\bar\xi_2}(\xi)\cdot|\xi-z|^2}\\
&-\frac{e_{-\lambda}(z)}{2\pi i}\frac{1}{\lambda}\int\limits_{|\xi_1|\le r}\mathrm{e}^{\lambda\xi_1-\bar\lambda\bar\xi_1}\chi_+F\partial\left(\frac{\det\left[\frac{\partial\bar P}{\partial\bar\xi}(\xi),\xi-z\right]\mathrm{d}\bar\xi_1}{\frac{\partial\bar P}{\partial\bar\xi_2}(\xi)\cdot|\xi-z|^2}\right)\\
&+e_{-\lambda}(z)\bar R_0(e_\lambda\chi_+f),
\end{aligned}
\tag{3.6}
$$

where R_0 is the operator from Proposition 1. Using the last expression for $u_+(z)$, together with the property $F\big|_{V_0}\in W^{1,p}(V_0)$ and Corollary 2, we obtain

$$
\|u_+\|_{L^\infty(V)}=O(1/\lambda)\|F\|_{\tilde W^{1,p}(V_0)}. \tag{3.7}
$$

This inequality together with (3.4) and statement (iii) proves the first part of statement (ii). Formula $u=R_\lambda f$ implies $\partial_z u=f-\lambda\mathrm{d}z_1 u-\mathcal{H}_\lambda f$. From this and from the already obtained estimates for u, we deduce the second part of statement (ii):

$$
\|\partial u\|_{L^{\tilde p}_{1,0}(V)}\le \mathrm{const}(V,p)\|\partial F\|_{L^p_{1,0}(V)}.
$$

(ii)$'$ In order to prove in this case the estimate for $u=R_\lambda f$ with $|\lambda|\le 1$, we combine the arguments above with Lemma 3, and obtain instead of (3.5) the following:

$$
\begin{aligned}
\|u-u(\infty_l)\|_{L^{2+\varepsilon}(V_l)}&\le\frac{1}{\varepsilon}O\left(r+\frac{1}{|\lambda|r}\right)\left(\|f_0\|_{\tilde W^{1,\tilde p}_{1,0}(V)}+\sum_{l=1}^{g}|c_l|\right)\\
\|\partial u\|_{L^{2\pm\varepsilon}_{1,0}(V)}&\le\frac{\lambda}{\varepsilon}O\left(r+\frac{1+r}{|\lambda|r}\right)\left(\|f_0\|_{\tilde W^{1,\tilde p}_{1,0}(V)}+\sum_{l=1}^{g}|c_l|\right)
\end{aligned}
\tag{3.5$'$}
$$

Putting in (3.5)$'$ $r=r_0/\sqrt{|\lambda|}$, we obtain the required estimate for $R_\lambda f$ with $|\lambda|\le 1$. To prove the estimate for $u=R_\lambda f$ with $|\lambda|\ge 1$, we use (3.6) and the Calderon–Zygmund $L^{2-\varepsilon}$-estimate for the singular integral on the right-hand side of (3.6).

In order to prove the statement concerning $H_\lambda f$, we just perform an integration by parts in the expression

$$
\mathcal{H}_\lambda f=e_{-\lambda}\mathcal{H}(e_\lambda f)=\sum_{l=1}^{g}e_{-\lambda}(z)\left(\int\limits_{\tilde V}e_\lambda(\xi)f(\xi)\wedge\overline{\omega_l(\xi)}\right)\omega_l(z),
$$

where $f=f_0+\sum_{l=1}^{g}c_l\hat R(\delta(z,a_l))$, and where $\{\omega_l,l=1,\dots,g\}$ is an orthonormal basis of holomorphic $(1,0)$-forms on $\tilde V$.

4 Faddeev-Type Green Function for $\bar{\partial}(\partial+\lambda \mathrm{d}z_1)u=\varphi$ and Further Results

Let $\hat{R}$ be the operator defined by formula (2.4) and let R_λ be the operator defined by formula (3.1).

Proposition 4. *Let $\varphi \in L^\infty_{1,1}(V)$ with support in $V_0=\{z\in V:\ |z_1|\le r_0\}$, where r_0 satisfies the condition of Sect. 1. Then for $u=G_\lambda\varphi \overset{\text{def}}{=} R_\lambda\circ\hat{R}\varphi$, where $\lambda\neq 0$, one has*

(i) $\bar{\partial}(\partial+\lambda \mathrm{d}z_1)u=\varphi+\bar{\lambda}\mathrm{d}\bar{z}_1\wedge\mathcal{H}_\lambda(\hat{R}\varphi)$ *on* V.
(ii) We have

$$\|u\|_{L^\infty(V)}\le \text{const}(V_0,\tilde{p})\cdot\min\left(1/\sqrt{|\lambda|},1/|\lambda|\right)\|\varphi\|_{L^\infty_{1,1}(V_0)},\quad \tilde{p}>2,$$
$$\|\partial u\|_{L^{\tilde{p}}_{1,0}(V)}\le \text{const}(V_0,\tilde{p})\,\|\varphi\|_{L^\infty_{1,1}(V_0)},\quad \tilde{p}>2.$$

Supplement. If we can write $\varphi=\varphi_0+\varphi_1$, where $\varphi_0\in L^\infty_{1,1}(V)$, $\operatorname{supp}\varphi_0\subset V_0$, and $\varphi_1=\sum\limits_{l=1}^{g} ic_l\delta(z,a_l)$, with $a_l\in V_{j(l)}\cap\tilde{V}_0$, then instead of (i)–(ii) we have (i) and the following conclusion:

(ii)$'$ We have

$$\|u-u(\infty_l)\|_{L^{2+\varepsilon}(V_l)}\le \text{const}(V,\varepsilon)\cdot\min\left(|\lambda|^{-1/2},|\lambda|^{-1}\right)\left(\|\varphi_0\|_{L^\infty_{1,1}(V_0)}+\sum_{j=1}^{g}|c_j|\right),$$
$$\|\partial u\|_{L^{2\pm\varepsilon}_{1,0}(V)}\le \text{const}(V,\varepsilon)\left(\|\varphi_0\|_{L^\infty_{1,1}(V_0)}+\sum_{l=1}^{g}|c_l|\right),$$

where $0<\varepsilon<1/2$.

Proof. By Proposition 2 we have

$$f=F\mathrm{d}z_1=\hat{R}\varphi\in\tilde{W}^{1,\tilde{p}}_{1,0}(V)\ \forall\tilde{p}\in(2,\infty),\quad F\big|_{V_0}\in W^{1,p}(V_0)\ \forall p\in(1,2).$$

Propositions 2 and 3 imply that $u=R_\lambda\circ\hat{R}\varphi\in\tilde{W}^{1,\tilde{p}}(V)$. Let us now verify statement (i) of Proposition 4. From Proposition 3(i), we obtain

$$(\partial+\lambda\mathrm{d}z_1)u=(\partial+\lambda\mathrm{d}z_1)R_\lambda\circ\hat{R}\varphi=\hat{R}\varphi+\mathcal{H}_\lambda(\hat{R}\varphi),\ \text{where}$$
$$\mathcal{H}_\lambda(\hat{R}\varphi)=e_{-\lambda}\mathcal{H}(e_\lambda\hat{R}\varphi). \tag{4.1}$$

From (4.1) and Proposition 2 we obtain

$$\bar{\partial}(\partial+\lambda \mathrm{d}z_1)u = \varphi + \bar{\partial}(\mathcal{H}_\lambda(\hat{R}\varphi)) = \varphi + \bar{\lambda}\mathrm{d}\bar{z}_1 \wedge \mathcal{H}_\lambda(\hat{R}\varphi),$$

where we have used that $\mathcal{H}(\hat{R}\varphi) \in H_{1,0}(\tilde{V})$.

Property 4(ii) follows from Proposition 3(ii), (iii). The supplement to Proposition 4 follows from the supplement to Proposition 3.

Definition 1. We define the Faddeev-type Green function for $\bar{\partial}(\partial+\lambda \mathrm{d}z_1)$ on V as the kernel $g_\lambda(z,\xi)$ of the integral operator $R_\lambda \circ \hat{R}$.

Definition 2. Let $q \in C_{1,1}(\tilde{V})$ be a form with supp$q$ contained in V_0, and let g denote the genus of $\tilde{V}$. The function $\psi(z,\lambda)$, $z \in V$, $\lambda \in \mathbb{C}$, will be called the Faddeev-type function associated with the potential q and the points $a_1,\ldots,a_g \in V\backslash\bar{V}_0$ if $\forall\lambda \in \mathbb{C}\backslash E$, where E is closed nowhere dense subset in $\mathbb{C}$, the function $\mu = \psi(z,\lambda)\mathrm{e}^{-\lambda z_1}$ has the following properties:

$$\bar{\partial}(\partial+\lambda \mathrm{d}z_1)\mu = \frac{i}{2}q\mu + i\sum_{l=1}^{g} c_l\delta(z,a_l) \text{ and } \lim_{\substack{z\to\infty \\ z\in V_1}} \mu(z,\lambda) = 1,$$
$$(\mu-\mu(\infty_j))\big|_{V_j} \in L^{\tilde{p}}(V_j),\ \ \tilde{p}>2,\ j=1,\ldots,d,$$

where $\delta(z,a_l)$ is the Dirac measure concentrated at the point a_l.

Based on the Faddeev-type Green function $g_\lambda(z,\xi)$, and on Proposition 4, we have in [HM] extended the Novikov reconstruction scheme from the case $X \subset \mathbb{C}$ to the case of a bordered Riemann surface $X \subset V$.

Definition 3. Let $\{\omega_j\}$ be an orthonormal basis for the holomorphic forms on $\tilde{V}$. An effective divisor $\{a_1,\ldots,a_g\}$ on V will be called generic if

$$\det\left[\frac{\omega_j}{\mathrm{d}z_1}(a_k)\big|_{j,k=1,2,\ldots,g}\right] \neq 0.$$

Lemma 4. *Let $\{a_j\}$ be a generic divisor on V. Put*

$$\Delta(\lambda) = \det\left[\int_{\xi\in V} \hat{R}(\delta(\xi,a_j))\wedge\bar{\omega}_l(\xi)\mathrm{e}^{\lambda\xi_1-\bar{\lambda}\bar{\xi}_1}\right]\Big|_{j,l=1,2,\ldots,g},$$
$$|\Delta(\lambda)|_\varepsilon = \sup_{|\lambda'-\lambda|<\varepsilon} |\Delta(\lambda')|,$$

where $\hat{R}$ is the operator from Proposition 2. Then for all $\varepsilon>0$, one has $\overline{\lim}_{\lambda\to\infty}|\lambda^g\cdot\Delta(\lambda)|<\infty$, $\underline{\lim}_{\lambda\to\infty}|\lambda^g\cdot\Delta(\lambda)|_\varepsilon>0$, and the set

$$E = \{\lambda\in\mathbb{C}:\ \Delta(\lambda)=0\} \text{ is a closed nowhere dense subset of } \mathbb{C}. \qquad (*)$$

The following is a corrected version of the main results from [HM]:

1. Let X be a domain with smooth boundary on V such that $X \supset \bar{V}_0$, $\bar{X} \subset Y \subset V$. Let $\sigma \in C^{(2)}(V)$, $\sigma > 0$ on V, and $\sigma = 1$ on $V \backslash X$. Let $a_1, \ldots, a_g$ be a generic divisor on $Y \backslash \bar{X}$, satisfying condition $(*)$. Then for all $\lambda \in \mathbb{C} \backslash E$, there exists a unique Faddeev-type function $\psi(z,\lambda) = \mu(z,\lambda) \mathrm{e}^{\lambda z_1}$ associated with the potential $q = \frac{dd^c \sqrt{\sigma}}{\sqrt{\sigma}}$ and the divisor $\{a_j\}$. Such a function can be found (together with constants $\{c_l\}$) from the integral equation
$$\mu(z,\lambda) = 1 + \frac{i}{2} \int\limits_{\xi \in X} g_\lambda(z,\xi)\mu(\xi,\lambda)q(\xi) + i\sum_{l=1}^{g} c_l(\lambda) g_\lambda(z,a_l), \tag{4.2}$$
where
$$\begin{gathered} \frac{1}{2}\mathcal{H}_\lambda(\hat{R}(q\mu)) = \sum_{l=1}^{g} c_l \mathcal{H}_\lambda(\hat{R}(z,a_l)), \\ \mu(z,\lambda) \to 1,\ z \in V_1,\ z \to \infty,\ \lambda \in \mathbb{C} \backslash E. \end{gathered} \tag{4.3}$$
The relation (4.3) is equivalent to the system of equations
$$\begin{aligned} &2\sum_{l=1}^{g} c_l(\lambda) \mathrm{e}^{\lambda a_{j,1} - \bar{\lambda}\bar{a}_{j,1}} \frac{\bar{\omega}_k}{\mathrm{d}\bar{z}_1}(a_j) = \\ &-\int\limits_{z \in X} \mathrm{e}^{\lambda z_1 - \bar{\lambda}\bar{z}_1} \left(\frac{dd^c\sqrt{\sigma}}{\sqrt{\sigma}} - 2i\partial \ln\sqrt{\sigma} \wedge \bar{\partial} \ln \sqrt{\sigma} \right) \mu(z,\lambda) \frac{\bar{\omega}_k}{\mathrm{d}\bar{z}_1}(z), \end{aligned}$$
where $k = 1, \ldots, g$ and $\{\omega_j\}$ is an orthonormal basis of holomorphic forms on $\tilde{V}$.
2. For all $\lambda \in \mathbb{C} \backslash E$, the restriction of $\mu = \mathrm{e}^{-\lambda z_1} \psi(z,\lambda)$ to bX can be found through Dirichlet-to-Neumann data for μ on bX by the Fredholm integral equation
$$\mu(z,\lambda)\big|_{bX} + \int\limits_{\xi \in bX} g_\lambda(z,\xi)(\bar{\partial}\mu(\xi,\lambda) - \bar{\partial}\mu_0(\xi,\lambda)) = I + i\sum_{j=1}^{g} c_j g_\lambda(z,a_j), \tag{4.4}$$
where $-I = \lim\limits_{\varepsilon \to 0} \int\limits_{|\xi - z| \geq \varepsilon} \mu_0 \bar{\partial}(\partial + \lambda d\xi_1) g_\lambda + \lim\limits_{R \to \infty} \int\limits_{|\xi_1| = R} [\bar{\partial} g_\lambda \mu_0 + g_\lambda(\partial + \lambda d\xi_1)\mu_0]$
$$-i\sum_{j=1}^{g} (a_{j,1})^{-k} c_j = \int\limits_{z \in bX} z_1^{-k}(\partial + \lambda \mathrm{d}z_1)\mu = 0, \quad k = 2, \ldots, g+1, \tag{4.5}$$
and μ_0 is the solution of the Dirichlet problem
$$\bar{\partial}(\partial + \lambda \mathrm{d}z_1)\mu_0\big|_X = 0,\ \ \mu_0\big|_{V\backslash X} = \mu\big|_{V \backslash X}.$$

The parameters $\{a_{j,1}\}$ (the first coordinates of $\{a_j\}$) are supposed to be distinct.

Equations (4.4) and (4.5) are solvable simultaneously with (4.2), (4.3).

The relations (4.5) are equivalent to the equality

$$\bar{\partial}(\partial + \lambda \mathrm{d}z_1)\mu\big|_{V\setminus X} = i\sum_{j=1}^{g} c_j \delta(z, a_j).$$

3. The Faddeev-type function $\mu = \psi(z,\lambda)\mathrm{e}^{-\lambda z_1}$ satisfies the Bers–Vekua-type $\bar{\partial}$-equation with respect to $\lambda \in \mathbb{C}\setminus E$:

$$\frac{\partial \mu(z,\lambda)}{\partial \bar{\lambda}} = b(\lambda)\bar{\mu}(z,\lambda)\mathrm{e}^{\bar{\lambda}\bar{z}_1 - \lambda z_1}, \tag{4.6}$$

where

$$b(\lambda) \overset{\mathrm{def}}{=} \lim_{\substack{z\to\infty\\ z\in V_l}} \frac{\bar{z}_1}{\bar{\lambda}} \mathrm{e}^{\lambda z_1 - \bar{\lambda}\bar{z}_1} \frac{\partial \mu}{\partial \bar{z}_1}(z,\lambda) \Big/ \lim_{\substack{z\to\infty\\ z\in V_l}} \overline{\mu(z,\lambda)},$$

with $l = 1,\dots,d$. The function $b(\lambda)$, referred to as nonphysical scattering data, can be found by (4.6) through $\mu\big|_{bX}$.

In addition, the following important formulas for the data $b(\lambda)$ are valid:

$$d\cdot\bar{\lambda}\cdot b(\lambda) = -\frac{1}{2\pi i}\int_{z\in bY} \mathrm{e}^{\lambda z_1 - \bar{\lambda}\bar{z}_1}\bar{\partial}\mu = \frac{1}{2\pi i}\int_{z\in X} \frac{i}{2}\mathrm{e}^{\lambda z_1 - \bar{\lambda}\bar{z}_1} q\mu + i\sum_{j=1}^{g} c_j \mathrm{e}^{\lambda a_{j,1} - \bar{\lambda}\bar{a}_{j,1}}, \tag{4.7}$$

where $\lambda \in \mathbb{C}\setminus E$.

On the basis of (4.3), (4.7), and Proposition 3, one can derive the estimate

$$|\lambda\cdot b(\lambda)| \le \mathrm{const}(V,\sigma)(1+|\lambda|)^{-g}|\Delta(\lambda)|^{-1}, \ \lambda \in \mathbb{C}\setminus E. \tag{4.8}$$

4. Let us suppose now that the divisor $\{a_1,\dots,a_g\}$ on $Y\setminus X$ is such that the exceptional set E in $\mathbb{C}$ consists of isolated points $\lambda_1,\dots,\lambda_N$, $N \le \infty$, and

$$|\Delta(\lambda)| \ge \mathrm{const}(V)\mathrm{dist}(\lambda, E) \ \text{ if } \ \mathrm{dist}(\lambda,E) \le \mathrm{const}. \tag{4.9}$$

Then the reconstruction procedure for $\mu\big|_{X\times\mathbb{C}}$ and $\sigma\big|_X$ through scattering data $b\big|_{\mathbb{C}}$ can be done in the following way.

The relations (4.2), (4.3), combined with the inequalities (4.8), (4.9), imply that the $\bar{\partial}$-equation (4.6) can be replaced by the singular integral equation

$$(\mu - 1) + \frac{1}{2\pi i}\lim_{\delta\to 0}\int_{\mathbb{C}\setminus\cup\{|\xi-\lambda_l|\le\delta\}} b(\xi)\mathrm{e}^{\bar{\xi}\bar{z}_1 - \xi z_1}\overline{(\mu-1)}\frac{\mathrm{d}\bar{\xi}\wedge\mathrm{d}\xi}{\xi - \lambda} + \frac{1}{2\pi i}\sum_{l=1}^{N}\frac{\mu_l}{\lambda_l - \lambda}$$

$$= -\frac{1}{2\pi i}\int\limits_{\mathbb{C}} b(\xi)e^{\bar{\xi}\bar{z}_1-\xi z_1}\frac{d\bar{\xi}\wedge d\xi}{\xi-\lambda}, \text{ where}$$

$$\mu_l = \lim_{\delta\to 0}\int\limits_{|\xi-\lambda_l|\le\delta} b\bar{\mu}e^{\bar{\xi}\bar{z}_1-\xi z_1}d\bar{\xi}\wedge d\xi = \lim_{\delta\to 0}\int\limits_{|\xi-\lambda_l|=\delta}\mu d\xi = O_z(1),$$

$$l = 1,2,\ldots,N,\ \lambda\in\mathbb{C}\backslash E. \tag{4.10}$$

This equation is of Fredholm–Noether type in the space of functions

$$\lambda\mapsto(\mu(\cdot,\lambda)-1):\ |\mu-1|\cdot|\Delta(\lambda)|(1+|\lambda|^g)\in L^{\tilde{p}}(\mathbb{C}),\ \tilde{p}>2.$$

In contrast to the planar case, when $d=1$, $g=0$, (4.10) does not necessarily have a unique solution. This makes it possible to find a basis of independent solutions of (4.10) for almost all $z_1\in\mathbb{C}$:

$$\lambda\mapsto\mu_k(z_1,\lambda),\ k=1,2,\ldots,\tilde{d},\ \lambda\in\mathbb{C},\ \tilde{d}\ge d.$$

Put

$$\mu(z_1,z_2,\lambda)=\mu(z_1,z_{2,j}(z_1),\lambda)=\sum_{k=1}^{\tilde{d}}\gamma_{j,k}(z_1)\mu_k(z_1,\lambda),$$

where $(z_1,z_2)=(z_1,z_{2,j}(z_1))\in V$, $j=1,2,\ldots,\tilde{d}$. The condition for the form $\mu^{-1}\bar{\partial}(\partial+\lambda dz_1)\mu$ to be independent of λ allows us to find (maybe not uniquely) the coefficients $\gamma_{j,k}(z)$ in the expression for $\mu(z_1,z_2,\lambda)$. The equalities

$$\frac{i}{2}\frac{dd^c\sqrt{\sigma}}{\sqrt{\sigma}}\Big|_X = q\big|_X = \mu^{-1}\bar{\partial}(\partial+\lambda dz_1)\mu\big|_X$$

finally permit us to find all q and σ with given scattering data $b\big|_{\mathbb{C}}$.

The uniqueness of the reconstruction of $\mu\big|_{X\times\mathbb{C}}$ and $\sigma\big|_X$ from the data b on $\mathbb{C}\setminus E$ is plausible but still unknown. Nevertheless, the unique reconstruction of $\sigma\big|_X$ from Dirichlet-to-Neumann data of the equation $d(\sigma d^cU)\big|_X=0$ can be obtained using Dirichlet-to-Neumann data not just for a single function, but for a family of Faddeev-type functions depending on a parameter θ:

$$\psi_\theta(z,\lambda)=e^{\lambda(z_1+\theta z_2)}\mu_\theta(z_1,z_2,\lambda),\ \text{ where}$$

$$\bar{\partial}(\partial+\lambda(dz_1+\theta dz_2))\mu_\theta=\frac{i}{2}q\mu_\theta+i\sum_{l=1}^{g}c_l\delta(z,a_l)\ \text{ and }\ \lim_{\substack{z\to\infty\\ z\in V_1}}\mu_\theta(z,\lambda)=1,$$

$$(\mu_\theta-\mu_\theta(\infty_j))\big|_{V_j}\in L^{\tilde{p}}(V_j),\ \tilde{p}>2,\ \lambda\in\mathbb{C}\backslash E_\theta,\ j=1,\ldots,d;$$

see [HN].

References

[BC1] Beals R., Coifman R., Multidimensional inverse scattering and nonlinear partial differential equations, Proc. Symp. Pure Math. **43** (1985), AMS, Providence, RI, 45–70.

[BC2] Beals R., Coifman R., The spectral problem for the Davey–Stewartson and Ishimori hierarchies. In: "Nonlinear Evolution Equations: Integrability and Spectral Methods," Proc. Workshop, Como, Italy 1988, Proc. Nonlinear Sci., 15–23, 1990.

[F1] Faddeev L. D., Increasing solutions of the Schrödinger equation, Dokl. Akad. Nauk SSSR **165** (1965), 514–517 (in Russian); Sov. Phys. Dokl. **10** (1966), 1033–1035.

[F2] Faddeev L. D., The inverse problem in the quantum theory of scattering, II, Current Problems in Math., **3**, 93–180, VINITI, Moscow, 1974 (in Russian); J. Soviet Math. **5** (1976), 334–396.

[GN] Grinevich P. G., Novikov S. P., Two-dimensional "inverse scattering problem" for negative energies and generalized analytic functions, Funct. Anal. Appl. **22** (1988), 19–27.

[H] Henkin G. M., Uniform estimate for a solution of the $\bar{\partial}$-equation in Weil domain, Uspekhi Mat. Nauk **26** (1971), 211–212.

[HM] Henkin G. M., Michel V., Inverse conductivity problem on Riemann surfaces, J. Geom. Anal. **18**(4) (2008) 1033–1052.

[HN] Henkin G. M., Novikov R. G., On the reconstruction of conductivity of bordered two-dimensional surface in $\mathbb{R}^3$ from electrical currents measurements on its boundary, J. Geom. Anal. **21** (2011), 543–887

[HP] Henkin G. M., Polyakov P. L., Homotopy formulas for the $\bar{\partial}$-operator on $\mathbb{C}P^n$ and the Radon–Penrose transform, Math. USSR Izvestiya **28** (1987), 555–587.

[Ho] Hodge W., The Theory and Applications of Harmonic Integrals, Cambridge University Press, Cambridge, 1952.

[N1] Novikov R., Multidimensional inverse spectral problem for the equation $-\Delta\psi+(v-Eu)\psi=0$, Funkt. Anal. i Pril. **22** (1988), 11–22 (in Russian); Funct. Anal Appl. **22** (1988), 263–278.

[N2] Novikov R., The inverse scattering problem on a fixed energy level for the two-dimensional Schrödinger operator, J. Funct. Anal. **103** (1992), 409–463.

[Na] Nachman A., Global uniqueness for a two-dimensional inverse boundary problem, Ann. Math. **143** (1996), 71–96.

[P1] Pompeiu D., Sur la représentation des fonctions analytiques par des intégrales définies, C. R. Acad. Sc. Paris **149** (1909), 1355–1357.

[P2] Pompeiu D., Sur une classe de fonctions d'une variable complexe et sur certaines équations intégrales, Rend. del. Circolo Matem. di Palermo **35** (1913), 277–281.

[Po] Polyakov P., The Cauchy–Weil formula for differential forms, Mat. Sb. **85** (1971), 388–402.

[S] Serre J.-P., Un théorème de dualité, Comm. Math. Helv. **29** (1955), 9–26.

[V] Vekua I. N., Generalized analytic functions, Pergamon Press; Addison-Wesley, 1962.

[W] Weil A., Variétés kähleriennes, Hermann, Paris, 1957.

Remarks on Khovanov Homology and the Potts Model

Louis H. Kauffman

Dedicated to Oleg Viro on his 60th birthday

Abstract In the paper we explore how the Potts model in statistical mechanics is related to Khovanov homology. This exploration is made possible because the underlying combinatorics for the bracket state sum for the Jones polynomial are shared by the Potts model for planar graphs. We show that Euler characteristics of Khovanov homology figure in the computation of the Potts model at certain imaginary temperatures and that these aspects of the Potts model can be reformulated as physical quantum amplitudes via Wick rotation. The paper concludes with a new conceptually transparent quantum algorithm for the Jones polynomial and with many further questions about Khovanov homology.

Keywords Khovanov homology • Potts model • Jones polynomial • Bracket state sum • Dichromatic polynomial

1 Introduction

This paper is about Khovanov homology and its relationships with finite combinatorial statistical mechanics models such as the Ising model and the Potts model.

Partition functions in statistical mechanics take the form

$$Z_G = \sum_{\sigma} \mathrm{e}^{(-1/kT)E(\sigma)},$$

L.H. Kauffman (✉)
Math UIC, 851 South Morgan Street, Chicago, Illinois 60607-7045
e-mail: kauffman@uic.edu

I. Itenberg et al. (eds.), *Perspectives in Analysis, Geometry, and Topology: On the Occasion of the 60th Birthday of Oleg Viro*, Progress in Mathematics 296,
DOI 10.1007/978-0-8176-8277-4_11,

where σ runs over the different physical states of a system G, and $E(\sigma)$ is the energy of the state σ. The probability of the system being in the state σ is taken to be

$$\mathrm{prob}(\sigma) = \mathrm{e}^{(-1/kT)E(\sigma)}/Z_G.$$

Since Onsager's work showing that the partition function of the Ising model has a phase transition, it has been a significant subject in mathematical physics to study the properties of partition functions for simply defined models based on graphs G. The underlying physical system is modeled by a graph G. and the states σ are certain discrete labelings of G. The reader can consult Baxter's book [4] for many beautiful examples. The Potts model, discussed below, is a generalization of the Ising model, and it is an example of a statistical-mechanical model that is intimately related to knot theory and to the Jones polynomial [8]. Since there are many connections between statistical mechanics and the Jones polynomial, it is remarkable that there have not been many connections between statistical mechanics and categorifications of the Jones polynomial. The reader will be find other points of view in [5, 6].

The partition function for the Potts model is given by the formula

$$P_G(Q,T) = \sum_{\sigma} \mathrm{e}^{(J/kT)E(\sigma)} = \sum_{\sigma} \mathrm{e}^{KE(\sigma)},$$

where σ is an assignment of one element of the set $\{1,2,\ldots,Q\}$ to each node of the graph G, and $E(\sigma)$ denotes the number of edges of a graph G whose end nodes receive the same assignment from σ. In this model, σ is regarded as a *physical state* of the Potts system, and $E(\sigma)$ is the *energy* of this state. Here $K = J\frac{1}{kT}$, where J is ± 1 (ferromagnetic and antiferromagnetic cases), k is Boltzmann's constant, and T is the temperature. The Potts partition function can be expressed in terms of the dichromatic polynomial of the graph G. Letting

$$v = e^K - 1,$$

it is shown in [12, 13] that the dichromatic polynomial $Z[G](v,Q)$ for a plane graph can be expressed in terms of a bracket state summation of the form

$$\{\times\} = \{\asymp\} + Q^{-\frac{1}{2}}v\{)(\},$$

with

$$\{\bigcirc\} = Q^{\frac{1}{2}}.$$

Then the Potts partition function is given by the formula

$$P_G = Q^{N/2}\{K(G)\},$$

where $K(G)$ is an alternating link diagram associated with the plane graph G such that the projection of $K(G)$ to the plane is a medial diagram for the graph. This translation of the Potts model in terms of a bracket expansion makes it possible to examine how the Khovanov homology of the states of the bracket is related to the evaluation of the partition function.

There are five sections in this paper beyond the introduction. In Sect. 2, we review the definition of Khovanov homology and observe, in parallel with [27], that it is very natural to begin by defining the Khovanov chain complex via enhanced states of the bracket polynomial model for the Jones polynomial. In fact, we begin with Khovanov's rewrite of the bracket state sum in the form

$$\langle \times \rangle = \langle \asymp \rangle - q \langle)(\rangle,$$

with $\langle \bigcirc \rangle = (q+q^{-1})$. We rewrite the state sum formula for this version of the bracket in terms of enhanced states (each loop is labeled $+1$ or -1 corresponding to q^{+1} and q^{-1} respectively) and show how, by collecting terms, the formula for the state sum has the form of a graded sum of Euler characteristics

$$\langle K \rangle = \sum_j q^j \sum_i (-1)^i \dim(\mathcal{C}^{ij}) = \sum_j q^j \chi(\mathcal{C}^{\bullet j}) = \sum_j q^j \chi(\mathcal{H}^{\bullet j}),$$

where $\mathcal{C}^{ij}$ is a module with basis the set of enhanced states s with i smoothings of type B (see Sect. 2 for definitions), $\mathcal{H}^{\bullet j}$ is its homology, and $j = j(s)$, where

$$j(s) = n_B(s) + \lambda(s),$$

where $n_B(s)$ denotes the number of B-smoothings in s and $\lambda(s)$ denotes the number of loops with positive label minus the number of loops with negative label in s. This formula suggests that there should be differentials $\partial : \mathcal{C}^{i,j} \longrightarrow \mathcal{C}^{i+1,j}$ such that j is preserved under the differential. We show that the restriction $j(s) = j(\partial(s))$ uniquely determines the differential in the complex, and how this leads to the Frobenius algebra structure that Khovanov used to define the differential. This part of our remarks is well known, but I believe that the method by which we arrive at the graded Euler characteristic is particularly useful for our subsequent discussion. We see clearly here that one should look for subcomplexes on which Euler characteristics can be defined, and one should attempt to shape the state summation so that these characteristics appear in the state sum. This happens miraculously for the bracket state sum and makes the combinatorics of that model dovetail with the Khovanov homology of its states, so that the bracket polynomial is seen as a graded Euler characteristic of the homology theory. The section ends with a discussion of how Grassmann algebra can be used, in analogy with de Rahm cohomology, to define the integral Khovanov chain complex. We further note that this analogy with de Rahm cohomology leads to other possibilities for chain complexes associated with the bracket states. These complexes will be the subject of a separate paper.

In Sect. 3 we recall the definition of the dichromatic polynomial and the fact [12] that the dichromatic polynomial for a planar graph G can be expressed as a special bracket state summation on an associated alternating knot $K(G)$. We recall that for special values of its two parameters, the Potts model can be expressed as a dichromatic polynomial. This sets the stage for examining the role of Khovanov homology in the evaluation of the Potts model and the dichromatic polynomial. In both cases (dichromatic polynomial in general and the Potts model in particular), one finds that the state summation does not so easily rearrange itself as a sum over Euler characteristics, as we have explained in Sect. 2. The states in the bracket summation for the dichromatic polynomial of $K(G)$ are the same as the states for the bracket polynomial of $K(G)$, so the Khovanov chain complex is present at the level of the states.

We clarify this relationship using a two-variable bracket expansion, the ρ-*bracket*, that reduces to the Khovanov version of the bracket as a function of q when ρ is equal to 1:

$$[\diagup\!\!\!\!\diagdown] = [\asymp] - q\rho[\,)(\,]$$

with

$$[\bigcirc] = q + q^{-1}.$$

We can regard this expansion as an intermediary between the Potts model (dichromatic polynomial) and the topological bracket. The ρ-bracket can be rewritten in the following form:

$$[K] = \sum_{i,j}(-\rho)^i q^j \dim(\mathcal{C}^{ij}) = \sum_j q^j \sum_i (-\rho)^i \dim(\mathcal{C}^{ij}) = \sum_j q^j \chi_\rho(\mathcal{C}^{\bullet j}),$$

where we define $\mathcal{C}^{ij}$ to be the linear span of the set of enhanced states with $n_B(s) = i$ and $j(s) = j$. Then the number of such states is the dimension $\dim(\mathcal{C}^{ij})$. We have expressed this ρ-bracket expansion in terms of generalized Euler characteristics of the complexes $\mathcal{C}^{\bullet j}$:

$$\chi_\rho(\mathcal{C}^{\bullet j}) = \sum_i (-\rho)^i \dim(\mathcal{C}^{ij}).$$

These generalized Euler characteristics become classical Euler characteristics when $\rho = 1$, and in that case, are the same as the Euler characteristic of the homology. With ρ not equal to 1, we do not have direct access to the homology. In that case, the polynomial is expressed in terms of ranks of chain modules, but not in terms of ranks of corresponding homology groups.

In Sect. 3 we analyze those cases of the Potts model in which $\rho = 1$ (so that Euler characteristics of Khovanov homology appear as coefficients in the Potts partition function), and we find that at criticality this requires $Q = 4$ and $e^K = -1$ ($K = J/kT$

as in the second paragraph of this introduction), hence an imaginary value of the temperature. When we simply require that $\rho = 1$, not necessarily at criticality, then we find that for $Q = 2$ we have $e^K = \pm$i. For $Q = 3$, we have $e^K = \frac{-1\pm\sqrt{3}i}{2}$. For $Q = 4$ we have $e^K = -1$. For $Q > 4$ it is easy to verify that e^K is real and negative. Thus in all cases of $\rho = 1$ we find that the Potts model has complex temperature values. Further work is called for to see how the evaluations of the Potts model at complex values influence its behavior for real temperature values in relation to the Khovanov homology. Such relationships between complex and real temperature evaluations are already known for the accumulation of the zeros of the partition function (the Lee–Yang zeros [20]). In Sect. 5 we return to the matter of imaginary temperature. This section is described below.

In Sect. 4 we discuss Stosic's categorification of the dichromatic polynomial, which involves using a differential motivated directly by the graphical structure and gives a homology theory distinct from Khovanov homology. We examine Stosic's categorification in relation to the Potts model and again show that it will work (in the sense that the coefficients of a partition function are Euler characteristics of the homology) when the temperature is imaginary. Specifically, we find this behavior for $K = i\pi + \ln(q + q^2 + \cdots + q^n)$, and again the challenge is to discover the influence of this graph homology on the Potts model at real temperatures. The results of this section do not require planarity of the graph G.

More generally, we see that in the case of this model, the differential in the homology is related to the combinatorial structure of the physical model. In that model, a given state has graphical regions of constant spin (calling the discrete assignments $\{1, 2, \ldots, Q\}$ the *spins*). We interpret the partial differentials in this model as taking two such regions and making them into a single region, by adding an edge that joins them, and reassigning a spin to the new region that is a combination of the spins of the two formerly separate regions. This form of partial differential is a way to think about relating the different states of the physical system. It is, however, not directly related to a classical physical process, since it invokes a global change in the spin configuration of the model. (One might consider such global transitions in quantum processes.)

The Stosic homology measures such global changes, and it is probably the nonclassical physicality of this patterning that makes it manifest at imaginary temperatures. The direct physical transitions from state to state that are relevant to the classical physics of the model may indeed involve changes of the form of the regions of constant spin, but this will happen by a local change, so that two disjoint regions with the same spin join into a single region, or a single region of constant spin bifurcates into two regions with this spin. These are the sorts of local-classical-time transitions that one works with in a statistical-mechanics model. Such transitions are part of the larger and more global transitions that are described by the differentials in our interpretation of the Stosic homology for the Potts model. Obviously, much more work needs to be done in this field. We have made some first steps in this paper.

In Sect. 5 we formulate a version of Wick rotation for the Potts model so that it is seen (for imaginary temperature) as a quantum amplitude. In this way, for those cases in which the temperature is pure imaginary, we obtain a quantum-physical interpretation of the Potts model and hence a relationship between Khovanov homology and quantum amplitudes at the special values discussed in Sect. 3.

In Sect. 6, we remark that the bracket state sum itself can be given a quantum-statistical interpretation, by choosing a Hilbert space whose basis is the set of enhanced states of a diagram K. We use the evaluation of the bracket at each enhanced state as a matrix element for a linear transformation on the Hilbert space. This transformation is unitary when the bracket variable (here denoted as above by q) is on the unit circle. In this way, using the Hadamard test, *we obtain a new quantum algorithm for the Jones polynomial* at all values of the Laurent polynomial variable that lie on the unit circle in the complex plane. This is not an efficient quantum algorithm, but it is conceptually transparent and it will allow us in subsequent work to analyze relationships between Khovanov homology and quantum computation.

2 Khovanov Homology

In this section, we describe Khovanov homology along the lines of [2, 19], and we tell the story in a way that allows the gradings and the structure of the differential to emerge in a natural way. This approach to motivating the Khovanov homology uses elements of Khovanov's original approach, Viro's use of enhanced states for the bracket polynomial [27], and Bar-Natan's emphasis on tangle cobordisms [3]. We use similar considerations in our paper [17].

Two key motivating ideas are involved in finding the Khovanov invariant. First of all, one would like to *categorify* a link polynomial such as $\langle K\rangle$. There are many meanings to the term categorify, but here the quest is to find a way to express the link polynomial as a *graded Euler characteristic* $\langle K\rangle = \chi_q\langle\mathcal{H}(K)\rangle$ for some homology theory associated with $\langle K\rangle$.

The bracket polynomial [11] model for the Jones polynomial [8–10, 28] is usually described by the expansion

and we have

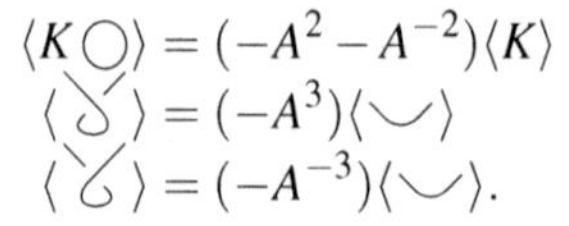

Letting $c(K)$ denote the number of crossings in the diagram K, if we replace $\langle K \rangle$ by $A^{-c(K)}\langle K \rangle$ and then replace A^2 by $-q^{-1}$, the bracket will be rewritten in the following form:

$$\langle \times \rangle = \langle \asymp \rangle - q \langle)(\rangle$$

with $\langle \bigcirc \rangle = (q + q^{-1})$. It is useful to use this form of the bracket state sum for the sake of the grading in the Khovanov homology (to be described below). We shall continue to refer to the smoothings labeled q (or A^{-1} in the original bracket formulation) as *B-smoothings*. We should further note that we use the well-known convention of *enhanced states*, where an enhanced state has a label of 1 or X on each of its component loops. We then regard the value of the loop $q + q^{-1}$ as the value of a circle labeled with a 1 (the value is q) added to the value of a circle labeled with an X (the value is q^{-1}). We could have chosen the more neutral labels of $+1$ and -1 so that

$$q^{+1} \Longleftrightarrow +1 \Longleftrightarrow 1$$

and

$$q^{-1} \Longleftrightarrow -1 \Longleftrightarrow X,$$

but since an algebra involving 1 and X naturally appears later, we take this form of labeling from the beginning.

To see how the Khovanov grading arises, consider the form of the expansion of this version of the bracket polynomial in enhanced states. We have the formula as a sum over enhanced states s:

$$\langle K \rangle = \sum_s (-1)^{n_B(s)} q^{j(s)},$$

where $n_B(s)$ is the number of B-type smoothings in s, $\lambda(s)$ is the number of loops in s labeled 1 minus the number of loops labeled X, and $j(s) = n_B(s) + \lambda(s)$. This can be rewritten in the following form:

$$\langle K \rangle = \sum_{i,j} (-1)^i q^j \dim(\mathcal{C}^{ij}),$$

where we define $\mathcal{C}^{ij}$ to be the linear span (over $k = Z/2Z$, since we will work with coefficients modulo 2) of the set of enhanced states with $n_B(s) = i$ and $j(s) = j$. Then the number of such states is the dimension $\dim(\mathcal{C}^{ij})$.

We would like to have a bigraded complex comprising the $\mathcal{C}^{ij}$ with a differential

$$\partial : \mathcal{C}^{ij} \longrightarrow \mathcal{C}^{i+1\,j}.$$

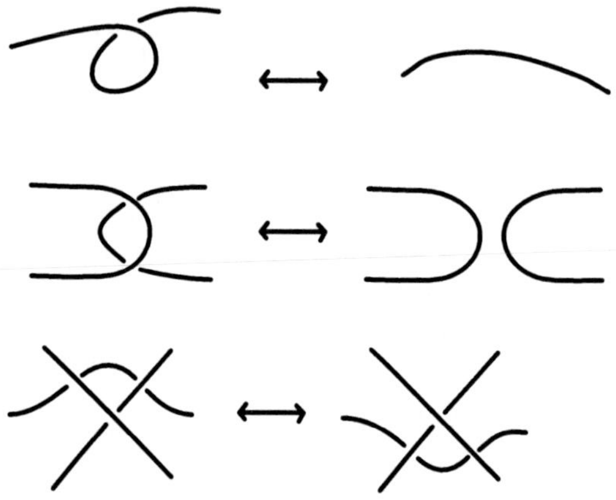

Fig. 1 Reidemeister moves

The differential should increase the *homological grading* i by 1 and preserve the *quantum grading* j. Then we could write

$$\langle K\rangle = \sum_j q^j \sum_i (-1)^i \dim(\mathcal{C}^{ij}) = \sum_j q^j \chi(\mathcal{C}^{\bullet j}),$$

where $\chi(\mathcal{C}^{\bullet j})$ is the Euler characteristic of the subcomplex $\mathcal{C}^{\bullet j}$ for a fixed value of j.

This formula would constitute a categorification of the bracket polynomial. Below, we shall see how *the original Khovanov differential ∂ is uniquely determined by the restriction that $j(\partial s) = j(s)$ for each enhanced state s.* Since j is preserved by the differential, these subcomplexes $\mathcal{C}^{\bullet j}$ have their own Euler characteristics and homology. We have

$$\chi(H(\mathcal{C}^{\bullet j})) = \chi(\mathcal{C}^{\bullet j}),$$

where $H(\mathcal{C}^{\bullet j})$ denotes the homology of the complex $\mathcal{C}^{\bullet j}$. We can write

$$\langle K\rangle = \sum_j q^j \chi(H(\mathcal{C}^{\bullet j})).$$

This last formula expresses the bracket polynomial as a *graded Euler characteristic* of a homology theory associated with the enhanced states of the bracket state summation. This is the categorification of the bracket polynomial. Khovanov proves that this homology theory is an invariant of knots and links (via the Reidemeister moves of Fig. 1), creating a new invariant that is stronger than the original Jones polynomial.

We will construct the differential in this complex first for mod-2 coefficients. The differential is based on regarding two states as *adjacent* if one differs from the other by a single smoothing at some site. Thus if (s,τ) denotes a pair consisting of an enhanced state s and site τ of that state with τ of type A, then we consider all enhanced states s' obtained from s by smoothing at τ and relabeling only those loops that are affected by the resmoothing. Call this set of enhanced states $S'[s,\tau]$.

Then we shall define the *partial differential* $\partial_\tau(s)$ as a sum over certain elements in $S'[s,\tau]$ and the differential by the formula

$$\partial(s) = \sum_\tau \partial_\tau(s)$$

with the sum over all type-A sites τ in s. It then remains to determine the possibilities for $\partial_\tau(s)$ for which $j(s)$ is preserved.

Note that if $s' \in S'[s,\tau]$, then $n_B(s') = n_B(s) + 1$. Thus

$$j(s') = n_B(s') + \lambda(s') = 1 + n_B(s) + \lambda(s').$$

From this we conclude that $j(s) = j(s')$ if and only if $\lambda(s') = \lambda(s) - 1$. Recall that

$$\lambda(s) = [s:+] - [s:-],$$

where $[s:+]$ is the number of loops in s labeled $+1$, and $[s:-]$ is the number of loops labeled -1 (same as labeled with X) and $j(s) = n_B(s) + \lambda(s)$.

Proposition. *The partial differentials $\partial_\tau(s)$ are uniquely determined by the condition $j(s') = j(s)$ for all s' involved in the action of the partial differential on the enhanced state s. This unique form of the partial differential can be described by the following structures of multiplication and comultiplication on the algebra $\mathcal{A} = k[X]/(X^2)$, where $k = Z/2Z$ for mod-2 coefficients, or $k = Z$ for integral coefficients:*

1. *The element 1 is a multiplicative unit and $X^2 = 0$.*
2. *$\Delta(1) = 1 \otimes X + X \otimes 1$ and $\Delta(X) = X \otimes X$.*

These rules describe the local relabeling process for loops in a state. Multiplication corresponds to the case that two loops merge to a single loop, while comultiplication corresponds to the case in which one loop bifurcates into two loops.

Proof. Using the above description of the differential, suppose that there are two loops at τ that merge in the resmoothing. If both loops are labeled 1 in s, then the local contribution to $\lambda(s)$ is 2. Let s' denote a resmoothing in $S[s,\tau]$. In order for the local λ contribution to become 1, we see that the merged loop must be labeled 1. Similarly, if the two loops are labeled 1 and X, then the merged loop must be labeled X, so that the local contribution for λ goes from 0 to -1. Finally, if the two loops are labeled X and X, then there is no label available for a single loop that will give -3, so we define ∂ to be zero in this case. We can summarize the result by saying that there is a multiplicative structure m such that $m(1,1) = 1, m(1,X) = m(X,1) = x, m(X,X) = 0$, and this multiplication describes the structure of the partial differential when two loops merge. Since this is the multiplicative structure of the algebra $\mathcal{A} = k[X]/(X^2)$, we take this algebra as summarizing the differential.

Now consider the case that s has a single loop at the site τ. Resmoothing produces two loops. If the single loop is labeled X, then we must label each of the two loops by X in order to make λ decrease by 1. If the single loop is labeled 1, then we can label the two loops by X and 1 in either order. In this second case we take the partial differential of s to be the sum of these two labeled states. This structure can be described by taking a coproduct structure with $\Delta(X) = X \otimes X$ and $\Delta(1) = 1 \otimes X + X \otimes 1$. We now have the algebra $\mathcal{A} = k[X]/(X^2)$ with product $m : \mathcal{A} \otimes \mathcal{A} \longrightarrow \mathcal{A}$ and coproduct $\Delta : \mathcal{A} \longrightarrow \mathcal{A} \otimes \mathcal{A}$, describing the differential completely. This completes the proof. □

Partial differentials are defined on each enhanced state s and a site τ of type A in that state. We consider states obtained from the given state by resmoothing the given site τ. The result of resmoothing τ is to produce a new state s' with one more site of type B than s. Forming s' from s, we either amalgamate two loops to a single loop at τ or divide a loop at τ into two distinct loops. In the case of amalgamation, the new state s acquires the label on the amalgamated circle that is the product of the labels on the two circles that are its ancestors in s. This case of the partial differential is described by the multiplication in the algebra. If one circle becomes two circles, then we apply the coproduct. Thus if the circle is labeled X, then the resultant two circles are each labeled X corresponding to $\Delta(X) = X \otimes X$. If the original circle is labeled 1, then we take the partial boundary to be a sum of two enhanced states with labels 1 and X in one case, and labels X and 1 in the other case, on the respective circles. This corresponds to $\Delta(1) = 1 \otimes X + X \otimes 1$. Modulo two, the boundary of an enhanced state is the sum, over all sites of type A in the state, of the partial boundaries at these sites. It is not hard to verify directly that the square of the boundary mapping is zero (this is the identity of mixed partials!) and that it behaves as advertised, keeping $j(s)$ constant. There is more to say about the nature of this construction with respect to Frobenius algebras and tangle cobordisms. In Figs. 2 and 3 we illustrate how the partial boundaries can be conceptualized in terms of surface cobordisms. The equality of mixed partials corresponds to topological equivalence of the corresponding surface cobordisms, and to the relationships between Frobenius algebras and the surface cobordism category. The proof of invariance of Khovanov homology with respect to the Reidemeister moves (respecting grading changes) will not be given here. See [2, 3, 19]. It is remarkable that this version of Khovanov homology is uniquely specified by natural ideas about adjacency of states in the bracket polynomial.

Remark on integral differentials. Choose an ordering for the crossings in the link diagram K and denote them by $1, 2, \ldots, n$. Let s be any enhanced state of K and let $\partial_i(s)$ denote the chain obtained from s by applying a partial boundary at the ith site of s. If the ith site is a smoothing of type A^{-1}, then $\partial_i(s) = 0$. If the ith site is a smoothing of type A, then $\partial_i(s)$ is given by the rules discussed above (with the same signs). The compatibility conditions that we have discussed show that partials commute in the sense that $\partial_i(\partial_j(s)) = \partial_j(\partial_i(s))$ for all i and j. One then defines signed boundary formulas in the usual way of algebraic topology. One way to think of this regards the complex as the analogue of a complex in de Rahm

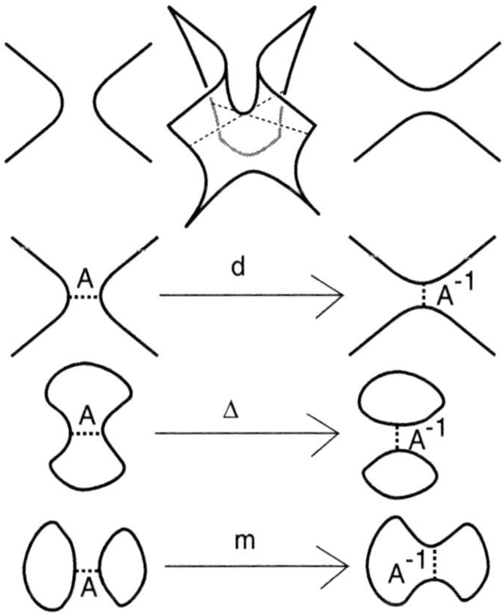

Fig. 2 Saddle points and state smoothings

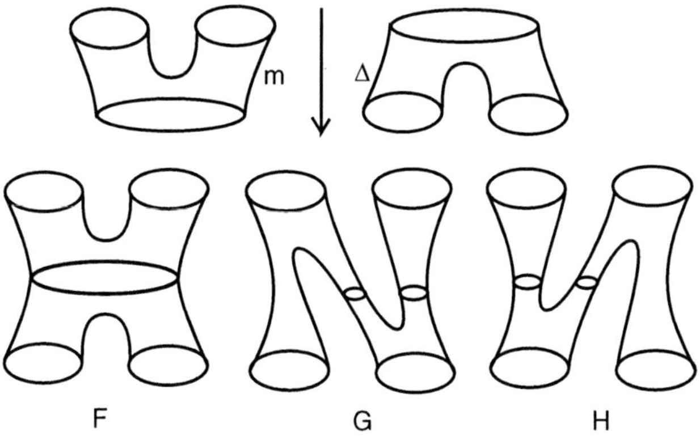

Fig. 3 Surface cobordisms

cohomology. Let $\{dx_1, dx_2, \ldots, dx_n\}$ be a formal basis for a Grassmann algebra such that $dx_i \wedge dx_j = -dx_i \wedge dx_j$. Starting with enhanced states s in $C^0(K)$ (that is, states with all A-type smoothings), define formally $d_i(s) = \partial_i(s)dx_i$ and regard $d_i(s)$ as identical with $\partial_i(s)$, as we have previously regarded it in $C^1(K)$. In general, given an enhanced state s in $C^k(K)$ with B-smoothings at locations $i_1 < i_2 < \cdots < i_k$, we represent this chain as $s\,dx_{i_1} \wedge \cdots \wedge dx_{i_k}$ and define

$$\partial(s\,dx_{i_1} \wedge \cdots \wedge dx_{i_k}) = \sum_{j=1}^{n} \partial_j(s)\,dx_j \wedge dx_{i_1} \wedge \cdots \wedge dx_{i_k},$$

just as in a de Rahm complex. The Grassmann algebra automatically computes the correct signs in the chain complex, and this boundary formula gives the original boundary formula when we take coefficients modulo two. Note that in this formalism, partial differentials ∂_i of enhanced states with a B-smoothing at the site i are zero due to the fact that $\mathrm{d}x_i \wedge \mathrm{d}x_i = 0$ in the Grassmann algebra. There is more to discuss about the use of Grassmann algebras in this context. For example, this approach clarifies parts of the construction in [18].

It of interest to examine this analogy between the Khovanov (co)homology and de Rahm cohomology. In that analogy the enhanced states correspond to the differentiable functions on a manifold. The Khovanov complex $C^k(K)$ is generated by elements of the form $s\,\mathrm{d}x_{i_1} \wedge \cdots \wedge \mathrm{d}x_{i_k}$, where the enhanced state s has B-smoothings at exactly the sites $i_1, \ldots, i_k$. If we were to follow the analogy with de Rahm cohomology literally, we would define a new complex $DR(K)$, where $DR^k(K)$ is generated by elements $s\,\mathrm{d}x_{i_1} \wedge \cdots \wedge \mathrm{d}x_{i_k}$, where s is *any* enhanced state of the link K. The partial boundaries are defined in the same way as before, and the global boundary formula is just as we have written it above. This gives a *new* chain complex associated with the link K. Whether its homology contains new topological information about the link K will be the subject of a subsequent paper.

A further remark on de Rham cohomology. There is another deep relationship with the de Rham complex: In [21], it was observed that Khovanov homology is related to Hochschild homology, and Hochschild homology is thought to be an algebraic version of de Rham chain complex (cyclic cohomology corresponds to de Rham cohomology); compare [22].

3 The Dichromatic Polynomial and the Potts Model

We define the *dichromatic polynomial* as follows:

$$Z[G](v,Q) = Z[G'](v,Q) + vZ[G''](v,Q),$$
$$Z[\bullet \sqcup G] = QZ[G],$$

where G' is the result of deleting an edge from G, while G'' is the result of contracting that same edge so that its end nodes have been collapsed to a single node. In the second equation, $\bullet$ represents a graph with one node and no edges, and $\bullet \sqcup G$ represents the disjoint union of the single-node graph with the graph G.

In [12, 13] it is shown that the dichromatic polynomial $Z[G](v,Q)$ for a plane graph can be expressed in terms of a bracket state summation of the form

$$\{\times\} = \{\asymp\} + Q^{-\frac{1}{2}} v \{)(\}$$

with

$$\{\bigcirc\} = Q^{\frac{1}{2}}.$$

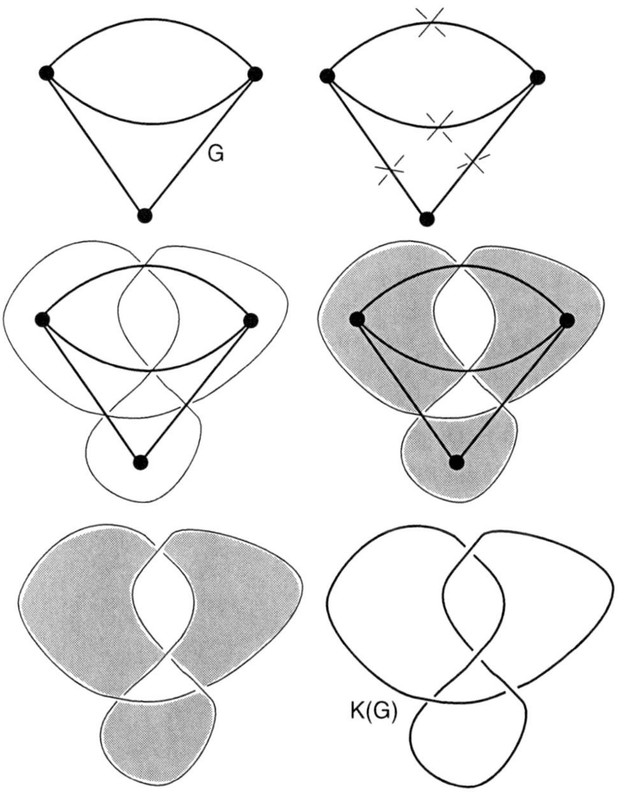

Fig. 4 Medial graph, Tait checkerboard graph, and $K(G)$

Here

$$Z[G](v,Q) = Q^{N/2}\{K(G)\},$$

where $K(G)$ is an alternating link diagram associated with the plane graph G in such a way that the projection of $K(G)$ to the plane is a medial diagram for the graph. Here we use the opposite convention from [13] in associating crossings to edges in the graph. We set $K(G)$ so that smoothing $K(G)$ along edges of the graph gives rise to B-smoothings of $K(G)$. See Fig. 4. The formula above, in bracket expansion form, is derived from the graphical contraction–deletion formula by translating first to the medial graph as indicated in the formulas below:

$$Z[\times] = Z[\asymp] + vZ[)(],$$
$$Z[R \sqcup K] = QZ[K].$$

Here the shaded medial graph is indicated by the shaded glyphs in these formulas. The medial graph is obtained by placing a crossing at each edge of G and then connecting all these crossings around each face of G as shown in Fig. 4. The medial can be checkerboard-shaded in relation to the original graph G (this is usually called the Tait checkerboard graph after Peter Guthrie Tait, who introduced these ideas into graph theory), and encoded with a crossing structure so that it represents a link diagram. Here R denotes a connected shaded region in the shaded medial graph. Such a region corresponds to a collection of nodes in the original graph, all labeled with the same color. The proof of the formula $Z[G] = Q^{N/2}\{K(G)\}$ then involves re-counting boundaries of regions in correspondence with the loops in the link diagram. The advantage of the bracket expansion of the dichromatic polynomial is that it shows that this graph invariant is part of a family of polynomials that includes the Jones polynomial, and it shows how the dichromatic polynomial for a graph whose medial is a braid closure can be expressed in terms of the Temperley–Lieb algebra. This, in turn, reflects on the structure of the Potts model for planar graphs, as we remark below.

It is well known that the partition function $P_G(Q,T)$ for the Q-state Potts model in statistical mechanics on a graph G is equal to the dichromatic polynomial when

$$v = \mathrm{e}^{J\frac{1}{kT}} - 1,$$

where T is the temperature for the model and k is Boltzmann's constant. Here $J = \pm 1$ according to whether we work with the ferromagnetic or antiferromagnetic model (see [4, Chap. 12]). For simplicity, we define

$$K = J\frac{1}{kT},$$

so that

$$v = e^K - 1.$$

We have the identity

$$P_G(Q,T) = Z[G](e^K - 1, Q).$$

The partition function is given by the formula

$$P_G(Q,T) = \sum_{\sigma} e^{KE(\sigma)},$$

where σ is an assignment of one element of the set $\{1,2,\ldots,Q\}$ to each node of the graph G, and $E(\sigma)$ denotes the number of edges of the graph G whose end nodes receive the same assignment from σ. In this model, σ is regarded as a *physical state* of the Potts system, and $E(\sigma)$ is the *energy* of that state. Thus we have a link diagrammatic formulation for the Potts partition function for planar graphs G:

$$P_G(Q,T) = Q^{N/2}\{K(G)\}(Q, v = e^K - 1),$$

where N is the number of nodes in the graph G.

This bracket expansion for the Potts model is very useful in thinking about the physical structure of the model. For example, since the bracket expansion can be expressed in terms of the Temperley–Lieb algebra, one can use this formalism to express the expansion of the Potts model in terms of the Temperley–Lieb algebra. This method clarifies the fundamental relationship between the Potts model and the algebra of Temperley and Lieb. Furthermore, the conjectured critical temperature for the Potts model occurs for T when $Q^{-\frac{1}{2}}v = 1$. We see clearly in the bracket expansion that this value of T corresponds to a point of symmetry of the model where the value of the partition function does not depend on the designation of over- and undercrossings in the associated knot or link. This corresponds to a symmetry between the plane graph G and its dual.

We first analyze how our heuristics leading to the Khovanov homology look when generalized to the context of the dichromatic polynomial. (This approach to the question is different from the methods of Stosic [26] and [7, 24], but see the next section for a discussion of Stosic's approach to categorifying the dichromatic polynomial.) We then ask questions about the relationship between Khovanov homology and the Potts model. It is natural to ask such questions, since the adjacency of states in the Khovanov homology corresponds to an adjacency for energetic states of the physical system described by the Potts model, as we shall describe below.

For this purpose we now adopt yet another bracket expansion as indicated below. We call this two-variable bracket expansion the ρ-*bracket.* It reduces to the Khovanov version of the bracket as a function of q when ρ is equal to one:

with

$$[\bigcirc] = q + q^{-1}.$$

We can regard this expansion as an intermediary between the Potts model (dichromatic polynomial) and the topological bracket. When $\rho = 1$, we have the topological bracket expansion in Khovanov form. When

$$-q\rho = Q^{-\frac{1}{2}}v$$

and

$$q + q^{-1} = Q^{\frac{1}{2}},$$

we have the Potts model. We shall return to these parameterizations shortly.

Just as in the last section, we have

$$[K] = \sum_{s} (-\rho)^{n_B(s)} q^{j(s)},$$

where $n_B(s)$ is the number of B-type smoothings in s, $\lambda(s)$ is the number of loops in s labeled 1 minus the number of loops labeled X, and $j(s) = n_B(s) + \lambda(s)$. This can be rewritten in the following form:

$$[K] = \sum_{i,j} (-\rho)^i q^j \dim(\mathcal{C}^{ij}) = \sum_{j} q^j \sum_{i} (-\rho)^i \dim(\mathcal{C}^{ij}) = \sum_{j} q^j \chi_\rho(\mathcal{C}^{\bullet j}),$$

where we define $\mathcal{C}^{ij}$ to be the linear span of the set of enhanced states with $n_B(s) = i$ and $j(s) = j$. Then the number of such states is the dimension $\dim(\mathcal{C}^{ij})$. Now we have expressed this general bracket expansion in terms of generalized Euler characteristics of the complexes:

$$\chi_\rho(\mathcal{C}^{\bullet j}) = \sum_{i} (-\rho)^i \dim(\mathcal{C}^{ij}).$$

These generalized Euler characteristics become classical Euler characteristics when $\rho = 1$, and in that case are the same as the Euler characteristic of the homology. With ρ not equal to 1, we do not have direct access to the homology.

Nevertheless, I believe that this raises a significant question about the relationship between $[K](q,\rho)$ and Khovanov homology. We get the Khovanov version of the bracket polynomial for $\rho = 1$ such that for $\rho = 1$, we have

$$[K](q, \rho = 1) = \sum_{j} q^j \chi_\rho(\mathcal{C}^{\bullet j}) = \sum_{j} q^j \chi(H(\mathcal{C}^{\bullet j})).$$

Away from $\rho = 1$ one can inquire into the influence of the homology groups on the coefficients of the expansion of $[K](q,\rho)$ and the corresponding questions about the Potts model. This is a way to generalize questions about the relationship between the Jones polynomial and the Potts model. In the case of the Khovanov formalism, we have the same structure of the states and the same homology theory for the states in both cases, but in the case of the Jones polynomial (ρ-bracket expansion with $\rho = 1$), we have expressions for the coefficients of the Jones polynomial in terms of ranks of the Khovanov homology groups. Only the ranks of the chain complexes figure in the Potts model itself. Thus we are suggesting here that it is worth asking about the relationship between the Khovanov homology and the dichromatic polynomial, the ρ-bracket and the Potts model, without changing the definition of the homology groups or chain spaces. This also raises the question of the relationship of the Khovanov homology to those constructions that have been made (e.g., [26]) where the homology has been adjusted to fit directly with the dichromatic polynomial. We will take up this comparison in the next section.

We now look more closely at the Potts model by writing a translation between the variables q,ρ and Q,v. We have

$$-q\rho = Q^{-\frac{1}{2}}v$$

and

$$q+q^{-1} = Q^{\frac{1}{2}},$$

and from this we conclude that

$$q^2 - \sqrt{Q}q + 1 = 0,$$

whence

$$q = \frac{\sqrt{Q} \pm \sqrt{Q-4}}{2}$$

and

$$\frac{1}{q} = \frac{\sqrt{Q} \mp \sqrt{Q-4}}{2}.$$

Thus

$$\rho = -\frac{v}{\sqrt{Q}q} = v\left(\frac{-1 \pm \sqrt{1-4/Q}}{2}\right).$$

For physical applications, Q is a positive integer greater than or equal to 2. Let us begin by analyzing the Potts model at criticality (see discussion above), where $-\rho q = 1$. Then

$$\rho = -\frac{1}{q} = \frac{-\sqrt{Q} \pm \sqrt{Q-4}}{2}.$$

For the Khovanov homology (its Euler characteristics) to appear directly in the partition function, we want

$$\rho = 1.$$

Thus we want

$$2 = -\sqrt{Q} \pm \sqrt{Q-4}.$$

Squaring both sides and collecting terms, we find that $4 - Q = \mp\sqrt{Q}\sqrt{Q-4}$. Squaring once more and collecting terms, we find that the only possibility for $\rho = 1$ is $Q = 4$. Returning to the equation for ρ, we see that this will be satisfied when we take $\sqrt{4} = -2$. This can be done in the parameterization, and then the partition function will have Khovanov topological terms. However, note that with this choice, $q = -1$, and so $v/\sqrt{Q} = -\rho q = 1$ implies that $v = -2$. Thus $e^K - 1 = -2$, and so

$$e^K = -1.$$

From this we see that in order to have $\rho = 1$ at criticality, we need a four-state Potts model with imaginary temperature variable $K = (2n+1)i\pi$. It is worthwhile considering the Potts models at imaginary temperature values. For example, the Lee–Yang theorem [20] shows that under certain circumstances, the zeros on the partition function are on the unit circle in the complex plane. We take the present calculation as an indication of the need for further investigation of the Potts model with real and complex values for its parameters.

Now we go back and consider $\rho = 1$ without insisting on criticality. Then we have $1 = -v/(q\sqrt{Q})$, so that

$$v = -q\sqrt{Q} = \frac{-Q \mp \sqrt{Q}\sqrt{Q-4}}{2}.$$

From this we see that

$$e^K = 1 + v = \frac{2 - Q \mp \sqrt{Q}\sqrt{Q-4}}{2}.$$

From this we get the following formulas for e^K: For $Q = 2$, we have $e^K = \pm \mathrm{i}$. For $Q = 3$, we have $e^K = \frac{-1 \pm \sqrt{3}i}{2}$. For $Q = 4$, we have $e^K = -1$. For $Q > 4$, it is easy to verify that e^K is real and negative. Thus in all cases of $\rho = 1$, we find that the Potts model has complex temperature values. In a subsequent paper, we shall attempt to analyze the influence of the Khovanov homology at these complex values on the behavior of the model for real temperatures.

4 The Potts Model and Stosic's Categorification of the Dichromatic Polynomial

In [26], Stosic gives a categorification for certain specializations of the dichromatic polynomial. In this section we describe this categorification, and discuss its relationship to the Potts model. The reader should also note that the relationships between dichromatic (and chromatic) homology and Khovanov homology are observed in [7, Theorem 24]. In this paper we use the Stosic formulation for our analysis.

For this purpose, we define, as in the previous section, the dichromatic polynomial through the formulas

$$Z[G](v,Q) = Z[G'](v,Q) + vZ[G''](v,Q)$$

and

$$Z[\bullet \sqcup G] = QZ[G],$$

where G' is the result of deleting an edge from G, while G'' is the result of contracting that same edge so that its end nodes have been collapsed to a single node. In the second equation, $\bullet$ represents a graph with one node and no edges, and $\bullet \sqcup G$ represents the disjoint union of the single-node graph with the graph G. The graph G is an arbitrary finite (multi)graph. This formulation of the dichromatic polynomial reveals its origins as a generalization of the chromatic polynomial for a graph G. The case $v = -1$ is that of the chromatic polynomial. In that case, the first equation asserts that the number of proper colorings of the nodes of G using Q colors is equal to the number of colorings of the deleted graph G' minus the number of colorings of the contracted graph G''. This statement is a tautology, since a proper coloring demands that nodes connected by an edge be colored with distinct colors, whence the deleted graph allows all colors, while the contracted graph allows only colorings where the nodes at the original edge receive the same color. The difference is then equal to the number of colorings that are proper at the given edge.

We reformulate this recursion for the dichromatic polynomial as follows: Instead of contracting an edge of the graph to a point in the second term of the formula, simply label that edge (say with the letter x) so that we know that it has been used in the recursion. For thinking of colorings from the set $\{1,2,\ldots,Q\}$ when Q is a positive integer, regard an edge marked with x as indicating that the colors on its two nodes are the same. This rule coincides with our interpretation of the coloring polynomial in the last paragraph. Then G'' in the deletion–contraction formula above denotes the labeling of the edge by the letter x. We then see that we can write the following formula for the dichromatic polynomial:

$$Z[G] = \sum_{H \subset G} Q^{|H|} v^{e(H)},$$

where H is a subgraph of G, $|H|$ is the number of components of H, and $e(H)$ is the number of edges of H. The subgraphs H correspond to the graphs generated by the new interpretation of the deletion–contraction formula, where contraction is replaced by edge labeling.

Moving now in the direction of Euler characteristics, we let $w = -v$, so that

$$Z[G] = Z[G'] - wZ[G''],$$

$$Z[\bullet \sqcup G] = QZ[G],$$

and

$$Z[G] = \sum_{H \subset G} (-1)^{e(H)} Q^{|H|} w^{e(H)}.$$

This suggests that differentials should increase the number of edges on the subgraphs, and that the terms $Q^{|H|}w^{e(H)}$ should not change under the application of the (partial) differentials. Stosic's solution to this requirement is to take

$$Q = q^n$$

and

$$w = 1 + q + q^2 + \cdots + q^n$$

so that

$$Z[G] = \sum_{H \subset G} (-1)^{e(H)} q^{n|H|} (1 + q + a^2 + \cdots + q^n)^{e(H)}.$$

To see how this works, we first rewrite this state sum (over states H that are subgraphs of G) as a sum over *enhanced states* h, where we define enhanced states h for a graph G to be *labeled subgraphs* h, where a labeling of h consists in an assignment of one of the elements of the set $S = \{1, X, X^2, \ldots, X^n\}$ to each component of h. Regard the elements of S as generators of the ring $R = Z[X]/(X^{n+1})$. Define the *degree* of X^i by the formula $\deg(X^i) = n - i$ and let

$$j(h) = n|h| + \sum_{\gamma \in C(h)} \deg(label(\gamma)),$$

where the sum goes over all γ in $C(h)$, the set of components of h (each component is labeled from S and $|h|$ denotes the number of components in h). Then it is easy to see that

$$Z[G] = \sum_{h \in S(G)} (-1)^{e(h)} q^{j(h)},$$

where $S(G)$ denotes the set of enhanced states of G.

We now define a chain complex for a corresponding homology theory. Let $C_i(G)$ be the module generated by the enhanced states of G with i edges. Partial boundaries applied to an enhanced state h simply add new edges between nodes, or from a node to itself. If A and B are components of an enhanced state that are joined by a partial boundary to form a new component C, then C is assigned the label X^{i+j} when A and B have respective labels X^i and X^j. This partial boundary does not change the labels on other components of h. It may happen that a component A is transformed by adding an edge to itself to form a new component A'. In this case, if A has label 1, we assign label X^n to A' and otherwise take the partial boundary to be zero if the label of A is not equal to 1. It is then easy to check that the partial boundaries defined in this way preserve $j(h)$ as defined in the last paragraph, and are compatible so that the composition of the boundary with itself is zero. We have described Stosic's homology theory for a specialization of the dichromatic polynomial. We have, as in the first section of this paper,

$$Z[G] = \sum_j q^j \chi(C^{\bullet j}(G)) = \sum_j q^j \chi(H^{\bullet j}(G)),$$

where $\chi(C^{\bullet j}(G))$ denotes the complex defined above generated by enhanced states h with $j = j(h)$, and correspondingly for the homology.

Now let us turn to a discussion of the Stosic homology in relation to the Potts model. In the Potts model we have that $Q = q^n$ is the number of spins in the model. Thus we can take any q such that q^n is a natural number greater than or equal to 2. For example, we could take q to be an nth root of 2, and then this would be a two-state Potts model. On the other hand, we have $w = -v = 1 - e^K$ as in our previous analysis for the Potts model. Thus we have

$$-e^K = q + q^2 + \cdots + q^n.$$

With q real and positive, we can take

$$K = \mathrm{i}\pi + \ln(q + q^2 + \cdots + q^n),$$

arriving at an imaginary temperature for the values of the Potts model where the partition function is expressed in terms of the homology. If we take $q = (1+n)^{1/n}$, then the model will have $Q = n + 1$ states, and so in this case, we can identify the enhanced states of this model as corresponding to the spin assignments of $\{1, X, X^2, \ldots, X^n\}$ to the subgraphs, interpreted as regions of constant spin. The partial boundaries for this homology theory describe particular (global) ways to transit between spin-labeled regions where the regions themselves change locally. Usually in thinking about the dynamics of a model in statistical physics, one looks for evolutions that are strictly local. In the case of the partial differentials, we change the configuration of regions at a single bond (edge in the graph), but we make a global change in the spin-labeling (for example, from X^i and X^j on two separate regions to X^{i+j} on the joined region). It is likely that the reason we see the results of this cohomology in the partition function only at imaginary temperature is related to this nonlocal structure. Nevertheless, the categorified homology is seen in direct relation to the Potts partition, function and this connection deserves further examination.

5 Imaginary Temperature, Real Time, and Quantum Statistics

The purpose of this section is to discuss the nature of imaginary temperature in the Potts model from the point of view of quantum mechanics. We have seen that for certain values of imaginary temperature, the Potts model can be expressed in terms of Euler characteristics of Khovanov homology. The suggests looking at the analytic continuation of the partition function to relate these complex values to real values of the temperature. However, it is also useful to consider reformulating the models so that *imaginary temperature is replaced with real time*, and the context of the models is shifted to quantum mechanics. To see how this works, let us recall again the general form of a partition function in statistical mechanics. The partition function is given by the formula

$$Z_G(Q,T) = \sum_{\sigma} \mathrm{e}^{(-1/kT)E(\sigma)},$$

where the sum runs over all states σ of the physical system, T is the temperature, k is Boltzmann's constant, and $E(\sigma)$ is the energy of the state σ. In the Potts model, the underlying structure of the physical system is modeled by a graph G, and the energy has the combinatorial form that we have discussed in previous sections.

A quantum amplitude analogous to the partition function takes the form

$$A_G(Q,t) = \sum_{\sigma} \mathrm{e}^{(it/\hbar)E(\sigma)},$$

where t denotes the *time* parameter in the quantum model. We shall make precise the Hilbert space for this model below. But note that the correspondence of form between the amplitude $A_G(Q,t)$ and the partition function $Z_G(Q,T)$ suggests that we make the substitution

$$-1/kT = it/\hbar$$

or equivalently that

$$t = (\hbar/k)(1/iT).$$

Time is, up to a factor of proportionality, inverse imaginary temperature. With this substitution, we see that when one evaluates the Potts model at imaginary temperature, it can be interpreted as an evaluation of a quantum amplitude at the corresponding time given by the formula above. Thus we obtain a quantum-statistical interpretation of those places where the Potts model can be expressed directly in terms of Khovanov homology. In the process, we have given a quantum-statistical interpretation of the Khovanov homology.

To complete this section, we define the associated states and Hilbert space for the quantum amplitude $A_G(Q,t)$. Let $\mathcal{H}$ denote the vector space over the complex numbers with orthonormal basis $\{|\sigma\rangle\}$, where σ runs over the states of the Potts model for the graph G. Define a unitary operator $U(t) = \mathrm{e}^{(it/\hbar)H}$ by the formula on the basis elements

$$U(t)|\sigma\rangle = \mathrm{e}^{(it/\hbar)E(\sigma)}|\sigma\rangle.$$

The operator $U(t)$ implicitly defines the Hamiltonian for this physical system. Let

$$|\psi\rangle = \sum_{\sigma} |\sigma\rangle$$

denote an initial state and note that

$$U(t)|\psi\rangle = \sum_{\sigma} \mathrm{e}^{(it/\hbar)E(\sigma)}|\sigma\rangle$$

and

$$\langle\psi|U(t)|\psi\rangle = \sum_{\sigma} \mathrm{e}^{(it/\hbar)E(\sigma)} = A_G(Q,t).$$

Thus the Potts amplitude is the quantum-mechanical amplitude for the state $|\psi\rangle$ to evolve to the state $U(t)|\psi\rangle$. With this we have given a quantum-mechanical interpretation of the Potts model at imaginary temperature.

Note that if in the Potts model, we write $v = e^K - 1$, then in the quantum model we would write

$$e^K = \mathrm{e}^{it/\hbar}.$$

Thus we can take $t = -\hbar K i$, and if K is pure imaginary, then the time will be real in the quantum model.

Returning now to our results in Sect. 3, we recall that in the Potts model we have the following formulas for e^K: For $Q = 2$ we have $e^K = \pm \mathrm{i}$. For $Q = 3$, we have $e^K = \frac{-1 \pm \sqrt{3}i}{2}$. For $Q = 4$ we have $e^K = -1$. It is at these values that we can interpret the Potts model in terms of a quantum model at a real time value. Thus we have these interpretations for $Q = 2$, $t = \hbar\pi/2$; $Q = 3$, $t = \hbar\pi/6$; and $Q = 4$, $t = \hbar\pi$. At these values the amplitude for the quantum model is $A_G(Q,t) = \sum_{\sigma} \mathrm{e}^{(it/\hbar)E(\sigma)}$, and it is given by the formula

$$A_G(Q,t) = Q^{N/2}\{K(G)\},$$

where

$$\{K(G)\} = \sum_{j} q^j \chi(H^{\bullet j}(K(G))),$$

where $H^{\bullet j}(K(G))$ denotes the Khovanov homology of the link $K(G)$ associated with the planar graph G and $q = (1 - \mathrm{e}^{it/\hbar})/\sqrt{Q}$. At these special values, the Potts partition function in its quantum form is expressed directly in terms of the Khovanov homology and is, up to normalization, an isotopy invariant of the link $K(G)$.

6 Quantum Statistics and the Jones Polynomial

In this section we apply the point of view of the last section directly to the bracket polynomial. In keeping with the formalism of this paper, we will use the bracket in the form

with $\langle \bigcirc \rangle = (q+q^{-1})$. We have the formula for the bracket as a sum over enhanced states s:

$$\langle K \rangle = \sum_{s} (-1)^{n_B(s)} q^{j(s)},$$

where $n_B(s)$ is the number of B-type smoothings in s, $\lambda(s)$ is the number of loops in s labeled 1 minus the number of loops labeled -1, and $j(s) = n_B(s) + \lambda(s)$. In analogy to the last section, we define a Hilbert space $\mathcal{H}(\mathcal{K})$ with orthonormal basis $\{|s\rangle\}$ in 1-to-1 correspondence with the set of enhanced states of K. Then for $q = \mathrm{e}^{\mathrm{i}\theta}$, define the unitary transformation $U : \mathcal{H}(\mathcal{K}) \longrightarrow \mathcal{H}(\mathcal{K})$ by its action on the basis elements:

$$U|s\rangle = (-1)^{n_B(s)} q^{j(s)} |s\rangle.$$

Setting $|\psi\rangle = \sum_s |s\rangle$, we conclude that

$$\langle K \rangle = \langle \psi | U | \psi \rangle.$$

Thus we can express the value of the bracket polynomial (and by normalization, the Jones polynomial) as a quantum amplitude when the polynomial variable is on the unit circle in the complex plane.

There are several conclusions that we can draw from this formula. First of all, this formulation constitutes a quantum algorithm for the computation of the bracket polynomial (and hence the Jones polynomial) at any specialization where the variable is on the unit circle. We have defined a unitary transformation U and then shown that the bracket is an evaluation of the form $\langle \psi | U | \psi \rangle$. This evaluation can be computed via the Hadamard test [23], and this gives the desired quantum algorithm. Once the unitary transformation is given as a physical construction, the algorithm will be as efficient as any application of the Hadamard test. This algorithm requires an exponentially increasing complexity of construction for the associated unitary transformation, since the dimension of the Hilbert space is equal to, $2^{c(K)}$, where $c(K)$ is the number of crossings in the diagram K.

Nevertheless, it is significant that the Jones polynomial can be formulated in such a direct way in terms of a quantum algorithm. By the same token, we can take the basic result of Khovanov homology that says that the bracket is a graded Euler characteristic of the Khovanov homology as telling us that we are taking a step in the direction of a quantum algorithm for the Khovanov homology itself. This will be the subject of a separate paper. For more information about quantum algorithms for the Jones polynomial, see [1, 14, 15, 25]. The form of this knot amplitude is also related to our research on quantum knots. See [16].

Acknowlegement It gives the author of this paper great pleasure to acknowledge a helpful conversation with John Baez.

References

1. D. Aharonov, V. Jones, and Z. Landau. A polynomial quantum algorithm for approximating the Jones polynomial, quant-ph/0511096. *Algorithmica* **55**, no. 3 (2009), pp. 395–421.
2. D. Bar-Natan (2002). On Khovanov's categorification of the Jones polynomial, *Algebraic and Geometric Topology*, **2**(16), pp. 337–370.
3. D. Bar-Natan (2005). Khovanov's homology for tangles and cobordisms, *Geometry and Topology*, **9–33**, pp. 1465–1499. `arXiv:mat.GT/0410495`.
4. R.J. Baxter. Exactly solved models in statistical mechanics. Academic Press, London 24–28 Oval Rd London NWITDX (1982).
5. R. Dijkgraaf, D. Orland, and S. Reffert. Dimer models, free fermions and super quantum mechanics, `arXiv:0705.1645`. *Adv. Theo. Math. Phys.* **13**, no. 5 (2009), pp. 1255–1315.
6. S. Gukov. Surface operators and knot homologies, `ArXiv:0706.2369`.
7. L. Helme-Guizon, J.H. Przytycki, and Y. Rong. Torsion in Graph Homology, *Fundamenta Mathematicae*, **190**; June 2006, 139–177.
8. V.F.R. Jones. A polynomial invariant for links via von Neumann algebras, *Bull. Amer. Math. Soc.* **129** (1985), 103–112.
9. V.F.R. Jones. Hecke algebra representations of braid groups and link polynomials. *Ann. Math.* **126** (1987), pp. 335–338.
10. V.F.R. Jones. On knot invariants related to some statistical mechanics models. *Pacific J. Math.* **137**, no. 2 (1989), pp. 311–334.
11. L.H. Kauffman. State models and the Jones polynomial. *Topology* **26** (1987), 395–407.
12. L.H. Kauffman. Statistical mechanics and the Jones polynomial. *AMS Contemp. Math. Series* **78** (1989), 263–297.
13. L.H. Kauffman. Knots and physics, World Scientific Publishers (1991), Second Edition (1993), Third Edition (2002).
14. L.H. Kauffman. Quantum computing and the Jones polynomial, in Quantum Computation and Information, S. Lomonaco, Jr. (ed.), AMS CONM/305, 2002, pp. 101–137, `math.QA/0105255`.
15. L.H. Kauffman and S. Lomonaco Jr. A Three-stranded quantum algorithm for the Jones polynonmial, in "Quantum Information and Quantum Computation V," Proceedings of Spie, April 2007, E.J. Donkor, A.R. Pirich and H.E. Brandt (eds.), pp. 65730T1-17, Intl Soc. Opt. Eng.
16. S.J. Lomonaco, Jr. and L.H. Kauffman. Quantum knots and mosaics. *J. Quan. Info. Proc.*. **7**, nos. 2–3 (2008), pp. 85–115, `arxiv.org/abs/0805.0339`.
17. H.A. Dye, L.H. Kauffman, and V.O. Manturov. On two categorifications of the arrow polynomial for virtual knots, `arXiv:0906.3408`. In M. Banagl and C. Vogel (eds.), "The Mathematics of Knots Theory and Application," Vol.1, "Contributions in Mathematical and Computational Sciences," pp. 95–124, Heidelberg University, Springer-Verlag (2011).
18. V.O. Manturov. Khovanov homology for virtual links with arbitrary coefficients, `math.GT/0601152`. *JKTR*. **16**, no. 3 (2007), pp. 345–377.
19. M. Khovanov (1997). A categorification of the Jones polynomial. *Duke Math. J.* **101**, no. 3, pp. 359–426.

20. T.D. Lee and C.N. Yang. (1952), Statistical theory of equations of state and phase transitions II., lattice gas and Ising model. *Phys. Rev. Lett.*, **87**, 410–419.
21. J.H. Przytycki. When the theories meet: Khovanov homology as Hochschild homology of links, arXiv:math.GT/0509334.
22. J-L. Loday. Cyclic Homology, Grund. Math. Wissen. Band 301, Springer, Berlin, 1992 (second edition, 1998).
23. M.A. Nielsen and I.L. Chuang. "Quantum Computation and Quantum Information," Cambridge University Press, Cambridge, (2000).
24. E.F. Jasso-Hernandez and Y. Rong. A categorification of the Tutte polynomial. *Alg. Geom. Topol.*, **6** 2006, 2031–2049.
25. Raimund Marx, Amr Fahmy, Louis Kauffman, Samuel Lomonaco, Andreas Sprl, Nikolas Pomplun, John Myers, and Steffen J. Glaser. NMR Quantum Calculations of the Jones Polynomial, arXiv:0909.1080. *Phy. Rev.* **81** (2010), 03239.
26. M. Stosic. Categorification of the dichromatic polynomial for graphs, arXiv:Math/0504239v2, *JKTR*. **17**, no. 1 (2008), pp. 31–45.
27. O. Viro (2004). Khovanov homology, its definitions and ramifications, *Fund. Math.*, **184** (2004), pp. 317–342.
28. E. Witten. Quantum Field Theory and the Jones Polynomial. *Comm. Math. Phys.* **121** (1989), 351–399.

Algebraic Equations and Convex Bodies

Kiumars Kaveh and Askold Khovanskii*

Dedicated to Oleg Yanovich Viro on the occasion of his sixtieth birthday

Abstract The well-known Bernstein–Kushnirenko theorem from the theory of Newton polyhedra relates algebraic geometry and the theory of mixed volumes. Recently, the authors have found a far-reaching generalization of this theorem to generic systems of algebraic equations on any algebraic variety. In the present note we review these results and their applications to algebraic geometry and convex geometry.

Keywords Bernstein–Kushnirenko theorem • Semigroup of integral points • Convex body • Mixed volume • Alexandrov–Fenchel inequality • Brunn–Minkowski inequality • Hodge index theorem • Intersection theory of Cartier divisors • Hilbert function

1 Introduction

The famous Bernstein–Kushnirenko theorem from the theory of Newton polyhedra relates algebraic geometry (mainly the theory of toric varieties) with the theory of mixed volumes in convex geometry. This relation is useful in both directions.

* The second author is partially supported by Canadian Grant N 156833-02.

K. Kaveh (✉)
Department of Mathematics, University of Pittsburgh, Pittsburgh, PA, 15260, USA
e-mail: kaveh@pitt.edu

A. Khovanskii
Department of Mathematics, University of Toronto, Toronto, ON M5S 2E4, Canada
e-mail: askold@math.utoronto.ca

I. Itenberg et al. (eds.), *Perspectives in Analysis, Geometry, and Topology: On the Occasion of the 60th Birthday of Oleg Viro*, Progress in Mathematics 296,
DOI 10.1007/978-0-8176-8277-4_12,

On the one hand it allows one to prove the Alexandrov–Fenchel inequality (the most important and hardest result in the theory of mixed volumes) using the Hodge inequality from the theory of algebraic surfaces. On the other hand, it suggests new inequalities in the intersection theory of Cartier divisors analogous to the known inequalities for mixed volumes (see [Teissier, Khovanskii-1]).

Recently, the authors found a far-reaching generalization of the Kushnirenko theorem in which instead of the complex torus $(\mathbb{C}^*)^n$, we consider any algebraic variety X, and instead of a finite-dimensional space of functions spanned by monomials in $(\mathbb{C}^*)^n$, we consider any finite-dimensional space of rational functions on X.

To this end, first we develop an intersection theory for finite-dimensional subspaces of rational functions on a variety. It can be considered a generalization of the intersection theory of Cartier divisors to general (not necessarily complete) varieties. We show that this intersection theory enjoys all the properties of the mixed volume [Kaveh–Khovanskii-2]. Then we introduce the *Newton-Okounkov body*, which is a far generalization of the Newton polyhedron of a Laurent polynomial. Our construction of the Newton-Okounkov body depends on the choice of a $\mathbb{Z}^n$-valued valuation on the field of rational functions on X. It associates a Newton-Okounkov body to any finite-dimensional space L of rational functions on X. We obtain a direct generalization of the Kushnirenko theorem in this setting (see Theorem 11.1).

This construction then allows us to give a proof of the Hodge inequality using elementary geometry of planar convex domains and (as a corollary) an elementary proof of the Alexandrov–Fenchel inequality. In general, our construction does not imply a generalization of the Bernstein theorem, although we also obtain a generalization of this theorem for some cases in which the variety X is equipped with a reductive group action.

In this paper we present a review of the results mentioned above. We have omitted most of the proofs in this short note. A preliminary version together with proofs can be found in [Kaveh–Khovanskii-1]. Refined and generalized versions appear in the authors' more recent preprints: [Kaveh–Khovanskii-2] is a detailed version of the first half of [Kaveh–Khovanskii-1] (mainly about the intersection index), and [Kaveh–Khovanskii-3] is a refinement and generalization of the results in the second half of [Kaveh–Khovanskii-1] (mainly about Newton-Okounkov bodies).

After these results had been posted on arXiv, we learned that we were not the only ones working in this direction. Firstly, A. Okounkov (in his interesting papers [Okounkov1,Okounkov-2]) was a pioneer in defining (in passing) an analogue of the Newton polyhedron in the general situation (although his case of interest is that in which X has a reductive group action). Secondly, R. Lazarsfeld and M. Mustata, based on Okounkov's previous works, and independently of our preprints, have come up with closely related results [Lazarsfeld–Mustata]. Recently, following [Lazarsfeld–Mustata], similar results and constructions have been obtained for line bundles on arithmetic surfaces [Yuan].

2 Mixed Volume

By a *convex body* we mean a convex compact subset of $\mathbb{R}^n$. There are two operations of addition and scalar multiplication on convex bodies: Let Δ_1, Δ_2 be convex bodies. Then their sum

$$\Delta_1 + \Delta_2 = \{x + y \mid x \in \Delta_1,\ y \in \Delta_2\}$$

is also a convex body, called the *Minkowski sum* of Δ_1, Δ_2. Also, for a convex body Δ and a scalar $\lambda \geq 0$,

$$\lambda\Delta = \{\lambda x \mid x \in \Delta\}$$

is a convex body.

Let Vol_n denote the n-dimensional volume in $\mathbb{R}^n$ with respect to the standard Euclidean metric. The function Vol_n is a homogeneous polynomial of degree n on the cone of convex bodies, i.e., its restriction to each finite-dimensional section of the cone is a homogeneous polynomial of degree n. More precisely, for any $k > 0$, let $\mathbb{R}^k_+$ be the positive octant in $\mathbb{R}^k$ consisting of all $\lambda = (\lambda_1, \ldots, \lambda_k)$ with $\lambda_1 \geq 0, \ldots, \lambda_k \geq 0$. The polynomiality of Vol_n means that for any choice of the convex bodies $\Delta_1, \ldots, \Delta_k$, the function $P_{\Delta_1,\ldots,\Delta_k}$ defined on $\mathbb{R}^k_+$ by

$$P_{\Delta_1,\ldots,\Delta_k}(\lambda_1, \ldots, \lambda_k) = \mathrm{Vol}_n(\lambda_1\Delta_1 + \cdots + \lambda_k\Delta_k),$$

is a homogeneous polynomial of degree n.

The coefficients of this homogeneous polynomial are obtained from the *mixed volumes* of all the possible n-tuples $\Delta_{i_1}, \ldots, \Delta_{i_n}$, of convex bodies for any choices of $i_1, \ldots, i_n \in \{1, \ldots, n\}$. By definition, the mixed volume of $V(\Delta_1, \ldots, \Delta_n)$ of an n-tuple $(\Delta_1, \ldots, \Delta_n)$ of convex bodies is the coefficient of the monomial $\lambda_1 \cdots \lambda_n$ in the polynomial $P_{\Delta_1,\ldots,\Delta_n}$ divided by $n!$.[1] Several important geometric invariants can be recovered as mixed volumes. For example, the $(n-1)$-dimensional volume of the boundary of an n-dimensional convex body Δ is equal to $(1/n)V(\Delta, \ldots, \Delta, B)$, where B is the n-dimensional unit ball. Indeed, it is easy to see that the $(n-1)$-dimensional volume of the boundary and the number $(1/n)V(\Delta, \ldots, \Delta, B)$ are both equal to the derivative $\partial/\partial\varepsilon \mathrm{Vol}_n(\Delta + \varepsilon B)$ evaluated at $\varepsilon = 0$. Many applications of the theory of mixed volumes can be found in the book [Burago–Zalgaller].

The definition of mixed volume implies that it is the *polarization* of the volume polynomial, i.e., it is the unique function on the n-tuples of convex bodies satisfying the following:

(i) (Symmetry) V is symmetric with respect to permuting the bodies $\Delta_1, \ldots, \Delta_n$.
(ii) (Multilinearity) It is linear in each argument with respect to the Minkowski sum. Linearity in the first argument means that for convex bodies Δ_1', Δ_1'', and $\Delta_2, \ldots, \Delta_n$, we have

$$V(\Delta_1' + \Delta_1'', \ldots, \Delta_n) = V(\Delta_1', \ldots, \Delta_n) + V(\Delta_1'', \ldots, \Delta_n).$$

[1] The notion of mixed volume was introduced by Hermann Minkowski (1864–1909).

(iii) (Relationship to volume) On the diagonal, it coincides with volume, i.e., if $\Delta_1 = \cdots = \Delta_n = \Delta$, then $V(\Delta_1,\ldots,\Delta_n) = \mathrm{Vol}_n(\Delta)$.

The above three properties characterize the mixed volume: it is the unique function satisfying (i)–(iii).

The following two inequalities are easy to verify:

1. Mixed volume is nonnegative. That is, for any n-tuple of convex bodies $\Delta_1,\ldots,\Delta_n$, we have
$$V(\Delta_1,\ldots,\Delta_n) \geq 0.$$
2. Mixed volume is monotone. That is, for two n-tuples of convex bodies $\Delta_1' \subset \Delta_1,\ldots,\Delta_n' \subset \Delta_n$, we have
$$V(\Delta_1,\ldots,\Delta_n) \geq V(\Delta_1',\ldots,\Delta_n').$$

The following inequality, attributed to Alexandrov and Fenchel, is important and very useful in convex geometry. All its previously known proofs are rather complicated. For a discussion of this inequality the reader can consult the book [Burago–Zalgaller] as well as the original three papers of A. D. Alexandrov cited therein.

Theorem 2.1 (Alexandrov–Fenchel). *Let $\Delta_1,\ldots,\Delta_n$ be convex bodies in $\mathbb{R}^n$. Then*
$$V(\Delta_1,\Delta_2,\ldots,\Delta_n)^2 \geq V(\Delta_1,\Delta_1,\Delta_3,\ldots,\Delta_n)V(\Delta_2,\Delta_2,\Delta_3,\ldots,\Delta_n).$$

Below, we mention a formal corollary of the Alexandrov–Fenchel inequality. First we need to introduce a notation for when we have repetition of convex bodies in the mixed volume. Let $2 \leq m \leq n$ be an integer and $k_1 + \cdots + k_r = m$ a partition of m with $k_i \in \mathbb{N}$. Denote by $V(k_1 * \Delta_1,\ldots,k_r * \Delta_r,\Delta_{m+1},\ldots,\Delta_n)$ the mixed volume of the Δ_i, where Δ_1 is repeated k_1 times, Δ_2 is repeated k_2 times, etc., and $\Delta_{m+1},\ldots,\Delta_n$ appear once.

Corollary 2.2. *With the notation as above, the following inequality holds:*
$$V^m(k_1 * \Delta_1,\ldots,k_r * \Delta_r,\Delta_{m+1},\ldots,\Delta_n) \geq \prod_{1\leq j\leq r} V^{k_j}(m * \Delta_j,\Delta_{m+1}\ldots,\Delta_n).$$

3 Brunn–Minkowski Inequality

The celebrated *Brunn–Minkowski inequality* concerns volumes of convex bodies in $\mathbb{R}^n$.

Theorem 3.1 (Brunn–Minkowski). *Let Δ_1, Δ_2 be convex bodies in $\mathbb{R}^n$. Then*
$$\mathrm{Vol}_n^{1/n}(\Delta_1) + \mathrm{Vol}_n^{1/n}(\Delta_2) \leq \mathrm{Vol}_n^{1/n}(\Delta_1 + \Delta_2).$$

The inequality was first found and proved by Brunn toward the end of nineteenth century in the following form.

Theorem 3.2. *Let $V_\Delta(h)$ be the n-dimensional volume of the section $x_{n+1} = h$ of a convex body $\Delta \subset \mathbb{R}^{n+1}$. Then $V_\Delta^{1/n}(h)$ is a concave function in h.*

To obtain Theorem 3.1 from Theorem 3.2, one takes $\Delta \subset \mathbb{R}^{n+1}$ to be the convex combination of Δ_1 and Δ_2, i.e.,

$$\Delta = \{(x,h) \mid 0 \leq h \leq 1,\ x \in h\Delta_1 + (1-h)\Delta_2\}.$$

The concavity of the function

$$V_\Delta(h)^{1/n} = \mathrm{Vol}_n^{1/n}(h\Delta_1 + (1-h)\Delta_2)$$

then readily implies Theorem 3.1.

For $n = 2$, Theorem 3.2 is equivalent to the Alexandrov–Fenchel inequality (see Theorem 4.1). Below we give a sketch of its proof in the general case.

Proof (Sketch of proof of Theorem 3.2).

(1) When the convex body $\Delta \subset \mathbb{R}^{n+1}$ is rotationally symmetric with respect to the x_{n+1}-axis, Theorem 3.2 is obvious: the section $x_{n+1} = h$ of the body Δ at level h is a ball (or empty), and $V_\Delta^{1/n}(h)$ is a constant times the radius, which is a concave function of h, since Δ is a convex body.

(2) Now suppose Δ is not rotationally symmetric. Fix a hyperplane H containing the x_{n+1}-axis. Then one can construct a new convex body Δ' that is symmetric with respect to the hyperplane H and such that the volume of sections of Δ' is the same as that of Δ. To do this, just think of Δ as the union of line segments perpendicular to the plane H. Then shift each segment along its line in such a way that its center lies on H. The resulting body is then symmetric with respect to H and has the same volume of sections as Δ. The above construction is called the *Steiner symmetrization process*.

(3) It will now be enough to show that by repeated application of Steiner symmetrization, we can make Δ as close as we wish to a rotationally symmetric body. This can be proved as follows: First, we show that given a non-rotationally symmetric body, there is always a Steiner symmetrization making it "more symmetric." Then we use a compactness argument on the collection of convex bodies inside a bounded closed domain to conclude the proof. □

4 Brunn–Minkowski and Alexandrov–Fenchel Inequalities

We recall the classical *isoperimetric inequality*, whose origins date back to antiquity. According to this inequality, if P is the perimeter of a simple closed curve in the plane and A is the area enclosed by that curve, then

$$4\pi A \leq P^2. \tag{1}$$

Equality is obtained when the curve is a circle. To prove (1), it is enough to prove it for convex regions. The Alexandrov–Fenchel inequality for $n = 2$ implies the isoperimetric inequality (1) as a particular case and hence has inherited the name.

Theorem 4.1 (Isoperimetric inequality). *If Δ_1 and Δ_2 are convex regions in the plane, then*

$$\mathrm{Area}(\Delta_1)\mathrm{Area}(\Delta_2) \leq A(\Delta_1,\Delta_2)^2,$$

where $A(\Delta_1,\Delta_2)$ is the mixed area.

When Δ_2 is the unit disk in the plane, $A(\Delta_1,\Delta_2)$ is one-half the perimeter of Δ_1. Thus the classical form (1) of the inequality (for convex regions) follows from Theorem 4.1.

Proof (Proof of Theorem 4.1). It is easy to verify that the isoperimetric inequality is equivalent to the Brunn–Minkowski inequality for $n = 2$. Let us check this in one direction, i.e., that the isoperimetric inequality follows from Brunn–Minkowski for $n = 2$:

$$\begin{aligned}
&\mathrm{Area}(\Delta_1) + 2A(\Delta_1,\Delta_2) + \mathrm{Area}(\Delta_2)\\
&\quad = \mathrm{Area}(\Delta_1 + \Delta_2)\\
&\quad \geq (\mathrm{Area}^{1/2}(\Delta_1) + \mathrm{Area}^{1/2}(\Delta_2))^2\\
&\quad = \mathrm{Area}(\Delta_1) + 2\mathrm{Area}(\Delta_1)^{1/2}\mathrm{Area}(\Delta_2)^{1/2} + \mathrm{Area}(\Delta_2),
\end{aligned}$$

which readily implies the isoperimetric inequality. □

The following generalization of the Brunn–Minkowski inequality is a corollary of the Alexandrov–Fenchel inequality.

Corollary 4.2. *(Generalized Brunn–Minkowski inequality) For any $0 < m \leq n$ and for any fixed convex bodies $\Delta_{m+1},\dots,\Delta_n$, the function F that assigns to a body Δ the number $F(\Delta) = V^{1/m}(m * \Delta,\Delta_{m+1},\dots,\Delta_n)$ is concave, i.e., for any two convex bodies Δ_1,Δ_2, we have*

$$F(\Delta_1) + F(\Delta_2) \leq F(\Delta_1 + \Delta_2).$$

On the other hand, the usual proof of the Alexandrov–Fenchel inequality deduces it from the Brunn–Minkowski inequality. But this deduction is the main part (and the most complicated part) of the proof (see [Burago–Zalgaller]). Interestingly, the main construction in the present paper (using algebraic geometry) allows us to obtain the Alexandrov–Fenchel inequality as an immediate corollary of the simplest case of the Brunn–Minkowski inequality, i.e., the isoperimetric inequality.

5 Generic Systems of Laurent Polynomial Equations in $(\mathbb{C}^*)^n$

In this section we recall the famous results due to Kushnirenko and Bernstein on the number of solutions of a generic system of Laurent polynomials in $(\mathbb{C}^*)^n$.

Let us identify the lattice $\mathbb{Z}^n$ with *Laurent monomials* in $(\mathbb{C}^*)^n$: to each integral point $k \in \mathbb{Z}^n$, $k = (k_1, \dots, k_n)$, we associate the monomial $z^k = z_1^{k_1} \cdots z_n^{k_n}$, where $z = (z_1, \dots, z_n)$. A *Laurent polynomial* $P = \sum_k c_k z^k$ is a finite linear combination of Laurent monomials with complex coefficients. The *support* $\mathrm{supp}(P)$ of a Laurent polynomial P is the set of exponents k for which $c_k \neq 0$. We denote the convex hull of a finite set $A \subset \mathbb{Z}^n$ by $\Delta_A \subset \mathbb{R}^n$. The *Newton polyhedron* $\Delta(P)$ of a Laurent polynomial P is the convex hull $\Delta_{\mathrm{supp}(P)}$ of its support. With each finite set $A \subset \mathbb{Z}^n$ one associates a vector space L_A of Laurent polynomials P with $\mathrm{supp}(P) \subset A$.

Definition 5.1. We say that a property holds for a *generic element* of a vector space L if there is a proper algebraic set Σ such that the property holds for all the elements in $L \setminus \Sigma$.

Definition 5.2. For a given n-tuple of finite sets $A_1, \dots, A_n \subset \mathbb{Z}^n$, the *intersection index* of the n-tuple of spaces $[L_{A_1}, \dots, L_{A_n}]$ is the number of solutions in $(\mathbb{C}^*)^n$ of a generic system of equations $P_1 = \cdots = P_n = 0$, where $P_1 \in L_{A_1}, \dots, P_n \in L_{A_n}$.

Problem: *Find the intersection index* $[L_{A_1}, \dots, L_{A_n}]$. That is, for a generic element $(P_1, \dots, P_n) \in L_{A_1} \times \cdots \times L_{A_n}$, find a formula for the number of solutions in $(\mathbb{C}^*)^n$ of the system of equations $P_1 = \cdots = P_n = 0$.

Kushnirenko found the following important result, which answers a particular case of the above problem [Kushnirenko].

Theorem 5.3. *When the convex hulls of the sets A_i are the same and equal to a polyhedron Δ, we have*

$$[L_{A_1}, \dots, L_{A_n}] = n!\mathrm{Vol}_n(\Delta),$$

where Vol_n *is the standard n-dimensional volume in* $\mathbb{R}^n$.

According to Theorem 5.3, if $P_1, \dots, P_n$ are sufficiently general Laurent polynomials with given Newton polyhedron Δ, the number of solutions in $(\mathbb{C}^*)^n$ of the system $P_1 = \cdots = P_n = 0$ is equal to $n!\mathrm{Vol}_n(\Delta)$.

The problem was solved by Bernstein in full generality [Bernstein]:

Theorem 5.4. *In the general case, i.e., for arbitrary finite subsets $A_1, \dots, A_n \subset \mathbb{Z}^n$, we have*

$$[L_{A_1}, \dots, L_{A_n}] = n!V(\Delta_{A_1}, \dots, \Delta_{A_n}),$$

where V is the mixed volume of convex bodies in $\mathbb{R}^n$.

According to Theorem 5.4, if $P_1,\dots,P_n$ are sufficiently general Laurent polynomials with Newton polyhedra $\Delta_1,\dots,\Delta_n$ respectively, then the number of solutions in $(\mathbb{C}^*)^n$ of the system $P_1 = \cdots = P_n = 0$ is equal to $n!V(\Delta_1,\dots,\Delta_n)$.

6 Convex Geometry and the Bernstein–Kushnirenko Theorem

Let us examine Theorem 5.4 (which we will call the Bernstein–Kushnirenko theorem) more closely. In the space of regular functions on $(\mathbb{C}^*)^n$, there is a natural class of finite-dimensional subspaces, namely the subspaces that are stable under the action of the multiplicative group $(\mathbb{C}^*)^n$. Each such subspace is of the form L_A for some finite set $A \subset \mathbb{Z}^n$ of monomials.

For two finite-dimensional subspaces L_1, L_2 of regular functions in $(\mathbb{C}^*)^n$, let us define the product L_1L_2 as the subspace spanned by the products fg, where $f \in L_1$, $g \in L_2$. Clearly, multiplication of monomials corresponds to the addition of their exponents, i.e., $z^{k_1}z^{k_2} = z^{k_1+k_2}$. This implies that $L_{A_1}L_{A_2} = L_{A_1+A_2}$.

The Bernstein–Kushnirenko theorem defines and computes the intersection index $[L_{A_1}, L_{A_2},\dots,L_{A_n}]$ of the n-tuples of subspaces L_{A_i} for finite subsets $A_i \subset \mathbb{Z}^n$. Since this intersection index is equal to the mixed volume, it enjoys the same properties, namely (1) positivity, (2) monotonicity, (3) multilinearity, and (4) the Alexandrov–Fenchel inequality and its corollaries. Moreover, if for a finite set $A \subset \mathbb{Z}^n$ we let $\overline{A} = \Delta_A \cap \mathbb{Z}^n$, then (5) the spaces L_A and $L_{\overline{A}}$ have the same intersection indices. That is, for any $(n-1)$-tuple of finite subsets $A_2,\dots,A_n \in \mathbb{Z}^n$, we have

$$[L_A, L_{A_2},\dots,L_{A_n}] = [L_{\overline{A}}, L_{A_2},\dots,L_{A_n}].$$

This means (surprisingly!) that enlarging $L_A \mapsto L_{\overline{A}}$ does not change any of the intersection indices we have considered. Hence in counting the number of solutions of a system, instead of support of a polynomial, its convex hull plays the main role. Let us denote the subspace $L_{\overline{A}}$ by $\overline{L}_A$ and call it the *completion of* L_A.

Since the semigroup of convex bodies with Minkowski sum has the cancellation property, we get the following cancellation property for the finite subsets of $\mathbb{Z}^n$: if for finite subsets $A,B,C \in \mathbb{Z}^n$ we have $\overline{\overline{A}+\overline{C}} = \overline{\overline{B}+\overline{C}}$, then $\overline{A} = \overline{B}$. And we have the same cancellation property for the corresponding semigroup of subspaces L_A. That is, if $\overline{\overline{L}_A\,\overline{L}_C} = \overline{\overline{L}_B\,\overline{L}_C}$, then $\overline{L}_A = \overline{L}_B$.

The Bernstein–Kushnirenko theorem relates the notion of mixed volume in convex geometry with that of intersection index in algebraic geometry. In algebraic geometry, the following inequality about intersection indices on a surface is well known:

Theorem 6.1 (Hodge inequality). *Let Γ_1,Γ_2 be algebraic curves on a smooth irreducible projective surface. Assume that Γ_1,Γ_2 have positive self-intersection indices. Then*

$$(\Gamma_1,\Gamma_2)^2 \geq (\Gamma_1,\Gamma_1)(\Gamma_2,\Gamma_2),$$

where (Γ_i,Γ_j) denotes the intersection index of the curves Γ_i and Γ_j.

On the one hand, Theorem 5.4 allows one to prove the Alexandrov–Fenchel inequality algebraically using Theorem 6.1 (see [Khovanskii-1, Teissier]). On the other hand, Theorem 5.4 suggests an analogy between the theory of mixed volumes and the intersection theory of Cartier divisors on a projective algebraic variety.

We will return to this discussion after stating our main theorem (Theorem 11.1) and its corollary, which is a version of the Hodge inequality.

7 An Extension of the Intersection Theory of Cartier Divisors

Now we discuss general results, inspired by the Bernstein–Kushnirenko theorem, that can be considered an analogue of the intersection theory of Cartier divisors for general (not necessarily complete) varieties [Kaveh–Khovanskii-2]. Instead of $(\mathbb{C}^*)^n$, we take any irreducible n-dimensional variety X, and instead of a finite-dimensional space of functions spanned by monomials, we take any finite-dimensional space of rational functions. For these spaces we define an intersection index and prove that it enjoys all the properties of the mixed volume of convex bodies.

Consider the collection $\mathbf{K}_{\text{rat}}(X)$ of all nonzero finite-dimensional subspaces of rational functions on X. The set $\mathbf{K}_{\text{rat}}(X)$ has a natural multiplication: the product L_1L_2 of two subspaces $L_1, L_2 \in \mathbf{K}_{\text{rat}}(X)$ is the subspace spanned by all the products fg, where $f \in L_1$, $g \in L_2$. With respect to this multiplication, $\mathbf{K}_{\text{rat}}(X)$ is a commutative semigroup.

Definition 7.1. The *intersection index* $[L_1, \dots, L_n]$ of $L_1, \dots, L_n \in \mathbf{K}_{\text{rat}}(X)$ is the number of solutions in X of a generic system of equations $f_1 = \cdots = f_n = 0$, where $f_1 \in L_1, \dots, f_n \in L_n$. In counting the solutions, we neglect the solutions x for which all the functions in some space L_i vanish as well as the solutions for which at least one function from some space L_i has a pole.

More precisely, let $\Sigma \subset X$ be a hypersurface that contains (1) all the singular points of X, (2) all the poles of functions from any of the L_i, (3) for any i, the set of common zeros of all the $f \in L_i$. Then for a generic choice of $(f_1, \dots, f_n) \in L_1 \times \cdots \times L_n$, the intersection index $[L_1, \dots, L_n]$ is equal to the number of solutions $\{x \in X \setminus \Sigma \mid f_1(x) = \cdots = f_n(x) = 0\}$.

Theorem 7.2. *The intersection index $[L_1, \dots, L_n]$ is well defined. That is, there is a Zariski-open subset U in the vector space $L_1 \times \cdots \times L_n$ such that for any $(f_1, \dots, f_n) \in U$, the number of solutions $x \in X \setminus \Sigma$ of the system $f_1(x) = \cdots = f_n(x) = 0$ is the same (and hence equal to $[L_1, \dots, L_n]$). Moreover, the above number of solutions is independent of the choice of Σ containing (1)–(3) above.*

The following properties of the intersection index are easy consequences of the definition:

Proposition 7.3. (1) $[L_1,\dots,L_n]$ *is a symmetric function of the n-tuples* $L_1,\dots,L_n \in \mathbf{K}_{\mathrm{rat}}(X)$ *(i.e., it takes the same value under a permutation of* $L_1,\dots,L_n$*).*

(2) *The intersection index is monotone (i.e., if* $L_1' \subseteq L_1,\dots,L_n' \subseteq L_n$*, then* $[L_1,\dots,L_n] \geq [L_1',\dots,L_n']$*.*

(3) *The intersection index is nonnegative (i.e.,* $[L_1,\dots,L_n] \geq 0$*).*

The next two theorems contain the main properties of the intersection index.

Theorem 7.4 (Multilinearity). (1) *Let* $L_1',L_1'',L_2,\dots,L_n \in \mathbf{K}_{\mathrm{rat}}(X)$ *and put* $L_1 = L_1'L_1''$*. Then*

$$[L_1,\dots,L_n] = [L_1',\dots,L_n] + [L_1'',\dots,L_n].$$

(2) *Let* $L_1,\dots,L_n \in \mathbf{K}_{\mathrm{rat}}(X)$ *and take 1-dimensional subspaces* $L_1',\dots,L_n' \in \mathbf{K}_{\mathrm{rat}}(X)$*. Then*

$$[L_1,\dots,L_n] = [L_1'L_1,\dots,L_n'L_n].$$

Let us say that $f \in \mathbb{C}(X)$ is *integral over a subspace* $L \in \mathbf{K}_{\mathrm{rat}}(X)$ if f satisfies an equation

$$f^m + a_1 f^{m-1} + \cdots + a_m = 0,$$

where $m > 0$ and $a_i \in L^i$ for each $i = 1,\dots,m$. It is well known that the collection $\overline{L}$ of all integral elements over L is a vector subspace containing L. Moreover, if L is finite-dimensional, then $\overline{L}$ is also finite-dimensional (see [Zariski–Samuel, Appendix 4]). It is called the *completion of* L. For two subspaces $L,M \in \mathbf{K}_{\mathrm{rat}}(X)$ we say that L is equivalent to M (written $L \sim M$) if there is $N \in \mathbf{K}_{\mathrm{rat}}(X)$ with $LN = MN$. One shows that the completion $\overline{L}$ is in fact the largest subspace in $\mathbf{K}_{\mathrm{rat}}(X)$ that is equivalent to L. The enlarging $L \to \overline{L}$ is analogous to the geometric operation $A \mapsto \Delta(A)$ that associates to a finite set A its convex hull $\Delta(A)$.

Theorem 7.5. (1) *Let* $L_1 \in K_{\mathrm{rat}}(X)$ *and let* $G_1 \in K_{\mathrm{rat}}(X)$ *be the subspace spanned by* L_1 *and a rational function g integral over* L_1*. Then for any* $(n-1)$*-tuple* $L_2,\dots,L_n \in K_{\mathrm{rat}}(X)$ *we have*

$$[L_1,L_2,\dots,L_n] = [G_1,L_2,\dots,L_n].$$

(2) *Let* $L_1 \in \mathbf{K}_{\mathrm{rat}}(X)$ *and let* $\overline{L}_1$ *be its completion as defined above. Then for any* $(n-1)$*-tuple* $L_2,\dots,L_n \in \mathbf{K}_{\mathrm{rat}}(X)$ *we have*

$$[L_1,L_2,\dots,L_n] = [\overline{L}_1,L_2,\dots,L_n].$$

The proof of Theorem 7.5 is not complicated. If X is a curve, statement (1) is obvious, and one can easily obtain the general case from the curve case (see [Kaveh–Khovanskii-2, Theorem 4.25]). Statement (2) follows from (1). Alternatively, statement (2) follows from the multilinearity of the intersection index and the fact that L and $\overline{L}$ are equivalent.

As with any other commutative semigroup, there corresponds a Grothendieck group to the semigroup $\mathbf{K}_{\text{rat}}(X)$. Let K be a commutative semigroup. The Grothendieck group $G(K)$ of K is defined as follows: two elements $x, y \in K$ are called *equivalent*, written $x \sim y$, if there is $z \in K$ with $xz = yz$. The Grothendieck group $G(K)$ is the collection of all formal fractions x_1/x_2, $x_1, x_2 \in K$, where two fractions x_1/x_2 and y_1/y_2 are considered equal if $x_1y_2 \sim y_1x_2$. There is a natural homomorphism $\phi : K \to G(K)$. The Grothendieck group has the following universal property: for any group G' and a homomorphism $\phi' : K \to G'$, there exists a unique homomorphism $\psi : G(K) \to G'$ such that $\phi' = \psi \circ \phi$.

From the multilinearity of the intersection index it follows that the intersection index extends to the the Grothendieck group of $\mathbf{K}_{\text{rat}}(X)$. The Grothendieck group of $\mathbf{K}_{\text{rat}}(X)$ can be considered an analogue (for a not necessarily complete variety X) of the group of Cartier divisors on a projective variety, and the intersection index on this Grothendieck group an analogue of the intersection index of Cartier divisors.

The intersection theory on the Grothendieck group of $\mathbf{K}_{\text{rat}}(X)$ enjoys all the properties of mixed volume. Some of those properties have already been discussed in the present section. The others will be discussed later (see Theorem 12.3 and Corollary 12.4 below).

8 Proof of the Bernstein–Kushnirenko Theorem Via the Hilbert Theorem

Let us recall the proof of the Bernstein–Kushnirenko theorem from [Khovanskii-2], which will be important for our generalization.

For each space $L \in \mathbf{K}_{\text{rat}}(X)$, let us define the Hilbert function H_L by $H_L(k) = \dim(L^k)$. For sufficiently large values of k, the function $H_L(k)$ is a polynomial in k, called the *Hilbert polynomial* of L.

With each space $L \in \mathbf{K}_{\text{rat}}(X)$, one associates a rational *Kodaira map* from X to $\mathbb{P}(L^*)$, the projectivization of the dual space L^*: to any $x \in X$ at which all the $f \in L$ are defined, there corresponds a functional in L^* that evaluates $f \in L$ at x. The Kodaira map sends x to the image of this functional in $\mathbb{P}(L^*)$. It is a rational map, i.e., defined on a Zariski-open subset in X. We denote by Y_L the closure of the image of X under the Kodaira map in $\mathbb{P}(L^*)$.

The following theorem is a version of the classical Hilbert theorem on the degree of a subvariety of the projective space.

Theorem 8.1 (Hilbert). *The degree m of the Hilbert polynomial of the space L is equal to the dimension of the variety Y_L, and its leading coefficient c is the degree of $Y_L \subset \mathbb{P}(L^*)$ divided by $m!$.*

Let A be a finite subset in $\mathbb{Z}^n$ with $\Delta(A)$ its convex hull. Denote by $k * A$ the sum $A + \cdots + A$ of k copies of the set A, and by $(k\Delta(A))_C$ the subset of $k\Delta(A)$ containing points whose distance to the boundary $\partial(k\Delta(A))$ is bigger than C. The following combinatorial theorem gives an estimate for the set $k * A$ in terms of the set of integral points in $k\Delta(A)$.

Theorem 8.2 ([Khovanskii-2]). (1) *One has* $k * A \subset k\Delta(A) \cap \mathbb{Z}^n$.
(2) *Assume that the differences* $a - b$ *for* $a, b \in A$ *generate the group* $\mathbb{Z}^n$. *Then there exists a constant* C *such that for any* $k \in \mathbb{N}$, *we have*

$$(k\Delta(A))_C \cap \mathbb{Z}^n \subset k * A.$$

Corollary 8.3. *Let* $A \subset \mathbb{Z}^n$ *be a finite subset satisfying the condition in Theorem 8.2(2). Then*

$$\lim_{k\to\infty} \frac{\#(k * A)}{k^n} = \mathrm{Vol}_n(\Delta(A)).$$

Corollary 8.3 together with the Hilbert theorem (Theorem 8.1) proves the Kushnirenko theorem for sets A such that the differences $a - b$ for $a, b \in A$ generate the group $\mathbb{Z}^n$. The Kushnirenko theorem for the general case easily follows from this. The Bernstein theorem, Theorem 5.4, follows from the Kushnirenko theorem (Theorem 5.3) and the identity $L_{A+B} = L_A L_B$.

9 Graded Semigroups in $\mathbb{N} \oplus \mathbb{Z}^n$ and the Newton-Okounkov Body

Let S be a subsemigroup of $\mathbb{N} \oplus \mathbb{Z}^n$. For any integer $k > 0$ we denote by S_k the section of S at level k, i.e., the set of elements $x \in \mathbb{Z}^n$ such that $(k, x) \in S$.

Definition 9.1. (1) A subsemigroup S of $\mathbb{N} \oplus \mathbb{Z}^n$ is called a *graded semigroup* if for any $k > 0$, S_k is finite and nonempty.
(2) Such a subsemigroup is called an *ample semigroup* if there is a natural m such that the set of all the differences $a - b$ for $a, b \in S_m$ generates the group $\mathbb{Z}^n$.
(3) Such a subsemigroup is called a *semigroup with restricted growth* if there is a constant C such that for any $k > 0$, we have $\#(S_k) \leq Ck^n$.

For a graded semigroup S, let $\mathrm{Con}(S)$ denote the closure of the convex hull of $S \cup \{0\}$. It is a cone in $\mathbb{R}^{n+1}$. Denote by $\tilde{S}$ the semigroup $\mathrm{Con}(S) \cap (\mathbb{N} \oplus \mathbb{Z}^n)$. The semigroup $\tilde{S}$ contains the semigroup S.

Definition 9.2. For a graded semigroup S, define the *Newton-Okounkov set* $\Delta(S)$ to be the section of the cone $\mathrm{Con}(S)$ at $k = 1$, i.e.,

$$\Delta(S) = \{x \mid (1, x) \in \mathrm{Con}(S)\}.$$

Theorem 9.3 (Asymptotics of graded semigroups). *Let S be an ample graded semigroup with restricted growth in $\mathbb{N} \oplus \mathbb{Z}^n$. Then:*

(1) *The cone* $\text{Con}(S)$ *is strictly convex, i.e., the Newton-Okounkov set* $\Delta(S)$ *is bounded.*
(2) *Let $d(k)$ denote the maximum distance of the points (k,x) from the boundary of* $\text{Con}(S)$ *for $x \in \tilde{S}_k \setminus S_k$. Then*

$$\lim_{k \to \infty} \frac{d(k)}{k} = 0.$$

Theorem 9.3 basically follows from Theorem 8.2. For a proof and generalizations, see [Kaveh–Khovanskii-3, Sect. 1.3].

Corollary 9.4. *Let S be an ample graded semigroup with restricted growth in $\mathbb{N} \oplus \mathbb{Z}^n$. Then*

$$\lim_{k \to \infty} \frac{\#(S_k)}{k^n} = \text{Vol}_n(\Delta(S)).$$

10 Valuations on the Field of Rational Functions

We start with the definition of a prevaluation. Let V be a vector space and let I be a set totally ordered with respect to some ordering $<$.

Definition 10.1. A *prevaluation* on V with values in I is a function $v : V \setminus \{0\} \to I$ satisfying the following:

(1) For all $f, g \in V \setminus \{0\}$, $v(f+g) \geq \min(v(f), v(g))$
(2) For all $f \in V \setminus \{0\}$ and $\lambda \neq 0$, $v(\lambda f) = v(f)$
(3) If for $f, g \in V \setminus \{0\}$ we have $v(f) = v(g)$ then there is $\lambda \neq 0$ such that $v(g - \lambda f) > v(g)$.

It is easy to verify that if $L \subset V$ is a finite-dimensional subspace, then $\dim(L)$ is equal to $\#v(L \setminus \{0\})$.

Example 10.2. Let V be a finite-dimensional vector space with basis $\{e_1, \ldots, e_n\}$, and let $I = \{1, \ldots, n\}$ with the usual ordering of numbers. For $f = \sum_i \lambda_i e_i$, define

$$v(f) = \min\{i \mid \lambda_i \neq 0\}.$$

Example 10.3 (Schubert cells in the Grassmannian). Let $\text{Gr}(n,k)$ be the Grassmannian of k-dimensional planes in $\mathbb{C}^n$. In Example 10.2, take $V = \mathbb{C}^n$ with the standard basis. Under the prevaluation v above, each k-dimensional subspace $L \subset \mathbb{C}^n$ goes to a subset $M \subset I$ containing k elements. The set of all the k-dimensional subspaces that are mapped onto M forms the *Schubert cell* X_M in the Grassmannian $\text{Gr}(n,k)$.

Similar to Example 10.3, the Schubert cells in the variety of complete flags can also be recovered from the prevaluation v above on $\mathbb{C}^n$.

Next we define the notion of a valuation with values in a totally ordered abelian group.

Definition 10.4. Let K be a field and Γ a totally ordered abelian group. A prevaluation $v : K \setminus \{0\} \to \Gamma$ is a *valuation* if it further satisfies the following: for any $f, g \in K \setminus \{0\}$ we have

$$v(fg) = v(f) + v(g).$$

The valuation v is called *faithful* if its image is all of Γ.

We will be concerned only with the field $\mathbb{C}(X)$ of rational functions on an n-dimensional irreducible variety X and $\mathbb{Z}^n$-valued valuations on it (with respect to some total order on $\mathbb{Z}^n$).

Example 10.5. Let X be an irreducible curve. Take the field of rational functions $\mathbb{C}(X)$ and $\Gamma = \mathbb{Z}$. Take a smooth point a on X. Then the map

$$v(f) = \mathrm{ord}_a(f)$$

defines a faithful valuation on $\mathbb{C}(X)$.

Example 10.6. Let X be an irreducible n-dimensional variety. Take a smooth point $a \in X$. Consider a local system of coordinates at a with analytic coordinate functions $x_1, \ldots, x_n$. Let $\Gamma = \mathbb{Z}^n_+$ be the semigroup in $\mathbb{Z}^n$ of points with nonnegative coordinates. Take any well-ordering $\prec$ that respects the addition, i.e., if $a \prec b$, then $a + c \prec b + c$. For a germ f at the point a of an analytic function in $x_1, \ldots, x_n$, let $cx^{\alpha(f)} = cx_1^{\alpha_1} \cdots x_n^{\alpha_n}$ be the term in the Taylor expansion of f with minimum exponent $\alpha(f) = (\alpha_1, \ldots, \alpha_n)$ with respect to the ordering $\prec$. For a germ F at the point a of a meromorphic function $F = f/g$, define $v(F)$ as $\alpha(f) - \alpha(g)$. This function v induces a faithful valuation on the field of rational functions $\mathbb{C}(X)$.

Example 10.7. Let X be an irreducible n-dimensional variety and Y any variety birationally isomorphic to X. Then the fields $\mathbb{C}(X)$ and $\mathbb{C}(Y)$ are isomorphic, and thus any faithful valuation on $\mathbb{C}(Y)$ gives a faithful valuation on $\mathbb{C}(X)$ as well.

11 Main Construction and Theorem

Let X be an irreducible n-dimensional variety. Fix a faithful valuation $v : \mathbb{C}(X) \setminus \{0\} \to \mathbb{Z}^n$, where $\mathbb{Z}^n$ is equipped with a total ordering respecting addition.

Let $L \in \mathbf{K}_{\mathrm{rat}}(X)$ be a finite-dimensional subspace of rational functions. Consider the semigroup $S(L)$ in $\mathbb{N} \oplus \mathbb{Z}^n$ defined by

$$S(L) = \bigcup_{k>0} \{(k, v(f)) \mid f \in L^k \setminus \{0\}\}.$$

It is easy to see that $S(L)$ is a graded semigroup. Moreover, by Hilbert's theorem, $S(L)$ is contained in a semigroup of restricted growth.

Definition 11.1 (Newton-Okounkov body for a subspace of rational functions). We define the *Newton-Okounkov body* for a subspace L to be the convex body $\Delta(S(L))$ associated to the semigroup $S(L)$.

Denote by $s(L)$ the index of the subgroup in $\mathbb{Z}^n$ generated by all the differences $a-b$ such that a, b belong to the same set $S_m(L)$ for some $m > 0$. Also let Y_L be the closure of the image of the variety X (in fact, the image of a Zariski-open subset of X) under the Kodaira rational map $\Phi_L : X \to \mathbb{P}(L^*)$. If $\dim(Y_L)$ is equal to $\dim(X)$, then the Kodaira map from X to Y_L has finite mapping degree. Denote this mapping degree by $d(L)$.

Theorem 11.1 (Main theorem). *Let X be an irreducible n-dimensional variety. Then:*

(1) *The complex dimension of the variety Y_L is equal to the real dimension of the Newton-Okounkov body $\Delta(S(L))$.*
(2) *If $\dim(Y_L) = n$, then*

$$[L, \ldots, L] = \frac{n!d(L)}{s(L)} \mathrm{Vol}_n(\Delta(S(L))).$$

(3) *In particular, if $\Phi_L : X \to Y_L$ is a birational isomorphism, then*

$$[L, \ldots, L] = n!\mathrm{Vol}_n(\Delta(S(L))).$$

(4) *For any two subspaces $L_1, L_2 \in \mathbf{K}_{\mathrm{rat}}(X)$ we have*

$$\Delta(S(L_1)) + \Delta(S(L_2)) \subseteq \Delta(S(L_1 L_2)).$$

The proof of the main theorem is based on Theorem 9.3 (which describes the asymptotic behavior of an ample graded semigroup with restricted growth) and Hilbert's theorem (Theorem 8.1). A sketch of a proof can be found in [Kaveh–Khovanskii-1]. For a complete proof as well as generalizations, see [Kaveh–Khovanskii-3, Sect. 4.5]. Also see [Kaveh–Khovanskii-3, Sect. 2.2] for an example of two spaces $L_1, L_2 \in \mathbf{K}_{\mathrm{rat}}(X)$ and a valuation on X such that the inclusion in (4) is not the identity.

12 Algebraic Analogue of the Alexandrov–Fenchel Inequality

Part (2) of the main theorem (Theorem 11.1) can be considered a far-reaching generalization of the Kushnirenko theorem, in which instead of $(\mathbb{C}^*)^n$, one takes any n-dimensional irreducible variety X, and instead of a finite-dimensional space generated by monomials, one takes any finite-dimensional space L of rational functions. The proof of Theorem 11.1 is an extension of the arguments used in [Khovanskii-2] to prove the Kushnirenko theorem (see also Sect. 8). As we mentioned, the Bernstein theorem (Theorem 5.4) follows immediately from the Kushnirenko theorem and the identity

$$L_{A+B} = L_A L_B.$$

Thus the Bernstein–Kushnirenko theorem is a corollary of our Theorem 11.1.

Note that although the Newton-Okounkov body $\Delta(S(L))$ depends on a choice of a faithful valuation, its volume depends on L only: after multiplication by $n!$, it equals the self-intersection index $[L, \ldots, L]$.

Our generalization of the Kushnirenko theorem does not imply the generalization of the Bernstein theorem. The point is that in general, we do not always have an equality $\Delta(S(L_1)) + \Delta(S(L_2)) = \Delta(S(L_1 L_2))$. In fact, by Theorem 11.1(4), what is always true is the inclusion

$$\Delta(S(L_1)) + \Delta(S(L_2)) \subseteq \Delta(S(L_1 L_2)).$$

This inclusion is sufficient for us to prove the following interesting corollary.

Let us call a subspace $L \in \mathbf{K}_{\text{rat}}(X)$ a *big subspace* if for some $m > 0$, the Kodaira rational map of the completion $\overline{L^m}$ is a birational isomorphism between X and its image. It is not hard to show that the product of two big subspaces is again a big subspace and that thus the big subspaces form a subsemigroup of $\mathbf{K}_{\text{rat}}(X)$.

Corollary 12.1 (Algebraic analogue of Brunn–Minkowski). *Assume that $L, G \in \mathbf{K}_{\text{rat}}(X)$ are big subspaces. Then*

$$[L, \ldots, L]^{1/n} + [G, \ldots, G]^{1/n} \leq [LG, \ldots, LG]^{1/n}.$$

Proof. Replacing L and G by $\overline{L^m}$ and $\overline{G^m}$, for any $m > 0$, does not change the inequality (see Theorems 7.4 and 7.5). Thus, without loss of generality, we can assume that the Kodaira maps of L and G are birational isomorphisms onto their images. From statement (4) in Theorem 11.1 we have $\Delta(S(L)) + \Delta(S(G)) \subseteq \Delta(S(LG))$. So $\text{Vol}_n(\Delta(S(L)) + \Delta(S(G))) \leq \text{Vol}_n(\Delta(S(LG)))$. Also, from statement (3) in the same theorem we have

$$[L, \ldots, L] = n!\text{Vol}_n(\Delta(S(L)),$$
$$[G, \ldots, G] = n!\text{Vol}_n(\Delta(S(G)),$$

$$[LG,\ldots,LG] = n!\mathrm{Vol}_n(\Delta(S(LG)).$$

To complete the proof, it is enough to use the Brunn–Minkowski inequality. □

In fact, it is shown in [Kaveh–Khovanskii-3, Theorem 4.23] that the above Brunn–Minkowski inequality holds without the assumption that L,G are big.

Corollary 12.2 (A version of the Hodge inequality). *If X is an algebraic surface and $L,G \in \mathbf{K}_{\mathrm{rat}}(X)$ are big, then*

$$[L,L][G,G] \le [L,G]^2.$$

Proof. From Corollary 12.1, for $n=2$, we have

$$\begin{aligned}[L,L]+2[L,G]+[G,G] &= [LG,LG] \ge ([L,L]^{1/2}+[G,G]^{1/2})^2\\ &= [L,L]+2[L,L]^{1/2}[G,G]^{1/2}+[G,G],\end{aligned}$$

which readily implies the Hodge inequality. □

Thus Theorem 11.1 immediately enables us to reduce the Hodge inequality to the isoperimetric inequality. In this way, we can easily prove an analogue of the Alexandrov–Fenchel inequality and its corollaries for intersection index:

Theorem 12.3 (Algebraic analogue of the Alexandrov–Fenchel inequality). *Let X be an irreducible n-dimensional variety and let $L_1,\ldots,L_n \in \mathbf{K}_{\mathrm{rat}}(X)$ be big subspaces. Then the following inequality holds:*

$$[L_1,L_2,L_3,\ldots,L_n]^2 \ge [L_1,L_1,L_3,\ldots,L_n][L_2,L_2,L_3,\ldots,L_n].$$

In fact, we can show that in Theorem 12.3, it is enough to assume only that $L_3,\ldots,L_n$ are big subspaces (see [Kaveh–Khovanskii-3, Theorem 4.27]).

Corollary 12.4 (Corollaries of the algebraic analogue of the Alexandrov–Fenchel inequality). *Let X be an irreducible n-dimensional variety.*

(1) *Let $2 \le m \le n$ and $k_1+\cdots+k_r = m$ with $k_i \in \mathbb{N}$. Take big subspaces of rational functions $L_1,\ldots,L_n \in \mathbf{K}_{\mathrm{rat}}(X)$. Then*

$$[k_1 * L_1,\ldots,k_r * L_r, L_{m+1},\ldots,L_n]^m \ge \prod_{1\le j\le r} [m * L_j, L_{m+1},\ldots,L_n]^{k_j}.$$

(1) *(Generalized Brunn–Minkowski inequality) For any fixed big subspaces $L_{m+1},\ldots,L_n \in \mathbf{K}_{\mathrm{rat}}(X)$, the function*

$$F : L \mapsto [m * L, L_{m+1},\ldots,L_n]^{1/m}$$

is a concave function on the semigroup of big subspaces.

As we saw above, the Bernstein–Kushnirenko theorem follows from the main theorem. Applying the algebraic analogue of the Alexandrov–Fenchel inequality to the situation considered in the Bernstein–Kushnirenko theorem, one can prove the Alexandrov–Fenchel inequality for convex polyhedra with integral vertices. Out of this one can easily complete the proof of Alexandrov–Fenchel for general convex bodies: the homogeneity gives the inequality for convex polyhedra with rational vertices. But each convex body can be approximated arbitrarily well with polyhedra with rational vertices. The statement now follows from the continuity of mixed volume.

Thus the Bernstein–Kushnirenko theorem and the Alexandrov–Fenchel inequality in algebra and in geometry can be considered corollaries of the main theorem (Theorem 11.1).

13 Additivity of the Newton-Okounkov Body for Varieties with a Reductive Group Action

In this section we announce some results that the authors are currently writing up. They will be presented in a forthcoming paper.[2]

While the additivity of the Newton-Okounkov body does not hold in general, we recall, as mentioned in Sect. 8, that it does hold for the subspaces L_A of Laurent polynomials on $(\mathbb{C}^*)^n$ spanned by monomials. The subspaces L_A are exactly the subspaces that are stable under the natural action of the multiplicative group $(\mathbb{C}^*)^n$ on Laurent polynomials (induced by the natural action of $(\mathbb{C}^*)^n$ on itself). It turns out that the additivity generalizes to some classes of varieties with a reductive group action.

Let G be a connected reductive algebraic group over $\mathbb{C}$ (in other words, the complexification of a connected compact real Lie group). Also let X be a G-variety, that is, a variety equipped with an algebraic action of G.

The group G naturally acts on $\mathbb{C}(X)$ by $(g \cdot f)(x) = f(g^{-1} \cdot x)$. A subspace $L \in \mathbf{K}_{\text{rat}}(X)$ is G-stable if for any $f \in L$ and $g \in G$ we have $g \cdot f \in L$.

A G-variety X is called *spherical* if a Borel subgroup of G has a dense orbit. Toric varieties, flag varieties, and group compactifications are well-known examples of spherical varieties.

Theorem 13.1. *Let X be an n-dimensional spherical G-variety. Then there is a naturally defined faithful valuation $v : \mathbb{C}(X) \setminus \{0\} \to \mathbb{Z}^n$ such that for any G-stable subspace $L \in \mathbf{K}_{\text{rat}}(X)$, the Newton-Okounkov body $\Delta(S(L))$ is in fact a polyhedron.*

Definition 13.2. Let V be a finite-dimensional representation of G. Let $v = v_1 + \cdots + v_k$ be a sum of highest-weight vectors in V. The closure of the G-orbit of v in V is called an *S-variety*.

---

[2]While the present volume was under preparation the related papers [Kaveh-Khovanskii-4] and [Kaveh-Khovanskii-5] appeared.

Affine toric varieties are S-varieties for $G = (\mathbb{C}^*)^n$. One can show that any S-variety is spherical.

Theorem 13.3. *Let X be an S-variety for one of the groups $G = \mathrm{SL}(n,\mathbb{C})$, $\mathrm{SO}(n,\mathbb{C})$, $\mathrm{SP}(2n,\mathbb{C})$, $(\mathbb{C}^*)^n$, or a direct product of them. Then for the valuation in Theorem 13.1 and for any choice of G-stable subspaces L_1, L_2 in $\mathbf{K}_{\mathrm{rat}}(X)$, we have*

$$\Delta(S(L_1L_2)) = \Delta(S(L_1)) + \Delta(S(L_2)).$$

Corollary 13.4 (Bernstein theorem for S-varieties). *Let X be an S-variety for one of the groups $G = \mathrm{SL}(n,\mathbb{C})$, $\mathrm{SO}(n,\mathbb{C})$, $\mathrm{SP}(2n,\mathbb{C})$, $(\mathbb{C}^*)^n$, or a direct product of them. Let $L_1,\ldots,L_n \in \mathbf{K}_{\mathrm{rat}}(X)$ be G-stable subspaces. Then for the valuation in Theorem 13.1, we have*

$$[L_1,\ldots,L_n] = n!V(\Delta(S(L_1)),\ldots,\Delta(S(L_n))),$$

where V is the mixed volume.

Another class of G-varieties for which the additivity of the Newton polyhedron holds is the class of symmetric homogeneous spaces.

Definition 13.5. Let σ be an involution of G, i.e., an order-2 algebraic automorphism. Let $H = G^{\sigma}$ be the fixed-point subgroup of σ. The homogeneous space G/H is called a *symmetric homogeneous space*.

Example 13.6. The map $M \mapsto (M^{-1})^t$ is an involution of $G = \mathrm{SL}(n,\mathbb{C})$ with the fixed-point subgroup $H = \mathrm{SO}(n,\mathbb{C})$. The symmetric homogeneous space G/H can be identified with the space of nondegenerate quadrics in $\mathbb{C}P^{n-1}$.

Any symmetric homogeneous space is an affine spherical G-variety (with the left G-action).

Under mild conditions on the L_i, analogues of Theorem 13.3 and Corollary 13.4 hold for symmetric varieties. Finally, the above theorems extend to subspaces of sections of G-line bundles.

Acknowledgements We would like to thank the referee for a careful reading of the manuscript and for helpful suggestions and comments that improved the presentation of the material in the paper.

References

Bernstein. Bernstein, D. N. *The number of roots of a system of equations*. English translation: Functional Anal. Appl. 9 (1975), no. 3, 183–185 (1976).

Burago–Zalgaller. Burago, Yu. D.; Zalgaller, V. A. *Geometric inequalities*. Translated from the Russian by A. B. Sosinskiĭ. Grundlehren der Mathematischen Wissenschaften, 285. Springer Series in Soviet Mathematics (1988).

Kaveh–Khovanskii-1. Kaveh, K.; Khovanskii, A. G. *Convex bodies and algebraic equations on affine varieties*. Preprint: arXiv:0804.4095v1.

Kaveh–Khovanskii-2. Kaveh, K.; Khovanskii, A. G. *Mixed volume and an extension of intersection theory of divisors*. Moscow Mathematical Journal, 10 (2010), no. 2, 343–375.

Kaveh–Khovanskii-3. Kaveh, K.; Khovanskii, A. G. *Newton-Okounkov bodies, semigroups of integral points, graded algebras and intersection theory*. Preprint: arXiv: 0904.3350v2.

Kaveh-Khovanskii-4. Kaveh, K.; Khovanskii, A. G. *Newton polytopes for horospherical varieties*. Moscow Mathematical Journal, 11 (2011) no. 2, 265–283.

Kaveh-Khovanskii-5. Kaveh, K.; Khovanskii, A. G. *Convex bodies associated to actions of reductive groups*. To appear in Moscow Mathematical Journal. Preprint: arXiv:1001.4830v1.

Khovanskii-1. Khovanskii, A. G. *Algebra and mixed volumes*. Appendix 3 in: Burago, Yu. D.; Zalgaller, V. A. *Geometric inequalities*. Translated from the Russian by A. B. Sosinskiĭ. Grundlehren der Mathematischen Wissenschaften, 285. Springer Series in Soviet Mathematics (1988).

Khovanskii-2. Khovanskii, A. G. *Sums of finite sets, orbits of commutative semigroups and Hilbert functions*. (Russian) Funktsional. Anal. i Prilozhen. 29 (1995), no. 2, 36–50, 95; translation in Funct. Anal. Appl. 29 (1995), no. 2, 102–112.

Kushnirenko. Kushnirenko, A. G. *Polyèdres de Newton et nombres de Milnor*. (French) Invent. Math. 32 (1976), no. 1, 1–31.

Lazarsfeld–Mustata. Lazarsfeld, R.; Mustata, M. *Convex bodies associated to linear series*. Ann. de l'ENS, 42 (2009), no. 5, 783–835.

Okounkov1. Okounkov, A. *Brunn–Minkowski inequality for multiplicities*. Invent. Math. 125 (1996), no. 3, 405–411.

Okounkov-2. Okounkov, A. *Why would multiplicities be log-concave?* The orbit method in geometry and physics (Marseille, 2000), 329–347, Progr. Math., 213, Birkhäuser Boston, Boston, MA, 2003.

Teissier. Teissier, B. *Du théorème de l'index de Hodge aux inégalités isopérimétriques*. C. R. Acad. Sci. Paris Sér. A-B 288 (1979), no. 4, A287–A289.

Yuan. Yuan, X. *On volumes of arithmetic line bundles*. Compos. Math. 145 (2009), no. 6, 1447–1464.

Zariski–Samuel. Zariski, O.; Samuel, P. *Commutative algebra*. Vol. II. Reprint of the 1960 edition. Graduate Texts in Mathematics, Vol. 29. Springer-Verlag, New York-Heidelberg, 1975.

Floer Homology on the Extended Moduli Space

Ciprian Manolescu and Christopher Woodward

Abstract Starting from a Heegaard splitting of a three-manifold, we use Lagrangian Floer homology to construct a three-manifold invariant in the form of a relatively $\mathbb{Z}/8\mathbb{Z}$-graded abelian group. Our motivation is to have a well-defined symplectic side of the Atiyah–Floer conjecture for arbitrary three-manifolds. The symplectic manifold used in the construction is the extended moduli space of flat $\mathrm{SU}(2)$-connections on the Heegaard surface. An open subset of this moduli space carries a symplectic form, and each of the two handlebodies in the decomposition gives rise to a Lagrangian inside the open set. In order to define their Floer homology, we compactify the open subset by symplectic cutting; the resulting manifold is only semipositive, but we show that one can still develop a version of Floer homology in this setting.

Keywords Floer homology • Three-manifold • Moduli space • Heegaard surface

1 Introduction

Floer's instanton homology [15] is an invariant of integral homology three-spheres Y that serves as target for the relative Donaldson invariants of four-manifolds with boundary; see [13]. It is defined from a complex whose generators are (suitably perturbed) irreducible flat connections in a trivial $\mathrm{SU}(2)$-bundle over Y, and whose differentials arise from counting anti-self-dual $\mathrm{SU}(2)$-connections on $Y \times \mathbb{R}$. There

C. Manolescu (✉)
Department of Mathematics, UCLA, 520 Portola Plaza, Los Angeles, CA 90095
e-mail: cm@math.ucla.edu

C. Woodward
Mathematics-Hill Center, Rutgers University, 110 Frelinghuysen Road, Piscataway, NJ 08854
e-mail: ctw@math.rutgers.edu

I. Itenberg et al. (eds.), *Perspectives in Analysis, Geometry, and Topology: On the Occasion of the 60th Birthday of Oleg Viro*, Progress in Mathematics 296,
DOI 10.1007/978-0-8176-8277-4_13,

is also a version of instanton Floer homology using connections in U(2)-bundles with c_1 odd [9, 17], an equivariant version [4, 5], and several other variants that use both irreducible and reducible flat connections [13]. More recently, Kronheimer and Mrowka [29] have developed instanton homology for sutured manifolds; a particular case of their theory leads to a version of instanton homology that can be defined for arbitrary closed three-manifolds.

In another remarkable paper [16], Floer associated a homology theory to two Lagrangian submanifolds of a symplectic manifold, under suitable assumptions. This homology is defined from a complex whose generators are intersection points between the two Lagrangians, and whose differentials count pseudoholomorphic strips. The Atiyah–Floer conjecture [2] states that Floer's two constructions are related: for any decomposition of the homology sphere Y into two handlebodies glued along a Riemann surface Σ, instanton Floer homology should be the same as the Lagrangian Floer homology of the SU(2)-character varieties of the two handlebodies, viewed as subspaces of the character variety of Σ.

As stated, an obvious problem with the Atiyah–Floer conjecture is that the symplectic side is ill defined: due to the presence of reducible connections, the SU(2)-character variety of Σ is not smooth. One way of dealing with the singularities is to use a version of Lagrangian Floer homology defined via the symplectic vortex equations on the infinite-dimensional space of all connections. This approach was pursued by Salamon and Wehrheim, who obtained partial results toward the conjecture in this setup; see [49, 50, 56]. Another approach is to avoid reducibles altogether by using nontrivial PU(2)-bundles instead. This road was taken by Dostoglou and Salamon [14], who proved a variant of the conjecture for mapping tori.

The goal of this paper is to construct another candidate that could sit on the symplectic side of the (suitably modified) Atiyah–Floer conjecture.

Here is a short sketch of the construction. Let Σ be a Riemann surface of genus $h \geq 1$, and $z \in \Sigma$ a base point. The moduli space $\mathscr{M}(\Sigma)$ of flat connections in a trivial SU(2)-bundle over Σ can be identified with the character variety $\{\rho : \pi_1(\Sigma) \to \mathrm{SU}(2)\}/\mathrm{PU}(2)$. The moduli space $\mathscr{M}(\Sigma)$ is typically singular. However, Jeffrey [26] and independently Huebschmann [23], showed that $\mathscr{M}(\Sigma)$ is the symplectic quotient of a different space, called the extended moduli space, by a Hamiltonian PU(2)-action. The extended moduli space is naturally associated not to Σ, but to Σ', a surface with boundary obtained from Σ by deleting a small disk around z. The extended moduli space has an open smooth stratum, which Jeffrey and Huebschmann equip with a natural closed two-form. This form is nondegenerate on a certain open set $\mathscr{N}(\Sigma')$, which we take as our ambient symplectic manifold. In fact, $\mathscr{N}(\Sigma')$ can also be viewed as an open subset of the Cartesian product $\mathrm{SU}(2)^{2h} \cong \{\rho : \pi_1(\Sigma') \to \mathrm{SU}(2)\}$. More precisely, if we pick $2h$ generators for the free groups $\pi_1(\Sigma')$, we can describe this subset as

$$\mathscr{N}(\Sigma') = \left\{ (A_1, B_1, \ldots, A_h, B_h) \in \mathrm{SU}(2)^{2h} \mid \prod_{i=1}^{h} [A_i, B_i] \neq -I \right\}.$$

Consider a Heegaard decomposition of a three-manifold Y as $Y = H_0 \cup H_1$, where the handlebodies H_0 and H_1 are glued along their common boundary Σ. There are smooth Lagrangians $L_i = \{\pi_1(H_i) \to \mathrm{SU}(2)\} \subset \mathscr{N}(\Sigma')$ for $i = 0, 1$. In order to take the Lagrangian Floer homology of L_0 and L_1, care must be taken with holomorphic strips going out to infinity; indeed, the symplectic manifold $\mathscr{N}(\Sigma')$ is not weakly convex at infinity. Our remedy is to compactify $\mathscr{N}(\Sigma')$ by (nonabelian) symplectic cutting. The resulting manifold $\mathscr{N}^c(\Sigma')$ is the union of $\mathscr{N}(\Sigma')$ and a codimension-two submanifold R. A new problem shows up here, because the natural two-form $\tilde{\omega}$ on $\mathscr{N}^c(\Sigma')$ has degeneracies on R. Nevertheless, $(\mathscr{N}^c(\Sigma'), \tilde{\omega})$ is monotone, in a suitable sense. One can deform $\tilde{\omega}$ into a symplectic form ω, at the expense of losing monotonicity. We are thus led to develop a version of Lagrangian Floer theory on $\mathscr{N}^c(\Sigma')$ by making use of the interplay between the forms $\tilde{\omega}$ and ω. Our Floer complex uses only holomorphic disks lying in the open part $\mathscr{N}(\Sigma')$ of $\mathscr{N}^c(\Sigma')$. We show that while holomorphic strips with boundary on L_0 and L_1 can go to infinity in $\mathscr{N}(\Sigma')$, they do so only in high codimension, without affecting the Floer differential. The resulting Floer homology group is denoted by

$$HSI(\Sigma; H_0, H_1) = HF(L_0, L_1 \text{ in } \mathscr{N}(\Sigma')),$$

and it admits a relative $\mathbb{Z}/8\mathbb{Z}$-grading. We call it *symplectic instanton homology*.

Using the theory of Lagrangian correspondences and pseudoholomorphic quilts developed in Wehrheim–Woodward [59] and Lekili–Lipyanskiy [30], we prove the following theorem.

Theorem 1. *The relatively $\mathbb{Z}/8\mathbb{Z}$-graded group $HSI(Y) = HSI(\Sigma; H_0, H_1)$ is an invariant of the three-manifold Y.*

Strictly speaking, if we are interested in canonical isomorphisms, then the symplectic instanton homology also depends on the base point $z \in \Sigma \subset Y$: as z varies inside Y, the corresponding groups form a local system. However, we drop z from the notation for simplicity.

Let us explain how we expect $HSI(Y)$ to be related to the traditional instanton theory on 3-manifolds. We restrict our attention to the original setup for Floer's instanton theory $I(Y)$ from [15] when Y is an integral homology sphere. It is then decidedly not the case that $HSI(Y)$ coincides with Floer's theory; for example, we have $HSI(S^3) \cong \mathbb{Z}$, but $I(S^3) = 0$. Nevertheless, in [13, Sect. 7.3.3], Donaldson introduced a different version of instanton homology, a $\mathbb{Z}/8\mathbb{Z}$-graded vector field over $\mathbb{Q}$ denoted by $\widetilde{HF}$ that satisfies $\widetilde{HF}(S^3) \cong \mathbb{Q}$. (Floer's theory I is denoted by HF in [13].) We state the following variant of the Atiyah–Floer conjecture:

Conjecture 1. For every integral homology sphere Y, the symplectic instanton homology $HSI(Y) \otimes \mathbb{Q}$ and the Donaldson–Floer homology $\widetilde{HF}(Y)$ from [13] are isomorphic as relatively $\mathbb{Z}/8\mathbb{Z}$-graded vector spaces.

Alternatively, one could hope to relate HSI to the sutured version of instanton Floer homology developed by Kronheimer and Mrowka in [29]. More open questions, and speculations along these lines, are presented in Sect. 7.3.

2 Floer Homology

2.1 The Monotone, Nondegenerate Case

Lagrangian Floer homology was originally constructed in [16] under some restrictive conditions, and later generalized by various authors to many different settings. We review here its definition in the monotone case, due to Oh [39, 41], together with a discussion of orientations following Fukaya–Oh–Ohta–Ono [18].

Let (M,ω) be a compact connected symplectic manifold. We denote by $\mathcal{J}(M,\omega)$ the space of compatible almost complex structures on (M,ω), and by $\mathfrak{J}_t(M,\omega) = C^\infty([0,1],\mathcal{J}(M,\omega))$ the space of *time-dependent* compatible almost complex structures. Any compatible almost complex structure J defines a complex structure on the tangent bundle TM. Since $\mathcal{J}(M,\omega)$ is contractible, the first Chern class $c_1(TM) \in H^2(M,\mathbb{Z})$ depends only on ω, not on J. The minimal Chern number N_M of M is defined as the positive generator of the image of $c_1(TM) : \pi_2(M) \to \mathbb{Z}$.

Definition 1. Let (M,ω) be a symplectic manifold. Then M is called *monotone* if there exists $\kappa > 0$ such that

$$[\omega] = \kappa \cdot c_1(TM).$$

In that case, κ is called the monotonicity constant.

Definition 2. A Lagrangian submanifold $L \subset (M,\omega)$ is called *monotone* if there exists a constant $\kappa > 0$ such that

$$2[\omega]|_{\pi_2(M,L)} = \kappa \cdot \mu_L,$$

where $\mu_L : \pi_2(M,L) \to \mathbb{Z}$ is the Maslov index.

Necessarily, if L is monotone, then M is monotone with the same monotonicity constant. The minimal Maslov number N_L of a monotone Lagrangian L is defined as the positive generator of the image of μ_L in $\mathbb{Z}$.

From now on, we will assume that M is monotone with monotonicity constant κ and that we are given two closed, simply connected Lagrangians $L_0, L_1 \subset M$. These conditions imply that L_0 and L_1 are monotone with the same monotonicity constant and

$$N_{L_0} = N_{L_1} = 2N_M.$$

We assume that $N_M > 1$ and set $N = 2N_M \geq 4$. We also assume that $w_2(L_0) = w_2(L_1) = 0$.

After a small Hamiltonian perturbation, we can arrange things so that the intersection $L_0 \cap L_1$ is transverse. Let $(J_t)_{0\leq t\leq 1} \in \mathfrak{J}_t(M,\omega)$. For any $x_\pm \in L_0 \cap L_1$, we denote by $\widetilde{\mathfrak{M}}(x_+,x_-)$ the space of *Floer trajectories* (or *J_t-holomorphic strips*) from x_+ to x_-, i.e., finite-energy solutions to Floer's equation

$$\begin{cases} u : \mathbb{R} \times [0,1] \to M, \\ u(s,j) \in L_j, \quad j = 0,1, \\ \partial_s u + J_t(u) \partial_t u = 0, \\ \lim_{s \to \pm\infty} u(s,\cdot) = x_{\pm}. \end{cases} \tag{2.1}$$

Let $\mathfrak{M}(x_+,x_-)$ denote the quotient of $\tilde{\mathfrak{M}}(x_+,x_-)$ by the translational action of $\mathbb{R}$. For $(J_t)_{0 \le t \le 1}$ chosen from a comeager[1] subset $\mathfrak{J}_t^{\mathrm{reg}}(L_0,L_1) \subset \mathfrak{J}_t(M,\omega)$ of (L_0,L_1)-*regular*, time-dependent compatible almost complex structures, $\mathfrak{M}(x_+,x_-)$ is a smooth, finite-dimensional manifold with dimension at a nonconstant $u \in \mathfrak{M}(x_+,x_-)$ given by $\dim T_u \mathfrak{M}(x_+,x_-) = I(u) - 1$. We denote by $\mathfrak{M}(x_+,x_-)_d$ the subset with $I(u) - 1 = d$ (note that $\mathfrak{M}(x_+,x_-)_{-1}$ is nonempty if $x_+ = x_-$). As explained in Oh [39], after shrinking $\mathfrak{J}_t^{\mathrm{reg}}(L_0,L_1)$ further, we may assume that $\mathfrak{M}(x_+,x_-)_0$ is finite and $\mathfrak{M}(x_+,x_-)_1$ is compact up to breaking of trajectories:

$$\partial \mathfrak{M}(x_+,x_-)_1 = \bigcup_{y \in L_0 \cap L_1} \mathfrak{M}(x_+,y)_0 \times \mathfrak{M}(y,x_-)_0. \tag{2.2}$$

The condition that the Lagrangians have vanishing w_2 is used in defining orientations on the moduli spaces, compatible with the identity (2.2). The Floer chain complex is then defined to be the free abelian group generated by the intersection points

$$CF(L_0,L_1) = \bigoplus_{x \in L_0 \cap L_1} \mathbb{Z}\langle x \rangle.$$

The Floer differential is

$$\partial \langle x_+ \rangle = \sum_{u \in \mathfrak{M}(x_+,x_-)_0} \varepsilon(u) \langle x_- \rangle,$$

where $\varepsilon(u) \in \{\pm 1\}$ is the sign comparing the orientation of the moduli space to the canonical orientation of a point; see, for example, [58].

Our assumptions allow one to define a relative Maslov index $I(x,y) \in \mathbb{Z}/N\mathbb{Z}$ for every $x,y \in L_0 \cap L_1$ such that $I(x,y) \equiv I(u) \pmod N$ for any $u \in \mathfrak{M}(x,y)$. The relative index satisfies $I(x,y) + I(y,z) = I(x,z)$, and it induces a relative $\mathbb{Z}/N\mathbb{Z}$-grading on the chain complex.

The Lagrangian Floer homology groups $HF(L_0,L_1)$ are the homology groups of $CF_*(L_0,L_1)$ with respect to the differential ∂. Equation (2.2) implies that $\partial^2 = 0$. An important property of the Floer homology groups $HF(L_0,L_1)$ is that they are

[1] A subset of a topological space is *comeager* if it is the intersection of countably many open dense subsets. Many authors use the term "Baire second category," which, however, denotes more generally subsets that are not meager, i.e., not the complement of a comeager subset. See, for example, [48, Chap. 7.8].

independent of the choice of path of almost complex structures, and invariant under Hamiltonian isotopies of both L_0 and L_1. Since $H_1(L_0) = H_1(L_1) = 0$, any isotopy of L_0 or L_1 through Lagrangians can be embedded in an ambient Hamiltonian isotopy; see, for example [44, Sect. 6.1] or the discussion in [52, Sect. 4(D)].

2.2 A Relative Version

Let $R \subset M$ denote a symplectic hypersurface disjoint from the Lagrangians L_0, L_1. Each pseudoholomorphic strip $u : \mathbb{R} \times [0,1] \to M$ meeting R in a finite number of points has a well-defined intersection number $u \cdot R$, defined by a signed count of intersection points of generic perturbations. The intersection numbers $u \cdot R$ depend only on the relative homology class of u, and are additive under concatenation of trajectories:

$$(u\#v) \cdot R = (u \cdot R) + (v \cdot R).$$

Let $\mathcal{J}(M,\omega,R)$ denote the space of compatible almost complex structures J for which R is a J-holomorphic submanifold. Let also $\mathfrak{J}_t(M,\omega,R) = C^\infty([0,1],\mathcal{J}(M,\omega,R))$ be the corresponding space of time-dependent almost complex structures. If $(J_t) \in \mathfrak{J}_t(M,\omega,R)$, then the intersection number of any J_t-holomorphic strip with R is a finite sum of positive local intersection numbers; see, for example, Cieliebak–Mohnke [11, Proposition 7.1]. In particular, if a J_t-holomorphic strip has trivial intersection number with R, it must be disjoint from R.

One can show that $\mathcal{J}_t^{\mathrm{reg}}(L_0,L_1,R) = \mathfrak{J}_t^{\mathrm{reg}}(L_0,L_1) \cap \mathfrak{J}_t(M,\omega,R)$ is comeager in $\mathfrak{J}_t(M,\omega,R)$. Since L_0, L_1 are disjoint from R, Floer homology may be defined using $J_t \in \mathfrak{J}_t(M,\omega,R)$. Moreover, for $J \in \mathcal{J}_t^{\mathrm{reg}}(L_0,L_1,R)$, the Floer differential decomposes as the sum

$$\partial = \sum_{m \geq 0} \partial_m,$$

where ∂_m counts the trajectories with intersection number m with R. By additivity of the intersection numbers, the square of the Floer differential satisfies the refined equality

$$\sum_{i+j=m} \partial_i \partial_j = 0.$$

In particular, $\partial_0^2 = 0$. Let $HF(L_0,L_1;R)$ denote the homology of ∂_0, counting Floer trajectories disjoint from R. We call $HF(L_0,L_1;R)$ the *Lagrangian Floer homology of L_0, L_1 relative to the hypersurface R*. This kind of construction has previously appeared in the literature in various guises; see, for example, Seidel's deformation of the Fukaya category [51, p. 8] or the hat version of Heegaard Floer homology [43]. Note that $HF(L_0,L_1;R)$ admits a relative $\mathbb{Z}/N'\mathbb{Z}$-grading, where $N' = 2N_{M\setminus R}$ is a positive multiple of N.

A standard continuation argument then shows that $HF(L_0,L_1;R)$ is independent of the choice of $J_t \in \mathcal{J}_t^{\mathrm{reg}}(L_0,L_1,R)$. Indeed, any two such compatible almost complex structures can be joined by a path $J_{t,\rho}, \rho \in [0,1]$, which equips the fiber bundle $\mathbb{R} \times [0,1] \times M \to \mathbb{R} \times [0,1]$ with an almost complex structure. The part of the continuation map counting pseudoholomorphic sections with zero intersection number with the almost complex submanifold $\mathbb{R} \times [0,1] \times R$ defines an isomorphism from the two Floer homology groups.

In fact, we may assume that all Floer trajectories are transverse to R by the following argument, which holds for not necessarily monotone M.

For any $k \in \mathbb{N}$, we denote by $\mathfrak{M}(x_+,x_-;k)$ the subset of $\mathfrak{M}(x_+,x_-)$ with a tangency of order exactly k to R. Given an open subset $\mathcal{W} \subset M$ containing L_0 and L_1 with closure disjoint from R and a $\tilde{J} \in \mathcal{J}(M,\omega,R)$, we denote by $\mathfrak{J}_t(M,\omega,\mathcal{W},\tilde{J})$ the space of compatible almost complex structures that agree with $\tilde{J}$ outside $\mathcal{W}$.

Lemma 1. *There exists a comeager subset $\mathfrak{J}_t^{\mathrm{reg}}(L_0,L_1,\mathcal{W},\tilde{J})$ of $\mathfrak{J}_t(M,\omega,\mathcal{W},\tilde{J})$ contained in $\mathfrak{J}_t^{\mathrm{reg}}(L_0,L_1,R)$ such that for any $(J_t) \in \mathfrak{J}_t^{\mathrm{reg}}(L_0,L_1,\mathcal{W},\tilde{J})$, the corresponding moduli space $\mathfrak{M}(x_+,x_-)$ is a smooth manifold, and for every $k \in \mathbb{N}$ and $x_\pm \in L_0 \cap L_1$, $\mathfrak{M}(x_+,x_-;k)$ is a smooth submanifold of $\mathfrak{M}(x_+,x_-)$ of codimension $2k$.*

Proof. For the closed case, see Cieliebak–Mohnke [11, Proposition 6.9]. The proof for Floer trajectories is the same, since the Lagrangians are disjoint from R. Note that [11] uses tamed almost complex structures; however, the arguments apply equally well to compatible almost complex structures; see [33, p. 47].

Corollary 1. *If $(J_t) \in \mathfrak{J}_t^{\mathrm{reg}}(L_0,L_1,\mathcal{W},\tilde{J})$, then for every element of $\mathfrak{M}(x_+,x_-)_0$ and $\mathfrak{M}(x_+,x_-)_1$, the intersection with R is transversal, and the number of intersection points equals the intersection pairing with R.*

2.3 *Floer Homology on Semipositive Manifolds*

In this section, we extend the definition of Floer homology to a semipositive setting. More precisely, we assume the following:

Assumption 2.1. (i) *(M,ω) is a compact symplectic manifold.*
(ii) *$\tilde{\omega}$ is a closed two-form on M.*
(iii) *The degeneracy locus $R \subset M$ of $\tilde{\omega}$ is a symplectic hypersurface with respect to ω.*
(iv) *$\tilde{\omega}$ is monotone, i.e., $[\tilde{\omega}] = \kappa \cdot c_1(TM)$ for some $\kappa > 0$.*
(v) *The restrictions of $\tilde{\omega}$ and ω to $M \setminus R$ have the same cohomology class in $H^2(M \setminus R)$.*
(vi) *The forms $\tilde{\omega}$ and ω themselves coincide on an open subset $\mathcal{W} \subset M \setminus R$.*
(vii) *We are given two closed submanifolds $L_0, L_1 \subset \mathcal{W}$ that are Lagrangian with respect to ω (hence Lagrangians with respect to $\tilde{\omega}$ as well).*
(viii) *L_0 and L_1 intersect transversely.*
(ix) *$\pi_1(L_0) = \pi_1(L_1) = 1$ and $w_2(L_0) = w_2(L_1) = 0$.*

(x) *The minimal Chern number* $N_{M\setminus R}$ *(with respect to* ω*) is at least* 2*, so that* $N = 2N_{M\setminus R} \geq 4$.

(xi) *There exists an almost complex structure that is compatible with respect to* ω *on* M *and compatible with respect to* $\tilde{\omega}$ *on* $M \setminus R$*, and for which* R *is an almost complex submanifold. We fix such a* $\tilde{J}$*, which we call the* base almost complex structure.

(xii) *Any* $\tilde{J}$*-holomorphic sphere in* M *of index zero (necessarily contained in* R*) has intersection number with* R *equal to a negative multiple of* 2.

Let us remark that because $\tilde{J}$ is compatible with respect to $\tilde{\omega}$ on $M \setminus R$, by continuity it follows that $\tilde{J}$ is semipositive with respect to $\tilde{\omega}$ on all of M; i.e., $\tilde{\omega}(v, \tilde{J}v) \geq 0$ for any $m \in M$ and $v \in T_m M$.

Our goal is to define a relatively $\mathbb{Z}/N\mathbb{Z}$-graded Floer homology group $HF(L_0, L_1, \tilde{J}; R)$ using Floer trajectories away from R and a path of almost complex structures that are small perturbations of $\tilde{J}$ supported in a neighborhood of $L_0 \cup L_1$. The construction is similar to the one in Sect. 2.2, but a priori it depends on $\tilde{J}$.

Definition 3. (a) We say that $J \in \mathcal{J}(M, \omega)$ is *spherically semipositive* if every J-holomorphic sphere has nonnegative Chern number $c_1(TM)[u] \geq 0$.

(b) We say that $J \in \mathcal{J}(M, \omega)$ is *hemispherically semipositive* if J is spherically semipositive and every J-holomorphic map $(D^2, \partial D^2) \to (M, L_i)$, $i \in 0, 1$, has nonnegative Maslov index $I(u)$, and further, if $I(u) = 0$, then u is constant.

Given a continuous map $u : (D^2, \partial D^2) \to (M, L_i)$, $i = 0, 1$, we define the *canonical area* of u by

$$\tilde{A}(u) := \frac{[\tilde{\omega}](u)}{\kappa}.$$

Lemma 2. *We have* $I(u) = \tilde{A}(u)$ *for any* $u : (D^2, \partial D^2) \to (M, L_i)$.

Proof. Since L_i is simply connected, we can find a disk v contained in L_i with boundary equal to that of u, but with reversed orientation. Let $u\#v : S^2 \to M$ be the map formed by gluing. By additivity of the Maslov index,

$$I(u) = I(u) + I(v) = I(u\#v) = 2\frac{[\tilde{\omega}](u\#v)}{\kappa} = 2\frac{[\tilde{\omega}](u)}{\kappa},$$

since both the index and the area of v are trivial.

We define a *strip with decay near the ends* to be a continuous map

$$u : (\mathbb{R} \times [0,1], \mathbb{R} \times \{0\}, \mathbb{R} \times \{1\}) \to (M, L_0, L_1) \tag{2.3}$$

such that $\lim_{s \to \infty} u(s,t)$, $\lim_{s \to -\infty} u(s,t) \in L_0 \cap L_1$ exist. Every strip with decay near the ends admits a relative homology class in $H_2(M, L_0 \cup L_1)$, and therefore has a well-defined canonical area

$$\tilde{A}(u) := \frac{[\tilde{\omega}](u)}{\kappa}$$

and a Maslov index $I(u)$.

The following lemma is [39, Proposition 2.7].

Lemma 3. *Strips* (2.3) *satisfy an index-action relation*

$$I(u) = \tilde{A}(u) + C$$

for some constant C depending only on the endpoints of u.

Proof (Sketch). Pick u_0 a reference strip with the same endpoints as u. Using the fact that $\pi_1(L_0) = 1$, we can find a map $v : D^2 \to L_0$ such that half of its boundary is taken to the image of $u_0(\mathbb{R} \times \{0\})$ and the other half to the image of $u(\mathbb{R} \times \{0\})$. By adjoining v to u and u_0 (the latter taken with reversed orientation), we obtain a disk $(-u_0)\#v\#u$ with boundary in L_1. Applying Lemma 2 to this disk and using the additivity of the index and canonical action under gluing, we obtain

$$I(u) - I(u_0) = \tilde{A}(u) - \tilde{A}(u_0).$$

We then take $C = I(u_0) - \tilde{A}(u_0)$.

As in Sect. 2.2, $\mathcal{J}(M,\omega,\mathcal{W},\tilde{J})$ denotes the space of compatible almost complex structures agreeing with $\tilde{J}$ outside $\mathcal{W}$. We let $\mathfrak{J}_t(M,\omega,\mathcal{W},\tilde{J}) = C^\infty([0,1], \mathcal{J}(M,\omega,\mathcal{W},\tilde{J})$.

Lemma 4. *Every J in $\mathcal{J}(M,\omega,\mathcal{W},\tilde{J})$ is hemispherically semipositive.*

Proof. Since $\tilde{\omega}$ agrees with ω on $\mathcal{W}$, we have $\tilde{\omega}(v,Jv) \geq 0$ for every $v \in T_mM$, where $m \in \mathcal{W}$. Since J agrees with $\tilde{J}$ outside $\mathcal{W}$, we in fact have $\tilde{\omega}(v,Jv) \geq 0$ everywhere. Nonnegativity of I then follows from the monotonicity of $\tilde{\omega}$ (for spheres) and Lemma 2 for disks. If a J-holomorphic disk u has $I(u) = 0$, its canonical area must be zero. Since J is compatible with respect to $\tilde{\omega}$ on $M \setminus R$, the disk should be contained in R. However, this is impossible, because the disk has boundary on a Lagrangian L_i with $L_i \cap R = \emptyset$. (By contrast, we could have $I(u) = 0$ for nonconstant $\tilde{J}$-holomorphic spheres contained in R.)

Let $\mathfrak{J}_t^{\text{reg}}(L_0,L_1,\mathcal{W},\tilde{J}) \subset \mathfrak{J}_t(M,\omega,\mathcal{W},\tilde{J}) \cap \mathfrak{J}_t^{\text{reg}}(L_0,L_1,R)$ be as in Lemma 1.

Proposition 1. *Let $M, L_0, L_1, \tilde{\omega}, \omega, \tilde{J}$ satisfy Assumption 2.1. If we choose $(J_t) \in \mathfrak{J}_t^{\text{reg}}(L_0,L_1, \mathcal{W},\tilde{J})$, then the relative Floer differential counting trajectories disjoint from R is finite and satisfies $\partial_0^2 = 0$. The resulting (relatively $\mathbb{Z}/N\mathbb{Z}$-graded) Floer homology groups $HF_*(L_0,L_1,\tilde{J};R)$ are independent of the choice of path (J_t), and are preserved under Hamiltonian isotopies of either Lagrangian, as long as Assumption 2.1 is still satisfied.*

Proof. Using parts (v) and (vi) of Assumption 2.1, we see that on the complement of R we have $\omega - \tilde{\omega} = da$, for some $a \in \Omega^1(M \setminus R)$ satisfying $da = 0$ in the neighborhood $\mathcal{W}$ of $L_0 \cup L_1$. Let u be a pseudoholomorphic strip whose image is contained in $M \setminus R$. Then

$$E(u) - \kappa\tilde{A}(u) = \int_{\mathbb{R}\times[0,1]} u^*(\omega - \tilde{\omega}) = \int_{\mathbb{R}\times[0,1]} \mathrm{d}(u^*a) = \int_{\gamma_0} u^*a - \int_{\gamma_1} u^*a,$$

where γ_i is a path in the Lagrangian L_i joining the endpoints of u. Since $\mathrm{d}a = 0$ on L_i, Stokes' theorem implies that $\int_\gamma u^*a$ is independent of γ; it depends only on the endpoints. Therefore, $E(u) - \kappa\tilde{A}(u)$ depends only on the endpoints of u. Together with Lemma 3 this gives an energy index relation as follows: for any u in $M \setminus R$, we have

$$I(u) = E(u)/\kappa + C',$$

where C' is a constant depending on the endpoints of u. Since there are only finitely many possibilities for these endpoints, it follows that there exists a constant $K > 0$ such that the energy of any such trajectory u is bounded above by K.

Let $(J_t) \in \mathcal{J}_t^{\mathrm{reg}}(L_0, L_1, \mathcal{W}, \tilde{J})$. By Proposition 4 each J_t is hemispherically semipositive. We define the Floer differential by counting J_t-holomorphic strips in $M \setminus R$. By Lemma 1, a sequence of such strips cannot converge to a strip that intersects R, unless further bubbling occurs.

We seek to rule out sphere bubbles and disk bubbles in the boundary of the zero- and one-dimensional moduli spaces of such strips (i.e., those of index 1 or 2). Assume that we have a sequence (u_ν) of pseudoholomorphic strips of index 1 or 2. Because of the energy bound, a subsequence Gromov converges to a limiting configuration consisting of a broken trajectory and a collection of disk and sphere bubbles. Since the J_t's are hemispherically semipositive, it follows that the indices of the bubbles are nonnegative. Further, by part (x) of Assumption 2.1, the index of each bubble is a multiple of 4. Since we started with a configuration of index at most 2, all bubbles have index zero. By the definition of hemispherical semipositivity, the index-zero disks are constant.

By item (xii) of Assumption 2.1, each index sphere bubble contributes a multiple of two to the intersection number with R. By Lemma 1, the intersection number of the limiting trajectory u_∞ (with sphere bubbles removed) is given by the number of intersection points, and each of these is transverse. Hence at most half of the intersection points with R have sphere bubbles attached. In particular, there exists a point $z \in \mathbb{R} \times [0,1]$ such that $u_\infty(z) \in R$ is a transverse intersection point but z is not in the bubbling set. Since the intersection points are stable under perturbation, it follows that u_∞ cannot be a limit of Floer trajectories disjoint from R. Indeed, by definition this convergence is uniform in all derivatives on the complement of the bubbling set, and in particular, on an open subset containing z. Since there are no sphere bubbles, there cannot be any disk bubbles either, since at least one disk bubble would have to be nonconstant. Hence the limit is a (possibly broken) Floer trajectory.

The rest of the argument is then as in the monotone case. In particular, the statement about the invariance of $HF(L_0, L_1, \tilde{J}; R)$ follows from the usual continuation arguments in Floer theory.

Remark 1. If $M, L_0, L_1, \tilde{\omega}, \omega, \tilde{J}$ satisfy Assumption 2.1, we can define $HF_*(L_0, L_1, \tilde{J}; R)$ even if L_0 and L_1 do not intersect transversely: one can simply isotope one of the Lagrangians to achieve transversality, and take the resulting Floer homology.

Remark 2. A priori, the construction of the Floer homologies $HF(L_0, L_1, \tilde{J}; R)$ depends on the open set $\mathcal{W}$, because (J_t) is chosen from the corresponding set $\mathcal{J}_t^{\text{reg}}(L_0, L_1, \mathcal{W}, \tilde{J})$. However, suppose we have another open set $\mathcal{W}' \subset M \setminus R$ satisfying $L_0 \cap L_1 \subset \mathcal{W}'$ and $\omega = \tilde{\omega}$ on $\mathcal{W}'$. Note that

$$\mathcal{J}_t^{\text{reg}}(L_0, L_1, \mathcal{W}, \tilde{J}) \cap \mathcal{J}_t^{\text{reg}}(L_0, L_1, \mathcal{W}', \tilde{J}) = \mathcal{J}_t^{\text{reg}}(L_0, L_1, \mathcal{W} \cap \mathcal{W}', \tilde{J}), \qquad (2.4)$$

because the regularity condition in Lemma 1 is intrinsic for (J_t) (it boils down to the surjectivity of certain linear operators). It follows that by choosing (J_t) in the (necessarily nonempty) intersection (2.4), the Floer homologies $HF(L_0, L_1, \tilde{J}; R)$ defined from $\mathcal{W}$ and $\mathcal{W}'$ are isomorphic. Thus, we can safely drop $\mathcal{W}$ from the notation.

Remark 3. A smooth variation of the base almost complex structure $\tilde{J}$ induces an isomorphism between the respective Floer homologies $HF(L_0, L_1, \tilde{J}; R)$. However, if we are given only $\tilde{\omega}$ and ω, it is not clear whether the space of possible $\tilde{J}$'s is contractible. This justifies keeping $\tilde{J}$ in the notation $HF(L_0, L_1, \tilde{J}; R)$.

3 Moduli Spaces

3.1 Notation

Throughout the rest of the paper, G will denote the Lie group $\mathrm{SU}(2)$, and $G^{\text{ad}} = \mathrm{PU}(2) = \mathrm{SO}(3)$ the corresponding group of adjoint type. We identify the Lie algebra $\mathfrak{g} = \mathfrak{su}(2)$ with its dual $\mathfrak{g}^*$ using the basic invariant bilinear form

$$\langle \cdot, \cdot \rangle : \mathfrak{g} \times \mathfrak{g} \to \mathbb{R}, \;\; \langle A, B \rangle = -\mathrm{Tr}\,(AB).$$

The maximal torus $T \cong S^1 \subset G$ consists of the diagonal matrices diag $(e^{2\pi t i}, e^{-2\pi t i}), t \in \mathbb{R}$. We let $T^{\text{ad}} = T/(\mathbb{Z}/2\mathbb{Z}) \subset G^{\text{ad}}$ and identify their Lie algebra $\mathfrak{t}$ with $\mathbb{R}$ by sending $\mathrm{diag}(\mathrm{i}, -\mathrm{i})$ to 1. Under this identification, the restriction of the inner product $\langle \cdot, \cdot \rangle$ to $\mathfrak{t}$ is twice the Euclidean metric. We use this inner product to identify $\mathfrak{t}$ with $\mathfrak{t}^*$ as well. Finally, we let $\mathfrak{t}^{\perp}$ denote the orthocomplement of $\mathfrak{t}$ in $\mathfrak{g}$.

Conjugacy classes in $\mathfrak{g}$ (under the adjoint action of G) are parameterized by the positive Weyl chamber $\mathfrak{t}_+ = [0,\infty)$. Indeed, the adjoint quotient map

$$Q : \mathfrak{g} \to [0,\infty)$$

takes $\theta \in \mathfrak{g}$ to t such that θ is conjugate to $\mathrm{diag}(ti, -ti)$.

On the other hand, conjugacy classes in G are parameterized by the fundamental alcove $\mathfrak{A} = [0, 1/2]$. Indeed, for any $g \in G$, there is a unique $t \in [0, 1/2]$ such that g is conjugate to the diagonal matrix $\mathrm{diag}(\mathrm{e}^{2\pi ti}, \mathrm{e}^{-2\pi ti})$.

3.2 The Extended Moduli Space

We review here the construction of the *extended moduli space* [23, 26], mostly following Jeffrey's gauge-theoretic approach from [26].

Let Σ be a compact connected Riemann surface of genus $h \geq 1$. Fix some $z \in \Sigma$ and let Σ' denote the complement in Σ of a small disk around z, so that $S = \partial\Sigma'$ is a circle. Identify a neighborhood of S in Σ' with $[0, \varepsilon) \times S$, and let $s \in \mathbb{R}/2\pi\mathbb{Z}$ be the coordinate on the circle S.

Consider the space $\mathscr{A}(\Sigma') \cong \Omega^1(\Sigma') \otimes \mathfrak{g}$ of smooth connections on the trivial G-bundle over Σ', and set

$$\mathscr{A}^{\mathfrak{g}}(\Sigma') = \{A \in \mathscr{A}(\Sigma') \mid F_A = 0,\ A = \theta \mathrm{d}s \text{ on some neighborhood of } S \text{ for some } \theta \in \mathfrak{g}\}.$$

The space $\mathscr{A}^{\mathfrak{g}}(\Sigma')$ is acted on by the gauge group

$$\mathscr{G}^c(\Sigma') = \{f : \Sigma' \to G) \mid f = I \text{ on some neighborhood of } S\}.$$

The extended moduli space is then defined as

$$\mathscr{M}^{\mathfrak{g}}(\Sigma') = \mathscr{A}^{\mathfrak{g}}(\Sigma')/\mathscr{G}^c(\Sigma').$$

A more explicit description of the extended moduli space is obtained by fixing a collection of simple closed curves α_i, β_i $(i = 1, \ldots, h)$ on Σ', based at a point in S, such that $\pi_1(\Sigma')$ is generated by their equivalence classes and the class of a curve γ around S, with the relation $\prod_{i=1}^h [\alpha_i, \beta_j] = \gamma$.

To each connection on Σ' one can then associate the holonomies $A_i, B_i \in G$ around the loops α_i and β_i, respectively, $i = 1, \ldots, h$. This allows us to view the extended moduli space as

$$\mathscr{M}^{\mathfrak{g}}(\Sigma') = \left\{ (A_1, B_1, \ldots, A_h, B_h) \in G^{2h}, \theta \in \mathfrak{g} \mid \prod_{i=1}^{h} [A_i, B_i] = \exp(2\pi\theta) \right\}. \quad (3.1)$$

There is a proper map
$$\Phi : \mathscr{M}^{\mathfrak{g}}(\Sigma') \to \mathfrak{g}$$
that takes the class $[A]$ of a connection A to the value $\theta = \Phi(A)$ such that $A|_S = \theta \mathrm{d}s$. (This corresponds to the variable θ appearing in (3.1).) There is also a natural G-action on $\mathscr{M}^{\mathfrak{g}}(\Sigma')$ given by constant gauge transformations. With respect to the identification (3.1), it is

$$g \in G : (A_i, B_i, \theta) \to (gA_i g^{-1}, gB_i g^{-1}, \mathrm{Ad}(g)\theta). \tag{3.2}$$

Observe that this action factors through G^{ad}. The map Φ is equivariant with respect to this action on its domain, and the adjoint action on its target. Set

$$\tilde{\Phi} : \mathscr{M}^{\mathfrak{g}}(\Sigma') \to [0, \infty), \quad \tilde{\Phi} = Q \circ \Phi.$$

Now consider the subspace

$$\mathscr{M}^{\mathfrak{g}}_s(\Sigma') = \{x \in \mathscr{M}^{\mathfrak{g}}(\Sigma') \mid \tilde{\Phi}(x) \notin \mathbb{Z} \setminus \{0\}\}.$$

Proposition 2. (a) *The space $\mathscr{M}^{\mathfrak{g}}_s(\Sigma')$ is a smooth manifold of real dimension $6h$.*
(b) *Every nonzero element $\theta \in \mathfrak{g}$ is a regular value for the restriction of Φ to $\mathscr{M}^{\mathfrak{g}}_s(\Sigma')$.*

Proof. Part (a) is proved in [26, Theorem 2.7]. We copy the proof here, and explain how the same arguments can be used to deduce part (b) as well.

Consider the commutator map $c : G^{2h} \to G, c(A_1, B_1, \ldots, A_h, B_h) = \prod_{i=1}^{h} [A_i, B_i]$. For $\rho = (A_1, B_1, \ldots, A_h, B_h) \in G^{2h}$, we denote by $Z(\rho) \subset G$ its stabilizer (under the diagonal action by conjugation). Let $z(\rho) \subset \mathfrak{g}$ be the Lie algebra of $Z(\rho)$. Note that $Z(\rho) = \{\pm I\}$ unless $c(\rho) = I$.

The image of $\mathrm{d}c_\rho \cdot c(\rho)^{-1}$ is $z(\rho)^\perp$; see, for example, [19, proof of Proposition 3.7]. In particular, the differential $\mathrm{d}c_\rho$ is surjective whenever $c(\rho) \neq I$.

Define the maps

$$f_1 : G^{2h} \times \mathfrak{g} \to G, \quad f_1(\rho, \theta) = c(\rho) \cdot \exp(-2\pi\theta)$$

and

$$f_2 : G^{2h} \times \mathfrak{g} \to G \times \mathfrak{g}, \quad f_2(\rho, \theta) = (f_1(\rho, \theta), \theta).$$

On the extended moduli space $\mathscr{M}^{\mathfrak{g}}(\Sigma') = f_1^{-1}(I)$, we have

$$(\mathrm{d}f_1)_{(\rho,\theta)} = (\mathrm{d}c)_\rho \exp(-2\pi\theta) + 2\pi \exp(2\pi\theta)(\mathrm{d}\exp)_{-2\pi\theta}.$$

When $c(\rho) = \exp(2\pi\theta) \neq I$, we have that $(dc)_\rho$ is surjective, hence so is $(\mathrm{d}f_1)_{(\rho,\theta)}$. Also, when $\theta = 0$, $(\mathrm{d}\exp)_{-2\pi\theta}$ is just the identity, so again $(\mathrm{d}f_1)_{(\rho,\theta)}$ is surjective. Claim (a) follows.

Next, observe that

$$(\mathrm{d}f_2)_{(\rho,\theta)}(\alpha,\lambda) = ((\mathrm{d}f_1)_{(\rho,\theta)}(\alpha,\lambda),\lambda) = (\mathrm{d}c_\rho(\alpha)\cdot\exp(-2\pi\theta)+l(\lambda),\lambda),$$

where $l(\lambda)$ does not depend on α. Hence, when $c(\rho) = \exp(2\pi\theta) \neq I$, the differential $(\mathrm{d}f_2)_{(\rho,\theta)}$ is surjective. This implies that any $\theta \in \mathfrak{g}$ with $Q(\theta) \notin \mathbb{Z}$ is a regular value for $\Phi|_{\mathscr{M}_s^{\mathfrak{g}}(\Sigma')}$. Since the values $\theta \in \mathfrak{g}$ with $Q(\theta) \in \mathbb{Z}\setminus\{0\}$ are not in the image of $\Phi|_{\mathscr{M}_s^{\mathfrak{g}}(\Sigma')}$, they are automatically regular values, and claim (b) follows.

Consider also the subspace

$$\mathscr{N}(\Sigma') = \tilde{\Phi}^{-1}\big([0,1/2)\big) \subset \mathscr{M}_s^{\mathfrak{g}}(\Sigma').$$

Note that the restriction of the exponential map $\theta \to \exp(2\pi\theta)$ to $Q^{-1}\big[0,1/2\big)$ is a diffeomorphism onto its image $G\setminus\{-I\}$. Therefore, using the identification (3.1), we can describe $\mathscr{N}(\Sigma')$ as

$$\mathscr{N}(\Sigma') = \left\{ (A_1,B_1,\dots,A_h,B_h) \in G^{2h} \;\middle|\; \prod_{i=1}^{h}[A_i,B_i] \neq -I \right\}. \tag{3.3}$$

3.3 Hamiltonian Actions

Let K be a compact connected Lie group with Lie algebra $\mathfrak{k}$. We let K act on the dual Lie algebra $\mathfrak{k}^*$ by the coadjoint action.

A *presymplectic manifold* is a smooth manifold M together with a closed form $\omega \in \Omega^2(M)$, possibly degenerate. A *Hamiltonian presymplectic K-manifold* (M,ω,Φ) is a presymplectic manifold (M,ω) together with a K-equivariant smooth map $\Phi : M \to \mathfrak{k}^*$ such that for any $\xi \in \mathfrak{g}$, if X_ξ denotes the vector field on M generated by the one-parameter subgroup $\{\exp(-t\xi) \mid t \in \mathbb{R}\} \subset K$, we have

$$d(\langle\Phi,\xi\rangle) = -\iota(X_\xi)\omega.$$

Under these hypotheses, the K-action on M is called Hamiltonian, and Φ is called the *moment map*. The quotient

$$M/\!/K := \Phi^{-1}(0)/K$$

is called the *presymplectic quotient* of M by K. The following result is known as the reduction theorem [32],[36], [20, Theorem 5.1].

Theorem 2. *Let (M,ω,Φ) be a Hamiltonian presymplectic K-manifold. Suppose that the level set $\Phi^{-1}(0)$ is a smooth submanifold on which K acts freely. Let $i : \Phi^{-1}(0) \hookrightarrow M$ be the inclusion and $\pi : \Phi^{-1}(0) \to M/\!/K$ the projection. Then there* *exists a unique closed form ω_{red} on the smooth manifold $M//K$ with the property that $i^*\omega = \pi^*\omega_{\text{red}}$. The reduced form ω_{red} is nondegenerate on $M//K$ if and only if ω is nondegenerate on M at the points of $\Phi^{-1}(0)$.*

Furthermore, if M admits another Hamiltonian K'-action (for some compact Lie group K') that commutes with the K-action, then $(M//K, \omega_{\text{red}})$ has an induced Hamiltonian K'-action.

When the form ω is symplectic, (M, ω, Φ) is called simply a *Hamiltonian K manifold*. In this case we can drop the condition that $\Phi^{-1}(0)$ be smooth from the hypotheses of Theorem 2; indeed, this condition is automatically implied by the assumption that K acts freely on $\Phi^{-1}(0)$.

3.4 A Closed Two-Form on the Extended Moduli Space

According to [26, (2.7)], the tangent space to the smooth stratum $\mathscr{M}_s^{\mathfrak{g}}(\Sigma') \subset \mathscr{M}^{\mathfrak{g}}(\Sigma')$ at some class $[A]$ can be naturally identified with

$$T_{[A]}\mathscr{M}_s^{\mathfrak{g}}(\Sigma') = \frac{\operatorname{Ker}(\mathrm{d}_A : \Omega^{1,\mathfrak{g}}(\Sigma') \to \Omega_c^2(\Sigma') \otimes \mathfrak{g})}{\operatorname{Im}(\mathrm{d}_A : \Omega_c^0(\Sigma') \otimes \mathfrak{g} \to \Omega^{1,\mathfrak{g}}(\Sigma'))}, \tag{3.4}$$

where $\Omega_c^p(\Sigma')$ denotes the space of p-forms compactly supported in the interior of Σ', and $\Omega^{1,\mathfrak{g}}(\Sigma')$ denotes the space of 1-forms A such that $A = \theta \mathrm{d}s$ near $S = \partial\Sigma'$ for some $\theta \in \mathfrak{g}$.

Define a bilinear form ω on $\Omega^{1,\mathfrak{g}}(\Sigma')$ by

$$\omega(a,b) = -\int_{\Sigma'} \langle a \wedge b \rangle,$$

where we combine the usual exterior product on forms with the inner product on $\mathfrak{g}$. Stokes' theorem implies that ω descends to a bilinear form on the tangent space to $\mathscr{M}_s^{\mathfrak{g}}(\Sigma')$ described in (3.4) above. Thus we can think of ω as a two-form on $\mathscr{M}_s^{\mathfrak{g}}(\Sigma')$.

Theorem 3 (Huebschmann–Jeffrey). *The two-form $\omega \in \Omega^2(\mathscr{M}_s^{\mathfrak{g}}(\Sigma'))$ is closed. It is nondegenerate when restricted to $\mathscr{N}(\Sigma') \subset \mathscr{M}_s^{\mathfrak{g}}(\Sigma')$. Moreover, the restriction of the G^{ad}-action* (3.2) *to $\mathscr{M}_s^{\mathfrak{g}}(\Sigma')$ is Hamiltonian with respect to ω. Its moment map is the restriction of Φ to $\mathscr{N}(\Sigma')$, which we henceforth also denote by Φ.*

For the proof, we refer to Jeffrey [26]; see also [34].

Theorem 3 says that $(\mathscr{M}_s^{\mathfrak{g}}(\Sigma'), \omega, \Phi)$ is a Hamiltonian presymplectic G^{ad}-manifold in the sense of Sect. 3.3, and that its subset $(\mathscr{N}(\Sigma'), \omega, \Phi)$ is a (symplectic) Hamiltonian G^{ad}-manifold. The symplectic quotient

$$\mathscr{N}(\Sigma')//G^{\text{ad}} = \Phi^{-1}(0)/G^{\text{ad}} = \mathscr{M}(\Sigma)$$

is the usual moduli space of flat G-connections on Σ, with the symplectic form (on its smooth stratum) being the one constructed by Atiyah and Bott [3]. If Σ is given a complex structure, $\mathscr{M}(\Sigma)$ can also be viewed as the moduli space of semistable bundles of rank two on Σ with trivial determinant, cf. [38].

For an alternative (group-theoretic) description of the form ω on $\mathscr{N}(\Sigma')$, see [27], [23], or [24].

Let us mention two results about the two-form ω. The first is proved in [35].

Theorem 4 (Meinrenken–Woodward). $(\mathscr{N}(\Sigma'),\omega)$ *is a monotone symplectic manifold, with monotonicity constant* $1/4$.

The second result is given in the following lemma.

Lemma 5. *The cohomology class of the symplectic form* $\omega \in \Omega^2(\mathscr{N}(\Sigma'))$ *is integral.*

Proof. The extended moduli space $\mathscr{M}^{\mathfrak{g}}(\Sigma')$ embeds in the moduli space $\mathscr{M}(\Sigma')$ of all flat connections on Σ'. The latter is an infinite-dimensional Banach manifold with a natural symplectic form that restricts to ω on $\mathscr{M}^{\mathfrak{g}}_s(\Sigma')$. Moreover, Donaldson [12] showed that $\mathscr{M}(\Sigma')$ has the structure of a Hamiltonian LG-manifold, where $LG = \mathrm{Map}(S^1, G)$ is the loop group of G.

Recall that a prequantum line bundle E for a symplectic manifold (M,ω) is a Hermitian line bundle equipped with an invariant connection ∇ whose curvature is $-2\pi\mathrm{i}$ times the symplectic form. If M is finite-dimensional, this implies that $[\omega] = c_1(E) \in H^2(M;\mathbb{Z})$. In our situation, a prequantum line bundle on $M = \mathscr{N}(\Sigma')$ can be obtained by restricting the well-known LG-equivariant prequantum line bundle on the infinite-dimensional symplectic manifold $\mathscr{M}(\Sigma')$. We refer the reader to [37], [46], and [60] for the construction of the latter; see also [34].

Corollary 2. *The minimal Chern number of the symplectic manifold* $\mathscr{N}(\Sigma')$ *is a positive multiple of* 4.

Proof. Use Theorem 4 and Lemma 5.

3.5 Other Versions

Although our main interest lies in the extended moduli space $\mathscr{M}^{\mathfrak{g}}(\Sigma')$ and its open subset $\mathscr{N}(\Sigma')$, in order to understand them better, we need to introduce two other moduli spaces. Both of them appeared in [26], where their main properties are spelled out. An alternative viewpoint on them is given in [34, Sect. 3.4.2], where they are interpreted as cross-sections of the full moduli space $\mathscr{M}(\Sigma')$.

The first auxiliary space that we consider is the *toroidal extended moduli space*

$$\mathscr{M}^{\mathfrak{t}}(\Sigma') = \Phi^{-1}(\mathfrak{t}) \subset \mathscr{M}^{\mathfrak{g}}(\Sigma').$$

It has a smooth stratum

$$\mathscr{M}^{\mathfrak{t}}_s(\Sigma') = \{x \in \mathscr{M}^{\mathfrak{t}}(\Sigma') \mid \tilde{\Phi}(x) \notin \mathbb{Z}\}.$$

The restrictions of ω and Φ to $\mathscr{M}^{\mathfrak{t}}_s(\Sigma')$ turn it into a Hamiltonian presymplectic T^{ad}-manifold. On the open subset $\mathscr{M}^{\mathfrak{t}}(\Sigma') \cap \tilde{\Phi}^{-1}(0,1/2)$, the two-form is nondegenerate.

The second space is the *twisted extended moduli space* from [26, Sect. 5.3]. In terms of coordinates, it is

$$\mathscr{M}^{\mathfrak{g}}_{\mathrm{tw}}(\Sigma') = \Big\{(A_1,B_1,\dots,A_h,B_h) \in G^{2h}, \theta \in \mathfrak{g} \Big| \prod_{i=1}^{h}[A_i,B_i] = -\exp(2\pi\theta)\Big\}.$$

This space admits a G^{ad}-action just like that on $\mathscr{M}^{\mathfrak{g}}(\Sigma')$ and also a natural projection $\Phi_{\mathrm{tw}} : \mathscr{M}^{\mathfrak{g}}_{\mathrm{tw}} \to \mathfrak{g}$. Set $\tilde{\Phi}_{\mathrm{tw}} = Q \circ \Phi_{\mathrm{tw}}$. The smooth stratum of $\mathscr{M}^{\mathfrak{g}}_{\mathrm{tw}}(\Sigma')$ is

$$\mathscr{M}^{\mathfrak{g}}_{\mathrm{tw},s}(\Sigma') = \Big\{x \in \mathscr{M}^{\mathfrak{g}}_{\mathrm{tw}}(\Sigma') \Big| \tilde{\Phi}_{\mathrm{tw}}(x) \notin \mathbb{Z} + \frac{1}{2}\Big\}.$$

Furthermore, $\mathscr{M}^{\mathfrak{g}}_{\mathrm{tw},s}(\Sigma')$ admits a natural two-form ω_{tw}, which turns it into a Hamiltonian presymplectic G^{ad}-manifold, with moment map Φ_{tw}. The restriction of ω_{tw} to the subspace

$$\mathscr{N}_{\mathrm{tw}}(\Sigma') = \tilde{\Phi}^{-1}_{\mathrm{tw}}\big([0,1/2)\big)$$

is nondegenerate.

Observe that the subspace $\Phi^{-1}_{\mathrm{tw}}(\mathfrak{t}) \subset \mathscr{M}^{\mathfrak{g}}_{\mathrm{tw}}(\Sigma')$ can be identified with the toroidal extended moduli space $\mathscr{M}^{\mathfrak{t}}(\Sigma')$, via the map

$$(A_1,B_1,\dots,A_h,B_h,t) \to (A_1,B_1,\dots,A_h,B_h,1/2-t).$$

This map is a diffeomorphism of the smooth strata and is compatible with the restrictions of the presymplectic forms ω and ω_{tw}.

3.6 The Structure of Degeneracies of $\mathscr{M}^{\mathfrak{g}}_s(\Sigma')$

Recall from Theorem 3 that the degeneracy locus of the presymplectic manifold $\mathscr{M}^{\mathfrak{g}}_s(\Sigma')$ is contained in the preimage $\tilde{\Phi}^{-1}(1/2)$. We seek to understand the structure of the degeneracies.

Let $\mu = \mathrm{diag}(\mathrm{i}/2, -\mathrm{i}/2)$. Note that the stabilizer G^{ad} of $\exp(2\pi\mu) = -I$ is bigger than the stabilizer $T^{\mathrm{ad}} = S^1$ of μ. Thus, we have an obvious diffeomorphism

$$\tilde{\Phi}^{-1}(1/2) \cong \mathcal{O}_\mu \times \Phi^{-1}(\mu),$$

where $\mathcal{O}_\mu$ denotes the coadjoint orbit of μ. The first factor $\mathcal{O}_\mu$ is diffeomorphic to the flag variety $G^{\mathrm{ad}}/T^{\mathrm{ad}} \cong \mathbb{P}^1$. The second factor $\Phi^{-1}(\mu)$ is smooth by Proposition 2(b).

There is a residual T^{ad}-action on the space $\Phi^{-1}(\mu)$. Thus $\Phi^{-1}(\mu)$ is an S^1-bundle over

$$\mathscr{M}_\mu(\Sigma') = \Phi^{-1}(\mu)/T^{\mathrm{ad}}.$$

Finally, $\mathscr{M}_\mu(\Sigma')$ is a $\mathbb{P}^1$-bundle over

$$\mathscr{M}_{-I}(\Sigma') = \Big\{(A_1,B_1,\dots,A_h,B_h) \in G^{2h} \Big| \prod_{i=1}^{h}[A_i,B_i] = -I\Big\}/G^{\mathrm{ad}}.$$

This last space $\mathscr{M}_{-I}(\Sigma')$ can be identified with the moduli space $\mathscr{M}_{\mathrm{tw}}(\Sigma)$ of projectively flat connections on $\mathfrak{E}$ with fixed central curvature, where $\mathfrak{E}$ is a U(2)-bundle of odd degree over the closed surface $\Sigma = \Sigma' \cup D^2$. Alternatively, it is the moduli space of rank-two stable bundles on Σ having fixed determinant of odd degree; cf. [3, 38]. It can also be viewed as the symplectic quotient of the twisted extended moduli space from Sect. 3.5:

$$\mathscr{M}_{\mathrm{tw}}(\Sigma) = \mathscr{N}_{\mathrm{tw}}(\Sigma')//G^{\mathrm{ad}} = \Phi_{\mathrm{tw}}^{-1}(0)/G^{\mathrm{ad}}.$$

We have described a string of fibrations that gives a clue to the structure of the space $\tilde{\Phi}^{-1}(1/2)$. Let us now reshuffle these fibrations and view $\tilde{\Phi}^{-1}(1/2)$ as a G^{ad}-bundle over the space $\mathcal{O}_\mu \times \mathscr{M}_{-I}(\Sigma')$. Its fiberwise tangent space (at any point) is $\mathfrak{g}$, which can be decomposed as $\mathfrak{t} \oplus \mathfrak{t}^\perp$, with $\mathfrak{t}^\perp \cong \mathbb{C}$.

Proposition 3. *Let $x \in \tilde{\Phi}^{-1}(1/2) \subset \mathscr{M}_s^{\mathfrak{g}}(\Sigma')$. The null space of the form ω at x consists of the fiber directions corresponding to $\mathfrak{t}^\perp \subset \mathfrak{g}$.*

Proof. Our strategy for proving Proposition 3 is to reduce it to a similar statement for the toroidal extended moduli space $\mathscr{M}^{\mathfrak{t}}(\Sigma')$, and then study the latter via its embedding into the twisted extended moduli space $\mathscr{M}^{\mathfrak{g}}_{\mathrm{tw}}(\Sigma')$.

First, note that by G^{ad}-invariance, we can assume without loss of generality that $\Phi(x) = \mu$. The symplectic cross-section theorem [21] says that near $\Phi^{-1}(\mu)$, the two-form on $\mathscr{M}^{\mathfrak{g}}(\Sigma')$ is obtained from the one on $\mathscr{M}^{\mathfrak{t}}(\Sigma') = \Phi^{-1}(\mathfrak{t})$ by a procedure called symplectic induction. (Strictly speaking, symplectic induction is described in [21] for nondegenerate forms; however, it applies to the Hamiltonian presymplectic case as well.) More concretely, we have a (noncanonical) decomposition

$$T_x\mathscr{M}^{\mathfrak{g}}(\Sigma') = T_x\mathscr{M}^{\mathfrak{t}}(\Sigma') \oplus T_\mu(\mathcal{O}_\mu) \tag{3.5}$$

such that $\omega|_x$ is the direct sum of its restriction to the first summand in (3.5) with the canonical symplectic form on the second summand.

Recall that we are viewing $\tilde{\Phi}^{-1}(1/2)$ as a G-bundle over $\mathcal{O}_\mu \times \mathscr{M}_{-I}(\Sigma')$. Its intersection with $\mathscr{M}^{\mathfrak{t}}(\Sigma')$ is $\Phi^{-1}(\mu)$, which is the part of the G^{ad}-bundle that lies over $\{\mu\} \times \mathscr{M}_{-I}(\Sigma')$. The decomposition (3.5) implies that in order to prove the final claim about the null space of $\omega|_x$, it suffices to show that the null space of $\omega|_{\mathscr{M}^{\mathfrak{t}}(\Sigma')}$ at x consists of the fiber directions corresponding to $\mathfrak{t}^{\perp} \subset \mathfrak{g}$.

Let us use the observation in the last paragraph of Sect. 3.5, and view $\mathscr{M}^{\mathfrak{t}}(\Sigma')$ as $\Phi_{\text{tw}}^{-1}(\mathfrak{t}) \subset \mathscr{M}^{\mathfrak{g}}_{\text{tw}}(\Sigma')$. The point x now lies in $\Phi_{\text{tw}}^{-1}(0)$.

Recall from Sect. 3.5 that the two-form $\mathscr{M}^{\mathfrak{g}}_{\text{tw}}(\Sigma')$ is nondegenerate near $\Phi_{\text{tw}}^{-1}(0)$. Further, it is easy to check that the action of G^{ad} on $\Phi_{\text{tw}}^{-1}(0)$ is free. This action is Hamiltonian; hence, the quotient $\Phi_{\text{tw}}^{-1}(0)/G^{\text{ad}} = \mathscr{M}_{-I}(\Sigma')$ is smooth, and the reduced two-form on it is nondegenerate. Further, there is a (noncanonical) decomposition

$$T_x\mathscr{M}^{\mathfrak{g}}_{\text{tw}}(\Sigma') \cong \pi^* T_{\pi(x)}\mathscr{M}_{-I}(\Sigma') \oplus \mathfrak{g} \oplus \mathfrak{g}^*, \tag{3.6}$$

where $\pi : \Phi_{\text{tw}}^{-1}(0) \to \mathscr{M}_{-I}(\Sigma')$ is the quotient map. (See, for example, [20, (5.6)].) The two-form ω_{tw} at x is the direct summand of the reduced form at $\pi(x)$ and the natural pairing of the two last factors in (3.6).

With respect to the decomposition (3.6), the subspace $T_x\mathscr{M}^{\mathfrak{t}}(\Sigma') \subset T_x\mathscr{M}^{\mathfrak{g}}_{\text{tw}}(\Sigma')$ corresponds to

$$T_x\mathscr{M}^{\mathfrak{t}}(\Sigma') \cong \pi^* T_{\pi(x)}\mathscr{M}_{-I}(\Sigma') \oplus \mathfrak{g} \oplus \mathfrak{t}^*.$$

Therefore, the null space of ω_{tw} on $T_x\mathscr{M}^{\mathfrak{t}}(\Sigma')$ is the null space of the restriction of the natural pairing on $\mathfrak{g} \oplus \mathfrak{g}^*$ to $\mathfrak{g} \oplus \mathfrak{t}^*$. This is $\mathfrak{g}/\mathfrak{t} \cong \mathfrak{t}^{\perp}$, as claimed.

4 Symplectic Cutting

4.1 Abelian Symplectic Cutting

We review here Lerman's definition of (abelian) symplectic cutting, following [31]. Consider a symplectic manifold (M, ω) with a Hamiltonian S^1-action and moment map $\Phi : M \to \mathbb{R}$. Pick some $\lambda \in \mathbb{R}$. The diagonal S^1-action on the space $M \times \mathbb{C}^-$ (endowed with the standard product symplectic structure, where $\mathbb{C}^-$ is $\mathbb{C}$ with negative the usual area form) is Hamiltonian with respect to the moment map

$$\Psi : M \times \mathbb{C}^- \to \mathbb{R}, \quad \Psi(m,z) = \Phi(m) + \frac{1}{2}|z|^2 - \lambda.$$

The symplectic quotient

$$M_{\leq\lambda} := \Psi^{-1}(0)/S^1 \cong \Phi^{-1}(\lambda)/S^1 \cup \Phi^{-1}(-\infty, \lambda)$$

is called the *symplectic cut* of M at λ. If the action of S^1 on $\Phi^{-1}(\lambda)$ is free, then $M_{\leq\lambda}$ is a symplectic manifold, and it contains $\Phi^{-1}(\lambda)/S^1$ (with its reduced form) as a symplectic hypersurface, i.e., a symplectic submanifold of real codimension two.

Remark 4. The normal bundle to $\Phi^{-1}(\lambda)/S^1$ in $M_{\leq\lambda}$ is the complex line bundle whose associated circle bundle is $\Phi^{-1}(\lambda) \to \Phi^{-1}(\lambda)/S^1$.

Remark 5. Symplectic cutting is a local construction. In particular, if (M,ω) is symplectic and $\Phi : M \to \mathbb{R}$ is a continuous map that induces a smooth Hamiltonian S^1-action on an open set $\mathcal{U} \subset M$ containing $\Phi^{-1}(\lambda)$, then we can still define $M_{\leq\lambda}$ as the union $(M \setminus \mathcal{U}) \cup \mathcal{U}_{\leq\lambda}$.

Remark 6. If M has an additional Hamiltonian K-action (for some other compact group K) commuting with that of S^1, then $M_{\leq\lambda}$ has an induced Hamiltonian K-action. This follows from a similar statement for symplectic reduction; cf. Theorem 2.

4.2 Nonabelian Symplectic Cutting

An analogue of symplectic cutting for nonabelian Hamiltonian actions was defined in [61]. We explain here the case of Hamiltonian PU(2)-actions, since this is all we need for our purposes.

We keep the notation from Sect. 3.1, with $G = \mathrm{SU}(2)$ and $G^{\mathrm{ad}} = \mathrm{PU}(2)$. Let (M,ω,Φ) be a Hamiltonian G^{ad}-manifold. Since $\mathfrak{g}$ and $\mathfrak{g}^*$ are identified using the bilinear form, from now on we will view the moment map Φ as taking values in $\mathfrak{g}$. Recall that

$$Q : \mathfrak{g} \to \mathfrak{g}/G^{\mathrm{ad}} \cong [0,\infty)$$

denotes the adjoint quotient map. The map Q is continuous and is smooth outside $Q^{-1}(0)$. Set

$$\tilde{\Phi} = Q \circ \Phi.$$

On the complement $\mathcal{U} = \mathcal{U}$ of $\Phi^{-1}(0)$ in M, the map $\tilde{\Phi}$ induces a Hamiltonian S^1-action. Explicitly, $u \in S^1 = \mathbb{R}/2\pi\mathbb{Z}$ acts on $m \in \mathcal{U}$ by

$$m \to \exp\left(u \cdot \frac{\Phi(m)}{2\tilde{\Phi}(m)}\right) \cdot m. \tag{4.1}$$

This action is well defined because $\exp(\pi H) = I$ in G^{ad}. We can describe it alternatively as follows: on $\Phi^{-1}(\mathfrak{t}) \subset M$, it coincides with the action of $T^{\mathrm{ad}} \subset G^{\mathrm{ad}}$; then it is extended to all of M in a G^{ad}-equivariant manner.

Fix $\lambda > 0$. Using the local version (from Remark 5) of abelian symplectic cutting for the action (4.1), we define the *nonabelian symplectic cut of M at λ* to be

$$M_{\leq\lambda} = \Phi^{-1}(0) \cup \mathcal{U}_{\leq\lambda} = M_{<\lambda} \cup R,$$

where

$$M_{<\lambda} = \Phi_1^{-1}([0,\lambda)), \;\; R_\lambda = \tilde{\Phi}^{-1}(\lambda)/S^1.$$

If S^1 acts freely on $\tilde{\Phi}^{-1}(\lambda)$, then $M_{\leq\lambda}$ is a smooth manifold. It can be naturally equipped with a symplectic form $\omega_{\leq\lambda}$, coming from the symplectic form ω on M. In fact, $M_{\leq\lambda}$ is a Hamiltonian G^{ad}-manifold; cf. Remark 6. With respect to the form $\omega_{\leq\lambda}$, R is a symplectic hypersurface in $M_{\leq\lambda}$.

4.3 Monotonicity

We aim to find a condition that guarantees that a nonabelian symplectic cut is monotone. As a toy model for our future results, we start with a general fact about symplectic reduction:

Lemma 6. *Let K be a Lie group with $H^2(K;\mathbb{R}) = 0$, and let (M,ω,Φ) be a Hamiltonian K-manifold that is monotone, with monotonicity constant κ. Assume that the moment map Φ is proper, and the K-action on $\Phi^{-1}(0)$ is free. Then the symplectic quotient $M//K = \Phi^{-1}(0)/K$ (with the reduced symplectic form ω^{red}) is also monotone, with the same monotonicity constant κ.*

Proof. Consider the Kirwan map from [28]:

$$H^2_K(M;\mathbb{R}) \to H^2(M//K;\mathbb{R}),$$

which is obtained by composing the map $H^2_K(M;\mathbb{R}) \to H^2_K(\Phi^{-1}(0);\mathbb{R})$ (induced by the inclusion) with the Cartan isomorphism $H^2_K(\Phi^{-1}(0);\mathbb{R}) \cong H^2(M//K;\mathbb{R})$. The Kirwan map takes the first equivariant Chern class $c_1^K(TM)$ to $c_1(T(M//K))$, and the equivariant two-form $\tilde{\omega} = \omega - \Phi$ to ω^{red}. Since $H^2_K(M;\mathbb{R}) \cong H^2(M;\mathbb{R})$, with c_1^K corresponding to c_1 and $[\tilde{\omega}]$ to $[\omega]$), the conclusion follows.

Let us now specialize to the case $K = G^{\mathrm{ad}} = \mathrm{PU}(2)$. For $\lambda \in (0,\infty)$, let $\mathcal{O}_\lambda \cong \mathbb{P}^1$ be the coadjoint orbit of $\mathrm{diag}(\mathrm{i}\lambda, -\mathrm{i}\lambda)$, endowed with the Kostant–Kirillov–Souriau form $\omega_{KKS}(\lambda)$. It has a Hamiltonian G^{ad}-action with moment map the inclusion $\iota : \mathcal{O}_\lambda \to \mathfrak{g}$. Let $\gamma =$ P.D.(pt) denote the generator of $H^2(\mathcal{O}_\lambda;\mathbb{Z}) \subset H^2(\mathcal{O}_\lambda;\mathbb{R})$, so that $c_1(\mathcal{O}_\lambda) = 2\gamma$. Then $c_1(\mathcal{O}_\lambda) = [\omega_{KKS}(1)]$ [6, Sects. 7.5 and 7.6], and so $[\omega_{KKS}(\lambda)] = 2\lambda\gamma$.

If (M,ω,Φ) is a Hamiltonian G^{ad}-manifold, let $M \times \mathcal{O}_\lambda{}^-$ denote the Hamiltonian manifold $(M \times \mathcal{O}_\lambda, \omega \times -\omega_{KKS}(\lambda), \Phi - \iota)$. The *reduction of M with respect to $\mathcal{O}_\lambda$* is defined as

$$M_\lambda = (M \times \mathcal{O}_\lambda{}^-)//G^{\mathrm{ad}} = \Phi^{-1}(\mathcal{O}_\lambda)/G^{\mathrm{ad}}.$$

If the G^{ad}-action on $\Phi^{-1}(\mathcal{O}_\lambda)$ is free, the quotient M_λ is smooth and admits a natural symplectic form ω_λ. It can be viewed as $\Phi^{-1}(\mathrm{diag}(\mathrm{i}\lambda, -\mathrm{i}\lambda))/T^{\mathrm{ad}}$; we let E_λ denote the complex line bundle on M_λ associated with the respective T^{ad}-fibration.

Lemma 7. *Let (M,ω,Φ) be a Hamiltonian G^{ad}-manifold such that the moment map Φ is proper and the action of G^{ad} is free outside $\Phi^{-1}(0)$. Assume that M*

is monotone, with monotonicity constant κ. Then the cohomology class of the reduced form ω_λ is given by the formula

$$[\omega_\lambda] = \kappa \cdot c_1(TM_\lambda) + (\lambda - \kappa) \cdot c_1(E_\lambda).$$

Proof. First, note that for any Hamiltonian G^{ad}-manifold M, we have $H^2_{G^{\mathrm{ad}}}(M;\mathbb{R}) \cong H^2(M;\mathbb{R})$, because $H^i(BG^{\mathrm{ad}};\mathbb{R}) = 0$ for $i = 1,2$. Thus the Kirwan map can viewed as going from $H^2(M;\mathbb{R})$ into $H^2(M//G^{\mathrm{ad}};\mathbb{R})$.

Let us consider the Kirwan map for the manifold $M \times \mathcal{O}_\lambda{}^-$, whose symplectic reduction is M_λ. By abuse of notation, we denote classes in $H^2(M)$ or $H^2(\mathcal{O}_\lambda{}^-)$ in the same way as their pullbacks to $H^2(M \times \mathcal{O}_\lambda{}^-)$.

Just as in the proof of Lemma 6, we get that the Kirwan map takes $[\omega] - [\omega_{KKS}(\lambda)] = \kappa c_1(TM) - 2\lambda\gamma$ to the reduced form $[\omega_\lambda]$, and $c_1(TM) - c_1(T\mathcal{O}_\lambda) = c_1(TM) - 2\gamma$ to the reduced Chern class $c_1(TM_\lambda)$. Note also that the image of $c_1(T\mathcal{O}_\lambda{}^-) = -2\gamma$ under the Kirwan map is $c_1(E_\lambda)$. Hence

$$[\omega_\lambda] - \kappa \cdot c_1(TM_\lambda) = (\lambda - \kappa) \cdot c_1(E_\lambda),$$

as desired.

We are now ready to study monotonicity for nonabelian cuts.

Proposition 4. *Let $G^{\mathrm{ad}} = \mathrm{PU}(2)$, and let (M, ω, Φ) be a Hamiltonian G^{ad}-manifold that is monotone with monotonicity constant $\kappa > 0$. Assume that the moment map Φ is proper, and that G^{ad} acts freely outside $\Phi^{-1}(0)$. Then the symplectic cut $M_{\leq\lambda}$ at the value $\lambda = 2\kappa \in (0,\infty)$ is also monotone, with the same monotonicity constant κ.*

Proof. Recall that the symplectic cut $M_{\leq\lambda}$ is the union of the open piece $M_{<\lambda}$ and the hypersurface $R_\lambda = \Phi^{-1}(\mathcal{O}_\lambda)/S^1$. Note that there is a natural symplectomorphism

$$R_\lambda \xrightarrow{\cong} \mathcal{O}_\lambda \times M_\lambda, \quad m \mapsto \big(\Phi(m), [m]\big). \tag{4.2}$$

The inverse to this symplectomorphism is given by the map $([g],[m]) \mapsto [gm]$.

By Remark 4, the normal bundle to R_λ is the line bundle associated with the defining T^{ad}-bundle on R_λ. We denote this T^{ad}-bundle by N_λ; it is the product of $G^{\mathrm{ad}} \to G^{\mathrm{ad}}/T^{\mathrm{ad}} \cong \mathcal{O}_\lambda$ on the $\mathcal{O}_\lambda$ factor and the circle bundle of E_λ on the M_λ factor.

Let $\nu(R_\lambda)$ be a regular neighborhood of R_λ, so that the intersection $M_{<\lambda} \cap \nu(R_\lambda)$ admits a deformation retract into a copy of N_λ.

We have a Mayer–Vietoris sequence

$$\ldots \to H^1(M_{<\lambda}) \oplus H^1(\nu(R_\lambda)) \to H^1(N_\lambda) \to H^2(M_{\leq\lambda}) \to H^2(M_{<\lambda}) \oplus H^2(\nu(R_\lambda)) \to \ldots.$$

Note that the first Chern class of the bundle $N_\lambda \to R_\lambda$ is nontorsion in $H^2(R_\lambda)$, because it is so on the $\mathcal{O}_\lambda$ factor. Hence, the map $H^1(\nu(R_\lambda);\mathbb{R}) \to H^1(N_\lambda;\mathbb{R})$ is onto. The Mayer–Vietoris sequence then tells us that the map

$$H^2(M_{\leq\lambda};\mathbb{R}) \to H^2(M_{<\lambda};\mathbb{R}) \oplus H^2(\nu(R_\lambda);\mathbb{R})$$

is injective. Therefore, in order to check the monotonicity of $M_{\leq\lambda}$, it suffices to check it on $M_{<\lambda}$ and $\nu(R_\lambda)$.

Since $M_{<\lambda}$ is symplectomorphic to a subset of M, monotonicity is satisfied there by assumption. Let us check it on $\nu(R_\lambda)$, or equivalently, on its deformation retract R_λ. We will use the symplectomorphism (4.2), and by abuse of notation, we will denote the objects on $\mathcal{O}_\lambda$ or M_λ in the same way as we denote their pullbacks to R_λ. Let γ be the generator of $H^2(\mathcal{O}_\lambda;\mathbb{Z})$ as in the proof of Lemma 7. By the result of that lemma, we have

$$[\omega_{\leq\lambda}|_{R_\lambda}] = 2\lambda\gamma + \kappa c_1(TM_\lambda) + (\lambda-\kappa)c_1(E_\lambda). \tag{4.3}$$

On the other hand, the tangent space to $M_{\leq\lambda}$ at a point of R_λ decomposes into the tangent and normal bundles to R_λ. Therefore,

$$c_1(TM_{\leq\lambda}|_{R_\lambda}) = c_1(TR_\lambda) + 2\gamma + c_1(E_\lambda) = 4\gamma + c_1(TM_\lambda) + c_1(E_\lambda).$$

Taking into account (4.3), for $\lambda = 2\kappa$ we conclude that $[\omega_{\leq\lambda}|_{R_\lambda}] = \kappa \cdot c_1(TM_{\leq\lambda}|_{R_\lambda})$.

4.4 Extensions to Presymplectic Manifolds

Abelian cutting and nonabelian cutting are simply particular instances of symplectic reduction. Since the latter can be extended to the presymplectic setting, one can also define abelian and nonabelian cutting for Hamiltonian presymplectic manifolds.

In general, one cannot define $c_1(TM)$ (and the notion of monotonicity) for presymplectic manifolds, because there is no good notion of compatible almost complex structure. In order to fix that, we introduce the following definition.

Definition 4. An *ϵ-symplectic manifold* $(M,\{\omega_t\})$ is a smooth manifold M together with a smooth family of closed two-forms $\omega_t \in \Omega^2(M)$, $t \in [0,\epsilon]$, for some $\epsilon > 0$, such that ω_t is symplectic for all $t \in (0,\epsilon]$.

One should think of an ϵ-symplectic manifold $(M,\{\omega_t\})$ as the presymplectic manifold (M,ω_0) together with some additional data given by the other ω_t's. In particular, by the *degeneracy locus of* $(M,\{\omega_t\})$ we mean the degeneracy locus of ω_0, i.e.,

$$R(\omega_0) = \{m \in M \mid \omega_0 \text{ is degenerate on } T_mM\}.$$

If $(M,\{\omega_t\})$ is any ϵ-symplectic manifold, we can define its first Chern class $c_1(TM) \in H^2(M;\mathbb{Z})$ by giving TM an almost complex structure compatible with some ω_t for $t > 0$. (Note that the resulting $c_1(TM)$ does not depend on t.) Thus, we can define the minimal Chern number of an ϵ-symplectic manifold just as we did for symplectic manifolds. Moreover, we can talk about monotonicity:

Definition 5. The ϵ-symplectic manifold $(M,\{\omega_t\})$ is called *monotone* (with monotonicity constant $\kappa > 0$) if

$$[\omega_0] = \kappa \cdot c_1(TM).$$

One source of ϵ-symplectic manifolds is symplectic reduction. Indeed, suppose we have a Hamiltonian presymplectic S^1-manifold (M,ω,Φ) with the moment map $\Phi : M \to \mathbb{R}$ proper. The form ω may have some degeneracies on $\Phi^{-1}(0)$; however, we assume that it is nondegenerate on $\Phi^{-1}((0,\epsilon])$ for some $\epsilon > 0$. Assume also that S^1 acts freely on $\Phi^{-1}([0,\epsilon])$ (hence any $t \in (0,\epsilon]$ is a regular value for Φ), and further, 0 is a regular value for Φ as well. Then the presymplectic quotients $M_t = \Phi^{-1}(t)/S^1$ for $t \in [0,\epsilon]$ form a smooth fibration over the interval $[0,\epsilon]$. By choosing a connection for this fiber bundle, we can find a smooth family of diffeomorphisms $\phi_t : M_0 \to M_t$, $t \in [0,\epsilon]$, with $\phi_0 = \mathrm{id}_{M_0}$. We can then put a structure of an ϵ-symplectic manifold on M_0 by using the forms $\phi_t^* \omega_t$, $t \in [0,\epsilon]$, where ω_t is the reduced form on M_t. Note that the space of choices involved in this construction (i.e., connections) is contractible. Therefore, whether $(M_0, \phi_t^* \omega_t)$ is monotone is independent of these choices.

Since abelian cutting and nonabelian cutting are instances of (pre)symplectic reduction, one can also turn presymplectic cuts into ϵ-symplectic manifolds in an essentially canonical way, provided that the form is nondegenerate on the nearby cuts. (By "nearby" we implicitly assume that we have chosen a preferred side for approximating the cut value: either from above or from below.) In this context, we have the following analogue of Proposition 4:

Proposition 5. *Let $G^{\mathrm{ad}} = \mathrm{PU}(2)$, and let (M,ω,Φ) be a Hamiltonian presymplectic G^{ad}-manifold. Set $\tilde{\Phi} = Q \circ \Phi : M \to [0,\infty)$ as usual. Assume that the following hold:*

- *The moment map Φ is proper.*
- *The form ω is nondegenerate on the open subset $M_{<\lambda} = \tilde{\Phi}^{-1}([0,\lambda))$, for some value $\lambda \in (0,\infty)$.*
- *G^{ad} acts freely on $\tilde{\Phi}^{-1}((0,\lambda])$ (hence any $t \in (0,\lambda)$ is a regular value for $\tilde{\Phi}$).*
- *λ is also a regular value for $\tilde{\Phi}$.*
- *As a symplectic manifold, $M_{<\lambda}$ is monotone, with monotonicity constant $\kappa = \lambda/2$.*

Fix some $\epsilon \in (0,\lambda)$ and view the presymplectic cut $M_{\le\lambda}$ as an ϵ-symplectic manifold with respect to forms $\phi_t^* \omega_{\le\lambda-t}$ for a smooth family of diffeomorphisms $\phi_t : M_{\le\lambda} \to M_{\le\lambda-t}$, $t \in [0,\epsilon]$, $\phi_0 = id$.

Then, $M_{\le\lambda}$ is monotone, with the same monotonicity constant $\kappa = \lambda/2$.

Proof. We can run the same arguments as in the proof of Proposition 4, as long as we apply them to the Hamiltonian manifold $M_{<\lambda}$, where ω is nondegenerate. This gives us the corresponding formulas for the cohomology classes $[\omega_{\le\lambda-t}]$ and $c_1(TM_{\le\lambda-t})$, for $t \in (0,\epsilon)$. In the limit $t \to 0$, we get monotonicity.

4.5 Cutting the Extended Moduli Space

Recall from Sect. 3.4 that the smooth part $\mathscr{M}_s^{\mathfrak{g}}(\Sigma')$ of the extended moduli space is a Hamiltonian presymplectic G^{ad}-manifold. Let us consider its nonabelian cut at the value $\lambda = 1/2$:

$$\mathscr{N}^c(\Sigma') = \mathscr{M}_s^{\mathfrak{g}}(\Sigma')_{\leq 1/2}.$$

The notation $\mathscr{N}^c(\Sigma')$ indicates that this space is a compactification of $\mathscr{N}(\Sigma') = \mathscr{M}_s^{\mathfrak{g}}(\Sigma')_{<1/2}$. Indeed, we have

$$\mathscr{N}^c(\Sigma') = \mathscr{N}(\Sigma') \cup R,$$

where

$$R \cong \left\{ (A_1, B_1, \dots, A_h, B_h, \theta) \in G^{2h} \times \mathfrak{g} \,\middle|\, \prod_{i=1}^{h} [A_i, B_i] = \exp(2\pi\theta) = -1 \right\} / S^1. \quad (4.4)$$

Here $u \in S^1 = \mathbb{R}/2\pi\mathbb{Z}$ acts by conjugating each A_i and B_i by $\exp(u\theta)$ and preserving θ.

The G^{ad}-action on $\tilde{\Phi}^{-1}\big((0,1/2]\big) \subset \mathscr{M}_s^{\mathfrak{g}}(\Sigma')$ is free. Since ω is nondegenerate on $\tilde{\Phi}^{-1}\big((0,1/2]\big)$ by Theorem 3, this implies that any $\theta \in \mathfrak{g}$ with $Q(\theta) \in (0,1/2]$ is a regular value for Φ. The last statement also follows from Proposition 2(b), which further says that the values $\theta \in \mathfrak{g}$ with $Q(\theta) = 1/2$ are also regular. Hence, any $t \in (0,1/2]$ is a regular value for $\tilde{\Phi}$. Lastly, note that Theorem 4 says that $\tilde{\Phi}^{-1}\big([0,1/2)\big)$ is monotone, with monotonicity constant $\kappa = 1/4 = \lambda/2$. We conclude that the hypotheses of Proposition 5 are satisfied.

Proposition 6. *Fix $\epsilon \in (0,1/2)$. Endow $\mathscr{N}^c(\Sigma')$ with the structure of an ϵ-symplectic manifold, using the forms $\phi_t^* \omega_{\leq 1/2-t}$, coming from a smooth family of diffeomorphisms*

$$\phi_t : \mathscr{N}^c(\Sigma') = M_{\leq 1/2} \to M_{\leq 1/2-t}, \quad t \in [0,\epsilon],\ \phi_0 = id.$$

Then $\mathscr{N}^c(\Sigma')$ is monotone with monotonicity constant $1/4$.

Thus, we have succeeded in compactifying the symplectic manifold $\mathscr{N}(\Sigma')$ while preserving monotonicity. The downside is that $\mathscr{N}^c(\Sigma')$ is only presymplectic. The resulting two-form has degeneracies on R.

Lemma 8. *Let us view $R = \tilde{\Phi}^{-1}(1/2)/S^1$ as a $\mathbb{P}^1$-bundle over the space $\mathcal{O}_\mu \times \mathscr{M}_{-I}(\Sigma')$; cf. Sect. 3.6. Then the null space of the form $\omega_{\leq 1/2}$ at $x \in R$ consists of the fiber directions. Furthermore, the intersection number (inside $\mathscr{N}^c(\Sigma)$) of R with any $\mathbb{P}^1$ fiber of R is -2.*

Proof. The first claim follows from Proposition 3. The second holds because the normal bundle is the associated bundle to $\mathfrak{t}^{\perp}$, which is a weight space with weight -2.

In a family of forms that make $\mathcal{N}^c(\Sigma')$ into an ϵ-symplectic manifold (as in Proposition 6), the degenerate form $\omega_{\leq 1/2}$ always corresponds to $t = 0$. Hence from now on, we will denote it by ω_0.

Proposition 7. *In addition to the degenerate form ω_0 coming from the cut, the space $\mathcal{N}^c(\Sigma') = \mathcal{N}(\Sigma') \cup R$ also admits a symplectic form ω_ϵ with the following properties:*

(i) R is a symplectic hypersurface with respect to ω_ϵ.
(ii) The restrictions of ω_0 and ω_ϵ to $\mathcal{N}(\Sigma')$ have the same cohomology class in $H^2(\mathcal{N}(\Sigma');\mathbb{R})$.
(iii) The forms ω_0 and ω_ϵ themselves coincide on the open subset $\mathcal{W} = \tilde{\Phi}^{-1}\big([0,1/4)\big) \subset \mathcal{N}(\Sigma')$.
(iv) There exists an almost complex structure $\tilde{J}$ on $\mathcal{N}^c(\Sigma')$ that preserves R, is compatible with respect to ω_ϵ on $\mathcal{N}^c(\Sigma')$, and compatible with respect to ω_0 on $\mathcal{N}(\Sigma')$, and for which any $\tilde{J}$-holomorphic sphere of index zero has intersection number with R a negative multiple of two.

Proof. As the name suggests, the form ω_ϵ will be part of a family $(\omega_t), t \in [0,\epsilon]$ of the type used to turn $\mathcal{N}^c(\Sigma')$ into an ϵ-symplectic manifold. In fact, it is easy to find such a form that satisfies conditions (i)–(iii) above. One needs to choose $\epsilon < 1/4$ and a smooth family of diffeomorphisms $\phi_t : \mathcal{N}^c(\Sigma') = M_{\leq 1/2} \to M_{\leq 1/2-t}$, $t \in [0,\epsilon]$, $\phi_0 = id$, such that $\phi_t = id$ on $\mathcal{W}$ and ϕ_t takes R to $R_{1/2-t} = \tilde{\Phi}^{-1}(1/2-t)/S^1$. Then set $\omega_\epsilon = \phi_\epsilon^* \omega_0$. Note that condition (ii) is automatic from (iii), because $\mathcal{W}$ is a deformation retract of $\mathcal{N}(\Sigma')$.

However, in order to ensure that condition (iv) is satisfied, more care is needed in choosing the diffeomorphisms above. We will construct only $\phi = \phi_\epsilon$, since this is all we need for our purposes; however, it will be easy to see that one could interpolate between ϕ and the identity.

The strategy for constructing ϕ and $\tilde{J}$ is the same as in the proofs of Proposition 3 and Lemma 8: we construct a diffeomorphism and an almost complex structure on the toroidal extended moduli space $\mathcal{M}^{\mathfrak{t}}(\Sigma')$, by looking at it as a subset of the twisted extended moduli space $\mathcal{M}^{\mathfrak{g}}_{\mathrm{tw}}(\Sigma')$; then we lift them to $\mathcal{M}^{\mathfrak{g}}(\Sigma')$; finally, we show how they descend to the cut.

Let $\mu = \mathrm{diag}(\mathrm{i}/2, -\mathrm{i}/2)$ as in Sect. 3.6. We start by carefully examining the restriction of the form ω to $\mathcal{M}^{\mathfrak{t}}(\Sigma')$, in a neighborhood of $\Phi^{-1}(\mu)$. By the remark at the end of Sect. 3.5, this is the same as looking at the restriction of ω_{tw} to $\Phi_{\mathrm{tw}}^{-1}(\mathfrak{t}^*)$ in a neighborhood of $\Phi_{\mathrm{tw}}^{-1}(0)$.

The zero set Z of the moment map Φ_{tw} on (the smooth, symplectic part of) $\mathcal{M}^{\mathfrak{g}}_{\mathrm{tw}}(\Sigma')$ is a coisotropic submanifold. Let $\omega_{\mathrm{tw},0}$ be the reduced form on $Z/G^{\mathrm{ad}} = \mathcal{M}_{-I}(\Sigma')$. Pick a connection form $\alpha \in \Omega^1(Z) \otimes \mathfrak{g}$ for the G^{ad}-action on Z. By the equivariant coisotropic embedding theorem [21, Proposition 39.2], we can find a G^{ad}-equivariant diffeomorphism between a neighborhood of $Z = \Phi_{\mathrm{tw}}^{-1}(0)$ in $\mathcal{M}^{\mathfrak{g}}_{\mathrm{tw}}(\Sigma')$ and a neighborhood of $Z \times \{0\}$ in $Z \times \mathfrak{g}^*$ such that the form ω_{tw} looks like

$$\omega_{\mathrm{tw}} = \pi_1^* \omega_{\mathrm{tw},0} + \mathrm{d}(\alpha, \pi_2),$$

where $\pi_1 : Z \times \mathfrak{g} \to Z \to Z/G^{\mathrm{ad}}$ and $\pi_2 : Z \times \mathfrak{g}^* \to \mathfrak{g}^*$ are projections. We can assume that π_2 corresponds to the moment map.

Restricting this diffeomorphism to $\Phi_{\mathrm{tw}}^{-1}(\mathfrak{t}^*)$, we obtain a local model $Z \times \mathfrak{t}^*$ for that space. This implies that locally near Z, we get a decomposition of its tangent spaces into several (nontrivial) bundles

$$T(\Phi_{\mathrm{tw}}^{-1}(\mathfrak{t}^*)) \cong T(Z/S^1) \oplus \mathfrak{g} \oplus \mathfrak{t}^* \cong T(\mathscr{M}_{-I}(\Sigma')) \oplus \mathfrak{t}^{\perp} \oplus (\mathfrak{t} \oplus \mathfrak{t}^*). \tag{4.5}$$

(We omitted the pullback symbols from the notation for simplicity.)

The restriction of ω_{tw} to $\Phi_{\mathrm{tw}}^{-1}(\mathfrak{t}^*)$ is nondegenerate in the horizontal directions $T\mathscr{M}_{-I}(\Sigma')$ as well as on $\mathfrak{t} \oplus \mathfrak{t}^*$. Let us compute it on the subbundle $\mathfrak{t}^{\perp} \subset \mathfrak{g}$. For a point x with $\Phi_{\mathrm{tw}}(x) = t\mu \in \mathfrak{t}^*$, and for $\xi_1, \xi_2 \in \mathfrak{t}^{\perp} \subset T_x\Phi_{\mathrm{tw}}^{-1}(\mathfrak{t}^*)$, we have

$$\omega_{\mathrm{tw}}(\xi_1, \xi_2) = (\mathrm{d}\alpha(\xi_1, \xi_2), t\mu) = -\frac{t}{2}\langle [\xi_1, \xi_2], \mu \rangle. \tag{4.6}$$

Thus the restriction of the form to $\mathfrak{t}^{\perp}$ is nondegenerate as long as $t \neq 0$. (For $t = 0$, we already knew that it was degenerate from the proof of Proposition 3.)

We construct a G^{ad}-equivariant almost complex structure J in a neighborhood of Z in $\Phi_{\mathrm{tw}}^{-1}(\mathfrak{t}^*)$ such that J is split with respect to the decomposition (4.5) and is compatible with ω_{tw} "as much as possible." More precisely, we choose G^{ad}-equivariant complex structures J_1, J_3 on each of the subbundles $T(\mathscr{M}_{-I}(\Sigma'))$ and $\mathfrak{t} \oplus \mathfrak{t}^*$ that are compatible with respect to the restriction of ω_{tw} on the respective subbundle. We also choose a G^{ad}-equivariant complex structure J_2 on $\mathfrak{t}^{\perp}$ that is compatible with respect to the form σ given by

$$\sigma(\xi_1, \xi_2) = -\langle [\xi_1, \xi_2], \mu \rangle.$$

By (4.6), we have $\omega_{\mathrm{tw}} = t\sigma/2$; hence J_2 is compatible with respect to ω_{tw} away from $t = 0$. We then let $J = J_1 \oplus J_2 \oplus J_3$ be the almost complex structure on $\Phi_{\mathrm{tw}}^{-1}(\mathfrak{t}^*)$ near Z.

Choose $\epsilon \in (0, 1/8)$ sufficiently small that $Z \times (-3\epsilon, 3\epsilon)$ is part of the local model for $\Phi_{\mathrm{tw}}^{-1}(\mathfrak{t}^*)$ described above. Pick a smooth function $f : \mathbb{R} \to \mathbb{R}$ with the following properties: $f(t) = t + \epsilon$ for t in a neighborhood of 0; $f(t) = t$ for $|t| \geq 2\epsilon$; and $f'(t) > 0$ everywhere. This induces a G^{ad}-equivariant self-diffeomorphism of the open subset $Z \times (-3\epsilon, 3\epsilon) \subset \Phi_{\mathrm{tw}}^{-1}(\mathfrak{t}^*)$, given by $(z, t) \to (z, f(t))$. Note that this diffeomorphism preserves J, it is the identity near the boundary, and it takes $Z \times [0, 2\epsilon)$ to $Z \times [\epsilon, 2\epsilon)$.

Now let us look at the constructions we have made in light of the identification between $\Phi_{\mathrm{tw}}^{-1}(\mathfrak{t}^*)$ and $\mathscr{M}^{\mathfrak{t}}(\Sigma') = \Phi^{-1}(\mathfrak{t}) \subset \mathscr{M}^{\mathfrak{g}}(\Sigma')$. We have obtained a local model $Z \times (-3\epsilon, 3\epsilon)$ for the neighborhood $N = \Phi^{-1}(-3\epsilon\mu, 3\epsilon\mu)$ of $\Phi^{-1}(\mu)$ in $\mathscr{M}^{\mathfrak{t}}(\Sigma')$, an almost complex structure on N, and a self-diffeomorphism of N.

The symplectic cross-section theorem [21] says that locally near $\tilde{\Phi}^{-1}(1/2)$, the extended moduli space $\mathscr{M}^{\mathfrak{g}}(\Sigma')$ looks like $G \times_T \mathscr{M}^{\mathfrak{t}}(\Sigma')$. Thus, we can lift the local model for $\mathscr{M}^{\mathfrak{t}}(\Sigma')$ and obtain a G^{ad}-equivariant local model $(G \times_T Z) \times (-3\epsilon, 3\epsilon)$

for $\mathscr{M}^{\mathfrak{g}}(\Sigma')$. Projection on the second factor corresponds to the map $1/2 - \tilde{\Phi}$. Further, locally we can decompose the tangent bundle to $\mathscr{M}^{\mathfrak{g}}(\Sigma')$ as in (3.5). The form ω is nondegenerate when restricted to $T_\mu(\mathscr{O}_\mu)$. Let us choose a G^{ad}-equivariant complex structure on this subbundle that is compatible with the restriction of ω there. By combining it with J, we obtain an equivariant almost complex structure $\tilde{J}$ on

$$\tilde{N} = \tilde{\Phi}^{-1}(1/2 - 3\epsilon, 1/2 + 3\epsilon) \subset \mathscr{M}^{\mathfrak{g}}(\Sigma').$$

We can also lift the self-diffeomorphism of $N \subset \mathscr{M}^{\mathfrak{t}}(\Sigma')$ to $\tilde{N} = G \times_T N$ in an equivariant manner. Since this self-diffeomorphism is the identity near the boundary, we can extend it by the identity to all of $\mathscr{M}^{\mathfrak{g}}_s(\Sigma')$. The result is a G^{ad}-equivariant diffeomorphism

$$\mathscr{M}^{\mathfrak{g}}_s(\Sigma') \to \mathscr{M}^{\mathfrak{g}}_s(\Sigma')$$

that preserves $\tilde{J}$ on $\tilde{N}$, takes $\tilde{\Phi}^{-1}(1/2)$ to $\tilde{\Phi}^{-1}(1/2 - \epsilon)$, and is the identity on $\tilde{\Phi}^{-1}\big([0, 1/2 - 2\epsilon)\big)$. This diffeomorphism descends to one between the corresponding cut spaces:

$$\phi : \mathscr{N}^c(\Sigma') = \mathscr{M}^{\mathfrak{g}}_s(\Sigma')_{\leq 1/2} \to \mathscr{M}^{\mathfrak{g}}_s(\Sigma')_{\leq 1/2 - \epsilon}.$$

We set $\omega_\epsilon = \phi^* \omega_0$. Note that ω_0 and ω_ϵ coincide on the subset $\tilde{\Phi}^{-1}\big([0, 1/2 - 2\epsilon)\big)$. Since we chose $2\epsilon < 1/4$, the latter subset contains $\mathscr{W} = \tilde{\Phi}^{-1}\big([0, 1/4)\big)$.

The almost complex structure $\tilde{J}$ on $\tilde{N}$ descends to the cut $\tilde{N}_{\leq 1/2}$ as well. Indeed, if $\mathfrak{t} \subset T\tilde{N}$ denotes the line bundle in the direction of the T^{ad}-action used for cutting, by construction we have $\tilde{J}\mathfrak{t} \cap T\big(\tilde{\Phi}^{-1}(1/2)\big) = 0$. Since $\tilde{J}$ equivariance, it is easy to see that it induces an almost complex structure on the cut, which we still denote by $\tilde{J}$. We extend $\tilde{J}$ to $\tilde{\Phi}^{-1}\big([0, 1/2 - 2\epsilon)\big)$ by choosing it to be compatible with $\omega_0 = \omega_\epsilon$ there.

It is easy to see that the resulting $\tilde{J}$ and ω_ϵ satisfy the required conditions (i)–(iv). With respect to the last claim in (iv), note that any $\tilde{J}$-holomorphic sphere of index zero is necessarily a multiple cover of one of the fibers of the $\mathbb{P}^1$-bundle $R \to \mathscr{M}_{-I}(\Sigma')$. Hence it has intersection number with R a positive multiple of the intersection number of the fiber, which by Lemma 8 is -2.

Remark 7. There were several choices made in the construction of ω_ϵ and $\tilde{J}$ in Proposition 7, for example, the connection α, the structures J_1, J_2, J_3, and the function f. The space of all these choices is contractible.

5 Symplectic Instanton Homology

5.1 *Lagrangians from Handlebodies*

Let H be a handlebody of genus $h \geq 1$ whose boundary is the compact Riemann surface Σ. We view Σ' and Σ as subsets of H, with $\Sigma' = \Sigma \setminus D^2$.

Let $\mathscr{A}^{\mathfrak{g}}(\Sigma'|H) \subset \mathscr{A}^{\mathfrak{g}}(\Sigma')$ be the subspace of connections that extend to flat connections on the trivial G-bundle over H. Consider also $\mathscr{A}(H)$, the space of flat connections on H, which is acted on by the based gauge group $\mathscr{G}_0(H) = \{f : H \to G \mid f(z) = I\}$. Since $\pi_1(G) = 1$ and Σ' has the homotopy type of a wedge of spheres, every map $\Sigma' \to G$ must be null homotopic. This implies that $\mathscr{G}^c(\Sigma')$ preserves $\mathscr{A}^{\mathfrak{g}}(\Sigma'|H)$, and furthermore, the natural map

$$\mathscr{A}(H)/\mathscr{G}_0(H) \longrightarrow \mathscr{A}^{\mathfrak{g}}(\Sigma'|H)/\mathscr{G}^c(\Sigma') \tag{5.1}$$

is a diffeomorphism.

Set

$$L(H) = \mathscr{A}(H)/\mathscr{G}_0(H) \cong \mathscr{A}^{\mathfrak{g}}(\Sigma'|H)/\mathscr{G}^c(\Sigma') \subset \mathscr{M}^{\mathfrak{g}}(\Sigma') = \mathscr{A}^{\mathfrak{g}}(\Sigma')/\mathscr{G}^c(\Sigma').$$

The left-hand side of (5.1) is the moduli space of flat connections on H. After a set of h simple closed curves $\alpha_1, \ldots, \alpha_h$ on H whose classes generate $\pi_1(H)$ has been chosen, the space $\mathscr{A}(H)/\mathscr{G}(H)$ can be identified with the space of homomorphisms $\pi_1(H) \to G$ or, alternatively, with the Cartesian product G^h.

In fact, if the curves $\alpha_1, \ldots, \alpha_h$ are the same as those chosen on Σ' for the identification (3.1), so that the remaining curves β_i are null homotopic in H, then with respect to the identification (3.3), we have

$$L(H) \cong \{(A_1, B_1, \ldots, A_h, B_h) \in G^{2h} \mid B_i = I, \quad i = 1, \ldots, h\} \subset \mathscr{N}(\Sigma'). \tag{5.2}$$

Let us now view $L(H)$ as $\mathscr{A}^{\mathfrak{g}}(\Sigma'|H)/\mathscr{G}^c(\Sigma')$ via (5.1). Note that connections A that extend to H extend in particular to Σ, which means that the value $\theta \in \mathfrak{g}$ such that $A|_S = \theta \mathrm{d}s$ is zero. In other words, $L(H)$ lies in $\Phi^{-1}(0) \subset \mathscr{N}(\Sigma')$.

Lemma 9. *With respect to the Huebschmann–Jeffrey symplectic form ω from Sect. 3.4, $L(H)$ is a Lagrangian submanifold of $\mathscr{N}(\Sigma')$.*

Proof. Let $\tilde{A}$ be a flat connection on H and A its restriction to Σ'. With respect to the description (3.4) of $T_{[A]}\mathscr{N}(\Sigma')$, the tangent space to $L(H)$ at A consists of equivalence classes of d_A-closed forms $a \in \Omega^{1,\mathfrak{g}}(\Sigma')$ that extend to $\mathrm{d}_{\tilde{A}}$-closed forms $\tilde{a} \in \Omega^1(H) \otimes \mathfrak{g}$. Let a, b be two such forms and $\tilde{a}, \tilde{b}$ their extensions to H. We have $a|_S = b|_S = 0$. Furthermore, by the Poincaré lemma for connections, on the disk D^2 that is the complement of Σ' in Σ there exists $\lambda \in \Omega^0(D^2; \mathfrak{g})$ such that $\mathrm{d}_{\tilde{A}}\lambda = \tilde{a}|_{D^2}$. By Stokes' theorem,

$$\int_{D^2} \langle a \wedge b \rangle = \int_S \langle \lambda \wedge b \rangle = 0.$$

Another application of Stokes' theorem gives

$$\int_{\Sigma'} \langle a \wedge b \rangle = \int_{\Sigma} \langle \tilde{a} \wedge \tilde{b} \rangle = \int_H \langle \mathrm{d}_{\tilde{A}}(\tilde{a} \wedge \tilde{b}) \rangle = 0.$$

This shows that ω vanishes on the tangent space to $L(H) \cong G^h$, which is half-dimensional.

5.2 Symplectic Instanton Homology

Let $Y = H_0 \cup H_1$ be a Heegaard decomposition of a three-manifold Y, where H_0 and H_1 are handlebodies of genus h, with $\partial H_0 = -\partial H_1 = \Sigma$. Let $L_0 = L(H_0)$ and $L_1 = L(H_1) \subset \mathscr{N}(\Sigma')$ be the Lagrangians associated respectively with H_0 and H_1, as in 5.1. View $\mathscr{N}(\Sigma')$ as an open subset of the compactified space $\mathscr{N}^c(\Sigma')$, as in Sect. 4.5, with R being its complement.

In Sect. 4.5 we gave $\mathscr{N}^c(\Sigma')$ the structure of an ϵ-symplectic manifold. By Lemma 8, its degeneracy locus is exactly R. Using the variant of Floer homology described in Sect. 2.3 and letting $\tilde{\omega} = \omega_0$, $\omega = \omega_\epsilon$, and $\tilde{J}$ be as in Proposition 7, we define

$$HSI(\Sigma'; H_0, H_1) = HF(L_0, L_1, \tilde{J}; R).$$

In order to ensure that the Floer homology is well defined, we should check that the hypotheses (i)–(ix) listed at the beginning of Sect. 2.3 are satisfied. Indeed, (i), (ii), (iii), (v), and (x) are subsumed in Proposition 7, while (iv), (v), and (ix) follow respectively from Proposition 6, Lemma 9, and Corollary 2. For (viii), the Lagrangians are simply connected and spin because they are diffeomorphic to G^h. By Theorem 4 and Lemma 5, the minimal Chern number of the open subset $\mathscr{N}(\Sigma)$ is a multiple of 4; therefore, the Floer groups admit a relative $\mathbb{Z}/8\mathbb{Z}$-grading.

A priori, the Floer homology depends on $\tilde{J}$. However, the set of choices used in the construction of $\tilde{J}$ is contractible; cf. Remark 7. By the usual continuation arguments in Floer theory, if we change $\tilde{J}$, the corresponding Floer homology groups are canonically isomorphic.

5.3 Dependence on the Base Point

Recall that the surface Σ' is obtained from a closed surface Σ by deleting a disk around some base point $z \in \Sigma$. Let $z_0, z_1 \in \Sigma$ be two choices of base point. Any choice of path $\gamma : [0,1] \to \Sigma$, $j \mapsto z_j$, $j = 0,1$, induces an identification of fundamental groups $\Sigma'_0 \to \Sigma'_1$ and equivariant presymplectomorphisms T_γ : $\mathscr{N}^c(\Sigma'_0) \to \mathscr{N}^c(\Sigma'_1)$ preserving the cut locus R. The pullbacks of the form ω and the almost complex structure $\tilde{J}$ from Proposition 7 (applied to $\mathscr{N}^c(\Sigma'_1)$) can act as the corresponding form and almost complex structure in Proposition 7 applied to $\mathscr{N}^c(\Sigma'_0)$. Moreover, if H_0, H_1 are handlebodies, the symplectomorphism T_γ preserves the corresponding Lagrangians L_0, L_1, since the vanishing holonomy condition is invariant under conjugation by paths. Therefore, the continuation arguments in Floer theory show that T_γ induces an isomorphism

$$HSI(\Sigma'_0; H_0, H_1) \to HSI(\Sigma'_1; H_0, H_1).$$

This isomorphism depends only on the homotopy class of γ relative to its endpoints. We conclude that the symplectic instanton homology groups naturally form a flat bundle over Σ. In particular, there is a natural action of $\pi_1(\Sigma, z_0)$ on $HSI(\Sigma_0'; H_0, H_1)$.

When we care about the Floer homology group only up to isomorphism (not canonical isomorphism), we drop the base point from the notation and write $HSI(\Sigma'; H_0, H_1) = HSI(\Sigma; H_0, H_1)$, as in the introduction.

6 Invariance

We prove here that the groups $HSI(\Sigma; H_0, H_1)$ are invariants of the 3-manifold $Y = H_0 \cup H_1$. The proof is based on the theory of Lagrangian correspondences in Floer theory; cf. [59]. We start by reviewing this theory.

6.1 Quilted Floer Homology

Let M_0, M_1 be compact symplectic manifolds. A *Lagrangian correspondence* from M_0 to M_1 is a Lagrangian submanifold $L_{01} \subset M_0^- \times M_1$. (The minus superscript means that we are considering the same manifold equipped with the negative of the given symplectic form.) Given Lagrangian correspondences $L_{01} \subset M_0^- \times M_1$, $L_{12} \subset M_1^- \times M_2$, their *composition* is the subset of $M_0^- \times M_2$ defined by

$$L_{01} \circ L_{12} = \pi_{02}(L_{01} \times_{M_1} L_{12}),$$

where $\pi_{02} : M_0^- \times M_1 \times M_1^- \times M_2 \to M_0^- \times M_2$ is the projection. If the intersection

$$L_{01} \times_{M_1} L_{12} = (L_{01} \times L_{12}) \cap (M_0^- \times \Delta_{M_1} \times M_2)$$

is transverse (hence smooth) in $M_0^- \times M_1 \times M_1^- \times M_2$, and the projection $\pi_{02} : L_{01} \times_{M_1} L_{12} \to L_{01} \circ L_{12}$ is embedded, we say that the composition $L_{02} = L_{01} \circ L_{12}$ is *embedded.* An embedded composition L_{02} is a smooth Lagrangian correspondence from M_0 to M_2.

Suppose now that M_0, M_1, M_2 are compact symplectic manifolds, monotone with the same monotonicity constant, and with minimal Chern number at least 2. Suppose that $L_0 \subset M_0, L_{01} \subset M_0^- \times M_1, L_{12} \subset M_1^- \times M_2, L_2 \subset M_2$ are simply connected Lagrangian submanifolds. (This implies that their minimal Maslov numbers are at least 4.)

Define

$$HF(L_0, L_{12}, L_{12}, L_2) := HF(L_0 \times L_{12}, L_{01} \times L_2)$$

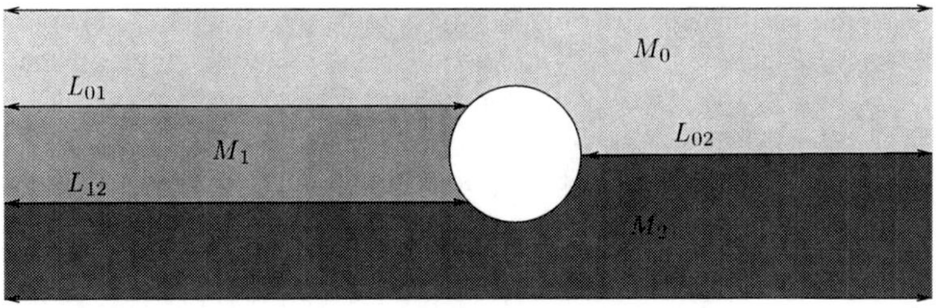

Fig. 1 Geometric composition via a quilt count of Y-maps

and

$$HF(L_0, L_{02}, L_2) := HF(L_0 \times L_2, L_{01} \circ L_{12}).$$

The main theorem of [59] implies the following.

Theorem 5. *With $M_0, M_1, M_2, L_0, L_{01}, L_{12}, L_2$ monotone as above, if $L_{02} := L_{01} \circ L_{12}$ is embedded, then there exists a canonical isomorphism of Lagrangian Floer homology groups*

$$HF(L_0, L_{01}, L_{12}, L_2) \to HF(L_0, L_{02}, L_2). \tag{6.1}$$

In Wehrheim–Woodward [59], an isomorphism is defined using pseudoholomorphic quilts, i.e., in this case, triples of strips in M_0, M_1, M_2 with boundary conditions in L_0, L_{01}, L_{12}, and L_2. The count of such quilts is used in the left-hand side of (6.1). In the limit whereby the width δ of the middle strip goes to 0, the same count produces the right-hand side. An alternative proof was given in Lekili–Lipyanskiy [30] using a count of *Y-maps*. This approach is better suited for the semipositive case in which we will need it, so we review the construction. Given $x_- \in (L_0 \times L_{12}) \cap (L_{01} \times L_2), x_+ \in (L_0 \times L_2) \cap L_{02}$, let $\mathfrak{M}(x_-, x_+)$ denote the set of holomorphic quilts with two striplike ends and one cylindrical end as shown in Fig. 1, with finite energy and limits $x_\pm$. The authors show that for a comeager subset of the space of point-dependent compatible almost complex structures, the moduli space $\mathfrak{M}(x_-, x_+)$ of Y-maps has the structure of a finite-dimensional manifold, and counting the zero-dimensional component $\mathfrak{M}(x_-, x_+)_0$ defines a cochain map

$$\Phi : CF(L_0, L_{01}, L_{12}, L_2) \to CF(L_0, L_{02}, L_2), \quad \langle x_- \rangle \mapsto \sum_{u \in \mathfrak{M}(x_-, x_+)_0} \varepsilon(u) \langle x_+ \rangle .$$

Here, in the case of integer coefficients, the map

$$\varepsilon : \mathfrak{M}(x_-, x_+)_0 \to \{\pm 1\}$$

is defined by comparing the orientations constructed in [58] with the canonical orientation of a point.

Counting Y-maps in the opposite direction defines a chain map

$$\Psi : CF(L_0, L_{02}, L_2) \to CF(L_0, L_{01}, L_{12}, L_2).$$

Lekili and Lipyanskiy [30] prove that the monotonicity constant for these Y-maps is the same as the monotonicity constant for Floer trajectories. They then show that Φ and Ψ induce isomorphisms on homology.

6.2 Relative Quilted Floer Homology in Semipositive Manifolds

We wish to have a version of the quilted Floer homology and composition theorem, Theorem 5, that holds for Floer homology *relative to hypersurfaces* in *semipositive* manifolds, as in Sect. 2.3. Suppose that R_0, R_1 are symplectic hypersurfaces in M_0, M_1. From them we obtain two hypersurfaces $\tilde{R}_0 = R_0^- \times M_1$, $\tilde{R}_1 = M_0^- \times R_1$ in $M_0^- \times M_1$. Let $\tilde{N}_{R_0}$, $\tilde{N}_{R_1}$ denote their normal bundles N_{R_0}, N_{R_1}, that is, the pullbacks of N_{R_0}, N_{R_1} to $\tilde{R}_0$, $\tilde{R}_1$. Because R_0, R_1 are symplectic, N_{R_0}, N_{R_1} are *oriented* rank-2 bundles, or equivalently up to homotopy, rank-1 complex line bundles. As we will see below, the following definition gives sufficient conditions for a sort of combined intersection number with R_0, R_1 to be well defined and given by the usual geometric formulas.

Definition 6. A simply connected Lagrangian correspondence $L_{01} \subset M_0^- \times M_1$ is called *compatible* with the pair (R_0, R_1) if

$$(R_0 \times M_1) \cap L_{01} = (M_0 \times R_1) \cap L_{01} = (R_0 \times R_1) \cap L_{01}$$

and there exist an isomorphism

$$\varphi : (\tilde{N}_{R_0})|_{(R_0 \times R_1) \cap L_{01}} \cong (\tilde{N}_{R_1})|_{(R_0 \times R_1) \cap L_{01}}$$

and tubular neighborhoods

$$\tau_0 : N_{R_0} \to M_0, \quad \tau_1 : N_{R_1} \to M_1$$

of R_0 and R_1 respectively such that $(\tau_0 \times \tau_1)^{-1}(L_{01}) \subset N_{R_0} \times N_{R_1} = \tilde{N}_{R_0} \times_{M_0 \times M_1} \tilde{N}_{R_1}$ is equal to the graph of φ.

To explain the conditions in the definition, note that the existence of φ implies that any map of a compact oriented surface with boundary to M with boundary conditions in L_{01} has a well-defined intersection number with $\tilde{R}_0 \cup \tilde{R}_1$. For example, suppose that $u : (D, \partial D) \to (M_0 \times M_1, L_{01})$ is a disk with Lagrangian boundary conditions. The sum of dual classes $[\tilde{R}_0]^\vee + [\tilde{R}_1]^\vee$ has trivial restriction to $H^2(L_{01}; \mathbb{Z})$. If L_{01} is simply connected, then $H^2(M, L_{01})$ is the kernel of $H^2(M) \to H^2(L_{01})$, and so we may consider $[\tilde{R}_0]^\vee + [\tilde{R}_1]^\vee$ as a class in $H^2(M, L_{01})$. Then the intersection number of a map $u : (D, \partial D) \to (M_0 \times M_1, L_{01})$ with $[\tilde{R}_0] + [\tilde{R}_1]$ is well defined and denoted by $u \cdot R$.

The existence of the tubular neighborhoods τ_0, τ_1 implies that $u \cdot R$ is given by a geometric count of intersection points. Indeed, we may identify a neighborhood of R_j with the normal bundle $\pi_j : N_j \to R_j$ via the tubular neighborhood τ_j. Then $\pi_j^* N_j$ is trivial on the complement of R_j, since the map π_j gives a nonvanishing section, and extends to a bundle L_{R_j} on M_j trivial on the complement of R_j. Then the dual class $[R_j]^\vee$ is given by a Thom class in the tubular neighborhood of R_j, and hence equals the Euler class of L_{R_j}. The bundles L_{R_0} and L_{R_1} are isomorphic on ∂D via φ, and so glue together to a bundle denoted by $u^* L_R$ over $S^2 = D \cup_{\partial D} D$. The intersection number is then the Euler number of $u^* L_R$, that is,

$$u \cdot R = ([S^2], \mathrm{Eul}(u^* L_R)).$$

The compatibility condition on the maps τ_j implies that the maps u_0, u_1 considered as sections of L_{R_j} near R_j glue together to a section of $u^* L_R$, which by abuse of notation we denote by u. If each u_j meets R_j in a finite number of points, then $u \cdot R$ is a sum of local intersection numbers $(u \cdot R)_z$, given by the image of a small loop around each intersection point z in $H_1(u^* L_R|_V - 0, \mathbb{Z}) \cong \mathbb{Z}$ in a small neighborhood V of z. Note that since we have constructed $u^* L_R$ only as a *topological* (or rather, piecewise smooth) bundle, such a loop will be only piecewise smooth if $z \in \partial D$.

Our examples will arise as follows:

Example 1. Suppose $\iota : C \to M_1$ is a fibered coisotropic submanifold of M_1 with structure group C, the fibration being $\pi : C \to M_0$. Then $(\pi \times \iota) : C \to M_0^- \times M_1$ defines a Lagrangian correspondence; cf. [59, Example 2.0.3(b)]. Suppose further that M_1 is a Hamiltonian U(1)-manifold with moment map Φ_1 and C is U(1)-invariant and meets $\Phi_0^{-1}(\lambda)$ transversely. Then the symplectic cut $M_{1,\leq\lambda}$ contains the closure $C_{\leq\lambda}$ of the image C as a fibered coisotropic submanifold, whose graph is a Lagrangian correspondence in $M_{0,\leq\lambda} \times M_{1,\leq\lambda}$. Furthermore, the submanifolds $R_0 := M_{0,\lambda}, R_1 := M_{1,\lambda}$ are symplectic submanifolds with the properties described in Definition 6. Indeed, any tubular neighborhood $N_{R_1} \to M_{1,\leq\lambda}$ of R_1 that is U(1)-invariant, maps $N_{R_1}|_{C_{\leq\lambda}}$ to $C_{\leq\lambda}$, and maps fibers to fibers induces a tubular neighborhood $N_{R_0} \to M_{0,\leq\lambda}$ with the required properties.

The intersection numbers described above are well defined more generally for quilted strips, as we now explain. Given symplectic manifolds $M_1, \ldots, M_k$, Lagrangian submanifolds $L_1 \subset M_1, L_k \subset M_k$ disjoint respectively from R_1 and R_k, and Lagrangian correspondences

$$L_{12} \subset M_1 \times M_2, \ldots, L_{(k-1)k} \subset M_{k-1}^- \times M_k$$

compatible with hypersurfaces $\underline{R} = (R_j \subset M_j)_{j=1,\ldots,k}$, the intersection number $\underline{u} \cdot \underline{R}$ of a quilted Floer trajectory

$$\underline{u} = (u_j : \mathbb{R} \times [0,1] \to M_j)_{j=1}^k$$

is the pairing of $\underline{u}$ with the sum of the dual classes $[R_j]^\vee$ to R_j.

If the intersection of $\underline{u}$ with $\underline{R}$ is finite, then the intersection number is the sum of local intersection numbers defined as follows. By assumption, there exists an isomorphism

$$N_{j-1}|_{L_{(j-1)j} \cap (R_{j-1} \times R_j)} \xrightarrow{\cong} N_j|_{L_{(j-1)j} \cap (R_{j-1} \times R_j)},$$

and this extends to an isomorphism of $\tilde{N}_{R_j}|_{L_{(j-1)j}}$ and $\tilde{N}_{R_{j-1}}|_{L_{(j-1)j}}$ by the assumption about the tubular neighborhoods. Thus the pullback bundles $u_j^* \tilde{N}_{R_j}$ patch together to a bundle on the quilted surface $\underline{S} = \cup_j S_j$, which we denote by $\underline{u}^* \tilde{N}_{\underline{R}}$. The intersection number is then the relative Euler number of $\underline{u}^* \underline{N}_{\underline{R}} \to \underline{S}$, that is, the pairing of the relative Euler class with the generator of $H^2(cl(\underline{S}), \partial cl(\underline{S}))$, where $cl(\underline{S})$ is the closed disk obtained by adding points at $\pm\infty$. The map $\underline{u}$ then provides a section of $\underline{u}^* \underline{N}_{\underline{R}}$, by the compatibility conditions in Definition 6. If the intersection is finite, then

$$\underline{u} \cdot \underline{R} = \sum_{\{z \in \underline{S} | \underline{u}(z) \in \underline{R}\}} (\underline{u} \cdot \underline{R})_z, \tag{6.2}$$

where $(\underline{u} \cdot \underline{R})_z \in \mathbb{Z}$ is, as in the case of disks discussed before, the image of a small loop around z in the complement $\underline{u}^* \underline{N}_{\underline{R}} - 0$ of the zero section, as a multiple of the generator of the first homology of the fiber, and the condition $\underline{u}(z) \in \underline{R}$ means that if z lies in the component S_j, then $u_j(z_j) \in R_j$. Note in particular that these local intersection numbers are topologically continuous, that is, given any loop in the domain of the quilt, the sum of the local intersection numbers is constant in any continuous family as long as none of the intersection points cross the loop.

If the intersection is not only finite but transverse, and the hypersurfaces $\underline{R}$ are almost complex, then the intersection number is the usual one counted with weight $1/2$ for the seam points:

Lemma 10. *Suppose that L_0 respectively L_k is disjoint from R_0 respectively R_k and each $L_{(j-1)j}$ is compatible with (R_{j-1}, R_j). Suppose that the almost complex structure on $M_0 \times \cdots \times M_k$ is of product form $J_0 \times \cdots \times J_k$ near each $\tilde{R}_j$, so that each R_j is an almost complex submanifold of M_j with respect to J_j. Let $\underline{u} : \underline{S} \to \underline{M}$ be a quilted Floer trajectory with Lagrangian boundary and seam conditions in $\underline{L}$ meeting each $\tilde{R}_j$ transversally. Then*

$$\underline{u} \cdot \underline{R} = \sum_{j=0}^{k} \#\{z_j \in \operatorname{int}(S_j) | u_j(z_j) \in R_j\} + \frac{1}{2} \#\{z_j \in \partial S_j | u_j(z_j) \in R_j\}.$$

Proof. The local intersection number in (6.2) at a transversal point of intersection $z \in \underline{S}$ is the homology class of the image of a small loop around z, considered as an element of $H_1(\underline{N}_z) \cong \mathbb{Z}$. We consider only the case of an intersection point z on the seam; the loop is divided into two loops, one coming from each component of the quilt, and is only piecewise smooth. The case of an interior intersection is easier and left to the reader.

Suppose z is on the seam $L_{(j-1)j}$ where the components u_{j-1} and u_j of the quilt meet. For $l = j-1$ or j, let us view u_l as a section of a piecewise smooth line bundle. Using a local trivialization of the bundle and a coordinate chart for $\underline{S}$ centered at z, we have that u_l near z (now viewed as a map to $\mathbb{C}$) is given approximately by its linearization at z:

$$|u_l(r\exp(it)) - (Du_l(z))r\exp(it)| < Cr^2.$$

We use here that since R_l is almost complex, the linearization Du_l is complex linear. Fix $\epsilon > 0$. For r sufficiently small, we have

$$|\arg(u_l(r\exp(it))) - \arg(Du_l(z)r\exp(it))| < \epsilon.$$

This implies that

$$\left|\int_0^1 u_l^* \mathrm{d}\theta - \pi\right| < \epsilon, \quad l = j-1, j,$$

and so

$$\int_0^1 u_{j-1}^* \mathrm{d}\theta + \int_0^1 u_j^* \mathrm{d}\theta \in (2\pi - 2\epsilon, 2\pi + 2\epsilon).$$

Since the integral must be an integer multiple of 2π (and ϵ can be chosen arbitrarily small), the integral must in fact equal 2π. It follows that the two paths patch together to a positive generator of $H_1(\mathbb{C}^*, \mathbb{Z})$, as claimed.

We can now define relative quilted Floer homology in semipositive manifolds.

Theorem 6. *Suppose that $\underline{M} = (M_i)_{i=0}^k$ are semipositive manifolds as in the first six items of Assumption 2.1, with a collection of open sets $\underline{\mathcal{W}} = (\mathcal{W}_i)_{i=0}^k$ on which the respective forms ω_i and $\tilde{\omega}_i$ coincide. Suppose the manifolds M_i come equipped with almost complex structures $\tilde{J}_i$, so that the degeneracy loci R_i of the forms $\tilde{\omega}_i$ are almost complex hypersurfaces in M_i, disjoint from $\mathcal{W}_i$. We denote by $\mathfrak{J}_t(\underline{M}, \underline{\mathcal{W}}, \tilde{J})$ the space of time-dependent almost complex structures on $M_0 \times \cdots \times M_k$ that agree with $\tilde{J} = \tilde{J}_0 \times \cdots \times \tilde{J}_k$ on $\mathcal{W} := \prod_{i=0}^k \mathcal{W}_i$.*

We are also given simply connected Lagrangians $L_0 \subset M_0, L_{01} \subset M_0^- \times M_1, \ldots, L_{(k-1)k} \subset M_{k-1}^- \times M_k, L_k \subset M_k$ such that the seam conditions $L_{(i-1)i}$ are compatible with (R_{i-1}, R_i), and L_0 and L_k are contained in $\mathcal{W}_0$ respectively $\mathcal{W}_k$. Also, we assume that

$$(L_0 \times L_{12} \times \cdots) \cap (L_{01} \times L_{23} \times \cdots) \subset \mathcal{W}_0 \times \cdots \times \mathcal{W}_k. \tag{6.3}$$

Suppose further that any holomorphic disk with boundary in $L_{(i-1)i}$, $i = 1, \ldots, k$, or holomorphic sphere with zero canonical area has intersection number with $\underline{R}$ given by a negative multiple of 2.

Then there exists a comeager subset $\mathcal{J}_t^{\rm reg}(\underline{L},\underline{\mathcal{W}},\tilde{J})$ *of* $\mathcal{J}_t(\underline{M},\underline{\mathcal{W}},\tilde{J})$ *such that if the almost complex structure* (J_t) *is chosen from* $\mathcal{J}_t^{\rm reg}(\underline{L},\underline{\mathcal{W}},\tilde{J})$, *then the part of the Floer differential of* $CF(\underline{L}) = CF(L_0 \times L_{12} \times \cdots, L_{01} \times L_{23} \times \cdots)$ *counting trajectories disjoint from* R_i, $i = 1,\dots,m$, *is finite and squares to zero. We denote by*

$$HF(\underline{L};\underline{R}) := HF(\underline{L},\tilde{J};\underline{R})$$

the resulting Floer homology group; it is independent up to isomorphism of all choices except possibly the base almost complex structures $\tilde{J}_i$.

Proof. First, note that the condition (6.3) implies that the endpoints of any holomorphic quilt are contained in $\mathcal{W} = \mathcal{W}_0 \times \cdots \times \mathcal{W}_k$. Hence, every quilt component u_i contains a point in the respective open set $\mathcal{W}_i$. This implies that the usual transversality arguments for holomorphic quilts apply, even when we restrict to almost complex structures J_t that are required to agree with $\tilde{J}$ on $\mathcal{W}$.

Next, we discuss compactness. We must rule out sphere and disk bubbling in the zero- and one-dimensional moduli spaces. For a suitable comeager subset of almost complex structures agreeing with the given $\tilde{J}_i$, the trajectories are transverse to the R_j in the zero- and one-dimensional moduli spaces, by the same argument we gave previously for the unquilted case (Corollary 1).

Suppose that $\underline{u}_\infty$ is the limit of a sequence of trajectories of index 1 or 2 disjoint from $\underline{R}$. By the assumption on the intersection number, any sphere bubble or disk bubble with boundary in some $L_{(j-1)j}$ contributes at least -2 to the intersection number with $\underline{R}$. It follows that at least one intersection point does not have a bubble attached. But then, since the intersection point is transverse, $\underline{u}_\infty$ cannot be the limit of a sequence of trajectories disjoint from $\underline{R}$, since transverse intersection points persist under deformation. Hence there is no such bubbling, and the limit is a (possibly broken) trajectory, as desired. Independence of the choice of almost complex structures is proved by the usual continuation argument, ruling out disk bubbles of index one and sphere bubbles by the same reasoning.

Remark 8. If the Lagrangian correspondences above are associated to fibered coisotropics, then the almost complex structures may be taken of split form, that is, products of the almost complex structures on $M_0,\dots,M_k$. This will be the case in our application.

Theorem 7. *Suppose that* $\underline{M} = (M_0,M_1,M_2)$ *and* $\underline{L} = (L_0,L_{01},L_{12},L_2)$ *satisfy the assumptions of Theorem 6. Suppose further that* $L_{01} \circ L_{12}$ *is an embedded composition, is simply connected, and is compatible with* (R_0,R_2), *and that all holomorphic quilted cylinders with seams in* $L_{01},L_{12},L_{01} \circ L_{12}$ *with zero canonical area have intersection number equal to a negative multiple of* 2. *Then the relative Lagrangian Floer homology groups* $HF(L_0,L_{01},L_{12},L_2;R_0,R_1,R_2)$, $HF(L_0,L_{01} \circ L_{12},L_2;R_0,R_2)$ *are isomorphic. Similar statements hold for the composition of any two adjacent pairs, as long as the compositions are smooth and embedded.*

Proof. If the Lagrangian correspondences had been monotone, the result would have been a slight extension of Theorem 5 in [59], by counting only those trajectories disjoint from R_i; indeed, since the intersection numbers are homotopy invariants, they do not change on taking the limit $\delta \to 0$.

In the semipositive case at hand, one can rule out disk and sphere bubbling as in the proof of Proposition 1, but not the figure-eight bubbles mentioned in [59, Sect. 5.3]. Indeed, removal of singularities, transversality, and Fredholm theory for figure-eight bubbles have not yet been developed. For this reason, we use instead the approach of Lekili–Lipyanskiy [30].

First, one checks that for a comeager subset of compatible almost complex structures, the ends of the cylinders of Y-maps will not map to R in the zero- and one-dimensional components of the moduli space, since this is a codimension-2 condition. Indeed, an examination of the weighted Sobolev space construction of the moduli space of Y-maps in [30] shows that the evaluation map at the end of the cylinder is smooth; indeed, it projects onto the factor of asymptotically constant maps in the Banach manifolds in which the moduli space of Y-maps is locally embedded: $W^{1,p,\varepsilon}(S;\underline{u}^*T\underline{M},\underline{u}^*T\underline{L}) \oplus T_{(\underline{u})_{02}(\infty)}L_{02}$, where the former is the space of from S with Lagrangian boundary conditions with finite ε-weighted Sobolev norm of class $(1,p)$, and the latter is the intersection of the linearized Lagrangian boundary conditions at infinity on the cylindrical end.

As a result, the intersection number $\underline{u} \cdot \underline{R}$ of any Y-map $\underline{u}$ of index zero and one with the collection $\underline{R}$ is well defined and given by the formula (10). (More generally, one could make the intersection number with *any* Y-map well defined by imposing the compatibility condition $\varphi_{01} \circ \varphi_{12} = \varphi_{02}$, so that the bundle $\underline{u}^* L_{\underline{R}}$ is well defined. But we will not need this.) In the zero- and one-dimensional moduli spaces, all intersections with the manifolds R_j are transverse for $\underline{J}$ chosen from a comeager subset of the space of compatible almost complex structures making R_j almost complex, by standard arguments [11, Sect. 6].

A Gromov compactness argument shows that finite-energy Y-maps have as limits configurations consisting of a (possibly broken) Y-map together with some sphere bubbles, disk bubbles, and cylinder bubbles. The cylinder bubbles may form when there is an accumulation of energy at the Y-end.

In the case at hand, sphere and disk bubbles are ruled out as in the proof of Theorem 6: any sphere or disk bubble appearing in the limit configuration $\underline{u}_\infty$ must have index zero, and therefore intersection number at most -2 with $\underline{R}$. By (10), any intersection point contributes at most 1 to the intersection number, and therefore at some intersection point with $\underline{R}$ it is not attached to a bubble. But then $\underline{u}_\infty$ cannot be the limit of a sequence of trajectories disjoint from $\underline{R}$, since the local intersection number of $\underline{u}_\infty$ is nonzero.

It remains to rule out cylinder bubbles. Since no trajectory of index zero or one maps the end of the cylinder to $\underline{R}$, any quilted cylinder bubble must capture positive canonical area. But then, for index reasons explained in Lekili–Lipyanskiy [30], the cylinder bubble must capture at least index two, so the index of the remaining Y-map is at most -1. (Here working with Y-maps, rather than strip-shrinking, provides an advantage: by exponential decay for holomorphic strips with boundary values in Lagrangians intersecting cleanly, one knows that these cylinder bubbles connect to a point outside of $\underline{R}$, whereas for figure-eight bubbles, such exponential decay estimates are missing.) But such a trajectory does not exist, since transversality is achieved for the chosen $\underline{J}$.

It follows that the moduli spaces of Y-maps of dimension zero and one that are disjoint from $\underline{R}$ are compact up to breaking off trajectories disjoint from $\underline{R}$. Furthermore, for these trajectories and Y-maps we have the same relationship as in [30], since the complements of $\underline{R}$ are monotone. The rest of the argument now goes as in [30, Sect. 3.1].

6.3 Proof of Invariance

Returning to topology, let Σ_0, Σ_1 be Riemann surfaces of genus h and $h+1$ respectively. Let H_{01} be a compression body with boundary $\Sigma_0^- \times \Sigma_1$, that is, a cobordism consisting of attaching a single handle of index one. Associated to H_{01} we have a Lagrangian correspondence

$$L_{01} \subset \mathscr{N}(\Sigma_0')^- \times \mathscr{N}(\Sigma_1')$$

defined as follows. Suppose that γ is a path from the base points z_0 to z_1, equipped with a framing of the normal bundle. Let H_{01}' denote the noncompact surface obtained from H_{01} by removing a regular neighborhood of γ. The boundary of H_{01}' then consists of Σ_0', Σ_1', and a cylinder $S \times [0,1]$. Let $\mathscr{N}(H_{01}')$ denote the moduli space of flat connections on H_{01} of the form $\theta \mathrm{d}s$ near $S \times [0,1]$ (where s is the coordinate on the circle S), for some $\theta \in \mathfrak{g}$, modulo gauge transformations equal to the identity in a neighborhood of $S \times [0,1]$. The same arguments as in the proof of Lemma 9 show that L_{01} is a Lagrangian correspondence.

The Lagrangian correspondence L_{01} has the following explicit description in terms of holonomies, similar to (3.3) and (5.2). Suppose that H_{01} consists in attaching a one-handle whose meridian is the generator B_{h+1} of $\pi_1(\Sigma_1)$. We have the following lemma.

Lemma 11. *The Lagrangian correspondence L_{01} is given by*

$$L_{01} = \{((A_1,\dots,B_h) \in \mathscr{N}(\Sigma_0'), (A_1,\dots,B_h,A_{h+1},B_{h+1}) \in \mathscr{N}(\Sigma_1')) \mid B_{h+1} = I\}.$$

Proof. H_{01}' has the homotopy type of the wedge product of Σ_0' with a circle, corresponding to a single additional generator a_{h+1}. Thus $\pi_1(H_{01}')$ is freely generated by $(a_1,\dots,b_h,a_{h+1})$, and the lemma follows.

Recall from Sect. 4.5 that $\mathscr{N}(\Sigma_0')$ admits a compactification $\mathscr{N}^c(\Sigma_0') = \mathscr{N}(\Sigma_0') \cup R_0$. We equip $\mathscr{N}^c(\Sigma')$ with the (nonmonotone) symplectic form constructed in Proposition 7, which we denote by $\omega_{\epsilon,0}$. Then R_0 is a symplectic hypersurface. Similarly, we have a symplectic form $\omega_{\epsilon,1}$ on $\mathcal{N}^c(\Sigma_1') = \mathcal{N}(\Sigma_1') \cup R_1$. Let L_{01}^c denote the closure of L_{01} in the compactification $\mathcal{N}^c(\Sigma_0')^- \times \mathcal{N}^c(\Sigma_1')$.

Lemma 12. *The Lagrangian correspondence L_{01}^c is compatible with the pair (R_0, R_1). Furthermore, any disk bubble with boundary in L_{01}^c with index zero has intersection number with (R_0, R_1) a negative multiple of* 2.

Proof. View L_{01}^c as a coisotropic submanifold of $\mathcal{N}^c(\Sigma_1')$, fibered over $\mathcal{N}^c(\Sigma_0')$ with fiber G. We are then exactly in the setting of Example 1. To prove the claim on the intersection number, note that any fiber of R_1 that intersects L_{01} is mapped symplectomorphically onto the corresponding fiber of R_0 via the projection of the fibered coisotropic $B_{h+1} = I$. Hence the patches of any such disk bubble, after projection to $\mathcal{N}^c(\Sigma_0')$, glue together to a sphere bubble in the $\mathbb{P}^1$-fiber of R_0. Furthermore, the projection induces an isomorphism of normal bundles by assumption, so the intersection number is equal to the intersection number of the sphere with R_0, which is a negative multiple of 2 as claimed.

Lemma 13. *Let $L_0 \subset \mathcal{N}^c(\Sigma_0')$, respectively $L_1 \subset \mathcal{N}^c(\Sigma_1')$, be the Lagrangian for the handlebody given by contracting the cycle $b_1, \ldots, b_h$, respectively $b_1, \ldots, b_{h+1}$. Then the composition $L_0 \circ L_{01}^c$ is embedded, and equals L_1.*

Proof. Immediate from Lemma 11 and the fact that L_0 does not meet the hypersurface R_0.

Lemma 14. *Let $L_{01}^c \subset \mathcal{N}^c(\Sigma_0')^- \times \mathcal{N}^c(\Sigma_1')$ be the Lagrangian correspondence for attaching a handle corresponding to adding the cycle a_{h+1}, and let $L_{10}^c \subset \mathcal{N}^c(\Sigma_1')^- \times \mathcal{N}^c(\Sigma_0')$ be the Lagrangian correspondence corresponding to contracting the cycle b_{h+1}. Then the composition $L_{01}^c \circ L_{10}^c$ is embedded, and it equals the diagonal $\Delta_0 \subset \mathcal{N}^c(\Sigma_0')^- \times \mathcal{N}^c(\Sigma_0')$. Furthermore, any quilted cylinder with seams in $L_{10}^c, L_{01}^c, \Delta_0$ with index zero has intersection number with (R_0, R_1, R_0) a negative multiple of* 2.

Proof. The first claim is immediate from Lemma 11. To see the assertion on the quilted cylinders, note that any quilted cylinder of index zero has zero canonical area, and so each component is contained in the corresponding R_j and maps onto a single fiber of the degeneracy locus. As in the proof of Lemma 12, the three holomorphic strips patch together to an orientation-preserving map of a sphere to a fiber of R_0, which must have intersection number a positive multiple of the intersection number of the fiber, which is -2.

Proof (Proof of Theorem 1). We seek to show that the Floer homology groups

$$HSI(\Sigma'; H_0, H_1) = HF(L_0, L_1; R)$$

are independent of the choice of Heegaard splitting of the 3-manifold Y.

By the Reidemeister–Singer theorem [47, 53], any two Heegaard splittings $Y = H_0 \cup_{\Sigma_0} H_1$, $Y = H_0' \cup_{\Sigma_1} H_1'$, are related by a sequence of stabilizations and destabilizations. Therefore, it suffices to consider the case that H_0', H_1' are obtained from H_0, H_1 by stabilization. That is,

$$H_0' = H_0 \cup_{\Sigma_0} H_{01}, \quad H_1' = H_1 \cup_{\Sigma_0} (-H_{10}),$$

where H_{01}, H_{10} are the compression bodies corresponding to adding the cycle a_{h+1}, respectively contracting b_{h+1}. Then, after three applications of Theorem 7, and taking into account Lemmas 13, 14, we have

$$\begin{aligned} HF(L_0, L_1; R_0) &\cong HF(L_0, \Delta_0, L_1; R_0, R_0) \\ &\cong HF(L_0, L_{01}^c, L_{10}^c, L_1; R_0, R_1, R_0) \\ &\cong HF(L_0 \circ L_{01}^c, L_{10}^c \circ L_1; R_1, R_1) \\ &= HF(L_0', L_1'; R_1). \end{aligned}$$

Remark 9. The symplectic instanton homology groups $HSI(Y,z)$ depend on the choice of base point $z \in \Sigma \subset Y$; cf. Sect. 5.3. As z varies, the groups naturally form a flat bundle over Y. Still, we usually drop z from the notation and denote them by $HSI(Y)$.

7 Properties and Examples

7.1 The Euler Characteristic

In general, the Euler characteristic of Lagrangian Floer homology is the intersection number of the two Lagrangians. In our situation, the corresponding intersection number is computed (up to a sign) in [1, Proposition 1.1(a),(b)]:

$$\chi\big(HSI(Y)\big) = [L_0] \cdot [L_1] = \begin{cases} \pm |H_1(Y;\mathbb{Z})| & \text{if } b_1(Y) = 0; \\ 0 & \text{otherwise.} \end{cases} \tag{7.1}$$

7.2 Examples

Proposition 8. *We have an isomorphism*

$$HSI(S^3) \cong \mathbb{Z}.$$

Proof. Let $\mathcal{H}_h$ denote the Heegaard decomposition $S^3 = H_0 \cup_\Sigma H_1$ of genus $h \geq 1$ such that there is a system of $2h$ curves α_i, β_i on Σ' as in Sect. 3.2 with the property that the β_i are null homotopic in H_0 and the α_i's are null homotopic in H_1.

With respect to the identification (3.3), the Lagrangians corresponding to H_0 and H_1 are given by

$$L_0 = \{(A_1, B_1, \dots, A_h, B_h) \in G^{2h} \mid B_i = I,\ i = 1, \dots, h\},$$
$$L_1 = \{(A_1, B_1, \dots, A_h, B_h) \in G^{2h} \mid A_i = I,\ i = 1, \dots, h\}.$$

These have exactly one intersection point, the reducible $A_i = B_i = I$. Clearly L_0 and L_1 intersect transversely in $\mathscr{N}(\Sigma') \subset G^{2h}$ at that point. It is somewhat counterintuitive that L_0 and L_1 can intersect transversely at I, because they both live in the subspace $\Phi^{-1}(0)$ of codimension three in $\mathscr{N}(\Sigma')$. However, that subspace is not smooth, so there is no contradiction. We conclude that the Floer chain group has one generator; hence so does the homology.

Proposition 9. *For $h \geq 1$, we have an isomorphism*

$$HSI(\#^h(S^1 \times S^2)) \cong \left(H_*(S^3; \mathbb{Z}/2\mathbb{Z})\right)^{\otimes h},$$

where the grading of the latter vector space is collapsed modulo 8.

Proof. Let $\mathcal{H}'_h$ be the Heegaard splitting of genus $h \geq 1$ for $\#^h(S^1 \times S^2)$. Since $L_0 = L_1 \cong G^h \cong (S^3)^h$, the cohomology ring of L_0 is generated by its degree-d ($d = 3$) part. Under the monotonicity assumptions that are satisfied in our setting, Oh [40] constructed a spectral sequence whose E^1 term is $H_*(L_0; \mathbb{Z}/2\mathbb{Z})$ and that converges to $HF_*(L_0, L_0; \mathbb{Z}/2\mathbb{Z})$. This sequence is multiplicative by the results of Buhovski [10] and Biran–Cornea [7, 8]. A consequence of multiplicativity is that the spectral sequence collapses at the E_1 stage, provided that $N_L > d + 1$; see, for example, [8, Theorem 1.2.2]. This is satisfied in our case because $N_{L_0} = N \geq 8$. Hence $HF_*(L_0, L_0; \mathbb{Z}/2\mathbb{Z}) \cong H_*(G^h; \mathbb{Z}/2\mathbb{Z})$.

Note that the results of Oh, Buhovski, and Biran–Cornea were originally formulated for monotone symplectic manifolds, i.e., in the setting of Section 2.1. However, they also apply to the Floer homology groups defined in Sect. 2.3. Indeed, the arguments in the proof of Proposition 1 about the finiteness of the Floer differential and the fact that $\partial^2 = 0$ apply equally well to the "string of pearls" complex used in [7, 8].

Proposition 10. *For a lens space $L(p,q)$, with* $\gcd(p,q) = 1$, *the symplectic instanton homology $HSI(L(p,q))$ is a free abelian group of rank p.*

Proof. Denote by $\mathcal{H}(p,q)$ the genus-one Heegaard splitting of $L(p,q)$. In terms of the coordinates $A = A_1$ and $B = B_1$, the two Lagrangians are given by $L_0 = B = 1$ and $L_1 = A^p B^{-q} = 1$. Their intersection consists of the space of representations $\pi_1(L(p,q)) \cong \mathbb{Z}/p \to \mathrm{SU}(2)$, which has several components: when p is odd, there are the reducible point ($A = B = I$) and $(p-1)/2$ copies of S^2; when p is even, there are two reducibles ($A = B = I$ and $A = -I, B = I$) and $(p-2)/2$ copies of S^2. It is straightforward to check that each component is a clean intersection in the sense of Poźniak [45]. Therefore, there exists a spectral sequence that starts at $H_*(L_0 \cap L_1) \cong \mathbb{Z}^p$ and converges to $HF(L_0, L_1)$; cf. [45]. Since the Euler characteristic of $HF(L_0, L_1)$ is p by (7.1), the sequence must collapse at the first stage.

Remark 10. More generally, whenever we have a Heegaard decomposition $\mathcal{H}$ of a three-manifold Y with $H^1(Y) = 0$, the two Lagrangians L_0 and L_1 will intersect transversely at the reducible I; cf. [1, Proposition 1.1(c)]. We could then fix an absolute $\mathbb{Z}/8\mathbb{Z}$-grading on $HSI(\mathcal{H})$ by requiring that the $\mathbb{Z}$ summand corresponding to I lie in grading zero.

7.3 Comparison with Other Approaches

Let $Y = H_0 \cup_\Sigma H_1$ be a Heegaard splitting of a 3-manifold, with Σ of genus h. Recall that the Lagrangians $L_0 = L(H_0)$ and $L_1 = L(H_1)$ live inside the subspace

$$\Phi^{-1}(0) = \Big\{(A_1, B_1, \ldots, A_h, B_h) \in G^{2h} \,\Big|\, \prod_{i=1}^{h} [A_i, B_i] = I\Big\} \subset \mathscr{N}(\Sigma').$$

There is an alternative way of embedding $\Phi^{-1}(0)$ inside a symplectic manifold of dimension $6h$. Namely, let Σ_+ be the closed surface (of genus $h+1$) obtained by gluing a copy of $T^2 \setminus D^2$ onto the boundary of $\Sigma' = \Sigma \setminus D^2$. Consider the moduli space $\mathscr{M}_{\mathrm{tw}}(\Sigma_+)$ of projectively flat connections (with fixed central curvature) in an odd-degree $\mathrm{U}(2)$-bundle over Σ_+, as in Sect. 3.6:

$$\mathscr{M}_{\mathrm{tw}}(\Sigma_+) = \Big\{(A_1, B_1, \ldots, A_{h+1}, B_{h+1}) \in G^{2h+2} \,\Big|\, \prod_{i=1}^{h+1} [A_i, B_i] = -I\Big\}/G.$$

Pick two particular matrices $X, Y \in G$ with the property that $[X, Y] = -I$. Then we can embed $\Phi^{-1}(0)$ into $\mathscr{M}_{\mathrm{tw}}(\Sigma_+)$ by the map

$$(A_1, B_1, \ldots, A_h, B_h) \to [(A_1, B_1, \ldots, A_h, B_h, X, Y)].$$

With respect to the natural symplectic form on $\mathscr{M}_{\mathrm{tw}}(\Sigma_+)$, the spaces $L_0, L_1 \subset \Phi^{-1}(0)$ are still Lagrangians. One can take their Floer homology and obtain a $\mathbb{Z}/4\mathbb{Z}$-graded abelian group. This was studied in [57, Sect. 4.1], where it is shown that it is a 3-manifold invariant. It is not obvious how this invariant relates to HSI.

The advantage of using $\mathscr{M}_{\mathrm{tw}}(\Sigma_+)$ instead of $\mathscr{N}(\Sigma')$ is that the former is already compact (and monotone); therefore, the definition of Floer homology is less technical, and this allows one to prove invariance. Nevertheless, the construction presented in this paper (using $\mathscr{N}(\Sigma')$) has certain advantages as well: first, the resulting groups are $\mathbb{Z}/8\mathbb{Z}$-graded rather than $\mathbb{Z}/4\mathbb{Z}$-graded. Second, it is better suited to defining an equivariant version of symplectic instanton homology. Indeed, unlike $\mathscr{M}_{\mathrm{tw}}(\Sigma_+)$, the space $\mathscr{N}(\Sigma')$ comes with a natural action of G that preserves the symplectic form and the Lagrangians. Following the ideas of Viterbo from [54, 55], we expect that one should be able to use this action to define equivariant Floer groups $HSI_*^G(Y)$ in the form of $H^*(BG)$-modules. For integral homology spheres, a suitable Atiyah–Floer conjecture would relate these to the equivariant instanton homology of Austin and Braam [5].

In a different direction, it would be interesting to study the connection between our construction and the Heegaard Floer homology groups $\widehat{HF}, HF^+$ of Ozsváth and Szabó [42, 43]. In particular, we ask the following question:

Question 7.1. For an arbitrary 3-manifold Y, are the total ranks of $HSI(Y) \otimes \mathbb{Q}$ and $\widehat{HF}(Y) \otimes \mathbb{Q}$ equal?

Finally, we remark that Jacobsson and Rubinsztein [25] have recently described a construction similar to the one in this paper, but for the case of knots in S^3 rather than 3-manifolds. Given a representation of a knot as a braid closure, they define two Lagrangians inside a certain symplectic manifold; this manifold was first constructed in [22] and is a version of the extended moduli space. Conjecturally, one should be able to take the Floer homology of the two Lagrangians and obtain a knot invariant.

Acknowledgments We would like to thank Yasha Eliashberg, Peter Kronheimer, Peter Ozsváth, Tim Perutz, and Michael Thaddeus for some very helpful discussions during the preparation of this paper. Especially, we would like to thank Ryszard Rubinsztein for pointing out an important mistake in an earlier version of this paper (in which topological invariance was stated as a conjecture).

The first author was partially supported by NSF grant DMS-0852439 and a Clay Research Fellowship. The second author was partially supported by the NSF grants DMS-060509 and DMS-0904358.

References

1. S. Akbulut and J. D. McCarthy. *Casson's invariant for oriented homology* 3*-spheres*, volume 36 of *Mathematical Notes*. Princeton University Press, Princeton, NJ, 1990.
2. M. Atiyah. New invariants of 3- and 4-dimensional manifolds. In *The mathematical heritage of Hermann Weyl (Durham, NC, 1987)*, volume 48 of *Proc. Sympos. Pure Math.*, pages 285–299. Amer. Math. Soc., Providence, RI, 1988.
3. M. F. Atiyah and R. Bott. The Yang–Mills equations over Riemann surfaces. *Philos. Trans. Roy. Soc. London Ser. A*, 308(1505):523–615, 1983.
4. D. M. Austin and P. J. Braam. Morse–Bott theory and equivariant cohomology. In H. Hofer, C. H. Taubes, A. Weinstein, and E. Zehnder, editors, *The Floer Memorial Volume*, number 133 in Progress in Mathematics, pages 123–183. Birkhäuser, 1995.
5. D. M. Austin and P. J. Braam. Equivariant Floer theory and gluing Donaldson polynomials. *Topology*, 35(1):167–200, 1996.

6. N. Berline, E. Getzler, and M. Vergne. *Heat kernels and Dirac operators*, volume 298 of *Grundlehren der Mathematischen Wissenschaften*. Springer, Berlin, 1992.
7. P. Biran and O. Cornea. Quantum structures for Lagrangian submanifolds. Preprint, arXiv:0708.4221.
8. P. Biran and O. Cornea. Rigidity and uniruling for Lagrangian submanifolds. *Geom. Topol.*, 13(5):2881–2989, 2009.
9. P. Braam and S. K. Donaldson. Floer's work on instanton homology, knots, and surgery. In H. Hofer, C. H. Taubes, A. Weinstein, and E. Zehnder, editors, *The Floer Memorial Volume*, number 133 in Progress in Mathematics, pages 195–256. Birkhäuser, 1995.
10. L. Buhovski. Multiplicative structures in Lagrangian Floer homology. Preprint, arXiv:math/0608063.
11. K. Cieliebak and K. Mohnke. Symplectic hypersurfaces and transversality in Gromov–Witten theory. *J. Symplectic Geom.*, 5(3):281–356, 2007.
12. S. K. Donaldson. Boundary value problems for Yang–Mills fields. *J. Geom. Phys.*, 8(1-4): 89–122, 1992.
13. S. K. Donaldson. *Floer homology groups in Yang–Mills theory*, volume 147 of *Cambridge Tracts in Mathematics*. Cambridge University Press, Cambridge, 2002. With the assistance of M. Furuta and D. Kotschick.
14. S. Dostoglou and D. Salamon. Self-dual instantons and holomorphic curves. *Ann. Math.*, 2(139):581–640, 1994.
15. A. Floer. An instanton-invariant for 3-manifolds. *Comm. Math. Phys.*, 119:215–240, 1988.
16. A. Floer. Morse theory for Lagrangian intersections. *J. Differential Geom.*, 28(3):513–547, 1988.
17. A. Floer. Instanton homology and Dehn surgery. In H. Hofer, C. H. Taubes, A. Weinstein, and E. Zehnder, editors, *The Floer Memorial Volume*, number 133 in Progress in Mathematics, pages 77–97. Birkhäuser, 1995.
18. K. Fukaya, Y.-G. Oh, K. Ono, and H. Ohta. *Lagrangian intersection Floer theory—anomaly and obstruction*. Kyoto University, 2000.
19. W. M. Goldman. The symplectic nature of fundamental groups of surfaces. *Adv. Math.*, 54(2):200–225, 1984.
20. V. Guillemin, V. Ginzburg, and Y. Karshon. *Moment maps, cobordisms, and Hamiltonian group actions*, volume 98 of *Mathematical Surveys and Monographs*. American Mathematical Society, Providence, RI, 2002. Appendix J by Maxim Braverman.
21. V. Guillemin and S. Sternberg. *Symplectic Techniques in Physics*. Cambridge University Press, Cambridge, second edition, 1990.
22. K. Guruprasad, J. Huebschmann, L. Jeffrey, and A. Weinstein. Group systems, groupoids, and moduli spaces of parabolic bundles. *Duke Math. J.*, 89(2):377–412, 1997.
23. J. Huebschmann. Symplectic and Poisson structures of certain moduli spaces. I. *Duke Math. J.*, 80(3):737–756, 1995.
24. J. Huebschmann and L. C. Jeffrey. Group cohomology construction of symplectic forms on certain moduli spaces. *Internat. Math. Res. Notices*, (6):245 ff., approx. 5 pp. (electronic), 1994.
25. M. Jacobsson and R. L. Rubinsztein. Symplectic topology of SU(2)-representation varieties and link homology, I: Symplectic braid action and the first Chern class. Preprint, arXiv:0806.2902.
26. L. C. Jeffrey. Extended moduli spaces of flat connections on Riemann surfaces. *Math. Ann.*, 298(4):667–692, 1994.
27. L. C. Jeffrey. Symplectic forms on moduli spaces of flat connections on 2-manifolds. In *Geometric topology (Athens, GA, 1993)*, volume 2 of *AMS/IP Stud. Adv. Math.*, pages 268–281. Amer. Math. Soc., Providence, RI, 1997.
28. F. C. Kirwan. *Cohomology of quotients in symplectic and algebraic geometry*, volume 31 of *Mathematical Notes*. Princeton University Press, Princeton, NJ, 1984.
29. P. B. Kronheimer and T. S. Mrowka. Knots, sutures and excision. *J. Differential Geom.*, 84(2):301–364, 2010.

30. Y. Lekili and M. Lipyanskiy. Geometric composition in quilted Floer theory. Preprint, arXiv:1003.4493.
31. E. Lerman. Symplectic cuts. *Math. Res. Lett.*, 2(3):247–258, 1995.
32. J. Marsden and A. Weinstein. Reduction of symplectic manifolds with symmetry. *Rep. Mathematical Phys.*, 5(1):121–130, 1974.
33. D. McDuff and D. Salamon. *J-holomorphic curves and symplectic topology*, volume 52 of *American Mathematical Society Colloquium Publications*. American Mathematical Society, Providence, RI, 2004.
34. E. Meinrenken and C. Woodward. Hamiltonian loop group actions and Verlinde factorization. *J. Differential Geom.*, 50(3):417–469, 1998.
35. E. Meinrenken and C. Woodward. Canonical bundles for Hamiltonian loop group manifolds. *Pacific J. Math.*, 198(2):477–487, 2001.
36. K. R. Meyer. Symmetries and integrals in mechanics. In *Dynamical systems (Proc. Sympos., Univ. Bahia, Salvador, 1971)*, pages 259–272. Academic, New York, 1973.
37. J. Mickelsson. String quantization on group manifolds and the holomorphic geometry of $\mathrm{Diff} S^1/S^1$. *Comm. Math. Phys.*, 112(4):653–661, 1987.
38. M. S. Narasimhan and C. S. Seshadri. Stable and unitary vector bundles on a compact Riemann surface. *Ann. Math. (2)*, 82:540–567, 1965.
39. Y.-G. Oh. Floer cohomology of Lagrangian intersections and pseudo-holomorphic disks. I. *Comm. Pure Appl. Math.*, 46(7):949–993, 1993.
40. Y.-G. Oh. Floer cohomology, spectral sequences, and the Maslov class of Lagrangian embeddings. *Internat. Math. Res. Notices*, (7):305–346, 1996.
41. Y.-G. Oh. Fredholm theory of holomorphic discs under the perturbation of boundary conditions. *Math. Z*, 222(3):505–520, 1996.
42. P. S. Ozsváth and Z. Szabó. Holomorphic disks and three-manifold invariants: properties and applications. *Ann. Math. (2)*, 159(3):1159–1245, 2004.
43. P. S. Ozsváth and Z. Szabó. Holomorphic disks and topological invariants for closed three-manifolds. *Ann. Math. (2)*, 159(3):1027–1158, 2004.
44. L. Polterovich. *The geometry of the group of symplectic diffeomorphisms*. Lectures in Mathematics ETH Zürich. Birkhäuser Verlag, Basel, 2001.
45. M. Poźniak. Floer homology, Novikov rings and clean intersections. In *Northern California Symplectic Geometry Seminar*, volume 196 of *Amer. Math. Soc. Transl. Ser. 2*, pages 119–181. Am. Math. Soc., Providence, RI, 1999.
46. T. R. Ramadas, I. M. Singer, and J. Weitsman. Some comments on Chern–Simons gauge theory. *Comm. Math. Phys.*, 126(2):409–420, 1989.
47. K. Reidemeister. Zur dreidimensionalen Topologie. *Abh. Math. Sem. Univ. Hamburg*, (9): 189–194, 1933.
48. H. L. Royden. *Real Analysis*. Macmillan, New York, 1963.
49. D. Salamon. *Lagrangian intersections, 3-manifolds with boundary, and the Atiyah–Floer conjecture*, pages 526–536. Birkhäuser, 1994.
50. D. Salamon and K. Wehrheim. Instanton Floer homology with Lagrangian boundary conditions. *Geom. Topol.*, 12(2):747–918, 2008.
51. P. Seidel. Vanishing cycles and mutation. In *European Congress of Mathematics, Vol. II (Barcelona, 2000)*, volume 202 of *Progr. Math.*, pages 65–85. Birkhäuser, Basel, 2001.
52. P. Seidel and I. Smith. A link invariant from the symplectic geometry of nilpotent slices. *Duke Math. J.*, 134(3):453–514, 2006.
53. J. Singer. Three-dimensional manifolds and their Heegaard diagrams. *Trans. Amer. Math. Soc.*, (35):88–111, 1933.
54. C. Viterbo. Functors and computations in Floer cohomology. Part II. Preprint, online at http://www.math.polytechnique.fr/cmat/viterbo/Prepublications.html.
55. C. Viterbo. Functors and computations in Floer homology with applications. I. *Geom. Funct. Anal.*, 9(5):985–1033, 1999.

56. K. Wehrheim. Lagrangian boundary conditions for anti-self-dual instantons and the Atiyah–Floer conjecture. *J. Symplectic Geom.*, 3(4):703–747, 2005. Conference on Symplectic Topology.
57. K. Wehrheim and C. Woodward. Floer field theory. Preprint, 2008.
58. K. Wehrheim and C. Woodward. Orientations for pseudoholomorphic quilts. Preprint.
59. K. Wehrheim and C. T. Woodward. Functoriality for Lagrangian correspondences in Floer theory. *Quantum Topol.*, 1(2):129–170, 2010.
60. E. Witten. Two-dimensional gauge theories revisited. *J. Geom. Phys.*, 9(4):303–368, 1992.
61. C. Woodward. The classification of transversal multiplicity-free group actions. *Ann. Global Anal. Geom.*, 14(1):3–42, 1996.

Projective Algebraicity of Minimal Compactifications of Complex-Hyperbolic Space Forms of Finite Volume

Ngaiming Mok†

Dedicated to Professor Oleg Viro on the joyous occasion of his sixtieth birthday

Abstract Let Ω be a bounded symmetric domain and $\Gamma \subset \mathrm{Aut}(\Omega)$ be an irreducible nonuniform torsion-free discrete subgroup. When Ω is of rank ≥ 2, Γ is necessarily arithmetic, and $X := \Omega/\Gamma$ admits a Satake-Baily-Borel compactification. When Ω is of rank 1, i.e., the complex unit ball B^n of dimension $n \geq 1$, Γ may be nonarithmetic. When $n \geq 2$, by a general result of Siu and Yau, X is pseudoconcave and it can be compactified to a Moishezon space by adding a finite number of normal isolated singularities. In this article we show that for $X := B^n/\Gamma$ the latter compactification is in fact projective-algebraic. We do this by showing that, just as in the arithmetic case of rank-1, X admits a smooth toroidal compactification $\overline{X}_M$ obtained by adjoining an Abelian variety to each of its finitely many ends, and $\overline{X}_M$ can be blown down to a normal projective-algebraic variety $\overline{X}_{\min}$ by solving $\overline{\partial}$ with L^2-estimates with respect to the canonical Kähler-Einstein metric and by normalization. As an application, we give an alternative proof of results of Koziarz-Mok on the submersion problem in the case of complex-hyperbolic space forms of finite volume by adapting the cohomological arguments in the compact case to general hyperplane sections of the minimal projective-algebraic compactifications which avoid the isolated singularities.

Keywords Minimal compactification • L^2-method • Hyperbolic space form • Kähler–Einstein metric

† Research partially supported by the CERG grant HKU7034/04P of the HKRGC, Hong Kong.

N. Mok (✉)
The University of Hong Kong, Pokfulam Road, Hong Kong
e-mail: nmok@hku.hk

I. Itenberg et al. (eds.), *Perspectives in Analysis, Geometry, and Topology: On the Occasion of the 60th Birthday of Oleg Viro*, Progress in Mathematics 296,
DOI 10.1007/978-0-8176-8277-4_14, © Springer Science+Business Media, LLC 2012

1 Introduction

Quotients X of bounded symmetric domains Ω with respect to torsion-free arithmetic lattices Γ have been well studied. In particular, the Satake–Borel–Baily compactifications (Satake [Sat60]; Borel-Baily [BB66]) give in general highly singular compactifications $X \subset \overline{X}_{\min}$ which are minimal in the sense that given any normal compactification $X \hookrightarrow \overline{X}$, the identity map on X extends to a holomorphic map $\overline{X} \to \overline{X}_{\min}$. The minimal compactifications are constructed using modular forms arising from Poincaré series, and for their construction, arithmeticity is used in an essential way.

When $X = \Omega/\Gamma$ is irreducible, by Margulis [Mar77] Γ is always arithmetic except in the case where Ω is of rank 1, i.e., in the case where Ω is isomorphic to the complex unit ball B^n, $n \geq 1$. When $n = 1$, the problem of compactifying Riemann surfaces of finite volume with respect to the Poincaré metric is classical and long understood, while in the case of higher-dimensional complex-hyperbolic space forms, i.e., quotients B^n/Γ, where $n \geq 2$ and $\Gamma \subset \mathrm{Aut}(B^n)$ are torsion-free lattices, minimal compactifications have not been described sufficiently explicitly in the literature.

It follows from the work of Siu–Yau [SY82] that X can be compactified by adding a finite number of normal isolated singularities. The proof in [SY82] is primarily differential-geometric in nature with a proof that applies to any complete Kähler manifold of finite volume with sectional curvature bounded between two negative constants. By the method of L^2-estimates of $\overline{\partial}$ it was proved in particular that $X = B^n/\Gamma$ is biholomorphic to a quasiprojective manifold. It leaves open the question whether the minimal compactification thus defined is projective-algebraic as in the case of arithmetic quotients.

In this article we give first of all a description of the structure near infinity of complex-hyperbolic space forms of dimension ≥ 2 which are not necessarily arithmetic quotients. We show that the picture of Mumford compactifications (smooth toroidal compactifications) obtained by adding an Abelian variety to each of the finitely many infinite ends remains valid (Ash–Mumford–Rappoport–Tai [AMRT75]). Each of these Abelian varieties has negative normal bundle and can be blown down to an isolated normal singularity, giving therefore a realization of the minimal compactification as proven in [SY82]. More importantly, we show that the minimal compactification is projective-algebraic.

Instead of using Poincaré series, we use the analytic method of solving $\overline{\partial}$ with L^2-estimates. The latter method originated from works of Andreotti–Vesentini [AV65] and Hörmander [Hör65], and the application of such estimates to the context of constructing holomorphic sections of Hermitian holomorphic line bundles on complete Kähler manifolds was initiated by Siu–Yau [SY77] (see also Mok [Mk90, Sects. 3 and 4] for a survey involving such methods). In our situation, from the knowledge of the asymptotic behavior with respect to a smooth toroidal compactification of the volume form of the canonical Kähler–Einstein metric, using L^2-estimates of $\overline{\partial}$ we construct logarithmic pluricanonical sections which are nowhere vanishing on given Abelian varieties at infinity when the logarithmic canonical line bundle is considered as a holomorphic line bundle over the Mumford compactification.

Using such sections and solving again the $\overline{\partial}$-equation with L^2-estimates with respect to appropriate singular weight functions (cf. Siu–Yau [SY77]), we construct a canonical map associated with certain positive powers of the logarithmic canonical bundle, showing that they are base-point-free. Thus, as opposed to the general case treated in [SY77], where the holomorphic map is defined only on the complete Kähler manifold X of finite volume, in the case of a ball quotient, our construction yields a holomorphic map defined on the Mumford compactification. It gives a holomorphic embedding of X onto a quasiprojective variety which admits a projective-algebraic compactification obtained by collapsing each Abelian variety at infinity to an isolated singularity.

The extension of the standard description of Mumford compactifications to the case of nonarithmetic higher-dimensional complex-hyperbolic space forms X of finite volume was known to the author but never published, and such a description was used in the proof of rigidity theorems for local biholomorphisms between such space forms in the context of Hermitian metric rigidity (Mok Mk89). A description of the asymptotic behavior of the canonical Kähler–Einstein metric with respect to Mumford compactifications also enters into play in the generalization of the immersion problem on compact complex hyperbolic space forms (Cao–Mok [CM90]) to the case of finite volume (To [To93]).

More recently, interest in the nature of minimal compactifications for nonarithmetic lattices in the rank-1 case was rekindled in connection with rigidity problems on holomorphic submersions between complex-hyperbolic space forms of finite volume (Koziarz–Mok [KM08]). There it was proved that any holomorphic submersion between compact complex-hyperbolic space forms must be a covering map, and a generalization was obtained also for the finite-volume case. Since the method of proof in [KM08] is cohomological, the most natural proof for a generalization to the finite-volume case can be obtained by compactifying such space forms by adding isolated singularities and by slicing such minimal compactifications by hyperplane sections, provided that it is known that the minimal compactifications are projective-algebraic. The proof of projective algebraicity by methods of partial differential equations and hence its validity also for the nonarithmetic case is the raison d'être of the current article.

In line with the purpose of bringing together analysis, geometry, and topology and establishing relationships among them, the substance of the current article makes use of a variety of results and techniques in these fields. To make the article accessible to a larger audience, in the exposition we have provided more details than is absolutely necessary. Especially, in regard to the technique of proving projective algebraicity by means of L^2-estimates of $\overline{\partial}$ we have included details to make the arguments as self-contained as possible for a nonspecialist.

2 Mumford Compactifications for Finite-Volume Complex-Hyperbolic Space Forms

2.1 Description of Mumford Compactifications for $X = B^n/\Gamma$ Arithmetic

Let B^n be the complex unit ball of complex dimension $n \geq 2$ and let $\Gamma \subset \mathrm{Aut}(B^n)$ be a torsion-free arithmetic subgroup. Let $E \subset \partial B^n$ be the set of boundary points b such that for the normalizer $N_b = \{\nu \in \mathrm{Aut}(B^n) : \nu(b) = b\}$, $\Gamma \cap N_b$ is an arithmetic subgroup of N_b. (Observe that every $\nu \in \mathrm{Aut}(B^n)$ extends to a real-analytic map from $\overline{B}^n$ to $\overline{B}^n$. We use the same notation ν to denote this extension.) The points $b \in E$ are the rational boundary components in the sense of Satake [Sat60] and Baily–Borel [BB66]. Modulo the action of Γ, those authors showed (in the general case of arithmetic quotients of bounded symmetric domains) that there are only a finitely many equivalence classes of rational boundary components. In the case of arithmetic quotients of the ball, the Satake–Baily–Borel compactification $\overline{X}_{\min}$ of X is set-theoretically obtained by adjoining a finite number of points, each corresponding to an equivalence class of rational boundary components. We fix a rational boundary component $b \in E$ and consider the Siegel domain presentation S_n of B^n with $b \in \partial B^n$ corresponding to infinity (Pyatetskii-Shapiro [Pya69]). In other words, we consider an inverse Cayley transform $\Phi : B^n \to S_n := \{(z_1, \ldots, z_n) \in \mathbb{C}^n : \mathrm{Im}\, z_n > |z_1|^2 + \cdots + |z_{n-1}|^2\}$ such that Φ extends real-analytically to $B^n - \{b\}$ and $\Phi|_{\partial B^n - \{b\}} \to \partial S_n$ is a real-analytic diffeomorphism. To simplify notation, we will write S for S_n. From now on, we will identify B^n with S via Φ and write $X = S/\Gamma$. Write $z' = (z_1, \ldots, z_{n-1}); z = (z'; z_n)$.

Let W_b be the unipotent radial of N_b. In terms of the Siegel domain presentation

$$\begin{aligned} W_b &= \{\nu \in N_b : \nu(z'; z_n) = (z' + a'; z_n + 2i\overline{a'} \cdot z' + i\|a'\|^2 + t); \\ a' &= (a_1, \ldots, a_{n-1}) \in \mathbb{C}^{n-1}, \quad t \in \mathbb{R}\}, \end{aligned} \tag{1}$$

where $\overline{a'} \cdot z' = \sum_{i=1}^{n-1} \overline{a}_i z_i$, W_b is a nilpotent group such that $U_b := [W_b, W_b]$ is real 1-dimensional, corresponding to the real one-parameter group of translations $\lambda_t, t \in \mathbb{R}$, given by $\lambda_t(z', z) = (z', z + t)$. Since $b \in \partial B^n$ is a rational boundary component, $\Gamma \cap W_b \subset W_b$ is a lattice, and in particular $\Gamma \cap W_b$ is Zariski dense in the real-algebraic group W_b. It follows that $[\Gamma \cap W_b, \Gamma \cap W_b] \subset \Gamma \cap U_b \subset U_b \cong \mathbb{R}$ must be nontrivial, otherwise $\Gamma \cap W_b$ and its Zariski closure W_b would be commutative, a plain contradiction. As a consequence, $\Gamma \cap U_b \subset U_b \cong \mathbb{R}$ must be a nontrivial discrete subgroup. Write $\lambda_\tau \in \Gamma \cap U_b$ for a generator of $\Gamma \cap U_b \cong \mathbb{Z}$. For any nonnegative integer N, define

$$S^{(N)} = \{(z'; z_n) \in \mathbb{C}^n : \mathrm{Im}\, z_n > \|z'\|^2 + N\} \subset S. \tag{2}$$

Consider the holomorphic map $\Psi : \mathbb{C}^{n-1} \times \mathbb{C} \to \mathbb{C}^{n-1} \times \mathbb{C}^*$ given by

$$\Psi(z';z_n) = (z', e^{\frac{2\pi i z_n}{\tau}}) := (w';w_n); \quad w' = (w_1,\dots,w_{n-1}), \tag{3}$$

which realizes $\mathbb{C}^{n-1} \times \mathbb{C}$ as the universal covering space of $\mathbb{C}^{n-1} \times \mathbb{C}^*$. Write $G = \Psi(S)$, and for any nonnegative integer N, write $G^{(N)} = \Psi(S^{(N)})$. Then G and each $G^{(N)}$ are total spaces of a family of punctured disks over $\mathbb{C}^{n-1}$. Define $\widehat{G} \subset \mathbb{C}^{n-1} \times \mathbb{C}$ by adding the "zero section" to G (i.e., by including the points $(w',0)$ for $w' \in \mathbb{C}^{n-1}$. Likewise, for each nonnegative integer N, define $\widehat{G}^{(N)} \subset \mathbb{C}^{n-1} \times \mathbb{C}$ by adding the "zero section" to $G^{(N)}$. We have

$$\begin{aligned} \widehat{G} &= \left\{(w';w_n) \in \mathbb{C} : |w_n|^2 < e^{\frac{-4\pi}{\tau}\|w'\|^2}\right\}; \\ \widehat{G}^{(N)} &= \left\{(w';w_n) \in \mathbb{C} : |w_n|^2 < e^{\frac{-4\pi N}{\tau}} \cdot e^{\frac{-4\pi}{\tau}\|w'\|^2}\right\}. \end{aligned} \tag{4}$$

The group $\Gamma \cap W_b$ acts as a discrete group of automorphisms on S. With respect to this action, any $\gamma \in \Gamma \cap W_b$ commutes with any element of $\Gamma \cap U_b$, which is generated by the translation λ_τ. Thus, $\Gamma \cap U_b \subset \Gamma \cap W_b$ is a normal subgroup, and the action of $\Gamma \cap W_b$ descends from S to $S/(\Gamma \cap U_b) \cong \Psi(S) = G$. Thus, there is a group homomorphism $\pi : \Gamma \cap W_b \to \mathrm{Aut}(G)$ such that $\Psi \circ \nu = \pi(\nu) \circ \Psi$ for any $\nu \cap \Gamma \cap W_b$. More precisely, given $\nu \in \Gamma \cap W_b$ of the form

$$\nu(z';z_n) = \left(z' + a'; z_n + 2i\overline{a'} \cdot z' + i\|a'\|^2 + k\tau\right) \quad \text{for some } a' \in \mathbb{C}^{n-1},\ k \in \mathbb{Z}, \tag{5}$$

we have

$$\pi(\nu)(w',w_n) = \left(w' + a', e^{-\frac{4\pi}{\tau}\overline{a'}\cdot w' - \frac{2\pi}{\tau}\|a'\|^2} \cdot w_n\right). \tag{6}$$

Then $S/(\Gamma \cap W_b)$ can be identified with $G/\pi(\Gamma \cap W_b)$. Since the action of W_b on S preserves ∂S, it follows readily from the definition of $\nu(z';z_n)$ that W_b preserves the domains $S^{(N)}$, so that $G^{(N)} \cong S^{(N)}/(\Gamma \cap U_b)$ is invariant under $\pi(\Gamma \cap W_b)$. The action of $\pi(\Gamma \cap W_b)$ extends to $\widehat{G}$. In fact, the action of $\pi(\Gamma \cap W_b)$ on the "zero section" $\mathbb{C}^{n-1} \times \{0\}$ is free, so that $\pi(\Gamma \cap W_b)$ acts as a torsion-free discrete group of automorphisms of $\widehat{G}$. Moreover, the action of $\pi(\Gamma \cap W_b)$ on $\mathbb{C}^{n-1} \times \{0\}$ is given by a lattice of translations Λ_b. Denoting the compact complex torus $(\mathbb{C}^{n-1} \times \{0\})/\Lambda_b$ by T_b, the Mumford compactification $\overline{X}_M$ of X is set-theoretically given by

$$\overline{X}_M = X\,(T_b), \tag{7}$$

where the disjoint union T_b is taken over the set of Γ-equivalence classes of rational boundary components $b \in E$. Define

$$\Omega_b^{(N)} = \widehat{G}^{(N)}/\pi(\Gamma \cap W_b) \supset G^{(N)}/\pi(\Gamma \cap W_b) \cong S^{(N)}/(\Gamma \cap W_b). \tag{8}$$

Then the natural map $G^{(N)}/\pi(\Gamma \cap W_b) = \Omega_b^{(N)} - T_b \hookrightarrow S/\Gamma = X$ is an open embedding for N sufficiently large, say $N \geq N_0$. Choose N_0 such that the latter statement is valid for every rational boundary component $b \in E$. As a complex manifold, $\overline{X}_M$ can be defined by

$$\overline{X}_M = X \quad (\Omega_b^{(N)})/\sim, \qquad \text{for any } N \geq N_0, \tag{9}$$

where $\sim$ is the equivalence relation which identifies points of X and $\Omega_b^{(N)}$ when they correspond to the same point of X (via the open embeddings $\Omega_b^{(N)} - T_b \hookrightarrow X$). For N sufficiently large, we may further assume that the images of $\Omega_b^{(N)} - T_b$ in X do not overlap. Thus, $\overline{X}_M$ is a complex manifold, and identifying $\Omega_b^{(N)}$, $N \geq N_0$, as open subsets of $\overline{X}_M, \{\Omega_b^{(N)}\}_{N \geq N_0}$ furnishes a fundamental system of neighborhoods of T_b in $\overline{X}_M$. It is possible to see from the preceding description of $\overline{X}_M$ that each compactifying divisor T_b can be blown down to a point. To see this, it suffices by the criterion of Grauert [Gra62] to show that the normal bundle of T_b in $\Omega_b^{(N)}(N \geq N_0)$ is negative. Actually, we are going to identify each $\Omega_b^{(N)}$ with a tubular neighborhood of the zero section of some negative holomorphic line bundle L over T_b. Recall that $\Omega_n^{(N)} = \widehat{G}^{(N)}/\pi(\Gamma \cap W_b)$, where by (6), $\pi(v)(w';w_n) = \left(w' + a'; \mathrm{e}^{-\frac{4\pi}{\tau}\overline{a'}\cdot w' - \frac{2\pi}{\tau}\|a'\|^2} \cdot w_n\right)$. Here $a' = a'(v)$ belongs to a lattice $\Lambda_b \subset \mathbb{C}^{n-1}$. Clearly, the nowhere-zero holomorphic functions $\Phi_{a'}(w') := \{\mathrm{e}^{-\frac{4\pi}{\tau}\overline{a'}\cdot w' - \frac{2\pi}{\tau}\|a'\|^2} : a' \in \Lambda_b\}$ on $\mathbb{C}^{n-1}$ constitute a system of factors of automorphy, i.e., they satisfy the composition rule $\Phi_{a'_2+a'_1}(w') = \Phi_{a'_2}(w' + a'_1) \cdot \Phi_{a'_1}(w')$. Extending the action of $\pi(\Gamma \cap W_b)$ to $\mathbb{C}^{n-1} \times \mathbb{C} \supset \widehat{G}$ yields that $(\mathbb{C}^{n-1} \times \mathbb{C})/\pi(\Gamma \cap W_b)$ is the total space of a holomorphic line bundle L over $T_b = (\mathbb{C}^{n-1} \times \{0\})/\Lambda_b$.

We introduce a Hermitian metric μ on the trivial line bundle $\mathbb{C}^{n-1} \times \mathbb{C}$ over $\mathbb{C}^{n-1}$. Namely, for $w = (w';w_n) \in \mathbb{C}^{n-1} \times \mathbb{C}$, we define

$$\mu(w;w) = \mathrm{e}^{\frac{4\pi}{\tau}\|w'\|^2} \cdot |w_n|^2. \tag{10}$$

The curvature form of μ is given by

$$-\sqrt{-1}\partial\overline{\partial}\log\mu = -\frac{4\pi}{\tau}\sqrt{-1}\partial\overline{\partial}\|w'\|^2, \tag{11}$$

which is a negative definite $(1,1)$-form on $\mathbb{C}^{n-1}$. For each N, the set $\widehat{G}^{(N)} = \{(w';w_n) \in \mathbb{C} : |w_n|^2 < \mathrm{e}^{\frac{-4\pi}{\tau}} \cdot \mathrm{e}^{\frac{-4\pi N}{\tau}\|w'\|^2}\}$ is nothing but the set of vectors of length not exceeding $\mathrm{e}^{\frac{-2\pi N}{\tau}}$ with respect to μ. Since $\widehat{G}^{(N)}$ is invariant under the action of $\pi(\Gamma \cap W_b)$, the latter must act as holomorphic isometries of the Hermitian line bundle $(\mathbb{C}^{n-1} \times \mathbb{C}; \mu)$. It follows that $\Omega_b^{(N)} = \widehat{G}^{(N)}/\pi(\Gamma \cap W_b)$ is the set of vectors on L of length less than $\mathrm{e}^{\frac{-2\pi N}{\tau}}$ on the Hermitian holomorphic line bundle $(L;\overline{\mu})$ over T_b, where $\overline{\mu}$ is the induced Hermitian metric on L. As a consequence, the normal bundle of T_b in $\Omega_b^{(N)}$ (being isomorphic to L) is negative, so that by the criterion of Grauert [Gra62], there exist a normal complex space Y and a holomorphic map $\sigma : \overline{X}_M \to Y$ such that $\sigma|_X$ is a biholomorphism onto $\sigma(X)$, and $\sigma(T_b)$ is a single point for each $b \in E$. In this way, one recovers the Satake–Baily–Borel compactification $Y = \overline{X}_{\min}$ from the toroidal compactification of Mumford.

2.2 *Description of the Canonical Kähler–Einstein Metric Near the Compactifying Divisors*

Fix a rational boundary component $b \in E$ and consider the tubular neighborhood $\Omega_b = \Omega_b^{(N)}$ of the compact complex torus T_b for some sufficiently large N. (T_b is in fact an Abelian variety because of the existence of the negative line bundle L.) Regard Ω_b as an open subset of the total space of the negative line bundle $(L;\overline{\mu})$ over T_b. One can now give on Ω_b an explicit description of the canonical Kähler–Einstein metric of X. For any $v \in L$ write $\|v\|^2$ for $\overline{\mu}(v;v)$ as defined toward the end of Sect. 2.1. Recall that on the Siegel domain $S \cong B^n$, the canonical Kähler–Einstein metric is defined by the Kähler form

$$\omega = \sqrt{-1}\partial\overline{\partial}\left(-\log\left(\operatorname{Im} z_n - \|z'\|^2\right)\right). \tag{1}$$

On the domain $\widehat{G} = \{(w';w_n) \in \mathbb{C}^{n-1} \times \mathbb{C}^* : |w_n| < e^{-\frac{2\pi}{\tau}\|w'\|^2}\}$, we have

$$|w_n| = e^{-\frac{2\pi}{\tau}}\operatorname{Im} z_n\,; \quad \text{i.e.,} \quad \operatorname{Im} z_n = -\frac{\tau}{2\pi}\log|w_n|, \tag{2}$$

so that the Kähler form of the canonical Kähler–Einstein metric on $G \cong S/(\Gamma \cap U_b)$ is given by

$$\omega = \sqrt{-1}\partial\overline{\partial}\left(-\log\left(-\frac{\tau}{2\pi}\log|w_n| - \|w'\|^2\right)\right). \tag{3}$$

From Sect. 2.1, (10), for a vector $w = (w';w_n) \in \mathbb{C}^{n-1} \times \mathbb{C}$ we have

$$\|w\| = \left(\mu(w,w)\right)^{\frac{1}{2}} = e^{\frac{2\pi}{\tau}\|w'\|^2} \cdot |w_n|. \tag{4}$$

It follows that

$$-\frac{\tau}{2\pi}\log|w_n| - \|w'\|^2 = -\frac{\tau}{2\pi}\left(\frac{2\pi}{\tau}\|w'\|^2 + \log|w_n|\right) = -\frac{\tau}{2\pi}\left(\log\|w\|\right), \tag{5}$$

and hence

$$\omega = \sqrt{-1}\partial\overline{\partial}\left(-\log\left(-\frac{\tau}{2\pi}\log\|w\|\right)\right) = \sqrt{-1}\partial\overline{\partial}\left(-\log(-\log\|w\|)\right). \tag{6}$$

Identifying Ω_b with an open tubular neighborhood of T_b in L, the same formula is valid on Ω_b with w replaced by a vector $v \in \Omega_b \subset L$. Then on Ω_b,

$$\omega = \frac{\sqrt{-1}\partial\overline{\partial}\log\|v\|}{-\log\|v\|} + \frac{\sqrt{-1}\partial(-\log\|v\|)\wedge\overline{\partial}(-\log\|v\|)}{(-\log\|v\|)^2}. \tag{7}$$

Write θ for minus the curvature form of the line bundle $(L, \overline{\mu})$. Note that θ is positive definite on T_b. Denote by π the natural projection of L onto T_b. Then

$$\omega = \frac{\pi^*\theta}{-2\log\|v\|} + \frac{\sqrt{-1}\partial\|v\|\wedge\overline{\partial}\|v\|}{\|v\|^2(-\log\|v\|)^2}. \tag{8}$$

In particular, we have the following result.

Proposition 1. *Denote by $\delta(x)$ the distance from $x \in \Omega_b$ to T_b in terms of any fixed Riemannian metric on $\overline{X}_M$. Let $\mathrm{d}V$ be a smooth volume form on $\overline{X}_M$. Then in terms of δ and $\mathrm{d}V$ and assuming that $\delta \leq \frac{1}{2}$ on Ω_b, the volume form $\mathrm{d}V_g$ of the canonical Kähler–Einstein metric g, given by $\mathrm{d}V_g = \frac{\omega^n}{n!}$ in terms of the Kähler form ω of (X, g), satisfies on Ω_b the estimate*

$$\frac{C_1}{\delta^2(-\log\delta)^{n+1}}\cdot \mathrm{d}V \leq \mathrm{d}V_g \leq \frac{C_2}{\delta^2(-\log\delta)^{n+1}}\cdot \mathrm{d}V$$

for some real constants $C_1, C_2 > 0$.

Proof. The estimate follows immediately by computing

$$\omega^n = \frac{n}{\|v\|^2(-\log\|v\|)^{n+1}}\cdot\left(\frac{\pi^*\theta}{2}\right)^{n-1}\wedge\sqrt{-1}\partial\|v\|\wedge\overline{\partial}\|v\|. \qquad \square$$

2.3 Extending the Construction of Smooth Toroidal Compactifications to Nonarithmetic Γ

Let $\Gamma \subset \mathrm{Aut}(B^n)$ be a torsion-free discrete subgroup such that $X = B^n/\Gamma$ is of finite volume with respect to the canonical Kähler–Einstein metric. According to the differential-geometric results of Siu–Yau [SY82], X can be compactified to a compact normal complex space by adding a finite number of points. We will now describe the structure of ends in differential-geometric terms according to Siu–Yau [SY82], which applies to any complete Kähler manifold Y of finite volume and of strictly negative Riemannian sectional curvature bounded between two negative constants, in which one considers the universal covering space $\rho : M \to Y$ and the Martin compactification $\overline{M}$, and we adapt the differential-geometric description to the special case where Y is a complex hyperbolic space form X of finite volume, i.e., $X = B^n/\Gamma$ for some torsion-free lattice Γ of automorphisms (hence necessarily isometries with respect to the canonical Kähler–Einstein metric). In the latter case, the Martin compactification of B^n is homeomorphic to the closure $\overline{B^n}$ with respect to the Euclidean topology, and we have knowledge of the stabilizers at a point $b \in \partial B^n$.

Let M be a simply connected complete Riemannian manifold of sectional curvature bounded between two negative constants. We remark that M can be compactified topologically by adding equivalence classes $M(\infty)$ of geodesic rays. Here two geodesic rays $\gamma_1(t), \gamma_2(t)$, $t > 0$, are equivalent if and only if the geodesic distance $d(\gamma_1(t), \gamma_2(t))$ is bounded independently of t. A topology (the cone topology) can be given such that the Martin compactification $\overline{M} = M \cup M(\infty)$ is homeomorphic to the closed Euclidean unit ball and every isometry of M extends to a homeomorphism of $\overline{M}$. There is a trichotomy of nontrivial isometries φ of M into the classes of elliptic, hyperbolic, and parabolic isometries. An isometry φ is elliptic whenever it has interior fixed points, φ is hyperbolic if it fixes exactly two points on the Martin boundary $M(\infty)$, and φ is parabolic if it fixes exactly one point on the boundary.

We briefly recall the scheme of arguments of [SY82] for the structure of ends, stated in terms of the special case of $X = B^n/\Gamma$ under consideration. Let $b \in \partial B^n$ and let $\Gamma_b' \subset \Gamma$ be the set of parabolic elements fixing b. A hyperbolic element of Γ and a parabolic element of Γ cannot share a common fixed point (cf. Eberlein–O'Neill [EO73]). Since Γ is torsion-free, it follows that either Γ_b' is empty or $\Gamma_b = \{id\} \cup \Gamma_b'$ is equal to the subgroup of Γ fixing b. By a result of Gromov [Gro78], there exists a positive constant ϵ (depending on Γ) such that the inequality $d(x, \gamma x) < \epsilon$ for some $x \in B^n$ implies that either γ is the identity or it is a parabolic element. For each $b_i \in \partial B^n$ (which corresponds to x_i in the notation of [SY82, following Lemma 2, p. 368]) such that $\Gamma_{b_i} \neq \{id\}$, define

$$A_i = \left\{ x \in B^n : \min_{\gamma \in \Gamma_{b_i}} d(x, \gamma x) < \epsilon \right\} \tag{1}$$

and

$$E = \left\{ x \in B^n : \min_{\gamma \in \Gamma} d(x, \gamma x) \geq \epsilon \right\}. \tag{2}$$

By the cited result of Gromov [Gro78], $B^n = E \cup (\cup A_i)$. In the present situation, the holomorphic parabolic isometries of B^n fixing b_i, together with the identity element, constitute precisely the unipotent radical W_{b_i} of the stabilizer N_{b_i} of b_i, as described here in Sect. 2.1, (1). Thus automatically, $\Gamma_{b_i} \subset W_{b_i}$ is nilpotent. It follows that the arguments of Siu–Yau [SY82, Lemma 3, p. 369] apply. Thus, denoting by $p : B^n \to X$ the canonical projection and shrinking ϵ if necessary, we have either $p(\overline{A}_i) = p(\overline{A}_j)$ or $p(\overline{A}_i) \cap p(\overline{A}_j) = \emptyset$. Using the finiteness of the volume, it was proved that there are only finitely many distinct ends $p(A_i)$, $1 \leq i \leq m$ [SY82, Lemma 4, p. 369] and that $p(E)$ is compact [SY82, preceding Lemma 3, p. 368]. Thus, we have the decomposition

$$X = p(E) \cup \Big(\bigcup_{1 \leq i \leq m} p(A_i) \Big). \tag{3}$$

Moreover, by [SY82, preceding Lemma 5, p. 370], the open sets A_i are connected.

Fix any i, $1 \le i \le m$, and write b for b_i. In order to show that the construction of the Mumford compactification extends to the present situation, it suffices to prove the following statements:

(I) $\Gamma \cap U_b$ is nontrivial, generated by $(z'; z_n) \to (z'; z_n + \tau)$ for some $\tau > 0$.
(II) There exists a lattice $A_b \subset \mathbb{C}^{n-1}$ such that $\Gamma_b = \Gamma \cap W_b$ can be written as

$$\Gamma_b = \left\{ \nu \in W_b : \nu(z'; z_n) = (z' + a'; z_n + 2i\overline{a'} \cdot z' + i\|a\|^2 + k\tau); a' \in \Lambda_b, k \in \mathbb{Z} \right\}.$$

(III) One can take A_i to contain $S^{(N)}$ for N sufficiently large. Here

$$S^{(N)} = \left\{ (z', z_n) \in \mathbb{C}^n : \mathrm{Im}\, z_n > \|z'\|^2 + N \right\} \subset S$$

in terms of the Siegel domain presentation S of B^n sending b to infinity (cf. Sect. 2.1, (2)).

We show first that (I) implies (II) and (III). First of all, (III) follows from (I) and the explicit form of the canonical Kähler–Einstein metric. In fact, the Kähler form is given by

$$\omega = \sqrt{-1}\partial\overline{\partial}\left(-\log\left(\mathrm{Im}\, z_n - \|z'\|^2\right)\right). \tag{4}$$

The restriction of ω to each upper half-plane $H_{z'_\circ} = \{(z'_\circ; z_n) : \mathrm{Im}\, z_n \ge |z'_\circ|^2\}$ is just the Poincaré metric on $H_{z'_\circ}$ with Kähler form

$$\omega\big|_{H_{z'_\circ}} = \frac{\sqrt{-1} dz_n \wedge d\overline{z}_n}{(\mathrm{Im}\, z_n - \|z'\|^2)^2}. \tag{5}$$

It follows immediately that for N sufficiently large and for χ the transformation $\chi(z'; z_n) = (z'; z_n + \tau)$, we have

$$d(z; \chi z) < \epsilon \text{ for all } z = (z'; z_n) \text{ with } \mathrm{Im}\, z_n > \|z'\|^2 + N. \tag{6}$$

Here d is the geodesic distance on S. Thus $S^{(N)} \subset A_i$ for N sufficiently large, proving that (I) implies (III).

We are now going to show that (I) implies (II). Write V_b for the group of translations of $\mathbb{C}^{n-1}$, and denote by $\rho : W_b \to W_b/U_b \cong V_b$ the canonical projection. We assert first of all that $\rho(\Gamma_b)$ is discrete in V_b. Suppose otherwise. Then there exists a sequence of $\gamma_j \in \Gamma_b = \Gamma \cap W_b$ such that $\rho(\gamma_j)$ are distinct and have an accumulation point in V_b. Say, $\rho(\gamma_j)(z') = z' + a'_j$ with $a'_j \to a'$. Then

$$\gamma_j(0; i) = (a'_j; i + i|a'_j|^2 + k_j\tau), \quad k_j \in \mathbb{Z}. \tag{7}$$

Thus,

$$\chi_j^{-k_j} \circ \gamma_j(0; i) = \left(a'_j; i + i|a'_j|^2\right) \to \left(a'; i + i\|a'\|^2\right). \tag{8}$$

Given (I), this contradicts the fact that Γ_b is discrete in W_b. Hence, (I) implies that $\rho(\Gamma_b)$ is discrete in V_b.

Next, we have to show that $\rho(\Gamma_b)$ is in fact a lattice, given (I). In order to do this, we need additional information about geodesic rays on A_i. By [SY82, Lemma 5, p. 371], for each $x \in A_i$ there is exactly one geodesic ray $\sigma(t), t \geq 0$ issuing from x and lying on $\overline{A_i}$. Namely, it is the ray joining x to $b = b_i$. Moreover, the geodesic $\sigma(t), -\infty < t < \infty$, must intersect E. Let Σ be the family of geodesic rays lying on $\overline{A}_i$ issuing from $\partial A_i \subset \partial E$. Then Σ is compact modulo the action of Γ_b in the sense that for every sequence (σ_j) of such geodesic rays there exists $\gamma_j \in \Gamma_b$ such that the family $(\gamma_j \circ \sigma_j)$ converges to a geodesic ray σ lying on A_i and issuing from ∂A_i. In fact, the set of equivalence classes Σ mod Γ_b is in one-to-one correspondence with $p(\partial A_i) \subset p(\partial E) \subset p(E)$. Since $p(E)$ is compact, for each sequence (σ_j) of geodesic rays issuing from ∂A_i there exists $\gamma_j \in \Gamma_b$ such that $\gamma_j \circ \sigma_j$ is convergent, given again by a geodesic ray σ. Since ∂A_i is closed, we must have $\sigma(0) = \lim_{j\to\infty} \gamma_j(\sigma_j(0)) \in \partial A_i$, proving the claim. In order to show that $\rho(\Gamma_b)$ is a lattice, given (I), it suffices to show that $V_b/\rho(\Gamma_b)$ is compact. In the Siegel domain presentation S, the geodesic ray from $z = (z'; z_n) \in S^{(N)}$ to b (located at "infinity") is given by the line segment $\{(z', z_n + it) : t \geq 0\}$. (Here t does not denote the geodesic length.) It follows that the family of geodesic rays in $\overline{A}_i$ issuing from ∂A_i is parameterized by $V_b \times \mathbb{R} = \mathbb{C}^{n-1} \times \mathbb{R}$, with the factor $\mathbb{R}$ corresponding to $\mathrm{Re}\, z_n$. Here Γ_b acts on $V_b \times \mathbb{R}$ in an obvious way. Modulo Γ_b, such geodesic rays are parameterized by $(V_b \times \mathbb{R})/\Gamma_b$, which is diffeomorphically a circle bundle over $V_b/\rho(\Gamma_b)$ (with fiber isomorphic to $\mathbb{R}/\mathbb{Z}\tau$). By the compactness of the family of geodesic rays Σ mod Γ_b, it follows that $V_b/\rho(\Gamma_b)$ must be compact, showing that (I) implies (II).

Finally, we have to justify (I), i.e., $\Gamma \cap U_b$ is nontrivial. Since $[W_b, W_b] = U_b$, in order for $\Gamma \cap U_b$ to be nontrivial it suffices to find two noncommuting elements of $\Gamma \cap W_b$. Take $\gamma_1, \gamma_2 \in \Gamma \cap W_b$ given by

$$\gamma_j(z'; z_n) = \left(z' + a'_j; z_n + 2i\overline{a'_j} \cdot z' + i\|a'_j\|^2 + t_j\right); \quad j = 1, 2. \tag{9}$$

Then

$$\gamma_k \circ \gamma_j(z'; z_n) = \Big(z' + a'_k + a'_j; z_n + 2i\overline{a'_k}(2 + a'_j) + 2i\overline{a'_j} \cdot z' + i\left(\|a'_k\|^2 + \|a'_j\|^2\right) + (t_k + t_j)\Big), \tag{10}$$

so that

$$\gamma_2 \circ \gamma_1 (z'; z_n) = \gamma_1 \circ \gamma_2 (z'; z_n) + 2i\left(\overline{a'_2} \cdot a'_1 - \overline{a'_1} \cdot a'_2\right), \tag{11}$$

(i.e.,)

$$\gamma_1^{-1} \circ \gamma_2^{-1} \circ \gamma_1 \circ \gamma_2 = \left(z'; z_n + 2i\left(\overline{a'_2} \cdot a'_1 - \overline{a'_1} \cdot a'_2\right)\right). \tag{12}$$

Therefore, two elements $\gamma_1, \gamma_2 \in \Gamma \cap W_b$ commute with each other if and only if $\overline{a'_1} \cdot a'_2$ is real, in other words, $a'_2 = ca'_1 + e'_2$ for some real number c and for some e'_2 orthogonal to a'_1. Assume that $\Gamma \cap U_b$ is trivial. Then necessarily $\Gamma \cap W_b$ is Abelian.

It follows readily that one can make a unitary transformation in the $(n-1)$ complex variables $z' = (z_n, \ldots, z_{n-1})$ such that any $\gamma \in \Gamma \cap W_b$ is of the form

$$\gamma(z'; z_n) = \left(z' + a'; z_n + 2i\overline{a'} \cdot z' + i\|a'\|^2 + t\right), \quad a' \in \mathbb{R}^{n-1}, t \in \mathbb{R}. \tag{13}$$

We argue that this would contradict the fact that Σ mod Γ_b is compact for the family of geodesic rays Σ issuing from ∂A_i. Consider the projection map $\theta(z'; z_n) = (z', \mathrm{Re} z_n)$. We assert first of all that $\theta : A_i \to \mathbb{C}^{n-1} \times \mathbb{R} = V_b \times \mathbb{R}$ is surjective. In fact, by [SY82, Appendix, p. 377 ff.], for any $z \in D, \sigma(t), t \geq 0$, a geodesic ray in D joining z to the infinity point b, and γ a parabolic isometry of D fixing b, $d(\sigma(t), \gamma \circ \sigma(t))$ decreases monotonically to 0 as $t \to +\infty$. It follows from the definition of A_i that for any $z = (z'; z_n) \in S$, we have $(z'; z_n + iy) \in A_i$ for y sufficiently large. Since θ is surjective, as in the last paragraph, the set Σ mod Γ_b of geodesic rays issuing from ∂A_i is now parameterized by $(V_b \times \mathbb{R})/\Gamma_b$. There is a natural map $(V_b \times \mathbb{R})/\Gamma_b \to V_b/\rho(\Gamma_b)$. If $\Gamma \cap W_b$ were commutative, then $\rho(\Gamma_b) \subset V_b \cong \mathbb{C}^{n-1}$ would be a discrete group of rank at most $n-1$, and hence $V_b/\rho(\Gamma_b)$ would be noncompact, in contradiction to the compactness of Σ mod $\Gamma_b \cong (V_b \times \mathbb{R})/\Gamma_b$. Thus, we have proved by contradiction that $\Gamma \cap W_b$ is noncommutative, so that $\Gamma \cap U_b \supset [\Gamma \cap W_b, \Gamma \cap W_b]$ is nontrivial, proving (I).

The extension of the construction of Mumford compactifications to nonarithmetic quotients $X = B^n/\Gamma$ of finite volume is completed. To summarize, we have proved the following result.

Theorem 1. *Let X be a complex hyperbolic space form of the finite volume $X = B^n/\Gamma$, where $\Gamma \subset \mathrm{Aut}(X)$ is a torsion-free lattice which is not necessarily arithmetic. Then X admits a smooth compactification $X \subset \overline{X}_M$ obtained by adding a finite number of Abelian varieties D_i such that each $D_i \subset \overline{X}_M$ is an exceptional divisor. As a consequence, X admits a normal compactification $\overline{X}_{\min}$, to be called the minimal compactification, by blowing down each exceptional divisor $D_i \subset \overline{X}_M$ to a normal isolated singularity. Moreover, the description of the volume form of the canonical Kähler–Einstein metric on X as given in Sect. 2.2, Proposition 1, remains valid also in the nonarithmetic case.*

3 Projective Algebraicity of Minimal Compactification of Finite-Volume Complex-Hyperbolic Space Forms

3.1 L^2-Estimates of $\overline{\partial}$ on Complete Kähler Manifolds

We are going to prove the projective algebraicity of minimal compactifications of complex-hyperbolic space forms of finite volume by means of the method of L^2-estimates of $\overline{\partial}$ over complete Kähler manifolds. To start with, we have the following standard existence theorem due to Andreotti–Vesentini [AV65] in combination with Hörmander [Hör65].

Theorem 2 (Andreotti–Vesentini [AV65], Hörmander [Hör65]). *Let (X,ω) be a complete Kähler manifold, where ω stands for the Kähler form of the underlying complete Kähler metric. Let (Λ,h) be a Hermitian holomorphic line bundle with curvature form $\Theta(\Lambda,h)$ and denote by* $\mathrm{Ric}(\omega)$ *the Ricci form of (X,ω). Let φ be a smooth function on X. Suppose c is a continuous positive function on X such that $\Theta(\Lambda,h)+\mathrm{Ric}(\omega)+\sqrt{-1}\partial\overline{\partial}\varphi \geq c\omega$ everywhere on X. Let f be a $\overline{\partial}$-closed square-integrable Λ-valued $(0,1)$-form on X such that $\int_X \frac{\|f\|^2}{c} < \infty$, where here and hereinafter, $\|\cdot\|$ denotes norms measured against natural metrics induced from h and ω. Then there exists a square-integrable Λ-valued section u solving $\overline{\partial}u = f$ and satisfying the estimate*

$$\int_X \|u\|^2 e^{-\varphi} \leq \int_X \frac{\|f\|^2}{c} e^{-\varphi} < \infty.$$

Furthermore, u can be taken to be smooth whenever f is smooth.

Siu–Yau proved in [SY82, Sect. 3] that a complete Kähler manifold of finite volume of sectional curvature bounded between two negative constants is biholomorphic to a quasiprojective manifold. Assuming without loss of generality that the complete Kähler manifold X under consideration is of complex dimension at least 2, they proved that there exists a projective manifold Z such that X is biholomorphic to a Zariski-open subset X' of Z such that, identifying X with X', $Z-X$ is an exceptional set of Z that can be blown down to a finite number of points. Their proof proceeds in fact by showing, using methods of complex differential geometry, that X is pseudoconcave and can be compactified by adding a finite number of points. Then using Theorem 1 and introducing appropriate singular weight functions as in [SY77], they showed that any pair of distinct points on X can be separated by pluricanonical sections, i.e., holomorphic sections of positive powers of the canonical line bundle K_X, and that furthermore, at every point $x \in X$ there exist some positive integer $\ell_0 > 0$ and $n+1$ holomorphic sections $s_0, s_1, \ldots, s_n$ of $K_X^{\ell_0}$, $n = \dim_{\mathbb{C}}(X)$, such that $s_0(x) \neq 0$ and such that $[s_0, s_1, \ldots, s_n]$ defines a holomorphic immersion into $\mathbb{P}^n$ in a neighborhood of x. Given this, and using the pseudoconcavity of X, together with a result of Andreotti–Tomassini [AT70, p. 97, Theorem 2], there exist some integer $\ell > 0$ and finitely many holomorphic sections of K_X^{ℓ} which embed X as a quasiprojective manifold.

3.2 Projective Algebraicity via L^2-Estimates of $\overline{\partial}$

Let $n \geq 1$ be a positive integer and let $\Gamma \subset \mathrm{Aut}(B^n)$ be a torsion-free discrete subgroup, $\Gamma \subset \mathrm{Aut}(B^n)$ being not necessarily arithmetic. Write $X = B^n/\Gamma$. As in Sect. 2, we write $\overline{X}_M$ for the Mumford compactification of X obtained by adding a finite number of Abelian varieties D_i and let $\overline{X}_{\min}$ be the minimal compactification of X obtained by blowing down each Abelian variety D_i at infinity to a normal isolated singularity. We are going to prove that $\overline{X}_{\min}$ is projective-algebraic. Here and in what follows, for a complex manifold Q we denote by K_Q its (holomorphic) canonical line bundle. We have the following theorem.

Main Theorem. *For a complex-hyperbolic space form $X = B^n/\Gamma$ of finite volume with Mumford compactification $\overline{X}_M$, write $\overline{X}_M - X = D$ for the divisor D at infinity. Write $D = D_1 \cup \cdots \cup D_m$ for the decomposition of D into connected components D_i, $1 \le i \le m$, each being biholomorphic to an Abelian variety. Write $E = K_{\overline{X}_M} \otimes [D]$ on $\overline{X}_M$. Then for a sufficiently large positive integer $\ell > 0$ and for each $i \in \{1,\dots,m\}$, there exists a holomorphic section $\sigma_i \in \Gamma(\overline{X}_M, E^\ell)$ such that $\sigma_i|_{D_i}$ is a nowhere-vanishing holomorphic section of $E^\ell|_{D_i} \cong \mathcal{O}_{D_i}$ and $\sigma_i|_{D_k} = 0$ for $1 \le k \le m, k \ne i$. Moreover, the complex vector space $\Gamma(\overline{X}_M, E^\ell)$ is finite-dimensional, and choosing a basis $s_0,\dots,s_{N_\ell}$, we have the canonical map $\Phi_\ell : X_M \to \mathbb{P}^{N_\ell}$, uniquely defined up to a projective-linear transformation on the target projective space, such that $s_0,\dots,s_{N_\ell}$ have no common zeros on $\overline{X}_M$ and such that the holomorphic map Φ_ℓ maps $\overline{X}_M$ onto a projective variety $Z \subset \mathbb{P}^{N_\ell}$ with m isolated singularities $\zeta_1,\dots,\zeta_m$ and restricts to a biholomorphism of X onto the complement $Z^0 := Z - \{\zeta_1,\dots,\zeta_m\}$. In particular, the isomorphism $\Phi_\ell|_X : X \xrightarrow{\cong} Z^0$ extends holomorphically to $\nu : \overline{X}_{\min} \to Z$, which is a normalization of the projective variety Z, and $\overline{X}_{\min}$ is projective-algebraic.*

Proof. We start with some generalities. For a holomorphic line bundle $\tau : L \to S$ over a complex manifold S, to avoid notational confusion we write $\mathfrak{L}$ (in place of L) for its total space. We have $K_{\mathfrak{L}} \cong \tau^*(L^{-1} \otimes K_S)$. Moreover, the zero section $O(L)$ of $\tau : L \to S$ defines a divisor line bundle $[O(L)]$ on $\mathfrak{L}$ isomorphic to $\tau^* L$.

Returning to the situation of the main theorem, we claim that the holomorphic line bundle $E = K_{\overline{X}_M} \otimes [D]$ is holomorphically trivial over a neighborhood of $D = D_1 \cup \cdots \cup D_m$. For $1 \le i \le m$ we denote by $\pi_i : N_i \to D_i$ the holomorphic normal bundle of D_i in the Mumford compactification $\overline{X}_M$. Then by construction, for each $i \in \{1,\dots,m\}$ there is some open neighborhood Ω_i of D_i in $\overline{X}_M$ on which there exists a biholomorphism $\nu_i : \Omega_i \xrightarrow{\cong} W_i \subset N_i$ of Ω_i onto some open neighborhood W_i of the zero section $O(N_i)$ of $\pi_i : N_i \to D_i$ such that ν_i restricts to a biholomorphism $\nu_i|_{D_i} : D_i \xrightarrow{\cong} O(N_i)$ of D_i onto the zero-section $O(N_i)$. Moreover, $\Omega_1,\dots,\Omega_m$ are mutually disjoint. On the total space $\mathfrak{N}_i$ of $\pi_i : N_i \to D_i$ as an $(n+1)$-dimensional complex manifold, the canonical line bundle $K_{\mathfrak{N}_i}$ is given by $K_{\mathfrak{N}_i} \cong \pi_i^*(N_i^{-1} \otimes K_{O(N_i)})$. Since $O(N_i) \cong D_i$ is an Abelian variety, its canonical line bundle $K_{O(N_i)}$ is holomorphically trivial, so that $K_{\mathfrak{N}_i} \cong \pi_i^* N_i^{-1}$. Denote by $\rho_i : \Omega_i \to D_i$ the holomorphic projection map corresponding to the canonical projection map $\pi_i : N_i \to D_i$. Restricting to W_i and transporting to $\Omega_i \subset \overline{X}_M$ by means of $\nu_i^{-1} : W_i \cong \Omega_i$, we have $K_{\Omega_i} \cong \rho_i^* N^{-1}_{D_i|\overline{X}_M}$. Here $N_{D_i|\overline{X}_M}$ denotes the holomorphic normal bundle of D_i in $\overline{X}_M$, and over $\Omega_i \subset \overline{X}_M$, the holomorphic line bundle $\rho_i^* N^{-1}_{D_i|\overline{X}_M}$ is biholomorphically isomorphic to the divisor line bundle $[D_i]^{-1}$. Thus, $K_{\Omega_i} \otimes [D_i]$ is holomorphically trivial over Ω_i, i.e., $K_{\overline{X}_M} \otimes [D]$ is holomorphically trivial over an open neighborhood of D which is the disjoint union $\Omega_1 \cup \cdots \cup \Omega_m$, proving the claim.

Base-Point Freeness on Divisors at Infinity. Fix i, $1 \le i \le m$. In the notation of Sect. 2, $\Omega_i \cong \widehat{G}_i/\Gamma_i$, and the isomorphism is realized by the uniformization map $\rho : S \to S/\Gamma \supset \widehat{G}_i/\Gamma_i$. At a point $x \in D_i$ we can use the Euclidean coordinates $w = (w'; w_n)$ as local holomorphic coordinates $w' = (w_1, \ldots, w_{n-1})$ on some open neighborhood $\Omega_x \Subset \Omega_i$. Without loss of generality, we assume that $|w_n| < \frac{1}{2}$ on Ω_x. Denote by dV_e the Euclidean volume form on Ω_x with respect to the standard Euclidean metric in the w-coordinates. By Proposition 1, the volume form dV_g of the canonical Kähler–Einstein metric satisfies on Ω_x the estimate

$$\frac{C_1}{|w_n|^2(-\log|w_n|)^{n+1}} \cdot dV_e \le dV_g \le \frac{C_2}{|w_n|^2(-\log|w_n|)^{n+1}} \cdot dV_e, \tag{1}$$

for some constants $C_1, C_2 > 0$, in which the constants may be different from those in Proposition 1 denoted by the same symbols. From the preceding paragraphs, the holomorphic line bundle $E = K_{\overline{X}_M} \otimes [D]^{-1}$ is holomorphically trivial on Ω_i, and from the proof it follows readily that a holomorphic basis of E over Ω_i can be chosen such that it corresponds to a meromorphic n-form ν_0 which is holomorphic and everywhere nonzero on $\Omega_i - D_i$, has precisely simple poles along D_i, and lifts to $\nu := \frac{dw^1 \wedge \cdots \wedge dw^n}{w_n}$ on $\widehat{G}_i$. Let $q > 0$ be an arbitrary positive integer. We have $\nu^q \in \Gamma(\Omega_i, E^q)$, and its restriction $\nu^q|_{\Omega_i - D_i}$ is a holomorphic section in $\Gamma(\Omega_i - D_i, K_X^q)$. Denote by h the Hermitian metric on K_X induced by the volume form dV_g. We assert that ν^q is not square-integrable when K_X^q is equipped with the Hermitian metric h^q. For $r > 0$, denote by $\Delta^n(r)$ the polydisk in $\mathbb{C}^n$ with coordinates $w = (w'; w_n)$ of polyradii $(r, \ldots, r)$ centered at the origin 0. Then for some $\delta > 0$, we have

$$\begin{aligned}
\int_{\Omega_x} \|\nu^q\|^2 dV_g &\ge C_1 \int_{\Delta^n(\delta)} \frac{1}{|w_n|^{2q}} |w_n|^{2q} (\log|w_n|)^{q(n+1)} \frac{1}{|w_n|^2(\log|w_n|)^{n+1}} \cdot dV_e \\
&= C_1 \int_{\Delta^n(\delta)} \frac{(\log|w_n|)^{(q-1)(n+1)}}{|w_n|^2} \cdot dV_e = \infty.
\end{aligned} \tag{2}$$

For any $k \in \{1, \ldots, m\}$, let Ω_k^0 be an open neighborhood of D_k such that $\Omega_k^0 \Subset \Omega_k$. For any i, $1 \le i \le m$, fixed as in the above, there exists a smooth function χ_i on $\overline{X}_M$ such that $\chi_i|_{\Omega_i^0} = 1$ and such that χ_i is identically 0 on some open neighborhood of $\overline{X}_M - \Omega_i$. Then on Ω_i, the smooth section $\chi_i \nu \in \mathcal{C}^\infty(\Omega_i, E)$ is of compact support, and it extends by zeros to a smooth section $\eta_i \in \mathcal{C}^\infty(\overline{X}_M, E)$. We have $\mathrm{Supp}(\overline{\partial}\eta_i) \subset \Omega_i - \Omega_i^0 \Subset X$. In particular, we have

$$\int_X \|\overline{\partial}\eta_i\|^2 < \infty. \tag{3}$$

Thus, η_i is not square-integrable, while $\overline{\partial}\eta_i$ is square-integrable. Regarding $\overline{\partial}\eta_i$ as a $\overline{\partial}$-closed K_X^q-valued smooth $(0,1)$-form, and noting that for $q \ge 2$ we have

$$\Theta(K_X^q, h^q) + \mathrm{Ric}(\omega_g) = (q-1)\omega_g \ge \omega_g, \tag{4}$$

by Theorem 1 there exists a smooth solution u_i of the inhomogeneous Cauchy–Riemann equation $\overline{\partial} u_i = \overline{\partial}\eta_i$ satisfying the estimate

$$\int_X \|u_i\|^2 \mathrm{d}V_g \le \int_X \frac{\|\overline{\partial}\eta_i\|^2}{q-1}\, \mathrm{d}V_g < \infty. \tag{5}$$

For each $k \in \{1,\ldots,m\}$, we have $\overline{\partial}\eta_i \equiv 0$ on $\Omega_k^0 - D_k$, so that u_i is holomorphic on each $\Omega_k^0 - D_k$. In what follows, k is arbitrary and fixed. In terms of the Euclidean coordinates $(w_1,\ldots,w_n)$ as used in (1), on $\Omega_k^0 - D_k$ we have $u_i = f\mathrm{d}w^1 \wedge \cdots \mathrm{d}w^n$, where f is a holomorphic function. Using the estimate of the volume form $\mathrm{d}V_g$ as given in Proposition 1, from the integral estimate (5) and the mean-value inequality for holomorphic functions, one deduces readily a pointwise estimate for f which implies that $|w_n^e f|$ is uniformly bounded on $\Omega_k^0 - D_k$ for some positive integer e. It follows that f is meromorphic on Ω_k, and hence $u_i|_{\Omega_k^0 - D_k}$ extends meromorphically to Ω_k. (Since k is arbitrary, u_i extends to a meromorphic section of $K_{\overline{X}_M}$.) As such, either u_i has removable singularities along D_k, or it has a pole of order p_k at a general point $x_k \in D_k$ for some positive integer p_k. In the former case, we will define p_k to be $-r_k$, where r_k is the vanishing order of the extended holomorphic section u_i at a general point of D_k. If $p_k \ge q$, then from the computation of integrals in (2), it follows readily that u_i cannot be square-integrable, which is a contradiction. So, either u_i has removable singularities along the divisor D_k, or it has poles of order $p_k < q$ at a general point of the divisor D_k. On the other hand, if we regard u_i rather as a holomorphic section of $E^q = K_{\overline{X}_M}^q \otimes [D]^q$ over each $\Omega_k^0 - D_k$, then u_i extends to Ω_k^0 as a holomorphic section with zeros of order $q - p_k > 0$. Define now $\sigma_i = \eta_i - u_i$. Then $\overline{\partial}\sigma_i = \overline{\partial}\eta_i - \overline{\partial} u_i = 0$ on $\overline{X}_M$ and $\sigma_i \in \Gamma(\overline{X}_M, E^q)$. Now, $\eta_i|_{D_i}$ is nowhere vanishing as a holomorphic section of the trivial holomorphic line bundle $E^q|_{D_i}$ over D_i, while $u_i|_{D_i}$ vanishes as a section in $\Gamma(D_i, E^q)$, so that $\sigma_i|_{D_i} = \eta_i|_{D_i}$ and σ_i is nowhere vanishing on D_i as a section of the trivial holomorphic line bundle $E^q|_{D_i}$. For $k \ne i$, we have $\eta_i|_{D_k} = 0$ by construction and $u_i|_{D_k} = 0$, where for the latter, one follows the same arguments as in the case $k = i$ in the above. Thus $\sigma_i|_{D_k} = 0$ for $k \ne i$. This proves the first statement of the main theorem. We proceed to prove the rest of the main theorem on the canonical maps Φ_ℓ in separate steps leading to the projective algebraicity of the minimal compactification $\overline{X}_{\min}$.

Base-Point Freeness on Mumford Compactifications. Fix any integer $q \ge 2$. From the preceding discussion, we have a finite number of holomorphic sections $\sigma_i \in \Gamma(\overline{X}_M, E^q)$, $1 \le i \le m$, whose common zero set $A = Z(\sigma_1,\ldots,\sigma_m)$ is disjoint from $D = D_1 \cup \cdots \cup D_m$. Thus, $A \subset X$ is a compact complex-analytic subvariety. We claim that for a positive and sufficiently large integer ℓ, the following holds: for each $x \in A$ there exists a holomorphic section $s \in \Gamma(\overline{X}_M, E^\ell)$ such that $s(x) \ne 0$. To prove the claim, let $(z_1,\ldots,z_n)$ be local holomorphic coordinates on a neighborhood U of x such that the base point x corresponds to the origin with respect to (z_j). Let χ be a smooth function of compact support on U such that $\chi \equiv 1$ on a neighborhood of x. Then $\varphi_\epsilon := n\chi\left(\log\left(\sum |z_j|^2 + \epsilon\right)\right)$ on U extends by zeros to a function on X, to be denoted by the same symbol. Since φ_ϵ is plurisubharmonic on some neighborhood of x and it vanishes outside a compact set (and hence $\sqrt{-1}\partial\overline{\partial}\varphi_\epsilon$ vanishes outside a compact set), there exists a positive real number C_ϵ such that

$$\sqrt{-1}\partial\overline{\partial}\varphi_\epsilon + C_\epsilon\omega \geq \omega. \tag{6}$$

As ϵ decreases to 0, the functions φ_ϵ converge monotonically to the function φ given by $n\chi\left(\log\left(\sum|z_j|^2\right)\right)$ on U and given by 0 on $X-U$. There exists a compact subset $Q \Subset U-\{x\}$ such that φ_ϵ is plurisubharmonic on $U-Q$ for each $\epsilon > 0$. Noting that $\log\left(\sum|z_i|^2\right)$ is smooth on Q, we see that (6) holds with C_ϵ replaced by some $C > 0$ independent of ϵ, provided that we require that $\epsilon \leq 1$, say. Letting ϵ converge to 0, we have also in the sense of currents the inequality

$$\sqrt{-1}\partial\overline{\partial}\varphi + C\omega \geq \omega. \tag{7}$$

In what follows, we are going to justify the solution of $\overline{\partial}$ with L^2-estimates for the singular weight function φ. Let ℓ be an integer such that $\ell \geq C+1$. Then we have

$$\sqrt{-1}\partial\overline{\partial}\varphi_\epsilon + \Theta(K_X^\ell, h^\ell) + \mathrm{Ric}(X,\omega) \geq \omega. \tag{8}$$

Let e be a holomorphic basis of the canonical line bundle K_U and consider the $\overline{\partial}$-exact K_U^ℓ-valued $(0,1)$-form $\overline{\partial}(\chi e^\ell)$, which will be regarded as a $\overline{\partial}$-exact (hence $\overline{\partial}$-closed) K_X^ℓ-valued $(0,1)$-form on X. Then Theorem 1 applies to give a solution to $\overline{\partial}u_\epsilon = \overline{\partial}(\chi e^\ell)$ satisfying the estimates

$$\int_X \|u_\epsilon\|^2 e^{-\varphi_\epsilon} \leq \int_X \|\overline{\partial}(\chi e^\ell)\|^2 e^{-\varphi_\epsilon} \leq M < \infty, \tag{9}$$

where M is a constant independent of ϵ, where we note that $\mathrm{Supp}(\overline{\partial}(\chi e^\ell))$ lies in a compact subset of X not containing x. From standard arguments involving Montel's theorem, choosing $\epsilon = \frac{1}{n}$, there exists a subsequence $\left(u_{\frac{1}{\sigma(n)}}\right)$ of $\left(u_{\frac{1}{n}}\right)$ which converges uniformly on compact subsets to a smooth solution of $\overline{\partial}u = \overline{\partial}(\chi e^\ell)$ satisfying the estimates

$$\int_X \|u\|^2 e^{-\varphi} \leq \int_X \|\overline{\partial}(\chi e^\ell)\|^2 e^{-\varphi} < \infty. \tag{10}$$

Define now $s = \chi e^\ell - u$. Then $\overline{\partial}s = \overline{\partial}(\chi e^\ell) - \overline{\partial}u = 0$, so that $s \in \Gamma(X, K_X^\ell)$. Now

$$e^{-\varphi} = \frac{1}{\left(\sum|z_j|^2\right)^n} = \frac{1}{r^{2n}} \tag{11}$$

in terms of the polar radius $r = \left(\sum |z_j|^2\right)^{\frac{1}{2}}$. Since the Euclidean volume form dV_e equals $r^{2n-1}dr \cdot dS$, where dS is the volume form of the unit sphere, it follows from (11) that $e^{-\varphi} = \frac{1}{r}dr \cdot dS$ is not integrable at $z = 0$. Then the estimate (10), according to which the solution u of $\overline{\partial} u = \overline{\partial}(\chi e^\ell)$ obtained must be integrable at 0, implies that we must have $u(x) = 0$. As a consequence, $s(x) = e^\ell \neq 0$. From the L^2-estimates (9) it follows that s is square-integrable with respect to the canonical Kähler–Einstein metric g on X and the Hermitian metrics h^ℓ on K_X^ℓ induced by g. From the volume estimates (2) it follows that s extends to a meromorphic section on $\overline{X}_M$ with at worst poles of order $\ell - 1$ along each of the divisors D_i, $1 \le i \le m$. Since A is compact, there exists a finite number of coordinate open sets U_α on X whose union covers A. Making use of these charts, it follows readily that there exists some positive integer ℓ_0 such that for $\ell \ge \ell_0$, the preceding arguments for producing $s \in \Gamma(X, K_X^\ell)$ apply for any $x \in A$. Let $\ell = pq$ be a multiple of q such that $\ell \ge \ell_0$. Here and in what follows, by a multiple of a positive integer q we will mean a product pq, where p is a positive integer. Further conditions will be imposed on ℓ later on. For the complex projective space $\mathbb{P}(V)$ associated to a finite-dimensional complex vector space V and for a positive integer e, we denote by $\nu_e : \mathbb{P}(V) \to \mathbb{P}(S^e V)$ the Veronese embedding defined by $\nu_e([\eta]) = [\otimes^e \eta] \in \mathbb{P}(S^e V)$. Then for the map $\Phi_q : \overline{X}_M \to \mathbb{P}^{N_q}$, the base locus of $\nu_p \circ \Phi_q : \overline{X}_M \to \mathbb{P}(S^p(\mathbb{C}^{N_q+1}))$ lies on A. If $\ell := pq$ is furthermore chosen such that $\ell \ge \ell_0$, then for any $x \in A$ there exists, moreover, $s \in \Gamma(\overline{X}_M, E^\ell)$ such that $s(x) \neq 0$, so that $\Gamma(\overline{X}_M, E^\ell)$ has no base locus, and hence $\Phi_\ell : \overline{X}_M \to \mathbb{P}^{N_\ell}$ is holomorphic.

Blowing Down Divisors at Infinity. For $\ell = pq$ as chosen, we denote by $\sigma_i^\ell \in \Gamma(\overline{X}_M, E^\ell)$ a holomorphic section in $\Gamma(\overline{X}_M, E^\ell)$ such that σ_i^ℓ is nowhere 0 on the divisor D_i, and $\sigma_i^\ell|_{D_k} = 0$ on any other irreducible divisor $D_k, k \neq i$ at infinity. (The notation σ_i used in earlier paragraphs is the same as σ_i^q, and we may take $\sigma_i^\ell = (\sigma_i^q)^p$.) Since $\sigma_i^\ell|_{D_i}$ is nowhere zero, for any section $s \in \Gamma(\overline{X}_M, E^\ell)$, $\frac{s}{\sigma_i^\ell}|_{D_i}$ is a holomorphic function on the irreducible divisor D_i; hence $\frac{s}{\sigma_i^\ell}\big|_{D_i}$ is some constant λ; i.e., $s = \lambda \sigma_i^\ell$ on D_i. It follows that the holomorphic mapping Φ_ℓ must be a constant map on each D_i, so that $\Phi_\ell(D_i)$ is a point on $\mathbb{P}^{N_\ell}$, to be denoted by ζ_i. Moreover, for $i_1 \neq i_2$, $1 \le i_1, i_2 \le m$, $\sigma_{i_1}^\ell\big|_{D_{i_1}}$ is nowhere vanishing, while $\sigma_{i_1}^\ell\big|_{D_{i_2}} = 0$ and $\sigma_{i_2}^\ell\big|_{D_{i_2}}$ is nowhere vanishing, while $\sigma_{i_2}^\ell\big|_{D_{i_1}} = 0$, implying that $\zeta_{i_1} \neq \zeta_{i_2}$. In other words, the points ζ_i, $1 \le i \le m$, are distinct.

Removing Ramified Points. It remains to show that for some choice of $\ell = pq$ the holomorphic mapping $\Phi_\ell : \overline{X}_M \to \mathbb{P}^{N_\ell}$ is a holomorphic embedding on X and that ζ_i is an isolated singularity of $Z := \Phi_\ell(\overline{X}_M)$. We start with showing that $\ell^\flat = p^\flat q$ can be chosen so that there are no ramified points on X, i.e., that $\Phi_{\ell^\flat}$ is a holomorphic immersion on X. Choose $\ell_1 = p_1 q$ to be a multiple of q such that the preceding arguments work for $\ell = \ell_1$. Let $S \subset \overline{X}$ be the subset where Φ_{ℓ_1} fails to be an immersion, to be called the ramification locus on X of Φ_{ℓ_1}. Clearly, $S \cup D \subset \overline{X}_M$ is a (compact) complex-analytic subvariety, so that $\overline{R} \subset X_M$

is also a (compact) complex-analytic subvariety. Then we have a decomposition $R = R_1 \cup \cdots \cup R_r$ into a finite number of irreducible components, so that writing $\overline{R}_i$ for the topological closure of R_i in $\overline{X}_M$, we have the decomposition $\overline{R} = \overline{R}_1 \cup \cdots \cup \overline{R}_r$ of the compact complex subvariety $\overline{R} \subset \overline{X}_M$ into a finite number of irreducible components. Suppose $\dim_{\mathbb{C}} R = r$. We are going to show that if we choose $\ell_2 = p_2 q$, where $p_2 = t_1 p_1$ is a sufficiently large multiple of p_1, then the ramification locus on X of Φ_{ℓ_2} is of dimension $\leq r-1$. Given this, by induction and taking ℓ to be an appropriate multiple of q, we will be able to prove that $\Phi_{\ell^\flat}|_X$ is a holomorphic immersion for some multiple $\ell^\flat = p^\flat q$ of q.

To reduce the ramification locus on X, for each R_j of dimension r we pick a point $x_j \in R_j$, and we are going to show that if $\ell_2 = p_2 q = t_1 p_1 q = t_1 \ell_1$ is sufficiently large, then there exists $s_j \in \Gamma(\overline{X}_M, E^{\ell_1})$ such that $s_j(x_j) \neq 0$. Since ℓ_2 is a multiple of ℓ_1, the ramification locus $R(\ell_2)$ on X of Φ_{ℓ_2} is contained in the ramification $R(\ell_1) = R$ on X of Φ_{ℓ_1}, and $R(\ell_2)$ does not contain any of the r-dimensional irreducible components of $R(\ell_1)$, it will follow that $\dim_{\mathbb{C}} R(\ell_2) \leq r-1$, as desired. To produce $s_j \in \Gamma(\overline{X}_M, E^{\ell_2})$, we use Theorem 1 with a slight modification, as follows. Recall that $z = (z_1, \ldots, z_n)$ are local holomorphic coordinates in a neighborhood U of x, where x corresponds to the origin in z. For the same cut-off function χ with $\mathrm{Supp}(\chi) \Subset U$ as above, and for $1 \leq k \leq n$, we solve the Cauchy–Riemann equation $\overline{\partial} u_k = \overline{\partial}(\chi z_k e^{\ell_2})$ with a more singular plurisubharmonic weight function $\psi = (n+1)\chi \log\left(\sum |z_k|^2\right)$. We choose $\ell_2 = t_1 p_1 q$ sufficiently large that

$$\sqrt{-1}\partial\overline{\partial}\psi + \Theta\left(K_X^{\ell_2}, h_2^{\ell_2}\right) + \mathrm{Ric}(X, \omega) \geq \omega \tag{12}$$

in the sense of currents. In analogy to (10), we obtain smooth solutions u_k on X to the equation $\overline{\partial} u_k = \overline{\partial}(\chi z_k e^{\ell_2})$ satisfying the L^2-estimate

$$\int_X \|u_k\|^2 e^{-\psi} \mathrm{d}V_g \leq \int_X \|\overline{\partial}\left(\chi z_k e^{\ell}\right)\|^2 e^{-\psi} \mathrm{d}V_g < \infty. \tag{13}$$

Since $e^{-\psi}\mathrm{d}V_e = \frac{1}{r^{2n+2}}\mathrm{d}V_e = \frac{1}{r^3}\mathrm{d}r \cdot \mathrm{d}S$, it follows from the integrability of $\|u_k\|^2 e^{-\psi}$ that we must have $u_k(x) = 0$ and also $\mathrm{d}u_k(x) = 0$. As explained above, by induction we have proven that there exists some multiple $\ell^\flat = p^\flat q$ such that $\Phi_{\ell^\flat}$ is a holomorphic immersion.

Separation of Points. To separate points, we are going to choose $\ell^\sharp = t\ell^\flat = tpq$ which is a multiple of $\ell^\flat$. For any positive integer k that is a multiple of q, we denote by $B^{(k)} \subset X \times X$ the subset of all pairs of points $(x_1, x_2) \in X \times X$ such that $\Phi_k(x_1) = \Phi_k(x_2)$. Clearly, $B^{(k)}$ contains the diagonal $\mathrm{Diag}(X \times X)$ as an irreducible component, which we will denote by B_0. Note that if k' is a multiple of k, then $B^{(k')} \subset B^{(k)}$ by the argument using Veronese embeddings as in the paragraph on removing ramification points. Since Φ_k is defined as a holomorphic map on $\overline{X}_M$, $\overline{B^{(k)}}$ has only a finite number of irreducible components; hence $B^{(k)}$ has only a finite number of irreducible components. Let now $k = \ell^\flat$ and write $B^{(\ell^\flat)} = B_0 \cup B_1 \cup \cdots \cup B_e$ for the decomposition of B into irreducible components. Let b be the maximum of the complex dimensions of $B_1,\ldots,B_e$. We are going to find a multiple $\ell^\sharp$ of $\ell^\flat$ for which the following holds. For each irreducible component B_c of complex dimension b we are going to find an ordered pair $(x_1,x_2) \in B_c - \mathrm{Diag}(X \times X)$ and holomorphic sections $s_c,t_c \in \Gamma(\overline{X}_M,E^\ell)$ such that $s_c(x_1) \neq 0, s_c(x_2) = 0$, while $t_c(x_1) = 0, t_c(x_2) \neq 0$. Given this, by the same reduction argument as in the above (in the paragraph for removing ramified points on X), by choosing $\ell^\sharp$ to be a multiple of $\ell^\flat$, we will have proven that $B^{(\ell^\sharp)}$ consists only of the diagonal $\mathrm{Diag}(X \times X)$, proving that $\Phi_{\ell^\sharp}$ separates points on X. To find the positive integral multiple ℓ of $\ell^\flat$ and a section $s = s_c$ in $\Gamma(\overline{X}_M,E^\ell)$ such that $s(x_1) \neq s(x_2)$, we choose holomorphic coordinate neighborhoods U_1 of x_1 respectively U_2 of x_2 such that $U_1 \cap U_2 = \emptyset$. For $i = 1,2$, denote by $z^{(i)} = \left(z_1^{(i)},\ldots,z_n^{(i)}\right)$ holomorphic coordinates in a neighborhood of x_i with respect to which the origin stands for the point x_i. Let χ_i, $i = 1,2$, be a smooth cut-off function such that χ_i is constant in a neighborhood of x_i, $i = 1,2$, and such that $\mathrm{Supp}(\chi_i) \Subset U_i$, so that in particular, $\mathrm{Supp}(\chi_1) \cap \mathrm{Supp}(\chi_2) = \emptyset$. Now we consider the weight function $\rho = n\chi_1 \log(\Sigma\,|z_i^{(1)}|^2) + n\chi_2 \log(\Sigma\,|z_i^{(2)}|^2)$. Let k be a positive integer. The smooth section $\chi_1 e^k$ of K_X^k over U_1 with compact support extends by zeros to a smooth section, to be denoted again by $\chi_1 e^k$, of K_X^k over X, so that $\chi_1 e^k|_{U_1^0} = e^k$ on some neighborhood $U_1^0 \Subset U_1$ and $\mathrm{Supp}(\eta) \Subset U_1$. Since $U_1 \cap U_2 = \emptyset$, we have in particular $\eta|_{U_2} = 0$. Our aim is to solve $\overline{\partial} u = \overline{\partial}(\chi_1 e^k)$ using Theorem 1. In analogy to (8) and using the same smoothing process as in preceding paragraphs, we have to find k such that

$$\sqrt{-1}\partial\overline{\partial}\rho + \Theta(K_X^k,h^k) + \mathrm{Ric}(X,\omega) \geq \omega \tag{14}$$

in the sense of currents. By exactly the same argument as in (8)–(10), the inequality (14) is satisfied for k sufficiently large. Applying Theorem 1, we have a smooth solution of $\overline{\partial} u = \overline{\partial}(\chi_1 e^k)$, where u satisfies the L^2-estimates

$$\int_X \|u\|^2 e^{-\rho} \leq \int_X \|\overline{\partial}(\chi_1 e^k)\|^2 e^{-\rho} < \infty, \tag{15}$$

so that $u(x_1) = u(x_2) = 0$ because of the choice of singularities of ρ at both x_1 and x_2. As a consequence, the smooth section $s := u - \chi_1 e^k$ of K_X^k over X satisfies $s(x_1) = e^k$, $s(x_2) = 0$, and s is a holomorphic section, since $\overline{\partial} s = \overline{\partial} u - \overline{\partial}(\chi_1 e^k) = 0$ on X. Interchanging x_1 and x_2, we obtain another holomorphic section $t \in \Gamma(X,K_X^k)$ such that $t(x_1) = 0$, while $t(x_2) = e^k$. Given this, taking $k = \ell^\sharp$ to be a sufficiently large multiple of $\ell^\flat$, the canonical map $\Phi_{\ell^\sharp} : X_M \to \mathbb{P}^{N_{\ell^\sharp}}$ is base-point-free (hence holomorphic) and a holomorphic immersion on X, and it furthermore separates points on X.

Blowing Down to Isolated Singularities. The map $\Phi_{\ell^\sharp} : X_M \to \mathbb{P}^{N_{\ell^\sharp}}$ sends each divisor D_i at infinity to a point $\zeta_i^\sharp \in P^{N_{\ell^\sharp}}$ such that $\zeta_i^\sharp$, $1 \leq i \leq m$, are distinct. We have, however, not ruled out the possibility that $\Phi_{\ell^\sharp}(x) = \Phi_{\ell^\sharp}(\zeta_i^\sharp)$ for some point $x \in X$ and some $i, 1 \leq i \leq m$. Since $\Phi_{\ell^\sharp}$ separate points on X, only a finite number of such pairs $(x_i, \zeta_i^\sharp)$ can actually occur. We claim that for a large enough multiple

ℓ of $\ell^\sharp$, we have $\Phi_\ell(x_i) \neq \zeta_i = \Phi_\ell(D_i)$. For this purpose it suffices to produce a holomorphic section $t \in \Gamma(X, K^{\ell}_{\overline{X}_M})$ such that $t|_{D_i}$ is nowhere vanishing whereas $t(x_i) = 0$. For this it suffices to solve the equation $\overline{\partial} u_i = \overline{\partial}\eta_i$ as in (3)–(5), choosing η_i to be 0 on some neighborhood of x_i, replacing q by ℓ and requiring at the same time that $u_i(x) = 0$. The latter requirement can be guaranteed by introducing a weight function φ as in (7) satisfying for ℓ sufficiently large the inequality

$$\sqrt{-1}\partial\overline{\partial}\varphi + \Theta(K^{\ell}_X, h^{\ell}) + \mathrm{Ric}(\omega_g) \geq \omega. \tag{16}$$

Thus the argument in (6)–(11) for the base-point freeness on $\overline{X}_M$ can be adapted here to yield the required sections t_i. Hence, we have proven that for some sufficiently large positive integer ℓ, the canonical map $\Phi_\ell : \overline{X}_M \to \mathbb{P}^{N_\ell}$ is holomorphic, blows down each divisor D_i at infinity to an isolated singularity ζ_i of $Z = \Phi_\ell(\overline{X}_M)$, and restricts to a holomorphic embedding on $X = \overline{X}_M - D$ into $Z^0 = Z - \{\zeta_1, \ldots, \zeta_m\}$.

End of Proof of Main Theorem. By definition, the minimal compactification $\overline{X}_{\min}$ of X is a normal complex space obtained by adding a finite number of isolated singularities μ_i, $1 \leq i \leq m$. Since $\Phi_\ell|_X : X \xrightarrow{\cong} Z^0 = Z - \{\zeta_1, \ldots, \zeta_m\}$ is a biholomorphism, for each $i \in \{1, \ldots, m\}$ there exist an open neighborhood V_i of μ_i in $\overline{X}_{\min}$ and an open neighborhood W_i of ζ_i in Z such that the biholomorphism $\Phi_\ell|_X$ restricts to a biholomorphism $\Phi_\ell|_{W_i} : W_i \xrightarrow{\cong} V_i$ and such that $\lim_{x \to \mu_i} \Phi_\ell(x) = \zeta_i$; Φ_ℓ extends to a continuous map $\widehat{\Phi}_\ell : \overline{X}_{\min} \to Z$ by defining $\widehat{\Phi}_\ell|_X = \Phi_\ell$ and $\widehat{\Phi}_\ell(\mu_i) = \zeta_i$. Since $\overline{X}_{\min}$ is normal, $\widehat{\Phi}_\ell$ is holomorphic, so that $\widehat{\Phi}_\ell : \overline{X}_{\min} \to Z$ is a normalization of Z. Finally, since $Z \subset \mathbb{P}^{N_\ell}$ is projective-algebraic, its normalization $\overline{X}_{\min}$ is projective-algebraic. The proof of Theorem 2 is complete. □

Remarks. (1) From the proof of the main theorem regarding base-point freeness on divisors at infinity, it follows that $\dim_{\mathbb{C}} \Gamma(\overline{X}_M, E^2) \geq m$, where m is the number of connected (equivalently irreducible) components of the divisor D at infinity. In other words, there are on X at least m linearly independent holomorphic 2-canonical sections of logarithmic growth (with respect to the Mumford compactification $X \hookrightarrow \overline{X}_M$ and hence with respect to any smooth compactification with normal-crossing divisors at infinity).

(2) For the statement and proof of the main theorem, it is not essential that the images $\Phi_\ell(D_i)$ be isolated singularities. We include this statement because the proof is more or less the same as in the steps yielding a holomorphic embedding Φ_ℓ on X.

3.3 *An Application of Projective Algebraicity of the Minimal Compactification to the Submersion Problem*

In relation to the submersion problem on complex-hyperbolic space forms, i.e., the study of holomorphic submersions between complex-hyperbolic space forms, Koziarz and Mok [KM08] proved the following rigidity result.

Theorem 3 (Koziarz-Mok [KM08]). *Let $\Gamma \subset \mathrm{Aut}(B^n)$ be a lattice of biholomorphic automorphisms. Let $\Phi : \Gamma \to \mathrm{Aut}(B^m)$ be a homomorphism and let $F : B^n \to B^m$ be a holomorphic submersion equivariant with respect to Φ. Suppose that $m \geq 2$ or $\Gamma \subset \mathrm{Aut}(B^n)$ is cocompact. Then $m = n$ and $F \in \mathrm{Aut}(B^n)$.*

The proof of Theorem 3 can be easily reduced to the case where $\Gamma \subset \mathrm{Aut}(B^n)$ is torsion-free, so that $X = B^n/\Gamma$ is a complex-hyperbolic space form of finite volume. One of the motivations to present a proof of the projective algebraicity of finite-volume complex-hyperbolic space forms arising from not necessarily arithmetic lattices is to give a deduction of the noncompact (finite-volume) case of Theorem 3 from the cohomological arguments in the compact case.

An Alternative Proof of Theorem 3 in the Case of Finite-Volume Quotients. Without loss of generality, assume that $\Gamma \subset \mathrm{Aut}(B^n)$ is torsion-free. We outline the arguments in the case where the complex-hyperbolic space form $X := B^n/\Gamma$ is compact. Write ω_X for the Kähler form of the canonical Kähler–Einstein metric on X of constant holomorphic sectional curvature -4π. Denote by $\overline{\omega}_{B^m}$ the closed $(1,1)$-form on B^m, by $\overline{\omega_m}$ the closed $(1,1)$-form on X induced by the Γ-invariant closed $(1,1)$-form $F^*\omega_{B^m}$, and by $[\cdots]$ the de Rham cohomology class on X of a closed differential form. Denote by $\mathcal{F}$ the holomorphic foliation on X induced by the Γ-equivariant foliation whose leaves are given by the level sets of the Γ-equivariant map $F : B^n \to B^m$, by $T_{\mathcal{F}}$ the associated holomorphic distribution on X, and by $N_{\mathcal{F}} := T_X/T_{\mathcal{F}}$ the holomorphic normal bundle of the foliation $\mathcal{F}$. In the case where the complex–hyperbolic space form X is compact, by an algebraic identity of Feder [Fed65] (cf. Koziarz–Mok [KM08, Lemma 1]) applied to Chern classes of the short exact sequence $0 \to T_{\mathcal{F}} \to T_X \to N_{\mathcal{F}} \to 0$ on X (which we will call the tangent sequence induced by $\mathcal{F}$ on X), and by the Hirzebruch proportionality principle, we have $[\omega_X - \overline{\omega_m}]^{n-m+1} = 0$. By the Schwarz lemma we have $\omega_X - \overline{\omega_m} \geq 0$ as a smooth $(1,1)$-form, and the identity on cohomology classes $[\omega_X - \overline{\omega_m}]^{n-m+1} = 0$ forces $(\omega_X - \overline{\omega_m})^{n-m+1} = 0$ everywhere on X, which implies that there are at least m zero eigenvalues of the nonnegative $(1,1)$-form on $\nu = \omega_X - \overline{\omega_m}$ on X. Since ν agrees with ω_X on the leaves of $\mathcal{F}$, we conclude that there are exactly m zero eigenvalues of ν everywhere on X. Thus $\mathcal{F} : B^n \to B^m$ is an isometric submersion in the sense of Riemannian geometry, and this leads to a contradiction.

In the noncompact case we need the extra condition $m \geq 2$. Since $X = B^n/\Gamma$ is of finite volume, by the main theorem, X admits the minimal compactification $\overline{X}_{\min}$ obtained by adding a finite number of normal isolated singularities ζ_i, $1 \leq i \leq m$, and moreover, $\overline{X}_{\min}$ is projective-algebraic. Embedding $\overline{X}_{\min} \subset \mathbb{P}^N$ as a projective-algebraic subvariety, a general section $H \cap \overline{X}_{\min}$ by a hyperplane $H \subset \mathbb{P}^N$ is smooth, and it avoids the finitely many isolated singularities ζ_i, $1 \leq i \leq m$. Write $X_H := H \cap \overline{X}_{\min}$, $X_H \subset X$. Restricting the short exact sequence $0 \to T_{\mathcal{F}} \to T_X \to N_{\mathcal{F}} \to 0$ to X_H, we conclude that $[\omega_X - \overline{\omega_m}]^{n-m+1} = 0$ as cohomology classes. Note that X_H is not a complex-hyperbolic space form, and we are not considering the tangent sequence of a holomorphic foliation induced by $F|_{\pi^{-1}(X_H)}$. (In fact, the restriction

$F|_{\pi^{-1}(X_H)}$ need not even have constant rank $m-1$.) In its place, we are considering the restriction of the tangent sequence induced by $\mathcal{F}$ on X to the compact complex submanifold X_H, and the cohomological identity $[\omega_X - \overline{\omega_m}]^{n-m+1} = 0$ results simply from the restriction of a cohomological identity on X as explained in the previous paragraph. We have $\dim_{\mathbb{C}}(X_H) = n-1$ and $n-m+1 \leq n-1$, since $m \geq 2$. From the cohomological identity $[\omega_X - \overline{\omega_m}]^{n-m+1} = 0$ on X_H and the inequality $\nu = \omega_X - \overline{\omega_m} \geq 0$ as $(1,1)$-forms, we conclude that there are at least $m-1$ zero eigenvalues of $\nu|_{X_H}$ everywhere on X_H. Thus, for every $z \subset B^n$ and for a general hyperplane $V \subset T_z(B^n)$, we have $\dim_{\mathbb{C}}(V \cap \mathrm{Ker}(\nu)) = m-1$. Since $\dim_{\mathbb{C}}(\mathrm{Ker}(\nu)) \leq m$, it follows that we must have $\dim_{\mathbb{C}}(\mathrm{Ker}(\nu)) = m$, and $F : B^n \to B^m$ is in fact a holomorphic submersion. This gives rise to a contradiction exactly as in the compact case. □

Acknowledgments The author would like to thank the organizers of the conference "Perspectives in Analysis, Geometry and Topology" in honor of Professor Oleg Viro on May 19–26, 2008, held in Stockholm, especially Professor Jöricke and Professor Passare, for inviting him to give a lecture at the conference. He also wishes to thank Vincent Koziarz for discussions that provided a motivation for finding an analytic proof of projective algebraicity of minimal compactifications of complex-hyperbolic space forms of finite volume, making it applicable to the nonarithmetic case.

References

[AMRT75] Ash, A., Mumford, D., Rapoport, M., Tai, Y.-S.: Smooth Compactification of Locally Symmetric Varieties, Lie Groups: History, Frontier and Applications. **4**, Math. Sci. Press, Brookline, MA (1975).

[AV65] Andreotti, A., Vesentini, E.: Carleman estimates for the Laplace–Beltrami operator on complex manifolds. Inst. Hautes Études Sci. Publ. Math., **25**, 81–130 (1965).

[AT70] Andreotti, A., Tomassini, G.: Some remarks on pseudoconcave manifolds, Essays on Topology and Related Topics. Dedicated to G. de Rham, ed. by A. Haefliger and R. Narasimhan, pp. 85–104, Springer, Berlin (1970).

[BB66] Baily, W.L., Jr., Borel, A.: Compactification of arithmetic quotients of bounded symmetric domains. Ann. Math., **84**, 442–528 (1966).

[CM90] Cao, H.-D., Mok, N.: Holomorphic immersions between compact hyperbolic space forms. Invent. Math., **100**, 49–61 (1990).

[EO73] Eberlin, P., O'Neill, B.: Visibility manifolds. Pacific J. Math., **46**, 45–109 (1973).

[Fed65] Feder, S.: Immersions and embeddings in complex projective spaces. Topology, **4**, 143–158 (1965).

[Gra62] Grauert, H.: Über Modifikationen und exzeptionelle analytische Menge. Math. Ann., **146**, 331–368 (1962).

[Gro78] Gromov, M.: Manifolds of negative curvature. J. Diff. Geom., **13**, 223–230 (1978).

[Hör65] Hörmander, L.: L^2-estimates and the existence theorems for the $\bar{\partial}$-operator. Acta. Math., **114**, 89–152 (1965).

[KM08] Koziarz, V., Mok, N.: Nonexistence of holomorphic submersians between complex unit balls equivariant with respect to a cocompact lattice. Amer. J. Math. **132** (2010), 1347–1363.

[Mar72] Margulis, R.A.: On connections between metric and topological properties of manifolds of nonpositive curvature. Proceedings of the VI topological Conference, Tbilisi (Russian) p. 83 (1972).

[Mar77] Margulis, G.A.: Discrete groups of motion of manifolds of nonpositive curvature. AMS Transl., (2) **109**, 33–45 (1977).

[Mok89] Mok, N.: Metric Rigidity Theorems on Hermitian Locally Symmetric Manifolds. Series in Pure Mathematics, World Scientific, Singapore **6** (1989).

[Mok90] Mok, N.: Topics in complex differential geometry, in Recent Topics in Differential and Analytic geometry pp. 1–141, Adv. Stud. Pure Math., 18-I, Academic, Boston, MA (1990).

[Pya69] Pyatetskii-Shapiro, I.I.: Automorphic Functions and the Geometry of Classical Domains. Gordon & Breach Science, New York (1969).

[Sat60] Satake, I.: On compactifications of the quotient spaces for arithmetically defined discontinuous groups. Ann. Math., **72**, 555–580 (1960).

[SY77] Siu, Y.-T., Yau, S.-T.: Complete Kähler manifolds with non-positive curvature of faster than quadratic decay. Ann. Math., **105**, 225–264 (1977).

[SY82] Siu, Y.-T., Yau, S.-T.: Compactification of negatively curved complete Kähler manifolds of finite volume. in Seminar on Differential Geometry, ed. Yau, S.-T., Ann. Math. Studies, **102**, pp. 363–380, Princeton University Press, Princeton, NJ (1982).

[To93] To, W.-K.: Total geodesy of proper holomorphic immersions between complex hyperbolic space forms of finite volume. Math. Ann., **297**, 59–84 (1993).

Some Examples of Real Algebraic and Real Pseudoholomorphic Curves

Stepan Yu. Orevkov

To Oleg Viro

Abstract In this paper we construct several examples (series of examples) of real algebraic and real pseudoholomorphic curves in $\mathbb{RP}^2$ in which we tried to maximize different characteristics among curves of a given degree. In Sect. 2, this is the number of nonempty ovals; in Sect. 4, the number of ovals of the maximal depth; in Sect. 5, the number n such that the curve has an A_n singularity. In the pseudoholomorphic case, the questions of Sects. 4 and 5 are equivalent to the same problem about braids, which is studied in Sect. 3. In Sect. 6.1, we construct a real algebraic M-curve of degree $4d+1$ with four nests of depth d (which shows that the congruence mod 8 proven in a joint paper with Viro is "nonempty"). In Sect. 6.2, we generalize this construction. In Sect. 7, we construct real algebraic M-curves of degree 9 with a single exterior oval, and we classify such curves up to isotopy.

Keywords Isotopy • M-curve • oval • Pseudoholomorphic curve • Real algebaric curve

1 Introductory Remarks

Let $\alpha = \limsup(\alpha_m/m^2)$, where α_m is twice the maximal number n such that there exists an algebraic curve in $\mathbb{CP}^2$ of degree m with an A_n singularity. Similarly, let $\beta = \limsup(\beta_k/k^2)$, where $\beta_k = \max l_{k-2}(A)$, where $l_{k-2}(A)$ is the number of ovals of A of depth $k-1$ and the maximum is taken over all real algebraic curves in $\mathbb{RP}^2$

S.Yu. Orevkov (✉)
IMT, Université Paul Sabatier, 118 route de Narbonne, Toulouse, France

Steklov Mathematical Institute, Gubkina 8, Moscow, Russia
e-mail: orevkov@math.ups-tlse.fr

I. Itenberg et al. (eds.), *Perspectives in Analysis, Geometry, and Topology: On the Occasion of the 60th Birthday of Oleg Viro*, Progress in Mathematics 296,
DOI 10.1007/978-0-8176-8277-4_15,

of degree $2k$. Let α_{ph} and β_{ph} be the same numbers for pseudoholomorphic curves. In the following table we summarize all known estimates for these numbers (LB/UB stand for lower/upper bound).

1	Evident LB for α, β, α_{ph}, β_{ph}
$28/27$, $8-4\sqrt{3}$	LB for α from [4, 14]
$9/8$	LB for β proved in Sect. 3.3
$7/6$	LB for α proved in Sect. 5
$4/3$	LB for α_{ph} and β_{ph} proved in Sects. 2–4
$3/2$	UB for α, β, α_{ph}, β_{ph} coming from signature estimates
2	Evident UB for α, β, α_{ph}, β_{ph}

2 Iteration of Wiman's Construction

Wiman [35] proposed a method to construct real algebraic M-curves in $\mathbb{RP}^2$ that have many nests. Here we use Wiman's construction to obtain curves with many nonempty ovals. As is shown in [16], the number I_d of isotopy types realizable by real algebraic curves of degree d in $\mathbb{RP}^2$ has the asymptotics $\log I_d = Cd^2 + o(d^2)$ for some positive constant C, and the only known upper bounds for C come from the fact that $C \leq \limsup f(L_d/d^2)$, where f is a certain effectively computable monotone function and L_d is the maximal number of nonempty ovals that a curve of degree d may have. All known upper bounds for L_d are of the form $d^2/4 + O(d)$. Here we construct real algebraic and real pseudoholomorphic curves, in particular M-curves, with as many nonempty ovals as we can. The best asymptotic that we can achieve for pseudoholomorphic curves is only $d^2/6 + o(d^2)$. In the algebraic case, the obtained asymptotics are yet worse.

Let us recall Wiman's construction. We start with an M-curve C of even degree d given by an equation $F = 0$. We double C and then perturb it, i.e., consider a curve $C' = \{F^2 - \varepsilon G = 0\}$, $|\varepsilon| \ll 1$, where G is some polynomial of degree $2d$. Suppose that the curve $G = 0$ meets C transversally. Then each arc of C where $G > 0$ provides an oval of C' (obtained by doubling the arc and joining the ends). In the same way, each oval of C where $G > 0$ provides a pair of nested ovals of C'. If we are lucky to find G such that it has $2d^2$ zeros on one oval of C and is positive on all other ovals, then we obtain an M-curve that has $O(d^2)$ nested pairs of ovals. This can be attained, for example, if we start with an M-curve C one of whose ovals maximally intersects a line.

In speaking of Wiman's construction, the divisor of G on C will be called the *branching divisor*.

If we work with real pseudoholomorphic curves, then we need not concern ourselves whether it is possible to place correctly the branching divisor. Perturbing if necessary the almost complex structure, we may place it wherever we want. The only restriction is the total degree and the parity of the number of points at each branch of C.

We say that an arrangement of embedded circles on $\mathbb{RP}^2$ is realizable by a real pseudoholomorphic curve if there exists a real pseudoholomorphic curve in $\mathbb{CP}^2$ whose set of real points is isotopic to the given arrangement.

Recall that a *nest of depth d* is a union of d ovals $V_1 \cup \cdots \cup V_d$ such that V_{i+1} is surrounded by V_i, $i = 1, \dots, n-1$. We say that a nest N of a curve C is *simple* if there exists an embedded disk $D \subset \mathbb{RP}^2$ such that $N = D \cap C$.

We shall use the encoding of isotopy types of smooth embedded curves in $\mathbb{RP}^2$ proposed by Viro. Namely, n denotes n ovals outside each other; $A \sqcup B$ denotes a union of two curves encoded by A and B respectively if there exist disjoint embedded disks containing them; $1\langle A\rangle$ denotes an oval surrounding a curve encoded by A; $n\langle A\rangle = 1\langle A\rangle \sqcup \cdots \sqcup 1\langle A\rangle$ (n times).

We extend this encoding as follows. Let $1\langle\langle d\rangle\rangle$ denote a simple nest of depth d and let $n\langle\langle d\rangle\rangle = 1\langle\langle d\rangle\rangle \sqcup \cdots \sqcup 1\langle\langle d\rangle\rangle$ (n times). Also, if S encodes the isotopy type of a curve A, and A' is obtained from A by replacing each component by k parallel copies, then we denote the isotopy type of A' by $\langle S\rangle^k$ or just by S^k in the case that S is of the form $n\langle S_1\rangle$. For example, $2\langle\langle 3\rangle\rangle = \langle 2\rangle^3 = 2\langle\,\langle 1\rangle^2\,\rangle = 2\langle 1\langle 1\rangle\rangle = 1\langle 1\langle 1\rangle\rangle \sqcup 1\langle 1\langle 1\rangle\rangle$ denotes .

Proposition 2.1. *(a) For any positive integers m and k there exists a real pseudoholomorphic M-curve $C_{m,k}$ in $\mathbb{RP}^2$ of degree $d = 2^k m$ realizing the isotopy type*

$$\frac{m^2-3m+2}{2}\langle\langle 2^k\rangle\rangle \sqcup \left(\bigsqcup_{j=1}^{k-1}(4^{j-1}m^2-1)\langle\langle 2^{k-j}\rangle\rangle\right) \sqcup 4^{k-1}m^2. \tag{1}$$

The number of nonempty ovals of this curve is $\frac{1}{6}(4^k-1)m^2 - \frac{3}{2}(2^k-1)m + k = \frac{1}{6}\,(d^2-m^2) - \frac{3}{2}(d-m) + k$. So for each series $\{C_{m,k}\}_{k\geq 0}$ with a fixed m, these numbers have the asymptotics $\frac{1}{6}d^2 + O(d)$.

(b) If $k \leq 3$, then for any m, the M-curve $C_{m,k}$ can be realized algebraically. The number of nonempty ovals of $C_{m,3}$ is $\frac{21}{2}(m^2-m)+3 = \frac{21}{128}d^2 + O(d)$.

(c) For any $k > 1$ there exists an algebraic curve $C'_{2,k}$ of degree $d = 2^{k+1}$ realizing the isotopy type

$$3\langle\langle 2^{k-1}\rangle\rangle \sqcup \left(\bigsqcup_{j=2}^{k-1}(4^j-2^{j-2})\langle\langle 2^{k-j}\rangle\rangle\right) \sqcup 4^k. \tag{2}$$

The number of ovals of $C'_{2,k}$ is $\frac{1}{2}d^2 - (\frac{k}{8}-1)d$, i.e., it is an $(M-r)$-curve for $r = (k-4)2^{k-2}+2 = O(d\log d)$.

The number of nonempty ovals of $C'_{2,k}$ is $\frac{1}{6}d^2 - \frac{k+7}{8}d + \frac{4}{3} = \frac{1}{6}d^2 + O(d\log d)$.

Proof. All these curves are obtained by iterating Wiman's construction.

(a) We start with Harnack's curve $C_{m,0}$ of degree m and apply Wiman's construction to it k times. At each step, we place the branching divisor on one empty exterior oval (see Fig. 1a–c) except at the first step, when we place it on the nonempty oval (for even m) or on the odd branch (for odd m).

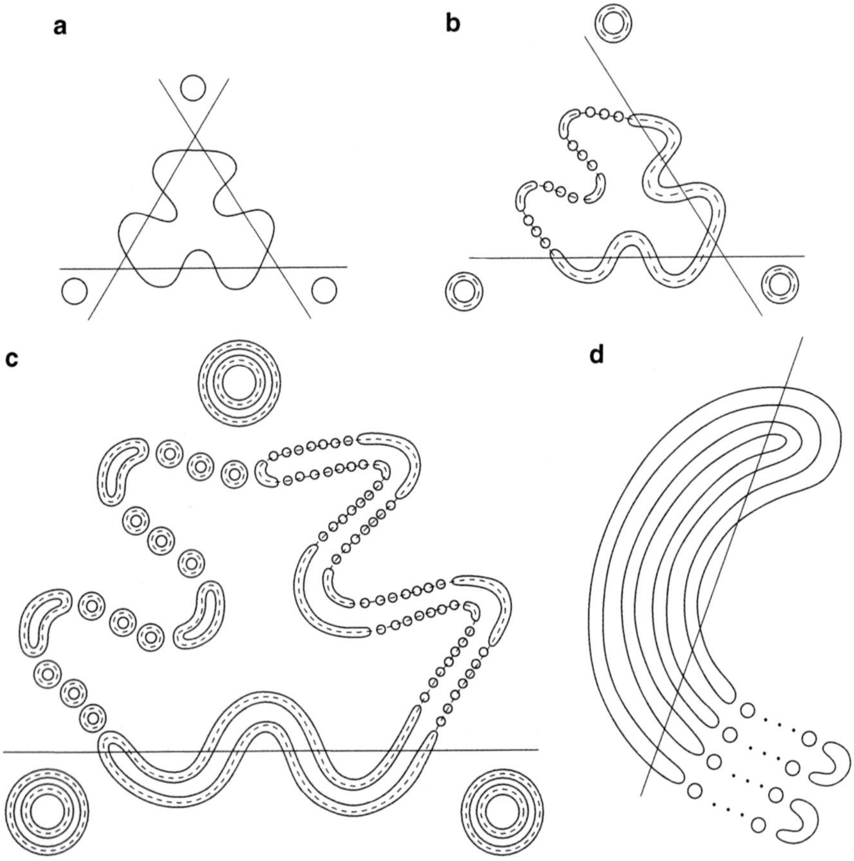

Fig. 1 (**a**) The curve $C_{4,0}$. (**b**) The curve $C_{4,1}$. (**c**) The curve $C_{4,2}$. (**d**) A part of $C'_{2,3}$

(b) The first three steps of this construction can be performed algebraically if the initial curve is arranged with respect to some three lines as in Fig. 1a. It means that there are three disjoint arcs on the nonempty oval (on the odd branch for odd m) meeting three lines at m points that lie on the arcs in the same order as on the lines. In classical terminology, such arcs are called *bases*.

(c) To continue iterations of Wiman's construction, we need more bases. By Mikhalkin's theorem [18], an M-curve of degree $d \geq 3$ cannot have more than three bases. So we start with $d = 2$. Choose a conic $C'_{2,0}$, disjoint arcs $\alpha_1, \cdots, \alpha_k$ on it, and lines $L_1, \cdots, L_k$ such that L_i cuts α_i at two points. Let $C'_{2,k+1}$ be obtained from $C'_{2,k}$ by Wiman's construction using the line L_k. It happens, however, that it is not enough to have many bases on the initial curve. The construction produces M-curves for $k \leq 3$ because the line L_k meets only one oval of $C'_{2,k-1}$, $k = 1, 2, 3$. Unfortunately, starting with $k = 4$, the line L_k meets

Fig. 2 Construction of the chart in the proof of Lemma 2.2

more than one oval (see Fig. 1d, where we depicted L_4 and the part of $C'_{2,3}$ obtained from that oval of $C'_{2,2}$ that meets L_3). It is easy to see that L_k meets 2^{k-3} ovals for $k \geq 3$. Using this fact, the result can be easily proven by induction. □

Lemma 2.2. *Let A be a real pseudoholomorphic curve of degree $d = 2k$. Suppose that an empty oval V of A has a tangency of order d with a line L. Let S be the isotopy type of $A \setminus V$. Then there exists a pseudoholomorphic curve A' of degree $2d$ one of whose empty ovals has a tangency of order $2d$ with L, and the isotopy type of A' is $S^2 \sqcup d^2$. In particular, if A is an M-curve, then A' is an M-curve also.*

Proof. Let p be the tangency point. We apply Wiman's construction in two steps. First, we perturb A so that the perturbed curve A'' has a tangency with A at p of order d and has $d^2 - d$ more intersection points, all lying on V. We may assume that $A \cup A''$ is holomorphic in some neighborhood of p and is defined by the equation $(y - ax^d)(y - bx^d) = 0$, $0 < a < b$. Then we perturb $A \cup A''$ by gluing at p the chart $(y - P(x))y + \varepsilon x^{2d}$ where roots of P are real negative (see Fig. 2). □

Corollary 2.3. *For any d there exists a real pseudoholomorphic M-curve A_d on $\mathbb{RP}^2$ of degree d that has at least $L_d = \frac{1}{6}d^2 - \frac{7}{54}(3d)^{4/3} + O(d)$ nonempty ovals.*

Proof. Let $k = [\frac{1}{3}\log_2(3d)]$ and $d = 2^k m + r$, $0 \leq r < 2^k$. Let $C = C_{m,k}$ be as in Proposition 2.1. By Lemma 2.2, we may suppose that C has a maximal tangency with some line. So let A be obtained from C by applying Harnack's construction r times.

Then A is an M-curve, and the number of its nonempty ovals is at least $L_d = \frac{1}{6}(d_1^2 - m^2) - \frac{3}{2}(d_1 - m) + k$, where $d_1 = 2^k m = \deg C$. Note that (x, r), $x = 2^k$, satisfies

$$(3d)^{1/3} \leq 2x \leq 2 \times (3d)^{1/3}, \qquad 0 \leq r \leq x - 1, \tag{3}$$

and $L_d = \frac{1}{6}f(2^k, r) + k$, where $f(x,r) = (d-r)^2(1 - x^{-2}) - 9(d-r)(1 - x^{-1})$. It is an easy calculus exercise to find the minimum of f under the constraints (3). □

Remark. It seems that the term $O(d^{4/3})$ in Corollary 2.3 is not optimal. Perhaps using a more careful construction (like that in Sect. 3) it can be replaced by $O(d)$.

In contrast, it is not clear at all how to construct real *algebraic* curves of *any* degree d with $\frac{1}{6}d^2 + o(d^2)$ nonempty ovals. Proposition 2.1(c) gives an example with these asymptotics for the sequence of degrees $d_k = 2^k$, but is it possible to do the same for, say, $d_k = 2^k - 1$?

3 When the Braid $\sigma_1^{-N}\Delta^n$ Is Quasipositive

The purpose of this section is, for given n and k, to find N as large as possible such that the braid $\sigma_1^{-N}\Delta_k^n$ is quasipositive (see Sect. 3.1 for definitions and see Sects. 4 and 5 for motivations). We propose here a recursive construction based on the binary decomposition of k. The best value of N obtained by this construction is presented in Theorem 3.13 (see also Corollary 3.15) in Sect. 3.6. We cannot prove that the obtained value of N is optimal.

3.1 *Quasipositive Braids*

Let B_n be the group of braids with n strings (n-braids). It is generated by $\sigma_1, \cdots, \sigma_{n-1}$, subject to relations $\sigma_i\sigma_j = \sigma_j\sigma_i$ for $j-i>1$ and $\sigma_i\sigma_j\sigma_i = \sigma_j\sigma_i\sigma_j$ for $j-i=1$. We suppose that $\{1\} = B_1 \subset B_2 \subset B_3 \subset \cdots$ by identifying σ_i of B_k with σ_i of B_n. We set $B_\infty = \bigcup_m B_n$. Let Δ_n be the *Garside element* of B_n. It is defined by

$$\Delta_0 = \Delta_1 = 1, \qquad \Delta_{n+1} = \sigma_1\sigma_2\cdots\sigma_n\Delta_n. \tag{1}$$

Let Q_n be the submonoid of B_n generated by $\{a^{-1}\sigma_i a \,|\, a \in B_n, 1 \le i < n\}$. The elements of Q_n are called *quasipositive braids* (this term was introduced by Lee Rudolph in [25]). Theorem 3.1 in Sect. 3.3 shows that $Q_{k+1} \cap B_k = Q_k$, i.e., the notion of quasipositivity is compatible with the convention that $B_k \subset B_{k+1}$.

We introduce a partial order on B_n by setting $a \le b$ if $ab^{-1} \in Q_n$. Then $Q_n = \{x \in B_n \,| x \ge 1\}$. Since Q_n is invariant under conjugation, this order is left and right invariant, i.e., $b' \le b$ implies $ab'c \le abc$. Indeed, if $b'b^{-1} \in Q_n$, then $(ab'c)(abc)^{-1} = a(b'b^{-1})a^{-1} \in Q_n$.

We write $a \sim b$ if a and b are conjugate. Note that $a \sim b \ge c$ does not imply $a \ge c$. Indeed, for $n=3$ we have $\sigma_2 \sim \sigma_1 \ge \sigma_1\sigma_2^{-1}$, but the assertion $\sigma_2 \ge \sigma_1\sigma_2^{-1}$ is wrong because $\sigma_2(\sigma_1\sigma_2^{-1})^{-1} = \sigma_2^2\sigma_1^{-1} \notin QP_3$ (see, e.g., [20] or [23]). However, $b_1 \sim b_2 \ge b_3 \sim b_4 \ge \cdots \sim b_{2n} \ge 1$ does imply $b_1 \ge 1$.

3.2 *Shifts and Cablings*

Let $s_m, c_m : B_\infty \to B_\infty$ be the group homomorphisms of m-shift and m-cabling defined respectively by $s_m(\sigma_i) = \sigma_{i+m}$ and

$$c_m(\sigma_i) = (\sigma_{mi}\sigma_{mi+1}\cdots\sigma_{mi+m-1})(\sigma_{mi-1}\cdots\sigma_{mi+m-2})\cdots(\sigma_{mi-m+1}\cdots\sigma_{mi})$$

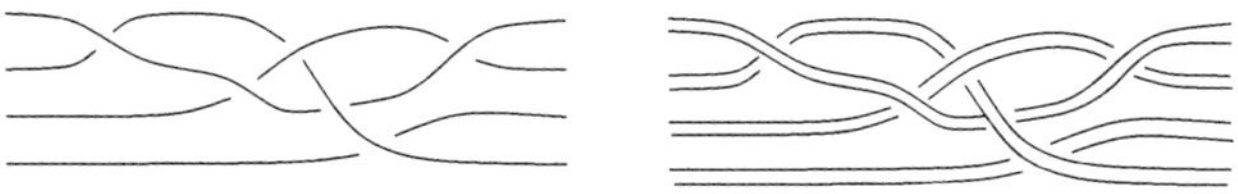

Fig. 3 Example of 2-cabling: $c(\sigma_3\sigma_2\sigma_3^{-1}\sigma_2\sigma_1\sigma_3)$

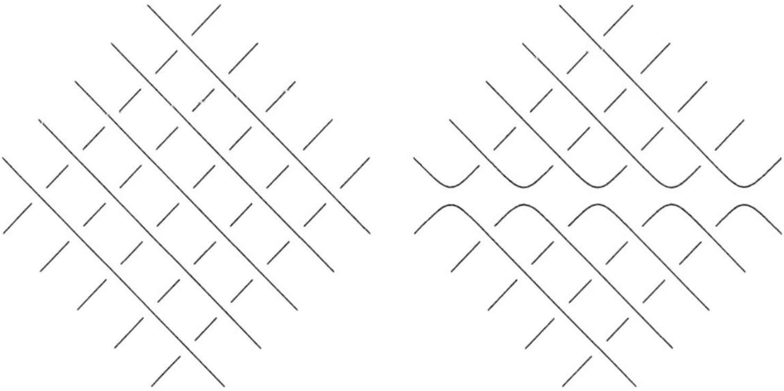

Fig. 4 $c_k(\sigma_1) \geq \Delta_k\tilde{\Delta}_k$ $(k = 5)$

(see the left-hand side of Fig. 4). We set $c = c_2$ (Fig. 3), $c^d = c_{2^d}$, and $s^d = s_{2^d}$. Then

$$c^d = c \circ \cdots \circ c \quad (d \text{ times}), \qquad c(\sigma_i) = \sigma_{2i}\sigma_{2i-1}\sigma_{2i+1}\sigma_{2i}.$$

Let $r_m : B_m \to B_m$ be the index-reversing homomorphism: $r_m(\sigma_j) = \sigma_{m-j}$. Let $\tilde{\Delta}_n = s_n(\Delta_n)$. Then we have

$$b\Delta_m = \Delta_m r_m(b), \quad b \in B_m; \qquad r_m(\Delta_m) = \Delta_m, \tag{2}$$

$$\tilde{\Delta}_k\Delta_{2k} = \Delta_{2k}\Delta_k, \qquad \Delta_k\Delta_{2k} = \Delta_{2k}\tilde{\Delta}_k, \tag{3}$$

$$\tilde{\Delta}_k c_k(\sigma_1) = c_k(\sigma_1)\Delta_k, \quad \Delta_k c_k(\sigma_1) = c_k(\sigma_1)\tilde{\Delta}_k, \tag{4}$$

$$s_{ki}(\Delta_k)s_{kl}(\Delta_k) = s_{kl}(\Delta_k)s_{ki}(\Delta_k), \tag{5}$$

$$\Delta_{2k} = \Delta_k\tilde{\Delta}_k c_k(\sigma_1) = \Delta_k c_k(\sigma_1)\Delta_k. \tag{6}$$

The last identity is the specialization for $a = 2$ of

$$\Delta_{ak} = c_k(\Delta_a)\prod_{j=0}^{a-1} s_{jk}(\Delta_k). \tag{7}$$

All these identities easily follow, for instance, from the characterization of Δ_k in [9].

Combining (3)–(4), we obtain

$$\Delta_{2k}^2 = \tilde{\Delta}_k^2\Delta_k^2 c_k(\sigma_1^2). \tag{8}$$

We have $c_k(\sigma_1) \geq \Delta_k \tilde{\Delta}_k$ (see Fig. 4). Combining this with (3), we obtain

$$c_k(\sigma_1) \geq \Delta_k^a \tilde{\Delta}_k^b \qquad \text{for any } a,b \text{ such that } a+b=2. \tag{9}$$

Indeed, $c_k(\sigma_1) \overset{(6)}{=} \Delta_k^{a-1} c_k(\sigma_1) \tilde{\Delta}_k^{1-a} \overset{\text{Fig. 4}}{\geq} \Delta_k^{a-1} (\Delta_k \tilde{\Delta}_k) \tilde{\Delta}_k^{1-a} = \Delta_k^a \tilde{\Delta}_k^{2-a}$.

Combining (9) and (6), we obtain also

$$\Delta_{2k} = \Delta_k c_k(\sigma_1) \Delta_k \geq \Delta_k^4. \tag{10}$$

3.3 *Quasipositivity and Stabilizations*

In this section we show that the quasipositivity is stable under two kinds of stabilizations: the inclusion $B_n \subset B_{n+1}$ and *positive* Markov moves.

3.3.1 Stability Under the Inclusion $B_n \subset B_{n+1}$

Theorem 3.1. $Q_{n+1} \cap B_n = Q_n$.

This is a specialization for $k = 1$ of the following fact.

Theorem 3.2. *Let $a \in B_k$, $b \in B_n$, and $c = s_n(a)b \in B_{n+k}$. Suppose that $c \in Q_{n+k}$. Then $a \in Q_k$ and $b \in Q_n$.*

Proof. Let D be the unit disk in $\mathbb{C}$. By Rudolph's theorem [25], a braid is quasipositive if and only if it is cut on $(\partial D) \times \mathbb{C}$ by an algebraic curve in $D \times \mathbb{C}$ that has no vertical asymptote.

Let L_a, L_b, and L_c be the links in the 3-sphere represented by a, b, and c. Let A_c be the algebraic curve bounded by L_c. The fact that $c = s_n(a)b$ means that $L_c = L_a \cup L_b$ and the sublinks L_a, L_b are separated by an embedded sphere. Then, by Eroshkin's theorem [10], A_c is a disjoint union of curves A_a and A_b bounded by L_a and L_b respectively. Hence, a and b are quasipositive. □

This proof of Theorem 3.2 relies on analytic methods (the filling disk technique is the main tool in [10]). However, Theorem 3.1 has a purely combinatorial proof based on Dehornoy's results [8] completed by Burckel–Laver's theorem [3, 17].

We say that a braid $b \in B_n$ is *Dehornoy i-positive,*[1] $i = 1, \cdots, n-1$, if there exist braids $b_0, \cdots, b_k \in B_{n-i}$, $k \geq 1$, such that $b = b_0 \prod_{j=0}^{k} (\sigma_{n-i} b_j)$. We say that b is *Dehornoy positive* if it is i-positive for some $i = 1, \ldots, n-1$. Let P_i be the set of $(n+1-i)$-positive braids and $\bar{P}_i = \bigcup_{j=1}^{i} P_j$.

[1]Our definitions differ from those in [8] only in the reversing of the string numbering.

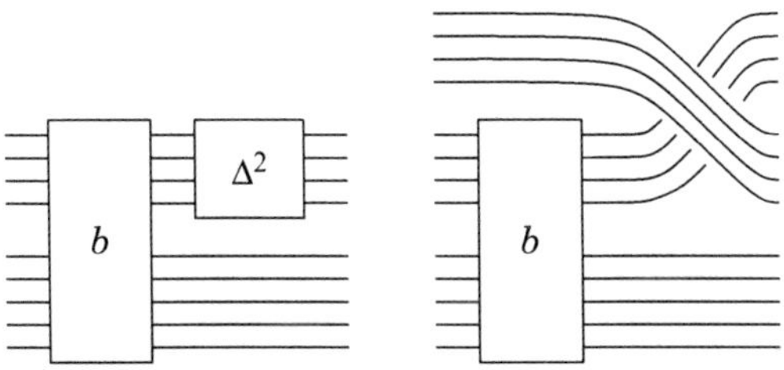

Fig. 5 The braids b' (*on the left*) and b'' (*on the right*)

In this notation, Dehornoy's theorem [8] (see also [11] for another proof) states that (i) B_n is a disjoint union $\{1\} \cup \bar{P}_n \cup \bar{P}_n^{-1}$. (ii) $\bar{P}_n$ is a disjoint union $P_2 \cup \ldots \cup P_n$. (iii) P_i and $\bar{P}_i$, $2 \leq i \leq n$, are subsemigroups of B_n. Burckel–Laver's theorem [3, 17] (see also [20] or [34] for another proof) states that (iv) $Q_n \subset \bar{P}_n$.

Combinatorial Proof of Theorem 3.1 The inclusion $Q_n \subset Q_{n+1} \cap B_n$ is evident. Let us show that $Q_{n+1} \cap B_n \subset Q_n$. Let $b \in Q_{n+1} \cap B_n$. Then $b = x_1 \cdots x_k$, each x_j being a conjugate of σ_1 in B_{n+1}. By (iv), we have $x_j \in \bar{P}_{n+1}$, $j = 1, \ldots, k$. If $x_j \in P_{n+1}$ for some j, then $b \in P_{n+1}$ by the definition of i-positivity. By (ii), this contradicts $b \in B_n$. Hence, each x_j is in P_n.

Thus, it remains to show that if x is a conjugate of σ_1 in B_{n+1}, then x is a conjugate of σ_1 in B_n. This follows from the fact that any conjugate of σ_1 can be presented *in a unique way* as $x = ca_{i,j}c^{-1}$, $i < j$, where $a_{i,j}$ is so-called band-generator (i.e., $a_{i,j} = a\sigma_i a^{-1}$ for $a = \sigma_{j-1}\sigma_{j-2} \cdots \sigma_{i+1}$) and c is in the kernel of the pure braid group homomorphism of forgetting the ith string. The latter fact can be easily proved using the braid combing theory. □

3.3.2 Stability Under Positive Markov Moves

Theorem 3.3. *Let $b \in B_n$. Then $b \in Q_n$ if and only if $b\sigma_n \in Q_{n+1}$.*

This fact is reduced in [21] to Gromov's theorem on pseudoholomorphic curves. The reduction given in [21] is rather cumbersome, but Michel Boileau observed that it can be considerably simplified using the arguments from our joint paper [2] (unfortunately, this observation was made when [2] had already been published). Indeed, it is proved (though not stated explicitly) in [2] that *if L is the boundary link of an analytic curve in $B^4 \subset \mathbb{C}^2$, and L is transversally isotopic*[2] *to a closed braid b, then b is quasipositive.* To deduce Theorem 3.3 from this fact, we note that $b\sigma_n$ bounds an analytic curve (by Rudolph's theorem [25]), and b is transversally isotopic to $b\sigma_n$ (an easy exercise; see, e.g., [25, Lemma 1]).

Corollary 3.4. *Let $b \in B_n$ and $k \leq n$. Then $b' = b\,s_{n-k}(\Delta_k^2)$ is quasipositive if and only if $b'' = b\,s_{n-k}(c_k(\sigma_1))$ is quasipositive; see Fig. 5.*

---

[2] In the sense of contact geometry.

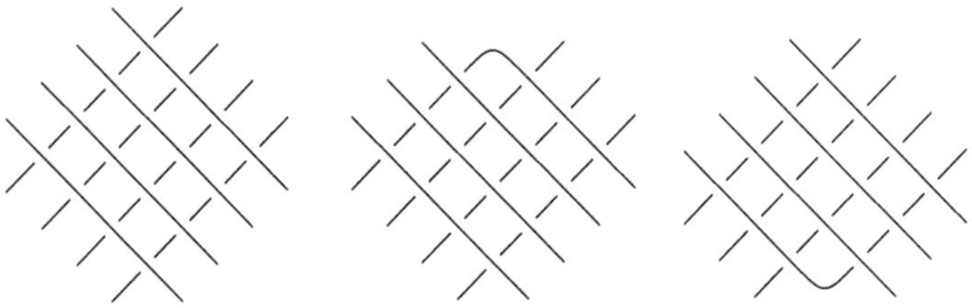

Fig. 6 $c_{k+1}(\sigma_1)\overset{Mm}{\to}\cdots = (\sigma_k\cdots\sigma_2\sigma_1)\times s_1(c_k(\sigma_1))\times(\sigma_1\sigma_2\cdots\sigma_k)$

Proof. We say that b_1b_2 is obtained from b_0 by a positive Markov move (and we write $b_0\overset{Mm}{\to}b_1b_2$) if $b_1,b_2\in B_n$ and $b_0=b_1\sigma_n b_2$. By Theorem 3.3, it is enough to prove that $b''\overset{Mm}{\to}\cdots\overset{Mm}{\to}b'$. If $k=0$, this is trivial. Suppose that this statement has been proved for k. Then

$$\begin{aligned} c_{k+1}(\sigma_1) &\overset{Mm}{\to}(\sigma_k\cdots\sigma_1)s_1(c_k(\sigma_1))(\sigma_1\cdots\sigma_k) && \text{(see Fig. 6)}\\ &\overset{Mm}{\to}(\sigma_k\cdots\sigma_1)s_1(\Delta_k^2)(\sigma_1\cdots\sigma_k) && \text{(by the induction hypothesis)}\\ &= r_{k+1}(\sigma_1\cdots\sigma_k\Delta_k^2\sigma_k\cdots\sigma_1)\overset{(4)}{=}\Delta_{k+1}^2. && \square \end{aligned}$$

3.4 The Subgroup A_∞ of B_∞

For an integer $d\geq 1$, let $X_d=\{s_{k2^d}(\Delta_{2^d})\mid k\geq 0, k\in\mathbb{Z}\}$ and let A_d be the subgroup of B_∞ generated by X_d. It is a free abelian group freely generated by X_d. For example, A_1 is the subgroup of B_∞ generated by $\sigma_1,\sigma_3,\sigma_5,\ldots$.

Let A_∞ be the subgroup of B_∞ generated by $\bigcup X_d$, i.e., the product of all the subgroups A_d. This product is semidirect in the sense that $A_1\ldots A_d$ is a normal subgroup of A_∞, and for any d,e, the subgroup A_e is a normal in A_eA_d if $e\leq d$. In the latter case, the action of A_d on A_e by conjugation is very easy to describe. Let $x\in X_e$, $y\in X_d$, $e\leq d$. Let P_x (respectively P_y) be the set of strings permuted by x (respectively by y). Only two cases are possible: either P_x and P_y are disjoint and then x and y commute, or $P_x\subset P_y$ and then y acts on x as in (5).

In particular, each element x of $A_1\ldots A_d$ can be uniquely presented in the form

$$x=x_1\ldots x_d,\qquad x_e\in A_e.$$

Let $\chi_d : A_d\to\mathbb{Z}$ be the homomorphism that takes each element of X_d to 1, and let $A_d^m=\chi_d^{-1}(m)$. Since A_∞ is a semidirect product of A_d's, the characters χ_d extend in a unique way to a homomorphism $\chi : A_\infty\to\bigoplus_{d=1}^\infty\mathbb{Z}$ such that $\chi(x_1\ldots x_d)=(\chi_1(x_1),\ldots,\chi_d(x_d))$ if $x_e\in A_e$ for $e=1,\ldots,d$ (here and below, we truncate the tail of zeros).

The above discussion implies also the following two easy facts:

Lemma 3.5. *Let $0 < r < 2^d$ and $m = 2^d q + r$. Then $A_\infty \cap B_m$ is the direct product of its subgroups $A_\infty \cap B_{m-r}$ and $s_{m-r}(A_\infty \cap B_r)$.* □

Lemma 3.6. *Let $B = B_{2^d}$, $\tilde{B} = s^d(B)$. Let $x \in A_\infty \cap B_{2^{d+1}}$ and $n = (n_1,\dots,n_d) = \chi(x)$. Then for any decomposition $n = n' + n'' + \tilde{n}' + \tilde{n}''$, there exist $x', x'' \in B$ and $\tilde{x}', \tilde{x}'' \in \tilde{B}$ such that $\chi(x') = n'$, $\chi(x'') = n''$, $\chi(\tilde{x}') = \tilde{n}'$, $\chi(\tilde{x}'') = \tilde{n}''$, and*

$$x\Delta_{2^{d+1}}^{2n+1} \sim x'\tilde{x}'\Delta_{2^{d+1}}^{2n+1}x''\tilde{x}''. \tag{11}$$

Proof. (The notation should be self-explanatory)

$$x\Delta_{2^{d+1}}^{2n+1} = abc\tilde{u}\tilde{v}\tilde{w}\Delta_{2^{d+1}}^{2n+1} = a\tilde{u}\Delta_{2^{d+1}}^{2n+1}vw\tilde{b}\tilde{c} \sim wa\tilde{c}\tilde{u}\Delta_{2^{d+1}}^{2n+1}v\tilde{b}. \quad \square$$

3.5 The Case in Which the Number of Strings Is a Power of 2

For any $d \geq 0$, we set

$$S_d = 1 + 4 + 4^2 + \cdots + 4^{d-1} = (4^d - 1)/3.$$

So $(S_0, S_1, \dots) = (0, 1, 5, 21, 85, 341, 1365, \dots)$. We have the recurrences

$$S_d - 4S_{d-1} = 1, \qquad S_d - 5S_{d-1} + 4S_{d-2} = 0. \tag{12}$$

Lemma 3.7. *Let $x \in A_\infty \cap B_{2^d}$, $\chi(x) = (n_1,\dots,n_d)$. If $d = 1$, we suppose only that $n_1 \geq 0$. If $d \geq 2$, we suppose that*

$$\sum_{e=k+1}^{d} (n_e S_{e-k} - \varepsilon_e) \geq 0, \qquad k = 0,\cdots,d-1, \tag{13}$$

where

$$\varepsilon_1 = 1, \qquad \varepsilon_d = \frac{3 + (-1)^{n_d}}{2}, \qquad \varepsilon_e = \frac{5 - (-1)^{n_e}}{2}, \quad 1 < e < d, \tag{14}$$

i.e., $n_d \geq \varepsilon_d$, $5n_d + n_{d-1} \geq \varepsilon_d + \varepsilon_{d-1}, \dots, S_d n_d + \cdots + 5n_2 + n_1 \geq \varepsilon_d + \cdots + \varepsilon_1$.
Then x is quasipositive.

Proof. Induction on d. If $d = 1$, then the statement is trivial because in this case, $x = \sigma_1^{n_1}$. So, let us assume that the statement is true for $d - 1$ and let us prove it for d.

Let $\Delta = \Delta_{2^{d-1}}$, $\tilde{\Delta} = \tilde{\Delta}_{2^{d-1}} = s^{d-1}(\Delta)$, $\delta_k = s_{(k-1)2^{d-2}}(\Delta_{2^{d-2}})$, $\hat{\sigma}_k = c^{d-2}(\sigma_k)$. The notation δ_{12}^a is an abbreviation for $\delta_1^{a'}\delta_2^{a-a'}$ when the value of a' is not important. In this notation, (3)–(6) and (7) specialize to

$$\Delta\Delta_{2^d} = \Delta_{2^d}\tilde{\Delta}, \qquad \delta_1\Delta = \Delta\delta_2, \qquad \delta_3\tilde{\Delta} = \tilde{\Delta}\delta_4, \tag{6'}$$

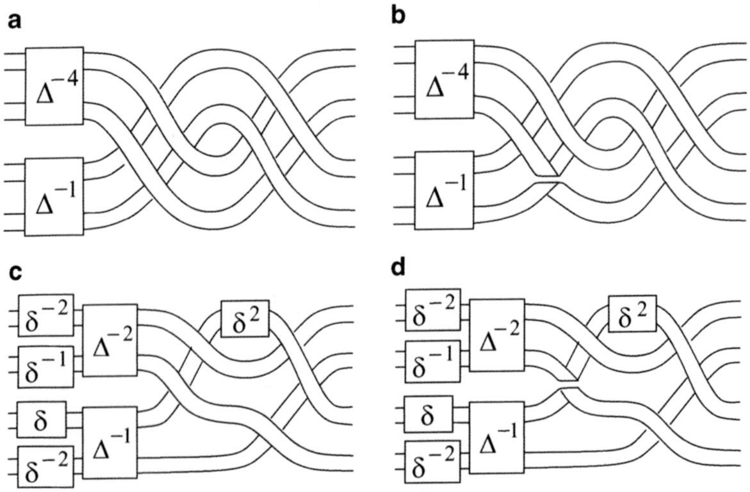

Fig. 7 Proof of (19)

$$\hat\sigma_i\delta_i=\delta_{i+1}\hat\sigma_i,\qquad \hat\sigma_i\delta_{i+1}=\delta_i\hat\sigma_i,\qquad \hat\sigma_i\delta_k=\delta_k\hat\sigma_i,\quad k\notin\{i,i+1\}, \tag{7'}$$

$$\delta_i\delta_l=\delta_l\delta_i,\qquad \Delta\tilde\Delta=\tilde\Delta\Delta, \tag{8'}$$

$$\Delta=\hat\sigma_1\delta_1\delta_2,\qquad \tilde\Delta=\hat\sigma_3\delta_3\delta_4, \tag{9'}$$

$$\forall a\in\mathbb{Z},\quad \hat\sigma_k\ge\delta_k^a\delta_{k+1}^{2-a}. \tag{12'}$$

Combining (9′) and (6), we obtain

$$\hat\sigma_1\hat\sigma_2\overset{(12)}{\ge}\hat\sigma_1\delta_2^2\overset{(9)}{=}\Delta\delta_1^{-1}\delta_2=\Delta\delta_{12}^0. \tag{15}$$

Let us show that

$$\tilde\Delta^{-6}\Delta^{-3}\Delta_{2d}^2\ge\delta_1^{-2}\delta_4^{-4}\hat\sigma_2 \tag{16}$$

(this is the heart of the proof). Indeed (see Fig. 7a–d)

$$\begin{aligned}\tilde\Delta^{-6}\Delta^{-3}\Delta_{2d}^2 &\overset{(11)}{=}\tilde\Delta^{-6}\Delta^{-3}(\Delta^2\tilde\Delta^2c^{d-1}(\sigma_1^2))=\tilde\Delta^{-4}\Delta^{-1}(\hat\sigma_2\hat\sigma_1\hat\sigma_3\hat\sigma_2)^2\\ &\overset{(12)}{\ge}\tilde\Delta^{-4}\Delta^{-1}\hat\sigma_2(\delta_1^{2-a}\delta_2^a)\hat\sigma_3\hat\sigma_2^2\hat\sigma_3\hat\sigma_1\hat\sigma_2\overset{(7)}{=}\tilde\Delta^{-4}\Delta^{-1}(\hat\sigma_2\hat\sigma_3\hat\sigma_2^2)\hat\sigma_3\hat\sigma_1\delta_1^a\delta_2^{2-a}\hat\sigma_2\\ &=\tilde\Delta^{-4}\Delta^{-1}\hat\sigma_3^2\sigma_2\hat\sigma_3^2\hat\sigma_1\delta_1^a\delta_2^{2-a}\hat\sigma_2\overset{(12)}{\ge}\tilde\Delta^{-4}\Delta^{-1}\hat\sigma_3^2(\delta_2^b\delta_3^{2-b})\hat\sigma_3^2\hat\sigma_1\delta_1^a\delta_2^{2-a}\hat\sigma_2\\ &\overset{(7)}{=}\tilde\Delta^{-4}\Delta^{-1}\hat\sigma_3^4\hat\sigma_1\delta_1^{a+b}\delta_2^{2-a}\delta_3^{2-b}\hat\sigma_2\overset{(9)}{=}(\delta_1\delta_2)^{-1}(\delta_3\delta_4)^{-4}\delta_1^{a+b}\delta_2^{2-a}\delta_3^{2-b}\hat\sigma_2,\end{aligned}$$

and we obtain (16) by setting $a = 1$, $b = -2$. We have also

$$\Delta_{2d} \geq \hat{\sigma}_1 \hat{\sigma}_2 \Delta^2 \delta_{12}^4. \tag{17}$$

Indeed,

$$\Delta_{2d} \overset{(9)}{=} \Delta c^{d-1}(\sigma_1)\Delta = \Delta \hat{\sigma}_2 \hat{\sigma}_1 \hat{\sigma}_3 \hat{\sigma}_2 \Delta \overset{(12)}{\geq} \Delta \hat{\sigma}_2 \hat{\sigma}_1 (\delta_3^2)(\delta_2^5 \delta_3^{-3})\Delta$$
$$\overset{(9)}{=} \hat{\sigma}_1 \delta_1 \delta_2 \hat{\sigma}_2 \hat{\sigma}_1 \delta_2^5 \delta_3^{-1} \Delta \overset{(7)}{=} \hat{\sigma}_1 \hat{\sigma}_2 \hat{\sigma}_1 \delta_2^6 \Delta \overset{(9)}{=} \hat{\sigma}_1 \hat{\sigma}_2 (\Delta \delta_1^{-1} \delta_2^{-1}) \delta_2^6 \Delta \overset{(9)}{=} \hat{\sigma}_1 \hat{\sigma}_2 \Delta^2 \delta_{12}^4.$$

We set $n_d = 2n+1+r$, $r \in \{0,1\}$. Let $m_{d-1} = n_{d-1} + 10n + 4r$, $m_{d-2} = n_{d-2} - 8n$,

$$n'_{d-1} = m_{d-1} + 3 = \; n_{d-1} + 5n_d - r - 2 = n_{d-1} + 5n_d - \varepsilon_d - 1,$$
$$n'_{d-2} = m_{d-2} + 4 = n_{d-2} - 4n_d + 4r + 8 = n_{d-2} - 4n_d + 4\varepsilon_d + 4,$$

and $n'_e = n_e$ for $e = 1, \ldots, d-3$. In the following computation we assume that $y_1, y_2, z, x' \in A_\infty \cap B_{2d-1}$ and $\chi(y_1) = \chi(y_2) = (n_1, \cdots, n_{d-2}, m_{d-1})$, $\chi(z) = (n_1, \cdots, n_{d-3}, m_{d-2}, m_{d-1})$, $\chi(x') = (n'_1, \cdots, n'_{d-1})$. Let $x = x_1 \Delta_{2d}^{n_d}$ with $x_1 \in (A_1 \cdots A_{d-1}) \cap B_{2d}$. So we have

$$x = x_1 \cdots x_{d-1} \Delta_{2d}^{n_d} \overset{(13)}{\geq} x_1 \ldots x_{d-1} \Delta^{4r} \Delta_{2d}^{2n+1} \overset{(14)}{\sim} y_1 \Delta^{-3n} \tilde{\Delta}^{-6n} \Delta_{2d}^{2n} \Delta_{2d} \Delta^{-n}$$
$$= y_1 (\Delta^{-3} \tilde{\Delta}^{-6} \Delta_{2d}^2)^n \Delta_{2d} \Delta^{-n} \overset{(9)}{\geq} y_1 \delta_1^{-2n} \delta_4^{-4n} \hat{\sigma}_2^n \Delta_{2d} \Delta^{-n}$$
$$= y_1 \delta_1^{-2n} \hat{\sigma}_2^n \Delta_{2d} \delta_1^{-4n} \Delta^{-n} \sim y_2 \delta_{12}^{-6n} \hat{\sigma}_2^n \Delta_{2d} \Delta^{-n} \overset{(9)}{=} y_2 \delta_{12}^{-6n} \hat{\sigma}_2^n \Delta_{2d} \hat{\sigma}_1^{-n} \delta_{12}^{-2n}$$
$$\sim z \hat{\sigma}_2^n \Delta_{2d} \hat{\sigma}_1^{-n} \overset{(20)}{\geq} z \hat{\sigma}_2^n \hat{\sigma}_1 \hat{\sigma}_2 \delta_{12}^4 \Delta^2 \hat{\sigma}_1^{-n} = z \hat{\sigma}_1 \hat{\sigma}_2 \delta_{12}^4 \Delta^2 \overset{(18)}{\geq} z \delta_{12}^4 \Delta^3 = x'.$$

It remains to check that the induction conditions are satisfied for x' and $d-1$. If $d = 2$, then $n'_1 = n_1 + 5n_2 - \varepsilon_2 - 1 = (n_1 S_1 - \varepsilon_1) + (n_2 S_2 - \varepsilon_2) \geq 0$, and we are done.

Suppose that $d > 2$. Let (13′) and (14′) refer to the formulas (13), (14), where $d-1$, n'_e, and ε'_e replace d, n_e, and ε_e. So we define $\varepsilon'_1, \ldots, \varepsilon'_{d-1}$ by (14′) and we have to check the inequalities (13′) for $k = 0, \ldots, d-2$. Indeed, we have $n'_e = n_e$ for $e < d-2$; $n'_{d-2} - n_{d-2} = -8n + 4$ is even, and $n'_{d-1} - n_{d-1} = 10n + 4r + 3$ is odd. Hence, $\varepsilon'_e = \varepsilon_e$ for $e \leq 2$, and

$$\varepsilon'_{d-1} = (3 + (-1)^{n'_{d-1}})/2 = (3 - (-1)^{n_{d-1}})/2 = (5 - (-1)^{n_{d-1}})/2 - 1 = \varepsilon_{d-1} - 1,$$

and we obtain for any $k = 0, \ldots, d-2$,

$$\sum_{e=k+1}^{d} \varepsilon_e - \sum_{e=k+1}^{d-1} \varepsilon'_e = \varepsilon_{d-1} + \varepsilon_d - \varepsilon'_{d-1} = \varepsilon_d + 1.$$

Since $n'_e = n_e$ for $e < d-2$, and $S_0 = 0$, we have for any $k = d - p \le d-2$,

$$\sum_{e=k+1}^{d} n_e S_{e-k} - \sum_{e=k+1}^{d-1} n'_e S_{e-k} = (n_{d-2} - n'_{d-2})S_{p-2} + (n_{d-1} - n'_{d-1})S_{p-1} + n_d S_p$$
$$= (4n_d - 4\varepsilon_d - 4)S_{p-2} + (-5n_d + \varepsilon_d + 1)S_{p-1} + n_d S_p$$
$$= (S_p - 5S_{p-1} + 4S_{p-2})n_d + (S_{p-1} - 4S_{p-2})(\varepsilon_d + 1) \overset{(15)}{=} \varepsilon_d + 1.$$

Thus, $(13')$ is equivalent to (13). □

Let us emphasize some particular cases of Lemma 3.7:

Corollary 3.8. *Let $x \in A_\infty \cap B_{2d}$, $d \ge 2$, $\chi(x) = (n_1, \dots, n_d)$, and let $\varepsilon_1, \dots, \varepsilon_d$ be as in (14).*

(a). *If $n_d > 0$, $n_e \ge 0$ for $e = 2, \dots, d-1$, and (13) holds for $k = 0$, i.e., $\sum_e (n_e S_e - \varepsilon_e) \ge 0$, then x is quasipositive.*

(b). *In particular, if $n_2, \dots, n_d$ are even and nonnegative, n_d is positive, and*

$$n_1 + 5n_2 + 21n_3 + \cdots + S_d n_d \ge 2d - 1, \tag{18}$$

then x is quasipositive.

Proof. (a) It is enough to check (13) for $k = 1, \dots, d-1$. First, note that (13) for $k = d-1$ is just $n_d \ge \varepsilon_d$, which is equivalent to $n_d > 0$. So let $1 \le k \le d-2$. For any $m \ge 1$ we have $3(m-1) \le S_m - 1$. Hence, $\varepsilon_{k+1} + \cdots + \varepsilon_{d-1} \le 3 + \cdots + 3 = 3(d-k-1) \le S_{d-k} - 1 \le n_d(S_{d-k} - 1)$. Thus,

$$\sum_{e=k+1}^{d} (n_e S_{e-k} - \varepsilon_e) = \left(n_d(S_{d-k} - 1) - \sum_{e=k+1}^{d-1} \varepsilon_e \right) + (n_d - \varepsilon_d) + \sum_{e=k+1}^{d-1} S_{e-k} n_e \ge 0.$$

(b) Immediate from (a). □

Corollary 3.9. *For positive integers d, n, if $N \le (4^d - 1)n/3 - 2d + (3 - (-1)^n)/2$, then $\sigma_1^{-N} \Delta_{2d}^n \ge 0$.*

Proof. $\chi(\sigma_1^{-N} \Delta_{2d}^n) = (-N, 0, \dots, 0, n)$, so we may apply Corollary 3.8. □

Remark. Corollary 3.8 combined with arguments similar to those in the proof of Corollary 2.3 allows us to show that for any k, the braid $\sigma_1^{-N} \Delta_k$ is quasipositive for $N = 1/3k^2 + O(k^{4/3})$. However, in the next subsection we give a better estimate for N of the form $1/3k^2 + O(k)$.

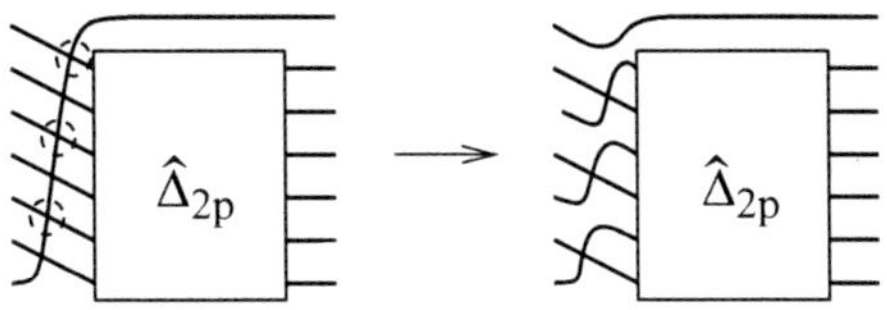

Fig. 8 Illustration to the proof of Lemma 3.10 ($p = 3$)

3.6 The General Case

Lemma 3.10. *Let $p, d > 0$, $m' = 2^d p$, $m = m' + 2^{d-1} = (2p+1)2^{d-1}$, and $x \in A_\infty \cap B_m$. Then $x\Delta_m \geq x'\Delta_{m'}$ for some $x' \in A_\infty \cap B_{m'}$ such that $\chi_{d-1}(x') = \chi_{d-1}(x) + 1$, $\chi_d(x') = \chi_d(x) + p$, and $\chi_e(x') = \chi_e(x)$ for $e \notin \{d-1, d\}$.*

Proof. By Lemma 3.5, we may write $x = y\tilde{y}$ with $y \in A_\infty \cap B_{m'}$, and $\tilde{y} \in A_\infty \cap s_{m'}(B_{2^{d-1}})$. Let $\delta_k = s_{2^{d-1}(k-1)}(\Delta_{2^{d-1}})$, $\Delta = \Delta_{2^k}$. We denote here $c^{d-1}(\alpha)$ by $\hat{\alpha}$ for any braid α.

Let $z = \Delta_m \tilde{y} \Delta_m^{-1}$ and $w = \Delta_{m'} z \Delta_{m'}^{-1}$. Then by (2), we have $z, w \in A_\infty \cap B_{m'}$ and $\chi(w) = \chi(z) = \chi(y)$. In the following computation, the "wild card character" δ^a stands for any product of the form $\delta_1^{a_1} \ldots \delta_{2p}^{a_{2p}}$ (no δ_{2p+1}) with $a_1 + \cdots + a_{2p} = a$ when the explicit values of the a_j are not important. In other words, δ^a stands for any element of $X^a_{2^{d-1}} \cap B_{m'}$. Similarly, Δ^a stands for any element of $X^a_{2^d} \cap B_{m'}$. So we have (see Fig. 8)

$$\begin{aligned} x\Delta_m &= y\tilde{y}\Delta_m = y\Delta_m z \overset{(7)}{=} y\hat{\Delta}_{2p+1}\delta^{2p}\delta_{2p+1}z \overset{(2)}{=} y\delta_1\hat{\Delta}_{2p+1}\delta^{2p}z \\ &\overset{(1)}{=} y\delta_1\hat{\sigma}_1\ldots\hat{\sigma}_{2p}\hat{\Delta}_{2p}\delta^{2p}z \overset{(7)}{=} y\delta_1(\hat{\sigma}_1\ldots\hat{\sigma}_{2p})\Delta_{m'}\delta^0 z \\ &\overset{(9)}{\geq} y\delta^1(\hat{\sigma}_1\delta_2^2\hat{\sigma}_3\delta_4^2\ldots\hat{\sigma}_{2p-1}\delta_{2p}^2)\Delta_{m'}z = y\delta^{2p+1}\hat{\sigma}_1\hat{\sigma}_3\cdots\hat{\sigma}_{2p-1}w\Delta_{m'} \\ &\overset{(6)}{=} y\delta^1\Delta^p w\Delta_{m'}. \square \end{aligned}$$

Lemma 3.11. *Let $k \geq 2$. Consider the binary decomposition*

$$k = \sum_{i=0}^{d} a_i 2^i, \qquad a_i \in \{0,1\}, \qquad a_d = 1. \tag{19}$$

Let $x \in A_\infty \cap B_k$. Then there exists $y \in A_\infty \cap B_{2^d}$ such that $x\Delta_k \geq y$ and

$$\chi_i(y) - \chi_i(x) = a_i + a_{i-1}\sum_{j=i}^{d} a_j 2^{j-i}, \qquad i = 1, \ldots, d. \tag{20}$$

Proof. Induction by $\nu(k)$, the number of ones in the binary decomposition of k. If $\nu = 1$, then $k = 2^d$ and $a_0 = \cdots = a_{d-1} = 0$; hence (20) holds for $y = x\Delta_k = x\Delta_{2^d}$.

Assume that the statement is proved for all k' with $\nu(k') < \nu(k)$ and let us prove it for k. Let 2^{e-1} be the maximal power of 2 that divides k, i.e., $(a_0, \ldots, a_d)$

$= (0,\dots,0,1,a_e,\dots,a_d)$. Let $k' = k - 2^{e-1}$. Then $k' = \sum a'_i 2^i$, where $(a'_0,\dots,a'_d) = (0,\dots,0,0,a_e,\dots,a_d)$. By Lemma 3.10, there exists $x' \in A_\infty \cap B_{k'}$ such that $x\Delta_k \geq x'\Delta_{k'}$ and $\chi(x') - \chi(x) = (n_1,\dots,n_d) = (0,\dots,0,1,p,0,\dots,0)$, where $p = k'/2^e = \sum_{j=e}^d a_j 2^{j-e}$, $n_{e-1} = 1$, and $n_e = p$.

Since $\nu(k') = \nu(k) - 1$, there exists $y \in A_\infty \cap B_{2d}$ such that $x'\Delta_k \geq y$ and (20) holds with x and a_i replaced by x' and a'_i. Hence,

$$\chi_i(y) - \chi_i(x) = \big(\chi_i(x') - \chi_i(x)\big) + \big(\chi_i(y) - \chi_i(x')\big) = n_i + a'_i + a'_{i-1}\sum_{j=i}^{d} a'_j 2^{j-i}$$
$$= \begin{cases} 0 + a_i + a_{i-1}(a_i + 2a_{i+1} + \dots + 2^{d-i}a_d), & i \geq e+1, \\ p + 1 + 0, & i = e, \\ 1 + 0 + 0, & i = e-1, \\ 0 + 0 + 0, & i \leq e-2. \end{cases}$$

This is equal to the right-hand side of (20) in all four cases. □

We define arithmetic functions $f(k)$, $g(k)$ via the binary decomposition (19):

$$f(k) = \sum_{i=0}^{d} a_i + \sum_{0\leq i<j\leq d} a_i a_j 2^{j-i-1}, \quad g(k) = a_{d-1} - 1 + \sum_{i=2}^{d-1} a_i(1 - a_{i-1}). \quad (21)$$

Corollary 3.12. *Let k be as in Lemma 3.11. Then there exists $y \in A_\infty \cap B_{2^d}$, $\chi(y) = (n_1,\dots,n_d)$, such that $\Delta_k \geq y$ and*

$$(1 - (-1)^{n_i})/2 = a_i(1 - a_{i-1}), \qquad i = 1,\dots,d,$$
$$S_1 n_1 + \dots + S_d n_d = (k^2 - f(k))/3.$$

Proof. By (20) we have $n_i = a_i + a_{i-1}(a_i + 2a_{i+1} + \dots) \equiv a_i(1 - a_{i-1}) \mod 2$ and

$$3\sum_{i=1}^{d} S_i \chi_i(y) = \sum_{i=1}^{d}(4^i - 1)\left(a_i + a_{i-1}\sum_{j=i}^{d} a_j 2^{j-i}\right)$$
$$= \sum_{i=0}^{d} a_i(4^i - 1) + \sum_{i=1}^{d}(4^i - 1)a_{i-1}\sum_{j=i}^{d} a_j 2^{j-i}$$
$$= \sum_{i=0}^{d} a_i 4^i - \sum_{i=0}^{d} a_i + \sum_{0\leq i<j\leq d} a_i a_j (4^{i+1} - 1)2^{j-i-1}$$
$$= \sum_{i=0}^{d} a_i^2 4^i + 2\sum_{0\leq i<j\leq d} a_i a_j 2^{i+j} - f(k) = k^2 - f(k). \quad \square$$

Theorem 3.13. *Let $k \geq 2$, $n \geq 1$. Let f and g be as in (19), (21). We set $\varepsilon = (1 - (-1)^n)/2$, $d = [\log_2 n]$. Then $\sigma_1^{-N}\Delta_k^n$ is quasipositive for*

$$N = \frac{n(k^2 - f(k))}{3} - 2d + 1 - \varepsilon g(k) + \left[\frac{n}{4}\right] \max\Big(0, f(k) - g(2k) - 2d - 1\Big).$$

Proof. Let $E = f(k) - g(2k) - 2d - 1$. If $E \le 0$, then the result follows immediately from Corollaries 3.8 and 3.12. Consider the case $E > 0$. Let $q = [n/4]$, $r = n - 4q$. We set $x = \sigma_1^{-N_1}\Delta_k^r$, $y = \sigma_1^{-N_2}\Delta_{2k}$, and $z = \sigma_1^{-N_2}\Delta_k^4$, where $N_1 = r(k^2 - f(k))/3 - 2d + 1 - \varepsilon g(k)$ and $N_2 = \big((2k)^2 - f(2k)\big)/3 - 2d - 1 - g(2k)$. By Corollaries 3.8 and 3.12, we have $x \ge 1$ and $y \ge 1$. Combining $y \ge 1$ with Corollary 3.4, we obtain $z \ge 1$. Since $f(2k) = f(k)$, we have $N = N_1 + qN_2$. Thus, $\sigma_1^{-N} = xz^q \ge 1$. □

Proposition 3.14. *(a) We have $1 \le f(k) \le k$ for any k. Moreover, $f(k) = k$ iff $k = 2^{d+1} - 1$ and $f(k) = 1$ iff $k = 2^d$ for some $d \ge 0$.*

(b) We have $k - f(k) - 3g(2k) \ge 0$. Equality is attained iff either $k = 2^{d+2} - 1$ or $k = 2^{d+3} - 2^d - 1$ for some $d \ge 0$.

Proof. (a)

$$k - f(k) = \sum_{j=0}^{d} a_j \left(2^j - 1 - \sum_{i=0}^{j-1} a_i 2^{j-i-1}\right) \ge \sum_{j=0}^{d} a_j \left(2^j - 1 - \sum_{i=0}^{j-1} 2^{j-i-1}\right) = 0,$$

and we have equality iff $k = 2^d - 1$. It is evident that $f(k) = 1$ iff $k = 2^d$.

(b) Exercise. □

Corollary 3.15. *(a) If $N \le \frac{2}{3}(k^2 - k) - 2[\log_2 k] + 1$, then $\sigma_1^{-N}\Delta_k^2$ is quasipositive.*

(b) If $N \le \frac{4}{3}k^2 - \frac{1}{3}k - 2[\log_2 k] - 1$, then $\sigma_1^{-N}\Delta_{2k}$ is quasipositive. □

4 Curves with a Deep Nest and with Many Innermost Ovals

4.1 Real Pseudoholomorphic Curves

Let A be a real curve on $\mathbb{RP}^2$. We say that the *depth* of an oval of $\mathbb{R}A$ is q if it is surrounded by q ovals. Degtyarev, Itenberg, and Kharlamov [7] ask, how many ovals of depth $k-2$ may a curve of degree $2k$ have? Note that $k-2$ is the maximal possible depth of ovals of a nonhyperbolic curve (a curve of degree $2k$ is called *hyperbolic* if it has k nested ovals and hence, by Bézout's theorem, cannot have more ovals). This question arises in the study of the number of components of an intersection of three real quadrics in higher-dimensional spaces (see details in [7]).

Let us denote the number of ovals of depth q of a curve A by $l_q = l_q(A)$. The improved Petrovsky inequality implies $l_{k-2} \le \frac{3}{2}k^2 + O(k)$. On the other hand, Hilbert's construction provides curves with $l_{k-2} \ge k^2 + O(k)$. We improve this

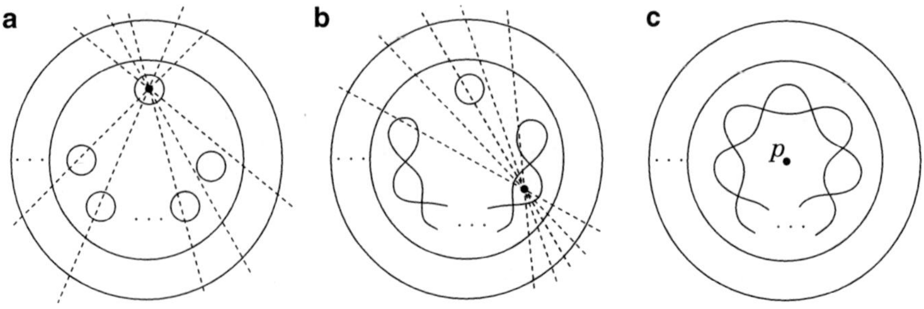

Fig. 9 Choice of pencils in the proof of Proposition 4.1

lower bound up to $9/8k^2$ for algebraic curves (see Proposition 4.3). The results of Sect. 3 (see Theorem 3.13 and Corollary 3.15(b)) provide a lower bound of the form $4/3k^2 + O(k)$ for real pseudoholomorphic curves because of the following fact.

Proposition 4.1. *The braid $\sigma_1^{-N}\Delta_{2k}$ is quasipositive if and only if there exists a real pseudoholomorphic curve A in $\mathbb{RP}^2$ of degree $2k$ such that $l_{k-2}(A) = N$.*

Proof. According to [22; Sect. 3.3], the fiberwise arrangement $[\supset_1 o_1^{N-1} \subset_1]$ is realizable by a real pseudoholomorphic curve of degree $2k$ if and only if the braid $x = \sigma_1^{-N}\Delta_{2k}$ is quasipositive. Thus, the quasipositivity of x implies the existence of a curve with $l_{k-2} = N$.

Suppose that there exists a pseudoholomorphic curve A of degree $2k$ with $l_{k-2} = N$. Let $v_1,\dots,v_N$ be the innermost ovals (i.e., the ovals of depth $k-2$). If some arrangement of embedded circles in $\mathbb{RP}^2$ is realizable by a real pseudoholomorphic curve and we erase an empty oval, then the new arrangement is also realizable by a real pseudoholomorphic curve. Thus, without loss of generality we may assume that A realizes the isotopy type $1\langle\cdots 1\langle N\rangle\cdots\rangle$. The arguments from [28] based on auxiliary conics through five innermost ovals prove that $v_1,\dots,v_N$ are in a convex position. Thus, choosing a pencil of lines centered at v_1, we see that $v_2,\dots,v_N$ form a single chain (see Fig. 9a); hence they can be replaced by a single branch B that has $N-2$ double points (see Fig. 9b). Choosing a pencil of lines as in Fig. 9b, we attach B to v_1 as in Fig. 9c. The braid corresponding to the arrangement of the obtained curve with respect to the pencil of lines centered at p (see Fig. 9c) is a conjugate of $\sigma_1^{-N}\Delta_{2k}$. □

Corollary 4.2. *For any integer $k \geq 2$, there exists a real pseudoholomorphic curve A on $\mathbb{RP}^2$ of degree $2k$ such that $l_{k-2}(A) \geq (4k^2 - f(k))/3 - 2[\log_2 k] - 1 - g(2k)$, where f,g are as in* (21)*, in particular, $l_{k-2}(A) \geq 4/3k^2 - 1/3k - 2[\log_2 k] - 1$.* □

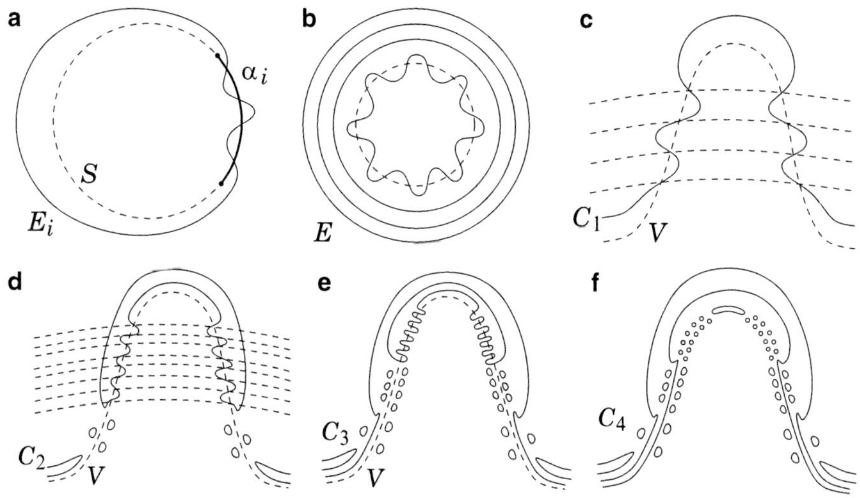

Fig. 10 Construction of an algebraic curve with many innermost ovals

4.2 *Real Algebraic Curves*

Proposition 4.3. *For any $k = 4p$ there exists a real algebraic curve of degree $2k$ in $\mathbb{RP}^2$ such that $l_{k-2} = 18p^2 - 2p = 9/8k^2 - 1/2k$.*

Proof. We fix an affine chart $\mathbb{R}^2$ on $\mathbb{RP}^2$. Let S be the unit circle and let $\alpha_1, \dots, \alpha_p$ be disjoint arcs of S. Let $E_1, \dots, E_p$ be ellipses such that E_i is arranged on $\mathbb{R}^2$ with respect to S and α_i as in Fig. 10a. Then $E_1 \cup \dots \cup E_p$ can be perturbed into a curve E of degree $2p$ consisting of a single nest of depth p (i.e., a *hyperbolic* curve), and the innermost oval V of E intersects S in k points that lie on S in the same order as on V (see Fig. 10b). Let $S_{\nu,1}, \dots, S_{\nu,\nu p}$, $\nu = 1, \dots, 4$, be concentric copies of S of increasing radii ($r_{1,1} < \dots < r_{1,p} < r_{2,1} < \dots < r_{2,2p} < r_{3,1} < \dots$) each of which intersects V at k points. Let

$$C_0 = 1, \quad C_\nu = EC_{\nu-1} + \varepsilon_\nu \prod_{i=1}^{\nu p} S_{\nu,i}, \quad \nu = 1, \dots, 4, \quad 0 < |\varepsilon_4| \ll \dots \ll |\varepsilon_1| \ll 1$$

(see Fig. 10c–f; we use the same notation for a curve and its defining polynomial). Then C_4 is the required curve. □

5 On A_N Singularity of a Plane Curve of a Given Degree

It is easy to see that the existence of a pseudoholomorphic curve of degree m that has a singular point of type A_n is equivalent to the quasipositivity of the braid $\sigma_1^{-(n+1)}\Delta_m^2$. Thus, Theorem 3.13 admits also the following interpretation.

Proposition 5.1. *For any m, there exists a pseudoholomorphic curve C_m in $\mathbb{CP}^2$ of degree m with a singularity of type A_n with $n = 2/3(m^2 - m) - 2[\log_2 k]$. Thus,* $\lim_{m\to\infty} 2n/m^2 = 4/3$. □

The question of the maximal $n = N(m)$ such that there exists an algebraic curve of degree m with an A_n singularity has been studied by several authors. Let $\alpha = \limsup 2N(m)/m^2$. Signature estimates for the double covering yield $\alpha \le 3/2$ (see [14]). An obvious example $(y+x^k)^2 - y^{2k} = 0$ yields $m = 2k$ and $n = 2k^2 - 1$, so $\alpha \ge 1$.

In a generic family of curves, the condition to have an A_n singularity defines a stratum of codimension n. Thus the so-called expected dimension of the variety of curves of degree m with a singularity A_n is equal to $m^2/2 - n + O(m)$, i.e., $\alpha > 1$ is "unexpected" from this point of view. Nevertheless, this is so. A series of examples providing $\alpha \ge 28/27$ was constructed by Gusein-Zade and Nekhoroshev in [14]. Cassou-Nogues and Luengo [4] improved this estimate up to $\alpha \ge 8 - 4\sqrt{3}$. Here we show that $\alpha \ge 7/6$. This follows from the following evident observation.

Proposition 5.2. *Let $F(X,Y)$ be a polynomial whose Newton polygon is contained in the triangle with vertices $(0,0)$, $(ac,0)$, and $(0,bc)$. Suppose that $F = 0$ has a singularity A_{k-1} at the origin, and $\operatorname{ord}_0 F(0,Y) = 2$. Then for any $p \ge b/a$, the curve $F(X^{pb}, Y^{pa} + X) = 0$ has a singularity A_n for $n = abkp^2 - 1$, and its degree is $m = abcp$. Hence $\alpha \ge \lim_{p\to\infty}(2n/m^2) = 2k/(abc^2)$.*

Proof. Indeed, $F_1(X,Y) = F(X^{pb},Y)$, $F_2(X,Y) = F_1(X,Y+X)$, and $F_3(X,Y) = F_2(X,Y^{pa})$ have singularities A_{bkp-1}, A_{bkp-1}, and A_{abkp^2-1} respectively. □

If we apply Proposition 5.2 to a sextic curve in $\mathbb{P}^2$ that has an A_{19} singularity ($a = b = 1$, $c = 6$, $k = 20$), then we obtain $\alpha \ge 10/9$. The existence of such a curve follows from the theory of K3 surfaces (see, e.g., [36]); an explicit equation is given in [5, Sect. 6].

If we apply Proposition 5.2 to $a = 2$, $b = 1$, $c = 4$, $k = 18$, then we obtain $\alpha \ge 9/8$. The existence of polynomials realizing this case can be proven using K3 surfaces (Alexander Degtyarev, private communication). Also, they can be written down explicitly:

$$\left(x^3 + 45x^4 + y - 2787x^2y + 60192y^2\right)^2 \\ + 12\left(x^8 + (1-87x)x^5y - (42-2943x)x^3y^2 + (288-36288x)xy^3 + 66816y^4\right)$$

or $\left(x^3 + y - 5x^2y\right)^2 - 4\left(2x^8 + 2x^5y + 9x^4y^2 + 3xy^3 + y^4\right)$ (the latter polynomial was found by Ignacio Luengo). To determine the singularity type at the origin, it is enough to compute the multiplicity at $x = 0$ of the discriminant with respect to y. Here is the corresponding Maple code for the second polynomial:

```
f := (x^3+y-5*x^2*y)^2 - 4*(2*x^8+2*x^5*y+9*x^4*y^2+3*x
*y^3+y^4); factor(discrim(f,y));
```

Finally, if we apply Proposition 5.2 to the case $a = 3$, $b = c = 2$, $k = 14$, then we obtain $\alpha \geq 7/6$. This case is realizable by the polynomial (also found by Ignacio Luengo)

$$\left(x^2 - 53x^3 + y - 60xy - \frac{2160}{7}y^2\right)^2 + \frac{4}{7}\left(5x^6 + 8x^4y + 3x^2y^2 + 41x^3y^2 + 27xy^3 + \frac{486}{7}y^4\right).$$

6 Odd-Degree Curves with Many Nests

6.1 Construction of Real Algebraic M-Curves of Degree $4d+1$ with Four Nests of Depth d

Let C be a nonsingular real pseudoholomorphic curve of odd degree $m = 2k+1$ in $\mathbb{RP}^2$. We say that an oval of C is *even* (respectively *odd*) if it is surrounded by an even (respectively *odd*) number of other ovals. Let us denote the number of even (respectively odd) ovals by p (respectively by n). In a joint note with Oleg Viro [32] we proved the following result.

Theorem 6.1. *If $k = 2d$ (i.e., $m = 4d+1$) and C has four disjoint nests of depth d, then:*

(*i*) *If C is an M-curve, then $p - n \equiv k^2 + k \mod 8$ (Gudkov–Rohlin congruence).*
(*ii*) *If C is an $(M-1)$-curve, then $p - n \pm 1 \equiv k^2 + k \mod 8$ (Kharlamov–Gudkov–Krakhnov congruence).*
(*iii*) *If C is an $(M-2)$-curve and $p - n + 4 \equiv k^2 + k \mod 8$, then C is of type I (Kharlamov congruence).*
(*iv*) *If C is of type I, then $p - n \equiv k^2 + k \mod 4$ (Arnold congruence).*

This is the first result of this kind for curves of odd degree. If $d = 1$, it is trivial. If $d = 2$, it was conjectured by Korchagin, who he constructed M-curves of degree 9 with four nests and observed the congruence mod 8. However, starting with $d = 3$, curves satisfying the hypothesis of Theorem 6.1 have not been known.

In this section we demonstrate the "nonemptiness" of Theorem 6.1 for any d for real *algebraic* curves.

Proposition 6.2. *For any integer $d \geq 1$, there exists a real algebraic M-curve of degree $m = 4d+1$ that has four disjoint nests of depth d. This curve realizes the isotopy type*

$$J \sqcup (4d^2 + 6d - 8) \sqcup 3\langle\langle d\rangle\rangle \sqcup \underbrace{1\langle \cdots 1\langle 1\langle 1\langle 1}_{d-1}\rangle \sqcup 8\rangle \sqcup 16\rangle \cdots \sqcup (8d-16)\rangle. \quad (1)$$

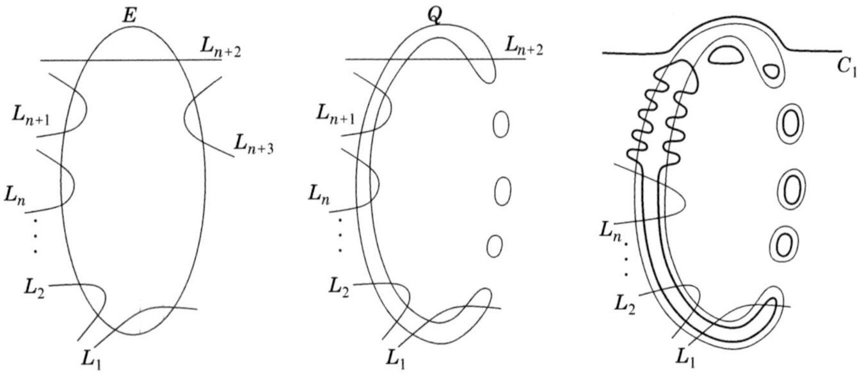

Fig. 11 Construction of Q, C_1, and $L_1,\dots,L_n$ in the proof of Proposition 6.2

The notation $3\langle\langle d\rangle\rangle$ is explained in Sect. 2.

Proof. The result follows immediately from the following statement $(\mathcal{H}_d)$, which we shall prove by induction:

$(\mathcal{H}_d)$. If $d \geq 1$, then for any $n > 0$ there exists a mutual arrangement of an M-quartic Q, an M-curve C_d of degree $m = 4d+1$, and n lines $L_1,\dots,L_n$ satisfying the following conditions:

- (i) The curve C_d belongs to the isotopy type (1).
- (ii) Each oval of Q (we denote them by $V_0,\dots,V_3$) surrounds a nest of C_d of depth d. the nests surrounded by V_1,V_2,V_3 are simple.
- (iii) One exterior empty oval of C_d (let us denote it by v) intersects V_0 at $4m$ distinct points all of which lie on V_0 in the same order as on v; so $(\operatorname{Int} V_0)\setminus(\operatorname{Int} v)$ is a disjoint union of $2m$ open disks (digons), which we denote by $D_1,\dots,D_{2m}$.
- (iv) $C_d \cap D_i = \varnothing$ for $i > 1$ and $C_d \cap D_1$ has the isotopy type $(8d-8) \sqcup S_d$, where S_d stands for the final part of the expression (1) starting with "$1\langle\dots$".
- (v) All the other exterior empty ovals are outside all the ovals of Q.
- (vi) There exist arcs $\alpha_1 \subset \dots \subset \alpha_n \subset V_0 \cap D_{m+1}$ such that for any $i = 1,\dots,n$, the line L_i intersects Q at four distinct points that lie on $\alpha_i \setminus \alpha_{i-1}$, two points on each connected component of $\alpha_i \setminus \alpha_{i-1}$ (here we assume that $\alpha_0 = \varnothing$).

Given a line L, we shall denote by $L^k(\varepsilon)$ a union of k generic lines depending on a real parameter ε such that each line tends to L as $\varepsilon \to 0$. We shall use the same notation for a curve and a polynomial that defines it. The notation $0 \ll \dots \ll \varepsilon_2 \ll \varepsilon_1 \ll 1$ means that we choose a small parameter ε_1, then we choose ε_2 that is small with respect to ε_1, and so on.

Let us prove $(\mathcal{H}_1)$. Let E be a conic and let $p_1,q_1,p_2,q_2,\dots,p_{n+3},q_{n+3}$ be points lying on E in this cyclic order. Let L_i be the line (p_iq_i) and let us set $Q = E^2 + \varepsilon_2 L^4_{n+3}(\varepsilon_1)$ and $C_1 = QL_{n+2} + \varepsilon_4 L^5_{n+1}(\varepsilon_3)$, where $0 \ll \varepsilon_4 \ll \dots \ll \varepsilon_1 \ll 1$. Then Q, C_1, and $L_1,\dots,L_n$ satisfy (i)–$(vi)_{d=1}$ for a suitable choice of signs of the equations (see Fig. 11).

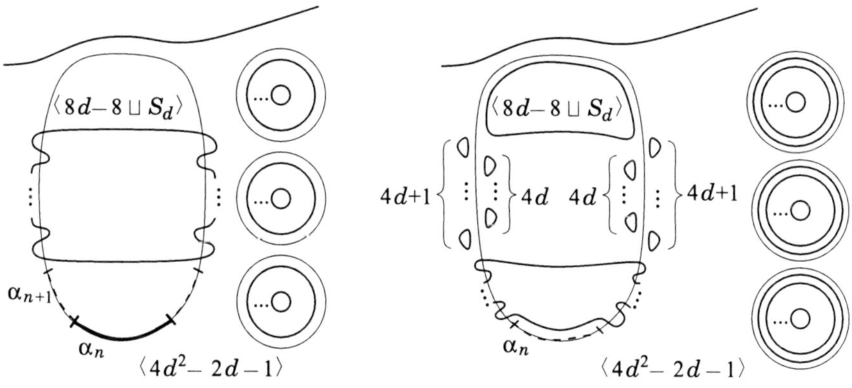

$$1\langle 8d-8\sqcup S_d\rangle = S_{d+1};$$
$$(4d^2-2d-1)+(8d+2) = 4(d+1)^2-2(d+1)-1.$$

Fig. 12 Induction step in the proof of Proposition 2.2

Now let us assume that $(\mathcal{H}_d)$ is true and let us prove $(\mathcal{H}_{d+1})$. Let Q, C_d, and $L_1,\cdots,L_{n+1}$ satisfy (i)–(vi) with $n+1$ instead of n and let us set $C_{d+1}=QC_d+\delta L_{n+1}^{4d+5}(\varepsilon)$ with $0\ll\delta\ll\varepsilon\ll 1$ (see Fig. 12). □

Remark. For the curve in Proposition 6.2, it is easy to check that $p-n=k^2+k$. Indeed, one sees in Fig. 12 that $p_{d+1}=n_d+4d^2+14d+6$ and $n_{d+1}=p_d-4d^2+2d$, whence $(p_{d+1}-n_{d+1})=-(p_d-n_d)+8d^2+12d+6$, i.e. the quantities p_d-n_d and $k^2+k=(2d)^2+2d$ satisfy the same recurrent relation. This gives another proof that the right-hand side of the congruences in Theorem 6.1 is correctly computed (it was computed in [32] via the Brown–van der Blij invariant of the Viro–Kharlamov quadratic form defined in [33]).

6.2 On M_d-Curves of Degree $2td+1$

Let A be a real algebraic (or real pseudoholomorphic) curve on $\mathbb{RP}^2$ of degree $m=2k+1$ with $k=td$. Recall that the *depth* of an oval is the number of ovals that surround it. Let V be an oval of A. We say that V is a *d-oval* of A if the depth of V is a multiple of d (perhaps zero) and V is the outermost oval of a nest of depth at least d (i.e., there are at least $d-1$ nested ovals inside V). We say that A is an M_d-curve if it is an M-curve of degree m and the number of its d-ovals is at least $2t^2-3t+2$.

For example, the curves discussed in Sect. 6.1 are M_d-curves of degree $4d+1$ (i.e., $t=2$).

Proposition 6.3. *(a) For any integers $t\geq 2$ and $d\geq 1$, there exist real pseudoholomorphic M_d-curves of degree $m=2td+1$.*

(b) For any integer $t\geq 2$, there exist real algebraic M_2-curves of degree $4t+1$. In particular:

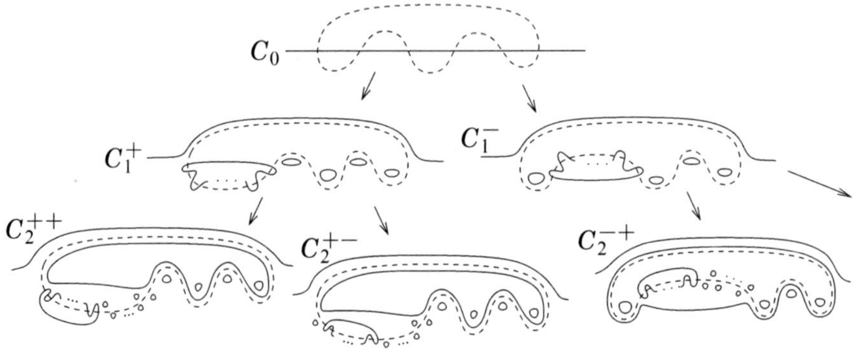

Fig. 13 Construction in Proposition 6.3(a)

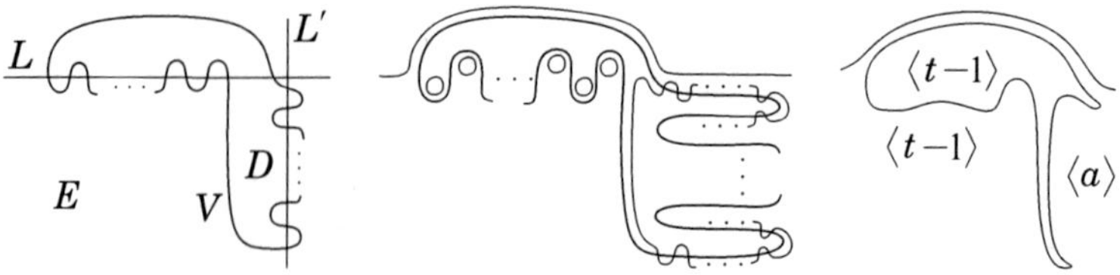

Fig. 14 Construction in Proposition 6.3(b)

(c) *For any integer $t \geq 2$ there exists a real algebraic M-curve of degree $m = 4t + 1$ realizing the isotopy type $J \sqcup g_{2t}\langle 1\rangle \sqcup 1\langle t-1\rangle \sqcup (4t^2+3t-2)$, where $g_{2t} = (t-1)(2t-1)$ is the genus of a curve of degree $2t$. So this curve has as many nests as the number of ovals of an M-curve of degree $2t$.*

Proof. (a) Let B be a real algebraic M-curve of degree $2t$ and let there be a line L satisfying the following conditions:

(i) An oval V of B has $2t$ intersections with L placed on V in the same order as on L.

(ii) $B \setminus V \subset E$, where E is the component of $\mathbb{RP}^2 \setminus (V \cup L)$ whose closure is nonorientable. Such a curve can be easily obtained by Harnack's method (see also the proof of (b)). We construct curves C_e of degrees $m_e = 2te + 1$, $e = 0, 1, 2, \cdots$, recursively (see Fig. 13). We set $C_0 = L$, and we define C_{e+1} as a small perturbation of $C_e \cup B$ such that C_{e+1} meets B at $2tm_e$ points all lying on an arc of B bounding a digon between B and C_e.

(b) For some curves B, the second step of the above construction can be realized in the class of algebraic curves. Suppose that B and L satisfy the conditions (i)–(ii), and moreover, V and L are arranged with respect to another line L' as shown in Fig. 14. Then we obtain the isotopy type

$$J \sqcup (a+t-1) \sqcup 1\langle t-1\rangle \sqcup S^2,$$

where $a = 2t(2t+1) - 1$ and S is the isotopy type of $B \setminus V$ (see Fig. 14).

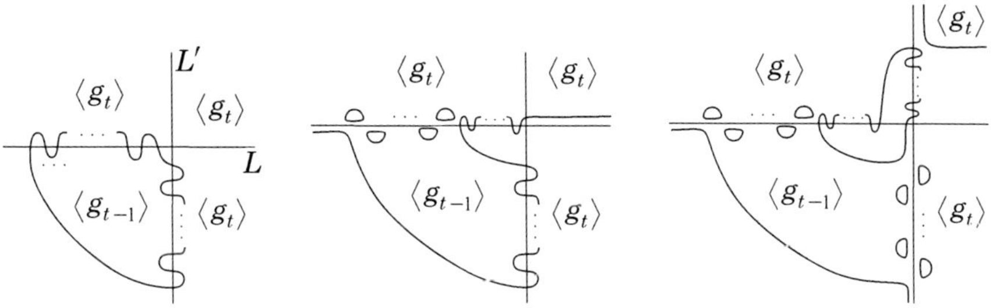

Fig. 15 Construction of the curve in Fig. 14 (on the left)

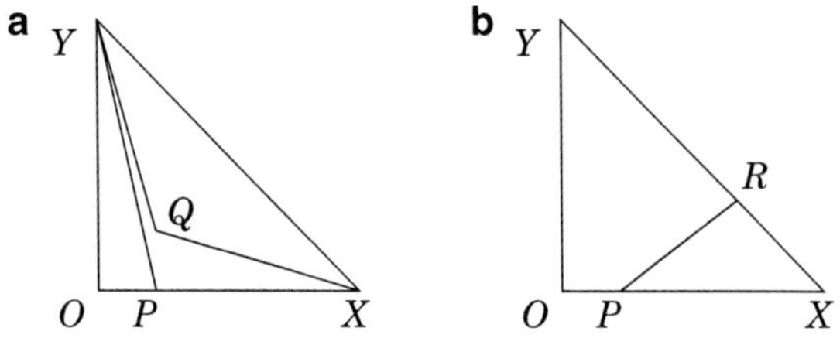

Fig. 16 Haas's zone decomposition

To construct the required arrangement of B, L, and L', we can start with a Harnack curve of degree $2t-2$ and proceed as shown in Fig. 15. Here $g_t = (t-1)(t-2)/2$ and $g_{t-1} = (t-2)(t-3)/2$.

The construction of (B,L,L') can be interpreted as Viro patchworking according to the Haas's zone decomposition (see [15]) of the triangle OXY into two triangles and one quadrangle OPY, XYQ, and $XPYQ$ (see Fig. 16b),where $O=(0,0)$, $X=(2t,0)$, $Y=(0,2t)$, $P=(1,0)$, and $Q=(1,1)$. This means that we choose any primitive triangulation that contains the edges XQ, QY, YP, and we define the sign distribution $\delta : (OXY)\cap\mathbb{Z}^2 \to \{\pm 1\}$,

$$\delta(x,y) = \begin{cases} (-1)^{(x+1)(y+1)}, & y>0, \\ -1, & y=0. \end{cases}$$

(c) Let B be the M-curve of degree $2t$ patchworked according to the Haas zone decomposition of OXY obtained by cutting it along the segment PR where O, X, Y, P are as above and $R=(2t-2,2)$ (see Fig. 16b). This means that we choose any primitive triangulation that contains the edge PR and we define the sign distribution $\delta : (OXY)\cap\mathbb{Z}^2 \to \{\pm 1\}$,

$$\delta(x,y) = \begin{cases} (-1)^{xy}, & (x.y)\in OPRY, \text{ i.e., } (2t-3)y \geq 2(x-1), \\ (-1)^{(x+1)y}, & (x,y)\in XPR, \text{ i.e., } (2t-3)y \leq 2(x-1). \end{cases}$$

Then B has an oval V that is arranged with respect to the lines L and L' (the axes Ox and Oy respectively) as in Fig. 14, but all other ovals of B are empty. Moreover, $(t-1)(t-2)/2$ empty ovals are in the domain D, and the other empty ovals are in the domain E. The rest of the construction is shown in Fig. 14. □

Remark. **1.** Let p and n be the numbers of positive and negative ovals of a curve C_d constructed in the proof of Proposition 6.3(a). It is easy to prove by induction that

$$p-n=\begin{cases} 2t(\pm m_1 \pm m_3 \pm \cdots \pm m_{d-1}), & d \text{ is even}, \\ 2t(1 \pm m_2 \pm m_4 \pm \cdots \pm m_{d-1}) + p_B - n_B - 2, & d \text{ is odd}, \end{cases}$$

where $m_e = 2te+1$, p_B (respectively n_B) is the number of positive (respectively negative) ovals of B, and the choice of signs is illustrated in Fig. 13. Thus it follows from the Gudkov–Rohlin congruence that for any choice of B satisfying (i) and (ii), we have

$$p-n\equiv\begin{cases} k^2+k & \text{mod } 8. \quad \text{if } t\equiv d\equiv 0 \mod 2, \\ k^2+k+t-2 & \text{mod } 8, \quad \text{if } t\equiv d+1\equiv 0 \mod 2, \\ k^2+k & \text{mod } 4, \quad \text{if } t+1\equiv d\equiv 0 \mod 2, \\ k^2+k+t-2 & \text{mod } 4, \quad \text{if } t\equiv d\equiv 1 \mod 2, \end{cases}$$

where $k = td$ (so $\deg C_d = 2k+1$). All values of $p-n$ satisfying these congruences are attained for pseudoholomorphic curves.

2. The slalgebraic curves constructed in the proof of Proposition 6.3(b,c) satisfy the congruence $p-n\equiv k^2+k \mod 8$. The first pseudoholomorphic curve constructed in Proposition 6.3(a) that does not satisfy this congruence is the curve of degree 13 ($t=3$, $d=2$) of isotopy type $J \sqcup 1 \sqcup 1\langle 44\rangle \sqcup 8\langle 1\rangle \sqcup 1\langle 1\langle 1\langle 1\rangle\rangle\rangle$ (the curve C_2^{-+} in Fig. 13 if Harnack's sextic is chosen for B). It would be of interest to study whether this curve is algebraically realizable.

7 *M*-Curves of Degree 9 with a Single Exterior Oval

Theorem 7.1. *(a) There exist real algebraic curves of degree* 9 *realizing the isotopy types*

$$J \sqcup 1\langle 2a \sqcup 1\langle 26-2a\rangle\rangle, \qquad 2 \le a \le 11. \tag{1}$$

(b) The isotopy type $J \sqcup 1\langle 24 \sqcup 1\langle 2\rangle\rangle$ *is unrealizable by real pseudoholomorphic (in particular, by real algebraic) curves of degree* 9.

Combined with the result of S. Fiedler–Le Touzé [12], Theorem 7.1 implies that among the isotopy types of the form $J \sqcup 1\langle b \sqcup 1\langle 26-b\rangle\rangle$, only the isotopy types in the list (1) are realizable by curves of degree 9.

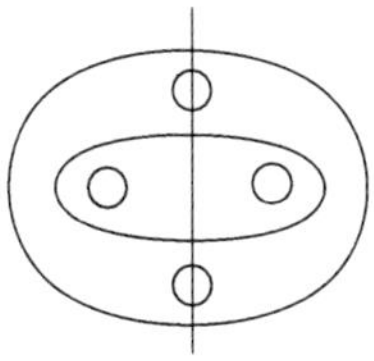

Fig. 17 O_1-jump

Following [12, Definition 1], we say that a curve of degree 9 has an *O_1-jump* if it has six ovals arranged with respect to some line as in Fig. 17. Theorem 7.1(b) follows immediately from [12, Theorem 2(2)] combined with the following fact:

Theorem 7.2. *Let A be an M-curve of degree 9 that realizes the isotopy type $J \sqcup 1\langle \beta \sqcup 1\langle \gamma \rangle \rangle$ with $\beta + \gamma = 26$. Then A has an O_1-jump.*

Theorem 7.1(a) is proven in Sect. 7.1; Theorem 7.2 is proven in Sect. 7.2.

Recall that an oval of a real algebraic plane curve is called *exterior* if it is not surrounded by another oval. We say that A is a *one-exterior-oval curve* (OEO curve) if it has exactly one exterior oval. Note that OEO M-curves of degree greater than three were previously unknown. It is evident that OEO M-curves do not exist in degree 4 and 5. The Petrovsky inequality excludes OEO M-curves of degree 6. Viro [28] (respectively Shustin [26]) excluded OEO M-curves of degree 7 (respectively 8). Using theta characteristics (the idea applied later in [7]), Kharlamov excluded OEO M-curves of odd degree of a very special form $J \sqcup 1\langle n \rangle$ (unfortunately, his proof still has not been written up). However, OEO M-curves of degree 9 do exist by Theorem 7.1(a).

It seems that OEO M-curves of even degree greater that 2 do not exist. Note that Hilbert's construction provides OEO $(M-r)$-curves of any even degree ≥ 6 for any $r \geq 1$.

7.1 Construction

Lemma 7.3. *For any $\alpha \in \{4,8,12,16,20\}$ and for any distinct real numbers $\lambda_1, \lambda_2, \lambda_3$, there exists a polynomial $g(x,y) = \sum_{i+9j \leq 27} g_{ij}x^i y^j$ such that the affine curve $g(x,y) = 0$ is as in Fig. 18 and $g^\Gamma = (y - \lambda_1 x^9)(y - \lambda_2 x^9)(y - \lambda_3 x^9)$, where g^Γ denotes the truncation of g to the edge $\Gamma = [(27,0),(0,3)]$ of the Newton polygon, i.e., $g^\Gamma = \sum_{i+9j=27} g_{ij}x^i y^j$*

Proof. The statement follows easily from the results of [29]. □

Proof of Theorem 7.1(a). All curves (1) are realizable as perturbations of the singular curve $F_3(F_3^2 + cF_2^3) = 0$, where $F_3 = 0$ is an M-cubic and $F_2 = 0$ is a conic that has maximal tangency with $F_3 = 0$ at a point p lying on the oval O_3 of the curve $F_3 = 0$.

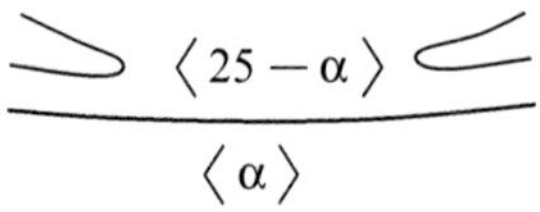

Fig. 18 $\alpha \in \{4,8,12,16,20\}$

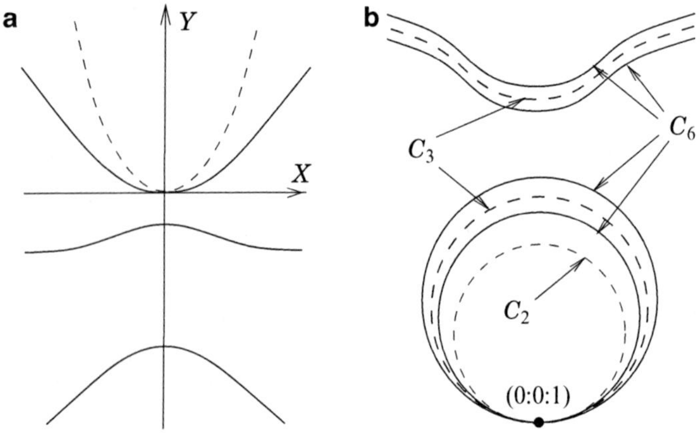

Fig. 19 (**a**) C_2 and C_3 in $0XY$ affine chart (**b**) C_2, C_3, C_6 on $\mathbb{RP}^2$

Let $F_2(X,Y) = Y - X^2$, $F_3(X,Y) = (Y - X^2)(1+3Y) + 2Y^3$, $F_6 = F_3^2 + cF_2^3$, $0 < c \ll 1$, and $F_9 = F_6F_3$. Let C_k be the curve $F_k = 0$, $k = 2,3,6,9$. Then C_2 has tangency of order 6 at the origin with C_3, and the mutual arrangement of C_2 and C_3 on $\mathbb{R}^2$ is as in Fig. 19a. Hence the arrangement of C_9 on $\mathbb{RP}^2$ is as in Fig. 19b. The curve C_9 has three smooth real local branches at the origin (two branches of C_6 and one of C_3) with pairwise tangencies of order 9.

We introduce local coordinates (x,y) at the origin $X = x$, $Y = y + \gamma(x)$, $\gamma(x) = x^2 - 2x^6 + 6x^8$. Let $f_k(x,y) = F_k(x, y+\gamma(x))$, $k = 2,3,6,9$, i.e., f_k is F_k rewritten in the coordinates (x,y). Then f_9 has the form $\sum_{i+9j\ge 27} a_{ij}x^iy^j$ and $f_9^\Gamma = y(y^2 - 8cx^{18})$, where f_9^Γ is the truncation of f_9 to Γ, i.e., $f_9^\Gamma = \sum_{i+9j=27} a_{ij}x^iy^j$. Here is the Mathematica code that checks it:

```
F2=Y-X^2; F3=F2(1+3Y)+2Y^3; F6=F3^2+c*F2^3; F9=F3*F6;
su={ X->x,Y->y+x^2-2x^6+6x^8} ; f9=Expand[F9//.su];
Table[Series[Coefficient[f9,y,j],{ x,0,27-9j}],{ j,0,3}]
```

We perturb the singularity of C_9 at the origin using the straightforward approach from [5]. Let $g(x,y)$ be as in Lemma 7.3, where we set $g^\Gamma = f_9^\Gamma$. We have $g_{18,1} = a_{18,1} = -8c \ne 0$; hence shifting if necessary the x-coordinate, we may assume that $g_{17,1} = 0$.

Let $\tilde F(X,Y) = \sum_{i+j\le 9} B_{ij}X^iY^j$ be a polynomial with indeterminate coefficients. We set $\tilde f(x,y) = \tilde F(x, y+\gamma(x)) = \sum_{i,j} b_{ij}x^iy^j$. Then the b_{ij} are linear functions of the B_{ij}. Let $\varphi(i,j) = 27 - i - 9j$. Solving a system of linear equations, we obtain

$B_{ij} = B_{ij}(t)$ such that

$$b_{ij} = g_{ij}t^{\varphi(i,j)} \qquad \text{for} \quad i+9j<27, \quad (i,j) \neq (17,1).$$

Substituting the solution into $b_{17,1}$, we see that $b_{17,1} = O(t^2)$:

```
ff=Expand[Sum[Sum[B[i,j]X^i Y^j,
  {i,0,9-j}],{j,0,9}]//.su];
Do[Do[b[i,j]=Coefficient[Coefficient[ff,x,i],y,j],
  {i,0,26-9j}],{j,0,2}];
var=eq={};
Do[Do[AppendTo[var,B[i,j]],{i,0,9-j}],{j,0,9}];
Do[Do[If[Not[i==17&&j==1],
   AppendTo[eq,b[i,j]==g[i,j]t^(27-9j-i)]],
  {i,0,26-9j}],{j,0,2}];
so=Solve[eq,var][[1]]; Factor[b[17,1]//.so]
```

Recall that $g_{17,1} = 0$. Thus, for any (i,j) such that $i+9j < 27$, we have $b_{ij} = g_{ij}t^{\varphi(i,j)} + O(t^{\varphi(i,j)+1})$. Therefore, the curve $F_9(X,Y) + \tilde{F}_t(X,Y) = 0$ for $0 < t \ll c$ is obtained from C_9 by Viro's patchworking by gluing the pattern in Fig. 18 into the singular point of C_9. We obtain in this way the isotopy types (1) with $a = 2,4,6,8,10$. Replacing $g(x,y)$ with $g(x,-y)$, we obtain those with $a = 3,5,7,9,11$. □

7.2 *Restrictions*

The main tool used in the proof of Theorem 7.2 is the analogue of the Murasugi–Tristram inequality for colored signatures obtained in [6, 13]. Given a *μ-colored oriented link*, i.e., an oriented link L in S^3 with a fixed decomposition $L = L_1 \sqcup \cdots \sqcup L_\mu$ into a disjoint union of sublinks, and a μ-tuple of complex numbers $\omega = (\omega_1, \cdots, \omega_\mu)$, $|\omega_i| = 1$, $\omega_i \neq 1$, V. Florens [13] defined the isotopy invariants *ω-signature* $\sigma_\omega(L)$ and *ω-nullity* $\eta_\omega(L)$. In [6], D. Cimasoni and V. Florens gave an efficient algorithm for the computation of σ_ω and n_ω via a generalized (colored) Seifert surface of L. This algorithm was used for the computations in the proof of Theorem 7.2. When $\mu = 1$, these invariants specialize to the usual Tristram signature and nullity. They satisfy the following analogue of the Murasugi–Tristram inequality.

We set $\mathbb{T}^1_* = \{z \in \mathbb{C};\, |z| = 1, z \neq 1\}$ and $\mathbb{T}^\mu_* = \mathbb{T}^1_* \times \cdots \times \mathbb{T}^1_*$ (μ times).

Theorem 7.4. (See [6, 13]). *Let $F_1,\dots,F_\mu$ be disjoint embedded oriented surfaces in the 4-ball B^4 transversal to the boundary $S^3 = \partial B^4$. Let $F = F_1 \cup \dots \cup F_\mu$. We consider the colored link $L = L_1 \sqcup \dots \sqcup L_\mu$, where $L_i = \partial F_i$, $i = 1,\dots,\mu$. Then for any $\omega \in \mathbb{T}_*^\mu$, we have*

$$\eta_\omega(L) \geq |\sigma_\omega(L)| + \chi(F), \tag{2}$$

where $\chi(F)$ is the Euler characteristic of F. □

Remark. In [30], Oleg Viro proposed another approach to defining η_ω, σ_ω and proving Theorem 7.4. This approach is based on [27].

To reduce the computations, we use the following fact, whose proof is very similar to that of [22; Proposition 3.3].

Proposition 7.5. *Let p,q be integers such that $0 < p < q$ and let L_0 and L_{2q} be two μ-colored links represented by braids b_0 and $b_{2q} = b_0\sigma_1^{2q}$ respectively. Let 1 and 2 be the colors of the first two strings in the part σ_1^{2q} of the braid b_{2q}. Let $t = (t_1,\cdots,t_\mu) \in \mathbb{T}_*^\mu$ be such that $t_1t_2 = \exp(2\pi i p/q)$. Let $t_j = \exp(2\pi i\theta_j)$, $0 < \theta_j < 1$, $j = 1,2$, and $\theta = \theta_1 + \theta_2$. Then $\eta_t(L_{2q}) = \eta_t(L_0)$ and $\sigma_t(L_{2q}) = \sigma_t(L_0) + (q-2p)\operatorname{sign}(1-\theta)$.* □

Corollary 7.6. *Let p,q be integers such that $0 < p < q$. Let $\{L_{2n}\}_{n\in\mathbb{Z}}$ be a family of μ-colored links such that L_{2n} is represented by the braid $b_{2n} = a_1\sigma_h^{2n}a_2\sigma_\ell^{-2n}a_3$ with some fixed braids a_1, a_2, a_3. Let j and k be the colors of the hth and the $(h+1)$th strings of the part σ_h^{2n} of b_{2n}. Suppose that the unordered pair of the colors of the ℓth and the $(\ell+1)$th strings of the part σ_ℓ^{-2n} of b_{2n} is also $\{j,k\}$ (we do not claim that $j \neq k$). Let $t = (t_1,\dots,t_\mu) \in \mathbb{T}_*^\mu$ be such that $t_jt_k = \exp(2\pi i p/q)$. Then $\eta_t(L_{2q}) = \eta_t(L_0)$ and $\sigma_t(L_{2q}) = \sigma_t(L_0)$.*

Proof. If $j = k$, the statement follows from [22; Proposition 3.3]. If $j \neq k$, it follows from Proposition 7.5. □

Proof of Theorem 7.2. Suppose that A has no O_1-jump. Then applying [22; Corollary 2.3] to a pencil of lines centered at a point inside an empty oval of depth 1, we may replace the group of the γ innermost ovals by a singular branch with $\gamma-1$ double points, as shown in Fig. 20. It follows from [12; proof of Theorem 2(2)] that if we choose p as in Fig. 20, then the fiberwise arrangement of the obtained curve with respect to $\mathcal{L}_p$ (the pencil of lines through p) is $[\times_2^{\gamma-2} \supset_2 o_3^{\beta_1} o_6^{\beta_2} o_3^{\beta_3} o_6^{\beta_4} \subset_7 \times_8]$ for some odd $\beta_1,\cdots,\beta_4$ such that $\beta_1 + \ldots + \beta_4 = \beta$; see [22; Sect. 3.2] for the notation of fiberwise arrangements.

Let b be the braid corresponding to $(\mathbb{R}A,\mathcal{L}_p)$. To fix the notation, we reproduce the definition of b from [19]. Let $\pi_p : \mathbb{CP}^2 \setminus p \to \mathbb{CP}^1$ be the linear projection from p. We fix complex orientations on $\mathbb{R}A$ and $\mathbb{RP}^1$. Let $A \setminus \mathbb{R}A = A_+ \sqcup A_-$ and $\mathbb{CP}^1 \setminus \mathbb{RP}^1 = \mathbb{CP}^1_+ \sqcup \mathbb{CP}^1_-$ be the corresponding partitions. Let H_+ be a closed disk in $\mathbb{CP}^1_+$ containing all nonreal critical values of $\pi_p|_A$. We define b as the closed braid corresponding to the braid monodromy of the curve A along the loop ∂H_+. We set also $F = \pi_p^{-1}(H_+) \cap A$, $F_\pm = F \cap A_\pm$, $L = \partial F$, and $L_\pm = \partial F_\pm$. Then L is the braid

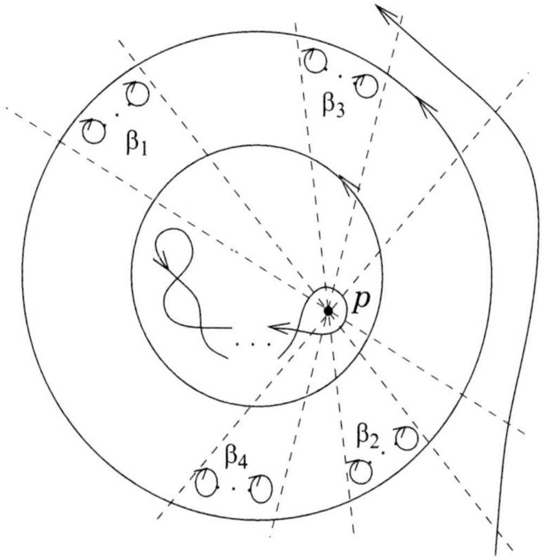

Fig. 20 Choice of the pencil for a curve without 0_1-jump

closure of b in the 3-sphere $\partial(\pi_p^{-1}(H_+) \setminus U_p)$, where U_p is a small ball centered at p. We have (see [22, Sect. 2.3])

$$b = \sigma_2^{-\gamma-1} \tau_{2,3} \sigma_3^{-\beta_1} \tau_{3,6} \sigma_6^{-\beta_2} \tau_{6,3} \sigma_3^{-\beta_3} \tau_{3,6} \sigma_6^{-\beta_4} \tau_{6,7} \sigma_8^{-1} \Delta_9, \tag{3}$$

where $\tau_{i,j} = \tau_{j,i}^{-1} = \left(\sigma_{i+1}^{-1} \cdots \sigma_j^{-1}\right)(\sigma_i \cdots \sigma_{j-1})$ for $i < j$. It follows from [12] that the complex orientation of $\mathbb{R}A$ is as in Fig. 20. Hence, in the braid (3), the strings 1, 8, 9 represent L_+, and the strings $2, \ldots, 7$ represent L_-.

To make the notation coherent with Theorem 7.4, we set $L_1 = L_+$, $L_2 = L_-$, $F_1 = F_+$, $F_2 = F_-$. The Riemann–Hurwitz formula for the projection $\pi_p|_F : F \to H_+$ yields $\chi(F) = 9 - e(b)$, where $e : B_9 \to \mathbb{Z}$ is the abelianization homomorphism, i.e., $e(b)$ is the number of branch points of the mapping $\pi_p|_F$. So we have $\chi(F) = 9 - 10 = -1$.

The result follows from the fact that for any choice of four odd numbers $\beta_1, \ldots, \beta_4$ with $\beta_1 + \cdots + \beta_4 \leq 24$, there exist $t = (t_1, t_2) \in \mathbb{T}_*^2$ such that the inequality (2) fails. To reduce the computations, we apply Corollary 7.6. Indeed, suppose that for some $\beta^{(0)} = (\beta_1^{(0)}, \ldots, \beta_4^{(0)})$ we find t such that $\operatorname{Arg} t_1 + \operatorname{Arg} t_2 \equiv 2\pi p/q \mod 2\pi$ and (2) fails. Then for any $\beta = (\beta_1, \ldots, \beta_4)$ such that $\beta \equiv \beta^{(0)} \mod 2q$, the inequality (2) also fails for the same t.

By chance, it happens that for any β there exists $t = (t_1, t_2)$ with $t_1 t_2 = -1$, so $q = 2$. Thus, it is enough to carry out the computations, for example, only when each of $\beta_1, \ldots, \beta_4$ is equal to 1 or 3. In all these 16 cases, the parameter choice $t_1 = -1/t_2 = \exp(2\pi i \theta_1)$, $\theta_1 \in]1/6, 7/40]$, provides $\eta_t(L) = 1$, $|\sigma_t(L)| = 4$, which contradicts

(2). When $\gamma \equiv 2 \mod 4$ (this is enough for Theorem 7.1), one can choose a larger interval $]1/6, 3/16]$ for θ_1. Note that the extremal value $\theta_1 = 1/6$ yields $\eta_t(L) = 2$, $|\sigma_t(L)| = 3$, which does not contradict (2). □

References

1. E. Artal, J. Carmona, J.I. Cogolludo *On sextic curves with big Milnor number*, in: Trends in Singularities (A. Libgober, M. Tibăr, eds) Trends Math., Birkhäuser Basel 2002 pp. 1–29.
2. M. Boileau, S.Yu. Orevkov *Quasipositivité d'une courbe analytique dans une boule pseudo-convexe*, C. R. Acad. Sci. Paris, Sér. I **332** (2001), 825–830.
3. S. Burckel *The wellordering on positive braids*, J. Pure Appl. Algebra **120** (1997) 1–17.
4. Pi. Cassou-Nogues, I. Luengo *On A_k singularities on plane curves of fixed degree*, Preprint, Oct. 31, 2000.
5. B. Chevallier *Four M-curves of degree* 8, Funct. Anal. Appl. **36** (2002) 76–78.
6. D. Cimasoni, V. Florens *Generalized Seifert surfaces and signatures of colored links*, Trans. Amer. Math. Soc. **360** (2008), 1223–1264.
7. A. Degtyarev, I. Itenberg, V. Kharlamov *On the number of components of a complete intersection of real quadrics*, arXiv:0806.4077.
8. P. Dehornoy Braid groups and left distributive operations Trans. AMS **345** (1994), 115–150.
9. E.A. El-Rifai, H.R. Morton *Algorithms for positive braids*, Quart. J. Math. Oxford (2) **45** (1994), 479–497.
10. O.G. Eroshkin *On a topological property of the boundary of an analytic subset of a strictly pseudoconvex domain in* $\mathbb{C}^2$ Mat. Zametki **49** (1991) no. 5 149–151 (Russian); English transl. Math. Notes **49** (1991) 546–547.
11. R. Fenn, M.T. Greene, D. Rolfsen, C. Rourke, B. Wiest *Ordering of the braid groups*. Pacific J. Math. **191** (1999), 49–74.
12. S. Fiedler-Le Touzé *M-curves of degree* 9 *with deep nests*, J. London Math. Soc, (2) **79** (2009), 649–662.
13. V. Florens *Signatures of colored links with application to real algebraic curves* J. Knot Theory Ramifications **14** (2005) 883–918.
14. S.M. Gusein-Zade, N.N. Nekhoroshev *On A_k singularities on plane curves of fixed degree*, Funk. Anal. i Prilozh. **34** (2000) no. 3, 69–70 (Russian); English transl., *On singularities of type A_k on plane curves of a chosen degree*, Funct. Anal. and Appl. **34** (2000), 214–215.
15. B. Haas *Real algebraic curves and combinatorial constructions* Ph.D. thesis, Basel Univ., 1997.
16. V.M. Kharlamov, S.Yu. Orevkov *The number of trees half of whose vertices are leaves and asymptotic enumeration of plane real algebraic curves*, J. of Combinatorial Theory, Ser. A **105** (2004) 127–142.
17. R. Laver *Braid group action on left distributive structures and well-ordering in the braid groups*, J. Pure Appl. Algebra **108** (1996) 81–98.
18. G. Mikhalkin *Real algebraic curves, the moment map and amoebas*, Ann. of Math. (2) **151** (2000), 309–326.
19. S.Yu. Orevkov *Link theory and oval arrangements of real algebraic curves*, Topology **38** (1999), 779–810.
20. S.Yu. Orevkov *Strong positivity in the right-invariant order on a braid group and quasipositivity*, Mat. Zametki **68** (2000), no. 5, 692–698 (Russian); English transl,. Math. Notes **68** (2000), 588–593.
21. S.Yu. Orevkov *Markov moves for quasipositive braids*, C. R. Acad. Sci. Paris, Sér. I **331** (2000), 557–562.

22. S.Yu. Orevkov, *Classification of flexible M-curves of degree 8 up to isotopy*, GAFA - Geom. and Funct. Anal. **12** (2002), 723–755.
23. S.Yu. Orevkov *Quasipositivity problem for 3-braids*, Turkish J. of Math. **28** (2004), 89–93.
24. S.Yu. Orevkov, V.V. Shevchishin *Markov theorem for transversal links*, J. Knot Theory and Ramifications **12** (2003), 905–913.
25. L. Rudolph *Algebraic functions and closed braids*, Topology **22** (1983) 191–202.
26. E.I. Shustin *New restrictions on the topology of real curves of degree a multiple of* 8 Izv. AN SSSR (Russian); English transl., Math. USSR-Izvestiya **37** (1991), 421–443.
27. O.Yu. Viro, *Signatures of links*, Tezisy VII Vsesoyuznoj topologicheskoj konferencii (1977), page 41 (Russian); English version, Available at http://www.pdmi.ras.ru/~olegviro/respapers.html.
28. O.Ya. Viro, *Plane real curves of degree 7 and 8: new restrictions*, Izv. AN SSSR (Russian); English transl., Math. USSR-Izvestiya **23** (1984) 409–422.
29. O.Ya. Viro, *Real algebraic plane curves: constructions with controlled topology*, Leningrad J. Math. **1** (1990), 1059–1134.
30. O.Ya. Viro, *Acyclicity of circle, twisting-untwisting and their applications*, Talk on the Conference "Géométrie et topologie en petite dimension (dédiée au 60ème anniversaire d'Oleg Viro)," CIRM, Luminy, November 17–21, 2008.
31. O. Viro, *Twisted acyclicity of a circle and signatures of a link*, J. Knot Theory and Ramifications, **6**(2009), 729–755.
32. O.Ya. Viro, S.Yu. Orevkov, *Congruence modulo 8 for real algebraic curves of degree 9*, Uspekhi Mat. Nauk **56:4** (2001), 137–138 (Russian): English transl., Russian Math. Surv. **56** (2001), 770–771.
33. O.Ya. Viro, S.Yu. Orevkov, *Congruence modulo 8 for real algebraic curves of degree 9*, Extended version. Available at http://picard.ups-tlse.fr/~orevkov.
34. B. Wiest *Dehornoy's ordering of the braid groups extends the subword ordering*. Pacific J. Math. **191** (1999) 183–188.
35. A.Wiman *Über die reellen Züge der ebenen algebraischen Kurven* Math. Ann. **90** (1923), 222–228
36. J.G. Yang *Sextic curves with simple singularities*, Tohoku Math J. (2) **48** (1996), 203–227.

Schur–Weyl-Type Duality for Quantized $\mathfrak{gl}(1|1)$, the Burau Representation of Braid Groups, and Invariants of Tangled Graphs

Nicolai Reshetikhin, Catharina Stroppel, and Ben Webster

Abstract We show that the Schur–Weyl-type duality between $\mathfrak{gl}(1|1)$ and GL_n gives a natural representation-theoretic setting for the relationship between reduced and introduced Burau representations.

Keywords Braid group • Burau representation • R-matrix • Tangled graph • Lie superalgebra

Introduction

The goal of this note is to clarify the relationship between reduced Burau representations of braid groups, nonreduced Burau representations, and the representation of the braid group defined by R-matrices related to $U_q(\mathfrak{gl}(1|1))$.

Much isknown about the relationships among the quantized universal enveloping algebra $U_q(\mathfrak{gl}(1|1)))$ of the Lie superalgebra of $\mathfrak{gl}(1|1)$, multivariable Alexander–

[1]Supported by the NSF grant DMS-0601912, and by DARPA.

[2]Supported by an NSF postdoctoral fellowship.

N. Reshetikhin (✉)
University of California at Berkeley and KDV Institute for Mathematics,
Universiteit van Amsterdam
e-mail: reshetik@math.berkeley.edu

C. Stroppel
Mathematik Zentrum, Universität Bonn

B. Webster
Northeastern University
e-mail: B.Webster@neu.edu

I. Itenberg et al. (eds.), *Perspectives in Analysis, Geometry, and Topology: On the Occasion of the 60th Birthday of Oleg Viro*, Progress in Mathematics 296,
DOI 10.1007/978-0-8176-8277-4_16,

Conway polynomials on links, and the Burau–Magnus representations of braid groups.

In this paper we show that the Schur–Weyl-type duality between $\mathfrak{gl}(1|1)$ and GL_n gives a natural representation-theoretic setting for the relationship between reduced and nonreduced Burau representations. We use this simple fact as an excuse to sum up some known (but partly folklore) facts about these representations and the invariants of knots.

In Sect. 1 we recall the definition of and basic facts about quantized $\mathfrak{gl}(1|1)$. Section 2 describes the duality between GL_n and $U_q(\mathfrak{gl}(1|1))$. In Sect. 3 we show how the Burau representation naturally reduces on the space of multiplicities. Section 4 relates the Alexander–Conway polynomial to the trace on the multiplicity space.

1 Quantum $\mathfrak{gl}(1|1)$ and Its Representations

1.1

Consider the Lie superalgebra $\mathfrak{g} = \mathfrak{gl}(1|1)$. Explicitly, this means that we consider the super vector space M of all complex 2×2 matrices with even part M_0 spanned by the matrix units $E_{1,1}$ and $E_{2,2}$, and odd part M_1 spanned by the matrix units $E_{1,2}$ and $E_{2,1}$, equipped with the Lie superalgebra structure given by the super commutator. The universal enveloping superalgebra $U(\mathfrak{g})$ has a quantum version $U_h(\mathfrak{gl}(1|1))$ defined as follows (see, e.g., [7]).

Let $\mathbb{C}[[h]]$ denote the ring of formal power series in h. The $\mathbb{C}[[h]]$-superalgebra $U_h(\mathfrak{gl}(1|1))$ is generated freely as a $\mathbb{C}[[h]]$-algebra by (odd) elements X, Y and (even) elements G, H modulo the defining relations

$$\{X,Y\} = e^{hH} - e^{-hH}, \quad X^2 = Y^2 = 0,$$

$$[G,X] = X, \ \ [G,Y] = -Y,$$

$$[H,X] = 0 \ \ [H,Y] = 0, \ \ [H,G] = 0$$

(using the common abbreviation $[A,B] := AB - BA$ and $\{A,B\} = AB + BA$).

Here $U_h(\mathfrak{gl}(1|1))$ is a Hopf superalgebra with comultiplication

$$\Delta X = X \otimes e^{\frac{hH}{2}} + e^{-\frac{hH}{2}} \otimes X, \ \ \Delta Y = Y \otimes e^{\frac{hH}{2}} + e^{-\frac{hH}{2}} \otimes Y,$$

$$\Delta H = H \otimes 1 + 1 \otimes H, \ \ \Delta G = G \otimes 1 + 1 \otimes G.$$

The Hopf superalgebra is quasitriangular with R-matrix

$$R = \exp(h(H \otimes G + G \otimes H)) \left(1 - e^{-\frac{hH}{2}} X \otimes e^{\frac{hH}{2}} Y\right). \tag{1}$$

That is, this element satisfies the following identities:

$$\Delta(a)^{op} = R\Delta(a)R^{-1}$$

and

$$(\Delta \otimes \mathrm{id})(R) = R_{13}R_{23}, \quad (\mathrm{id} \otimes \Delta)(R) = R_{13}R_{12}.$$

There is an integral form $U_q(\mathfrak{gl}(1|1)) \subset U_h(\mathfrak{gl}(1|1))$ that is generated by X, Y, G and the invertible element $t = e^{\frac{hH}{2}}$ as a $\mathbb{C}[e^h, e^{-h}]$-algebra. As usual, we write $q = e^h$.

1.2

Recall that there is up to isomorphism precisely one irreducible $\mathfrak{gl}(2)$-module of a fixed dimension n (for instance, the natural representation for $n = 2$). In contrast, the algebra $U_q(gl(1|1)$ has a 2-(complex)-parameter family of irreducible representations on $\mathbb{C}^{1|1}$. For $z \in \mathbb{C}^*$, $n \in \mathbb{C}$, denote by $V_{z,n}$ the irreducible two-dimensional representation $V_{z,n} = \mathbb{C}v \oplus \mathbb{C}u$ with v even and u odd such that

$$Xv = u, \quad Yv = 0, \quad Gv = nv, \quad tv = zv \tag{2}$$

(from which $Xu = 0$, $Gu = (n+1)u$, $Yu = (z^2 - z^{-2})v$, and $tu = zu$ follow). Obviously, one can also consider the representation $\Pi V_{z,n}$ with the parity of the elements reversed. The representation $\Pi V_{z,n}$ can be realized as $\epsilon \otimes V_{z,n}$, where ε is an odd one-dimensional representation. These representations and their tensor products will in fact be essentially the only $\mathfrak{gl}(1|1)$-representations of interest to us. For more details on the representation theory see, e.g., [13, Sect. 11].

1.3

Let V be a finite-dimensional representation of $U_q(\mathfrak{gl}(1|1))$. It decomposes into a direct sum of weight spaces for G,

$$V = \bigoplus_{n \in \mathbb{C}} V(n).$$

Note that we do not assume the weights to be integral. As usual, the elements X and Y act from one weight space to another,

$$X : V(n) \to V(n+1), \qquad Y : V(n) \to V(n-1),$$

and we have $X^2 = Y^2 = 0$. Hence, V can be viewed as a complex with two differentials acting in opposite directions. The de Rham complex of any Kähler manifold carries an action of $\mathfrak{gl}(1|1)$ such that the element H acts as the Laplace operator. Thus, the algebra $U_h(\mathfrak{gl}(1|1))$, and for the same reasons $U(\mathfrak{gl}(1|1))$, is, in a certain sense, an abstraction of the structures of Hodge theory.

These are, in fact, isomorphic as algebras; the difference between them lies in the action of the differential on $V \otimes W$: the usual, diagonal action for $U(\mathfrak{gl}(1|1))$; the comultiplication for $U_q(\mathfrak{gl}(1|1))$ gives another action.

Alternatively, any $\mathfrak{gl}(1|1)$-representation can be thought of as a matrix factorization with extra structure (primarily, an upgrade of the $\mathbb{Z}_2$-grading to a $\mathbb{Z}$-grading). The underlying super vector space remains unchanged, with $X+Y$ giving the differential, and with

$$(X+Y)^2 = \{X,Y\} = t^2 - t^{-2}$$

as the potential.

2 The Decomposition of the Tensor Product

2.1

Let Cl_N be the Clifford algebra (over $\mathbb{C}$) with $2N$ generators:

$$\mathrm{Cl}_N = \langle a_i, b_i, i = 1, \ldots, N | \{a_i, a_j\} = \{b_i, b_j\} = 0, \{a_i, b_j\} = \delta_{ij} \rangle$$

The algebra Cl_N has an irreducible 2^N-dimensional representation U_N generated by a cyclic vector v with $b_i v = 0$. We might identify the basis vectors with the set of $\{0,1\}$-sequences of length N such that $v = (0,0,0,\ldots,0)$ and a_i annihilates all basis vectors $S = (s_1, s_2, \ldots, s_N)$ with $s_i = 1$, and otherwise sends S to $(-1)^{\sum_{k=1}^{i-1} s_k} S'$, where S' differs from S exactly in the ith entry. If we consider the subspace $U = \mathrm{span}\langle a_1, \ldots, a_n \rangle$ of Cl_N, then there is a natural isomorphism of graded vector spaces:

$$U_N \longrightarrow \overset{\bullet}{\bigwedge} U \tag{3}$$

$$S = (s_1, s_2, \ldots, s_N) \longmapsto a_{j_1} \wedge a_{j_2} \wedge \cdots \wedge a_{j_k}, \tag{4}$$

where $s_{j_1}, s_{j_2}, \ldots, s_{j_k}$ are precisely the 1's appearing (in this order) in S. The action of a_i gets turned into $a_i \wedge \_$).

In case $N = 1$, U_N is two-dimensional, and the irreducible two-dimensional representation (2) is obtained by pulling back the Clifford algebra action to $U_q(\mathfrak{gl}(1|1))$ via the algebra homomorphism $U_q(\mathfrak{gl}(1|1)) \to \mathrm{Cl}_1$:

$$X \mapsto a_1,\ \ Y \mapsto (z-z^{-1})b_1,\ \ t \mapsto z,\ \ G \mapsto n + a_1 b_1.$$

This formalism can be extended to the N-fold tensor product (via the comultiplication Δ) of these representations:

Proposition 2.1. *Let $V(\mathbf{n},\mathbf{z}) = V_{z_1,n_1} \otimes \cdots \otimes V_{z_N,n_N}$. Then the mapping*

$$\begin{array}{ll} X \mapsto \sum_{i=1}^N z_1^{-1} \ldots z_{i-1}^{-1} z_{i+1} \ldots z_N a_i, & t \mapsto z_1 \ldots z_N, \\ Y \mapsto \sum_{i=1}^N z_1^{-1} \ldots z_{i-1}^{-1} (z_i^2 - z_i^{-2}) z_{i+1} \ldots z_N b_i & G \mapsto \sum_{i=1}^N (n_i + a_i b_i) \end{array}$$

defines uniquely an algebra homomorphism $\Phi_{\mathbf{n},\mathbf{z}} : U_q(\mathfrak{gl}(1|1)) \to \mathrm{Cl}_N$. Pulling back the representation U_N of Cl_N via this map gives the tensor product representation $V(\mathbf{z},\mathbf{n})$.

Proof. One easily verifies that the map is compatible with the relations of $U_q(\mathfrak{gl}(1|1))$. The second statement follows then also by explicit calculations. □

2.2

The vector $v_N = v \otimes \cdots \otimes v \in V(\mathbf{z},\mathbf{n})$ is a lowest-weight vector of lowest weight $\lambda = \sum_{i=1}^N n_i$, i.e., $Y v_N = 0$ and $G v_N = \lambda v_N$.

The subspaces $U = \mathrm{span}\langle a_1, \ldots, a_N \rangle$ and $U' = \mathrm{span}\langle b_1, \ldots, b_N \rangle$ of Cl_N can be paired via $U \otimes U' \to \mathbb{C}$, $a_j \otimes b_i \mapsto \delta_{i,j}$. Abbreviate $\Phi = \Phi_{\mathbf{n},\mathbf{z}}$ and let $W = (\mathbb{C}\Phi(Y))^{\perp}$ and $W' = (\mathbb{C}\Phi(X))^{\perp}$.

Lemma 2.2. *Let $z := z_1 z_2 \cdots z_N$. Assume $z^2 - z^{-2} \neq 0$. Then*

1. *$U = \mathbb{C}\Phi(X) \oplus W$ and $U' = \mathbb{C}\Phi(Y) \oplus W'$.*
2. *The subspaces W, W' generate*[3] *a subalgebra $C(X,Y)$ of Cl_N isomorphic to $\mathrm{Cl}_{\dim W}$, which is the supercommutant of the subalgebra generated by X and Y.*

Proof. The inclusion $U \subseteq \mathbb{C}\Phi(X) \oplus W$ holds by definition. For the inverse, it is enough to find (for $1 \leq i \leq N$) $\beta_i \in \mathbb{C}$ such that $a_i - \beta_i \Phi(X) \in W$. One easily verifies that $\beta_i = \frac{z^2(1-z_i^{-4})}{z^2 - z^{-2}}$ does the job. The sum is direct, since an element u in the intersection is of the form $u = \alpha \sum_{i=1}^N \gamma_i^2 a_i$, where $0 = \alpha \sum_{i=1}^N \gamma_i^2 (z^2 - z^{-2}) = 0$; hence with our assumption, $\alpha = 0$, and so $u = 0$. The argument for U' is similar. Part 1 follows.

Now, $C(X,Y)$ is clearly contained in the commutant of X and Y, since $\dim\langle X, Y \rangle = 4$, and the action of this subalgebra on U_N is semisimple, by 2^{N-1} copies of the unique two-dimensional irreducible representation of $\langle X, Y \rangle$. Thus, its

[3] Elements of these spaces are elements of the associative algebra Cl_N.

commutant is of dimension $2^{2(N-1)}$. Since $C(X,Y)$ has this dimension, it must be the entire commutant, obviously isomorphic to the Clifford algebra as claimed. □

In order to find the supercommutant not just of $\Phi(X)$ and $\Phi(Y)$ but of all of $U_q(\mathfrak{gl}(1|1))$, we must find the subalgebra that also commutes with $\Phi(G)$.

Proposition 2.3 (Schur–Weyl duality). *Let us still have $z^2 - z^{-2} \neq 0$. The subalgebra of* Cl_W *commuting with $\Phi(G)$ is that of Euler degree* 0*, i.e., that generated by elements of the form $W \cdot W'$. There is a natural map $U(\mathfrak{gl}(W)) \to \mathrm{Cl}_W \subset \mathrm{Cl}_N$ whose image is this subalgebra.*

Proof. The first statement is obvious. Recall that W and W' generate a Clifford algebra, say with generators a_i', b_i'. Note that $W \cdot W'$ forms a Lie subalgebra of Cl_W isomorphic to $\mathfrak{gl}(W)$ (by mapping $a_i'b_j'$ to the matrix unit $E_{i,j}$); hence this extends to an algebra map $U(\mathfrak{gl}(W)) \to \mathrm{Cl}_W \subset \mathrm{Cl}_N$. The image of this map is precisely the commutant, because by the PBW theorem for Clifford algebras, the subspace of Euler degree 0 is that of the form $\bigoplus_n W^n \cdot (W')^n = \bigoplus_n (W \cdot W')^n$. □

Under the action of Cl_W, U_N decomposes into two copies of $U_W = \bigwedge^\bullet W$, one with parity reversed. Thus, $V(\mathbf{z},\mathbf{n})$ is completely decomposable, and up to grading shift and parity reversal, the summands are precisely the two-dimensional simple modules from above. Of course, the highest-weight vector v_N generates a copy of $V_{z,\lambda}$, so all simple submodules must be of the form $V_{z,\lambda+k}$ for some k (possibly with parity reversed). Thus, we have the following.

Proposition 2.4 (Tensor space decomposition). *The multiplicity space of $V_{z,\lambda+k}$ in $V(\mathbf{n},\mathbf{z})$ is the space of weight k (for G) in U_W. That is,*

$$V(\mathbf{z},\mathbf{n}) \simeq \bigoplus_{k=0}^{N-1} \bigwedge^k W \otimes \Pi^m V_{z,\lambda+k}, \tag{5}$$

where Π is the shift of parity, $\Pi^2 = id$.

2.3

This decomposition of the tensor product can be made more explicit if we choose a basis c_i, $i = 1, \ldots, N-1$ in the subspace $W \subset U$ complementary to $\mathbb{C}\Phi(X)$, thereby fixing a decomposition $U = \mathbb{C}\Phi(X) \oplus_{i=1}^{N-1} \mathbb{C}c_i$. From now on, we will just write X, Y instead of $\Phi(X)$, $\Phi(Y)$.

Lemma 2.5. *We have the following formulas:*

$$Xc_{i_1}\ldots c_{i_k}w = (-1)^k c_{i_1}\ldots c_{i_k}Xw,$$

$$Yc_{i_1}\ldots c_{i_k}w = (-1)^k c_{i_1}\ldots c_{i_k}Yw + \sum_{j=1}^{k} y_{i_j}(-1)^{j-1}c_{i_1}\ldots \widehat{c_{i_j}}\ldots c_{i_k}w,$$

where $w \in U$ and the y_i's are defined by $Yc_i + c_iY = y_i$.

Proof. Obvious. □

For a vector $v \in U$, define

$$(v)_{i_1,\dots,i_k} = c_{i_1}\dots c_{i_k}v + (-1)^k \frac{1}{z-z^{-1}} \sum_{a=1}^{k} y_{i_a}(-1)^{a-1} c_{i_1}\dots \widehat{c_{i_a}} \dots c_{i_k} v.$$

Proposition 2.6. *The space*

$$V_{i_1,\dots,i_k} = \mathbb{C}(v_N)_{i_1,\dots,i_k} \oplus \mathbb{C}X(v_N)_{i_1,\dots,i_k},$$

where v_N is the highest-weight vector (see Sect. 2.2), is an irreducible submodule isomorphic to $V_{z,(\sum_{i=1}^N n_i)+k}$. This submodule corresponds to the monomial $c_{i_1} \wedge \dots \wedge c_{i_k}$ in the decomposition (5).

Proof. We have

$$X(v_N)_{i_1,\dots,i_k} = (Xv_N)_{i_1,\dots,i_k},$$

and since $Yv_N = 0$, we have

$$Y(v_N)_{i_1,\dots,i_k} = 0, \quad YX(v_N)_{i_1,\dots,i_k} = (z^2 - z - 2)(v_N)_{i_1,\dots,i_k}.$$

The statement follows directly from the action of t and G and (2). □

3 The Relationship to the Burau Representation

3.1

The action of the universal R-matrix (1) in the tensor product representation $V_{z_1,n_1} \otimes V_{z_2,n_2}$ can easily be computed explicitly. Namely, in terms of the weight basis (by abuse of language we use the basis $\{v, Xv\}$ for either module), this *right* action looks as follows:

$$\begin{aligned}
R(v\otimes v) &= z_1^{2n_2} z_2^{2n_1} v\otimes v,\\
R(v\otimes Xv) &= z_1^{2n_2+2} z_2^{2n_1} v\otimes Xv - z_1^{-1} z_2 (z_2^2 - z_2^{-2}) z_1^{2n_2} z_2^{2n_1+2} Xv\otimes v,\\
R(Xv\otimes v) &= z_1^{2n_2} z_2^{2n_1+2} Xv\otimes v,\\
R(Xv\otimes Xv) &= z_1^{2n_2+2} z_2^{2n_1+2} Xv\otimes Xv.
\end{aligned}$$

In the tensor product basis $v\otimes v, v\otimes Xv, Xv\otimes v, Xv\otimes Xv$, it produces the 4×4 matrix $R^{(z_1,z_2)} = z_1^{-2n_2} z_2^{-2n_1}(R)$,

$$R^{(z_1,z_2)} = \begin{bmatrix} 1 & 0 & 0 & 0 \\ 0 & z_1^2 & -(z_2^2 - z_2^{-2}) z_2^2 \frac{z_2}{z_1} & 0 \\ 0 & 0 & z_2^2 & 0 \\ 0 & 0 & 0 & z_1^2 z_2^2 \end{bmatrix}, \tag{6}$$

with

$$R^{(z_1,z_2)^{-1}} = \begin{bmatrix} 1 & 0 & 0 & 0 \\ 0 & z_1^{-2} & z_1^{-2}(z_2^2 - z_2^{-2}) \frac{z_2}{z_1} & 0 \\ 0 & 0 & z_2^{-2} & 0 \\ 0 & 0 & 0 & z_1^{-2} z_2^{-2} \end{bmatrix}. \tag{7}$$

3.2

Consider the groupoid of braids whose strands are labeled by elements in $\mathbb{C} \times \mathbb{C}^*$. Each N-braid with colors (z_i, n_i), $1 \le i \le N$, on its N strands defines a morphism from the tuple $(\mathbf{z}, \mathbf{n})$ to the permuted tuple $(\sigma\mathbf{z}, \sigma\mathbf{n})$ given by the braid. Assigning to a tuple $(\mathbf{z}, \mathbf{n})$ the representation $V(\mathbf{z}, \mathbf{n}) = V_{z_1,n_1} \otimes \cdots \otimes V_{z_N,n_N}$ and to the single (positive) braid β_i with strands colored by $a := (z_i, n_i)$ and $b := (z_{i+1}, n_{i+1})$, the mapping

$$\pi(\beta_i)(a,b): \quad V(\mathbf{z},\mathbf{n}) \to V(s_i\mathbf{z}, s_i\mathbf{n}),$$
$$\pi(\beta_i)(a,b) = P_{i,i+1} \circ \left(1 \otimes \cdots \otimes 1 \otimes R^{(z_1,z_2)} \otimes 1 \otimes \cdots \otimes 1\right),$$

defines a representation π of the colored braid groupoid. Here $R^{(z_1,z_2)}$ is as above; hence up to a multiple, the universal R-matrix (1) acting on $V_{n_{\sigma_1},z_{\sigma_1}} \otimes V_{n_{\sigma_2},z_{\sigma_2}}$, and $P_{i,i+1}$ is the flip map of simply swapping the two tensor factors as $x \otimes y \mapsto (-1)^{\overline{x}\overline{y}}$ $y \otimes x$. To verify the claim, note that the braid relations amount to the relations

$$\pi(\beta_i)(a,b) \circ \pi(\beta_{i+1})(a,c) \circ \pi(\beta_i)(b,c) = \pi(\beta_{i+1})(b,c) \circ \pi(\beta_i)(a,c) \circ \pi(\beta_{i+1})(a,b),$$
$$\pi(\beta_i)(a,b) \circ \pi(\beta_j)(c,d) = \pi(\beta_j)(c,d) \circ \pi(\beta_i)(a,b),$$

for $j \neq i-1, i, i+1$ and a,b,c,d arbitrary colors. These relations can easily be checked by direct calculations. In particular, the subgroup $\mathbb{B}_{\mathbf{z}}$ of the braid group that preserves $(\mathbf{z}, \mathbf{n})$ acts on $V(\mathbf{z}, \mathbf{n})$. Because the operators $\pi(\beta_i)$ commute with the action of $U_q(\mathfrak{gl}(1|1))$, the action is determined by the action on multiplicity spaces.

The first interesting multiplicity space is W considered as a subspace of U:

Proposition 3.1. *In the case $z_1 = \cdots = z_n$, this braid group representation on U is isomorphic to the Burau representation. Similarly, the action on W gives rise to the reduced Burau representation in the case $z_1 = \cdots = z_n$.*

Remark 3.2. In general, we obtain a representation that might be viewed as a colored version of the Magnus representation of $\mathbb{B}_{\mathbf{z}}$ obtained from an action on the free group on N generators (see [2, p. 102 ff] for the noncolored version, and for the colored version see [3, Sect. 4]).

Proof. Choose the basis $b_i = v \otimes v \otimes \cdots v \otimes Xv \otimes v \otimes \cdots \otimes v$, where X is applied to the ith factor. Then $\pi(\beta_i) = P_{i,i+1} \circ R_{i,i+1}$ acts on this basis as follows: matrix: $\pi(\beta_i)(b_j) = b_j$ for $j \neq i, i+1$, and on b_i, b_{i+1} by

$$\begin{pmatrix} 0 & z_2^2 \\ z_1^2 & -(z_2^2 - z_2^{-2}) z_2^2 \frac{z_2}{z_1} \end{pmatrix}. \tag{8}$$

If we scale the basis by $b_i' = z_2^2 z_3^2 \cdots z_i^2$, then $\pi(\beta_i)(b_j') = b_j'$ for $j \neq i, i+1$, and $\pi(\beta_i)$ acts on b_i', b_{i+1}' by

$$\begin{pmatrix} 0 & 1 \\ z_1^2 z_2^2 & -(z_2^2 - z_2^{-2}) z_2^2 \frac{z_2}{z_1} \end{pmatrix}. \tag{9}$$

In case $z_i = z_j$ for any i, j, we set we $t := z_i^{-4}$ and obtain the matrix A with its inverse A^{-1}:

$$A = \begin{pmatrix} 0 & 1 \\ t^{-1} & 1 - t^{-1} \end{pmatrix} \; A^{-1} = \begin{pmatrix} 1 - t & t \\ 1 & 0 \end{pmatrix}.$$

But this is exactly the form of Burau representation given (for example) in [1, p. 118 Example 3]. The invariant subspace is $\mathbb{C}Xv$. The reduced Magnus representation acts in the quotient space $W = U/\mathbb{C}Xv$. □

4 Multivariable Alexander–Conway Polynomial

In this section we will use Theorem 2.4 to obtain the Alexander–Conway polynomial of a knot in terms of R-matrices for quantum $\mathfrak{gl}(1|1)$. These results are very closely related to the results in [6] and [8].

4.1

To construct invariants of links and tangled graphs, let us start with the explicit decomposition of the two-fold tensor product. We abbreviate

$$\gamma := \left(z_1^2 z_2^2 - z_1^{-2} z_2^{-2}\right)^{-1}$$

(assuming from now on that this inverse exists). The following linear maps explicitly describe the decomposition of the tensor product of two generic irreducible two-dimensional representations:

$$\varphi : \quad V_{z_1 z_2, n+m} \oplus V_{z_1 z_2, n+m+1} \to V_{z_1, n} \otimes V_{z_2, m}$$

and

$$\psi: \quad V_{z_1,n} \otimes V_{z_2,m} \to V_{z_1z_2,n+m} \oplus V_{z_1z_2,n+m+1}.$$

We denote by w_1, Xw_1 (respectively w_2, Xw_2) the standard basis in $V_{z_1z_2,n+m+1}$ and in $V_{z_1z_2,n+m}$, and by v_1, Xv_1 (respectively v_2, Xv_2) the standard basis in $V_{z_1,n}$ and in $V(z_2,m)$. Then the maps are defined as follows:

$$\begin{aligned}
\varphi(w_1) &= v_1 \otimes v_2, \\
\varphi(Xw_1) &= z_2 Xv_1 \otimes v_2 + (-1)^n z_1^{-1} v_1 \otimes Xv_2, \\
\varphi(w_2) &= (-1)^{n+1} z_1^{-1}(z_2^2 - z_2^{-2})\gamma Xv_1 \otimes v_2 + z_2(z_1^2 - z_1^{-2})\gamma v_1 \otimes Xv_2, \\
\varphi(Xw_2) &= Xv_1 \otimes Xv_2,
\end{aligned}$$

and

$$\begin{aligned}
\psi(v_1 \otimes v_2) &= w_1, \\
\psi(Xv_1 \otimes v_2) &= z_2(z_1^2 - z_1^{-2})\gamma Xw_1 + (-1)^{n+1} z_1^{-1} w_2, \\
\psi(v_1 \otimes Xv_2) &= (-1)^n z_1^{-1}(z_2^2 - z_2^{-2})\gamma Xw_1 + z_2 w_2, \\
\psi(Xv_1 \otimes Xv_2) &= Xw_2.
\end{aligned}$$

One easily verifies that they are inverse to each other:

$$\psi \circ \varphi = \mathrm{id}_{V\otimes V}, \quad \varphi \circ \psi = \mathrm{id}_{V\otimes V}.$$

Let $P_0, P_1 \in \mathrm{End}(V_{z_1z_2,n+m} \oplus V_{z_1z_2,n+m+1})$ be the natural orthogonal projections to the first and the second summand respectively.

For any $A \in \mathrm{End}(M)$ with M an arbitrary superspace, define the supertrace $\mathrm{str}(A)$ to be the trace of A restricted to the even part of M minus the trace of A restricted to the odd part of M. For instance, if $M = V$, then $\mathrm{str}(A) = A_{v,v} - A_{Xv,Xv}$, where $v, Xv = u$ is the weight basis in V.

We have the following identities for the supertraces:

$$\begin{aligned}
\mathrm{str}_2(\phi P_0 \psi) &= z_2^2(z_1^2 - z_1^{-2})\gamma \mathrm{id}_{V(z_1,n)}, \\
\mathrm{str}_2(\phi P_1 \psi) &= -z_2^2(z_1^2 - z_1^{-2})\gamma \mathrm{id}_{V(z_1,n)}, \\
\mathrm{str}_1(\phi P_0 \psi) &= z_1^{-2}(z_2^2 - z_2^{-2})\gamma \mathrm{id}_{V(z_2,n)}, \\
\mathrm{str}_1(\phi P_1 \psi) &= -z_1^{-2}(z_2^2 - z_2^{-2})\gamma \mathrm{id}_{V(z_2,n)}.
\end{aligned} \tag{10}$$

Here $\mathrm{str}_{1,2}$ are partial supertraces:

$$\mathrm{str}_2(a \otimes b) = a\,\mathrm{str}(b), \quad \mathrm{str}_1(a \otimes b) = \mathrm{str}(a)b.$$

The matrix $PR^{(z,z)}$ has the spectral decomposition

$$PR^{(z,z)} = z^2 \phi P_0 \psi - z^{-2} \phi P_1 \psi.$$

We also have

$$\mathrm{str}_2(PR^{(z,z)}) = z^2\,\mathrm{id}, \quad \mathrm{str}_2((PR^{(z,z)})^{-1}) = z^2\,\mathrm{id}, \tag{11}$$

and it is easy to check that these identities agree with the spectral decomposition and supertrace identities above.

4.2

Let π be the representation from above and β a braid. The partial trace $\mathrm{tr}_{23\ldots N}(\pi(\beta))$ is the evaluation of a central element in $U_h(\mathfrak{gl}(1|1))$ in the irreducible representation V_{z_1,n_1}. This is a general fact about the construction of link invariants from quasi-triangular Hopf algebras [10], [14, Definition 2.1]. Therefore, this partial trace is proportional to the identity. We will write

$$\mathrm{str}_{23\ldots N}(\beta) = \langle \mathrm{str}_{23\ldots N}(\pi(\beta))\rangle I_1,$$

where I_1 is the identity operator in $V(z_1,n_1)$.

Theorem 4.1. *Abbreviating $z = z_1\ldots z_N$, the following holds:*

$$\langle \mathrm{str}_{23\ldots N}(\pi(\beta))\rangle = \frac{z_1^2 - z_1^{-2}}{z^2 - z^{-2}} z^2 \sum_{m=0}^{N-1} \mathrm{tr}_{\wedge^m W}(\pi_W(\beta)).$$

Proof. This theorem follows immediately from Proposition 2.4. The decomposition of the tensor product $V(\mathbf{z},\mathbf{n})$ defines linear maps $f_m : \wedge^m W \to \wedge^m W$ for each element $f \in \mathrm{End}(V(\mathbf{z},\mathbf{n}))$. Using the explicit formulas for the decomposition of two irreducible two-dimensional representations from the previous section and the formulas for partial traces $\mathrm{str}_a(\phi P_b \psi)$ from the previous subsection, we arrive at the identity

$$\langle \mathrm{str}_{23\ldots N}(f)\rangle = \sum_{m=0}^{N-1} \mathrm{tr}_{\wedge^n W}(f_m) \frac{z_1^2 - z_1^{-2}}{z^2 - z^{-2}} z^2. \qquad \square$$

Let $\widehat{\beta}$ be the link that is the closure of the braid β. We number its connected components by $1 \le i \le k$ and denote by $w_i(\beta)$ the winding number of the ith component of $\widehat{\beta}$. Then the following holds:

Theorem 4.2. *The function*

$$\tau(\beta) = \langle \mathrm{tr}_{23\ldots N}(\beta)\rangle z^{2\sum_{i=1}^k w_i(\beta)} \tag{12}$$

is an invariant of the link $\widehat{\beta}$.

Proof. We have to verify the invariance with respect to Markov moves. The invariance with respect to the first Markov move means that $\tau(\sigma\beta\sigma^{-1}) = \tau(\beta)$. But this identity follows immediately from the conjugation invariance of the ordinary trace and Theorem 4.1. The second Markov move means that $\tau(\beta s_{n-1}^{\pm 1}) = \tau(\beta)$, where β is a braid that has no factors $s_{n-1}^{\pm 1}$. But this identity follows immediately from the property (11) of R-matrices. □

As was shown in [8], the invariant (12) is the Alexander–Conway polynomial $\Delta_{z_1,\dots,z_N}$:

$$\langle \mathrm{tr}_{23\dots N}(\pi(\beta))\rangle = \frac{z_1^2 - z_1^{-2}}{z^2 - z^{-2}} z^2 \Delta_{z_1,\dots,z_N}(\widehat{\beta}).$$

Remark 4.3. For any $U_q(\mathfrak{gl}(1|1))$-linear map $f \in \mathrm{End}(V(z_1,n_1)\otimes V(z_2,n_2))$, we have the following identity:

$$z_1^{-2}\,\mathrm{str}_2(f) = z_2^2\,\mathrm{str}_1(f).$$

This follows immediately from (10). This property is a projective version of the ambidextrous [15] property of $U_q(\mathfrak{gl}(1|1))$-modules. Using this formula, the Alexander–Conway polynomial in terms of $U_q(\mathfrak{gl}(1|1))$ can be written as a state sum, and as a state sum, it can be extended to invariants of framed graphs (see [6, Sect. 3]).

Acknowledgment The authors are grateful for the hospitality shown them by the Mathematics Department of Aarhus University, where this work was completed, and for a Niels Bohr grant from the Danish National Research Foundation.

References

1. J. Birman, Braids, links, and mapping class groups, Ann Math Studies, No. 82. Princeton University Press (1974).
2. J. Birman, D. D. Long, J. Moody, Finite-dimensional representations of Artin's braid group. In: The Mathematical Legacy of Wilhelm Magnus: Groups, Geometry and Special Functions, Contemporary Math. 169, Amer. Math. Soc. (1994), pp. 123–132.
3. F. Constantinescu, F. Toppan, On the linearized Artin braid representation. J. Knot Theory Ramifications 2, no. 4 (1993), 399–412.
4. N. Geer, B. Patureau-Mirand, An invariant supertrace for the category of representations of Lie superalgebras Pacific J. Math, 238, no. 2 (2008), 331–348.
5. N. Geer, B. Patureau-Mirand, V. Turaev, Modified quantum dimensions and re-normalized link invariants, Compos. Math. 145 no. 1 (2009), 196–212.
6. L. H. Kauffman, H. Saleur, Free fermions and the Alexander–Conway polynomial, Comm. Math. Phys. 141, no. 2 (1991), 293–327.
7. P. P. Kulish, Quantum Lie superalgebras and supergroups, Problems of Modern Quantum Field Theory (Alushta, 1989) Springer, 1989, pp. 14–21.

8. J. Murakami, A state model for multi-variable Alexander polynomial, Pacific J. Math. 157, no. 1 (1993), 109–135.
9. N. Reshetikhin, Quantum Supergroups, Proceedings of the NATO advanced research workshop, Quantum Field Theory, Statistical Mechanics, Quantum Groups, and Topology. (Coral Gables, FL, 1991), 264–282.
10. N. Reshetikhin, V. Turaev, Ribbon graphs and their invariants derived from quantum groups, Comm. Math. Phys. 127, no. 1 (1990), 1–26.
11. M. Rosso, Alexander polynomial and Koszul resolution, Algebra Monpellier Announcements, 1999.
12. V. Turaev, Quantum invariants of knots and 3-manifolds. de Gruyter Studies in Mathematics, 18. Walter de Gruyter & Co., Berlin, 1994.
13. O. Viro, Quantum relatives of the Alexander polynomial, Algebra i Analiz 18:3 (2006) 63-157 (in Russian), St. Petersburg Math. J. 18 no. 3 (2007), 391–457 (in English).
14. N. Geer, B. Patureau-Mirand, An invariant supertrace for the category of representations of Lie superalgebras, Pacific J. Math, 238, no. 2 (2008), 331–348.
15. N. Geer, B. Patureau-Mirand, V. Turaev, Modified quantum dimensions and re-normalized link invariants, Compos. Math. 145, no. 1 (2009), 196–212.

Khovanov Homology Theories and Their Applications

Alexander Shumakovitch

To my teacher and advisor, Oleg Yanovich Viro,
on the occasion of his 60th birthday

Abstract This is an expository paper discussing various versions of Khovanov homology theories, interrelations between them, their properties, and their applications to other areas of knot theory and low-dimensional topology.

Keywords Categorification • Khovanov homology • Odd Khovanov homology • Homological thickness • Rasmussen invariant • Bounds on the Thurston–Bennequin number • Quasialternating knots

1 Introduction

Khovanov homology is a special case of *categorification*, a novel approach to construction of knot (or link) invariants that has been under active development over the last decade following a seminal paper [15] by Mikhail Khovanov. The idea of categorification is to replace a known polynomial knot (or link) invariant with a family of chain complexes, such that the coefficients of the original polynomial are the Euler characteristics of these complexes. Although the chain complexes themselves depend heavily on a diagram that represents the link, their homology depends on the isotopy class of the link only. Khovanov homology categorifies the Jones polynomial [13].

A. Shumakovitch (✉)
Department of Mathematics, The George Washington University, Monroe Hall 2115 G St. NW, Washington, DC 20052, U.S.A
e-mail: Shurik@gwu.edu

I. Itenberg et al. (eds.), *Perspectives in Analysis, Geometry, and Topology: On the Occasion of the 60th Birthday of Oleg Viro*, Progress in Mathematics 296, DOI 10.1007/978-0-8176-8277-4_17,

More specifically, let L be an oriented link in $\mathbb{R}^3$ represented by a planar diagram D and let $J_L(q)$ be a version of the Jones polynomial of L that satisfies the following identities (called the *Jones skein relation* and *normalization*):

$$-q^{-2}J_{\times_+}(q)+q^{2}J_{\times_-}(q)=(q-1/q)J_{)(_0}(q); \qquad J_{\bigcirc}(q)=q+1/q. \tag{1}$$

The skein relation should be understood as relating the Jones polynomials of three links whose planar diagrams are identical everywhere except in a small disk, where they are different as depicted in (1). The normalization fixes the value of the Jones polynomial on the trivial knot; $J_L(q)$ is a Laurent polynomial in q for every link L and is completely determined by its skein relation and normalization.

In [15], Mikhail Khovanov assigned to D a family of abelian groups $\mathscr{H}^{i,j}(L)$ whose isomorphism classes depend on the isotopy class of L only. These groups are defined as homology groups of an appropriate (graded) chain complex $\mathscr{C}^{i,j}(D)$ with integer coefficients. Groups $\mathscr{H}^{i,j}(L)$ are nontrivial for finitely many values of the pair (i,j) only. The gist of the categorification is that the graded Euler characteristic of the Khovanov chain complex equals $J_L(q)$:

$$J_L(q)=\sum_{i,j}(-1)^i q^j h^{i,j}(L), \tag{2}$$

where $h^{i,j}(L)=\mathrm{rk}(\mathscr{H}^{i,j}(L))$, the Betti numbers of $\mathscr{H}$. The reader is referred to Sect. 2 for a detailed treatment (see also [2, 15]).

In our paper we also make use of another version of the Jones polynomial, denoted by $\widetilde{J}_L(q)$, that satisfies the same skein relation (1) but is normalized to equal 1 on the trivial knot. For the sake of completeness, we also list the skein relation for the original Jones polynomial $V_L(t)$ from [13]:

$$t^{-1}V_{\times_+}(t)-tV_{\times_-}(t)=(t^{1/2}-t^{-1/2})V_{)(_0}(t); \qquad V_{\bigcirc}(t)=1. \tag{3}$$

We note that $J_L(q)\in\mathbb{Z}[q,q^{-1}]$, while $V_L(t)\in\mathbb{Z}[t^{1/2},t^{-1/2}]$. In fact, the terms of $V_L(t)$ have half-integer (respectively integer) exponents if L has an even (respectively odd) number of components. This is one of the main motivations for our convention (1) to be different from (3). We also want to ensure that the Jones polynomial of the trivial link has only positive coefficients. The different versions of the Jones polynomial are related as follows:

$$J_L(q)=(q+1/q)\widetilde{J}_L(q), \qquad \widetilde{J}_L(-t^{1/2})=V_L(t), \qquad V_L(q^2)=\widetilde{J}_L(q). \tag{4}$$

Another way to look at Khovanov's identity (2) is via the *Poincaré polynomial* of the Khovanov homology:

$$Kh_L(t,q)=\sum_{i,j}t^i q^j h^{i,j}(L). \tag{5}$$

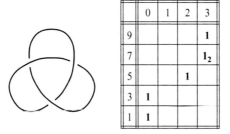

	0	1	2	3
9				**1**
7				$\mathbf{1_2}$
5			**1**	
3	**1**			
1	**1**			

Fig. 1 Right trefoil and its Khovanov homology

With this notation, we get

$$J_L(q) = Kh_L(-1,q). \tag{6}$$

Example 1.1. Consider the right trefoil K. Its nonzero homology groups are tabulated in Fig. 1, where the i-grading is represented horizontally and the j-grading vertically. The homology is nontrivial for odd j-grading only, and thus even rows are not shown in the table. A table entry of **1** or $\mathbf{1_2}$ means that the corresponding group is $\mathbb{Z}$ or $\mathbb{Z}_2$, respectively (one can find a more interesting example in Fig. 10). In general, an entry of the form $\mathbf{a}, \mathbf{b_2}$ would correspond to the group $\mathbb{Z}^a \oplus \mathbb{Z}_2^b$. For the trefoil K, we have that $\mathscr{H}^{0,1}(K) \simeq \mathscr{H}^{0,3}(K) \simeq \mathscr{H}^{2,5}(K) \simeq \mathscr{H}^{3,9}(K) \simeq \mathbb{Z}$ and $\mathscr{H}^{3,7}(K) \simeq \mathbb{Z}_2$. Therefore, $Kh_K(t,q) = q + q^3 + t^2q^5 + t^3q^9$. On the other hand, the Jones polynomial of K equals $V_K(t) = t + t^3 - t^4$. Relation (4) implies that $J_K(q) = (q+1/q)(q^2+q^6-q^8) = q+q^3+q^5-q^9 = Kh_K(-1,q)$.

Without going into details, we note that the initial categorification of the Jones polynomial by Khovanov was followed by a flurry of activity. Categorifications of the colored Jones polynomial [5, 17] and skein $\mathfrak{sl}(3)$ polynomial [18] were based on Khovanov's original construction. A matrix factorization technique was used to categorify the $\mathfrak{sl}(n)$ skein polynomials [21], the HOMFLY-PT polynomial [22], the Kauffman polynomial [23], and more recently, colored $\mathfrak{sl}(n)$ polynomials [47, 48]. Ozsváth, Szabó, and, independently, Rasmussen used a completely different method of Floer homology to categorify the Alexander polynomial [31, 34]. Ideas of categorification were successfully applied to tangles, virtual links, skein modules, and polynomial invariants of graphs.

One of the most important recent developments in the Khovanov homology theory was the introduction in 2007 of its *odd* version by Ozsváth, Rasmussen, and Szabó [29]. The odd Khovanov homology equals the original (even) one modulo 2, and in particular, categorifies the same Jones polynomial. On the other hand, the odd and even homology theories often have drastically different properties (see Sects. 2.4 and 3 for details). The odd Khovanov homology appears to be one of the connecting links between the Khovanov and Heegaard–Floer homology theories [32].

The importance of the Khovanov homology became apparent after a seminal result by Jacob Rasmussen [35], who used the Khovanov chain complex to give the first purely combinatorial proof of the Milnor conjecture. This conjecture states that the 4-dimensional (slice) genus (and hence the genus) of a (p,q)-torus knot equals $\frac{(p-1)(q-1)}{2}$. This was originally proved by Kronheimer and Mrowka [24] using gauge theory in 1993.

There are numerous other applications of Khovanov homology theories. They can be used to provide combinatorial proofs of the slice–Bennequin inequality and to give upper bounds on the Thurston–Bennequin number of Legendrian links, detect quasialternating links and find topologically locally flat slice knots that are not smoothly slice. We refer the reader to Sect. 4 for the details.

The goal of this paper is to give an overview of the current state of research in Khovanov homology. The exposition is mostly self-contained, and no advanced knowledge of the subject is required from the reader. We intentionally limit the scope of our paper to the categorifications of the Jones polynomial, so as to keep its size under control. The reader is referred to other expository papers on the subject [1, 20, 36] to learn more about the interrelations between different types of categorifications.

We also pay significant attention to experimental aspects of the Khovanov homology. As is often the case with new theories, the initial discovery was the result of experimentation. It is especially true for Khovanov homology, since it can be computed by hand for a very limited family of knots only. At the moment, there are two programs [4, 39] that compute Khovanov homology. The first one was written by Dror Bar-Natan and his student Jeremy Green in 2005 and implements the methods from [3]. It works significantly faster for knots with sufficiently many crossings (say, more than 15) than the older program `KhoHo` by the author. On the other hand, `KhoHo` can compute all the versions of the Khovanov homology that are mentioned in this paper. It is currently the only program that can deal with the odd Khovanov homology. Most of the experimental results that are referred to in this paper were obtained with `KhoHo` .

This paper is organized as follows. In Sect. 2 we give a quick overview of constructions involved in the definition of various Khovanov homology theories. We compare these theories with each other and list their basic properties in Sect. 3. Sect. 4 is devoted to some of the more important applications of the Khovanov homology to other areas of low-dimensional topology.

2 Definition of the Khovanov Homology

In this section we give a brief outline of various Khovanov homology theories starting with Khovanov's original construction. Our setting is slightly more general than that in the introduction, for we allow different coefficient rings, not only $\mathbb{Z}$.

2.1 Algebraic Preliminaries

Let R be a commutative ring with unit. In this paper, we are mainly interested in the cases $R = \mathbb{Z}$, $\mathbb{Q}$, and $\mathbb{Z}_2$.

Definition 2.1. A *$\mathbb{Z}$-graded* (or simply *graded*) R-module M is an R-module decomposed into a direct sum $M = \bigoplus_{j\in\mathbb{Z}} M_j$, where each M_j is an R-module itself. The summands M_j are called the *homogeneous components* of M, and elements of M_j are called the *homogeneous elements of degree j*.

Definition 2.2. Let $M = \bigoplus_{j\in\mathbb{Z}} M_j$ be a graded free R-module. The *graded dimension* of M is the power series $\dim_q(M) = \sum_{j\in\mathbb{Z}} q^j \dim(M_j)$ in the variable q. If $k \in \mathbb{Z}$, the *shifted module* $M\{k\}$ is defined as having homogeneous components $M\{k\}_j = M_{j-k}$.

Definition 2.3. Let M and N be two graded R-modules. A map $\varphi : M \to N$ is said to be *graded of degree k* if $\varphi(M_j) \subset N_{j+k}$ for each $j \in \mathbb{Z}$.

Note 2.4. It is an easy exercise to check that $\dim_q(M\{k\}) = q^k \dim_q(M)$, $\dim_q(M \oplus N) = \dim_q(M) + \dim_q(N)$, and $\dim_q(M \otimes_R N) = \dim_q(M)\dim_q(N)$, where M and N are graded R-modules. Moreover, if $\varphi : M \to N$ is a graded map of degree k', then the *shifted map* $\varphi : M \to N\{k\}$ is graded of degree $k' + k$. We slightly abuse notation here by denoting the shifted map in the same way as the map itself.

Definition 2.5. Let $(\mathscr{C}, d) = \cdots \longrightarrow \mathscr{C}^{i-1} \xrightarrow{d^{i-1}} \mathscr{C}^i \xrightarrow{d^i} \mathscr{C}^{i+1} \longrightarrow \cdots$ be a (co)chain complex of graded free R-modules with graded differentials d^i having degree 0 for all $i \in \mathbb{Z}$. Then the *graded Euler characteristic* of $\mathscr{C}$ is defined as $\chi_q(\mathscr{C}) = \sum_{i\in\mathbb{Z}} (-1)^i \dim_q(\mathscr{C}^i)$.

Remark. One can think of a graded (co)chain complex of R-modules as a *bigraded R-module* in which the homogeneous components are indexed by pairs of numbers $(i, j) \in \mathbb{Z}^2$.

Let $A = R[X]/X^2$ be the algebra of truncated polynomials. As an R-module, A is freely generated by 1 and X. We put a grading on A by specifying that $\deg(1) = 1$ and $\deg(X) = -1$ (we follow the original grading convention from [15] and [2] here; it is different by a sign from that of [1]). In other words, $A \simeq R\{1\} \oplus R\{-1\}$ and $\dim_q(A) = q + q^{-1}$. At the same time, A is a (graded) commutative algebra with unit 1 and multiplication $m : A \otimes A \to A$ given by

$$m(1 \otimes 1) = 1, \qquad m(1 \otimes X) = m(X \otimes 1) = X, \qquad m(X \otimes X) = 0. \tag{7}$$

The algebra A can also be equipped with a coalgebra structure with comultiplication $\Delta : A \to A \otimes A$ and counit $\varepsilon : A \to R$ defined as

$$\Delta(1) = 1 \otimes X + X \otimes 1, \qquad \Delta(X) = X \otimes X; \tag{8}$$

$$\varepsilon(1) = 0, \qquad \varepsilon(X) = 1. \tag{9}$$

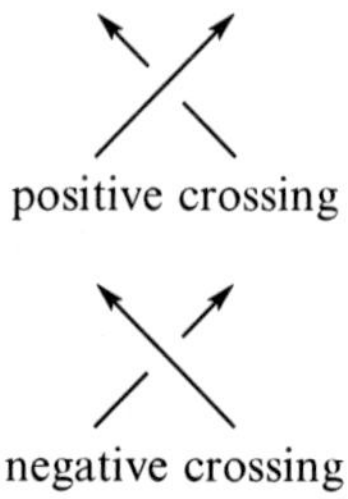

Fig. 2 Positive and negative crossings

The comultiplication Δ is coassociative and cocommutative and satisfies

$$(m \otimes \mathrm{id}_A) \circ (\mathrm{id}_A \otimes \Delta) = \Delta \circ m, \tag{10}$$
$$(\varepsilon \otimes \mathrm{id}_A) \circ \Delta = \mathrm{id}_A. \tag{11}$$

Together with the unit map $\iota : R \to A$ given by $\iota(1) = 1$, this makes A into a commutative Frobenius algebra over R [19].

It follows directly from the definitions that ι, ε, m, and Δ are graded maps with

$$\deg(\iota) = \deg(\varepsilon) = 1 \quad \text{and} \quad \deg(m) = \deg(\Delta) = -1. \tag{12}$$

2.2 *Khovanov Chain Complex*

Let L be an oriented link and D its planar diagram. We assign a number ± 1, called the *sign*, to every crossing of D according to the rule depicted in Fig. 2. The sum of these signs over all the crossings of D is called the *writhe number* of D and is denoted by $w(D)$.

Every crossing of D can be *resolved* in two different ways according to a choice of a *marker*, which can be either *positive* or *negative*, at this crossing (see Fig. 3). A collection of markers chosen at every crossing of a diagram D is called a *(Kauffman) state* of D. For a diagram with n crossings, there are, obviously, 2^n different states. Denote by $\sigma(s)$ the difference between the numbers of positive and negative markers in a given state s. Define

$$i(s) = \frac{w(D) - \sigma(s)}{2}, \qquad j(s) = \frac{3w(D) - \sigma(s)}{2}. \tag{13}$$

Since both $w(D)$ and $\sigma(s)$ are congruent to n modulo 2, $i(s)$ and $j(s)$ are always integers. For a given state s, the result of the resolution of D at each crossing according to s is a family D_s of disjointly embedded circles. Denote the number of these circles by $|D_s|$.

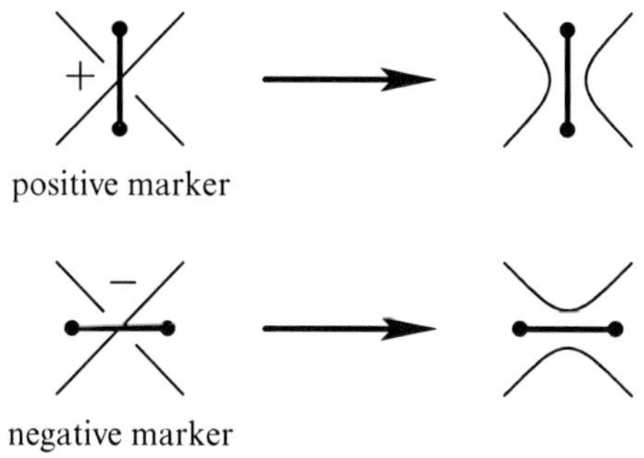

Fig. 3 Positive and negative markers and the corresponding resolutions of a diagram

For each state s of D, let $\mathscr{A}(s) = A^{\otimes|D_s|}\{j(s)\}$. One should understand this construction as assigning a copy of the algebra A to each circle from D_s, taking the tensor product of all of these copies, and shifting the grading of the result by $j(s)$. By construction, $\mathscr{A}(s)$ is a graded free R-module of graded dimension $\dim_q(\mathscr{A}(s)) = q^{j(s)}(q+q^{-1})^{|D_s|}$. Let $\mathscr{C}^i(D) = \bigoplus_{i(s)=i}\mathscr{A}(s)$ for each $i \in \mathbb{Z}$. In order to make $\mathscr{C}(D)$ into a graded complex, we need to define a (graded) differential $d^i : \mathscr{C}^i(D) \to \mathscr{C}^{i+1}(D)$ of degree 0. But even before the differential is defined, the (graded) Euler characteristic of $\mathscr{C}(D)$ makes sense.

Lemma 2.6. *The graded Euler characteristic of $\mathscr{C}(D)$ equals the Jones polynomial of the link L. That is, $\chi_q(\mathscr{C}(D)) = J_L(q)$.*

Proof.

$$\begin{aligned}
\chi_q(\mathscr{C}(D)) &= \sum_{i\in\mathbb{Z}}(-1)^i \dim_q(\mathscr{C}^i(D)) \\
&= \sum_{i\in\mathbb{Z}}(-1)^i \sum_{i(s)=i} \dim_q(\mathscr{A}(s)) \\
&= \sum_{s}(-1)^{i(s)} q^{j(s)}(q+q^{-1})^{|D_s|} \\
&= \sum_{s}(-1)^{\frac{w(D)-\sigma(s)}{2}} q^{\frac{3w(D)-\sigma(s)}{2}}(q+q^{-1})^{|D_s|}.
\end{aligned}$$

Let us forget for a moment that A denotes an algebra and (temporarily) use this letter for a variable. Substituting $(-A^{-2})$ instead of q and noticing that $w(D) \equiv \sigma(s)$ (mod 2), we arrive at

$$\chi_q(\mathscr{C}(D)) = (-A)^{-3w(D)}\sum_{s} A^{\sigma(s)}(-A^2-A^{-2})^{|D_s|} = (-A^2-A^{-2})\langle L\rangle_N,$$

where $\langle L\rangle_N$ is the normalized Kauffman bracket polynomial of L (see [14] for details). The normalized bracket polynomial of a link is related to the bracket polynomial of its diagram by $\langle L\rangle_N = (-A)^{-3w(D)}\langle D\rangle$. Kauffman proved in [14] that $\langle L\rangle_N$ equals the Jones polynomial $V_L(t)$ of L after substituting $t^{-1/4}$ for A. The relation (4) between $V_L(t)$ and $J_L(q)$ completes our proof. □

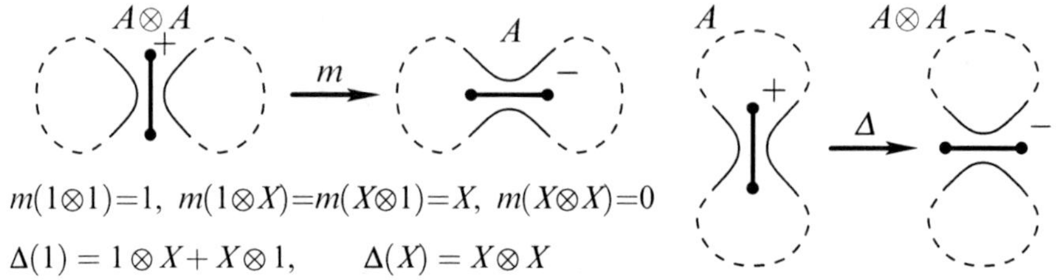

Fig. 4 Diagram resolutions corresponding to adjacent states and maps between the algebras assigned to the circles

Let s_+ and s_- be two states of D that differ at a single crossing, where s_+ has a positive marker, while s_- has a negative one. We call two such states *adjacent*. In this case, $\sigma(s_-) = \sigma(s_+) - 2$ and, consequently, $i(s_-) = i(s_+) + 1$ and $j(s_-) = j(s_+) + 1$. Consider now the resolutions of D corresponding to s_+ and s_-. One can readily see that D_{s_-} is obtained from D_{s_+} by either merging two circles into one or splitting one circle into two (see Fig. 4). All the circles that do not pass through the crossing at which s_+ and s_- differ remain unchanged. We define $d_{s_+:s_-} : \mathscr{A}(s_+) \to \mathscr{A}(s_-)$ as either $m \otimes \mathrm{id}$ or $\Delta \otimes \mathrm{id}$ depending on whether the circles merge or split. Here, the multiplication or comultiplication is performed on the copies of A that are assigned to the affected circles, as in Fig. 4, while $d_{s_+:s_-}$ acts as the identity on all the A's corresponding to the unaffected ones. The difference in grading shift between $\mathscr{A}(s_+)$ and $\mathscr{A}(s_-)$ and equalities (12) ensure that $\deg(d_{s_+:s_-}) = 0$ by Theorem 2.4.

We need one more ingredient in order to finish the definition of the differential on $\mathscr{C}(D)$, namely, an ordering of the crossings of D. For an adjacent pair of states (s_+, s_-), define $\xi(s_+, s_-)$ to be the number of the *negative* markers in s_+ (or s_-) that appear in the ordering of the crossings *after* the crossing at which s_+ and s_- differ. Finally, let $d^i = \sum_{(s_+,s_-)} (-1)^{\xi(s_+,s_-)} d_{s_+:s_-}$, where (s_+, s_-) runs over all adjacent pairs of states with $i(s_+) = i$. It is straightforward to verify [15] that $d^{i+1} \circ d^i = 0$ and hence $d : \mathscr{C}(D) \to \mathscr{C}(D)$ is indeed a differential.

Definition 2.7 (Khovanov, [15]). The resulting (co)chain complex $\mathscr{C}(D) = \cdots \longrightarrow \mathscr{C}^{i-1}(D) \xrightarrow{d^{i-1}} \mathscr{C}^i(D) \xrightarrow{d^i} \mathscr{C}^{i+1}(D) \longrightarrow \cdots$ is called the *Khovanov chain complex* of the diagram D. The homology of $\mathscr{C}(D)$ with respect to d is called the *Khovanov homology* of L and is denoted by $\mathscr{H}(L)$. We write $\mathscr{C}(D;R)$ and $\mathscr{H}(L;R)$ if we want to emphasize the ring of coefficients with which we work. If R is omitted from the notation, integer coefficients are assumed.

Theorem 2.8 (Khovanov, [15]; see also [2]). *The isomorphism class of $\mathscr{H}(L;R)$ depends on the isotopy class of L only, and hence is a link invariant. In particular, it does not depend on the ordering chosen for the crossings of D. The Khovanov homology $\mathscr{H}(L;R)$ categorifies $J_L(q)$, a version of the Jones polynomial defined by (1).*

Remark. One can think of $\mathscr{C}(D;R)$ as a bigraded (co)chain complex $\mathscr{C}^{i,j}(D;R)$ with a differential of bidegree $(1,0)$. In this case, i is the homological grading of

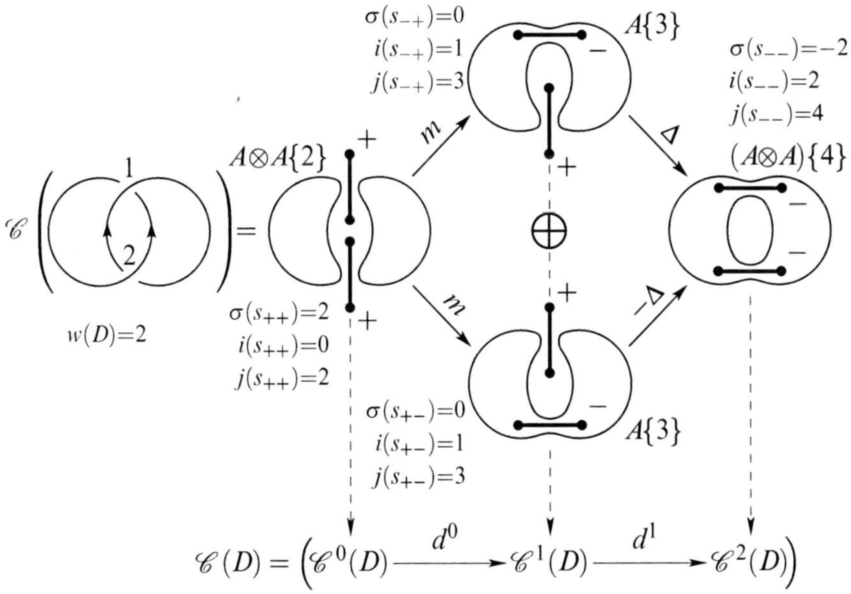

Fig. 5 Khovanov chain complex for the Hopf link

this complex, and j is its q-grading, also called the *Jones grading*. Consequently, $\mathscr{H}(L;R)$ can be considered to be a bigraded R-module as well.

Note 2.9. Let $\#L$ be the number of components of a link L. One can check that $j(s)+|D_s|$ is congruent modulo 2 to $\#L$ for every state s. It follows that $\mathscr{C}(D;R)$ has nontrivial homogeneous components only in the degrees that have the same parity as $\#L$. Consequently, $\mathscr{H}(L;R)$ is nontrivial only in the q-gradings with this parity (see Example 1.1).

Example 2.10. Figure 5 shows the Khovanov chain complex for the Hopf link with the indicated orientation. The diagram has two positive crossings, so its writhe number is 2. Let $s_{\pm\pm}$ be the four possible resolutions of this diagram, where each "+" or "−" describes the sign of the marker at the corresponding crossings. The chosen ordering of crossings is depicted by numbers placed next to them. By looking at Fig. 5, one easily computes that $\mathscr{A}(s_{++})=A^{\otimes 2}\{2\}$, $\mathscr{A}(s_{+-})=\mathscr{A}(s_{-+})=A\{3\}$, and $\mathscr{A}(s_{--})=A^{\otimes 2}\{4\}$. Consequently, $\mathscr{C}^0(D)=\mathscr{A}(s_{++})=A^{\otimes 2}\{2\}$, $\mathscr{C}^1(D)=\mathscr{A}(s_{+-})\oplus\mathscr{A}(s_{-+})=(A\oplus A)\{3\}$, and $\mathscr{C}^2(D)=\mathscr{A}(s_{--})=A^{\otimes 2}\{4\}$. It is convenient to arrange the four resolutions in the corners of a square placed in the plane in such a way that its diagonal from s_{++} to s_{--} is horizontal. Then the edges of this square correspond to the maps between the adjacent states (see Fig. 5). We observe that only one of the maps, namely the one corresponding to the edge from s_{+-} to s_{--}, comes with a negative sign.

In general, 2^n resolutions of a diagram D with n crossings can be arranged into an n-dimensional *cube of resolutions*, where vertices correspond to the 2^n states of D. The edges of this cube connect adjacent pairs of states and can be oriented from s_+ to s_-. Every edge is assigned either m or Δ with the sign $(-1)^{\xi(s_+,s_-)}$, as described above. It is easy to check that this makes each square (that is, a 2-dimensional face) of the cube anticommutative (all squares are commutative without the signs). Finally, the differential d^i restricted to each summand $\mathscr{A}(s)$ with $i(s) = i$ equals the sum of all the maps assigned to the edges that originate at s.

2.3 Reduced Khovanov Homology

Let, as before, D be a diagram of an oriented link L. Fix a base point on D that is different from all the crossings. For each state s, we define $\widetilde{\mathscr{A}}(s)$ in almost the same way as $\mathscr{A}(s)$, except that we assign XA instead of A to the circle from the resolution D_s of D that contains the base point. That is, $\widetilde{\mathscr{A}}(s) = ((XA) \otimes A^{\otimes(|D_s|-1)})\{j(s)\}$. We can now build the *reduced Khovanov chain complex* $\widetilde{\mathscr{C}}(D;R)$ in exactly the same way as $\mathscr{C}(D;R)$ by replacing $\mathscr{A}$ with $\widetilde{\mathscr{A}}$ everywhere. The grading shifts and differentials remain the same. It is easy to see that $\widetilde{\mathscr{C}}(D;R)$ is a subcomplex of $\mathscr{C}(D;R)$ of index 2. In fact, it is the image of the chain map $\mathscr{C}(D;R) \to \mathscr{C}(D;R)$ that acts by multiplying elements assigned to the circle containing the base point by X.

Definition 2.11 (Khovanov [16]; cf. Definition 2.7). The homology of $\widetilde{\mathscr{C}}(D;R)$ is called the *reduced Khovanov homology of L* and is denoted by $\widetilde{\mathscr{H}}(L;R)$. It is clear from the construction of $\widetilde{\mathscr{C}}(D;R)$ that its graded Euler characteristic equals $\widetilde{J}_L(q)$.

Theorem 2.12 (Khovanov [16]; cf. Theorem 2.8). *The isomorphism class of of the reduced Khovanov homology $\widetilde{\mathscr{H}}(L;R)$ is a link invariant that categorifies $\widetilde{J}_L(q)$, a version of the Jones polynomial defined by (1) and (4). Moreover, if two base points are chosen on the same component of L, then the corresponding reduced Khovanov homologies are isomorphic. On the other hand, $\widetilde{\mathscr{H}}(L;R)$ might depend on the component of L on which the base point is chosen.*

Although $\widetilde{\mathscr{C}}(D;R)$ can be determined from $\mathscr{C}(D;R)$, it is in general not clear how $\mathscr{H}(L;R)$ and $\widetilde{\mathscr{H}}(L;R)$ are related. There are several examples of pairs of knots (the first being 14^n_{9933} and $\overline{15}^n_{129763}$) that have the same rational Khovanov homology, but different rational reduced Khovanov homologies.[1] No such examples are known for homologies over $\mathbb{Z}$ among all prime knots with at most 15 crossings. On the other hand, it has been proved that $\mathscr{H}(L;\mathbb{Z}_2)$ and $\widetilde{\mathscr{H}}(L;\mathbb{Z}_2)$ determine each other completely.

[1] Here, 14^n_{9933} denotes the nonalternating knot number 9933 with 14 crossings from the Knotscape knot table [12], and $\overline{15}^n_{129763}$ is the mirror image of the knot 15^n_{129763}. See also Remark on page 417.

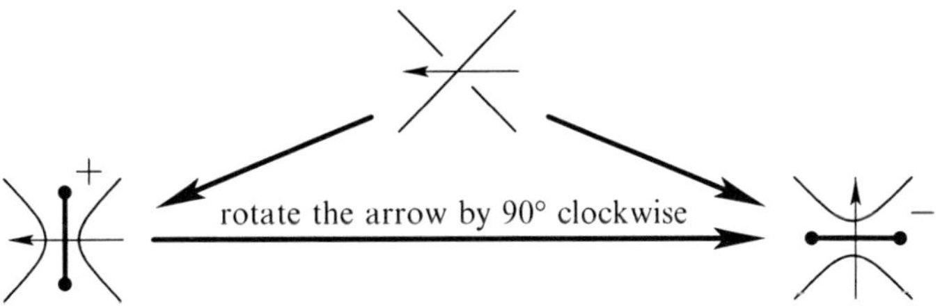

Fig. 6 Choice of arrows at the diagram crossings

Theorem 2.13 ([40]). *$\mathscr{H}(L;\mathbb{Z}_2) \simeq \widetilde{\mathscr{H}}(L;\mathbb{Z}_2) \otimes_{\mathbb{Z}_2} A_{\mathbb{Z}_2}$. In particular, $\widetilde{\mathscr{H}}(L;\mathbb{Z}_2)$ does not depend on the component on which the base point is chosen.*

Remark. $XA \simeq R\{0\}$ as a graded R-module. It follows that $\widetilde{\mathscr{C}}$ and $\widetilde{\mathscr{H}}$ are nontrivial only in the q-gradings with parity different from that of $\#L$, the number of components of L (cf. 2.9).

2.4 Odd Khovanov Homology

In 2007, Ozsváth, Rasmussen, and Szabó introduced [29] an *odd* version of the Khovanov homology. In their theory, the nilpotent variables X assigned to each circle in the resolutions of the link diagram (see Sect. 2.2) anticommute rather than commute. The odd Khovanov homology equals the original (*even*) one modulo 2, and in particular, categorifies the same Jones polynomial. In fact, the corresponding chain complexes are isomorphic as free bigraded R-modules, and their differentials differ only by sign. On the other hand, the resulting homology theories often have drastically different properties. We define the odd Khovanov homology below.

Let L be an oriented link and D its planar diagram. To each resolution s of D we assign a free graded R-module $\Lambda(s)$ as follows. Label all circles from the resolution D_s by some independent variables, say, $X_1^s, X_2^s, \dots, X_{|D_s|}^s$, and let $V_s = V(X_1^s, X_2^s, \dots, X_{|D_s|}^s)$ be a free R-module generated by them. We define $\Lambda(s) = \Lambda^*(V_s)$, the exterior algebra of V_s. Then $\Lambda(s) = \Lambda^0(V_s) \oplus \Lambda^1(V_s) \oplus \dots \oplus \Lambda^{|D_s|}(V_s)$, and we grade $\Lambda(s)$ by specifying $\Lambda(s)_{|D_s|-2k} = \Lambda^k(V_s)$ for each $0 \le k \le |D_s|$, where $\Lambda(s)_{|D_s|-2k}$ is the homogeneous component of $\Lambda(s)$ of degree $|D_s| - 2k$. It is an easy exercise for the reader to check that $\dim_q(\Lambda(s)) = \dim_q(A^{\otimes |D_s|})$.

Just as in the case of even Khovanov homology, these R-modules $\Lambda(s)$ can be arranged into an n-dimensional cube of resolutions. Let $\mathscr{C}^i_{\text{odd}}(D) = \bigoplus_{i(s)=i} \Lambda(s)\{j(s)\}$. Then, similarly to Lemma 2.6, we have that $\chi_q(\mathscr{C}_{\text{odd}}(D)) = J_L(q)$. In fact, $\mathscr{C}_{\text{odd}}(D) \simeq \mathscr{C}(D)$ as bigraded R-modules. In order to define the differential on $\mathscr{C}_{\text{odd}}$, we need to introduce an additional structure, a choice of an arrow at each crossing of D that is parallel to the *negative* marker at that crossing (see Fig. 6). There are obviously

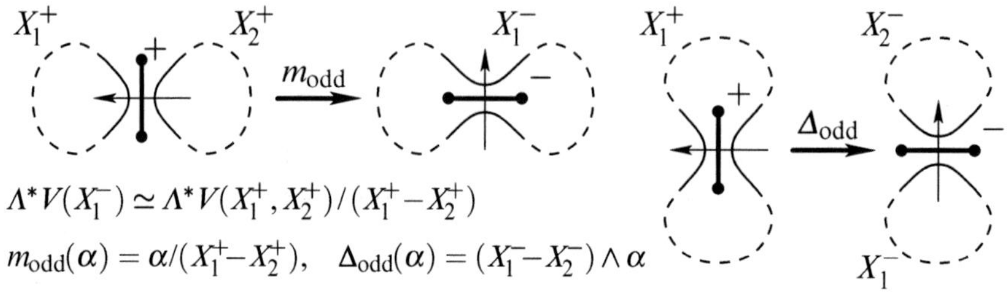

Fig. 7 Adjacent states and differentials in the odd Khovanov chain complex

2^n such choices. For every state s on D, we place arrows that connect two branches of D_s near each (former) crossing according to the rule from Fig. 6.

We now assign (graded) maps m_{odd} and Δ_{odd} to each edge of the cube of resolutions that connects adjacent states s_+ and s_-. If s_- is obtained from s_+ by merging two circles together, then $\Lambda(s_-) \simeq \Lambda(s_+)/(X_1^+ - X_2^+)$, where X_1^+ and X_2^+ are the generators of V_{s_+} corresponding to the two merging circles, as depicted in Fig. 7. We define $m_{\mathrm{odd}} : \Lambda(s_+) \to \Lambda(s_-)$ to be this isomorphism composed with the projection $\Lambda(s_+) \to \Lambda(s_+)/(X_1^+ - X_2^+)$.

The case in which one circle splits into two is more interesting. Let X_1^- and X_2^- be the generators of V_{s_-} corresponding to these two circles such that the arrow points from X_1^- to X_2^- (see Fig. 7). Now for each generator X_k^+ of V_{s_+}, we define $\Delta_{\mathrm{odd}}(X_k^+) = (X_1^- - X_2^-) \wedge X_{\eta(k)}^-$, where η is the correspondence between circles in D_{s_+} and D_{s_-}. While $\eta(1)$ can equal either 1 or 2, this choice does not affect $\Delta_{\mathrm{odd}}(X_1^+)$, since $(X_1^- - X_2^-) \wedge X_2^- = X_1^- \wedge X_2^- = -X_2^- \wedge X_1^- = (X_1^- - X_2^-) \wedge X_1^-$.

This definition makes each square in the cube of resolutions commutative, anticommutative, or both. The latter case means that both double composites corresponding to the square are trivial. This is a major departure from the situation that we had in the even case, in which each square was commutative. In particular, it makes the choice of signs on the edges of the cube much more involved.

Theorem 2.14 (Ozsváth–Rasmussen–Szabó [29]). *It is possible to assign a sign to each edge in this (odd) cube of resolutions in such a way that every square becomes anticommutative. This results in a graded (co)chain complex $\mathscr{C}_{\mathrm{odd}}(D;R)$. The homology $\mathscr{H}_{\mathrm{odd}}(L;R)$ of $\mathscr{C}_{\mathrm{odd}}(D;R)$ does not depend on the choice of arrows at the crossings, the choice of edge signs, and some other choices needed in the construction. Moreover, the isomorphism class of $\mathscr{H}_{\mathrm{odd}}(L;R)$ is a link invariant, called* odd Khovanov homology, *that categorifies $J_L(q)$.*

Remark. There is no explicit construction for assigning signs to the edges of the cube of resolutions in the case of the odd Khovanov chain complex. The theorem above only ensures that signs exist.

Note 2.15. By comparing the definitions of $\mathscr{C}_{\mathrm{odd}}(D;\mathbb{Z}_2)$ and $\mathscr{C}(D;\mathbb{Z}_2)$, it is easy to see that they are isomorphic as graded chain complexes (since the signs do not matter modulo 2). It follows that $\mathscr{H}_{\mathrm{odd}}(D;\mathbb{Z}_2) \simeq \mathscr{H}(D;\mathbb{Z}_2)$ as well.

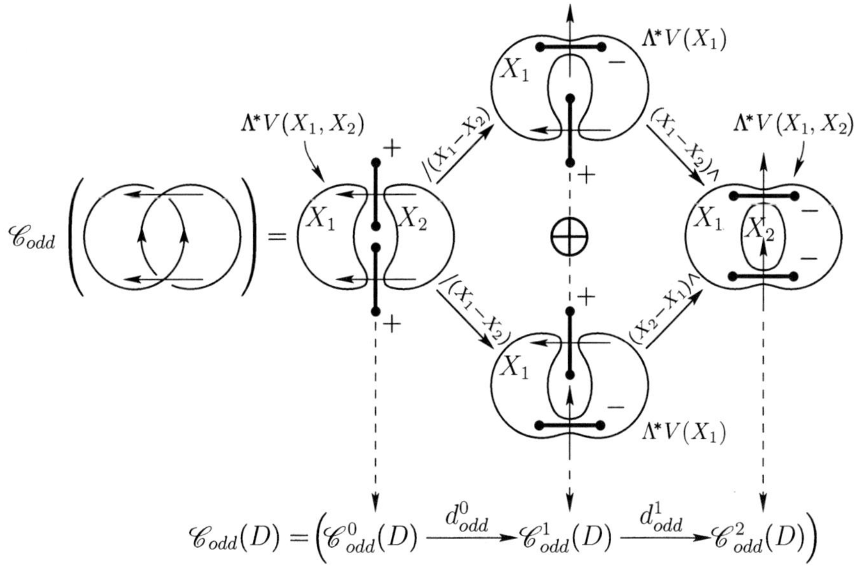

Fig. 8 Odd Khovanov chain complex for the Hopf link

Note 2.16. One can construct the *reduced* odd Khovanov chain complex $\widetilde{\mathscr{C}}_{\text{odd}}(D;R)$ and the *reduced* odd Khovanov homology $\widetilde{\mathscr{H}}_{\text{odd}}(L;R)$ using methods similar to those from Sect. 2.3. In this case, in contrast to the even situation, reduced and nonreduced odd Khovanov homologies determine each other completely (see [29]). Namely, $\mathscr{H}_{\text{odd}}(L;R) \simeq \widetilde{\mathscr{H}}_{\text{odd}}(L;R)\{1\} \oplus \widetilde{\mathscr{H}}_{\text{odd}}(L;R)\{-1\}$ (cf. Theorem 2.13). It is therefore enough to consider the reduced version of the odd Khovanov homology only.

Example 2.17. The odd Khovanov chain complex for the Hopf link is depicted in Fig. 8. All the grading shifts in this case are the same as in Example 2.10 and in Fig. 5, so we do not list them again. We observe that the resulting square of resolutions is anticommutative, so no adjustment of signs is needed.

The odd Khovanov homology should provide an insight into interrelations between the Khovanov and Heegaard–Floer [30] homology theories. Its definition was motivated by the following result.

Theorem 2.18 (Ozsváth–Szabó [32]). *For each link L with a diagram D, there exists a spectral sequence with $E^1 = \widetilde{\mathscr{C}}(D;\mathbb{Z}_2)$ and $E^2 = \widetilde{\mathscr{H}}(L;\mathbb{Z}_2)$ that converges to the $\mathbb{Z}_2$-Heegaard–Floer homology $\widehat{HF}(\Sigma(L);\mathbb{Z}_2)$ of the double branched cover $\Sigma(L)$ of S^3 along L.*

Conjecture 2.19. There exists a spectral sequence that starts with $\widetilde{\mathscr{C}}_{\text{odd}}(D;\mathbb{Z})$ and $\widetilde{\mathscr{H}}_{\text{odd}}(L;\mathbb{Z})$ and converges to $\widehat{HF}(\Sigma(L);\mathbb{Z})$.

3 Properties of the Khovanov Homology

In this section we summarize the main properties of the Khovanov homology and list related constructions. We emphasize similarities and differences in properties exhibited by different versions of the Khovanov homology. Some of them were already mentioned in the previous sections.

Note 3.1. Let L be an oriented link and D its planar diagram. Then

- $\widetilde{\mathscr{C}}(D;R)$ is a subcomplex of $\mathscr{C}(D;R)$ of index 2.
- $\mathscr{H}_{\text{odd}}(L;\mathbb{Z}_2) \simeq \mathscr{H}(L;\mathbb{Z}_2)$ and $\widetilde{\mathscr{H}}_{\text{odd}}(L;\mathbb{Z}_2) \simeq \widetilde{\mathscr{H}}(L;\mathbb{Z}_2)$.
- $\chi_q(\mathscr{H}(L;R)) = \chi_q(\mathscr{H}_{\text{odd}}(L;R)) = J_L(q)$ and $\chi_q(\widetilde{\mathscr{H}}(L;R)) = \chi_q(\widetilde{\mathscr{H}}_{\text{odd}}(L;R)) = \widetilde{J}_L(q)$.
- $\mathscr{H}(L;\mathbb{Z}_2) \simeq \widetilde{\mathscr{H}}(L;\mathbb{Z}_2) \otimes_{\mathbb{Z}_2} A_{\mathbb{Z}_2}$ [40] and $\mathscr{H}_{\text{odd}}(L;R) \simeq \widetilde{\mathscr{H}}_{\text{odd}}(L;R)\{1\} \oplus \widetilde{\mathscr{H}}_{\text{odd}}(L;R)\{-1\}$ [29]. On the other hand, $\mathscr{H}(L;\mathbb{Z})$ and $\mathscr{H}(L;\mathbb{Q})$ do not split in general.
- For links, $\widetilde{\mathscr{H}}(L;\mathbb{Z}_2)$ and $\widetilde{\mathscr{H}}_{\text{odd}}(L;R)$ do not depend on the choice of a component with the base point. This is, in general, not the case for $\widetilde{\mathscr{H}}(L;\mathbb{Z})$ and $\widetilde{\mathscr{H}}(L;\mathbb{Q})$.
- If L is a nonsplit alternating link, then $\mathscr{H}(L;\mathbb{Q})$, $\widetilde{\mathscr{H}}(L;R)$, and $\widetilde{\mathscr{H}}_{\text{odd}}(L;R)$ are completely determined by the Jones polynomial and signature of L [16,26,29].
- $\mathscr{H}(L;\mathbb{Z}_2)$ and $\mathscr{H}_{\text{odd}}(L;\mathbb{Z})$ are invariant under the component-preserving link mutations [6,46]. It is unclear whether the same holds for $\mathscr{H}(L;\mathbb{Z})$. On the other hand, $\mathscr{H}(L;\mathbb{Z})$ is known not to be preserved under a mutation that exchanges components of a link [45] and under a cabled mutation [8].
- $\mathscr{H}(L;\mathbb{Z})$ almost always has torsion (except for several special cases), but mostly of order 2. The first knot with 4-torsion is the $(4,5)$-torus knot, which has 15 crossings. The first known knot with 3-torsion is the $(5,6)$-torus knot with 24 crossings. On the other hand, $\mathscr{H}_{\text{odd}}(L;\mathbb{Z})$ was observed to have torsion of various orders even for knots with relatively few crossings (see Remark on page 426), although orders 2 and 3 are the most popular.
- $\widetilde{\mathscr{H}}(L;\mathbb{Z})$ has very little torsion. The first knot with torsion has 13 crossings. On the other hand, $\widetilde{\mathscr{H}}_{\text{odd}}(L;\mathbb{Z})$ has as much torsion as $\mathscr{H}_{\text{odd}}(L;\mathbb{Z})$.

Remark. The properties above show that $\widetilde{\mathscr{H}}_{\text{odd}}(L;\mathbb{Z})$ behaves similarly to $\widetilde{\mathscr{H}}(L;\mathbb{Z}_2)$ but not to $\widetilde{\mathscr{H}}(L;\mathbb{Z})$. This is by design (see Theorem 2.18 and Conjecture 2.19).

3.1 *Homological Thickness*

Definition 3.2. Let L be a link. The *homological width* of L over a ring R is the minimal number of adjacent diagonals $j-2i = \text{const}$ such that $\mathscr{H}(L;R)$ is zero outside of these diagonals. It is denoted by $\text{hw}_R(L)$. The *reduced homological width*, $\widetilde{\text{hw}}_R(L)$ of L, *odd homological width*, $\text{ohw}_R(L)$ of L, and *reduced odd homological width*, $\widetilde{\text{ohw}}_R(L)$ of L are defined similarly.

Note 3.3. It follows from Theorem 2.16 that $\widetilde{\mathrm{ohw}}_R(L) = \mathrm{ohw}_R(L) - 1$. The same holds in the case of the even Khovanov homology over $\mathbb{Q}$: $\widetilde{\mathrm{hw}}_{\mathbb{Q}}(L) = \mathrm{hw}_{\mathbb{Q}}(L) - 1$ (see [16]).

Definition 3.4. A link L is said to be *homologically thin* over a ring R, or simply RH-thin, if $\mathrm{hw}_R(L) = 2$; otherwise, L is *homologically thick*, or RH-thick. We define *odd-homologically* thin and thick, or simply ROH-thin and ROH-thick, links similarly.

Theorem 3.5 (Lee, Ozsváth–Rasmussen–Szabó, Manolescu–Ozsváth [26, 27, 29]). *Quasialternating links (see Sect. 4.3 for the definition) are RH-thin and ROH-thin for every ring R. In particular, this is true for nonsplit alternating links.*

Theorem 3.6 (Khovanov [16]). *Adequate links are RH-thick for every R.*

Note 3.7. Homological thickness of a link L often does not depend on the base ring. The first prime knot with $\mathrm{hw}_{\mathbb{Q}}(L) < \mathrm{hw}_{\mathbb{Z}_2}(L)$ and $\mathrm{hw}_{\mathbb{Q}}(L) < \mathrm{hw}_{\mathbb{Z}}(L)$ is 15^n_{41127}; it has 15 crossings (see Fig. 9). The first prime knot that is $\mathbb{Q}$H-thin but $\mathbb{Z}$H-thick, 16^n_{197566}, has 16 crossings (see Fig. 10). Its mirror image, $\overline{16}^n_{197566}$, is both $\mathbb{Q}$H- and $\mathbb{Z}$H-thin but is $\mathbb{Z}_2$H-thick with $\mathscr{H}^{-8,-21}(\overline{16}^n_{197566};\mathbb{Z}_2) \simeq \mathbb{Z}_2$, for example, because of the universal coefficient theorem. Also observe that $\mathscr{H}^{9,25}(16^n_{197566};\mathbb{Z})$ and $\mathscr{H}^{-8,-25}(\overline{16}^n_{197566};\mathbb{Z})$ have 4-torsion, indicated by a small box in the tables.

Remark. Throughout this paper we use the following notation for knots: knots with 10 crossings or fewer are numbered according to the Rolfsen's knot table [37], and knots with 11 crossings or more are numbered according to the knot table from Knotscape [12]. Mirror images of knots from either table are denoted with a bar on top. For example, $\overline{9}_{46}$ is the mirror image of the knot number 46 with 9 crossings from Rolfsen's table, and 16^n_{197566} is the nonalternating knot number 197566 with 16 crossings from Knotscape's table.

Note 3.8. Odd Khovanov homology is often thicker over $\mathbb{Z}$ than the corresponding even homology. This is crucial for applications (see Sect. 4). On the other hand, $\widetilde{\mathrm{ohw}}_{\mathbb{Q}}(L) \leq \widetilde{\mathrm{hw}}_{\mathbb{Q}}(L)$ for all but one prime knot with at most 15 crossings. The homology for this knot, 15^n_{41127}, is shown in Fig. 9. Please observe that $\widetilde{\mathscr{H}}_{\mathrm{odd}}(15^n_{41127})$ has 3-torsion (in gradings $(-2,-2)$ and $(-1,0)$), while $\widetilde{\mathscr{H}}(15^n_{41127})$ has none.

3.2 Lee Spectral Sequence and the Knight-Move Conjecture

In [26], Eun Soo Lee introduced a structure of a spectral sequence on the rational Khovanov chain complex $\mathscr{C}(D;\mathbb{Q})$ of a link diagram D. Namely, Lee defined a differential $d' : \mathscr{C}(D;\mathbb{Q}) \to \mathscr{C}(D;\mathbb{Q})$ of bidegree $(1,4)$ by setting

	-7	-6	-5	-4	-3	-2	-1	0	1	2	3
8											**1**
6										**1**	
4									**1**		
2							**1**	**2**			
0						**1**		$\mathbf{1}_2$			
-2					**1**	**1**					
-4				**2**	**1**						
-6			**1**								
-8		**1**									
-10	**1**										

$\mathrm{hw}_{\mathbb{Q}} = 3, \quad \widetilde{\mathrm{hw}}_{\mathbb{Q}} = 2, \quad \mathrm{hw}_{\mathbb{Z}} = 4, \quad \widetilde{\mathrm{hw}}_{\mathbb{Z}} = 3$

	-7	-6	-5	-4	-3	-2	-1	0	1	2	3
8											**1**
6										**1**	
4									**1**		
2								$\mathbf{1}, \mathbf{1}_2$			
0							$\boxed{\mathbf{1}_6}$	**1**			
-2						$\boxed{\mathbf{1}_6}$					
-4				**1**	$\mathbf{1}_2$						
-6			**1**								
-8		**1**									
-10	**1**										

$\mathrm{ohw}_{\mathbb{Q}} = 4, \quad \widetilde{\mathrm{ohw}}_{\mathbb{Q}} = 3, \quad \mathrm{ohw}_{\mathbb{Z}} = 4, \quad \widetilde{\mathrm{ohw}}_{\mathbb{Z}} = 3$

Fig. 9 Integral reduced even Khovanov homology (above) and odd Khovanov Homology (below) of the knot 15^n_{41127}

$$\begin{aligned} m' : A \otimes A \to A: &\quad m'(1 \otimes 1) = m'(1 \otimes X) = m'(X \otimes 1) = 0, \quad m'(X \otimes X) = 1 \\ \Delta' : A \to A \otimes A: &\quad \Delta'(1) = 0, \qquad \Delta'(X) = 1 \otimes 1. \end{aligned} \tag{14}$$

It is straightforward to verify that d' is indeed a differential and that it anticommutes with d, that is, $d \circ d' + d' \circ d = 0$. This makes $(\mathscr{C}(D;\mathbb{Q}), d, d')$ into a double complex. Let d'_* be the differential induced by d' on $\mathscr{H}(L;\mathbb{Q})$. Lee proved that d'_* is functorial, that is, it commutes with the isomorphisms induced on $\mathscr{H}(L;\mathbb{Q})$ by isotopies of L. It follows that there exists a spectral sequence with $(E_1, d_1) = (\mathscr{C}(D;\mathbb{Q}), d)$ and

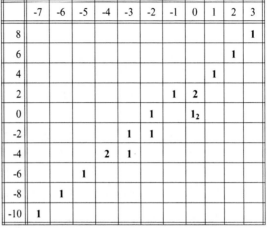

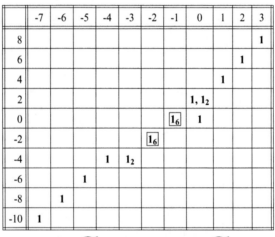

	-2	-1	0	1	2	3	4	5	6	7	8	9	10
29													**1**
27												**4**	$\mathbf{1_2}$
25											**7**	$\mathbf{1, 3_2 \boxed{1_4}}$	
23										**12**	$\mathbf{4, 8_2}$		
21									**15**	$\mathbf{7, 12_2}$	$\mathbf{1_2}$		
19								**17**	$\mathbf{12, 16_2}$				
17							**16**	$\mathbf{15, 18_2}$	$\mathbf{1_2}$				
15						**15**	$\mathbf{17, 16_2}$	$\mathbf{1_2}$					
13					**10**	$\mathbf{16, 15_2}$							
11				**6**	$\mathbf{15, 10_2}$								
9			**3**	$\mathbf{10, 6_2}$									
7		**1**	$\mathbf{7, 2_2}$										
5		$\mathbf{2, 1_2}$											
3	**1**												

The free part of $\mathscr{H}(16^n_{197566};\mathbb{Z})$ is supported on diagonals $j-2i=7$ and $j-2i=9$.
On the other hand, there is 2-torsion on the diagonal $j-2i=5$.
Therefore, 16^n_{197566} is $\mathbb{Q}$H-thin, but $\mathbb{Z}$H-thick and $\mathbb{Z}_2$H-thick.

	-10	-9	-8	-7	-6	-5	-4	-3	-2	-1	0	1	2
-3													**1**
-5												**2**	$\mathbf{1_2}$
-7											**7**	$\mathbf{1, 2_2}$	
-9										**10**	$\mathbf{3, 6_2}$		
-11									**15**	$\mathbf{6, 10_2}$			
-13								**16**	$\mathbf{10, 15_2}$				
-15							$\mathbf{17, 1_2}$	$\mathbf{15, 16_2}$					
-17						$\mathbf{15, 1_2}$	$\mathbf{16, 18_2}$						
-19					**12**	$\mathbf{17, 16_2}$							
-21				$\mathbf{7, 1_2}$	$\mathbf{15, 12_2}$								
-23			**4**	$\mathbf{12, 8_2}$									
-25		**1**	$\mathbf{7, 3_2 \boxed{1_4}}$										
-27		$\mathbf{4, 1_2}$											
-29	**1**												

$\mathscr{H}(\overline{16}^n_{197566};\mathbb{Z})$ is supported on diagonals $j-2i=-7$ and $j-2i=-9$.
But there is 2-torsion on the diagonal $j-2i=-7$.
Therefore, $\overline{16}^n_{197566}$ is $\mathbb{Q}$H-thin and $\mathbb{Z}$H-thin, but $\mathbb{Z}_2$H-thick.

Fig. 10 Integral Khovanov homology of the knots 16^n_{197566} and $\overline{16}^n_{197566}$

$(E_2, d_2) = (\mathscr{H}(L;\mathbb{Q}), d'_*)$ that converges to the homology of the total (filtered) complex of $\mathscr{C}(D;\mathbb{Q})$ with respect to the differential $d + d'$. It is called the *Lee spectral sequence*. The differentials d_n in this spectral sequence have bidegree $(1, 4(n-1))$.

Theorem 3.9 (Lee [26]). *If L is an oriented link with #L components, then the limit of the Lee spectral sequence,* $H(\mathrm{Total}(\mathscr{C}(D;\mathbb{Q})), d + d')$, *consists of* 2^{n-1} *copies of* $\mathbb{Q} \oplus \mathbb{Q}$, *each located in a specific homological grading that is explicitly defined by linking numbers of the components of L. In particular, if L is a knot, then the Lee spectral sequence converges to* $\mathbb{Q} \oplus \mathbb{Q}$ *localed in the homological grading* 0.

The following theorem is the cornerstone in the definition of the Rasmussen invariant, one of the main applications of the Khovanov homology (see Sect. 4.1).

Theorem 3.10 (Rasmussen [35]). *If L is a knot, then the two copies of* $\mathbb{Q}$ *in the limiting term of the Lee spectral sequence for L are "neighbors," that is, their q-gradings differ by* 2.

Corollary 3.11. *If the Lee spectral sequence for a link L collapses after the second page (that is,* $d_n = 0$ *for* $n \geq 3$*), then* $\mathscr{H}(L;\mathbb{Q})$ *consists of one "pawn-move" pair in homological grading* 0 *and multiple "knight-move" pairs, shown below, with appropriate grading shifts.*

Pawn-move pair:

$\mathbb{Q}$
$\mathbb{Q}$

Knight-move pair:

	$\mathbb{Q}$
$\mathbb{Q}$	

Corollary 3.12. *Since* d_3 *has bidegree* $(1, 8)$, *the Lee spectral sequence collapses after the second page for all knots with homological width* 2 *or* 3, *in particular, for all alternating and quasialternating knots. Hence Corollary 3.11 can be applied to such knots.*

Knight-Move Conjecture (Garoufalidis–Khovanov–Bar-Natan [2, 15]). The conclusion of Corollary 3.11 is true for every knot.

Remark. There are currently no known counterexamples to the knight-move conjecture. In fact, the Lee spectral sequence can be proved (in one way or another) to collapse after the second page for every known example of Khovanov homology.

Remark. While the Lee spectral sequence exists over any ring R, the statement of Theorem 3.9 does not hold over all rings. In particular, it is false over $\mathbb{Z}_2$. In this case, though, a similar theory was constructed by Paul Turner [43]. In fact, his construction works for reduced Khovanov homology as well because of Theorem 2.13. While Theorems 3.9 and 3.10 are still true over $\mathbb{Z}_p$ with odd prime p, the knight-move conjecture is known to be false over such rings [3].

Remark. The Lee spectral sequence has no analogue in the odd and reduced Khovanov homology theories (except over $\mathbb{Z}_2$, as noted above, where the two theories coincide). The knight-move conjecture has no analogue in these theories either.

3.3 Long Exact Sequence of the Khovanov Homology

One of the most useful tools in studying Khovanov homology is the long exact sequence that categorifies Kauffman's *unoriented* skein relation for the Jones polynomial [15]. If we forget about the grading, then it is clear from the construction from Sect. 2.2 that $\mathscr{C}(\asymp)$ is a subcomplex of $\mathscr{C}(\times)$ and $\mathscr{C}()() \simeq \mathscr{C}(\times)/\mathscr{C}(\asymp)$ (see also Fig. 5). Here, $\asymp$ and $)($ depict link diagrams where a single crossing $\times$ is resolved in a negative or, respectively, positive direction. This results in a short exact sequence of nongraded chain complexes:

$$0 \longrightarrow \mathscr{C}(\asymp) \xrightarrow{in} \mathscr{C}(\times) \xrightarrow{p} \mathscr{C}()() \longrightarrow 0, \tag{15}$$

where *in* is the inclusion and p is the projection.

In order to introduce grading into (15), we need to consider the cases in which the crossing to be resolved is either positive or negative. We get (see [36])

$$\begin{aligned} &0 \longrightarrow \mathscr{C}(\asymp)\{2+3\omega\}[1+\omega] \xrightarrow{in} \mathscr{C}(\overset{+}{\times}) \xrightarrow{p} \mathscr{C}()()\{1\} \longrightarrow 0,\\ &0 \longrightarrow \mathscr{C}()()\{-1\} \xrightarrow{in} \mathscr{C}\left(\overset{-}{\times}\right) \xrightarrow{p} \mathscr{C}(\asymp)\{1+3\omega\}[\omega] \longrightarrow 0, \end{aligned} \tag{16}$$

where ω is the difference between the numbers of negative crossings in the unoriented resolution $\asymp$ (it has to be oriented somehow in order to define its Khovanov chain complex) and in the original diagram. The notation $\mathscr{C}[k]$ is used to represent a shift in the homological grading of a complex $\mathscr{C}$ by k. The graded versions of *in* and p are both homogeneous, that is, have bidegree $(0,0)$.

By passing to homology in (16), we get the following result.

Theorem 3.13 (Khovanov, Viro, Rasmussen [15, 36, 44]). *The Khovanov homology is subject to the following long exact sequences:*

$$\begin{aligned} &\cdots \to \mathscr{H}()()\{1\} \xrightarrow{\partial} \mathscr{H}(\asymp)\{2+3\omega\}[1+\omega] \xrightarrow{in_*} \mathscr{H}(\overset{+}{\times}) \xrightarrow{\sqrt{}^*} \mathscr{H}()()\{\infty\} \to \cdots,\\ &\cdots \to \mathscr{H}()()\{-1\} \xrightarrow{in_*} \mathscr{H}(\overset{-}{\times}) \xrightarrow{\sqrt{}^*} \mathscr{H}(\asymp)\{\infty+\ni\omega\}[\omega] \xrightarrow{\partial} \mathscr{H}()()\{-\infty\} \to \cdots, \end{aligned} \tag{17}$$

where in_* and p_* are homogeneous and ∂ is the connecting differential and has bidegree $(1,0)$.

Remark. The long exact sequences (17) work equally well over any ring R and for every version of the Khovanov homology, including the odd one (see [29]). This is both a blessing and a curse. On the one hand, this means that all of the properties of the even Khovanov homology that are proved using these long exact sequences (and most of them are) are automatically true for the odd Khovanov homology as well. On the other hand, this makes it very hard to find explanations for many differences among these homology theories.

4 Applications of the Khovanov Homology

In this section we collect some of the more prominent applications of the Khovanov homology theories. This list is by no means complete; it is chosen to provide the reader with a broader view of the type of problems that can be solved with the help of the Khovanov homology. We make a special effort to compare the performance of different versions of the homology, where applicable.

4.1 Rasmussen Invariant and Bounds on the Slice Genus

One of the most important applications of Khovanov's construction was obtained by Jacob Rasmussen in 2004. In [35], he used the structure of the Lee spectral sequence to define a new invariant of knots that gives a lower bound on the slice genus. More specifically, for a knot L, its Rasmussen invariant $s(L)$ is defined as the mean q-grading of the two copies of $\mathbb{Q}$ that remain in the homological grading 0 of the limiting term of the Lee spectral sequence; see Theorem 3.10. Since the q-gradings of these $\mathbb{Q}$'s are odd and differ by 2, the Rasmussen invariant is an even integer.

Theorem 4.1 (Rasmussen [35]). *Let L be a knot and $s(L)$ its Rasmussen invariant. Then*

- $|s(L)| \leq 2g_s(L)$*, where $g_s(L)$ is the slice genus of L, that is, the smallest possible genus of a smoothly embedded surface in the 4-ball D^4 that has $L \subset S^3 = \partial D^4$ as its boundary;*
- $s(L) = \sigma(L)$ *for alternating L, where $\sigma(L)$ is the signature of L;*
- $s(L) = 2g_s(L) = 2g(L)$ *for a knot L that possesses a planar diagram with positive crossings only, where $g(L)$ is the genus of L;*
- *if knots L_- and L_+ have diagrams that are different at a single crossing in such a way that this crossing is negative in L_- and positive in L_+, then* $s(L_-) \leq s(L_+) \leq s(L_-) + 2$.

Corollary 4.2. $s(T_{p,q}) = (p-1)(q-1)$ *for $p, q > 0$, where $T_{p,q}$ is the (p,q)-torus knot. This implies the Milnor conjecture, first proved by Kronheimer and Mrowka in 1993 using gauge theory [24]. This conjecture states that the slice genus (and hence the genus) of $T_{p,q}$ is $\frac{1}{2}((p-1)(q-1))$. The upper bound on the slice genus is straightforward, so the lower bound provided by the Rasmussen invariant is sharp.*

Remark. Although the Rasmussen invariant was originally defined for knots only, its definition was later extended to the case of links by Anna Beliakova and Stephan Wehrli [5].

The Rasmussen invariant can be used to search for knots that are topologically locally flat slice but are not smoothly slice (see [41]). A knot is slice if its slice genus is 0. Theorem 4.1 implies that knots with nontrivial Rasmussen invariant are not smoothly slice. On the other hand, it was proved by Freedman [9] that knots with Alexander polynomial 1 are topologically locally flat slice. There are 82 knots with up to 16 crossings that possess these two properties [41]. Each such knot gives rise to a family of exotic $\mathbb{R}^4$ [10, Exercise 9.4.23]. It is worth noticing that most of these 82 examples were not previously known.

The Rasmussen invariant was also used [33, 41] to deduce the combinatorial proof of the slice–Bennequin inequality. This inequality states that

$$g_s(\widehat{\beta}) \leq \frac{1}{2}(w(\beta) - k + 1), \tag{18}$$

where β is a braid on k strands with the closure $\widehat{\beta}$ and $w(\beta)$ is its writhe number. The slice–Bennequin inequality provides one of the upper bounds for the Thurston–Bennequin number of Legendrian links (see below). It was originally proved by Lee Rudolph [38] using gauge theory. The approach via the Rasmussen invariant and Khovanov homology avoids gauge theory and symplectic Floer theory and results in a purely combinatorial proof.

Remark. Since Rasmussen's construction relies on the existence and convergence of the Lee spectral sequence, the Rasmussen invariant can be defined for only the even nonreduced Khovanov homology. In fact, a knot might not have any rational homology in the homological grading 0 of the odd Khovanov homology at all; see Fig. 12 below.

4.2 Bounds on the Thurston–Bennequin Number

Another useful application of the Khovanov homology is in finding upper bounds on the Thurston–Bennequin number of Legendrian links. Consider $\mathbb{R}^3$ equipped with the standard contact structure $dz - y\,dx$. A link $K \subset \mathbb{R}^3$ is said to be *Legendrian* if it is everywhere tangent to the 2-dimensional plane distribution defined as the kernel of this 1-form. Given a Legendrian link K, one defines its *Thurston–Bennequin number* $tb(K)$ as the linking number of K with its push-off K' obtained using a vector field that is tangent to the contact planes but orthogonal to the tangent vector field of K. Roughly speaking, $tb(K)$ measures the framing of the contact plane field around K. It is well known that the TB-number can be made arbitrarily small within the same class of topological links via stabilization but is bounded from above.

Definition 4.3. For a given *topological* link L, let $\overline{tb}(L)$, the *TB-bound* of L, be the maximal possible TB-number among all the Legendrian representatives of L. In other words, $\overline{tb}(L) = \max_K\{tb(K)\}$, where K runs over all the Legendrian links in $\mathbb{R}^3$ that are topologically isotopic to L.

Finding TB-bounds for links has attracted considerable interest lately, since such bounds can be used to demonstrate that certain contact structures on $\mathbb{R}^3$ are not isomorphic to the standard one. Such bounds can be obtained from the Bennequin and slice–Bennequin inequalities, degrees of HOMFLY-PT and Kauffman polynomials, knot Floer homology, and so on (see [28] for more details). The TB-bound coming from the Kauffman polynomial is usually one of the strongest, since most of the others incorporate another invariant of Legendrian links, the rotation number, into the inequality. In [28], Lenhard Ng used Khovanov homology to define a new bound on the TB-number.

Theorem 4.4 (Ng [28]). *Let L be an oriented link. Then*

$$\overline{tb}(L) \leq \min\Big\{k \Big| \bigoplus_{j-i=k} \mathscr{H}^{i,j}(L;R) \neq 0\Big\}. \tag{19}$$

Moreover, this bound is sharp for alternating links.

This Khovanov bound on the TB-number is often better than those that were known before. There are only two prime knots with up to 13 crossings for which the Khovanov bound is worse than the one coming from the Kauffman polynomial [28]. There are 45 such knots with at most 15 crossings.

Example 4.5. Fig. 11 shows computations of the Khovanov TB-bound for the $(4,-5)$-torus knot. The Khovanov homology groups in (19) can be used over any ring R, and this example shows that the bound coming from the integral homology is sometimes better than that from the rational homology, due to a strategically placed torsion. It is interesting to note that the integral Khovanov bound of -20 is computed incorrectly in [28]. In particular, this was one of the cases in which Ng thought that the Kauffman polynomial provided a better bound. In fact, the TB-bound of -20 is sharp for this knot.

The proof of Theorem 4.4 is based on the long exact sequences (17) and hence can be applied verbatim to the reduced as well as the odd Khovanov homology. By making appropriate adjustments to the grading, we immediately get [42] that

$$\overline{tb}(L) \leq -1 + \min\Big\{k \Big| \bigoplus_{j-i=k} \widetilde{\mathscr{H}}^{i,j}(L;R) \neq 0\Big\}, \tag{20}$$

$$\overline{tb}(L) \leq -1 + \min\Big\{k \Big| \bigoplus_{j-i=k} \widetilde{\mathscr{H}}^{i,j}_{\text{odd}}(L;R) \neq 0\Big\}. \tag{21}$$

As it turns out, the odd Khovanov TB-bound is often better than the even one. In fact, computations performed in [42] show that the odd Khovanov homology

	-9	-8	-7	-6	-5	-4	-3	-2	-1	0
-11										**1**
-13										**1**
-15								**1**		
-17						**1**		$\mathbf{1_2}$		
-19				**1**		**1**	**1**			
-21				$\mathbf{1, 1_2}$	**1**					
-23		**1**	**1**	$\mathbf{1_2}$	**1**					
-25		$\boxed{\mathbf{1_4}}$	**1**							
-27	$\mathbf{1, 1_2}$									
-29	$\mathbf{1_2}$									

j-i=-18

j-i=-20

TB-bound equals -18 over $\mathbb{Q}$ and -20 over $\mathbb{Z}$

Fig. 11 Khovanov TB-bound for the $(4, -5)$-torus knot

	0	1	2	3	4	5	6	7
13								**1**
11								$\mathbf{1_2}$
9						**1**	**1**	
7					**1**	$\mathbf{1_2}$		
5					$\mathbf{1, 1_2}$			
3			**1**	**1**				
1	**1**		$\mathbf{1_2}$					
-1	**1**	**1**						

(even) Khovanov homology
TB-bound equals -2

	0	1	2	3	4	5	6	7
12								**1**
10							**1**	
8						**1**		
6					**2**			
4				**1**				
2			$\mathbf{1}, \boxed{\mathbf{1_3}}$					
0		$\boxed{\mathbf{1_8}}$						
-2	$\boxed{\mathbf{1_3}}$							

odd reduced Khovanov homology
TB-bound equals -3

Fig. 12 Khovanov TB-bounds for the knot 12^n_{475}

provide the best upper bound on the TB-number among all currently known bounds for all prime knots with at most 15 crossings. In particular, the odd Khovanov TB-bound equals the Kauffman bound on all the 45 knots with at most 15 crossings where the latter is better than the even Khovanov TB-bound.

Example 4.6. The odd Khovanov TB-bound is better than the even Khovanov bound and equals the Kauffman bound for the knot 12^n_{475}, as shown in Fig. 12.

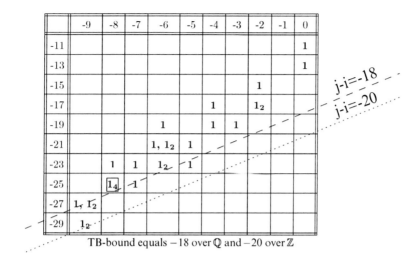

4.3 Finding Quasialternating Knots

Quasialternating links were introduced by Ozsváth and Szabó in [32] as a way to generalize the class of alternating links.

Definition 4.7. The class $\mathcal{Q}$ of quasialternating links is the smallest set of links such that

- the unknot belongs to $\mathcal{Q}$;
- if a link L has a planar diagram D such that the two resolutions of this diagram at one crossing represent two links, L_0 and L_1, with the properties that $L_0, L_1 \in \mathcal{Q}$ and $\det(L) = \det(L_0) + \det(L_1)$, then $L \in \mathcal{Q}$ as well.

Remark. It is well known that all nonsplit alternating links are quasialternating.

The main motivation for studying quasialternating links is the fact that the double branched covers of S^3 along such links are so-called L-spaces. A 3-manifold M is called an *L-space* if the order of its first homology group H_1 is finite and equals the rank of the Heegaard–Floer homology of M (see [32]). Unfortunately, due to the recursive style of Definition 4.7, it is often highly nontrivial to prove that a given link is quasialternating. It is equally challenging to show that it is not.

To determine that a link is not quasialternating, one usually employs the fact that such links have homologically thin Khovanov homology over $\mathbb{Z}$ and knot Floer homology over $\mathbb{Z}_2$ (see [27]). Thus, $\mathbb{Z}$H-thick knots are not quasialternating. There are 12 such knots with up to 10 crossings. Most of the others can be shown to be quasialternating by various constructions. After the work of Champanerkar and Kofman [7], there were only two knots left, 9_{46} and 10_{140}, for which it was not known whether they were quasialternating. Both of them have homologically thin Khovanov and knot Floer homologies.

As it turns out, odd Khovanov homology is much better at detecting quasialternating knots. The proof of the fact that such knots are $\mathbb{Z}$H-thin is based on the long exact sequences (17), and therefore, can be applied verbatim to the odd homology as well [29]. Computations show [42] that the knots 9_{46} and 10_{140} have homologically thick odd Khovanov homology and hence are not quasialternating; see Figs. 13 and 14.

Remark. It is worth mentioning that the knots 9_{46} and 10_{140} are $(3,3,-3)$- and $(3,4,-3)$-pretzel knots, respectively (see Fig. 15 for the definition). Computations show that $(n,n,-n)$- and $(n,n+1,-n)$-pretzel links for $n \leq 6$ all have torsion of order n outside of the main diagonal that supports the free part of the homology. This suggests a certain n-fold symmetry on the odd Khovanov chain complexes for these pretzel links that cannot be explained by the construction.

Remark. Joshua Greene has recently determined [11] all quasialternating pretzel links by considering 4-manifolds that are bounded by the branched double covers of the links. In particular, he found several knots that are not quasialternating, yet both H-thin and OH-thin. The smallest such knot is 11^n_{50}.

	-6	-5	-4	-3	-2	-1	0
0							**2**
-2						**1**	
-4					**1**		
-6				**2**			
-8			**1**				
-10		**1**					
-12	**1**						

(even) reduced Khovanov homology

	-6	-5	-4	-3	-2	-1	0
0							**2**
-2						**1**	$\boxed{\mathbf{1}_3}$
-4					**1**		
-6				**2**			
-8			**1**				
-10		**1**					
-12	**1**						

odd reduced Khovanov homology

Fig. 13 Khovanov homology of 9_{46}, the $(3,3,-3)$-pretzel knot

	-7	-6	-5	-4	-3	-2	-1	0
0								**1**
-2							**1**	
-4						**1**		
-6					**1**			
-8				**2**				
-10			**1**					
-12		**1**						
-14	**1**							

(even) reduced Khovanov homology

	-7	-6	-5	-4	-3	-2	-1	0
0								**1**
-2							**1**	
-4						**1**	$\boxed{\mathbf{1}_3}$	
-6					**1**			
-8				**2**				
-10			**1**					
-12		**1**						
-14	**1**							

odd reduced Khovanov homology

Fig. 14 Khovanov homology of 10_{140}, the $(3,4,-3)$-pretzel knot

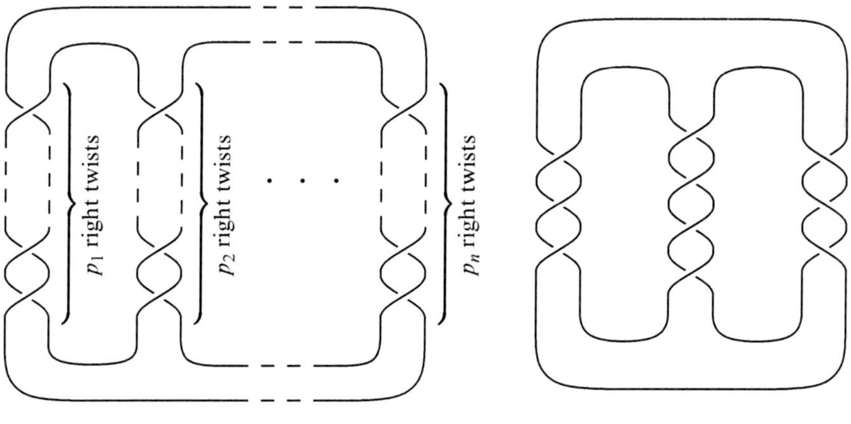

Fig. 15 $(p_1, p_2, \ldots, p_n)$-pretzel link and $(3,4,-3)$-pretzel knot 10_{140}

4.4 Detection of the Unknot

It was recently showed by Kronheimer and Mrowka [25] that Khovanov homology detects the unknot. More specifically, they proved that a knot is the unknot if and only if its reduced Khovanov homology has rank 1. This development is a major step toward proving a longstanding conjecture that the Jones polynomial itself detects the unknot.

Acknowledgments This paper was originally presented at the Marcus Wallenberg Symposium on Perspectives in Analysis, Geometry, and Topology at Stockholm University in May 2008. The author would like to thank all the organizers of the symposium for a very successful and productive meeting. He extends his special thanks to Ilia Itenberg, Burglind Jöricke, and Mikael Passare, the editors of these proceedings, for their patience with the author. The author is indebted to Mikhail Khovanov for much advice and enlightening discussions during the work on this paper. Finally, the author would like to express his deepest gratitude to Oleg Yanovich Viro for introducing him to the wonderful world of topology twenty years ago and for continuing to be his guide in this world ever since. The author is partially supported by NSF grant DMS–0707526.

References

1. M. Asaeda and M. Khovanov, *Notes on link homology*, in *Low dimensional topology*, 139–195, IAS/Park City Math. Ser., 15, Amer. Math. Soc., Providence, RI, 2009; arXiv:0804.1279.
2. D. Bar-Natan, *On Khovanov's categorification of the Jones polynomial*, Alg. Geom. Top., **2** (2002) 337–370; arXiv:math.QA/0201043.
3. D. Bar-Natan, *Fast Khovanov Homology Computations*, J. Knot Th. and Ramif. **16** (2007), no. 3, 243–255; arXiv:math.GT/0606318.
4. D. Bar-Natan and J. Green, `JavaKh` — a fast program for computing Khovanov homology, part of the `KnotTheory`' Mathematica Package, http://katlas.math.utoronto.ca/wiki/Khovanov_Homology
5. A. Beliakova and S. Wehrli, *Categorification of the colored Jones polynomial and Rasmussen invariant of links*, Canad. J. Math. **60** (2008), no. 6, 1240–1266; arXiv:math.GT/0510382.
6. J. Bloom, *Odd Khovanov homology is mutation invariant*, Math. Res. Lett. **17** (2010), no. 1, 1–10; arXiv:0903.3746.
7. A. Champanerkar and I. Kofman, *Twisting quasi-alternating links*, Proc. Amer. Math. Soc. **137** (2009), 2451–2458; arXiv:0712.2590.
8. N. Dunfield, S. Garoufalidis, A. Shumakovitch, and M. Thistlethwaite, *Behavior of knot invariants under genus 2 mutation*, New York J. Math. **16** (2010), 99–123; arXiv:math.GT/0607258.
9. M. Freedman, *A surgery sequence in dimension four; the relations with knot concordance*, Invent. Math. **68** (1982), no. 2, 195-226.
10. R. Gompf and A. Stipsicz, *4-manifolds and Kirby calculus*, Graduate Studies in Mathematics **20**, American Mathematical Society, Providence, RI, 1999.
11. J. Greene, *Homologically thin, non-quasi-alternating links*, Math. Res. Lett. **17** (2010), no. 1, 39–49; arXiv:0906.2222.
12. J. Hoste and M. Thistlethwaite, `Knotscape` — a program for studying knot theory and providing convenient access to tables of knots, http://www.math.utk.edu/~morwen/knotscape.html
13. V. Jones, *A polynomial invariant for knots via von Neumann algebras*, Bull. Amer. Math. Soc. **12** (1985), 103–111.

14. L. Kauffman, *State models and the Jones polynomial*, Topology **26** (1987), no. 3, 395–407.
15. M. Khovanov, *A categorification of the Jones polynomial*, Duke Math. J. **101** (2000), no. 3, 359–426; arXiv:math.QA/9908171.
16. M. Khovanov, *Patterns in knot cohomology I*, Experiment. Math. **12** (2003), no. 3, 365–374; arXiv:math.QA/0201306.
17. M. Khovanov, *Categorifications of the colored Jones polynomial*, J. Knot Th. Ramif. **14** (2005), no. 1, 111–130; arXiv:math.QA/0302060.
18. M. Khovanov, $\mathfrak{sl}(3)$ *link homology I*, Algebr. Geom. Topol. **4** (2004), 1045–1081; arXiv:math.QA/0304375.
19. M. Khovanov, *Link homology and Frobenius extensions*, Fundamenta Mathematicae, **190** (2006), 179–190; arXiv:math.QA/0411447.
20. M. Khovanov, *Link homology and categorification*, ICM–2006, Madrid, Vol. II, 989–999, Eur. Math. Soc., Zürich, 2006; arXiv:math/0605339.
21. M. Khovanov and L. Rozansky, *Matrix factorizations and link homology*, Fund. Math. **199** (2008), no. 1, 1–91; arXiv:math.QA/0401268.
22. M. Khovanov and L. Rozansky, *Matrix factorizations and link homology II*, Geom. Topol. **12** (2008), no. 3, 1387–1425; arXiv:math.QA/0505056.
23. M. Khovanov and L. Rozansky, *Virtual crossings, convolutions and a categorification of the* $SO(2N)$ *Kauffman polynomial*, J. Gökova Geom. Topol. GGT **1** (2007), 116–214; arXiv:math/0701333.
24. P. Kronheimer and T. Mrowka, *Gauge Theory for Embedded Surfaces I*, Topology **32** (1993), 773–826.
25. P. Kronheimer and T. Mrowka, *Khovanov homology is an unknot-detector*, Publ. Math. Inst. Hautes Études Sci. **113** (2011), 97–208; arXiv:1005.4346.
26. E. S. Lee, *An endomorphism of the Khovanov invariant*, Adv. Math. **197** (2005), no. 2, 554–586; arXiv:math.GT/0210213.
27. C. Manolescu, P. Ozsváth, *On the Khovanov and knot Floer homologies of quasi-alternating links* in Proc. of Gökova Geometry-Topology Conference 2007, 60–81, Gökova Geometry/Topology Conference (GGT), Gökova, 2008; arXiv:0708.3249.
28. L. Ng, *A Legendrian Thurston-Bennequin bound from Khovanov homology*, Algebr. Geom. Topol. **5** (2005) 1637–1653; arXiv:math.GT/0508649.
29. P. Ozsváth, J. Rasmussen, and Z. Szabó, *Odd Khovanov homology*, arXiv:0710.4300.
30. P. Ozsváth and Z. Szabó, *Holomorphic disks and topological invariants for closed three-manifolds*, Ann. of Math. **159** (2004) 1027–1158; arXiv:math.SG/0101206.
31. P. Ozsváth and Z. Szabó, *Holomorphic disks and knot invariants*, Adv. Math. **186** (2004), no. 1, 58–116; arXiv:math.GT/0209056.
32. P. Ozsváth and Z. Szabó, *On the Heegaard Floer homology of branched double-covers*, Adv. Math. **194** (2005), no. 1, 1–33; arXiv:math.GT/0309170.
33. O. Plamenevskaya, *Transverse knots and Khovanov homology*, Math. Res. Lett. **13** (2006), no. 4, 571–586; arXiv:math.GT/0412184.
34. J. Rasmussen, *Floer homology and knot complements*, Ph.D. Thesis, Harvard U.; arXiv:math.GT/0306378.
35. J. Rasmussen, *Khovanov homology and the slice genus*, Invent. Math. **182** (2010), no. 2, 419–447; arXiv:math.GT/0402131.
36. J. Rasmussen, *Knot polynomials and knot homologies*, Geometry and Topology of Manifolds (Boden et al eds.), Fields Institute Communications **47** (2005) 261–280, AMS; arXiv:math.GT/0504045.
37. D. Rolfsen, *Knots and Links*, Publish or Perish, Mathematics Lecture Series 7, Wilmington 1976.
38. L. Rudolph, *Quasipositivity as an obstruction to sliceness*, Bull Amer. Math. Soc. (N.S.) **29** (1993), no. 1, 51–59; arXiv:math.GT/9307233.
39. A. Shumakovitch, `KhoHo` — a program for computing and studying Khovanov homology, http://www.geometrie.ch/KhoHo

40. A. Shumakovitch, *Torsion of the Khovanov homology*, arXiv:math.GT/0405474; to appear in Fund. Math.
41. A. Shumakovitch, *Rasmussen invariant, slice–Bennequin inequality, and sliceness of knots*, J. Knot Th. and Ramif., **16** (2007), no. 10, 1403–1412; arXiv:math.GT/0411643.
42. A. Shumakovitch, *Patterns in odd Khovanov homology*, J. Knot Th. and Ramif. **20** (2011) no. 1, 203–222; arXiv:1101.5607.
43. P. Turner, *Calculating Bar-Natan's characteristic two Khovanov homology*, J. Knot Th. Ramif. **15** (2006), no. 10, 1335–1356; arXiv:math.GT/0411225.
44. O. Viro, *Khovanov homology, its definitions and ramifications*, Fund. Math. **184** (2004), 317–342; arXiv:math.GT/0202199.
45. S. Wehrli, *Khovanov Homology and Conway Mutation*, arXiv:math/0301312.
46. S. Wehrli, *Mutation invariance of Khovanov homology over* $\mathbb{F}_2$, Quantum Topol. **1** (2010), no. 2, 111–128; arXiv:0904.3401.
47. H. Wu, *A colored* $\mathfrak{sl}(N)$*-homology for links in* S^3, arXiv:0907.0695.
48. Y. Yonezawa, *Quantum* $(\mathfrak{sl}_n, \wedge V_n)$ *link invariant and matrix factorizations*, arXiv:0906.0220.

Tropical and Algebraic Curves with Multiple Points

Eugenii Shustin

To Oleg Yanovich Viro on the occasion of his 60th birthday

Abstract Patchworking theorems serve as a basic element of the correspondence between tropical and algebraic curves, which is a core area of tropical enumerative geometry. We present a new version of a patchworking theorem that relates plane tropical curves to complex and real algebraic curves having prescribed multiple points. It can be used to compute Welschinger invariants of nontoxic del Pezzo surfaces.

Keywords Patchworking • Tropical curve • Multiple point • Welschinger invariant • Del Pezzo surface

1 Introduction

The patchworking construction in the toric context originated in a method suggested by Viro in 1979–1980 for obtaining real algebraic hyperuricemia with a prescribed topology [19–21]. Later it was developed and applied to other problems, in particular to tropical geometry. Namely, it serves as in important step in the proof of a correspondence between tropical and algebraic curves, which in turn is central to enumerative applications of tropical geometry (see, e.g., Mikhalkin foundational work [9] and other versions and modifications in [10, 13, 15, 17]). We continue the latter line and present here a new patchworking theorem. The novelty of our version

E. Shustin (✉)
School of Mathematical Sciences, Raymond and Beverly Sackler Faculty of Exact Sciences, Tel Aviv University, Ramat Aviv, 69978 Tel Aviv, Israel
e-mail: shustin@post.tau.ac.il

I. Itenberg et al. (eds.), *Perspectives in Analysis, Geometry, and Topology: On the Occasion of the 60th Birthday of Oleg Viro*, Progress in Mathematics 296,
DOI 10.1007/978-0-8176-8277-4_18,

is that it allows one to patchwork algebraic curves with prescribed multiple points, whereas similar existing statements in tropical geometry apply only to nonsingular or nodal curves.[1]

The results cited are restricted to the case of curves in toric varieties (e.g., the plane blown up in at most three points). Since the consideration of curves on a blown-up surface is equivalent to the study of curves with fixed multiple points on the original surface, one can apply tropical enumerative geometry to count curves on the plane blown up at more than three points. This approach naturally leads to the following question: What are the plane tropical curves that correspond (as non-Archimedean amoebas or logarithmic limits) to algebraic curves with fixed generic multiple points on toric surfaces? The question appears to be more complicated than that resolved in [9, 13], and no general answer is known so far.

The goal of the present paper is to prove a patchworking theorem for a specific sort of plane tropical curves, i.e., we show that each tropical curve in the chosen class gives rise to an explicitly described set of algebraic curves on a given toric surface, in a given linear system, of a given genus, and with a given collection of fixed points with prescribed multiplicities (Theorem 2, Sect. 3). Furthermore, in the real situation, we compute the contribution of the constructed curves to the Welschinger invariant (Theorem 3, Sect. 3).

In fact, we do not know all the tropical curves that may give rise to the above-mentioned algebraic curves, and furthermore, we restrict our patchworking theorem to a statement that is sufficient to settle the following two problems:

- Prove recursive formulas of Caporaso–Harris type for the Welschinger invariants of $(\mathbb{P}^1)^2_{(0,2)}$, the quadric hyperboloid, blown up at two imaginary points, and for $\mathbb{P}^2_{(k,2l)}$, $k+2l\le 5$, $l\le 1$, the plane, blown up at k generic real points and at l pairs of conjugate imaginary points [6].
- Establish a new correspondence theorem between algebraic curves of a given genus in a given linear system on a toric surface and some tropical curves, and find new real tropical enumerative invariants of real toric surfaces [18].

We mention here an important consequence of the former result.

Theorem 1. ([6]) *Let Σ be one of the real del Pezzo surfaces $(\mathbb{P}^1)^2_{(0,2)}$, $\mathbb{P}^2_{(k,2l)}$, $k+2l\le 5$, $l\le 1$, and let $D\subset\Sigma$ be a real ample divisor. Then the Welschinger invariants $W_0(\Sigma,D)$ corresponding to the totally real configurations of points are positive, and they satisfy the asymptotic relation*

$$\lim_{n\to\infty}\frac{\log W_0(\Sigma,nD)}{n\log n}=\lim_{n\to\infty}\frac{\log GW_0(\Sigma,nD)}{n\log n}=-K_\Sigma D,$$

where $GW_0(\Sigma,D)$ are the genus-zero Gromov–Witten invariants.

[1] Rephrasing Selman Akbulut, who called Viro's disciples "little Viros," our contribution is a "little patchworking theorem" descended from "Viro's great patchworking theorem."

A similar statement for all the real toric del Pezzo surfaces except for $(\mathbb{P}^1)^2_{(0,2)}$ has been known previously [4,5].

Preliminary notation and definitions. If $P \subset \mathbb{R}^n$ is a pure-dimensional lattice polyhedral complex, $\dim P = d \le n$, by $|P|$ we denote the lattice volume of P, counted so that the lattice volume of a d-dimensional lattice polytope $\Delta \subset \mathbb{R}^n$ is the ratio of the Euclidean volume of Δ and the minimal Euclidean volume of a d-dimensional lattice simplex in the linear d-subspace of $\mathbb{R}^n$ parallel to the affine d-space spanned by Δ. In particular, $|P| = \#P$ if P is finite.

Given a lattice polyhedron Δ, by $\mathrm{Tor}_K(\Delta)$ we denote[2] the toric variety over a field K associated with Δ, and by $\mathcal{L}_\Delta$ we denote the tautological line bundle (i.e., the bundle generated by the monomials z^ω, $\omega \in \Delta$, as global sections). The divisors $\mathrm{Tor}_K(\sigma) \subset \mathrm{Tor}_K(\Delta)$ corresponding to the facets (faces of codimension 1) σ of Δ, we call *toric divisors*. By $\mathrm{Tor}_K(\partial\Delta)$ we denote the union of all the toric divisors in $\mathrm{Tor}_K(\Delta)$.

The main field we use is $\mathbb{K} = \bigcup_{m\ge 1} \mathbb{C}((t^{1/m}))$, the field of locally convergent complex Puiseux series possessing a non-Archimedean valuation

$$\mathrm{Val} : \mathbb{K}^* \to \mathbb{R}, \quad \mathrm{Val}\left(\sum_r a_r t^r\right) = -\min\{r \ : \ a_r \ne 0\}.$$

Define

$$\mathrm{ini}\left(\sum_r a_r t^r\right) = a_v, \quad \text{where} \quad v = -\mathrm{Val}\left(\sum_r a_r t^r\right).$$

The field $\mathbb{K}$ is algebraically closed and contains a closed real subfield $\mathbb{K}_\mathbb{R} = \mathrm{Fix}(\mathrm{Conj})$, $\mathrm{Conj}(\sum_r a_r t^r) = \sum_r \overline{a}_r t^r$.

We recall here the definition of Welschinger invariants [23], restricting ourselves to a particular situation. Let Σ be a real *unnodal* (i.e., without $(-n)$-curves, $n \ge 2$) del Pezzo surface with a connected real part $\mathbb{R}\Sigma$, and let $D \subset \Sigma$ be a real ample divisor. Consider a generic configuration ω of $c_1(\Sigma)\cdot D - 1$ distinct real points of Σ. The set $R(D,\omega)$ of real (i.e., complex-conjugation-invariant) rational curves $C \in |D|$ passing through the points of ω is finite, and all these curves are nodal and irreducible. Put

$$W(\Sigma, D, \omega) = \sum_{C \in R(D,\omega)} (-1)^{s(C)},$$

where $s(C)$ is the number of solitary nodes of C (i.e., real points where a local equation of the curve can be written over $\mathbb{R}$ in the form $x^2 + y^2 = 0$). By Welschinger's theorem [23], the number $W(\Sigma, D, \omega)$ does not depend on the choice of a generic configuration ω, and hence we simply write $W(\Sigma, D)$, omitting the configuration in the notation of this Welschinger invariant.

---

[2]We omit the subscript in the complex case, writing simply $\mathrm{Tor}(\Delta)$.

In what follows, we shall use a generalized definition of the Welschinger sign of a curve. Namely, let C be a real algebraic curve on a smooth real algebraic surface Σ, and let $\overline{p} \subset \Sigma$ be a conjugation-invariant finite subset. Assume that C has no singular local branches (i.e., is an immersed curve). Then we define the Welschinger sign

$$W_{\Sigma,\overline{p}}(C) = (-1)^{s(C,\Sigma,\overline{p})}, \quad \text{where} \quad s(C,\Sigma,\overline{p}) = \sum_{z \in \mathrm{Sing}(C')} s(C',z), \qquad (1)$$

C' being the strict transform of C under the blowup of Σ at $\overline{p}$, and $s(C',z)$ is the number of solitary nodes in a local δ-const deformation of the singular point z of C' into $\delta(C',z)$ nodes, where δ denotes the δ-invariant of singularity (i.e., the maximal possible number of nodes in its deformation). It is evident that $s(C',z)$ is correctly defined modulo 2, and hence $W_{\Sigma,\overline{p}}(C)$ is well defined.

Organization of the material. In Sect. 2, we set forth the geometry of plane tropical curves adapted to our purposes, finishing with the definition of weights of tropical curves that in the complex case, designate the number of algebraic curves associated with the given tropical curves in the further patchworking theorem, and in the real case designate the contribution of the real algebraic curves in the associated set to the Welschinger number. In Sect. 3, we provide two patchworking theorems, the complex theorem and the real theorem, in which we explicitly construct algebraic curves associated with the tropical curves under consideration.

2 Parameterized Plane Tropical Curves

For the reader's convenience, we recall here some basic definitions and facts about tropical curves that we shall use in the sequel. The details can be found in [8, 9, 12].

2.1 Definition

An *abstract tropical curve* is a compact graph $\overline{\Gamma}$ without divalent vertices and isolated points such that $\Gamma = \overline{\Gamma} \setminus \Gamma^0_\infty$, where Γ^0_∞ is the set of univalent vertices, is a metric graph whose closed edges are isometric to closed segments in $\mathbb{R}$, and nonclosed edges Γ – *ends* are isometric to rays in $\mathbb{R}$ or to $\mathbb{R}$ itself. Denote by $\overline{\Gamma}^0$, respectively Γ^0, the set of vertices of $\overline{\Gamma}$, respectively Γ, and split the set $\overline{\Gamma}^1$ of edges of $\overline{\Gamma}$ into Γ^1_∞, the set of Γ-ends, and Γ^1, the set of closed (finite-length) edges of Γ. The *genus* of Γ is $g = b_1(\Gamma) - b_0(\Gamma) + 1$.

A *plane parameterized tropical curve* (*PPT-curve* for short) is a pair $(\overline{\Gamma}, h)$, where $\overline{\Gamma}$ is an abstract tropical curve and $h : \Gamma \to \mathbb{R}^2$ is a continuous map whose restriction to any edge of Γ is a nonzero $\mathbb{Z}$-affine map and that satisfies the following *balancing* and *nondegeneracy* conditions at any vertex v of Γ: For each $v \in \Gamma^0$,

$$\sum_{v\in e,\ e\in\overline{\Gamma}^1} dh_v(\tau_v(e)) = 0, \tag{2}$$

and

$$\mathrm{Span}\left\{dh_v(\tau_v(e)),\quad v\in e,\ e\in\overline{\Gamma}^1\right\} = \mathbb{R}^2,$$

where $\tau_v(e)$ is the unit tangent vector to an edge e at the vertex v. The *degree* of a PPT-curve $(\overline{\Gamma},h)$ is the unordered multiset of vectors $\{dh(\tau(e))\ :\ e\in\Gamma^1_\infty\}$, where $\tau(e)$ denotes the unit tangent vector of a Γ-end e pointing to the univalent vertex.

Observe that

$$\sum_{e\in\Gamma^1_\infty} dh(\tau(e)) = 0, \tag{3}$$

which follows immediately from the balancing condition (2). We shall also use another form of the Γ-end-balancing condition. For each Γ-end e pick any point $\boldsymbol{x}_e\in h(e\setminus\Gamma^0_\infty)$. Then

$$\sum_{e\in\Gamma^1_\infty}\langle R_{\pi/2}(dh(\tau(e))),\boldsymbol{x}_e\rangle = 0, \tag{4}$$

where $R_{\pi/2}$ is the (positive) rotation by $\pi/2$. This is an elementary consequence of the material discussed in the next section: one can lift a PPT-curve to a plane algebraic curve over a non-Archimedean field, consider the defining polynomial, and then use the fact that the product of the roots of the (quasihomogeneous) truncations of this polynomial on the sides of its Newton polygon is 1. We leave the details to the reader.

Since $dh_v((\tau_v(e))\in\mathbb{Z}^2$, we have a well-defined positive weight function $w:\overline{\Gamma}^1\to\mathbb{Z}$ in the relation $dh_v(\tau_v(e)) = w(e)\boldsymbol{u}_v(e)$ with $\boldsymbol{u}_v(e)$ the primitive integral tangent vector to $h(e)$, emanating from $h(v)$. In the sequel, when modifying tropical curves, we speak of changes of edge weights, which in terms of h and $\overline{\Gamma}$ means that h remains unchanged, whereas the metric on the chosen edges is multiplied by a constant.

Observe that a connected component of $\overline{\Gamma}\setminus F$, where F is finite, naturally induces a new PPT-curve (further on referred to as *induced*) when one is making the metric on the nonclosed edges of that component complete and respectively correcting the map h on these edges. These induced curves and the unions of a few of them, coming from the same $\overline{\Gamma}\setminus F$, are called PPT-curves *subordinate* to $(\overline{\Gamma},h)$.

The *deformation space* $\mathcal{M}(\overline{\Gamma},h)$ of a PPT-curve $(\overline{\Gamma},h)$ is obtained by variation of the length of the finite edges of Γ and combining h with shifts. It can be identified with an open rational convex polyhedron in Euclidean space, and its closure $\overline{\mathcal{M}}(\overline{\Gamma},h)$ can be obtained by adding the boundary of that polyhedron that corresponds to PPT-curves with some edges $e\in\Gamma^1$ contracted to points.

Deformation-equivalent PPT-curves are often said to be of the same *combinatorial type*. The degree and the genus are invariants of the combinatorial type as well as the following characteristics. We call a PPT-curve $(\overline{\Gamma},h)$

- *Irreducible* if Γ is connected,
- *Simple* if Γ is trivalent and

- *Pseudosimple* if for any vertex $v \in \Gamma^0$ incident to $m > 3$ edges $e_1, e_2, \dots, e_m$, one has $\boldsymbol{u}_v(e_1) \neq \boldsymbol{u}_v(e_j)$, $1 < j \leq m$, and only two distinct vectors among $\boldsymbol{u}_v(e_2), \dots, \boldsymbol{u}_v(e_m)$.

In the latter case, an edge e_i emanating from a vertex $v \in \Gamma^0$ of valency $m > 3$ is called *simple* if $\boldsymbol{u}_v(e_i) \neq \boldsymbol{u}_v(e_j)$ for all $j \neq i$, and is called *multiple* otherwise.

2.2 Newton Polygon and Its Subdivision Dual to a Plane Tropical Curve

Given a PPT-curve $Q = (\Gamma, h)$, the image $T = h(\Gamma) \subset \mathbb{R}^2$ is a finite planar graph that supports an embedded plane tropical curve (*EPT*-curve for short) $h_*Q := (T, h_*w)$ with the (edge) weight function

$$h_*w : T^1 \to \mathbb{Z}, \quad h_*w(E) = \sum_{e \in \overline{\Gamma}^1,\ h(e) \supset E} w(e).$$

The respective balancing condition immediately follows from (2). Furthermore, there exist a convex lattice polygon $\Delta \subset \mathbb{R}^2$ (different from a point) and a convex piecewise linear function

$$f_T : \mathbb{R}^2 \to \mathbb{R}, \quad f(\boldsymbol{x}) = \max_{\omega \in \Delta \cap \mathbb{Z}^2} (\langle \omega, \boldsymbol{x} \rangle + c_\omega), \quad \boldsymbol{x} \in \mathbb{R}^2, \tag{5}$$

such that

- T is the corner locus of f_T
- For any two linearity domains D_1, D_2 of f_T corresponding to linear functions in formula (5) with gradients ω_1, ω_2, respectively, and having a common edge $E = D_1 \cap D_2$ of T, one has $\omega_2 - \omega_1 = h_*w(E) \cdot \boldsymbol{u}(E)$, where $\boldsymbol{u}(E)$ is the primitive integral vector orthogonal to E and directed from D_1 to D_2.

Here the polygon Δ, called the *Newton polygon* of Q, is defined uniquely up to a shift in $\mathbb{R}^2$, and f_T is defined uniquely up to addition of a linear affine function.

The Legendre function $\nu_T : \Delta \to \mathbb{R}$ dual to f_T is convex piecewise linear, and its linearity domains define a subdivision S_T of Δ into convex lattice subpolygons. This subdivision S_T is dual to the pair $(\mathbb{R}^2, T)$ in the following way: there is a 1-to-1 correspondence between the faces of subdivision of $\mathbb{R}^2$ determined by T and the faces of subdivision S_T such that (i) the sum of the dimensions of dual faces is 2, (ii) the correspondence inverts the incidence relation, (iii) the dual edges of T and S_T are orthogonal, and the weight of an edge of T equals the lattice length of the dual edge of S_T. In particular, if $V = (\alpha, \beta)$ is a vertex of T, then $\nabla \nu_T = (-\alpha, -\beta)$ along the dual polygon Δ_V of the subdivision S_T.

Furthermore, we can obtain extra information on the subdivision S_T out of the original PPT-curve Q. Namely,

- With each edge $e \in \overline{\Gamma}^1$ we associate a lattice segment σ_e that is orthogonal to $h(e)$ and satisfies $|\sigma_e| = w(e)$.
- With each vertex $v \in \Gamma^0$ we associate a convex lattice polygon Δ_v whose sides are suitable translates of the segments σ_e, $e \in \overline{\Gamma}^1$, $v \in e$. Denote by $\sigma_{v,e}$ the side of Δ_v that is a translate of σ_e and whose outward normal is $dh_v(\tau_v(e))$.

Let a polygon Δ_V of the subdivision S_T be dual to a vertex V of T. Then (up to a shift)

$$\Delta_V = \sum_{\substack{e \in \overline{\Gamma}^1 \\ \mathrm{Int}(e) \cap h^{-1}(V) \neq \emptyset}} \sigma_e + \sum_{\substack{v \in \Gamma^0 \\ h(v) = V}} \Delta_v. \tag{6}$$

In this connection, we can speak of $\nabla \nu_T$ along the polygons Δ_v appearing in (6).

A EPT curve T is called *nodal* if the dual subdivision S_T consists of triangles and parallelograms, i.e., if the nontrivalent vertices of T are locally intersections of two straight lines. A nodal EPT curve canonically lifts to a simple PPT curve when one resolves all nodes of the given curve.

2.3 Compactified Tropical Curves

For a given convex lattice polygon Δ different from a point, we define a compactification $\mathbb{R}^2_\Delta$ of $\mathbb{R}^2$ in the following way. If $\dim\Delta = 2$, we identify $\mathbb{R}^2$ with the positive orthant $(\mathbb{R}_{>0})^2$ by coordinatewise exponentiation, then identify $(\mathbb{R}_{>0})^2$ with the interior of $\mathbb{R}^2_\Delta := \mathrm{Tor}_\mathbb{R}(\Delta)_+ \simeq \Delta$, the nonnegative part of the real toric variety $\mathrm{Tor}_\mathbb{R}(\Delta)$, via the moment map

$$\mu(x) = \frac{\sum_{\omega \in \Delta \cap \mathbb{Z}^2} x^\omega \omega}{\sum_{\omega \in \Delta \cap \mathbb{Z}^2} x^\omega}, \quad x \in (\mathbb{R}_{>0})^2.$$

If Δ is a segment, then we take $\Delta' = \Delta \times \sigma$, σ being a transverse lattice segment, and define $\mathbb{R}^2_\Delta$ as the quotient of $\mathbb{R}^2_{\Delta'}$ by contracting the sides parallel to σ. We observe that the rays in $\mathbb{R}^2$ directed by an external normal $\boldsymbol{u}$ to a side σ of Δ and emanating from distinct points on a line transverse to σ close up at distinct points on the part of $\partial(\mathbb{R}^2_\Delta)$ corresponding to the interior of σ in the above construction.[3]

So we can naturally compactify a PPT-curve (Γ, h) into $(\overline{\Gamma}, \overline{h})$ by extending h up to a map $\overline{h} : \overline{\Gamma} \to \mathbb{R}^2_\Delta$.

[3]Clearly, the rays directed by vectors distinct from any exterior normal to sides on Δ close up at respective vertices of $\mathbb{R}^2_\Delta$.

2.4 Marked Tropical Curves

An abstract tropical curve with n marked points is a pair $(\overline{\Gamma}, G)$, where $\overline{\Gamma}$ is an abstract tropical curve and $G = (\gamma_1, \ldots, \gamma_n)$ is an ordered n-tuple of distinct points of $\overline{\Gamma}$. We say that a marked tropical curve $(\overline{\Gamma}, G)$ is *regular* if each connected component of $\overline{\Gamma} \setminus G$ is a tree containing precisely one vertex from Γ^0_∞. Furthermore, a marked tropical curve $(\overline{\Gamma}, G)$ is called

- *End-marked*, if $G \cap \Gamma^0 = \emptyset$ and the points of G lie on the ends of $\overline{\Gamma}$, one on each end.
- *Regularly end-marked* if $G \cap \Gamma^0 = \emptyset$, the points of G lie on the ends of $\overline{\Gamma}$, and $(\overline{\Gamma}, G)$ is regular.

A parameterization of a (compact) plane tropical curve with marked points is a triple $(\overline{\Gamma}, G, \overline{h})$, where $(\overline{\Gamma}, G)$ is a marked abstract tropical curve, and $(\overline{\Gamma}, \overline{h})$ is a PPT-curve. We define the deformation space $\mathcal{M}(\overline{\Gamma}, G, \overline{h})$ by fixing the combinatorial type of the pair $(\overline{\Gamma}, G)$ (G being an ordered sequence). It can be identified with a convex polyhedron in $\mathbb{R}^N$, where the coordinates designate the two coordinates of the image $\overline{h}(v)$ of a fixed vertex $v \in \Gamma^0$, the lengths of the edges $e \in \Gamma^1$, and the distances between the marked points lying inside edges of Γ to some fixed points inside these edges (chosen one on each edge); cf. [1]. Further on, the deformation type of a marked PPT-curve will be called a *combinatorial* type.

Lemma 1. *Let Δ be a convex lattice polygon, $X = (\boldsymbol{x}_1, \ldots, \boldsymbol{x}_n)$ a sequence of points in $\mathbb{R}^2_\Delta$ (not necessarily distinct). Then there exists at most one n-marked regular PPT-curve $(\overline{\Gamma}, G, \overline{h})$ with the Newton polygon Δ and with a fixed combinatorial type such that $\overline{h}(\gamma_i) = \boldsymbol{x}_i$, $\gamma_i \in G$, $i = 1, \ldots, n$.*

Proof. If such a marked PPT-curve exists, it is sufficient to uniquely restore each connected component of $\overline{\Gamma} \setminus G$, and hence the general situation reduces to the case of an irreducible rational PPT-curve (a subordinate curve defined by such a connected component) with $|\Gamma^0_\infty| - 1 = |\Gamma^1_\infty| - 1$ marked univalent vertices. We proceed by induction on $|\overline{\Gamma}^1|$. The base of induction, i.e., the case $|\overline{\Gamma}^1| = 1$, is evident. Assume that $|\overline{\Gamma}^1| > 1$.

If there are two Γ-ends e_1, e_2 with marked points that emanate from one vertex $v \in \Gamma^0$ and are mapped into the same straight line by $\overline{h}$, then either $\overline{h}(e_1) = \overline{h}(e_2)$, in which case we replace e_1, e_2 by one end of weight $w(e_1) + w(e_2)$ and respectively replace two marked points by one, thus reducing $|\overline{\Gamma}^1|$ by 1 and keeping the irreducibility and the rationality of the tropical curve, or $\overline{h}(e_1)$ and $\overline{h}(e_2)$ are the opposite rays emanating from $\overline{h}(v)$, in which case we remove the Γ-end with lesser weight, leaving the other with weight $|w(e_1) - w(e_2)|$, thus reducing $|\overline{\Gamma}^1|$ by 1 or 2.

If there are no Γ-ends as above, from

$$|\Gamma^0| - |\Gamma^1| = 1 \quad \text{and} \quad 3 \cdot |\Gamma^0| \leq 2 \cdot |\Gamma^1| + |\Gamma^1_\infty|$$

we deduce that $|\Gamma^0| \le |\Gamma^1_\infty| - 2$. Hence there are two nonparallel Γ-ends with marked points that merge to a common vertex $v \in \Gamma^0$, which thereby is determined uniquely. So we remove the above Γ-ends and the vertex v from $\overline{\Gamma}$, then extend the other edges of $\overline{\Gamma}$ coming to v up to new ends and mark on them the points mapped to $h(v)$. Thus, the induction assumption completes the proof. □

2.5 Tropically Generic Configurations of Points

Let Δ be a convex lattice polygon, $x = (\boldsymbol{x}_1,\dots,\boldsymbol{x}_k)$ a sequence of distinct points in $\mathbb{R}^2_\Delta$ such that $\boldsymbol{x}_i \in \sigma_i$, $1 \le i \le r$, where $\sigma_1,\dots,\sigma_r \subset \mathbb{R}^2_\Delta$ correspond to certain sides of Δ, and $\boldsymbol{x}_i \in \mathbb{R}^2 \subset \mathbb{R}^2_\Delta$, $r < i \le k$. Let $\overline{m} = (m_1,\dots,m_k)$ be a sequence of nonnegative integers, called weights of the points $\boldsymbol{x}_1,\dots,\boldsymbol{x}_k$, respectively. A subconfiguration of $(x,\overline{m})$ is a configuration $(x,\overline{m}')$ with $\overline{m}' \le \overline{m}$ (componentwise).

Let $\mathcal{C}$ be a combinatorial type of an irreducible end-marked PPT-curve with Newton polygon Δ, with $m = m_1 + \cdots + m_k$ Γ-ends and marked points $\gamma_1,\dots,\gamma_m$. A weighted configuration $(\overline{\boldsymbol{x}},\overline{m})$ is called $\mathcal{C}$*-generic* if there is no end-marked irreducible PPT-curve $(\overline{\Gamma},G,\overline{h})$ of type $\mathcal{C}$ such that $\overline{h}(G) = (\overline{\boldsymbol{x}},\overline{m})$, i.e.,

$$\overline{h}(\gamma_i) = \boldsymbol{x}_i,\ \sum_{j<i} m_j < i \le \sum_{j\le i} m_j, \quad i = 1,\dots,k.$$

A weighted configuration $(\overline{\boldsymbol{x}},\overline{m})$ is called Δ*-generic* if it together with all its subconfigurations is generic with respect to the combinatorial types of end-marked irreducible PPT-curves that have $m \le |\partial\Delta \cap \mathbb{Z}^2|$ Γ-ends and directing vectors of all edges orthogonal to integral segments in Δ. A (nonweighted) configuration $\overline{\boldsymbol{x}}$ is called Δ*-generic* if all possible weighted configurations $(\overline{\boldsymbol{x}},\overline{m})$ are Δ-generic.

Lemma 2. *The Δ-generic configurations with rational coordinates $\overline{\boldsymbol{x}} = (\boldsymbol{x}_1,\dots,\boldsymbol{x}_k) \subset \mathbb{R}^2_\Delta$ such that $\boldsymbol{x}_i \in \sigma_i$, $1 \le i \le r$, $\boldsymbol{x}_i \in \mathbb{R}^2$, $r < i \le k$, form a dense subset of $\sigma_1 \times \cdots \times \sigma_r \times (\mathbb{R}^2)^{k-r}$.*

Proof. Notice that there are only finitely many (up to the choice of edge weights) combinatorial types of end-marked irreducible PPT-curves under consideration and only finitely many weight collections $\overline{m}$ to consider. We shall prove that for any such combinatorial type $\mathcal{C}$ of end-marked irreducible PPT-curves, the image of the natural evaluation map $\mathrm{Ev} : \mathcal{M}(\mathcal{C}) \to \sigma_1 \times \cdots \times \sigma_r \times (\mathbb{R}^2)^{k-r}$ is nowhere dense, and hence is a finite polyhedral complex of positive codimension. This would suffice for the proof of the lemma due to the aforementioned finiteness.

Thus, assuming that an end-marked irreducible curve $(\overline{\Gamma},G,\overline{h})$ of type $\mathcal{C}$ matches a weighted rational configuration $(\overline{\boldsymbol{x}},\overline{m})$, we shall show that this imposes a nontrivial relation on the coordinates of the points of $\overline{\boldsymbol{x}}$ and hence complete the proof. This is, in fact, a direct consequence of (4). We explain this in detail, however, since formally one can assume that all the points of $\overline{\boldsymbol{x}}$ are multiple, and for each point $\boldsymbol{x}_i \in \overline{\boldsymbol{x}}$, the corresponding scalar products in (4) sum to zero.

Since any point $\boldsymbol{x}_i \in \overline{\boldsymbol{x}}$ lying on $\partial\mathbb{R}^2_\Delta$ is a univalent vertex for some ends of $\overline{\Gamma}$ whose $\overline{h}$-images lie on the same straight line, by pushing all the points of $\overline{\boldsymbol{x}} \cap \partial\mathbb{R}^2_\Delta$ along the corresponding lines, we can make $\overline{\boldsymbol{x}} \subset \mathbb{R}^2$. Take an irrational vector $a \in \mathbb{R}^2$ and choose the point $\boldsymbol{x}_i \in \overline{\boldsymbol{x}}$ with the maximal value of the functional $\langle a, \boldsymbol{x}\rangle$. Notice that there is no vertex $v \in \Gamma^0$ with $\langle a, h(v)\rangle \geq \langle a, \boldsymbol{x}_i\rangle$, since otherwise, due to the balancing condition (2), one would find an end e of Γ with $h(e)$ lying entirely in the half-plane $\langle a, \boldsymbol{x}\rangle > \langle a, \boldsymbol{x}_i\rangle$, contrary to our assumptions. Hence, for each end $e \in \Gamma^1_\infty$ with $h(e)$ passing through $\boldsymbol{x}_i$, we have $\langle a, \tau(e)\rangle > 0$, which yields that

$$\sum_{\substack{e \in \Gamma^1_\infty \\ \boldsymbol{x}_i \in h(e)}} m_e \cdot \mathrm{d}h(\tau(e)) \neq 0$$

for any positive integers m_e, and which finally implies that (some of) the coordinates of $\boldsymbol{x}_i$ enter relation (4) with nonzero coefficients. □

Lemma 3. *Let $\overline{\boldsymbol{x}}$ be a Δ-generic configuration of points, $Q = (\overline{\Gamma}, G, \overline{h})$ a marked regular PPT-curve with Newton polygon Δ that matches $\overline{\boldsymbol{x}}$. Then*

(i) $(\overline{h})^{-1}(\overline{\boldsymbol{x}}) = G$;
(ii) If K is a connected component of $\overline{\Gamma} \setminus G$, then its edges can be oriented in such a way that

  (a) The edges merging to marked points emanate from these points.
  (b) The unmarked Γ-end is oriented toward its univalent endpoint.
  (c) From any vertex $v \in K^0$ there emanates precisely one edge, and this edge is simple.

Remark 4. It follows from Lemma 3 that if an edge of Γ is multiple for both of its endpoints and contains a marked point inside that matches a point $\boldsymbol{x} \in \overline{\boldsymbol{x}}$, then all the other edges joining the same vertices contain marked points matching $\boldsymbol{x}$.

Proof of Lemma 3. (i) Assume that there is a point $\gamma \in (\overline{h})^{-1}(\overline{\boldsymbol{x}}) \setminus G$. It belongs to a component K of $\overline{\Gamma} \setminus G$, which is a tree due to the regularity of the considered marked tropical curve, and hence is cut by γ into two trees K_1, K_2, and only one of them, say K_1, contains a Γ-end free of marked points. Then marking the new point γ, we obtain that the irreducible (rational) PPT-curve induced by K_2 is end-marked and matches a subconfiguration of $\overline{\boldsymbol{x}}$, contrary to its Δ-genericity.

(ii) Observe that the image of the unmarked ray does not coincide with the image of any other edge of K, which follows immediately from the statement (i).

Next we notice that if p is a vertex of K, e a multiple edge merging to p, then the connected component $K(e)$ of $K \setminus \{p\}$, starting with the edge e, does not contain the unmarked K-end. Indeed, otherwise, we consider another edge e' of K merging to p so that $\boldsymbol{u}_p(e') = \boldsymbol{u}_p(e)$. Then we take the graph $K \setminus K(e)$, multiply the weights of the edges of the component $K(e')$ of $K \setminus \{p\}$ starting with e' by $w(e) + w(e')$, and multiply the weights of the edges in $K \setminus (K(e) \cup K(e'))$ by $w(e')$, where $w(e), w(e')$

are the weights of e and e' in K, in order to preserve the balancing condition at the vertex p, and thereby the newly weighted $K \setminus K(e)$ induces an end-marked (rational) PPT-curve matching the Δ-generic configuration $\overline{x}$, a contradiction.

It follows from the latter observation that K has no edge that is multiple for both of its endpoints. Indeed, otherwise we would have two vertices $v_1, v_2 \in K^0$, joined by an edge $e \in K^1$, multiple for both v_1 and v_2, and then would obtain that the unmarked K-end is contained either in the component of $K \setminus \{v_1\}$ starting with e, or in the component of $K \setminus \{v_2\}$ starting with e, contrary to the above conclusion.

Finally, we define an orientation of the edges of K, opposite to the required one. Start with the unmarked K-end and orient it toward its multivalent endpoint. In any other step, in arriving at a vertex $v \in K^0$ along some edge, we orient all other edges merging to v outward. Since K is a tree, the orientation smoothly extends to all of its edges. The preceding observations confirm that any edge e oriented in this manner toward a vertex $v \in E$ is simple for v. □

2.6 *Weights of Marked Pseudosimple Regular PPT-Curves*

In this section, $Q = (\overline{\Gamma}, G, \overline{h})$ is always a regular marked pseudosimple PPT-curve. Set $G_\infty = G \cap \Gamma^0_\infty$ and $G_0 = G \setminus G_\infty$, and put $\overline{x} = \overline{h}(G), \overline{x}^\infty = \overline{h}(G_\infty)$. Throughout this section we assume that

(T1) No edge of Γ is multiple for two vertices of Γ.
(T2) G_0 does not contain vertices of valency greater than 3.
(T3) $\overline{x}$ is Δ-generic.

In particular, by Lemma 3, we have that $(\overline{h})^{-1}(\overline{x}) = G$.

Complex weights. We define the *complex weight* of a PTT-curve $Q = (\overline{\Gamma}, G, \overline{h})$ as

$$M(Q) = \prod_{v \in \Gamma^0} M(Q,v) \cdot \prod_{e \in \Gamma^1} M(Q,e) \cdot \prod_{\gamma \in G} M(Q,\gamma), \tag{7}$$

where the values $M(Q,v), M(Q,e), M(Q,\gamma)$ are computed according to the following rules:

(M1) $M(Q,e) = w(e)$ for each edge $e \in \Gamma^1$.
(M2) $M(Q,\gamma) = 1$ for each $\gamma \in G \cap (\Gamma^0 \cup \Gamma^0_\infty)$, and $M(Q,\gamma) = w(e)$ for each $\gamma \in G \setminus (\Gamma^0 \cup \Gamma^0_\infty)$, $\gamma \in e \in \overline{\Gamma}^1$.

To define $M(Q,v)$, $v \in \Gamma^0$, we introduce some notation: We denote by Δ_v the lattice triangle whose boundary is composed of the vectors $dh(\tau_v(e))$, rotated clockwise by $\pi/2$, where e runs over all the edges of $\overline{\Gamma}$ emanating from v. Next, we put the following:

(M3) If $v \in \Gamma^0 \cap G$, then $M(Q,v) = |\Delta_v|$.

(M4) If $v \in \Gamma^0 \setminus G$ is trivalent, then it belongs to a connected component K of $\overline{\Gamma} \setminus G$, which we orient as in Lemma 3(ii) and thus define two edges $e_1, e_2 \in \overline{\Gamma}^1$ merging to v. In this case we put $M(Q,v) = |\Delta_v|(w(e_1)w(e_2))^{-1}$.

(M5) Let $v \in \Gamma^0$ be of valency $s+r+1 > 3$, where $1 \le r \le s$, $2 \le s$, and let e_i, $i = 1,\dots,s+r+1$, be all the edges with endpoint v, so that the edges e_i, $1 \le i \le s$, have a common directing vector $\boldsymbol{u}_v(e_1)$, the edges e_i, $s < i \le s+r$, have a common directing vector $\boldsymbol{u}_v(e_{s+1})$, and e_{s+r+1} is a simple edge emanating from v along the orientation of Lemma 3(ii). Consider a rational PPT-curve Q_v induced by the graph $\overline{\Gamma}_v = \{v\} \cup \bigcup_{i=1}^{s++r+1} e_i \subset \overline{\Gamma}$, pick auxiliary marked points $\gamma_i \in e_i \setminus \{v\}, i = 1,\dots,s+r$, in such a way that $\overline{h}(\gamma_i) = \boldsymbol{y}' \in \mathbb{R}^2$ as $1 \le i \le s$, and $\overline{h}(\gamma_i) = \boldsymbol{y}'' \in \mathbb{R}^2$ as $s < i \le s+r$. Then we replace $\boldsymbol{y}'$ (respectively $\boldsymbol{y}''$) by a generic set of distinct points $\boldsymbol{y}_1,\dots,\boldsymbol{y}_s$ close to $\boldsymbol{y}'$ (respectively distinct points $\boldsymbol{y}_{s+1},\dots,\boldsymbol{y}_{s+r}$ close to $\boldsymbol{y}''$), and take rational regularly end-marked PPT-curves of degree $\{dh(\tau_v(e_i))\}_{i=1,\dots,s+r+1}$ matching the configuration $\boldsymbol{y}_1,\dots,\boldsymbol{y}_{s+r}$, so that the $\overline{h}$-image of the Γ-end of weight $w(e_i)$ with the directing vector $\boldsymbol{u}_v(e_i)$ passes through the point $\boldsymbol{y}_i$, $i = 1,\dots,s+r$ (see Fig. 1).[4] By [9, Corollaries 2.24 and 4.12], the set $\mathcal{T}$ of these PPT-curves is finite, and they all are simple. Then put

$$M(Q,v) = \sum_{Q' \in \mathcal{T}} M(Q') \,, \tag{8}$$

where all terms $M(Q')$ are computed by formula (7) and the rules (M1)–(M4).

Remark 5. (1) We point out that the right-hand side of (8) does not depend on the choice of the configuration $(\boldsymbol{y}_i)_{i=1,\dots,s+r+1}$, which follows from [1, Theorem 4.8] (observe that the degree of the evaluation map as in [1, Definition 4.6] coincides with the right-hand side of (8) in our situation). Slightly modifying the Mikhalkin correspondence theorem [9, Theorem 1], one can deduce that $M(Q,v)$ as defined in (8) equals the number of complex rational curves C on the toric surface $\mathrm{Tor}(\Delta_v)$ such that

- C belongs to the tautological linear system $|\mathcal{L}_{\Delta_v}|$.
- For each side σ of Δ_v, the intersection points of C with toric divisor $\mathrm{Tor}(\sigma) \subset \mathrm{Tor}(\Delta_v)$ are in 1-to-1 correspondence with the Γ-ends of the tropical curves from $\mathcal{T}$ orthogonal to σ, C is nonsingular along $\mathrm{Tor}(\sigma)$, and the intersection multiplicities are respectively equal to the weights of the above Γ-ends.
- C passes through a generic configuration of $s+r+1$ points in $\mathrm{Tor}(\Delta_v)$.

Furthermore, $\mathcal{T}$ consists of just one curve since $r = 1$. Indeed, its dual subdivision of the Newton triangle Δ_v must be as described above with the order of segments dual to the parallel Γ-ends, which is determined uniquely by the disposition of the points $\boldsymbol{y}_1,\dots,\boldsymbol{y}_s$.

[4] Notice that by construction there is a canonical 1-to-1 correspondence between the ends of Q_v and the ends of any of the curves obtained in the deformation.

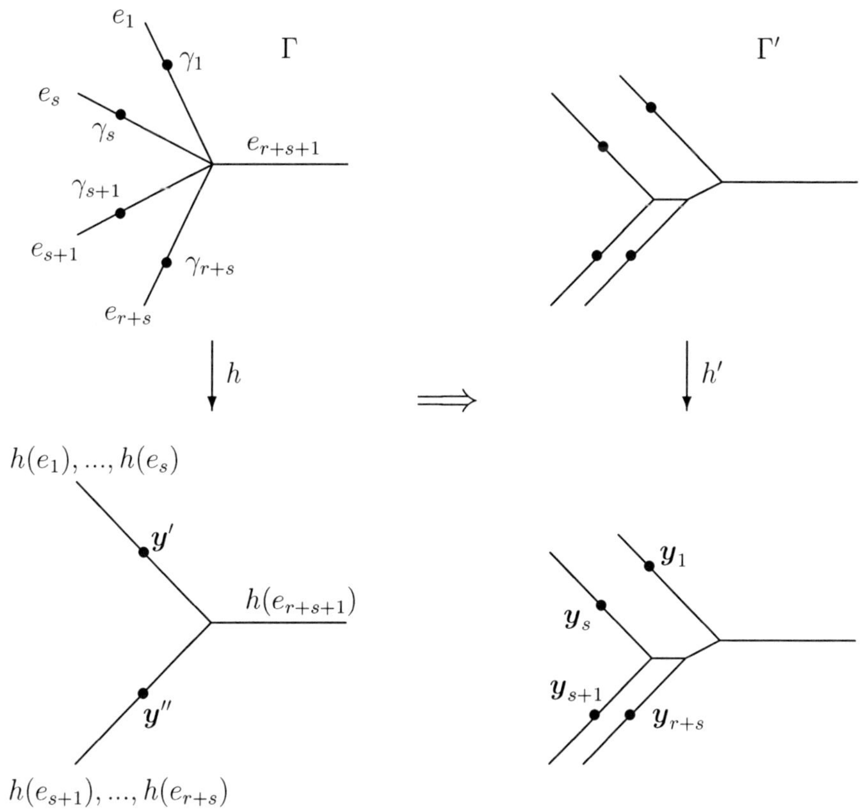

Fig. 1 Local deformation of a tropical curve in (M5)

(2) If Q is simple, i.e., all the vertices of Γ are trivalent, then (7) gives

$$M(Q) = \frac{\prod_{v\in\Gamma^0} |\Delta_v|}{\prod_{\gamma\in G_\infty,\ \gamma\in e\in\Gamma^1_\infty} w(e)}, \tag{9}$$

which generalizes Mikhalkin's weight introduced in [9, Definitions 2.16 and 4.15], and coincides with the multiplicity of a tropical curve from [1].

Real weights. A PPT-curves $Q = (\overline{\Gamma}, G, \overline{h})$ equipped with the additional structure of a continuous involution $c : (\overline{\Gamma}, G, \overline{h}) \to (\overline{\Gamma}, G, \overline{h})$ and a subdivision $G = \Re G \cup \Im G$ invariant with respect to c is called *real*.

Clearly, $\Im\overline{\Gamma} := \overline{\Gamma} \setminus \Re\overline{\Gamma}$, where $\Re\overline{\Gamma} = \text{Fix}\ c\big|_{\overline{\Gamma}}$ consists of two disjoint subsets $\Im\overline{\Gamma}'$, $\Im\overline{\Gamma}''$ interchanged by c.

Given a real PPT-curve Q, we can construct a (usual) PPT-curve $Q/c = (\overline{\Gamma}/c, G/c, \overline{h}/c)$. Notice that the weights of the edges obtained here by identifying

$\Im\overline{\Gamma}'$ and $\Im\overline{\Gamma}''$ are even. Conversely, given a (usual) PPT-curve $Q = (\overline{\Gamma}, G, \overline{h})$ and a set $I(\overline{\Gamma}^1) \subset \overline{\Gamma}^1$ that includes only edges of even weight, we construct a real PPT-curve $Q' = (\overline{\Gamma}', G', \overline{h}')$ as follows: (i) Put $K = \bigcup_{e \in I(\overline{\Gamma}^1)} e$ and obtain the graph $\overline{\Gamma}'$ by gluing up $\overline{\Gamma}$ with another copy K' of K at the vertices of Γ, common for K and the closure of $\overline{\Gamma} \setminus K$. (ii) The map $\overline{h}$ coincides on K and K', whereas the weights of the doubled edges are divided by 2 in order to keep the balancing condition. (iii) The points of $G \cap K$ are respectively doubled to K'. Finally, define an involution c on Q' interchanging K and K', and define a subdivision $G' = \Re G' \cup \Im G'$.

We shall consider only real PPT-curves with the following properties:

(R1) $\Re\overline{\Gamma}$ is nonempty and has no one-point connected component;
(R2) $\Im\overline{\Gamma}$ has only univalent and trivalent vertices (if nonempty);
(R3) The marked points $G_0 \cap \Im\overline{\Gamma}$ are not vertices of $\Im\overline{\Gamma}$;
(R4) $\Im G \setminus \Im\overline{\Gamma}$ is empty or consists of some trivalent vertices of Γ;
(R5) The closure of any component of $\Im\overline{\Gamma} \setminus G$ contains a point from $\Im G$.

Observe that the closure of $\Im\overline{\Gamma}$ joins $\Re\overline{\Gamma}$ at vertices of valency greater than 3 (which are not in G by condition (T2)).

The real weight of a real PPT-curve Q is defined as

$$W(Q) = (-1)^{\ell_1} 2^{\ell_2} \cdot \prod_{v \in \Gamma^0} W(Q,v) \cdot \prod_{e \in \overline{\Gamma}^1} W(Q,e) \cdot \prod_{\gamma \in G_0} W(Q,\gamma),$$
$$\ell_1 = \frac{|\Re G \cap \Im\overline{\Gamma}|}{2}, \quad \ell_2 = \frac{|\Im G \cap \Im\overline{\Gamma}| - b_0(\Im\overline{\Gamma})}{2}, \tag{10}$$

with $W(Q,v), W(Q,e), W(Q,\gamma)$ computed along the following rules:

(W1) For an edge $e \subset \Re\overline{\Gamma}$, put $W(Q,e) = 0$ or 1 according to whether $w(e)$ is even or odd. For an edge $e \in \Gamma^1$, $e \subset \Im\overline{\Gamma}$, put $W(Q,e)W(Q,c(e)) = w(e)$. For an edge $e \in \Gamma^1_\infty$, $e \subset \Im\overline{\Gamma}$, put $W(Q,e) = 1$.
(W2) For $\gamma \in G_0 \cap \Re\overline{\Gamma} \setminus \Gamma^0$, put $W(Q,\gamma) = 1$. For $\gamma \in \Re G \cap \Gamma^0$, put $W(Q,\gamma) = 1$. For $\gamma \in \Im G$, $\gamma = v \in \Gamma^0$, put $W(Q,\gamma) = |\Delta_v|$. For $\gamma \in G_0$, $\gamma \in e \subset \Im\overline{\Gamma}$, put $W(Q,\gamma)W(Q,c(\gamma)) = w(e)$.
(W3) For a vertex $v \in \Gamma^0 \cap \Im\overline{\Gamma}$, put $W(Q,v)W(Q,c(v)) = (-1)^{|\partial\Delta_v \cap \mathbb{Z}^2|} M(Q,v)$ (see condition (M4) for the definition of $M(Q,v)$). For a trivalent vertex $v \in \Gamma^0 \cap \Re\overline{\Gamma}$, put $W(Q,v) = (-1)^{|\mathrm{Int}(\Delta_v) \cap \mathbb{Z}^2|}$.
(W4) For a four-valent vertex $v \in \Gamma^0$ incident to two simple edges from $\Re\overline{\Gamma}$ and two multiple edges e', e'' from $\Im\overline{\Gamma}$, put $W(Q,v) = (-1)^{|\mathrm{Int}(\Delta_v) \cap \mathbb{Z}^2|} |\Delta_v| / (2w(e'))$.
(W5) Let $v \in \Gamma^0$ be of valency >3 incident to

- A simple edge $e_1 \subset \Re\overline{\Gamma}$,
- Edges $e_i \subset \Re\overline{\Gamma}$, $1 < i \le r_1 + 1$, and $e'_i \subset \Im\overline{\Gamma}'$, $e''_i \subset \Im\overline{\Gamma}''$, $1 \le i \le s_1$, for some nonnegative r_1, s_1, all with the same directing vector $\boldsymbol{u}' \ne \boldsymbol{u}_v(e_1)$ and
- Edges $e_i \subset \Re\overline{\Gamma}$, $r_1 + 1 < i \le r_1 + r_2 + 1$, and $e'_i \subset \Im\overline{\Gamma}'$, $e''_i \subset \Im\overline{\Gamma}''$, $s_1 < i \le s_1 + s_2$, for some nonnegative r_2, s_2 such that $r_2 + 2s_2 \ge 2$, all with the same directing vector $\boldsymbol{u}'' \ne \boldsymbol{u}_v(e_1), \boldsymbol{u}'$.

Take the real PPT-curve Q_v induced by v and the edges emanating from v, correspondingly restrict on Q_v the involution c, and introduce a finite c-invariant set of marked points G_v picking up one point on each edge emanating from v but e_1. Consider the PPT-curve Q_v/c and perform with it the deformation procedure described in (M5) (cf. Fig. 1), getting a finite set of simple rational regularly end-marked PPT-curves. We turn any curve $\widetilde{Q} = (\widetilde{\overline{\Gamma}}, \widetilde{G}, \widetilde{\overline{h}})$ from this set into a real PPT-curve. Namely, first we include in the set $I(\widetilde{\overline{\Gamma}}^1)$ all the Γ-ends that correspond to the Γ/c-ends of Q_v from $\Im\overline{\Gamma}_v/c$. Then we maximally extend the set $I(\widetilde{\overline{\Gamma}}^1)$ by the following inductive procedure: If two edges $f_1, f_2 \in I(\widetilde{\overline{\Gamma}}^1)$ merge to a vertex $p \in \widetilde{\Gamma}^0$, then the third edge f_3 emanating from p should be added to $I(\widetilde{\overline{\Gamma}}^1)$. Clearly, by construction, the weights of the edges $e \in I(\widetilde{\overline{\Gamma}}^1)$ are even; hence we can make a real PPT-curve $Q' = (\overline{\Gamma}', G', \overline{h}')$, letting $\Re G' = G' \cap \Re\overline{\Gamma}'$, $\Im G' = G' \cap \Im\overline{\Gamma}'$. Denoting the final set of real PPT-curves by $\mathcal{T}$ and observing that their real weight $W(Q')$ can be computed by the above rules (W1–(W4), we define

$$W(Q,v) = \sum_{Q' \in \mathcal{T}} W(Q').$$

The fact that the latter expression does not depend on the choice of the perturbation of the points $\boldsymbol{y}'$, $\boldsymbol{y}''$ (cf. construction in (M5) and Fig. 1) follows from a more general statement proven in [18].

Remark 6. (1) If $c = \mathrm{Id}$, $\Im G = \emptyset$, and Q is simple, we obtain the well-known formula $W(Q) = 0$ when $\overline{\Gamma}$ contains an even weight edge, and $W(Q) = (-1)^a$, $a = \sum_{v \in \Gamma^0} |\mathrm{Int}(\Delta_v) \cap \mathbb{Z}^2|$, when all the edge weights of $\overline{\Gamma}$ are odd (cf. [9, Definition 7.19] or [13, Proposition 6.1], where in addition, $\deg Q$ consists of only primitive integral vectors).

(2) If Q is rational, Q/c is simple, and $G_\infty = \Re G \cap \Gamma^0 = \Re G \cap \Im\overline{\Gamma} = \emptyset$, we obtain a generalization of [15, Formula (2.12)] (in the version at arXiv:math/0406099). Indeed, if $\Re\overline{\Gamma}$ contains an edge of even weight, we obtain $W(Q) = 0$ in (10) due to (W1), and accordingly we obtain $w(Q/c) = 0$ in [15, Sect. 2.5] (in the notation therein). If $\Re\overline{\Gamma}$ contains only edges of odd weight, then (under assumption that the weights of the ends of Γ are 1) [15, Formula (2.12)] reads

$$w(Q/c) = (-1)^{a+b} \prod_{v \in \Gamma^0 \cap \Im G} |\Delta_v| \cdot \prod_{v \in (\Gamma/c)^0 \cap \overline{(\Im\Gamma/c)}} \frac{|\Delta_v|}{2} \qquad (11)$$

with $a = \sum_{v \in (\Gamma/c)^0} |\mathrm{Int}(\Delta_v) \cap \mathbb{Z}^2|$, $b = |(\Gamma/c)^0 \cap (\Im\Gamma/c)|$, whereas in (10) we obtain $\ell_1 = 0$ by the assumption $\Re G \cap \Im\overline{\Gamma} = \emptyset$, $\ell_2 = |(\Gamma/c)^0 \cap (\Im\Gamma/c)|$ due to the rationality of Q and simplicity of Q/c, and furthermore, taking into account that $w(e) = 2w(e') = 2w(c(e'))$ for $e = (e' \cup c(e'))/c \in (\Gamma/c)^1$, $e' \in \Gamma^1$, $e' \subset \Im\Gamma$, we compute the other factors in (10):

$$\prod_{v\in\Gamma^0} W(Q,v) = \prod_{v\in\Gamma^0\setminus\overline{\Im\Gamma}} (-1)^{|\mathrm{Int}(\Delta_v)\cap\mathbb{Z}^2|} \cdot \prod_{\{v,c(v)\}\in(\Gamma/c)^0\cap(\Im\Gamma/c)} (-1)^{|\partial\Delta_v\cap\mathbb{Z}^2|} M(Q,v)$$
$$\times \prod_{\substack{v\in\Re\Gamma\cap\overline{\Im\Gamma/c}\\ v\in e\subset\overline{\Im\Gamma/c},\ e\in(\Gamma/c)^1}} (-1)^{|\mathrm{Int}(\Delta_v)\cap\mathbb{Z}^2|} \frac{|\Delta_v|}{w(e)}$$
$$= \prod_{v\in(\Gamma/c)^0} (-1)^{|\mathrm{Int}(\Delta_v)\cap\mathbb{Z}^2|} \cdot (-4)^{-|(\Gamma/c)^0\cap\Im\Gamma/c|} \cdot \prod_{v\in(\Gamma/c)^0\cap\overline{\Im\Gamma/c}} |\Delta_v|$$
$$\times 2^{-|\Re\Gamma\cap\overline{\Im\Gamma}|} \prod_{e\in(\Gamma/c)^1,\ e\subset\overline{\Im\Gamma/c}} \frac{2}{w(e)} \prod_{\substack{e\in(\Gamma/c)^1,\ e\subset\overline{\Im\Gamma/c}\\ e\cap G/c\neq\emptyset}} \frac{2}{w(e)},$$
$$\prod_{e\in\overline{\Gamma}^1} W(Q,e) = \prod_{e\in(\Gamma/c)^1,\ e\subset\overline{\Im\Gamma/c}} \frac{w(e)}{2},$$
$$\prod_{\gamma\in G_0} W(Q,\gamma) = \prod_{v\in\Im G\cap\Re\Gamma} |\Delta_v| \prod_{\substack{e\in(\Gamma/c)^1,\ e\subset\overline{\Im\Gamma/c}\\ e\cap G/c\neq\emptyset}} \frac{w(e)}{2},$$

which altogether gives (with a,b from (11))

$$W(Q) = (-1)^{a+b} \prod_{v\in\Gamma^0\cap\Im G} |\Delta_v| \cdot \prod_{v\in(\Gamma/c)^0\cap\overline{(\Im\Gamma/c)}} \frac{|\Delta_v|}{2} = w(Q/c).$$

3 Patchworking Theorem

3.1 Patchworking Data

Combinatorial-geometric part. In the notation of Sect. 2.6, let $Q = (\overline{\Gamma}, G, \overline{h})$ be a pseudosimple irreducible regular marked PPT-curve of genus g that has a nondegenerate Newton polygon Δ and that satisfies conditions (T1)–(T3) of Sect. 2.6.

Let G_0 split into disjoint subsets $G_0 = G_0^{(\mathrm{m})} \cup G_0^{(\mathrm{dm})}$ such that $G_0^{(\mathrm{m})} \cap \Gamma^0 = \emptyset$ and $h(G_0^{(\mathrm{m})}) \cap h(G_0^{(\mathrm{dm})}) = \emptyset$. We equip the points of G_0 with the following multiplicities:

- If $\gamma \in G^{(\mathrm{m})}$, put $\mathrm{mt}(\gamma) = 1$.
- If $\gamma \in G^{(\mathrm{dm})}$ is a (trivalent) vertex of Γ, put $\mathrm{mt}(\gamma) = (1,1)$.
- If $\gamma \in G^{(\mathrm{dm})}$ is not a vertex of Γ, put $\mathrm{mt}(\gamma) = (1,0)$ or $(0,1)$.

In the sequel, by $\hat{Q}$ we denote the PPT-curve Q equipped with the subdivision $G_0 = G_0^{(\mathrm{m})} \cup G_0^{(\mathrm{dm})}$ and the multiplicity function $\mathrm{mt}(\gamma)$, $\gamma \in G_0$, as above.

Definition 7. A pair γ, γ' of distinct points in G_0 is called *special* if $h(\gamma) = h(\gamma')$ and $\mathrm{mt}(\gamma) = \mathrm{mt}(\gamma')$. A pair of parallel multiple edges $e, e' \in \overline{\Gamma}^1$ emanating from a

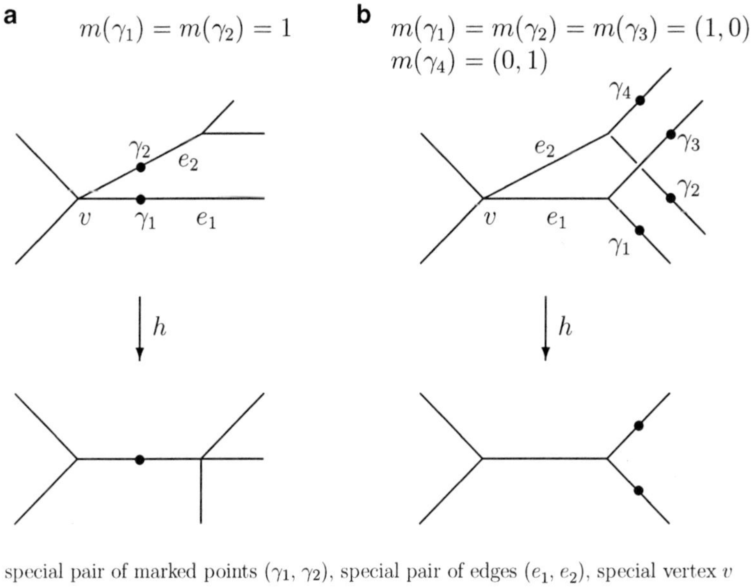

Fig. 2 Illustration to Definition 7

vertex $v \in \Gamma^0$ of valency greater than 3 is called *special* if there are disjoint open connected subsets K, K' of $\Gamma \setminus \{v\}$ and a special pair of points $\gamma \in K$, $\gamma' \in K'$ such that

- K contains the germ of e at v; K' contains the germ of e' at v
- There is a homeomorphism $\varphi : K \to K'$ satisfying $h\big|_K = h\big|_{K'} \circ \varphi$.

A vertex $v \in \Gamma^0$ incident to a special pair of edges is called *special*. See Fig. 2.

Then we assume the following:

(T4) The edges in special pairs have weight 1, and at least one of the simple edges emanating from a special vertex has weight 1.

(T5) Let e, e' be a special pair of edges emanating from a vertex $v \in \Gamma^0$, and let K, K' be disjoint connected subsets of $\Gamma \setminus \{v\}$ as in Definition 7; then $K \cup K'$ contains at most one special pair of points of G_0.

(T6) A special pair of edges cannot be a pair of Γ-ends and cannot be a pair of finite-length edges that end up at a special pair $\gamma, \gamma' \in \Gamma^0$ such that $h(\gamma) = h(\gamma')$ and $\mathrm{mt}(\gamma) = \mathrm{mt}(\gamma') = (1,1)$.

(T7) Let a vertex $v \in \Gamma^0$ be a special vertex, and let $\{e_1, \ldots, e_s\}$ be a maximal (with respect to inclusion) set of edges of Γ incident to v and such that

- Each edge e_i contains a point $\gamma_i \in G_0$, $1 \le i \le s$,
- $h(\gamma_1) = \cdots = h(\gamma_s)$ and $\mathrm{mt}(\gamma_1) = \cdots = \mathrm{mt}(\gamma_s)$.

Suppose that $\mathrm{dist}(v, v_i) \le \mathrm{dist}(v, v_{i+1})$, $1 \le i < s$, $v_i \in \overline{\Gamma}^0 \setminus \{v\}$ being the second vertex of e_i. Then we require

$$\mathrm{dist}(v, \gamma_1) > \sum_{1 \le i < s-1} \mathrm{dist}(\gamma_i, v_i) + 2 \cdot \mathrm{dist}(\gamma_{s-1}, v_{s-1}). \tag{12}$$

Notice that in condition (T7), at most one edge e_i is a Γ-end (cf. (T6)), and it must be e_s.

We introduce also the semigroup

$$\mathbb{Z}^\infty_{\ge 0} = \{\alpha = (\alpha_1, \alpha_2, \dots) \ : \ \alpha_i \in \mathbb{Z},\ \alpha_i \ge 0,\ i = 1, 2, \dots\ ,\ |\{i \ : \ \alpha_i > 0\}| < \infty\},$$

equipped with two norms

$$\|\alpha\|_0 = \sum_{i=1}^\infty \alpha_i, \quad \|\alpha\|_1 = \sum_{i=1}^\infty i\alpha_i\ ,$$

and the partial order

$$\alpha \ge \beta \quad \Leftrightarrow \quad \alpha - \beta \in \mathbb{Z}^\infty_{\ge 0}.$$

For each side σ of Δ, we introduce the vectors $\beta^\sigma \in \mathbb{Z}^\infty_{\ge 0}$ such that the coordinate β^σ_i of β^σ equals the number of the univalent vertices $v \in \Gamma^0_\infty$ such that $\overline{h}(v) \in \sigma \subset \mathbb{R}^2_\Delta$ and $w(e) = i$ for the Γ-end e merging to v, for all $i = 1, 2, \dots$.

Algebraic part. Let $\Sigma = \mathrm{Tor}_{\mathbb{K}}(\Delta)$. The coordinatewise valuation map $\mathrm{Val} : (\mathbb{K}^*)^2 \to \mathbb{R}^2$ naturally extends up to $\mathrm{Val} : \Sigma \to \mathbb{R}^2_\Delta$. Let $\overline{\boldsymbol{p}} \subset \Sigma := \mathrm{Tor}_{\mathbb{K}}(\Delta)$ be finite and satisfy $\mathrm{Val}(\overline{\boldsymbol{p}}) = \overline{\boldsymbol{x}} = \overline{h}(G)$.[5] Suppose that

(A1) Each point $\boldsymbol{x} \in h(G^{(\mathrm{m})}) \subset \overline{\boldsymbol{x}}$ has a unique preimage in $\overline{\boldsymbol{p}}$.
(A2) The preimage of each point $\boldsymbol{x} \in h(G^{(\mathrm{dm})}) \subset \overline{\boldsymbol{x}}$ consists of an ordered pair of points $\boldsymbol{p}_{1,\boldsymbol{x}}, \boldsymbol{p}_{2,\boldsymbol{x}} \in \overline{\boldsymbol{p}}$.
(A3) There is a bijection $\psi : \overline{\boldsymbol{p}}^\infty \to G_\infty$, where $\overline{\boldsymbol{p}}^\infty := \mathrm{Val}^{-1}(\overline{\boldsymbol{x}}^\infty)$, $\overline{\boldsymbol{x}}^\infty = \overline{h}(G_\infty)$, such that $\mathrm{Val}(\boldsymbol{p}) = \overline{h}(\psi(\boldsymbol{p}))$, $\boldsymbol{p} \in \overline{\boldsymbol{p}}^\infty$;
(A4) The sequence $\overline{\boldsymbol{p}}$ is generic among the sequences satisfying the above conditions.

Define the multiplicity function $\mu : \overline{\boldsymbol{p}} \cap (\mathbb{K}^*)^2 \to \mathbb{Z}_{>0}$ such that:

- For $\boldsymbol{p} \in \overline{\boldsymbol{p}} \cap (\mathbb{K}^*)^2$, $\mathrm{Val}(\boldsymbol{p}) = \overline{x} \in h(G^{\mathrm{m}})$, put

$$\mu(\boldsymbol{p}) = \sum_{\gamma \in G^{(\mathrm{m})},\ h(\gamma) = \boldsymbol{x}} \mathrm{mt}(\gamma). \tag{13}$$

[5] This means, in particular, that the points of $\overline{\boldsymbol{x}}$ have rational coordinates.

- For the points $\boldsymbol{p}_{1,\boldsymbol{x}},\boldsymbol{p}_{2,\boldsymbol{x}}$ where $\mathrm{Val}(\boldsymbol{p}_{1,\boldsymbol{x}})=\mathrm{Val}(\boldsymbol{p}_{2,\boldsymbol{x}})=\boldsymbol{x}\in h(G^{(\mathrm{dm})})$, put

$$\mu(\boldsymbol{p}_{1,\boldsymbol{x}})=m_1,\ \mu(\boldsymbol{p}_{2,\boldsymbol{x}})=m_2,\quad (m_1,m_2)=\sum_{\gamma\in G^{(\mathrm{dm})},\ h(\gamma)=\boldsymbol{x}}\mathrm{mt}(\gamma).\qquad(14)$$

From this definition and from the count of the Euler characteristic of $\overline{\Gamma}$, we derive

$$\sum_{\boldsymbol{p}\in\overline{\boldsymbol{p}}\cap(\mathbb{K}^*)^2}\mu(\boldsymbol{p})+|\overline{\boldsymbol{p}}^\infty|-|\Gamma^0_\infty|=g-1.\qquad(15)$$

Let $\Delta'\subset\mathbb{R}^2$ be a convex lattice polygon such that there is another lattice polygon (or segment, or point) Δ'' satisfying $\Delta'+\Delta''=\Delta$. Then we have a well-defined line bundle $\mathcal{L}_{\Delta'}$ on $\mathrm{Tor}_{\mathbb{K}}(\Delta)$. Let $\overline{\boldsymbol{p}}'\subset\overline{\boldsymbol{p}}$ and $\mu':\overline{\boldsymbol{p}}'\cap(\mathbb{K}^*)^2\to\mathbb{Z}_{>0}$ be such that $\mu'(\boldsymbol{p})\le\mu(\boldsymbol{p})$ for all $\boldsymbol{p}\in\overline{\boldsymbol{p}}'$. Let $(\beta^\sigma)'\in\mathbb{Z}^\infty_{\ge0}$, $\sigma\subset\partial\Delta$, be such that $(\beta^\sigma)'\le\beta^\sigma$ for all sides σ of Δ. We say that the tuple $(\Delta',g',\overline{\boldsymbol{p}}',\mu',\{(\beta^\sigma)'\}_{\sigma\subset\partial\Delta})$, where $g'\in\mathbb{Z}_{\ge0}$, is compatible if

- $\|(\beta^\sigma)'\|_1=\langle c_1(\mathcal{L}_{\Delta'}),\mathrm{Tor}_{\mathbb{K}}(\sigma)\rangle$ for all sides σ of Δ;
- $(\beta^\sigma)'_i\ge|\{\boldsymbol{p}\in\overline{\boldsymbol{p}}'\cap\mathrm{Tor}_{\mathbb{K}}(\sigma)\ :\ \psi(\boldsymbol{p})=\gamma\in e\in\Gamma^1_\infty,\ w(e)=i\}|$ for all sides σ of Δ and all $i=1,2,\ldots$;
- $g'\le|\mathrm{Int}(\Delta'\cap\mathbb{Z}^2)|$ and

$$\sum_{\boldsymbol{p}\in\overline{\boldsymbol{p}}'\cap(\mathbb{K}^*)^2}\mu'(\boldsymbol{p})+|\overline{\boldsymbol{p}}'\cap\overline{\boldsymbol{p}}^\infty|-\sum_{\sigma\subset\partial\Delta}\|(\beta^\sigma)'\|_0=g'-1.$$

In view of (15) and $|\Gamma^0_\infty|=\sum_{\sigma\subset\partial\Delta}\|\beta^\sigma\|_0$, the tuple $(\Delta,g,\overline{\boldsymbol{p}},\mu,\{\beta^\sigma\}_{\sigma\subset\partial\Delta})$ is compatible.

For any compatible tuple $(\Delta',g',\overline{\boldsymbol{p}}',\mu',\{(\beta^\sigma)'\}_{\sigma\subset\partial\Delta})$, we introduce the set $\mathcal{C}(\Delta',g',\overline{\boldsymbol{p}}',m',\{(\beta^\sigma)'\}_{\sigma\subset\partial\Delta})$ of reduced irreducible curves $C\in|\mathcal{L}_{\Delta'}|$ passing through $\overline{\boldsymbol{p}}'$ and such that

- The points $\boldsymbol{p}\in\overline{\boldsymbol{p}}'\cap\overline{\boldsymbol{p}}^\infty$ are nonsingular for C, and

$$(C\cdot\mathrm{Tor}_{\mathbb{K}}(\partial\Delta))_{\boldsymbol{p}}=w(e),$$

  where $\gamma=\psi(\boldsymbol{p})\in G_\infty$, and $e\in\Gamma^1_\infty$ merges to γ.
- The local branches of C centered at the points of $C\cap\mathrm{Tor}_{\mathbb{K}}(\partial\Delta)$ are smooth, and for each side σ of Δ and each $i=1,2,\ldots$, there are precisely $(\beta^\sigma)'_i$ local branches P of C centered at $C\cap\mathrm{Tor}_{\mathbb{K}}(\sigma)$ such that

$$(P\cdot\mathrm{Tor}_{\mathbb{K}}(\sigma))=i,\quad i=1,2,\ldots.$$

- C has genus $\le g'$.
- At each point $\boldsymbol{p}\in\overline{\boldsymbol{p}}'\cap(\mathbb{K}^*)^2$, the multiplicity of C is $\mathrm{mt}(C,\boldsymbol{p})\ge\mu'(\boldsymbol{p})$.

We now impose new conditions on the algebraic pathchworking data:

(A5) For any compatible tuple $(\Delta', g', \overline{\boldsymbol{p}}', m', \{(\beta^\sigma)'\}_{\sigma\subset\partial\Delta})$, the set $\mathcal{C}(\Delta', g', \overline{\boldsymbol{p}}', m', \{(\beta^\sigma)'\}_{\sigma\subset\partial\Delta})$ is finite, and all the curves $C\in\mathcal{C}(\Delta', g', \overline{\boldsymbol{p}}', m', \{(\beta^\sigma)'\}_{\sigma\subset\partial\Delta})$ are immersed and have genus g' and multiplicity $\mathrm{mt}(C,\boldsymbol{p}) = m'(\boldsymbol{p})$ at each point $\boldsymbol{p}\in\overline{\boldsymbol{p}}'\cap(\mathbb{K}^*)^2$; furthermore,

$$H^1(C^\nu, \mathcal{J}_Z(C^\nu)) = 0, \tag{16}$$

where C^ν is the normalization and $\mathcal{J}_Z(C^\nu)$ is the (twisted with C^ν) ideal sheaf of the zero-dimensional scheme $Z\subset C^\nu$ that contains the lift of $\overline{\boldsymbol{p}}$ and of the points of tangency of C and $\mathrm{Tor}_{\mathbb{K}}(\partial\Delta)$ on C^ν and that has length $(C\cdot\mathrm{Tor}_{\mathbb{K}}(\partial\Delta))_{\boldsymbol{p}}$ at the lift of $\boldsymbol{p}\in\overline{\boldsymbol{p}}\cap\mathrm{Tor}_{\mathbb{K}}(\partial\Delta)$, and length $(C\cdot\mathrm{Tor}_{\mathbb{K}}(\partial\Delta))_z - 1$ at the lift of each point $z\in C\cap\mathrm{Tor}_{\mathbb{K}}(\partial\Delta)\setminus\overline{\boldsymbol{p}}$.

Here we verify the condition (A5) for the versions of the patchworking theorem used in [6, 18].

Lemma 8. *Condition (A5) holds if*

- *Either* $\mu(\boldsymbol{p}) = 1$ *for all* $\boldsymbol{p}\in\overline{\boldsymbol{p}}\cap(\mathbb{K}^*)^2$,
- *or the surface* $\Sigma = \mathrm{Tor}_{\mathbb{K}}(\Delta)$ *is one of* $\mathbb{P}^2$, $\mathbb{P}^2_k$ *with* $1\le k\le 3$, $(\mathbb{P}^1)^2$, *the configuration* $\overline{\boldsymbol{p}}^\infty$ *is contained in one toric divisor* E *of* Σ, *and*

$$\left|\{\boldsymbol{p}\in\overline{\boldsymbol{p}}\cap(\mathbb{K}^*)^2 \,:\, \mu(\boldsymbol{p})>1\}\right| \le \begin{cases} 4, & \Sigma=\mathbb{P}^2,\\ 5-E^2-k, & \Sigma=\mathbb{P}^2_k,\\ 3 & \Sigma=(\mathbb{P}^1)^2.\end{cases}$$

Moreover, all the curves C in the considered sets are nonsingular along $\mathrm{Tor}_{\mathbb{K}}(\partial\Delta)$, *are nodal outside* $\overline{\boldsymbol{p}}$, *and have ordinary singularity of order* $m'(\boldsymbol{p})$ *at each point* $\boldsymbol{p}\in\overline{\boldsymbol{p}}'\cap(\mathbb{K}^*)^2$.

Proof. We prove the statement only for the original data $(\Delta, g, \overline{\boldsymbol{p}}, m, \{\beta^\sigma\}_{\sigma\subset\partial\Delta})$, since the other compatible tuples can be treated in the same way.

Observe that in the first case, each curve $C\in\mathcal{C}(\Delta, g, \overline{\boldsymbol{p}}, m, \{\beta^\sigma\}_{\sigma\subset\partial\Delta})$ satisfies

$$\sum_{\substack{\boldsymbol{p}\in\overline{\boldsymbol{p}}\cap(\mathbb{K}^*)^2\\ \mu(\boldsymbol{p})>1}} \mu(\boldsymbol{p}) < |C\cap\mathrm{Tor}(\partial\Delta)| - 1. \tag{17}$$

In the second situation, except for finitely many lines or conics (which, of course, satisfy (A5)), the other curves obey (17) by Bézout's theorem (just consider intersections with suitable lines or conics; we leave this to the reader as a simple exercise).

We proceed further under the condition (17). Let $\overline{\boldsymbol{p}}' = \{\boldsymbol{p}\in\overline{\boldsymbol{p}}\cap(\mathbb{K}^*)^2 \,:\, \mu(\boldsymbol{p})>1\}$. Consider the family $\mathcal{C}'$ of reduced irreducible curves $C'\in|\mathcal{L}_\Delta|$ of genus at most g that have multiplicity $\ge\mu(\boldsymbol{p})$ at each point $\boldsymbol{p}\in\overline{\boldsymbol{p}}'$, whose local branches centered along $\mathrm{Tor}_{\mathbb{K}}(\partial\Delta)$ are nonsingular, and for which the number of such branches crossing the toric divisor $\mathrm{Tor}_{\mathbb{K}}(\sigma) \subset \mathrm{Tor}_{\mathbb{K}}(\Delta)$ with multiplicity i is β_i^σ for all sides σ of Δ and all $i = 1, 2, \ldots$

By a classical deformation theory argument (see, for instance, [2, 3]), the Zariski tangent space to $\mathcal{C}'$ at $C \in \mathcal{C} := \mathcal{C}(\Delta, g, \overline{\boldsymbol{p}}, m, \{\beta^\sigma\}_{\sigma \subset \partial\Delta})$ is naturally isomorphic to $H^0(C^\nu, \mathcal{J}_Z(C^\nu))$, where C^ν, $\mathcal{J}_Z(C^\nu)$ are defined in (A5). So we have

$$\begin{aligned}\deg Z = \sum_{\boldsymbol{p} \in \overline{\boldsymbol{p}}'} \mu(\boldsymbol{p}) + C \cdot \mathrm{Tor}_{\mathbb{K}}(\partial\Delta) - |C \cap \mathrm{Tor}(\partial\Delta)| &= \sum_{\boldsymbol{p} \in \overline{\boldsymbol{p}}'} \mu(\boldsymbol{p}) \\ -CK_\Sigma - |C \cap \mathrm{Tor}(\partial\Delta)| < -CK_\Sigma - 1. &\end{aligned} \tag{18}$$

Hence (see [2, 3]) $H^1(C^\nu, \mathcal{J}_Z(C^\nu)) = 0$, which yields

$$\begin{aligned} h^0(C^\nu, \mathcal{J}_Z(C^\nu)) &= C^2 - 2\delta(C) - \deg Z - g(C) + 1 \\ &= -CK_\Sigma + 2g(C) - 2 - \deg Z - g(C) + 1 \\ &= g(C) - 1 + |C \cap \mathrm{Tor}(\partial\Delta)| - \sum_{\boldsymbol{p} \in \overline{\boldsymbol{p}}'} \mu(\boldsymbol{p}) \overset{(15)}{=} |\overline{\boldsymbol{p}} \setminus \overline{\boldsymbol{p}}'| - (g - g(C)). \end{aligned} \tag{19}$$

Since $\overline{\boldsymbol{p}} \setminus \overline{\boldsymbol{p}}'$ is a configuration of generic points (partly on $\mathrm{Tor}_{\mathbb{K}}(\partial\Delta)$), we derive that $g(C) = g$ and that $\mathcal{C}$ is finite.

For the rest of the required statement, we assume that a curve $C \in \mathcal{C}$ is either not nodal outside $\overline{\boldsymbol{p}}$ or has singularities on $\mathrm{Tor}_{\mathbb{K}}(\partial\Delta)$ or has at some point $\boldsymbol{p} \in \overline{\boldsymbol{p}} \cap (\mathbb{K}^*)^2$ a singularity more complicated than an ordinary point of order $\mu(\boldsymbol{p})$. Then (cf. the argument in the proof of [11, Proposition 2.4]) one can find a zero-dimensional scheme $Z \subset Z' \subset C^\nu$ of degree $\deg Z' = \deg Z + 1$ such that the Zariski tangent space to $\mathcal{C}'$ at C is contained in $H^0(C^\nu, \mathcal{J}_{Z'}(C^\nu))$. However, then one derives from (18) that $\deg Z' < -CK_\Sigma$, and hence again $H^1(C^\nu, \mathcal{J}_{Z'}(C^\nu)) = 0$, which in view of (19) will lead to

$$h^0(C^\nu, \mathcal{J}_{Z'}(C^\nu)) = |\overline{\boldsymbol{p}} \setminus \overline{\boldsymbol{p}}'| - 1,$$

which finally implies the emptiness of $\mathcal{C}$. □

3.2 Algebraic Curves over $\mathbb{K}$ and Tropical Curves

If $C \in |\mathcal{L}_\Delta|$ is a curve on the toric surface $\mathrm{Tor}_{\mathbb{K}}(\Delta)$, then the closure $\mathrm{Cl}(\mathrm{Val}(C \cap (\mathbb{K}^*)^2)) \subset \mathbb{R}^2$ supports an EPT-curve T with Newton polygon Δ (cf. Sect. 2.2) defined by a convex piecewise linear function (5) coming from a polynomial equation $F(z) = 0$ of C in $(\mathbb{K}^*)^2$:

$$F(z) = \sum_{\omega \in \Delta \cap \mathbb{Z}^2} A_\omega z^\omega, \quad A_\omega \in \mathbb{K},\ c_\omega = \mathrm{Val}(A_\omega), \quad z \in (\mathbb{K}^*)^2. \tag{20}$$

The EPT-curve obtained does not depend on the choice of the defining polynomial of C and will be denoted by $\mathrm{Trop}(C)$.

Observe also that the polynomial (20) can be written

$$F(z) = \sum_{\omega \in \Delta \cap \mathbb{Z}^2} (a_\omega + O(t^{>0}))\, t^{\nu_T(\omega)} z^\omega$$

with the convex piecewise linear function $\nu : \Delta \to \mathbb{R}$ as in Sect. 2.2 and the coefficients $a_\omega \in \mathbb{C}$ nonvanishing at the vertices of the subdivision S_T of Δ.

3.3 Patchworking Theorems

The algebraically closed version.

Theorem 2. *Given the patchworking data, a PPT-curve $\hat{Q}$, and a configuration $\overline{p}$ satisfying all the conditions of Sect. 3.1, there exists a subset $\mathcal{C}(\hat{Q}) \subset \mathcal{C}(\Delta, g, \overline{p}, m, \{\beta^\sigma\}_{\sigma \subset \partial\Delta})$ of $M(Q)$ curves C such that* $\mathrm{Trop}(C) = h_*Q$. *Furthermore, for any distinct (nonisomorphic) curves $\hat{Q}_1$ and $\hat{Q}_2$, the sets $\mathcal{C}(\hat{Q}_1)$ and $\mathcal{C}(\hat{Q}_2)$ are disjoint.*

Remark 9. We would like to underscore one useful consequence of Theorem 2: The PPT-curve Q and the multiplicities of its marked curves must satisfy the restrictions known for the respective algebraic curves with multiple points.

The real version. In addition to all the above hypotheses, we assume the following:

(R6) The configuration $\overline{p}$ is Conj-invariant, $\mathrm{Val}(\Re\overline{p}) \cap \mathrm{Val}(\Im\overline{p}) = \emptyset$, where $\Re\overline{p} := \mathrm{Fix}(\mathrm{Conj}|_{\overline{p}})$ and $\Im\overline{p} = \overline{p} \setminus (\Re\overline{p})$.

(R7) The PPT-curve Q possesses a real structure $c : Q \to Q$, $G = \Re G \cup \Im G$ such that

- (i) The bijection ψ from (A3) takes $G_\infty \cap \Re G$ into $\Re\overline{p} \cap \mathrm{Tor}_{\mathbb{K}}(\partial\Delta)$ and takes $G_\infty \cap \Im G$ into $\Im\overline{p} \cap \mathrm{Tor}_{\mathbb{K}}(\partial\Delta)$, respectively.
- (ii) $h(G_o \cap \Re G) \subset \mathrm{Val}(\Re\overline{p})$, $h(G_0 \cap \Im G \cap \Gamma^0) \subset \mathrm{Val}(\Im\overline{p})$.
- (iii) $\Re G \cap G^{(\mathrm{dm})} \cap \Im\overline{\Gamma} = \emptyset$.
- (iv) If $\gamma \in G_0 \cap \Im G \cap \Im\overline{\Gamma}$, $\mathrm{mt}(\gamma) = (1,0)$, then $\mathrm{mt}(c(\gamma)) = (0,1)$.
- (v) If $e \in \Gamma^1_\infty$, $e \subset \Re\overline{\Gamma}$, then $w(e)$ is odd.

Theorem 3. *In the notation and hypotheses of Theorem 2 and under assumptions (R1)–(R7), the following holds:*

$$\sum_{C \in \Re\mathcal{C}(\hat{Q})} W_{\Sigma,\overline{p}}(C) = W(Q), \tag{21}$$

where $\Re\mathcal{C}(\hat{Q})$ is the set of real curves in $\mathcal{C}(\hat{Q})$, and $W_{\Sigma,\overline{p}}(C)$ is as defined in (1).

3.4 Proof of Theorem 2

Our argument is as follows. First, we dissipate each multiple point $\boldsymbol{p} \in \overline{\boldsymbol{p}}$ of multiplicity $k > 1$ into k generic simple points (in a neighborhood of $\boldsymbol{p}$), and then, using the known patchworking theorems ([13, Theorem 5] and [16, Theorem 2.4])[6], we obtain $M(Q)$ curves $C \in |\mathcal{L}_\Delta|$ of genus g matching the deformed configuration $\overline{\boldsymbol{q}}$. After that, we specialize the configuration $\overline{\boldsymbol{q}}$ back into the original configuration $\overline{\boldsymbol{p}}$ and show that each of the constructed curves converges to a curve with multiple points and tangencies as asserted in Theorem 2.

Remark 10. The deformation part of our argument works well in a rather more general situation, whereas the degeneration part appears to be more problematic, and at the moment we do not have a unified approach to treating all possible degenerations that may lead to algebraic curves with multiple points.

Following [13, Sect. 3], we obtain the algebraic curves C over $\mathbb{K}$ as germs of one-parameter families of complex curves $C^{(t)}$, $t \in (\mathbb{C},0)$, with irreducible fibers $C^{(t)}$, $t \neq 0$, of genus g and a reducible central fiber $C^{(0)}$. The given data of Theorem 2 provide us with a collection of suitable central fibers $C^{(0)}$ out of which we restore the families using the patchworking statement [13, Theorem 5].

Step 1. We start with a simple particular case that later will serve as an element of the proof in the general situation. Assume that Q is a rational, simple, regularly end-marked PPT-curve, $h_*Q \subset \mathbb{R}^2_\Delta$ is a (compactified) nodal embedded plane tropical curve, $\overline{\boldsymbol{p}} \subset \mathrm{Tor}_{\mathbb{K}}(\partial\Delta)$, $G = G_\infty$, $\overline{\boldsymbol{x}} = \overline{\boldsymbol{x}}^\infty$, and $\overline{\boldsymbol{p}} \overset{\mathrm{Val}}{\to} \overline{\boldsymbol{x}} \overset{\overline{h}}{\leftarrow} G$ are bijections. Here $M(Q)$ is given by formula (9), and this number of required rational curves $C \subset \mathrm{Tor}_{\mathbb{K}}(\Delta)$ is obtained by a direct application of [13, Theorem 5].

The combinatorial part of the patchworking data for the construction of curves over $\mathbb{K}$ consists of the tropical curve Q that defines a piecewise linear function $\nu :$ $\Delta \to \mathbb{R}$ and a subdivision $S:\ \Delta = \Delta_1 \cup \cdots \cup \Delta_N$ (see Sect. 2.2). The algebraic part of the patchworking data includes the limit curves $C_k \subset \mathrm{Tor}(\Delta_k)$, the deformation patterns C_e associated with the (finite-length) edges $e \in \Gamma^1$ (see [13, Sect. 5.1] and [16, Sect. 2.1]), and the refined conditions to pass through the fixed points (see [13, Sect. 5.4] and [7, Sect. 2.5.9]).

First, we orient the edges of Γ as in Lemma 3(ii). Then we define complex polynomials f_e, $e \in \overline{\Gamma}^1$, and f_v, $v \in \Gamma^0$, by the following inductive procedure. At the very beginning, for the Γ-ends e with marked points γ, we define

$$f_e(x,y) = (\eta^q x^p - \xi^p y^q)^{w(e)}, \tag{22}$$

where $\boldsymbol{u}(e) = (p,q)$ and (ξ,η) are quasiprojective coordinates of the point $\mathrm{ini}(\boldsymbol{p})$ on $\mathrm{Tor}(e) \subset \mathrm{Tor}(\Delta)$ such that $\gamma = \psi(\boldsymbol{p})$ (here ψ is the bijection from condition (A3) above). We define a linear order on Γ^0 compatible with the orientation of Γ. On each

[6] The complete proof is provided in [16].

stage, we take the next vertex $v \in \Gamma^0$ and define f_v and f_e, where e is the edge emanation from v. Namely, the polynomials f_{e_1}, f_{e_2} associated with the two edges merging to v determine points $z_1 \in \mathrm{Tor}(\sigma_1)$, $z_2 \in \mathrm{Tor}(\sigma_2)$ on the surface $\mathrm{Tor}(\Delta_v)$, where σ_1, σ_2 are the sides of Δ_v orthogonal to $\overline{h}(e_1)$, $\overline{h}(e_2)$, respectively, and we construct a polynomial f_v with Newton polygon Δ_v that defines an irreducible rational curve $C_v \subset \mathrm{Tor}(\Delta_v)$, nonsingular along $\mathrm{Tor}(\partial\Delta_v)$, crossing $\mathrm{Tor}(e_i)$ at z_i, $i = 1, 2$, and crossing $\mathrm{Tor}(e)$ at one point z_0 (at which one has $(C_v \cdot \mathrm{Tor}(e))_{z_0} = w(e)$). By [13, Lemma 3.5], up to a constant factor there are $|\Delta_v|/(w(e_1)w(e_2)) = M(Q, v)$ choices for such a polynomial f_v. After that, we define $f_e(x, y)$ via (22) with ξ, η the (quasihomogeneous) coordinates of z_0 in $\mathrm{Tor}(\sigma)$, where σ is the side of Δ_v orthogonal to $\overline{h}(e)$. Thus the limit curves $C_k \subset \mathrm{Tor}(\Delta_k)$ are C_v for the triangles Δ_k dual to $h(v)$, and they are given by $f_{e_1} f_{e_2}$, where $e_1, e_2 \in \overline{\Gamma}^1$ appear in the decomposition (6) of a parallelogram Δ_k.

The set of limit curves is completed by a set of deformation patterns (see [13, Sects. 3.5 and 3.6]) as follows. Namely, for each edge $e \in \Gamma^1$ with $w(e) > 1$, the deformation pattern is an irreducible rational curve $C_e \subset \mathrm{Tor}(\Delta_e)$, where $\Delta_e := \mathrm{conv}\{(0,1), (0,-1), (w(e), 0)\}$, whose defining (Laurent) polynomial $f_e(x, y)$ has the zero coefficient of $x^{w(e)-1}$ and the truncations to the edges $[(0,1), (w(e), 0)]$ and $[(0,-1), (w(e), 0)]$ of Δ_e fitting the polynomials f_{v_1}, f_{v_2}, where v_1, v_2 are the endpoints of e (see the details in [13, Sects. 3.5 and 3.6]). Recall that by [13, Lemma 3.9], there are $w(e) = M(Q, e)$ suitable polynomials f_e.

The conditions to pass through a given configuration $\overline{\boldsymbol{p}}$ do not admit a refinement. Indeed, following [13, Sect. 5.4], we can turn a given fixed point $\boldsymbol{p}$ into $(\xi, 0)$, $\xi = \xi^0 + O(t^{>0}) \in \mathbb{K}$, by means of a suitable toric transformation. Then, in [13, Formula (6.4.26)], the term with the power $1/m$ will vanish.[7]

The above collections of limit curves and deformation patterns coincide with those considered in [13]; the transversality hypotheses of [13, Theorem 5] are verified in [13, Sect. 5.4]. Hence, each of the

$$\prod_{v \in \Gamma^0} M(Q, v) \cdot \prod_{e \in \Gamma^1} M(Q, e) = M(Q)$$

above patchworking data gives rise to a rational curve $C \subset \mathrm{Tor}_{\mathbb{K}}(\Delta)$ as asserted in Theorem 2. Notice that all these curves are nodal by construction.

Step 2. Now we return to the general situation and deform the given configuration $\overline{\boldsymbol{p}}$ into the following new configuration $\overline{\boldsymbol{q}}$.

We replace each point $\boldsymbol{p} = (\xi t^a + \cdots, \eta t^b + \cdots) \in \overline{\boldsymbol{p}} \cap (\mathbb{K}^*)^2$ with multiplicity $\mu(\boldsymbol{p}) > 1$ (defined by (13) or (14)) by $\mu(\boldsymbol{p})$ generic points in $(\mathbb{K}^*)^2$ with the same valuation image $\mathrm{Val}(\boldsymbol{p}) = (-a, -b)$ and the initial coefficients of the coordinates

[7] The mentioned term contains η_s^0, the initial coefficient of the second coordinate of $\boldsymbol{p}$, and not ξ_s^0 as appears in the published text. The correction is clear, since in the preceding formula for τ one has just η_s^0.

close to $\xi, \eta \in \mathbb{C}^*$, respectively. Furthermore, we extend the bijection ψ from (A3) up to a map $\psi : \overline{\boldsymbol{q}} \to G$ in such a way that

- $\mathrm{Val}\big|_{\overline{\boldsymbol{q}}} = \overline{h} \circ \psi$.
- Each point $\gamma \in G \setminus \Gamma^0$ has a unique preimage, and each point $\gamma \in G \cap \Gamma^0$ has precisely two preimages.
- If $\gamma \in G^{(\mathrm{dm})} \setminus \Gamma^0$, $h(\gamma) = \boldsymbol{x}$, then $\psi^{-1}(\gamma)$ is close to $\boldsymbol{p}_{1,\boldsymbol{x}}$ or to $\boldsymbol{p}_{2,\boldsymbol{x}}$ according to whether $\mathrm{mt}(\gamma) = (1,0)$ or $(0,1)$.
- If $\gamma \in G^{(\mathrm{dm})} \cap \Gamma^0$, $h(\gamma) = \boldsymbol{x}$, then $\psi^{-1}(\gamma)$ consists of two points, one close to $\boldsymbol{p}_{1,\boldsymbol{x}}$ and the other close to $\boldsymbol{p}_{2,\boldsymbol{x}}$.

Next we construct a set $\mathcal{C}' \subset \mathcal{C}(\Delta, g, \overline{\boldsymbol{q}}, 1, \{\beta^\sigma\}_{\sigma \subset \partial\Delta})$ of $M(Q)$ curves with the tropicalization h_*Q. By Lemma 8, they are irreducible, nodal, of genus g, and with specified tangency conditions along $\mathrm{Tor}_{\mathbb{K}}(\partial\Delta)$.

Step 3. Similarly to Step 1, we obtain the limit curves from a collection of polynomials in $\mathbb{C}[x,y]$ associated with the edges and vertices of the parameterizing graph Γ of Q:

(i) Let $\gamma \in G \setminus \Gamma^0$ lie on the edge $e \in \overline{\Gamma}^1$. Then we associate with the edge e a polynomial $f_e(x,y)$ given by (22) with the parameters described in Step 1.

(ii) Let $\gamma \in G_0$ be a (trivalent) vertex v of Γ. Then $f_v(x,y)$ is a polynomial with Newton triangle Δ_v (see Sect. 2.2) defining in $\mathrm{Tor}(\Delta_v)$ a rational curve $C_v \in |\mathcal{L}_{\Delta_v}|$ that crosses each toric divisor of $\mathrm{Tor}(\Delta_v)$ at one point, where it is nonsingular, and that passes through the two points $\mathrm{ini}(\psi^{-1}(\gamma))$. Observe that by [15, Lemma 2.4], up to a constant factor there are precisely $|\Delta_v|$ polynomials f_v as above. (Although the assertion and the proof of [15, Lemma 2.4] are restricted to the real case, the proof works well in the same manner in the complex case regardless of the parity of the side length of Δ_v.)

(iii) Edges emanating from a vertex $v \in \Gamma^0 \cap G_0$ do not contain any other point of G due to the Δ-general position, and we define polynomials f_e for them by formula (22), where ξ, η are the (quasihomogeneous) coordinates of the intersection point of C_v with $\mathrm{Tor}(\sigma)$, with σ the side of Δ_v orthogonal to $\overline{h}(e)$.

(iv) Pick a connected component K of $\overline{\Gamma} \setminus G$ and orient it as in Lemma 3(ii). Then we inductively define polynomials for the vertices and closed edges of K: In each stage we define polynomials f_v and f_e for a vertex v and a simple closed edge e emanating from v, whereas the polynomials $f_{e'}$ for all the edges e' of K merging to v are given. Each of the latter polynomials defines a point on $\mathrm{Tor}(\partial\Delta_v)$, and these points are distinct. We denote their set by X. Then we choose a polynomial $f_v(x,y)$ with Newton triangle Δ_v defining an irreducible rational curve $C_v \subset \mathrm{Tor}(\Delta_v)$ that

    - is nonsingular along $\mathrm{Tor}(\partial\Delta_v)$
    - crosses $\mathrm{Tor}(\partial\Delta_v)$ at each point $z \in X$ with multiplicity $w(e')$, where the edge $e' \in \overline{\Gamma}^1$ merging to v is associated with a polynomial $f_{e'}$ that determines the point z
    - crosses $\mathrm{Tor}(\partial\Delta_v) \setminus X$ at precisely one point z_0.

Notice that z_0 is the unique intersection point of C_v with the toric divisor $\mathrm{Tor}(\sigma) \subset \mathrm{Tor}(\Delta_v)$, where σ is orthogonal to $\overline{h}(e)$, and $(C_v \cdot \mathrm{Tor}(\sigma))_{z_0} = w(e)$. We claim that up to a constant factor there are precisely $M(Q,v)$ polynomials f_v, as required.

The case of a trivalent vertex v was considered in Step 1. In general, observe that the set of the required curves is finite, since we impose

$$(C_v \cdot \mathrm{Tor}(\partial\Delta_v)) - 1 = -C_v K_{\mathrm{Tor}(\Delta_v)} - 1$$

conditions on the rational curves $C_v \in |\mathcal{L}_{\Delta_v}|$, and the conditions are independent by Riemann–Roch. The cardinality of this set depends neither on the choice of a generic configuration of fixed points on $\mathrm{Tor}(\partial\Delta_v)$ nor on the choice of an algebraically closed ground field of characteristic zero. Thus, we consider the field $\mathbb{K}$ and pick the fixed points on $\mathrm{Tor}_{\mathbb{K}}(\partial\Delta_v)$ so that the valuation takes them injectively to a Δ_v-generic configuration in $\partial\mathbb{R}^2_{\Delta_v}$. Then the rule (M5) and the construction in Step 1 provide $M(Q,v)$ curves as required. The fact that there are no other curves under consideration follows from a slightly modified Mikhalkin's correspondence theorem (for details, see, for instance, [18]).

We then define $f_e(x,y)$ via (22) with ξ,η the (quasihomogeneous) coordinates of z_0 in $\mathrm{Tor}(\sigma)$.

Summarizing, we deduce that the number of choices of the curves C_v, $v \in \Gamma^0$, and C_e, $e \in \overline{\Gamma}^1$, is

$$\prod_{v\in\Gamma^0} M(Q,v).$$

Step 4. Now we define the limit curves, the deformation patterns, and the refined conditions to pass through fixed points.

For each polygon Δ_k of the subdivision S of Δ, the limit curve $C_k \subset \mathrm{Tor}(\Delta_k)$ is defined by the product of the polynomials f_v, f_e constructed above corresponding to the summands in the decomposition (6) of Δ_k.

The deformation pattern for each edge $e \in \Gamma^1$ such that $w(e) > 1$ is defined in the way described in Step 1.

Finally, the condition to pass through a given point $\boldsymbol{q} \in \overline{\boldsymbol{q}}$ such that $\gamma = \psi(\boldsymbol{q}) \in G$ lies in the interior of an edge $e \in \overline{\Gamma}^1$ with $w(e) > 1$ admits a refinement (see [13, Sects. 5.4] and [7, 2.5.9]) that in turn is defined up to the choice of a $w(e)$th root of unity, where $e \in \overline{\Gamma}^1$ contains γ.

So, the total number of choices we made up to now is

$$\prod_{v\in\Gamma^0} M(Q,v) \cdot \prod_{e\in\Gamma^1} M(Q,e) \cdot \prod_{\gamma\in G} M(Q,\gamma) = M(Q).$$

Step 5. Let us verify the hypotheses of the patchworking theorem from [13, 16].

First, the requirement on the limit curves (see [13], conditions (A), (B), (C) in Sect. 5.1, or [16], conditions (C1), (C2) in Sect. 2.1) is ensured by the generic choice of $\text{ini}(\boldsymbol{q})$, $\boldsymbol{q} \in \overline{\boldsymbol{q}}$. Namely, the limit curves do not contain multiple nonbinomial components (i.e., defined by polynomials with nondegenerate Newton polygons), any two distinct components of any limit curve $C_k \subset \text{Tor}(\Delta_k)$ intersect transversally at nonsingular points all of which lie in the big torus $(\mathbb{C}^*)^2 \subset \text{Tor}(\Delta_k)$, and finally, the intersection points of any component of a limit curve C_k with $\text{Tor}(\partial\Delta_k)$ are nonsingular.

The main requirement is the transversality condition for the limit curves and deformation patterns (see [13, Sect. 5.2] and [16, Sect. 2.2]), which is relative to the choice of an orientation of the edges of the underlying tropical curve. In [13, 16], one considers an orientation of edges of the embedded plane tropical curve (cf. Sect. 2.2), which in our setting is just $h_*(Q) \subset \mathbb{R}^2$. Here we consider the orientation of the edges of the connected components of $\overline{\Gamma} \setminus G$ as defined in Lemma 3(ii). Since this orientation does not define oriented cycles and since the intersection points of distinct components of any limit curve with toric divisors are distinct, the proof of [13, Theorem 5] and [16, Theorem 2.4] with the orientation of Γ is a word-for-word copy of the proof with the orientation of $h(\Gamma)$. Moreover, comparing with [13, 16], here we impose extra conditions to pass through the points $\text{ini}(\boldsymbol{q})$, $\boldsymbol{q} \in \overline{\boldsymbol{q}}$.

The deformation patterns are transversal in the sense of [13, Definition 5.2], due to [13, Lemma 5.5(ii)], where both inequalities hold, since the deformation patterns are nodal ([13, Lemma 3.9]) and thus do not contribute to the left-hand side of the inequalities, whereas their right-hand sides are positive.

The transversality of the limit curve $C_k \subset \Sigma_k := \text{Tor}(\Delta_k)$ in the sense of [13, Definition 5.1] means the triviality (i.e., zero-dimensionality) of the Zariski tangent space at C_k to the stratum in $|\mathcal{L}_{\Delta_k}|$ formed by the curves that split into the same number of rational components as C_k (i.e., the components of C_k do not glue up when one is deforming along such a stratum), each of them having the same number of intersection points with $\text{Tor}(\partial\Delta_k)$ as the respective component of C_k and with the same intersection number, and such that all but one of these intersection points are fixed. In other words, the conditions imposed on each of the components of C_k determine a stratum with the one-point Zariski tangent space. Indeed, the above fixation of intersection numbers of a component C' of C_k and all but one intersection point of C' with $\text{Tor}(\partial\Delta_k)$ has imposed $-C'K_{\Sigma_k} - 1$ conditions, all of which are independent due to Riemann–Roch on C'.

Thus, [13, Theorem 5] applies, and each of the $M(Q)$ refined patchworking data constructed above produces a curve $C \subset \text{Tor}_{\mathbb{K}}(\Delta)$ as asserted in Theorem 2.

Step 6. Now we specialize the configuration $\overline{\boldsymbol{q}}$ to $\overline{\boldsymbol{p}}$ and prove that each of the curves $C \in \mathcal{C}'$ constructed above tends (in an appropriate topology) to some curve $\hat{C} \in |\mathcal{L}_\Delta|$.

To obtain the required limits, we introduce a suitable topology. Since the variation of $\overline{\boldsymbol{q}}$ does not affect its valuation image, the same holds for the (variable) curves $C' \in \mathcal{C}'$, and hence one can fix once and for all the function $\nu : \Delta \to \mathbb{R}$. Then, writing each coordinate of any point $\boldsymbol{q} \in \overline{\boldsymbol{q}}$ as $X = t^{-\text{Val}(X)}\Psi_{q,X}(t)$ and each coefficient of the defining polynomials of C as $A_\omega = t^{\nu(\omega)}\Psi_\omega(t)$, and assuming (without loss of generality) that all the exponents of t in the above coordinates and coefficients are integral, we deal with the following topology in the space of the functions $\psi_\omega(t)$ holomorphic in a neighborhood of zero: Take the C^0 topology in each subspace consisting of the functions convergent in $|t| \le \varepsilon$ and then define the inductive limit topology in the whole space.

So we assume that the variation of $\overline{\boldsymbol{q}}$ reduces to only variation of $\mathrm{ini}(\boldsymbol{q})$, $\boldsymbol{q} \in \overline{\boldsymbol{q}} \cap (\mathbb{K}^*)^2$, whereas the reminders of the corresponding series in t are unchanged.

To show that the families of the curves $C \in \mathcal{C}'$ have limits, we recall that their coefficients appear as solutions to a system of analytic equations that is soluble by the implicit function theorem due to the transversality of the initial (refined) patchworking data (cf. [13, 16]). Thus, to confirm the existence of the limits of the curves $C \in \mathcal{C}'$, it is sufficient to show that the system of equations and the (refined) patchworking data have limits and that the latter limit is transverse. In particular, we shall obtain that in each coefficient $A_\omega = t^{\nu(\omega)}\Psi_\omega(t)$, $\omega \in \Delta \cap \mathbb{Z}^2$, the factor Ψ_ω converges uniformly in the family.

We start with analyzing the specialization of limit curves. Since the given tropical curve Q stays the same, we go through the curves C_v, $v \in \Gamma^0$, and C_e, $e \in \overline{\Gamma}^1$. Clearly, the curves C_e, $e \in \overline{\Gamma}^1$, keep their form (22) with the parameters ξ, η possibly changing as $\mathrm{ini}(\boldsymbol{q})$ tends to $\mathrm{ini}(\boldsymbol{p})$, $\boldsymbol{p} \in \overline{\boldsymbol{p}}$. Similarly, the curves C_v corresponding to the vertices $v \in \Gamma^0$ of valency 3 remain as described in Step 1, i.e., nodal nonsingular along $\mathrm{Tor}(\partial\Delta_v)$, and crossing each toric divisor at one point. Furthermore, the curves C_v corresponding to the nonspecial vertices $v \in \Gamma^0$ of valency greater than 3, remain as described in Step 3, paragraph (iv), since the intersection points of C_v with the toric divisors that correspond to the edges of $\overline{\Gamma}$, merging to v, do not collate and remain generic in the specialization, since they are not affected by possible collisions of the points $\mathrm{ini}(\boldsymbol{q})$, $\boldsymbol{q} \in \overline{\boldsymbol{q}}$. So let us consider the case of a special vertex $v \in \Gamma^0$. By (T4), C_v cannot split into proper components, and hence it specializes to an irreducible rational curve. Furthermore, the intersection points of C_v with toric divisors that correspond to the special edges may collate, forming singular points, centers of several smooth branches.

So finally, the transversality conditions for such a curve reduce to the fact that the Zariski tangent space at C_v to the stratum in $|\mathcal{L}_{\Delta_v}|$ consisting of rational curves with given intersection points along the two toric divisors that are related to the oriented edges of $\overline{\Gamma}$ merging to v is zero-dimensional. These are precisely the same stratum conditions as in Step 5, and the argument of Step 5 (Riemann–Roch on the rational curve C_v) shows that all the $-C_vK_{\mathrm{Tor}(\Delta_v)} - 1$ conditions defining the stratum in the Severi variety parameterizing the rational curves in $|\mathcal{L}_{\Delta_v}|$ are independent.

Next, we notice that by assumption (T4), the possible collision of intersection points of C_v with $\mathrm{Tor}(\partial\Delta_v)$ concerns only transverse intersection points (i.e., those that correspond to edges of weight 1), and hence affects neither the deformation patterns nor the refined conditions to pass through $\overline{\boldsymbol{p}}$. Thus, each of the curves $C \in \mathcal{C}'$ degenerates into some curve $\hat{C} \in |\mathcal{L}_\Delta|$ that is given by a polynomial with coefficients $A_\omega = t^{\nu(\omega)}\Psi_\omega(t)$, $\omega \in \Delta$, containing factors Ψ_ω convergent uniformly in some neighborhood of 0 in $\mathbb{C}$.

Remark 11. (1) Observe that the genus of $\hat{C}$ does not exceed the genus of C.

(2) Notice also that there is no need to study refinements of possible singular points appearing in the above collisions of the intersection points of C_v with $\mathrm{Tor}(\partial\Delta_v)$. Indeed, the number of transverse conditions we found equals the number of parameters; hence no extra ramification is possible.

Step 7. Next we show that at each point $\boldsymbol{p} \in \overline{\boldsymbol{p}}$ with $\mu(\boldsymbol{p}) > 1$, the obtained curve $\hat{C}$ has $\mu(\boldsymbol{p})$ local branches.

Considering the point $\boldsymbol{p} \in (\mathbb{K}^*)^2$ as a family of points $\boldsymbol{p}^{(t)} \in (\mathbb{C}^*)^2$, $t \neq 0$, we claim that the curves $\hat{C}^{(t)} \subset \mathrm{Tor}(\Delta)$ have $\mu(\boldsymbol{p})$ branches at $\boldsymbol{p}^{(t)}$, $t \neq 0$. Indeed, we will describe how to glue up the limit curves forming $\hat{C}^{(0)}$ when $\hat{C}^{(0)}$ deforms into $\hat{C}^{(t)}$, $t \neq 0$. Our approach is to compare the above gluing with the gluing of the limit curves in the deformation of $C^{(0)}$ into $C^{(t)}$, $t \neq 0$, where $C \in \mathcal{C}'$ passes through the configuration $\overline{\boldsymbol{q}}$, and this comparison heavily relies on the one-to-one correspondence between the limit curves of $\hat{C}$ and $C \in \mathcal{C}'$ established in Step 6.

Let $\boldsymbol{q}_1, \ldots, \boldsymbol{q}_s$ be all the points of the configuration $\overline{\boldsymbol{q}}$ that appear in the dissipation of the point $\boldsymbol{p} \in \overline{\boldsymbol{p}}$ (cf. Step 2), and let $\gamma_i = \psi(\boldsymbol{q}_i)$, $i = 1, \ldots, s$, be the corresponding marked points on Γ, so that $\gamma_i \in e_i \in \Gamma^1$, $i = 1, \ldots, s$. If the edges $h(e_i), h(e_j)$ intersect transversally at $V = h(\gamma_i) = h(\gamma_j)$, then V is a vertex of the plane tropical curve $h_*(Q)$ dual to a polygon Δ_V of the corresponding subdivision of Δ. The components $C_i, C_j \subset \mathrm{Tor}(\Delta_V)$ of the curve $C^{(0)}$ passing through $\mathrm{ini}(\boldsymbol{q}_i), \mathrm{ini}(\boldsymbol{q}_j) \in (\mathbb{C}^*)^2 \subset \mathrm{Tor}(\Delta_V)$, respectively, intersect transversally in $(\mathbb{C}^*)^2$, and their intersection points in $(\mathbb{C}^*)^2$ do not smooth up in the deformation $C^{(t)}$, $t \neq 0$, and the same holds for the corresponding components $\hat{C}_i, \hat{C}_j$ of $\hat{C}^{(0)}$ meeting at $\mathrm{ini}(\boldsymbol{p}) \in (\mathbb{C}^*)^2 \subset \mathrm{Tor}(\Delta_V)$, since the smoothing out of an intersection point $\mathrm{ini}(\boldsymbol{p})$ of $\hat{C}_i$ and $\hat{C}_j$ would raise the genus of $\hat{C}$ above the genus of C, contrary to Remark 11.

Suppose that in the above notation, $h(e_i)$ and $h(e_j)$ lie on the same straight line, but e_i, e_j have no vertex in common (see Fig. 3a). We consider the case of finite-length edges e_i, e_j; the case of ends can be treated similarly. Let v_i, v_i' be the vertices of e_i, and let v_j, v_j' be the vertices of e_j. Their dual polygons $\Delta_{v_i}, \Delta_{v_i'}, \Delta_{v_j}, \Delta_{v_j'}$ (see Sect. 2.2) have sides E_i, E_i', E_j, E_j' orthogonal to $h(e_i)$. In the deformation $C^{(0)} \to C^{(t)}$, $t \neq 0$, the limit curves $C_i \subset \mathrm{Tor}(\Delta_{v_i})$ and $C_i' \subset \mathrm{Tor}(\Delta_{v_i'})$ passing through $\mathrm{ini}(\boldsymbol{q}_i) \in \mathrm{Tor}(E_i) = \mathrm{Tor}(E_i')$ glue up to form a branch centered at $\boldsymbol{q}_i^{(t)}$, and similarly the limit curves $C_j \subset \mathrm{Tor}(\Delta_{v_j})$ and $C_j' \subset \mathrm{Tor}(\Delta_{v_j'})$ passing through $\mathrm{ini}(\boldsymbol{q}_j) \in \mathrm{Tor}(E_j) = \mathrm{Tor}(E_j')$ glue up to form a branch centered at $\boldsymbol{q}_j^{(t)}$. The same happens when C specializes to $\hat{C}$ and $\boldsymbol{q}_i, \boldsymbol{q}_j$ specialize to $\boldsymbol{p}$, since again the aforementioned restriction $g(\hat{C}) \leq g(C)$ does not allow the limit curves $\hat{C}_i, \hat{C}_i'$ to glue up with the limit curves $\hat{C}_j, \hat{C}_j'$.

The remaining case to study is given by the tropical data described in condition (T7), Sect. 3.1. Without loss of generality we can assume that all the edges $e_1, \ldots, e_s$ have a common vertex v and that their h-images lie on the same line (see an example in Fig. 3b). Applying an appropriate invertible integral–affine transformation, we

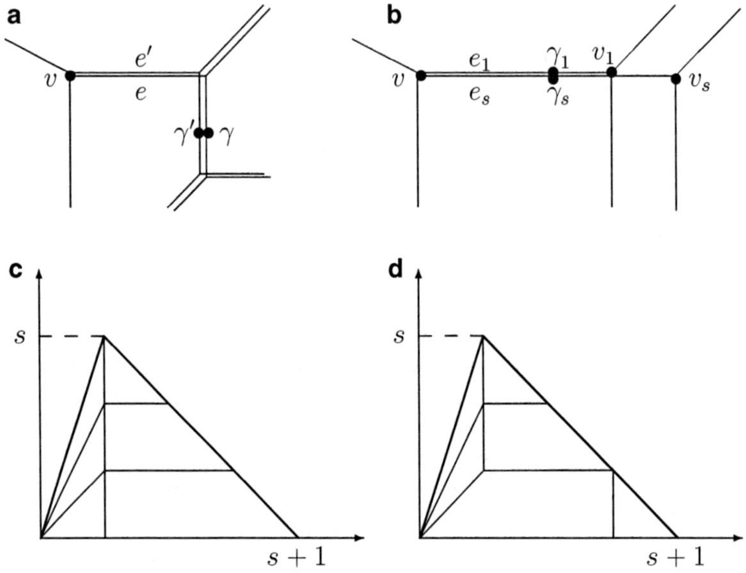

Fig. 3 Illustration to Step 7 of the proof of Theorem 2

can make the edges $e_1,\dots,e_s$ horizontal and the point $\boldsymbol{x} = \text{Val}(\boldsymbol{p}) \in \mathbb{R}^2$ to be the origin. Correspondingly, $v = (-\alpha, 0)$, $v_i = (\alpha_i, 0)$, $i = 1,\dots,s$, with $0 < \alpha_1 \le \dots \le \alpha_s \le \infty$ and

$$\alpha > \sum_{1 \le i < s-1} \alpha_i + 2\alpha_{s-1}. \tag{23}$$

In what follows we suppose that $\alpha_s < \infty$. The case $\alpha_s = \infty$ admits the same treatment as the case of finite $\alpha_s \gg \alpha$.

Let $\boldsymbol{q}_i = \psi^{-1}(\gamma_i)$, $1 \le i \le s$, be the points of the configuration $\overline{\boldsymbol{q}}$ that appear in the deformation of the point $\boldsymbol{p}$ described in Step 2. Our assumptions yield that

$$\boldsymbol{p} = (\xi + O(t^{>0}), \eta + O(t^{>0})), \quad \boldsymbol{q}_i = (\xi_i + O(t^{>0}), \eta_i + O(t^{>0})),\ i = 1,\dots,s,$$

with some $\xi, \eta \in \mathbb{C}^*$, ξ_i close to ξ, η_i close to η, $i = 1,\dots,s$. Furthermore, the triangles Δ_v and Δ_{v_i} dual to the vertices v and v_i, $1 \le i \le s$, respectively, have vertical edges $\sigma \subset \partial\Delta_v$ and $\sigma_i \subset \partial\Delta_{v_i}$, $1 \le i \le s$, along which the function ν (see Sect. 2.2) is constant. By assumptions (T4)–(T6), the limit curve $C_v \subset \text{Tor}(\Delta_v)$ crosses the toric divisor $\text{Tor}(\sigma)$ at the points $\eta_1,\dots,\eta_s$ with total intersection multiplicity s, and each of the limit curves $C_{v_i} \subset \text{Tor}(\Delta_{v_i})$, $1 \le i \le s$, crosses the toric divisor $\text{Tor}(\sigma_i)$ at the unique point η_i transversally, and the corresponding limit curve $\hat{C}_{v_i}$ crosses $\text{Tor}(\sigma_i)$ at the point η transversally, too.

Now we move the points $\boldsymbol{q}_1,\dots,\boldsymbol{q}_s$, keeping their x-coordinates and making $(\boldsymbol{q}_1)_y=\dots=(\boldsymbol{q}_s)_y=(\boldsymbol{p})_y$. As shown in Step 6, the curve C (depending on $\boldsymbol{q}_1,\dots,\boldsymbol{q}_s$) converges to a curve C' with the same Newton polygon, genus, and tropicalization, and the limit curves of C componentwise converge to limit curves of C'. Consider now the polynomial $\widetilde{F}(x,y):=F'(x,y+(\boldsymbol{p})_y)$, where the polynomial $F'(x,y)$ defines the curve C'. As in the refinement procedure described in [13, Sect. 3.4] or [7, Sect. 2.5.8], the subdivision of the Newton polygon $\widetilde{\Delta}$ of $\widetilde{F}$ contains the fragment bounded by the triangle $\delta=\operatorname{conv}\{(0,0),(1,s),(s+1,0)\}$ (see Fig. 3d), which matches the points $\boldsymbol{q}_1,\dots,\boldsymbol{q}_s$. The corresponding function $\widetilde{\nu}:\widetilde{\Delta}\to\mathbb{R}$ takes the values

$$\widetilde{\nu}(0,0)=\alpha,\quad \widetilde{\nu}(1,s)=0,\quad \widetilde{\nu}(k,s+1-k)=\sum_{1\le i<k}\alpha_i,\ k=2,\dots,s+1,$$

along the inclined part of $\partial\delta$. The tropical limit of $\widetilde{F}$ restricted to the above fragment consists of a subdivision of δ determined by some extension of the function $\widetilde{\nu}$ inside δ and of limit curves that must meet the following conditions:

- These limit curves glue up into a rational curve (with Newton triangle δ), since in the original tropical curve, the aforementioned fragment corresponds to a tree (see Fig. 3b).
- The intersection points $\boldsymbol{q}$ of the curve $C_\delta:=\{\widetilde{F}_\delta=0\}$ with the line $x=(\boldsymbol{p})_x$ such that $\operatorname{Val}(\boldsymbol{q})_y\le 0$ converge to $\boldsymbol{p}$ as $\boldsymbol{q}_1,\dots,\boldsymbol{q}_s$ tend to $\boldsymbol{p}$, where $\widetilde{F}_\delta$ is the sum of the monomials of $\widetilde{F}$ matching the set $\Delta\cap\mathbb{Z}^2$, and the convergence is understood in the topology of Step 6.
- The subdivision of δ contains a segment $\widetilde{\sigma}$ of length s lying inside the edge $[(0,0),(s+1,0)]$, along which the function $\widetilde{\nu}$ is constant and such that the corresponding toric divisor $\operatorname{Tor}(\widetilde{\sigma})$ intersects the limit curves at the points $\xi_1,\dots,\xi_s$.

These restrictions and inequality (23) leave only one possibility: the subdivision of δ shown in Fig. 3c,d (the subdivision (c) for the case $\alpha>\alpha_1+\dots+\alpha_s$, and the subdivision (d) for the case $\alpha<\alpha_1+\dots+\alpha_s$). The limit curve $C_{\delta'}\subset\operatorname{Tor}(\delta')$ for a triangle $\delta'\subset\delta$ having a horizontal base splits into $H(\delta)$ distinct straight lines (any of them crossing each toric divisor at one point), where $H(\delta')$ is the height. The limit curve $\mathbb{C}_{\delta'}\subset\operatorname{Tor}(\delta')$ for a trapeze $\delta'\subset\delta$ splits into $H(\delta')$ straight lines as above and a suitable number of straight lines $x=\mathrm{const}$ (which reflect the splitting of the trapeze into the Minkowski sum of a triangle with a horizontal segment). All the limit curves are uniquely defined by the intersections with the toric divisors $\operatorname{Tor}(\sigma')$ for inclined segments σ' (in our construction, these data are determined by the points $\overline{q}\setminus\{\boldsymbol{q}_1,\dots,\boldsymbol{q}_s\}$ and by the condition $(\boldsymbol{q}_1)_y=\dots=(\boldsymbol{q}_s)_y=0$) and by intersections with $\operatorname{Tor}(\widetilde{\sigma})$ introduced above. When $\boldsymbol{q}_1,\dots,\boldsymbol{q}_s$ tend to $\boldsymbol{p}$, the subdivision of δ remains unchanged, whereas the limit curves naturally converge componentwise. Then we immediately derive that components of the limit curves passing through $\operatorname{ini}(\boldsymbol{p})$ do not glue up together in the deformation $\hat{C}^{(t)}$, $t\in(\mathbb{C},0)$, since otherwise, the (geometric) genus of $\hat{C}^{(t)}$, $t\ne 0$, would jump above the genus of C, which is impossible (see Remark 11).

Step 8. By assumption (A5), Sect. 3.1, the curves $\hat{C} \subset \mathrm{Tor}_{\mathbb{K}}(\Delta)$ are immersed, irreducible, of genus g, have multiplicity $\mu(\boldsymbol{p})$ at each point $\boldsymbol{p} \in \overline{\boldsymbol{p}} \cap (\mathbb{K}^*)^2$, and satisfy the tangency conditions with $\mathrm{Tor}_{\mathbb{K}}(\partial\Delta)$ as specified in the assertion of Theorem 2. It remains to show that we have constructed precisely $M(Q)$ curves $\hat{C}$.

Indeed, condition (16) implies that for any dissipation of each point $\boldsymbol{p} \in \overline{\boldsymbol{p}} \cap (\mathbb{K}^*)^2$ into $\mu(\boldsymbol{p})$ distinct points there exists a unique deformation of $\hat{C}$ into a curve $C \in \mathcal{C}$ such that a priori prescribed branches of $\hat{C}$ at $\boldsymbol{p}$ will pass through prescribed points of the dissipation.

Finally, we notice that the sets $\mathcal{C}(\hat{Q}_1)$ and $\mathcal{C}(\hat{Q}_2)$ are disjoint for distinct (nonisomorphic) PPT-curves $\hat{Q}_1, \hat{Q}_2$. Indeed, the collections of limit curves as constructed in Steps 1 and 3 appear to be distinct for distinct curves $\hat{Q}_1$ and $\hat{Q}_2$ and the given configuration $\overline{\boldsymbol{p}}$. □

3.5 Proof of Theorem 3

The curves $C \in \Re\mathcal{C}(\hat{Q})$ constructed in the proof of Theorem 2 are immersed, and hence the formula (1) for the Welschinger weight applies. Thus the left-hand side of (21) is well defined.

Next we go through the proof of Theorem 2, counting the contribution to the right-hand side of (21).

First, we deform the configuration $\overline{\boldsymbol{p}}$ as described in Step 2, assuming that the deformed configuration $\overline{\boldsymbol{q}}$ is Conj-invariant and that the map $\psi : \overline{\boldsymbol{q}} \to G$ sends $\Re\overline{\boldsymbol{q}} = \overline{\boldsymbol{q}} \cap \mathrm{Fix}(\mathrm{Conj})$ to $\Re G$ and sends $\Im\overline{\boldsymbol{q}} = \overline{\boldsymbol{q}} \setminus \Re\overline{\boldsymbol{q}}$ to $\Im G$, respectively. In particular, if $\boldsymbol{p} \in \Re\overline{\boldsymbol{p}}$, and the points $\mathrm{Val}(\boldsymbol{p})$ is an image of r points of $\Re G \cap \Re\overline{\Gamma}$ and s pairs of points of $\Re G \cap \Im\overline{\Gamma}$, then $\boldsymbol{p}$ deforms into r real points and s pairs of imaginary conjugate points.

Notice that the replacement of $\overline{\boldsymbol{p}}$ by $\overline{\boldsymbol{q}}$ causes a change of sign in the left-hand side of (21) and a change in the quantity of the real curves in the count on the right-hand side of (21). We now explain the change of sign: The dissipation of a real point $\boldsymbol{p}$ as in the preceding paragraph means that for each curve $C \in \mathcal{C}'$, we count an additional r real solitary nodes in a neighborhood of $\boldsymbol{p}$, since in the nondeformed situation, the point $\boldsymbol{p}$ should be blown up for the computation of the Welschinger sign. This change is reflected in the sign $(-1)^{\ell_1}$, $\ell_1 = |\Re G \cap \Im\overline{\Gamma}|/2$, in the right-hand side of formula (10).

Next we follow the procedure in Steps 3 and 4 of the proof of Theorem 2 and construct Conj-invariant collections of limit curves, deformation patterns, and refined conditions to pass through fixed points:

- By [13, Proposition 8.1(i)], the existence of an even-weight edge $e \in \Gamma^1$, $e \subset \Re\overline{\Gamma}$, annihilates the contribution to the Welschinger number, and hence by (R7)(v), we can assume that all the edges $e \subset \Re\overline{\Gamma}$ have odd weight. In particular, with the finite-length edges $e \subset \Re\overline{\Gamma}$ one can associate a unique real deformation pattern with an even number of solitary nodes (cf. [15, Lemma 2.3]).

- The limit curves associated with the vertices of $\Re\overline{\Gamma}$ contribute as designated in rules (W2)–(W4) in Sect. 2.6 (cf. [15, Lemmas 2.3, 2.4, and 2.5]).
- The construction of limit curves and deformation patterns associated with the vertices and edges of $\Im\overline{\Gamma}'$ (a half of $\Im\overline{\Gamma}$) contributes as designated in rules (W1)–(W3) (cf. the complex formulas in the proof of Theorem 2 and [15, Sect. 2.5]). Accordingly, the data associated with $\Im\overline{\Gamma}''$ are obtained by conjugation.
- The refinement of the condition to pass through fixed points contributes as designated in rule (W2), since we have a unique refinement for $\gamma \in \Re\overline{\Gamma}$ and $w(e)$ refinements for $\gamma \in e \in \Im\overline{\Gamma}^1$.

The remaining step is to explain the factor 2^{ℓ_2}, $\ell_2 = (|\Im G \cap \Im\overline{\Gamma}| - b_0(\Im\overline{\Gamma}))/2$, in formula (10). Indeed, when constructing the limit curves associated with the vertices of $\Im\overline{\Gamma}'$, we start with the respective fixed points, which are all are imaginary in the configuration $\overline{\boldsymbol{q}}$, and thus we choose a point in each of the $|G \cap \Im\overline{\Gamma}'| = |G \cap \Im\overline{\Gamma}|/2$ pairs of the corresponding points in $\overline{\boldsymbol{q}}$.

Observe that in the degeneration $\overline{\boldsymbol{q}} \to \overline{\boldsymbol{p}}$, $|\Re G \cap \Im\overline{\Gamma}'|$ pairs ofs imaginary points of $\overline{\boldsymbol{q}}$ merge to real points in $\overline{\boldsymbol{p}}$, which leaves only $|\Im G \cap \Im\overline{\Gamma}|/2$ choices in the original configuration $\overline{\boldsymbol{p}}$. After all, we factorize by the interchange of the components of $\Im\overline{\Gamma}$, arriving at the required factor 2^{ℓ_2}. □

Acknowledgments The author was supported by the grant 465/04 from the Israel Science Foundation, a grant from the Higher Council for Scientific Cooperation between France and Israel, and a grant from Tel Aviv University. This work was completed during the author's stay at the Centre Interfacultaire Bernoulli, École Polytechnique Fédérale da Lausanne and at Laboratoire Émile Picard, Université Paul Sabatier, Toulouse. The author thanks the CIB-EPFL and UPS for the hospitality and excellent working conditions. Special thanks are due to I. Itenberg, who pointed out a mistake in the preliminary version of Theorem 3. Finally, I express my gratitude to the unknown referee for numerous remarks, corrections, and suggestions.

References

1. Gathmann, A., and Markwig, H.: The numbers of tropical plane curves through points in general position. *J. reine angew. Math.* **602** (2007), 155–177.
2. Greuel, G.-M., and Karras, U.: Families of varieties with prescribed singularities. *Compos. Math.* **69** (1989), no. 1, 83–110.
3. Greuel, G.-M., and Lossen, C.: Equianalytic and equisingular families of curves on surfaces. *Manuscripta Math.* **91** (1996), no. 3, 323–342.
4. Itenberg, I. V., Kharlamov, V. M., and Shustin, E. I.: Logarithmic equivalence of Welschinger and Gromov–Witten invariants. *Russian Math. Surveys* **59** (2004), no. 6, 1093–1116.
5. Itenberg, I. V., Kharlamov, V. M., and Shustin, E. I.: New cases of logarithmic equivalence of Welschinger and Gromov–Witten invariants. *Proc. Steklov Math. Inst.* **258** (2007), 65–73.
6. Itenberg, I., Kharlamov, V., and Shustin, E.: *Welschinger invariants of small non-toric del Pezzo surfaces*. Preprint at arXiv:1002.1399.
7. Itenberg, I., Mikhalkin, G., and Shustin, E.: *Tropical algebraic geometry* Oberwolfach seminars, vol. 35. Birkhäuser, 2007.

8. Mikhalkin, G.: Decomposition into pairs-of-pants for complex algebraic hypersurfaces. *Topology* **43** (2004), 1035–1065.
9. Mikhalkin, G.: Enumerative tropical algebraic geometry in $\mathbb{R}^2$. *J. Amer. Math. Soc.* **18** (2005), 313–377.
10. Nishinou, T., and Siebert, B.: Toric degenerations of toric varieties and tropical curves. *Duke Math. J.* **135** (2006), no. 1, 1–51.
11. Orevkov, S., and Shustin, E.: Pseudoholomorphic, algebraically unrealizable curves. *Moscow Math. J.* **3** (2003), no. 3, 1053–1083.
12. Richter-Gebert, J., Sturmfels, B., and Theobald, T.: First steps in tropical geometry. Idempotent mathematics and mathematical physics. *Contemp. Math.* 377, *Amer. Math. Soc.*, Providence, RI, 2005, pp. 289–317.
13. Shustin, E.: A tropical approach to enumerative geometry. *Algebra i Analiz* **17** (2005), no. 2, 170–214 (English translation: *St. Petersburg Math. J.* **17** (2006), 343–375).
14. Shustin, E.: On manifolds of singular algebraic curves. *Selecta Math. Sov.* **10**, no. 1, 27–37 (1991).
15. Shustin, E.: A tropical calculation of the Welschinger invariants of real toric del Pezzo surfaces. *J. Alg. Geom.* **15** (2006), no. 2, 285–322 (corrected version at arXiv:math/0406099).
16. Shustin, E.: Patchworking construction in the tropical enumerative geometry. *Singularities and Computer Algebra*, C. Lossen and G. Pfister, eds., Lond. Math. Soc. Lec. Notes Ser. **324**, Proceedings of Conference dedicated to the 60th birthday of G.-M. Greuel, Cambridge University Press, Cambridge, MA, 2006, pp. 273–300.
17. Shustin, E.: Welschinger invariants of toric del Pezzo surfaces with non-standard real structures. *Proc. Steklov Math. Inst.* **258** (2007), 219–247.
18. Shustin, E.: *New enumerative invariants and correspondence theorems for plane tropical curves*, Preprint, 2011.
19. Viro, O. Ya.: Gluing of plane real algebraic curves and construction of curves of degrees 6 and 7. *Lect. Notes Math.* **1060**, Springer, Berlin, 1984, pp. 187–200.
20. Viro, O. Ya.: Real algebraic plane curves: constructions with controlled topology. *Leningrad Math. J.* **1** (1990), 1059–1134.
21. Viro, O. Ya.: *Patchworking Real Algebraic Varieties*. Preprint at arXiv:math/0611382.
22. Viro, O.: Dequantization of real algebraic geometry on a logarithmic paper. *Proceedings of the 3rd European Congress of Mathematicians*, Birkhäuser, Progress in Math. **201** (2001), 135–146.
23. Welschinger, J.-Y.: Invariants of real symplectic 4-manifolds and lower bounds in real enumerative geometry. *Invent. Math.* **162** (2005), no. 1, 195–234.

CPSIA information can be obtained at www.ICGtesting.com
Printed in the USA
LVOW100240060112

262663LV00006B/26/P

9 780817 682767